DESIGN WITH OPERATIONAL AMPLIFIERS AND
ANALOG INTEGRATED CIRCUITS

McGraw-Hill Series in Electrical and Computer Engineering

SENIOR CONSULTING EDITOR

Stephen W. Director, University of Michigan, Ann Arbor

Circuits and Systems
Communications and Signal Processing
Computer Engineering
Control Theory and Robotics
Electromagnetics
Electronics and VLSI Circuits
Introductory
Power
Antennas, Microwaves, and Radar

PREVIOUS CONSULTING EDITORS

Ronald N. Bracewell, Colin Cherry, James F. Gibbons, Willis W. Harman, Hubert Heffner, Edward W. Herold, John G. Linvill, Simon Ramo, Ronald A. Rohrer, Anthony E. Siegman, Charles Susskind, Frederick E. Terman, John G. Truxal, Ernst Weber, and John R. Whinnery

RELATED TITLES

DeMicheli: *Synthesis and Optimization of Digital Circuits*
Hodges and Jackson: *Analysis and Design of Digital Integrated Circuits*
Kang and Leblebici: *CMOS Digital Integrated Circuits Analysis and Design*
Kasap: *Principles of Electronic Materials and Devices*
Kovacs: *Micromachined Transducers Sourcebook*
Jaeger: *Microelectronic Circuit Design*
Neamen: *Electronic Circuit Analysis and Design*
Neamen: *Semiconductor Physics and Devices*
Razavi: *Design of Analog CMOS Integrated Circuits*
Smith: *Modern Communication Circuits*
Tsividis: *Operation and Modeling of the MOS Transistor*

Also Available from McGraw-Hill

Schaum's Outline Series in Electronics & Electrical Engineering

Most Outlines include basic theory, definitions, and hundreds of example problems solved in step-by-step detail, and supplementary problems with answers.

Related titles on the current list include:

Analog & Digital Communications
Basic Circuit Analysis
Basic Electrical Engineering
Basic Electricity
Basic Mathematics for Electricity & Electronics
Digital Principles
Electric Circuits
Electric Machines & Electromechanics
Electric Power Systems
Electromagnetics
Electronic Communication
Electronic Devices & Circuits
Feedback & Control Systems
Introduction to Digital Systems
Microprocessor Fundamentals
Signals & Systems

Schaum's Electronic Tutors

A Schaum's Outline plus the power of Mathcad® software. Use your computer to learn the theory and solve problems—every number, formula, and graph can be changed and calculated on screen.

Related titles on the current list include:

Electric Circuits
Feedback & Control Systems
Electromagnetics
College Physics

Available at most college bookstores, or for a complete list of titles and prices, write to:

The McGraw-Hill Companies
Schaum's
11 West 19th Street
New York, New York 11011-4285
(212-337-4097)

DESIGN WITH OPERATIONAL AMPLIFIERS AND ANALOG INTEGRATED CIRCUITS

THIRD EDITION

Sergio Franco

San Francisco State University

Boston   Burr Ridge, IL   Dubuque, IA   Madison, WI   New York   San Francisco   St. Louis
Bangkok   Bogotá   Caracas   Kuala Lumpur   Lisbon   London   Madrid   Mexico City
Milan   Montreal   New Delhi   Santiago   Seoul   Singapore   Sydney   Taipei   Toronto

In Memory of My Parents
Luigia Braidotti and Luigi Franco

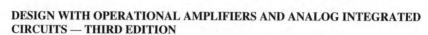

McGraw-Hill Higher Education
A Division of The McGraw-Hill Companies

DESIGN WITH OPERATIONAL AMPLIFIERS AND ANALOG INTEGRATED CIRCUITS — THIRD EDITION
International Edition 2002

10 09 08
20 09 08
CRC   BJE

Library of Congress Cataloging-in-Publication Data
Franco, Sergio.
   Design with operational amplifiers and analog integrated circuits / Sergio Franco.
—3rd ed.
      p.   cm. — (McGraw-Hill series in electrical and computer engineering)
   Includes bibliographical references and index.
   ISBN 0-07-232084-2
   1. Linear integrated circuits. 2. Operational amplifiers. I. Title. II. Series.
TK7874 .F677 2002
621.3815—dc21

                                                                    2001030803
                                                                    CIP

When ordering this title, use ISBN 0-07-120703-1

Printed in Singapore

www.mhhe.com

ABOUT THE AUTHOR

SERGIO FRANCO is Professor of Electrical Engineering at San Francisco State University. He was born in Friuli, Italy, and earned his Ph.D. from the University of Illinois at Urbana-Champaign. Prior to becoming professor, Dr. Franco had extensive industrial experience and has worked and published in such diverse areas as solid state physics, pattern recognition, electronic music, IC design, and medical, consumer, and automotive electronics. Dr. Franco is also the author of the textbook *Electric Circuit Fundamentals,* Oxford University Press, 1995. In addition to teaching, Dr. Franco consults for industry.

CONTENTS

Preface   xi

1 Operational Amplifier Fundamentals   1

   1.1 Amplifier Fundamentals   2
   1.2 The Operational Amplifier   5
   1.3 Basic Op Amp Configurations   8
   1.4 Ideal Op Amp Circuit Analysis   15
   1.5 Negative Feedback   23
   1.6 Feedback in Op Amp Circuits   29
   1.7 The Loop Gain   37
   1.8 Op Amp Powering   41
     Problems   47
     Bibliography   58
     Appendix 1A   Standard Resistor Values   58

2 Circuits with Resistive Feedback   60

   2.1 Current-to-Voltage Converters   61
   2.2 Voltage-to-Current Converters   63
   2.3 Current Amplifiers   71
   2.4 Difference Amplifiers   73
   2.5 Instrumentation Amplifiers   79
   2.6 Instrumentation Applications   86
   2.7 Transducer Bridge Amplifiers   91
     Problems   98
     References   105

3 Active Filters: Part I   106

   3.1 The Transfer Function   109
   3.2 First-Order Active Filters   115
   3.3 Audio Filter Applications   121
   3.4 Standard Second-Order Responses   126
   3.5 *KRC* Filters   133
   3.6 Multiple-Feedback Filters   141
   3.7 State-Variable and Biquad Filters   144
   3.8 Sensitivity   150
     Problems   153
     References   158

4 Active Filters: Part II   160

   4.1 Filter Approximations   161
   4.2 Cascade Design   166
   4.3 Generalized Impedance Converters   175
   4.4 Direct Design   181

4.5 The Switched Capacitor   187
4.6 Switched-Capacitor Filters   192
4.7 Universal SC Filters   198
    Problems   204
    References   210

5 Static Op Amp Limitations   211

5.1 Simplified Op Amp Circuit Diagram   212
5.2 Input Bias and Offset Currents   217
5.3 Low-Input-Bias-Current Op Amps   221
5.4 Input Offset Voltage   225
5.5 Low-Input-Offset-Voltage Op Amps   230
5.6 Input Offset-Error Compensation   235
5.7 Maximum Ratings   240
    Problems   244
    References   248
    Appendix 5A Data Sheets of the μA741 Op Amp   249

6 Dynamic Op Amp Limitations   258

6.1 Open-Loop Response   259
6.2 Closed-Loop Response   263
6.3 Input and Output Impedances   269
6.4 Transient Response   275
6.5 Effect of Finite GBP on Integrator Circuits   283
6.6 Effect of Finite GBP on Filters   289
6.7 Current-Feedback Amplifiers   293
    Problems   303
    References   309

7 Noise   311

7.1 Noise Properties   313
7.2 Noise Dynamics   317
7.3 Sources of Noise   322
7.4 Op Amp Noise   328
7.5 Noise in Photodiode Amplifiers   335
7.6 Low-Noise Op Amps   339
    Problems   342
    References   346

8 Stability   347

8.1 The Stability Problem   348
8.2 Stability in Constant-GBP Op Amp Circuits   356
8.3 Internal Frequency Compensation   365
8.4 External Frequency Compensation   374
8.5 Stability in CFA Circuits   381
8.6 Composite Amplifiers   384
    Problems   390
    References   396

9 Nonlinear Circuits   398

9.1 Voltage Comparators   399
9.2 Comparator Applications   407

9.3 Schmitt Triggers 416
9.4 Precision Rectifiers 422
9.5 Analog Switches 428
9.6 Peak Detectors 433
9.7 Sample-and-Hold Amplifiers 437
Problems 443
References 448

10 Signal Generators 449
10.1 Sine Wave Generators 451
10.2 Multivibrators 457
10.3 Monolithic Timers 465
10.4 Triangular Wave Generators 471
10.5 Sawtooth Wave Generators 476
10.6 Monolithic Waveform Generators 478
10.7 *V-F* and *F-V* Converters 486
Problems 492
References 497

11 Voltage References and Regulators 499
11.1 Performance Specifications 500
11.2 Voltage References 506
11.3 Voltage-Reference Applications 512
11.4 Linear Regulators 519
11.5 Linear-Regulator Applications 527
11.6 Switching Regulators 535
11.7 Monolithic Switching Regulators 544
Problems 551
References 557

12 D-A and A-D Converters 559
12.1 Performance Specifications 561
12.2 D-A Conversion Techniques 567
12.3 Multiplying DAC Applications 580
12.4 A-D Conversion Techniques 584
12.5 Oversampling Converters 595
Problems 603
References 606

13 Nonlinear Amplifiers and Phase-Locked Loops 607
13.1 Log/Antilog Amplifiers 608
13.2 Analog Multipliers 615
13.3 Operational Transconductance Amplifiers 620
13.4 Phase-Locked Loops 627
13.5 Monolithic PLLs 635
Problems 642
References 646

Index 647

PREFACE

During the last decades much has been prophesized that there will be little need for analog circuitry in the future because digital electronics is taking over. Far from having proven true, this contention has provoked controversial rebuttals, as epitomized by statements such as "If you cannot do it in digital, it's got to be done in analog." Add to this the common misconception that analog design, compared to digital design, seems to be more of a whimsical art than a systematic science, and what is the confused student to make of this controversy? Is it worth pursuing some coursework in analog electronics, or is it better to focus just on digital?

There is no doubt that many functions that were traditionally the domain of analog electronics are nowadays implemented in digital form, a popular example being offered by digital audio. Here, the analog signals produced by microphones and other acoustic transducers are suitably conditioned by means of amplifiers and filters, and are then converted to digital form for further processing, such as mixing, editing, and the creation of special effects, as well as for the more mundane but no less important tasks of transmission, storage, and retrieval. Finally, digital information is converted back to analog signals for playing through loudspeakers. One of the main reasons why it is desirable to perform as many functions as possible digitally is the generally superior reliability and flexibility of digital circuitry. However, *the physical world is inherently analog,* indicating that there will *always* be a need for analog circuitry to condition physical signals such as those associated with transducers, as well as to convert information from analog to digital for processing, and from digital back to analog for reuse in the physical world. Moreover, new applications continue to emerge, where considerations of speed and power make it more advantageous to use analog front ends; wireless communications provide a good example.

Indeed many applications today are best addressed by mixed-mode integrated circuits (mixed-mode ICs) and systems, which rely on analog circuitry to interface with the physical world, and digital circuitry for processing and control. Even though the analog circuitry may constitute only a small portion of the total chip area, it is often the most challenging part to design as well as the limiting factor on the performance of the entire system. In this respect, it is usually the analog designer who is called to devise ingenious solutions to the task of realizing analog functions in decidedly digital technologies; switched-capacitor techniques in filtering and sigma-delta techniques in data conversion are popular examples. In light of the above, the need for competent analog designers will continue to remain very strong. Even purely digital circuits, when pushed to their operational limits, exhibit analog behavior. Consequently, a solid grasp of analog design principles and techniques is a valuable asset in the design of any IC, not just purely digital or purely analog ICs.

THE BOOK

The goal of this book is the illustration of general analog principles and design methodologies using practical devices and applications. The book is intended as a textbook for undergraduate and graduate courses in design and applications with analog integrated circuits (analog ICs), as well as a reference book for practicing engineers. The reader is expected to have had an introductory course in electronics, to be conversant in frequency-domain analysis techniques, and to possess basic skills in the use of PSpice. Though the book contains enough material for a two-semester course, it can also serve as basis for a one-semester course after suitable selection of topics. The selection process is facilitated by the fact that the book as well as its individual chapters have generally been designed to proceed from the elementary to the complex.

At San Francisco State University we use the book for a sequence of two one-semester courses, one at the senior and the other at the graduate level. In the senior course we cover Chapters 1–3, Chapters 5 and 6, and most of Chapters 9 and 10; in the graduate course we cover all the rest. The senior course is taken concurrently with a course in analog IC fabrication and design. For an effective utilization of analog ICs, it is important that the user be cognizant of their internal workings, at least qualitatively. To serve this need, the book provides intuitive explanations of the technological and circuital factors intervening in a design decision.

The third edition retains the features that distinguished the second edition from the first: namely, considerable pedagogical enhancements, inclusion of PSpice simulations, and expanded subject coverage to include, among others, current-feedback amplifiers, switching regulators, sigma-delta converters, and phase-locked loops. In addition, negative-feedback concepts have been clarified and emphasized further, and the number of end-of-chapter problems has been increased by 10% to achieve a total of 579. Although many readers will undoubtedly prefer to use Windows versions of PSpice, it was decided after much deliberation to retain the netlist version of the second edition because of its pedagogical advantages. However, the interested reader can find Windows versions of the book's netlist examples in the accompanying Website at http://www.mhhe.com/franco.

The desire to address general and lasting principles in a manner that transcends the latest technological trend has motivated the choice of well-established and widely documented devices and technologies as vehicles to illustrate such principles. However, whenever necessary, the reader is made aware of more contemporary alternatives, as well as bibliographical sources and where to find them.

THE CONTENTS AT A GLANCE

Although not explicitly indicated, the book consists of three parts. Part I (Chapters 1–4) introduces fundamental concepts and applications based on the op amp as a predominantly ideal device. It is felt that the student needs to develop sufficient confidence with ideal (or near-ideal) op amp situations before tackling and assessing the consequences of practical device limitations. Limitations are the subject of Part II (Chapters 5–8), which covers the topic in more systematic detail than previous editions. Finally, Part III (Chapters 9–13) exploits the maturity and judgment developed by the reader in the first two parts to address a variety of design-oriented applications. Following is a brief chapter-by-chapter description of the material covered.

Chapter 1 reviews basic amplifier concepts, including negative feedback. Much emphasis is placed on the loop gain T as a gauge of circuit performance. The student is introduced to simple PSpice models, which will become more sophisticated as we progress through the book. Those instructors who find the loop-gain treatment overwhelming this early in the book, may skip it to return to it later, at a more suitable time. Coverage rearrangements of this sort are facilitated by the fact that individual sections and chapters have been designed to be as independent as possible from each other; moreover, the end-of-chapter problems are grouped by section.

Chapter 2 deals with I-V, V-I, and I-I converters, along with various instrumentation and transducer amplifiers. The chapter places much emphasis on feedback topologies and the role of the loop gain T.

Chapter 3 covers first-order filters, audio filters, and popular second-order filters such as the *KRC,* multiple-feedback, state-variable, and biquad topologies. The chapter emphasizes complex-plane systems concepts, and concludes with filter sensitivities.

The reader who wants to go deeper into the subject of filters will find Chapter 4 useful. This chapter covers higher-order filter synthesis using both the cascade and the direct approaches. Moreover, these approaches are presented for both the case of active RC filters and the case of switched-capacitor (SC) filters.

Chapter 5 addresses input-referrable op amp errors such as V_{OS}, I_B, I_{OS}, CMRR, PSRR, and drift, along with operating limits. The student is introduced to datasheet interpretation, PSpice macromodels, and also to different technologies and topologies.

Chapter 6 addresses dynamic limitations in both the frequency and time domains, and investigates their effect on the resistive circuits and the filters that were studied in Part I using mainly ideal op amp models. Voltage-feedback and current-feedback are compared in detail, and PSpice is used extensively to visualize both the frequency and transient responses of representative circuit examples. Having mastered the material of the first four chapters using ideal or nearly-ideal op amps, the student is now in a better position to appreciate and evaluate the consequences of practical device limitations.

The subject of ac noise, covered in Chapter 7, follows naturally since it combines the principles learned both in Chapters 5 and 6. Noise calculations and estimation represent another area in which PSpice proves a most useful tool.

Part II concludes with the subject of stability, in Chapter 8. The material has been arranged to facilitate topic selection, and puts much emphasis on a systems-oriented approach. Again, PSpice is used profusely to visualize the effect of the different frequency-compensation techniques presented.

Part III begins with nonlinear applications, in Chapter 9. Here, nonlinear behavior stems from either the lack of feedback (voltage comparators), or the presence of feedback, but of the positive type (Schmitt triggers), or the presence of negative feedback, but using nonlinear elements such as diodes and switches (precision rectifiers, peak detectors, track-and-hold amplifiers).

Chapter 10 covers signal generators, including Wien-bridge and quadrature oscillators, multivibrators, timers, function generators, and V-F and F-V converters.

Chapter 11 addresses regulation. It starts with voltage references, proceeds to linear voltage regulators, and concludes with switching regulators. The last topic has been at the center of much attention and industrial activity since the eighties, and entire books have been written on the subject. Of necessity, this chapter exposes the student only to the fundamentals of this most important area.

Chapter 12 deals with data conversion. Data-converter specifications are treated in systematic fashion, and various applications with multiplying DACs are presented. The chapter concludes with oversampling-conversion principles and sigma-delta converters. Much has been written also about this subject, so this chapter of necessity exposes the student only to the fundamentals.

Chapter 13 concludes the book with a variety of nonlinear circuits, such as log/antilog amplifiers, analog multipliers, and operational transconductance amplifiers with a brief exposure to g_m-C filters. The chapter culminates with an introduction to phase-locked loops, a subject that combines important materials addressed at various points in the preceding chapters.

THE WEBSITE

The book is accompanied by a website (http://www.mhhe.com/franco) containing a variety of Instructor and Student Resources as well as other useful links. A website icon has been placed in the margin in those places in the text where the website resources would prove most useful. The Instructor Resources consist of a Solutions Manual, Downloadable Software, Windows Version of PSpice Examples, and PageOut (a link to McGraw-Hill's course Website Development Tools). The Student Resources consist of Downloadable Software and Windows Version of PSpice Examples. The author welcomes feedback via email, at sfranco@sfsu.edu.

ACKNOWLEDGMENTS

Some of the changes in the third edition were made in response to feedback received from a number of readers in both industry and academia, and I am grateful to all who took the time to e-mail me. In addition, the following reviewers provided detailed commentaries on the previous edition as well as valuable suggestions for the current revision. All suggestions have been examined in detail, and if only a portion of them has been honored, it was not out of callousness, but because of production constraints or personal philosophy. To all reviewers, my sincere thanks: J. Alvin Connelly, Georgia Institute of Technology; Dragan Maksimovic, University of Colorado-Boulder; Philip C. Munro, Youngstown State University; Thomas G. Owen, University of North Carolina-Charlotte; Dr. Guillermo Rico, New Mexico State University; Mahmoud F. Wagdy, California State University-Long Beach; Subbaraya Yuvarajan, North Dakota State University. Richard C. Jaeger, Auburn University also gave input on the issue of PSpice.

I remain grateful to the reviewers of the previous editions: Stanley G. Burns, Iowa State University; Michael M. Cirovic, California Polytechnic State University-San Luis Obispo; J. Alvin Connelly, Georgia Institute of Technology; William J. Eccles, Rose-Hulman Institute of Technology; Amir Farhat, Northeastern University; Ward J. Helms, University of Washington; Frank H. Hielscher, Lehigh University; Richard C. Jaeger, Auburn University; Franco Maddaleno, Politecnico di Torino, Italy; Dragan Maksimovic, University of Colorado-Boulder; and Arthur B. Williams, Coherent Communications Systems Corporation. Finally, I wish to express my gratitude to Diana May, my wife, for her encouragement and steadfast support.

Sergio Franco
San Francisco, California, 2001

1

OPERATIONAL AMPLIFIER
FUNDAMENTALS

1.1 Amplifier Fundamentals
1.2 The Operational Amplifier
1.3 Basic Op Amp Configurations
1.4 Ideal Op Amp Circuit Analysis
1.5 Negative Feedback
1.6 Feedback in Op Amp Circuits
1.7 The Loop Gain
1.8 Op Amp Powering
    Problems
    Bibliography
    Appendix 1A

The term *operational amplifier*, or *op amp* for short, was coined in 1947 by John R. Ragazzini to denote a special type of amplifier that, by proper selection of its external components, could be configured for a variety of operations such as amplification, addition, subtraction, differentiation, and integration. The first applications of op amps were in analog computers. The ability to perform mathematical operations was the result of combining high gain with negative feedback.

Early op amps were implemented with vacuum tubes, so they were bulky, power-hungry, and expensive. The first dramatic miniaturization of the op amp came with the advent of the bipolar junction transistor (BJT), which led to a whole generation of op amp modules implemented with discrete BJTs. However, the real breakthrough occurred with the development of the integrated circuit (IC) op amp, whose elements are fabricated in monolithic form on a silicon chip the size of a pinhead. The first such device was developed by Robert J. Widlar at Fairchild Semiconductor Corporation in the early 1960s. In 1968 Fairchild introduced the op amp that was to become the industry standard, the popular μA741. Since then the number of op amp families and manufacturers has swollen considerably. Nevertheless, the 741 remains one of

1

the most popular types in spite of competition from devices of comparable cost but superior performance. Because of its enduring popularity and the fact that it is the most widely documented op amp in the literature, we shall use it as a vehicle to illustrate general op amp principles and also as a yardstick to assess the relative merits of other op amp families. However, we shall not hesitate to turn to other op amp types if they prove better suited to the application at hand.

Op amps have made lasting inroads into virtually every area of analog and mixed analog-digital electronics. Such widespread use has been aided by dramatic price drops. Today, the cost of an op amp that is purchased in volume quantities can be comparable to that of more traditional and less sophisticated components such as trimmers, quality capacitors, and precision resistors. In fact, the prevailing attitude is to regard the op amp as just another component, a viewpoint that has had a profound impact on the way we think of analog circuits and design them today.

The internal circuit diagram of the 741 op amp is shown in Fig. 5A.2 of the Appendix at the end of Chapter 5. The circuit may be intimidating, especially if your understanding of BJTs is not sufficiently deep. Be reassured, however, that it is possible to design a great number of op amp circuits without a detailed knowledge of the op amp's inner workings. Indeed, in spite of its internal complexity, the op amp lends itself to a black-box representation with a very simple relationship between output and input. We shall see that this simplified schematization is adequate for a great variety of situations. When it is not, we shall turn to the data sheets and predict circuit performance from specified data, again avoiding a detailed consideration of the inner workings.

To promote their products, op amp manufacturers maintain applications departments with the purpose of identifying areas of application for their products and publicizing them by means of application notes and articles in trade journals. You are encouraged to start building your own reference library of linear data books and application notes. Browse through them in your spare time, and you will be amazed by the wealth of information they provide. For your convenience, we maintain an updated list of the major op amp manufacturers. This list can be accessed by visiting the Web site at http://www.mhhe.com/franco.

This study of op amp principles should be corroborated by practical experimentation. You can either assemble your circuits on a protoboard and try them out in the lab, or you can simulate them with a personal computer using any of the various CAD/CAE packages available, such as SPICE. For best results, you may wish to do both.

After reviewing basic amplifier concepts, this chapter introduces the op amp as well as analytical techniques suitable for investigating a variety of basic op amp circuits. Central to the operation of these circuits is the concept of *negative feedback*. In particular, the reader is introduced to the concept of *loop gain* as the most important characteristic of negative-feedback circuits. The chapter concludes with some practical considerations, such as op amp powering, output saturation, and internal power dissipation.

1.1
AMPLIFIER FUNDAMENTALS

Before embarking on the study of the operational amplifier, it is worth reviewing the fundamental concepts of amplification and loading. Recall that an amplifier is a two-port device that accepts an externally applied signal, called *input*, and generates a signal called *output* such that *output* = *gain* × *input*, where *gain* is a suitable proportionality constant. A device conforming to this definition is called a *linear amplifier* to distinguish it from devices with nonlinear input-output relationships, such as quadratic and log/antilog amplifiers. Unless stated to the contrary, the term *amplifier* will here signify *linear amplifier*.

An amplifier receives its input from a *source* upstream and delivers its output to a *load* downstream. Depending on the nature of the input and output signals, we have different amplifier types. The most common is the *voltage amplifier*, whose input v_I and output v_O are voltages. Each port of the amplifier can be modeled with a Thévenin equivalent, consisting of a voltage source and a series resistance. The input port usually plays a purely passive role, so we model it with just a resistance R_i, called the *input resistance* of the amplifier. The output port is modeled with a voltage-controlled voltage source (VCVS) to signify the dependence of v_O on v_I, along with a series resistance R_o called the *output resistance*. The situation is depicted in Fig. 1.1, where A_{oc} is called the *voltage gain factor* and is expressed in volts per volt. Note that the input source is also modeled with a Thévenin equivalent consisting of the source v_S and series resistance R_s; the output load, playing a passive role, is modeled with a mere resistance R_L.

We now wish to derive an expression for v_O in terms of v_S. Applying the voltage divider formula at the output port yields

$$v_O = \frac{R_L}{R_o + R_L} A_{oc} v_I \qquad (1.1)$$

We note that in the absence of any load ($R_L = \infty$) we would have $v_O = A_{oc} v_I$. Hence, A_{oc} is called the *unloaded*, or *open-circuit*, voltage gain. Applying the voltage divider formula at the input port yields

$$v_I = \frac{R_i}{R_s + R_i} v_S \qquad (1.2)$$

Eliminating v_I and rearranging, we obtain the *source-to-load gain*,

$$\frac{v_O}{v_S} = \frac{R_i}{R_s + R_i} A_{oc} \frac{R_L}{R_o + R_L} \qquad (1.3)$$

As the signal progresses from source to load, it undergoes first some attenuation at the input port, then magnification by A_{oc} inside the amplifier, and finally additional attenuation at the output port. These attenuations are referred to as *loading*. It is apparent that because of loading, Eq. (1.3) gives $|v_O/v_S| \leq |A_{oc}|$.

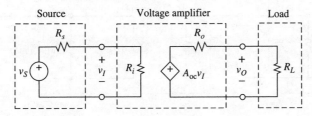

FIGURE 1.1
Voltage amplifier.

EXAMPLE 1.1. (*a*) An amplifier with $R_i = 100$ kΩ, $A_{oc} = 100$ V/V, and $R_o = 1$ Ω is driven by a source with $R_s = 25$ kΩ and drives a load $R_L = 3$ Ω. Calculate the overall gain as well as the amount of input and output loading. (*b*) Repeat, but for a source with $R_s = 50$ kΩ and a load $R_L = 4$ Ω. Compare.

Solution.

(*a*) By Eq. (1.3), the overall gain is $v_O/v_S = [100/(25 + 100)] \times 100 \times 3/(1 + 3) = 0.80 \times 100 \times 0.75 = 60$ V/V, which is less than 100 V/V because of loading. Input loading causes the source voltage to drop to 80% of its unloaded value; output loading introduces an additional drop to 75%.

(*b*) By the same equation, $v_O/v_S = 0.67 \times 100 \times 0.80 = 53.3$ V/V. We now have more loading at the input but less loading at the output. Moreover, the overall gain has changed from 60 V/V to 53.3 V/V.

Loading is generally undesirable because it makes the overall gain dependent on the particular input source and output load, not to mention gain reduction. The origin of loading is obvious: when the amplifier is connected to the input source, R_i draws current and causes R_s to drop some voltage. It is precisely this drop that, once subtracted from v_S, leads to a reduced voltage v_I. Likewise, at the output port the magnitude of v_O is less than the dependent-source voltage $A_{oc}v_I$ because of the voltage drop across R_o.

If loading could be eliminated altogether, we would have $v_O/v_S = A_{oc}$ regardless of the input source and the output load. To achieve this condition, the voltage drops across R_s and R_o must be zero regardless of R_s and R_L. The only way to achieve this is by requiring that our voltage amplifier have $R_i = \infty$ and $R_o = 0$. For obvious reasons such an amplifier is termed *ideal*. Though these conditions cannot be met in practice, an amplifier designer will strive to approximate them as closely as possible by ensuring that $R_i \gg R_s$ and $R_o \ll R_L$ for all input sources and output loads that the amplifier is likely to be connected to.

Another popular amplifier is the *current amplifier*. Since we are now dealing with currents, we model the input source and the amplifier with Norton equivalents, as in Fig. 1.2. The parameter A_{sc} of the current-controlled current source (CCCS) is called the *unloaded*, or *short-circuit, current gain*. Applying the current divider formula twice yields the source-to-load gain,

$$\frac{i_O}{i_S} = \frac{R_s}{R_s + R_i} A_{sc} \frac{R_o}{R_o + R_L} \tag{1.4}$$

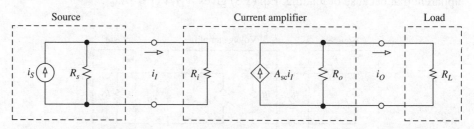

FIGURE 1.2
Current amplifier.

TABLE 1.1
Basic amplifiers and their ideal terminal resistances

Input	Output	Amplifier type	Gain	R_i	R_o
v_I	v_O	Voltage	V/V	∞	0
i_I	i_O	Current	A/A	0	∞
v_I	i_O	Transconductance	A/V	∞	∞
i_I	v_O	Transresistance	V/A	0	0

We again witness loading both at the input port, where part of i_S is lost through R_s, making i_I less than i_S, and at the output port, where part of $A_{sc}i_I$ is lost through R_o. Consequently, we always have $|i_O/i_S| \leq |A_{sc}|$. To eliminate loading, an *ideal* current amplifier has $R_i = 0$ and $R_o = \infty$, exactly the opposite of the ideal voltage amplifier.

An amplifier whose input is a voltage v_I and whose output is a current i_O is called a *transconductance amplifier* because its gain is in amperes per volt, the dimensions of conductance. The situation at the input port is the same as that of the voltage amplifier of Fig. 1.1; the situation at the output port is similar to that of the current amplifier of Fig. 1.2, except that the dependent source is now a voltage-controlled current source (VCCS) of value $A_g v_I$, with A_g in amperes per volt. To avoid loading, an ideal transconductance amplifier has $R_i = \infty$ and $R_o = \infty$.

Finally, an amplifier whose input is a current i_I and whose output is a voltage v_O is called a *transresistance amplifier,* and its gain is in volts per ampere. The input port appears as in Fig. 1.2, and the output port as in Fig. 1.1, except that we now have a current-controlled voltage source (CCVS) of value $A_r i_I$, with A_r in volts per ampere. Ideally, such an amplifier has $R_i = 0$ and $R_o = 0$, the opposite of the transconductance amplifier.

The four basic amplifier types, along with their ideal input and output resistances, are summarized in Table 1.1.

1.2
THE OPERATIONAL AMPLIFIER

The operational amplifier is a voltage amplifier with extremely high gain. For example, the popular 741 op amp has a typical gain of 200,000 V/V, also expressed as 200 V/mV. Gain is also expressed in decibels (dB) as $20 \log_{10} 200,000 = 106$ dB. The OP-77, a more recent type, has a gain of 12 million, or 12 V/μV, or $20 \log_{10}(12 \times 10^6) = 141.6$ dB. In fact, what distinguishes op amps from all other voltage amplifiers is the size of their gain. In the next sections we shall see that the higher the gain the better, or that an op amp would ideally have an infinitely large gain. Why one would want gain to be extremely large, let alone infinite, will become clearer as soon as we start analyzing our first op amp circuits.

Figure 1.3*a* shows the symbol of the op amp and the power-supply connections to make it work. The inputs, identified by the "$-$" and "$+$" symbols, are designated *inverting* and *noninverting*. Their voltages with respect to ground are denoted v_N and v_P, and the output voltage as v_O. The arrowhead signifies signal flow from the inputs to the output.

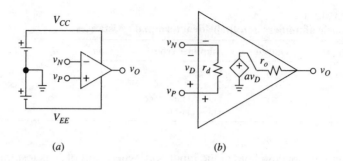

(a) (b)

FIGURE 1.3

(a) Op amp symbol and power-supply connections. (b) Equivalent circuit of a powered op amp. (The 741 op amp has typically $r_d = 2\ \text{M}\Omega$, $a = 200\ \text{V/mV}$, and $r_o = 75\ \Omega$.)

Op amps do not have a 0-V ground terminal. Ground reference is established externally by the power-supply common. The supply voltages are denoted V_{CC} and V_{EE}, and their values are typically ± 15 V, though other values are possible, as we shall see. To minimize cluttering in circuit diagrams, it is customary not to show the power-supply connections. However, when we try out an op amp in the lab, we must remember to apply power to make it function.

Figure 1.3b shows the equivalent circuit of a properly powered op amp. Though the op amp itself does not have a ground pin, the ground symbol inside its equivalent circuit models the power-supply common of Fig. 1.3a. The equivalent circuit includes the *differential* input resistance r_d, the *voltage gain a*, and the *output resistance r_o*. For reasons that will become clear in the next sections, r_d, a, and r_o are referred to as *open-loop* parameters and are symbolized by lowercase letters. The difference

$$v_D = v_P - v_N \tag{1.5}$$

is called the *differential input voltage,* and gain a is also called the *unloaded gain* because in the absence of output loading we have

$$v_O = a v_D = a(v_P - v_N) \tag{1.6}$$

Since both input terminals are allowed to attain independent potentials with respect to ground, the input port is said to be of the *double-ended* type. Contrast this with the output port, which is of the *single-ended* type. Equation (1.6) indicates that the op amp responds only to the difference between its input voltages, not to their individual values. Consequently, op amps are also called *difference amplifiers*.

Reversing Eq. (1.6), we obtain

$$v_D = \frac{v_O}{a} \tag{1.7}$$

which allows us to find the voltage v_D causing a given v_O. We again observe that this equation yields only the difference v_D, not the values of v_N and v_P themselves. Because of the large gain a in the denominator, v_D is bound to be very small. For instance, to sustain $v_O = 6$ V, an unloaded 741 op amp needs $v_D = 6/200{,}000 = 30\ \mu\text{V}$, quite a small voltage. An unloaded OP-77 would need $v_D = 6/(12 \times 10^6) = 0.5\ \mu\text{V}$, an even smaller value!

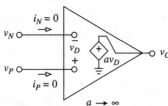

FIGURE 1.4
Ideal op amp model.

The Ideal Op Amp

We know that to minimize loading, a well-designed voltage amplifier must draw negligible (ideally zero) current from the input source and must present negligible (ideally zero) resistance to the output load. Op amps are no exception, so we define the ideal op amp as an ideal voltage amplifier with infinite open-loop gain:

$$a \rightarrow \infty \qquad (1.8a)$$

Its ideal terminal conditions are

$$r_d = \infty \qquad (1.8b)$$

$$r_o = 0 \qquad (1.8c)$$

$$i_P = i_N = 0 \qquad (1.8d)$$

where i_P and i_N are the currents drawn by the noninverting and inverting inputs. The ideal op amp model is shown in Fig. 1.4.

We observe that in the limit $a \rightarrow \infty$ we obtain $v_D \rightarrow v_O/\infty \rightarrow 0$! This result is often a source of puzzlement because it makes one wonder how an amplifier with zero input can sustain a nonzero output. Shouldn't the output also be zero by Eq. (1.6)? The answer lies in the fact that as gain a approaches infinity, v_D does indeed approach zero, but in such a way as to maintain the product av_D nonzero and equal to v_O.

Real-life op amps depart somewhat from the ideal, so the model of Fig. 1.4 is only a conceptualization. But during our initiation into the realm of op amp circuits, we shall use this model because it relieves us from worrying about loading effects so that we can concentrate on the role of the op amp itself. Once we have developed enough understanding and confidence, we shall backtrack and use the more realistic model of Fig. 1.3b to assess the validity of our results. We shall find that the results obtained with the ideal and with the real-life models are in much closer agreement than we might have suspected, corroborating the claim that the ideal model, though a conceptualization, is not that academic after all.

SPICE Simulation

Circuit simulation by computer has become a powerful and indispensable tool in both analysis and design. In this book we shall use the popular program known as PSpice to verify the results of our calculations. As we proceed, we shall develop op

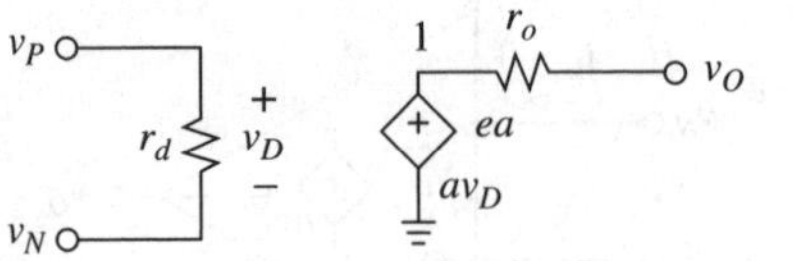

FIGURE 1.5
Simple op amp model for PSpice.

amp models of increasing complexity. We begin with the basic model depicted in Fig. 1.5. The following code reflects typical 741 op amp parameters at dc:

```
*Simple op amp model
.subckt OA vP vN vO
rd vP vN 2Meg        ;input resistance
ea 1 0 vP vN 200k    ;gain
ro 1 vO 75           ;output resistance
.ends OA
```

If a pseudo-ideal model is desired, then r_d is left open, r_o is shorted out, and the source value is increased from 200 kV/V to some huge value, say, 1 GV/V. (However, the reader is cautioned that too large a value may cause convergence problems.)

1.3
BASIC OP AMP CONFIGURATIONS

By connecting external components around an op amp, we obtain what we shall henceforth refer to as an *op amp circuit*. It is crucial that you understand the difference between an op amp circuit and a plain op amp. Think of the latter as a component of the former, just as the external components are. The most basic op amp circuits are the *inverting, noninverting,* and *buffer amplifiers.*

The Noninverting Amplifier

The circuit of Fig. 1.6a consists of an op amp and two external resistors. To understand its function, we need to find a relationship between v_O and v_I. To this end we redraw it as in Fig. 1.6b, where the op amp has been replaced by its equivalent model and the resistive network has been rearranged to emphasize its role in the circuit. We can find v_O via Eq. (1.6); however, we must first derive expressions for v_P and v_N. By inspection,

$$v_P = v_I \tag{1.9}$$

Using the voltage divider formula yields $v_N = [R_1/(R_1 + R_2)]v_O$, or

$$v_N = \frac{1}{1 + R_2/R_1}v_O \tag{1.10}$$

The voltage v_N represents the fraction of v_O that is being fed back to the inverting input. Consequently, the function of the resistive network is to create *negative*

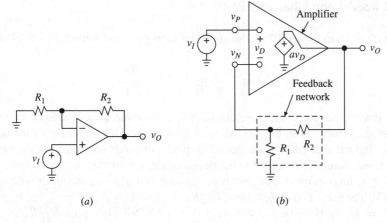

(a) (b)

FIGURE 1.6
Noninverting amplifier and circuit model for its analysis.

feedback around the op amp. Letting $v_O = a(v_P - v_N)$, we get

$$v_O = a\left(v_I - \frac{1}{1 + R_2/R_1}v_O\right) \qquad (1.11)$$

Collecting terms and solving for the ratio v_O/v_I, which we shall designate as A, yields, after minor rearrangement,

$$A = \frac{v_O}{v_I} = \left(1 + \frac{R_2}{R_1}\right)\frac{1}{1 + (1 + R_2/R_1)/a} \qquad (1.12)$$

This result reveals that the circuit of Fig. 1.6a, consisting of an op amp plus a resistor pair, is itself an amplifier, and that its gain is A. Since A is positive, the polarity of v_O is the same as that of v_I—hence the name *noninverting amplifier.*

The gain A of the op amp circuit and the gain a of the basic op amp are quite different. This is not surprising, as the two amplifiers, while sharing the same output v_O, have different inputs, namely, v_I for the former and v_D for the latter. To underscore this difference, a is referred to as the *open-loop gain,* and A as the *closed-loop gain,* the latter designation stemming from the fact that the op amp circuit contains a loop. In fact, starting from the inverting input in Fig. 1.6b, we can trace a clockwise loop through the op amp and then through the resistive network, which brings us back to the starting point.

EXAMPLE 1.2. In the circuit of Fig. 1.6a, let $v_I = 1$ V, $R_1 = 2$ kΩ, and $R_2 = 18$ kΩ. Find v_O if (a) $a = 10^2$ V/V, (b) $a = 10^4$ V/V, (c) $a = 10^6$ V/V. Comment on your findings.

Solution. Equation (1.12) gives $v_O/1 = (1 + 18/2)/(1 + 10/a)$, or $v_O = 10/(1 + 10/a)$. So

(a) $v_O = 10/(1 + 10/10^2) = 9.091$ V,
(b) $v_O = 9.990$ V,
(c) $v_O = 9.9999$ V.

The higher the gain a, the closer v_O is to 10.0 V.

Ideal Closed-Loop Characteristics

Letting $a \to \infty$ in Eq. (1.12) yields a closed-loop gain that we refer to as ideal:

$$A_{\text{ideal}} = \lim_{a \to \infty} A = 1 + \frac{R_2}{R_1} \qquad (1.13)$$

In this limit A becomes independent of a and its value is set exclusively by the *external resistance ratio*, R_2/R_1. We can now appreciate the reason for wanting $a \to \infty$. Indeed, a circuit whose closed-loop gain depends only on a resistance ratio offers tremendous advantages for the designer since it makes it easy to tailor gain to the application at hand. For instance, suppose you need an amplifier with a gain of 2 V/V. Then, by Eq. (1.13), pick $R_2/R_1 = A - 1 = 2 - 1 = 1$; for example, pick $R_1 = R_2 = 100$ kΩ. Do you want $A = 10$ V/V? Then pick $R_2/R_1 = 9$; for example, $R_1 = 20$ kΩ and $R_2 = 180$ kΩ. Do you want an amplifier with variable gain? Then make R_1 or R_2 variable by means of a potentiometer (pot). For example, if R_1 is a fixed 10-kΩ resistor and R_2 is a 100-kΩ pot configured as a variable resistance from 0 Ω to 100 kΩ, then Eq. (1.13) indicates that the gain can be varied over the range 1 V/V $\leq A \leq$ 11 V/V. No wonder it is desirable that $a \to \infty$. It leads to the simpler expression of Eq. (1.13) and it makes op amp circuit design a real snap!

Another advantage of Eq. (1.13) is that gain A can be made as accurate and stable as needed by using resistors of suitable quality. Actually it is not even necessary that the individual resistors be of high quality; it only suffices that their ratio be so. For example, using two resistances that track each other with temperature so as to maintain a constant ratio will make gain A temperature-independent. Contrast this with gain a that depends on the characteristics of the resistors, diodes, and transistors inside the op amp, and is therefore sensitive to thermal drift, aging, and production variations. This is a prime example of one of the most fascinating aspects of electronics, namely, the ability to implement high-performance circuits using inferior components!

The advantages afforded by Eq. (1.13) do not come for free. The price is the size of gain a needed to make this equation acceptable within a given degree of accuracy (more on this will follow). It is often said that we are in effect throwing away a good deal of open-loop gain for the sake of stabilizing the closed-loop gain. Considering the benefits, the price is well worth paying, especially with IC technology, which, in mass production, makes it possible to achieve high open-loop gains at extremely low cost.

Since the op amp circuit of Fig. 1.6 has proven to be an amplifier itself, besides gain A it must also present input and output resistances, which we shall designate as R_i and R_o and call the *closed-loop input* and *output resistances*. You may have noticed that to keep the distinction between the parameters of the basic op amp and those of the op amp circuit, we are using lowercase letters for the former and uppercase letters for the latter.

Though we shall have more to say about R_i and R_o from the viewpoint of negative feedback in Section 1.6, we presently use the simplified model of Fig. 1.6b to state that $R_i = \infty$ because the noninverting input terminal appears as an open circuit, and $R_o = 0$ because the output comes directly from the source av_D.

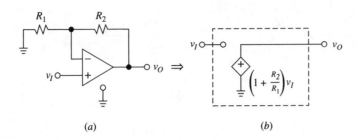

(a)   (b)

FIGURE 1.7
Noninverting amplifier and its ideal equivalent circuit.

In summary,

$$R_i = \infty \qquad R_o = 0 \qquad\qquad (1.14)$$

which, according to Table 1.1, represent the ideal terminal charactistics of a voltage amplifier. The equivalent circuit of the ideal noninverting amplifier is shown in Fig. 1.7.

The Voltage Follower

Letting $R_1 = \infty$ and $R_2 = 0$ in the noninverting amplifier turns it into the *unity-gain amplifier*, or *voltage follower*, of Fig. 1.8a. Note that the circuit consists of the op amp and a wire to feed the entire output back to the input. The closed-loop parameters are

$$A = 1 \text{ V/V} \qquad R_i = \infty \qquad R_o = 0 \qquad\qquad (1.15)$$

and the equivalent circuit is shown in Fig. 1.8b. As a voltage amplifier, the follower is not much of an achiever since its gain is only unity. Its specialty, however, is to act as a *resistance transformer*, since looking into its input we see an open circuit, but looking into its output we see a short circuit to a source of value $v_O = v_I$.

To appreciate this feature, consider a source v_S whose voltage we wish to apply across a load R_L. If the source were ideal, all we would need would be a plain wire to connect the two. However, if the source has nonzero output resistance R_s, as in Fig. 1.9a, then R_s and R_L will form a voltage divider and the magnitude of v_L will be less than that of v_S because of the voltage drop across R_s. Let us now replace

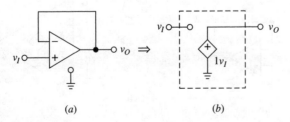

(a)   (b)

FIGURE 1.8
Voltage follower and its ideal equivalent circuit.

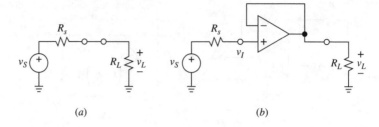

FIGURE 1.9
Source and load connected (*a*) directly, and (*b*) via a voltage follower to eliminate loading.

the wire by a voltage follower as in Fig. 1.9*b*. Since the follower has $R_i = \infty$, there is no loading at the input, so $v_I = v_S$. Moreover, since the follower has $R_o = 0$, loading is absent also from the output, so $v_L = v_I = v_S$, indicating that R_L now receives the full source voltage with no losses. The role of the follower is thus to act as a *buffer* between source and load.

We also observe that now the source delivers no current and hence no power, while in the circuit of Fig. 1.9*a* it did. The current and power drawn by R_L are now supplied by the op amp, which in turn takes them from its power supplies, not explicitly shown in the figure. Thus, besides restoring v_L to the full value of v_S, the follower relieves the source v_S from supplying any power. The need for a buffer arises so often in electronic design that special circuits are available whose performance has been optimized for this function. The BUF-03 (Analog Devices) is a popular example.

The Inverting Amplifier

Together with the noninverting amplifier, the inverting configuration of Fig. 1.10*a* constitutes a cornerstone of op amp applications. The inverting amplifier was invented before the noninverting amplifier because in their early days op amps had only one input, namely, the inverting one. Referring to the equivalent circuit of Fig. 1.10*b*,

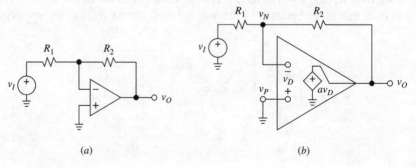

FIGURE 1.10
Inverting amplifier and circuit model for its analysis.

we have

$$v_P = 0 \tag{1.16}$$

Applying the superposition principle yields $v_N = [R_2/(R_1 + R_2)]v_I + [R_1/(R_1 + R_2)]v_O$, or

$$v_N = \frac{1}{1 + R_1/R_2}v_I + \frac{1}{1 + R_2/R_1}v_O \tag{1.17}$$

Letting $v_O = a(v_P - v_N)$ yields

$$v_O = a\left(-\frac{1}{1 + R_1/R_2}v_I - \frac{1}{1 + R_2/R_1}v_O\right) \tag{1.18}$$

Comparing with Eq. (1.11), we observe that the resistive network still feeds the portion $1/(1 + R_2/R_1)$ of v_O back to the inverting input, thus providing the same amount of negative feedback. Solving for the ratio v_O/v_I and rearranging, we obtain

$$A = \frac{v_O}{v_I} = \left(-\frac{R_2}{R_1}\right)\frac{1}{1 + (1 + R_2/R_1)/a} \tag{1.19}$$

Our circuit is again an *amplifier.* However, the gain A is now *negative,* indicating that the polarity of v_O will be opposite to that of v_I. This is not surprising, because we are now applying v_I to the inverting side of the op amp. Hence, the circuit is called an *inverting amplifier.* If the input is a sine wave, the circuit will introduce a *phase reversal,* or, equivalently, a *180° phase shift.*

Ideal Closed-Loop Characteristics

Letting $a \to \infty$ in Eq. (1.19), we obtain

$$A_{\text{ideal}} = \lim_{a \to \infty} A = -\frac{R_2}{R_1} \tag{1.20}$$

That is, the closed-loop gain again depends only on an external resistance ratio, yielding well-known advantages for the circuit designer. For instance, if we need an amplifier with a gain of -5 V/V, we pick two resistances in a $5\!:\!1$ ratio, such as $R_1 = 20$ kΩ and $R_2 = 100$ kΩ. If, on the other hand, R_1 is a fixed 20-kΩ resistor and R_2 is a 100-kΩ pot configured as a variable resistance, then the closed-loop gain can be varied anywhere over the range -5 V/V $\leq A \leq 0$. Note in particular that the magnitude of A can now be controlled all the way down to zero.

We now turn to the task of determining the closed-loop input and output resistances R_i and R_o. Since $v_D = v_O/a$ is vanishingly small because of the large size of a, it follows that v_N is very close to v_P, which is zero. In fact, in the limit $a \to \infty$, v_N would be zero exactly, and would be referred to as *virtual ground* because to an outside observer things appear as if the inverting input were permanently grounded. We conclude that the effective resistance seen by the input source is just R_1. Moreover, since the output comes directly from the source av_D, we have $R_o = 0$. In summary,

$$R_i = R_1 \qquad R_o = 0 \tag{1.21}$$

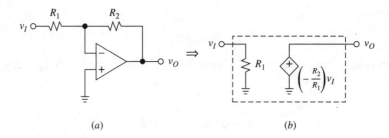

(a) (b)

FIGURE 1.11
Inverting amplifier and its ideal equivalent circuit.

The equivalent circuit of the inverting amplifier is shown in Fig. 1.11.

EXAMPLE 1.3. Use PSpice to verify the values of R_i, A, and R_o in the circuit of Fig. 1.11 if $R_1 = 10\,\text{k}\Omega$ and $R_2 = 100\,\text{k}\Omega$.

Solution. To obtain the transfer and terminal characteristics, we use the .tf statement. The PSpice input file is as follows.

```
Inverting amplifier
vi 1 0 ac 1V
R1 1 2 10k
R2 2 3 100k
eoa 3 0 2 0 1G
.tf v(3) vi
.end
```

After running PSpice, we get the following output-file printout:

```
v(3)/vi = -1.000E+01
Input resistance at vi = 1.000E+04
Output resistance at v(3) = 0.000E+00
```

This confirms that $R_i = 10\,\text{k}\Omega$, $A = -10\,\text{V/V}$, and $R_o = 0$.

Unlike its noninverting counterpart, the inverting amplifier will load down the input source if the source is nonideal. This is depicted in Fig. 1.12. Since in the limit $a \to \infty$, the op amp keeps $v_N \to 0\,\text{V}$ (virtual ground), we can apply the voltage divider formula and write

$$v_I = \frac{R_1}{R_s + R_1} v_S \tag{1.22}$$

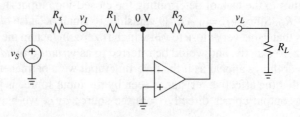

FIGURE 1.12
Input loading by the inverting amplifier.

indicating that $|v_I| \leq |v_S|$. Applying Eq. (1.20), $v_L/v_I = -R_2/R_1$. Eliminating v_I, we obtain

$$\frac{v_L}{v_S} = -\frac{R_2}{R_s + R_1} \qquad (1.23)$$

Because of loading at the input, the magnitude of the overall gain, $R_2/(R_s + R_1)$, is less than that of the amplifier alone, R_2/R_1. The amount of loading depends on the relative magnitudes of R_s and R_1, and only if $R_s \ll R_1$ can loading be ignored.

We can look at the above circuit also from another viewpoint. Namely, to find the gain v_L/v_S, we can still apply Eq. (1.20), provided, however, that we regard R_s and R_1 as a *single* resistance of value $R_s + R_1$. Thus, $v_L/v_S = -R_2/(R_s + R_1)$, the same as above.

1.4
IDEAL OP AMP CIRCUIT ANALYSIS

Considering the simplicity of the ideal closed-loop results of the previous section, we wonder whether there is not a simpler technique to derive them, bypassing some of the tedious algebra. Such a technique exists and is based on the fact that when the op amp is operated with negative feedback, in the limit $a \to \infty$ its input voltage $v_D = v_O/a$ approaches zero,

$$\lim_{a\to\infty} v_D = 0 \qquad (1.24)$$

or, since $v_N = v_P - v_D = v_P - v_O/a$, v_N approaches v_P,

$$\lim_{a\to\infty} v_N = v_P \qquad (1.25)$$

This property, referred to as the *input voltage constraint,* makes the input terminals appear as if they were shorted together, though in fact they are not. We also know that an ideal op amp draws no current at its input terminals, so this apparent short carries no current, a property referred to as the *input current constraint.* In other words, for voltage purposes the input port appears as a short circuit, but for current purposes it appears as an open circuit! Hence the designation *virtual short.* Summarizing, *when operated with negative feedback, an ideal op amp will output whatever voltage and current it takes to drive v_D to zero or, equivalently, to force v_N to track v_P, but without drawing any current at either input terminal.*

Note that it is v_N that follows v_P, not the other way around. The op amp controls v_N via the external feedback network. Without feedback, the op amp would be unable to influence v_N and the above equations would no longer hold.

To better understand the action of the op amp, consider the simple circuit of Fig. 1.13a, where we have, by inspection, $i = 0$, $v_1 = 0$, $v_2 = 6$ V, and $v_3 = 6$ V. If we now connect an op amp as in Fig. 1.13b, what will happen? As we know, the op amp will drive v_3 to whatever it takes to make $v_2 = v_1$. To find these voltages, we equate the current entering the 6-V source to that exiting it; or

$$\frac{0 - v_1}{10} = \frac{(v_1 + 6) - v_2}{30}$$

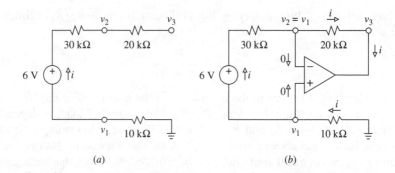

FIGURE 1.13
The effect of an op amp in a circuit.

Letting $v_2 = v_1$ and solving yields $v_1 = -2$ V. The current is

$$i = \frac{0 - v_1}{10} = \frac{2}{10} = 0.2 \text{ mA}$$

and the output voltage is

$$v_3 = v_2 - 20i = -2 - 20 \times 0.2 = -6 \text{ V}$$

Summarizing, as the op amp is inserted in the circuit, it swings v_3 from 6 V to -6 V because this is the voltage that makes $v_2 = v_1$. Consequently, v_1 is changed from 0 V to -2 V, and v_2 from 6 V to -2 V. The op amp also sinks a current of 0.2 mA at its output terminal, but without drawing any current at either input.

The Basic Amplifiers Revisited

It is instructive to derive the noninverting and inverting amplifier gains using the concept of the *virtual short*. In the circuit of Fig. 1.14a we exploit this concept to label the inverting-input voltage as v_I. Applying the voltage divider formula, we have $v_I = v_O/(1 + R_2/R_1)$, which is readily turned around to yield the familiar relationship $v_O = (1 + R_2/R_1)v_I$. In words, the noninverting amplifier provides the *inverse* function of the voltage divider: the divider attenuates v_O to yield v_I, whereas the amplifier magnifies v_I by the inverse amount to yield v_O. This action can be visualized via the lever analog depicted above the amplifier in the figure. The lever pivots around a point corresponding to ground. The lever segments correspond to resistances, and the swings correspond to voltages.

In the circuit of Fig. 1.14b we again exploit the virtual-short concept to label the inverting input as a virtual ground, or 0 V. Applying KCL, we have $(v_I - 0)/R_1 = (0 - v_O)/R_2$, which is readily solved for v_O to yield the familiar relationship $v_O = (-R_2/R_1)v_I$. This can be visualized via the mechanical analog shown above the amplifier. An upswing (downswing) at the input produces a downswing (upswing) at the output. By contrast, in Fig. 1.14a the output swings in the same direction as the input.

So far, we have only studied the basic op amp configurations. It is time to familiarize ourselves with other op amp circuits. These we shall study using the virtual-short concept.

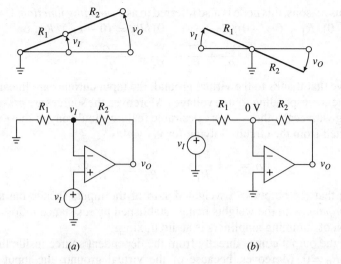

(a)   *(b)*

FIGURE 1.14
Mechanical analogies of the noninverting and the inverting amplifiers.

The Summing Amplifier

The summing amplifier has two or more inputs and one output. Though the example of Fig. 1.15 has three inputs, v_1, v_2, and v_3, the following analysis can readily be generalized to an arbitrary number of them. To obtain a relationship between output and inputs, we impose that the total current entering the virtual-ground node equal that exiting it, or

$$i_1 + i_2 + i_3 = i_F$$

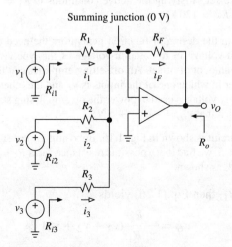

FIGURE 1.15
Summing amplifier.

For obvious reasons, this node is also referred to as a *summing junction*. Using Ohm's law, $(v_1 - 0)/R_1 + (v_2 - 0)/R_2 + (v_3 - 0)/R_3 = (0 - v_O)/R_F$, or

$$\frac{v_1}{R_1} + \frac{v_2}{R_2} + \frac{v_3}{R_3} = -\frac{v_O}{R_F}$$

We observe that thanks to the virtual ground, the input currents are linearly proportional to the corresponding source voltages. Moreover, the sources are prevented from interacting with each other—a very desirable feature should any of these sources be disconnected from the circuit. Solving for v_O yields

$$v_O = -\left(\frac{R_F}{R_1}v_1 + \frac{R_F}{R_2}v_2 + \frac{R_F}{R_3}v_3\right) \tag{1.26}$$

indicating that the output is a weighted sum of the inputs (hence the name *summing amplifier*), with the weights being established by resistance ratios. A popular application of summing amplifiers is audio mixing.

Since the output comes directly from the dependent source inside the op amp, we have $R_o = 0$. Moreover, because of the virtual ground, the input resistance R_{ik} ($k = 1, 2, 3$) seen by source v_k equals the corresponding resistance R_k. In summary,

$$R_{ik} = R_k \qquad k = 1, 2, 3$$
$$R_o = 0 \tag{1.27}$$

If the input sources are nonideal, the circuit will load them down, as in the case of the inverting amplifier. Equation (1.26) is still applicable provided we replace R_k by $R_{sk} + R_k$ in the denominators, where R_{sk} is the output resistance of the kth input source.

EXAMPLE 1.4. Using standard 5% resistances, design a circuit such that $v_O = -2(3v_1 + 4v_2 + 2v_3)$.

Solution. By Eq. (1.26) we have $R_F/R_1 = 6$, $R_F/R_2 = 8$, $R_F/R_3 = 4$. One possible standard resistance set satisfying the above conditions is $R_1 = 20\ \text{k}\Omega$, $R_2 = 15\ \text{k}\Omega$, $R_3 = 30\ \text{k}\Omega$, and $R_F = 120\ \text{k}\Omega$.

EXAMPLE 1.5. In the design of function generators the need arises to *offset* as well as *amplify* a given voltage v_I to obtain a voltage of the type $v_O = Av_I + V_O$, where V_O is the desired amount of offset. An offsetting amplifier can be implemented with a summing amplifier in which one of the inputs is v_I and the other is either V_{CC} or V_{EE}, the regulated supply voltages used to power the op amp. Using standard 5% resistances, design a circuit such that $v_O = -10v_I + 5$ V.

Solution. The circuit is shown in Fig. 1.16. Imposing $v_O = -(R_F/R_1)v_I - (R_F/R_2) (-15) = -10v_I + 5$, we find that a possible resistance set is $R_1 = 10\ \text{k}\Omega$, $R_2 = 300\ \text{k}\Omega$, and $R_F = 100\ \text{k}\Omega$, as shown.

If $R_3 = R_2 = R_1$, then Eq. (1.26) yields

$$v_O = -\frac{R_F}{R_1}(v_1 + v_2 + v_3) \tag{1.28}$$

that is, v_O is proportional to the *true sum* of the inputs. The proportionality constant $-R_F/R_1$ can be varied all the way down to zero by implementing R_F with a variable

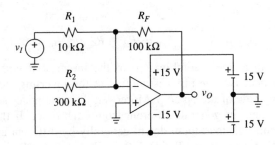

FIGURE 1.16
A dc-offsetting amplifier.

resistance. If all resistances are equal, the circuit yields the (inverted) sum of its inputs, $v_O = -(v_1 + v_2 + v_3)$.

The Difference Amplifier

As shown in Fig. 1.17, the difference amplifier has one output and two inputs, one of which is applied to the inverting side, the other to the noninverting side. We can find v_O via the superposition principle as $v_O = v_{O1} + v_{O2}$, where v_{O1} is the value of v_O with v_2 set to zero, and v_{O2} that with v_1 set to zero.

Letting $v_2 = 0$ yields $v_P = 0$, making the circuit act as an inverting amplifier with respect to v_1. So $v_{O1} = -(R_2/R_1)v_1$ and $R_{i1} = R_1$, where R_{i1} is the input resistance seen by the source v_1.

Letting $v_1 = 0$ makes the circuit act as a noninverting amplifier with respect to v_P. So $v_{O2} = (1 + R_2/R_1)v_P = (1 + R_2/R_1) \times [R_4/(R_3 + R_4)]v_2$ and $R_{i2} = R_3 + R_4$, where R_{i2} is the input resistance seen by the source v_2. Letting $v_O = v_{O1} + v_{O2}$ and rearranging yields

$$v_O = \frac{R_2}{R_1} \left(\frac{1 + R_1/R_2}{1 + R_3/R_4} v_2 - v_1 \right) \qquad (1.29)$$

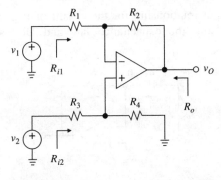

FIGURE 1.17
Difference amplifier.

Moreover,

$$R_{i1} = R_1 \qquad R_{i2} = R_3 + R_4 \qquad R_o = 0 \qquad (1.30)$$

The output is again a linear combination of the inputs, but with coefficients of opposite polarity because one input is applied to the inverting side and the other to the noninverting side of the op amp. Moreover, the resistances seen by the input sources are finite and, in general, different from each other. If these sources are nonideal, the circuit will load them down, generally by different amounts. Let the sources have output resistances R_{s1} and R_{s2}. Then Eq. (1.29) is still applicable provided we replace R_1 by $R_{s1} + R_1$ and R_3 by $R_{s2} + R_3$.

> **EXAMPLE 1.6.** Design a circuit such that $v_O = v_2 - 3v_1$ and $R_{i1} = R_{i2} = 100$ kΩ.
>
> **Solution.** By Eq. (1.30) we must have $R_1 = R_{i1} = 100$ kΩ. By Eq. (1.29) we must have $R_2/R_1 = 3$, so $R_2 = 300$ kΩ. By Eq. (1.30) $R_3 + R_4 = R_{i2} = 100$ kΩ. By Eq. (1.29), $3[(1 + 1/3)/(1 + R_3/R_4)] = 1$. Solving the last two equations for their two unknowns yields $R_3 = 75$ kΩ and $R_4 = 25$ kΩ.

An interesting case arises when the resistance pairs in Fig. 1.17 are in equal ratios:

$$\frac{R_3}{R_4} = \frac{R_1}{R_2} \qquad (1.31)$$

When this condition is met, the resistances are said to form a *balanced bridge*, and Eq. (1.29) simplifies to

$$v_O = \frac{R_2}{R_1}(v_2 - v_1) \qquad (1.32)$$

The output is now proportional to the *true difference* of the inputs—hence the name of the circuit. A popular application of the true difference amplifier is as a building block of instrumentation amplifiers, to be studied in the next chapter.

The Differentiator

To find the input-output relationship for the circuit of Fig. 1.18, we start out by imposing $i_C = i_R$. Using the capacitance law and Ohm's law, this becomes

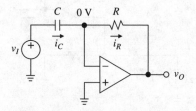

FIGURE 1.18
The op amp differentiator.

$Cd(v_I - 0)/dt = (0 - v_O)/R$, or

$$v_O(t) = -RC\frac{dv_I(t)}{dt} \qquad (1.33)$$

The circuit yields an output that is proportional to the *time derivative* of the input—hence the name. The proportionality constant is set by R and C, and its units are seconds (s).

If you try out the differentiator circuit in the lab, you will find that it tends to oscillate. Its stability problems stem from the open-loop gain rolloff with frequency, an issue that will be addressed in Chapter 8. Suffice it to say here that the circuit is usually stabilized by placing a suitable resistance R_s in series with C. After this modification the circuit will still provide the differentiation function, but only over a limited frequency range.

The Integrator

The analysis of the circuit of Fig. 1.19 mirrors that of Fig. 1.18. Imposing $i_R = i_C$, we now get $(v_I - 0)/R = C\,d(0 - v_O)/dt$, or $dv_O(t) = (-1/RC)v_I(t)\,dt$. Changing t to the dummy integration variable ξ and then integrating both sides from 0 to t yields

$$v_O(t) = -\frac{1}{RC}\int_0^t v_I(\xi)\,d\xi + v_O(0) \qquad (1.34)$$

where $v_O(0)$ is the value of the output at $t = 0$. This value depends on the charge initially stored in the capacitor. Equation (1.34) indicates that the output is proportional to the *time integral* of the input—hence the name. The proportionality constant is set by R and C, but its units are now s^{-1}. Mirroring the analysis of the inverting amplifier, you can readily verify that

$$R_i = R \qquad R_o = 0 \qquad (1.35)$$

Thus, if the driving source has an output resistance R_s, in order to apply Eq. (1.34) we must replace R with $R_s + R$.

The op amp integrator, also called a *precision integrator* because of the high degree of accuracy with which it can implement Eq. (1.34), is a workhorse of electronics. It finds wide application in function generators (triangle and sawtooth wave generators), active filters (state-variable and biquad filters, switched-capacitor

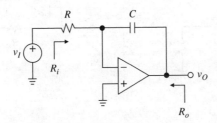

FIGURE 1.19
The op amp integrator.

filters), analog-to-digital converters (dual-slope converters, quantized-feedback converters), and analog controllers (PID controllers).

If $v_I(t) = 0$, Eq. (1.34) predicts that $v_O(t) = v_O(0) = $ constant. In practice, when the integrator circuit is tried out in the lab, it is found that its output will drift until it saturates at a value close to one of the supply voltages, even with v_I grounded. This is due to the so-called input offset error of the op amp, an issue to be discussed in Chapter 5. Suffice it to say here that a crude method of preventing saturation is to place a suitable resistance R_p in parallel with C. The resulting circuit, called a *lossy integrator,* will still provide the integration function, but only over a limited frequency range. Fortunately, in most applications integrators are placed inside a control loop designed to automatically keep the circuit away from saturation, at least under proper operating conditions, thus eliminating the need for the aforementioned parallel resistance.

The Negative-Resistance Converter (NIC)

We conclude by demonstrating another important op amp application beside signal processing, namely, *impedance transformation.* To illustrate, consider the plain resistance of Fig. 1.20a. To find its value experimentally, we apply a test source v, we measure the current i *out* of the source's positive terminal, and then we let $R_{eq} = v/i$, where R_{eq} is the value of the resistance as seen by the source. Clearly, in this simple case $R_{eq} = R$. Moreover, the test source releases power and the resistance absorbs power.

Suppose we now lift the lower terminal of R off ground and drive it with a noninverting amplifier with the input tied to the other terminal of R, as shown in Fig. 1.20b. The current is now $i = [v - (1 + R_2/R_1)v]/R = -R_2v/(R_1 R)$. Letting $R_{eq} = v/i$ yields

$$R_{eq} = -\frac{R_1}{R_2}R \qquad (1.36)$$

indicating that the circuit simulates a *negative resistance.* The meaning of the negative sign is that current is now actually flowing *into* the test source's positive

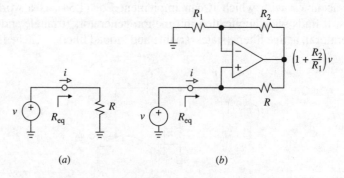

(a)   (b)

FIGURE 1.20
(a) Positive resistance: $R_{eq} = R$. (b) Negative-resistance converter: $R_{eq} = -(R_1/R_2)R$.

terminal, causing the source to absorb power. Consequently, a negative resistance *releases* power.

If $R_1 = R_2$, then $R_{eq} = -R$. In this case the test voltage v is amplified to $2v$ by the op amp, making R experience a net voltage v, positive at the right. Consequently $i = -v/R = v/(-R)$.

Negative resistances can be used to neutralize unwanted ordinary resistances, as in the design of current sources, or to control pole location, as in the design of active filters and oscillators.

Looking back at the circuits covered so far, note that by interconnecting suitable components around a high-gain amplifier we can configure it for a variety of *operations:* multiplication by a constant, summation, subtraction, differentiation, integration, and resistance conversion. This explains why it is called *operational!*

1.5
NEGATIVE FEEDBACK

Section 1.3 informally introduced the concept of negative feedback. Since the majority of op amp circuits employ this type of feedback, we shall now discuss it in a more systematic fashion.

Figure 1.21 shows the basic structure of a negative-feedback circuit. The arrows indicate signal flow, and the generic symbol x stands for either a voltage or a current signal. Besides the source and load, we identify the following basic blocks:

1. An *amplifier,* also called *error amplifier* in control theory, which accepts the signal x_d and yields the *output signal*

$$x_o = ax_d \qquad (1.37)$$

where a, the forward gain of the amplifier, is called the *open-loop gain* of the circuit.
2. A *feedback network,* which samples x_o and produces the *feedback signal*

$$x_f = \beta x_o \qquad (1.38)$$

where β, the gain of the feedback network, is called the *feedback factor* of the circuit.
3. A *summing network,* denoted Σ, which generates the difference

$$x_d = x_i - x_f \qquad (1.39)$$

also called the *error signal.* The designation *negative feedback* stems from the

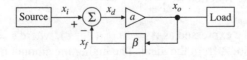

FIGURE 1.21
Block diagram of a negative-feedback system.

fact that we are in effect *feeding* a portion of x_o *back* to the input, where it is then *subtracted* from x_i to yield the reduced signal x_d. Were it added instead, the feedback would be positive. For reasons that will become clearer as we proceed, negative feedback is also said to be *degenerative*, and positive feedback *regenerative*.

Eliminating x_f and x_d from the above equations and solving for the ratio $A = x_o/x_i$ yields

$$A = \frac{a}{1 + a\beta} \tag{1.40}$$

where A is called the *closed-loop gain* of the circuit. Note that for feedback to be negative we must have $a\beta > 0$. Consequently, A will be smaller than a by the amount $1 + a\beta$, which is aptly called the *amount of feedback*.

As a signal propagates around the loop made up of the amplifier, feedback network, and summer, it experiences an overall gain of $a \times \beta \times (-1)$, or $-a\beta$. Its negative is called the *return ratio* or, more commonly, the *loop gain*,

$$T = a\beta \tag{1.41}$$

and it allows us to express Eq. (1.40) as $A = (1/\beta)T/(1 + T)$. Letting $T \to \infty$ yields the ideal situation

$$A_{\text{ideal}} = \lim_{T \to \infty} A = \frac{1}{\beta} \tag{1.42}$$

that is, A becomes independent of a and is set exclusively by the feedback network. By proper choice of the topology of this network as well as the quality of its components, we can tailor the circuit to a variety of different applications. For instance, specifying $0 < \beta < 1$ will cause x_o to be a magnified replica of x_i since $1/\beta > 1$. Conversely, implementing the feedback network with reactive elements such as capacitors will yield a frequency-selective circuit with the transfer function $H(jf) = 1/\beta(jf)$; filters and oscillators belong to this class of circuits.

Henceforth we shall express the closed-loop gain in the insightful form

$$A = A_{\text{ideal}} \times \frac{1}{1 + 1/T} \tag{1.43}$$

If we define

$$\text{Error function} = \frac{1}{1 + 1/T} = 1 - \frac{1}{1 + T} = 1 - \epsilon \tag{1.44}$$

then Eq. (1.43) can be expressed as $A = A_{\text{ideal}}(1 - \epsilon)$, where $\epsilon = 1/(1 + T)$ is the *fractional deviation* of A from the ideal. The higher the amount of feedback $1 + T$, the smaller is the fractional error ϵ, and the closer the error function is to unity. The *gain error* is the percentage deviation of A from the ideal. For $T \gg 1$ we have

$$\text{Gain error (\%)} = 100\frac{A - A_{\text{ideal}}}{A_{\text{ideal}}} \cong -\frac{100}{T} \tag{1.45}$$

EXAMPLE 1.7. (a) Find the loop gain needed to approximate A_{ideal} within 0.1%. (b) Find the open-loop gain needed to achieve $A = 100$ with the above degree of accuracy. (c) With the value of a found in (b), find β so that $A = 100$ exactly.

Solution.

(a) By Eq. (1.45), we want $100/T \leq 0.1$. Consequently, we need $T \geq 10^3$.
(b) For $A_{ideal} = 1/\beta = 100$ we need $\beta = 10^{-2}$. Then $a\beta \geq 10^3$ implies $a \geq 10^3/10^{-2} = 10^5$.
(c) Imposing $100 = 10^5/(1 + 10^5\beta)$ yields $\beta = 9.99 \times 10^{-3}$.

This example evidences the price for a tight closed-loop accuracy, namely, the need to start out with $a \gg A$. As we close the loop around the error amplifier, we are in effect throwing away a good deal of open-loop gain, namely, the amount $1 + T$. It is also evident that for a given a, *the smaller the closed-loop gain A, the smaller its percentage deviation from the ideal.*

It is instructive to examine the effect of negative feedback also on the signals x_d and x_f. We have $x_d = x_o/a = Ax_i/a = (A/a)x_i$, or

$$x_d = \frac{1}{1+T}x_i \tag{1.46}$$

Moreover, $x_f = \beta x_o = \beta Ax_i$, or

$$x_f = \frac{1}{1+1/T}x_i \tag{1.47}$$

As $T \to \infty$, the error signal x_d will approach zero, and the feedback signal x_f will track the input signal x_i. This is the familiar virtual short introduced in the previous section.

The most direct implementation of the feedback structure of Fig. 1.21 is the familiar noninverting amplifier. As shown in Fig. 1.22, the feedback signal is the inverting-input voltage $v_N = \beta v_O$, where $\beta = R_1/(R_1 + R_2)$. Moreover, since the op amp is a difference amplifier, the operation of subtraction of v_N from v_I is performed implicitly by the op amp itself.

Gain Desensitivity

We wish to investigate how variations in the open-loop gain a affect the closed-loop gain A. Differentiating Eq. (1.40) with respect to a and simplifying yields

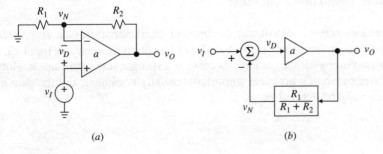

(a)   (b)

FIGURE 1.22
The noninverting amplifier as a negative-feedback system.

$dA/da = 1/(1 + a\beta)^2$. Since $1 + a\beta = a/A$, we can write, after rearranging,

$$\frac{dA}{A} = \frac{1}{1+T}\frac{da}{a} \tag{1.48}$$

Replacing differentials with finite increments and multiplying both sides by 100, we can approximate

$$100\frac{\Delta A}{A} \cong \frac{1}{1+T}\left(100\frac{\Delta a}{a}\right) \tag{1.49}$$

indicating that the effect of a given percentage change in a upon A is reduced by $1 + T$. For T sufficiently large, even a substantial change in a will cause an insignificant change in A. It is apparent that negative feedback desensitizes gain, and this is the reason why $1 + T$ is also called the *desensitivity factor*. The stabilization of A is highly desirable because the open-loop gain a of a practical amplifier is ill-defined due to process variations, thermal drift, aging, and power-supply variations.

Evaluating $dA/d\beta$ and proceeding in a similar fashion, we find that for T sufficiently large,

$$100\frac{\Delta A}{A} \cong -100\frac{\Delta\beta}{\beta} \tag{1.50}$$

A given increase (or decrease) in β will cause A to decrease (or increase) by the same amount, indicating that negative feedback does not stabilize A against variations in β. Hence the need to implement the feedback network with components of adequate quality and tracking capabilities.

EXAMPLE 1.8. A negative-feedback circuit has $a = 10^5$ and $\beta = 10^{-3}$. (a) Estimate the percentage change in A brought about by a $\pm10\%$ change in a. (b) Repeat if $\beta = 1$.

Solution.

(a) The desensitivity factor is $1 + T = 1 + 10^5 \times 10^{-3} = 101$. Thus, a $\pm10\%$ change in a will cause a percentage change in A 101 times as small; that is, A changes by $\pm10/101 \cong \pm0.1\%$.

(b) Now the desensitivity increases to $1 + 10^5 \times 1 \cong 10^5$. The percentage change in A is now $\pm10/10^5 = 0.0001\%$, or one part per million (1 ppm). We note that for a given a, the lower the value of A, the higher the desensitivity because $1 + T = a/A$.

Nonlinear Distortion Reduction

A convenient way of visualizing the transfer characteristic of an amplifier is by means of its *transfer curve,* that is, the plot of its output x_o versus its input x_d. Since a linear amplifier yields $x_o = ax_d$, its curve is a straight line with slope a. However, the transfer curve of a practical amplifier is usually nonlinear, and the gain a must more generally be defined as

$$a = \frac{dx_o}{dx_d} \tag{1.51}$$

The top curve in Fig. 1.23a is the voltage transfer curve (VTC) of an amplifier having the characteristic

$$v_O = V_o \tanh \frac{v_D}{V_d} \qquad (1.52)$$

where V_d and V_o are suitable input and output scaling voltages. In the present case $V_d = 10^{-4}$ V and $V_o = 10$ V. The curve is approximately linear near the origin, but as the operating point is moved toward the periphery, the slope decreases until the curve eventually flattens out and saturates at $\pm V_o = \pm 10$ V. As shown in the bottom curve in Fig. 1.23a, the slope, or gain a, is maximum at the origin, decreases away from the origin, and finally drops to zero in saturation. A nonlinear curve yields a distorted output even if the peak values are kept below the saturation limits. For instance, applying a sinusoidal input will result in a pseudosinusoidal output with the top and bottom flattened out because of the decreased gain away from the origin.

Consider now the effect of applying negative feedback around such an amplifier. In light of Eq. (1.42) we expect that as long as a is sufficiently large to make $T \gg 1$, A will be fairly constant and close to $1/\beta$ in spite of the decrease of a away from the origin. This is also confirmed by Eq. (1.49). Figure 1.23b shows the effect of applying feedback with $\beta = 0.1$ V/V. The closed-loop curve is much more linear than the open-loop curve, and it is so over a wider signal range. Of course, as we approach saturation, where a drops to zero, the linearizing effect of negative

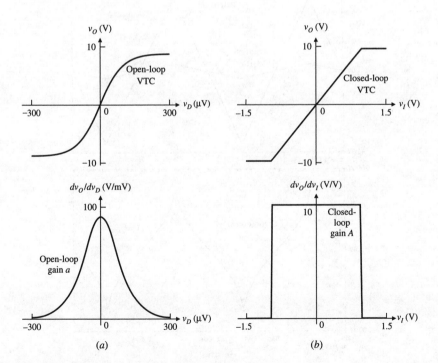

(a)   (b)

FIGURE 1.23
The linearizing effect of negative feedback.

feedback no longer applies because of the lack of loop gain there; hence, *A* itself drops to zero.

It is instructive to carry out a computer simulation to visualize the various waveforms in the circuit. Using the noninverting configuration of Fig. 1.22*a*, we write the following PSpice file:

```
Waveforms with a nonlinear open-loop VTC
*vO = 10 tanh (10,000*vD)
vI 1 0 pulse (-0.9V 0.9V -0.25ms 0.5ms 0.5ms 1us 1ms)
rd 1 2 1Meg              ;input resistance
R1 0 2 2k               ;beta = R1/(R1+R2)
R2 2 3 18k              ;beta = 0.1
eOA 3 0 value = {10*((exp(2E4*v(1,2))-1)/(exp(2E4*v(1,2))+1))}
.tran 100us 2ms 0ms 100us
.probe                  ;vI=v(1), vO=v(3), vD=v(1,2)
.end
```

The PSpice waveforms of Fig. 1.24 indicate that v_O is a magnified and fairly undistorted replica of v_I, as expected. Note, however, how distorted v_D is! It is the error amplifier itself that predistorts v_D, via the feedback network, in order to compensate for its own distorted VTC and thus yield an undistorted output.

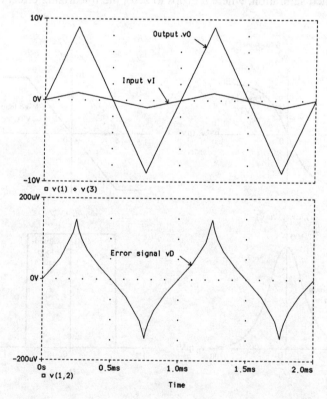

FIGURE 1.24
Effect of a nonlinear VTC on signal waveforms.

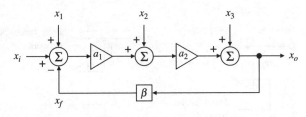

FIGURE 1.25
Investigating the effect of negative feedback on disturbances and noise.

Effect of Feedback on Disturbances and Noise

Negative feedback provides a means also for reducing circuit sensitivity to certain
types of disturbances. Figure 1.25 illustrates three types of disturbances: x_1, entering
the circuit at the input, might represent unwanted signals such as input offset errors
and input noise, both of which will be covered in later chapters; x_2, entering the
circuit at some intermediate point, might represent power-supply hum; x_3, entering
the circuit at the output, might represent output load changes.

To accommodate x_2, we break the amplifier into two stages with individual gains
a_1 and a_2. The overall forward gain is then $a = a_1 \times a_2$. The output is found as
$x_o = x_3 + a_2[x_2 + a_1(x_i - \beta x_o + x_1)]$, or

$$x_o = \frac{a_1 a_2}{1 + a_1 a_2 \beta} \left(x_i + x_1 + \frac{x_2}{a_1} + \frac{x_3}{a_1 a_2} \right) \tag{1.53}$$

We observe that x_1 undergoes no attenuation relative to x_i. However, x_2 and x_3
are attenuated by the forward gains from the input to the points of entry of the
disturbances themselves. This feature is widely exploited in the design of audio
amplifiers. The output stage of such an amplifier is a power stage that is usually
afflicted by an intolerable amount of hum. Preceding such a stage by a high-gain,
low-noise preamplifier stage and then closing a proper feedback loop around the
composite amplifier reduces hum by the first-stage gain.

For $a_1 a_2 \beta \gg 1$, Eq. (1.53) simplifies to $x_o = (1/\beta)(x_i + x_1 + x_2/a_1 + x_3/a_1 a_2)$.
The quantity $1/\beta$ is aptly called the *noise gain* because this is the gain with which
the circuit amplifies the input noise x_1.

1.6
FEEDBACK IN OP AMP CIRCUITS

We now wish to relate the concepts of the previous section to circuits based on op
amps. Figure 1.26 shows typical topologies for input summing and output sampling.
As we proceed, we shall make frequent references to these basic topologies.

In Fig. 1.26*a* we are summing voltages; since voltages are combined in series with each other, this is referred to as an *input-series* topology. By contrast, in
Fig. 1.26*b* we are summing currents, and this is an *input-shunt* topology. As a rule

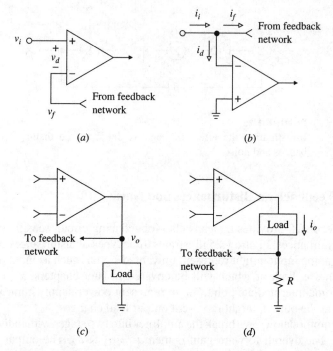

FIGURE 1.26
Negative-feedback topologies: (*a*) series at the input; (*b*) shunt
at the input; (*c*) shunt at the output; and (*d*) series at the output.

of thumb, if the input and the feedback signals enter the amplifier at different nodes, the input topology is of the series type; if they enter at the same node, it is of the shunt type.

In Fig. 1.26*c* we are sampling the load voltage, an operation that is performed in parallel, or shunt; hence this is an *output-shunt* topology. In Fig. 1.26*d* we are using a series resistance R to sample the load current; hence this is an *output-series* topology. As a rule of thumb, if we short (open) the output load and we still witness some feedback signal at the input, then we are not sampling a voltage (current).

Using intuitive arguments, we expect negative feedback to alter not only the gain but also the input and output resistances. Referring to Fig. 1.26*a*, we know that the op amp tends to reduce v_d. The current v_d/r_d drawn from the input source will thus be small, indicating that the input-series topology raises the input resistance. By contrast, the input-shunt topology of Fig. 1.26*b* lowers the input resistance because the voltage at the summing junction is forced to closely track the noninverting input voltage, which in this case is ground.

Turning next to the voltage-sampling topology of Fig. 1.26*c* we observe that a disturbance in the form of a load-current change will have a reduced effect on the output voltage, implying that the output-shunt topology lowers the output resistance. Conversely, the output-series topology raises the output resistance because a load-voltage change will have a reduced effect on the output current. In summary, whether at the input or at the output port, a *series* topology *raises* and a *shunt* topology *lowers* the corresponding port resistance. We will soon find that the amount of increase or decrease is given by the amount of feedback itself!

To get a feel for the effects of negative feedback, let us investigate the basic inverting and noninverting configurations, which are the workhorses of op amp applications. Specifically, let us derive expressions for the closed-loop parameters R_i, A, and R_o, but using the full-fledged op amp model of Fig. 1.3b; then let us compare the results with those of the ideal op amp of Section 1.4. This kind of approach will be applied to other circuits as well in subsequent chapters.

The Noninverting Configuration

Comparing Fig. 1.27 with Fig. 1.26a and c, we observe that the noninverting amplifier is of the input-series, output-shunt type, or *series-shunt* for short. To find its closed-loop gain, we sum currents at the nodes labeled v_N and v_O:

$$\frac{v_I - v_N}{r_d} - \frac{v_N}{R_1} + \frac{v_O - v_N}{R_2} = 0$$

$$\frac{v_N - v_O}{R_2} + \frac{a(v_I - v_N) - v_O}{r_o} = 0$$

where we have used $v_D = v_I - v_N$. Eliminating v_N and solving for the ratio $A = v_O/v_I$ yields

$$A = \frac{(1 + R_2/R_1)a + r_o/r_d}{1 + a + R_2/R_1 + (R_2 + r_o)/r_d + r_o/R_1} \tag{1.54}$$

In a well-designed amplifier the ratios r_o/r_d, $(R_2 + r_o)/r_d$, and r_o/R_1 are negligible compared to $1 + a$, so we can simplify as

$$A \cong \left(1 + \frac{R_2}{R_1}\right) \frac{1}{1 + 1/T} \tag{1.55}$$

where $T = a\beta$ is the loop gain and

$$\beta = \frac{R_1}{R_1 + R_2} \tag{1.56}$$

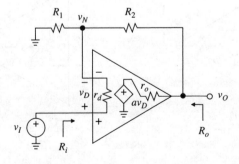

FIGURE 1.27
The noninverting configuration.

is the feedback factor. The quantity

$$\lim_{a \to 0} A = \frac{r_o/r_d}{1 + R_2/R_1 + (R_2 + r_o)/r_d + r_o/R_1} \tag{1.57}$$

is called the *feedthrough gain* because it refers to signal transmission from input to output via the feedback network. This unwanted term is negligible and will be ignored in future analysis.

To find R_i we apply a test voltage v as in Fig. 1.28a, we find the current i out of the test source's positive terminal, and then we let $R_i = v/i$. Summing currents at the node labeled v_N yields

$$\frac{v - v_N}{r_d} - \frac{v_N}{R_1} + \frac{a(v - v_N) - v_N}{R_2 + r_o} = 0$$

Letting $v_N = v - v_D = v - r_d i$, collecting terms, and solving for the ratio v/i yields

$$R_i = r_d \left(1 + \frac{a}{1 + (R_2 + r_o)/R_1} \right) + R_1 \| (R_2 + r_o) \tag{1.58}$$

For a sufficiently large gain a we can ignore the last term. Moreover, in a well-designed circuit we usually have $r_o \ll R_2$. Therefore, $R_i \cong r_d[1 + a/(1/\beta)]$, or

$$R_i \cong r_d(1 + T) \tag{1.59}$$

To find R_o, we suppress the input source v_I and again apply the test-voltage technique. With reference to Fig. 1.28b, we have, by the voltage divider formula,

$$v_N = \frac{R_1 \| r_d}{R_1 \| r_d + R_2} v$$

Summing currents at the output node,

$$i + \frac{v_N - v}{R_2} + \frac{-a v_N - v}{r_o} = 0$$

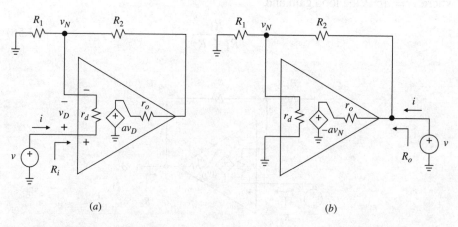

(a)   (b)

FIGURE 1.28
Finding R_i and R_o for the noninverting configuration.

Eliminating v_N and solving for the ratio $R_o = v/i$, we obtain

$$R_o = \frac{r_o}{1 + (a + r_o/R_1 + r_o/r_d)/(1 + R_2/R_1 + R_2/r_d)} \qquad (1.60)$$

Typically r_d is in the megohm range or greater, R_1 and R_2 are in the kilohm range, and r_o is on the order of 10^2 Ω. The terms r_o/R_1, r_o/r_d, and R_2/r_d can thus be ignored to yield

$$R_o \cong \frac{r_o}{1 + T} \qquad (1.61)$$

Looking back at Eqs. (1.55), (1.59), and (1.61), we observe that negative feedback, besides desensitizing gain by the factor $1 + T$, raises r_d and lowers r_o by the same factor. We find these features extremely helpful in our attempt to approximate the ideal voltage-amplifier terminal conditions of Table 1.1.

EXAMPLE 1.9. Let the op amp of Fig. 1.27 be the 741, for which $r_d = 2$ MΩ, $r_o = 75$ Ω, and $a = 200$ V/mV. Find the *exact*, the *approximated*, and the *ideal* values of A, R_i, and R_o if (a) $R_1 = 1$ kΩ and $R_2 = 999$ kΩ; (b) $R_1 = \infty$ and $R_2 = 0$. Confirm with PSpice.

Solution.

(a) Substituting the given parameter values into Eq. (1.54) gives $A = 995.022$ V/V; using Eq. (1.55) with $T = 200$ gives $A = 995.024$ V/V; moreover, $A_{ideal} = 1000$ V/V. Proceeding in similar fashion, we find $R_i = 401.97$ MΩ, 402.00 MΩ, ∞; $R_o = 373.32$ mΩ, 373.13 mΩ, 0 Ω.

(b) We now have $T = 200,000$. Because of this much larger value, we simply ignore the exact calculations and use only the approximations. We thus find $A = 0.999995$ V/V, 1 V/V; $R_i = 400$ GΩ, ∞; $R_o = 375\mu\Omega$, 0.

Using the subcircuit OA appearing in the PSpice code at the end of Section 1.2, we write the following circuit file for part (a):

```
Noninverting amplifier with A = 1 V/mV
vi 1 0 ac 1V          ;input source
R1 0 2 1k             ;resistance @ input
R2 2 3 999k           ;feedback resistance
X 1 2 3 OA            ;activates the op amp
.tf v(3) vi           ;xfer-function analysis
.end
```

The PSpice simulation yields

```
v(3)/vi = 9.950E+02
Input resistance at vi = 4.020E+08
Output resistance at v(3) = 3.733E-01
```

This confirms the results of hand calculations. A similar simulation confirms the results of part (b).

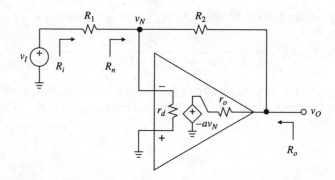

FIGURE 1.29
The inverting configuration.

The Inverting Configuration

To find the gain of the inverting configuration of Fig. 1.29, we proceed as in the noninverting case. Summing currents at the nodes labeled v_N and v_O, eliminating v_N, and then solving for the ratio $A = v_O/v_I$ yields

$$A = -\frac{aR_2 - r_o}{(1+a)R_1 + (R_2 + r_o)(1 + R_1/r_d)} \tag{1.62}$$

In a well-designed circuit we usually have $r_o \ll R_2$ and $R_1/r_d \ll 1$. Consequently, we can simplify as

$$A \cong \left(-\frac{R_2}{R_1}\right)\frac{1}{1 + 1/T} \tag{1.63}$$

where T, the loop gain, is given by

$$T = \frac{aR_1}{R_1 + R_2} \tag{1.64}$$

We observe that T is the same as in the noninverting configuration. However, the expressions for A_{ideal} are different, since the inputs are applied at different points of the same circuit. The *feedthrough gain* is now

$$\lim_{a \to 0} A = \frac{r_o}{R_1 + (R_2 + r_o)(1 + R_1/r_d)} \tag{1.65}$$

Though not as small as in the noninverting case, this gain will also be ignored in future calculations.

To find R_i, we first determine the equivalent resistance R_n of the inverting input. Then we let $R_i = R_1 + R_n$. To this end we apply a test current i, as in Fig. 1.30, we find the resulting voltage v, and then we let $R_n = v/i$. Comparing with Fig. 1.26b and c we observe that this is a *shunt-shunt* topology. Summing currents at node v and then solving for the ratio $R_n = v/i$, we get

$$R_n = \frac{R_2 + r_o}{1 + a + (R_2 + r_o)/r_d} \tag{1.66}$$

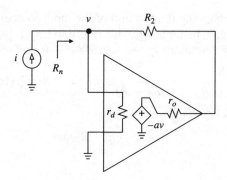

FIGURE 1.30
Finding the virtual-ground resistance R_n.

Ignoring the term $(R_2 + r_o)/r_d$ compared to a, we approximate R_n as

$$R_n \cong \frac{R_2 + r_o}{1 + a} \tag{1.67a}$$

For $r_o \ll R_2$, this simplifies further as

$$R_n \cong \frac{R_2}{1 + a} \tag{1.67b}$$

indicating that a negative-feedback amplifier's feedback resistance R_2 is divided by $(1 + a)$ when reflected to the input. This transformation is referred to as the *Miller effect,* and it holds also in the more general case in which the feedback element is an impedance. Because of the large gain a, we expect $R_n \ll R_1$. In fact, in the limit $a \to \infty$ we would get $R_n \to 0$, the condition for a *perfect virtual ground,* as we already know. Summarizing,

$$R_i = R_1 + R_n \cong R_1 \tag{1.68}$$

To find R_o in the circuit of Fig. 1.29 we suppress the input source v_I and apply a test voltage v at the output, ending up again with the situation of Fig. 1.28b. Consequently,

$$R_o \cong \frac{r_o}{1 + T} \tag{1.69}$$

where T is as given in Eq. (1.64).

EXAMPLE 1.10. Let the op amp of Fig. 1.29 be the 741 type. Find A, R_n, R_i, and R_o if (a) $R_1 = R_2 = 100\,\text{k}\Omega$; (b) $R_1 = 1\,\text{k}\Omega$ and $R_2 = 1\,\text{M}\Omega$.

Solution.

(a) $T = a\beta = 200{,}000 \times 100/(100 + 100) = 10^5$; $A = -1/(1 + 1/10^5) = -0.99999$ V/V; $R_n = (10^5 + 75)/(1 + 10^5) \cong 1\ \Omega$; $R_i = 10^5 + 1 \cong 100$ kΩ; $R_o = 75/(1 + 10^5) \cong 0.75$ mΩ.

(b) $T = 200{,}000/1001 = 199.8$; $A = -995.0$ V/V; $R_n \cong 5\ \Omega$; $R_i = 10^3 + 5 = 1.005$ kΩ; $R_o \cong 0.374\ \Omega$.

The reader is encouraged to verify the above results using PSpice.

It is intriguing that the expressions for A and R_i for the inverting and noninverting configurations are so different, even though one configuration can be derived from the other simply by changing the location of the input source; yet, those for T and R_o are the same. To gain better insight, we return to Eq. (1.18), which was derived under the simplifying conditions $r_d \to \infty$ and $r_o \to 0$, and rewrite as

$$v_o = -a\left(\frac{R_2}{R_1 + R_2}v_I + \frac{R_1}{R_1 + R_2}v_O\right) = -(R_1//R_2)a\left(\frac{v_I}{R_1} + \frac{v_O}{R_2}\right)$$

Letting $v_I/R_1 \to i_I$, $v_O/R_2 \to -i_F$, and

$$a_r = -(R_1//R_2)a$$

allows us to write

$$v_O = a_r\,(i_I - i_F)$$

confirming an input-shunt topology of the type of Fig. 1.26b. Even though the op amp is a voltage-type amplifier with gain a in V/V, when used in the inverting configuration it functions as a transresistance amplifier with gain a_r in V/A. Moreover, rewriting as $v_O = a_r(i_I - \beta_g v_O)$, with

$$\beta_g = -\frac{1}{R_2}$$

confirms a feedback factor β_g in A/V. The loop gain is $T = a_r \beta_g$, or

$$T = \frac{aR_1}{R_1 + R_2}$$

in agreement with Eq. (1.64). The *closed-loop transresistance gain*, defined as $A_r = v_O/i_I$, is, by Eqs. (1.42) and (1.43),

$$A_r = \frac{1}{\beta_g} \times \frac{1}{1 + 1/T} = -R_2 \times \frac{1}{1 + 1/T}$$

Finally, the *closed-loop voltage gain*, defined as $A = v_O/v_I$, is found as $A = [v_O/(v_I/R_1)]/R_1 = A_r/R_1$, or

$$A = -\frac{R_2}{R_1} \times \frac{1}{1 + 1/T}$$

in agreement with Eq. (1.63). Summarizing, we can state that the inverting amplifier, though commonly applied as a voltage-in, voltage-out circuit, when analyzed as a negative-feedback system is more properly treated as a current-in, voltage-out circuit, thus confirming the designation *shunt-shunt configuration*.

Concluding Remarks

The above examples confirm that A, R_i, R_o, and R_n are remarkably close to ideal. For a given value of a, the lower the closed-loop gain, the closer the results are to ideal. Even with closed-loop gains on the order of 10^3 V/V, which is about the upper limit of practical interest, the deviation from ideal is still quite small, at least for the value of a used in the examples. It seems therefore reasonable to assume *ideal* closed-loop parameters even if the open-loop parameters are those of a *nonideal* op amp, especially in view of the simplicity of the ideal closed-loop expressions and of the virtual-short concept. This is also justified by the fact that in a great many practical situations, accuracies within a few percentage points are adequate. Even in precision applications, where small deviations may matter, it is always convenient to start with the ideal op amp model in order to gain a quick, albeit approximate, understanding of what the circuit is supposed to do, and then refine the analysis in the course of a second pass. We shall see many examples of this.

Once again we reiterate that the benefits of negative feedback stem from the availability of a sufficiently large-loop gain T. Put another way, if you had to choose between an op amp with poor r_d and r_o but excellent a, and one with excellent r_d and r_o but poor a, go for the former! The large size of a will make up for its poor r_d and r_o specifications (see Problem 1.53).

1.7
THE LOOP GAIN

By now it is apparent that the loop gain T plays a central role in negative-feedback theory. The larger T is, the closer to ideal the closed-loop parameters are. In Chapter 8 we shall see that T also determines whether a circuit is stable as opposed to oscillatory.

As we know, the gain of an op amp circuit is generally found as

$$A = A_{\text{ideal}} \times \frac{1}{1 + 1/T} \tag{1.70}$$

where A_{ideal} is calculated using the ideal op amp model and, hence, the virtual-short technique. Moreover, the closed-loop terminal resistances are generally found as

$$R \cong r \times (1 + T)^{\pm 1} \tag{1.71}$$

where r is the open-loop resistance calculated in the limit $a \to 0$, and we use $+1$ for a series topology, -1 for a shunt topology.

Finding the Loop Gain T Directly

We can find T directly by suppressing all input sources, breaking the loop at some convenient point, and injecting a *test signal* v_T. As this signal propagates around the loop, it comes back as the *return signal* $v_R = a \times \beta \times (-1) \times v_T$, so $T = a\beta$ is found as

$$T = -\frac{v_R}{v_T}\bigg|_{x_I=0} \tag{1.72}$$

where we use the generic symbol x_I to denote the input source (or sources, in the case of multiple-input circuits such as summing and difference amplifiers). The procedure is illustrated in Fig. 1.31, where for completeness we have included also an output load R_L. This circuit could pertain to both the inverting and the noninverting configurations, as they are indistinguishable once the external source has been suppressed. In fact, the previous section has revealed that T depends only on the amplifier and its feedback network, regardless of where we apply the input signal. Breaking the loop right at the dependent source's output, as shown, yields the convenient result

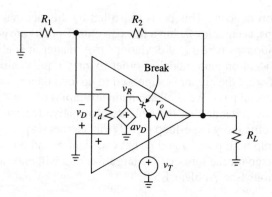

FIGURE 1.31
Determining the loop gain directly as
$T = -av_D/v_T$.

$v_R = av_D$. Using the voltage divider formula twice, we get

$$v_R = a\left(-\frac{R_1 \| r_d}{R_1 \| r_d + R_2} \times \frac{(R_1 \| r_d + R_2) \| R_L}{(R_1 \| r_d + R_2) \| R_L + r_o}\right)v_T$$

Expanding and then using Eq. (1.72) yields

$$T = a\left(\frac{1}{1 + R_2/R_1 + R_2/r_d} \times \frac{1}{1 + r_o/(R_1 \| r_d + R_2) + r_o/R_L}\right) \quad (1.73)$$

Note that for r_o sufficiently small the last term tends to unity, and for r_d sufficiently large the ratio R_2/r_d can be ignored, thus yielding the familiar result $T = a/(1 + R_2/R_1)$.

EXAMPLE 1.11. In Fig. 1.32*a* let $R_1 = R_2 = 1\,\text{M}\Omega$, $R_3 = 100\,\text{k}\Omega$, $R_4 = 1\,\text{k}\Omega$, and $R_L = 2\,\text{k}\Omega$. (*a*) Find the ideal gain A. (*b*) Find the actual gain if the op amp has $r_d = 1\,\text{M}\Omega$, $a = 10^5$ V/V, and $r_o = 100\,\Omega$. What is its percentage departure from the ideal?

Solution.

(*a*) If the op amp were ideal, we would have $v_N = 0$ and $v_1 = -(R_2/R_1)v_I$. Summing currents at node v_1 yields $-v_1/R_2 - v_1/R_4 + (v_O - v_1)/R_3$. Eliminating v_1 and solving for v_O/v_I,

$$A_{\text{ideal}} = -\frac{R_2}{R_1}\left(1 + \frac{R_3}{R_2} + \frac{R_3}{R_4}\right)$$

Substituting the given component values yields $A_{\text{ideal}} = -101.1$ V/V.

(*b*) Find T using the equivalent circuit of Fig. 1.32*b*. Let $R_A = R_1 \| r_d = 500\,\text{k}\Omega$, $R_B = R_A + R_2 = 1.5\,\text{M}\Omega$, $R_C = R_B \| R_4 \cong 1\,\text{k}\Omega$, $R_D = R_C + R_3 \cong 101\,\text{k}\Omega$, $R_E = R_D \| R_L = 1.961\,\text{k}\Omega$, and $R_F = R_E + r_o = 2.061\,\text{k}\Omega$. Applying the voltage divider formula repeatedly, we get $-v_D = (R_A/R_B)v_1 = v_1/3$, $v_1 = (R_C/R_D)v_O = v_O/101$, $v_O = (R_E/R_F)v_T = v_T/1.051$. Thus, $v_R = av_D = -10^5 v_T/(3 \times 101 \times 1.051) = -314v_T$. So $T = -v_R/v_T = 314$, and $A = -101.1/(1+1/T) = -100.8$ V/V. By Eq. (1.45), the deviation from the ideal is -0.32%.

The reader may find it instructive to verify the above results using PSpice. This is a practical circuit for realizing a large inverting gain while using a relatively large resistance R_1 to ensure high input resistance.

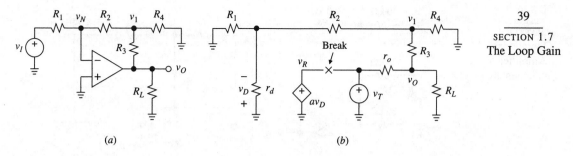

FIGURE 1.32
(a) Amplifier of Example 1.11; (b) circuit for finding its loop gain T.

Finding the Feedback Factor β

An alternative approach is to focus on the feedback circuitry to find the amount β of *voltage feedback* around the op amp, consistent with the fact that the op amp is a voltage-type amplifier, and then combine with data-sheet information about the *voltage gain a* to obtain the loop gain as $T = a\beta$. We shall follow this approach extensively when studying stability, in Chapter 8. To find β, we suppress all input sources, we disconnect the op amp, and we replace it with its terminal resistances r_d and r_o to retain the same loading conditions. Then we apply a test source v_T via r_o, we find the difference v_D across r_d, and we finally let

$$\beta = -\frac{v_D}{v_T}\bigg|_{x_I=0} \qquad (1.74)$$

This is illustrated in Fig. 1.33 for the circuit of Fig. 1.31. Using the voltage divider formula twice, we get

$$\beta = -\frac{v_D}{v_T} = \frac{v_N}{v_T} = \frac{R_1 \parallel r_d}{R_1 \parallel r_d + R_2} \times \frac{(R_1 \parallel r_d + R_2) \parallel R_L}{(R_1 \parallel r_d + R_2) \parallel R_L + r_o}$$

which is readily rearranged as

$$\beta = \frac{1}{1 + R_2/R_1 + R_2/r_d} \times \frac{1}{1 + r_o/(R_1 \parallel r_d + R_2) + r_o/R_L} \qquad (1.75)$$

in agreement with Eq. (1.73). This expression accounts for loading both of the output port by the feedback network and of the feedback network itself by the input port. Only in the limits $r_d \to \infty$ and $r_o \to 0$ does it tend to the simplified form $\beta = R_1/(R_1 + R_2) = 1/(1 + R_2/R_1)$ of Eq. (1.56).

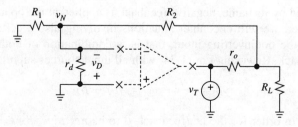

FIGURE 1.33
Finding the feedback factor β ($\times$ denotes a cut).

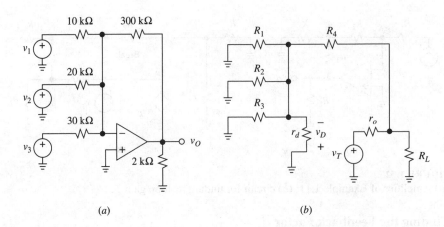

(a)   (b)

FIGURE 1.34
Summing amplifier and equivalent circuit for finding its feedback factor β.

Except for special cases such as heavy capacitive loading at the output, the external circuitry in a well-designed amplifier will cause negligible voltage loss across r_o. At the risk of a small error, we will often let $r_o \cong 0$ to simplify our calculations. This will yield slightly overestimated values for β and T.

EXAMPLE 1.12. Let the op amp of Fig. 1.34a have $r_d = 1$ MΩ, $a = 10^4$ V/V, and $r_o = 100$ Ω. (a) Find β and T. (b) Find the ideal as well as the actual transfer characteristic of the circuit.

Solution.

(a) After suppressing all input sources, replacing the op amp with its terminal resistances, and applying a test voltage v_T, we get the circuit of Fig. 1.34b. Let $R_A = R_1 \parallel R_2 \parallel R_3 \parallel r_d = 10 \parallel 20 \parallel 30 \parallel 1000 = 5.425$ kΩ, $R_B = R_A + R_4 = 305.4$ kΩ, $R_C = R_B \parallel R_L = 1.987$ kΩ, and $R_D = R_C + r_o = 2.087$ kΩ. Then $v_N = (R_A/R_B)v_O = v_O/56.23$, $v_O = (R_C/R_D)v_T = v_T/1.050$, and $\beta = v_N/v_T = 1/(56.23 \times 1.050) = 1/59.13$ V/V. The loop gain is $T = a\beta = 10^4/59.13 = 169.1$.

(b) Ideally, $v_O = -(300/10)v_1 - (300/20)v_2 - (300/30)v_3$, or

$$v_O = -(30v_1 + 15v_2 + 10v_3)$$

To find the actual characteristic, each coefficient must be multiplied by $1/(1 + 1/T) = 1/(1 + 1/169.1) = 0.9941$. Thus,

$$v_O = -(29.82v_1 + 14.91v_2 + 9.941v_3)$$

As implied by its name, negative feedback is applied at the op amp's inverting input. However, we will encounter situations involving also a certain amount of feedback via the noninverting input, that is, a combination of both negative and positive feedback. Rewriting Eq. (1.74) with all input sources suppressed as

$$\beta = \frac{v_N}{v_T} - \frac{v_P}{v_T} = \beta_N - \beta_P \tag{1.76}$$

indicates that in order for the net feedback β to be *negative*, $\beta_N \ (= v_N/v_T)$ must prevail over $\beta_P \ (= v_P/v_T)$. We shall see in Chapter 9 that if β_P prevails over β_N,

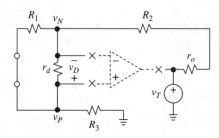

FIGURE 1.35
Finding β for the circuit of Fig. 1.13b.

then feedback is of the *positive type*, something that forces the op amp into saturation and causes it to operate as a Schmitt trigger. Unless stated to the contrary, henceforth we shall assume feedback to be always negative.

EXAMPLE 1.13. Find β in the circuit of Fig. 1.13b if $r_d = 100$ kΩ and $r_o = 100\ \Omega$.

Solution. After the necessary modifications, we end up with the circuit of Fig. 1.35. Applying the voltage divider formula twice, we obtain

$$\beta_N = \frac{(R_1//r_d) + R_3}{r_o + R_2 + (R_1//r_d) + R_3} = 0.622 \qquad \beta_P = \frac{R_3}{r_o + R_2 + (R_1//r_d) + R_3} = 0.188$$

so that $\beta = 0.622 - 0.188 = 0.434$ V/V. Since the amount (0.622) of negative feedback is greater than the amount (0.188) of positive feedback, the net feedback around the op amp is negative.

In Chapters 6 and 8 we shall have much more to say about the loop gain T.

1.8
OP AMP POWERING

In order to function, op amps need to be externally powered. Powering serves the twofold purpose of biasing the internal transistors and providing the power that the op amp must in turn supply to the output load and the feedback network. Figure 1.36 shows a recommended way of powering op amps. To prevent the ac noise usually present on the supply lines from interfering with the op amps, the supply pins of each IC must be bypassed to ground by means of low-inductance capacitors (0.1-μF ceramic capacitors are usually adequate). These decoupling capacitors also help neutralize any spurious feedback loops arising from the nonzero impedances of the supply and ground lines, or busses, which might pose stability problems. For this cure to be effective, the leads must be kept short to minimize their distributed inductance, which rises at the rate of about 1 nH/mm, and the capacitors must be mounted as close as possible to the op amp pins. A well-constructed circuit board will also include 10-μF polarized capacitors at the points of entry of the supply voltages to provide board-level bypass. Moreover, using wide ground traces will help maintain an electrically clean ground reference.

Typically V_{CC} and V_{EE} are generated with a dual ± 15-V regulated power supply. Though these values have long been the standard in analog systems, today's

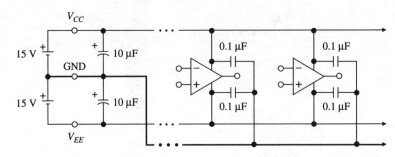

FIGURE 1.36
Op amp powering with bypass capacitors.

mixed-mode applications call for a single 5-V supply to power both digital and analog circuitry. In this case we have $V_{CC} = 5$ V and $V_{EE} = 0$ V. Unless otherwise specified, we assume $V_{CC} = 15$ V and $V_{EE} = -15$ V. Though the power-supply interconnections are normally omitted from circuit diagrams for the sake of simplicity, we must remember to power our op amps when we try them out in the lab. Some of the most frequent sources of frustration for the beginner are due to improper powering, such as faulty wire connections, interchanging V_{CC} and V_{EE}, or even forgetting to turn the power on! When troubleshooting, it is good practice to check the voltages right at the supply pins of the op amp.

Current Flow and Power Dissipation

Since virtually no current flows in or out of the input pins of an op amp, the only current-carrying terminals are the output and the supply pins. We shall designate their currents as i_O, i_{CC}, and i_{EE}. Since V_{CC} is the most positive and V_{EE} the most negative voltage in the circuit, under proper operation i_{CC} will always flow *into* and i_{EE} always *out of* the op amp. However, i_O may flow either out of or into the op amp, depending on circuit conditions. In the former case the op amp is said to be *sourcing* current, and in the latter it is *sinking* current. At all times, the three currents must satisfy KCL. So for an op amp sourcing current we have $i_{CC} = i_{EE} + i_O$, and for an op amp sinking current we have $i_{EE} = i_{CC} + i_O$.

In the special case in which $i_O = 0$, we have $i_{CC} = i_{EE} = I_Q$, where I_Q is called the *quiescent supply current*. This is the current that biases the internal transistors to keep them electrically alive. Its magnitude depends on the op amp type and, to a certain extent, on the supply voltages; typically, I_Q is in the milliampere range. Op amps intended for portable equipment applications may have I_Q in the microampere range and are therefore called *micropower op amps*.

Figure 1.37 shows current flow in the noninverting and inverting circuits, both for the case of a positive and a negative input. Trace each circuit in detail until you are fully convinced that the various currents flow as shown. Note that the output current consists of two components, one to feed the load and the other to feed the feedback network. Moreover, the flow of currents I_Q and i_O through the op amp causes *internal power dissipation*. This dissipation must never exceed the maximum rating specified in the data sheets.

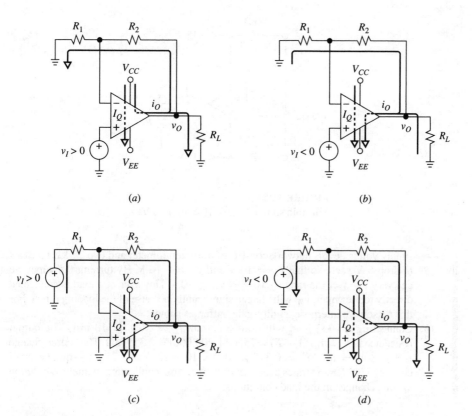

FIGURE 1.37
Current flow for the noninverting [(a) and (b)] and the inverting [(c) and (d)] amplifiers.

EXAMPLE 1.14.   An inverting amplifier with $R_1 = 10$ kΩ, $R_2 = 20$ kΩ, and $v_I = 3$ V drives a 2-kΩ load. (a) Assuming $I_Q = 0.5$ mA, find i_{CC}, i_{EE}, and i_O. (b) Find the power dissipated inside the op amp.

Solution.

(a) With reference to Fig. 1.37c, we have $v_O = -(20/10)3 = -6$ V. Denoting the currents through R_L, R_2, and R_1 as i_L, i_2, and i_1, we have $i_L = 6/2 = 3$ mA, and $i_2 = i_1 = 3/10 = 0.3$ mA. Thus, $i_O = i_2 + i_L = 0.3 + 3 = 3.3$ mA; $i_{CC} = I_Q = 0.5$ mA; $i_{EE} = i_{CC} + i_O = 0.5 + 3.3 = 3.8$ mA.

(b) Whenever a current i experiences a voltage drop v, the corresponding power is $p = vi$. Thus, $p_{OA} = (V_{CC} - V_{EE})I_Q + (v_O - V_{EE})i_O = 30 \times 0.5 + [-6 - (-15)] \times 3.3 = 44.7$ mW.

EXAMPLE 1.15.   When experimenting with op amps it is handy to have a variable source over the range -10 V $\leq v_S \leq 10$ V. (a) Design one such source using a 741 op amp and a 100-kΩ potentiometer. (b) If v_S is set to 10 V, how much does it change when we connect a 1-kΩ load to the source?

Solution.

(a) We first design a resistive network to produce an adjustable voltage over the range -10 V to $+10$ V. As shown in Fig. 1.38, where we use a concise notation for the

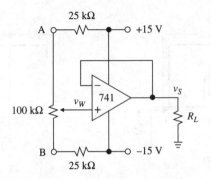

FIGURE 1.38
Variable source from -10 V to $+10$ V.

supply voltages, this network consists of the potentiometer and two 25-kΩ resistors to drop 5 V each, so that $v_A = 10$ V and $v_B = -10$ V. By turning the wiper we can vary v_W over the range -10 V $\leq v_W \leq 10$ V. However, if a load is connected directly to the wiper, v_W will change significantly because of the loading effect. For this reason we interpose a unity-gain buffer, as shown.

(b) Connecting a 1-kΩ load will draw a current $i_L = 10/1 = 10$ mA. The output resistance is $R_o = r_o/(1 + T) = 75/(1 + 200{,}000) = 0.375$ mΩ. The source change is thus $\Delta v_S = R_o\,\Delta i_L = 0.375 \times 10^{-3} \times 10 \times 10^{-3} = 3.75$ μV—quite a small change! This demonstrates a most important op amp application, namely, *regulation* against changes in the load conditions.

Output Saturation

The supply voltages V_{CC} and V_{EE} set upper and lower bounds on the output swing capability of the op amp. This is best visualized in terms of the VTC of Fig. 1.39, which reveals three different regions of operation.

In the *linear region* the curve is approximately straight and its slope represents the open-loop gain a. With a as large as 200,000 V/V, the curve is so steep that it practically coalesces with the vertical axis, unless we use different scales for the two axes. If we express v_O in volts and v_D in microvolts, as shown, then the slope becomes 0.2 V/μV. As we know, op amp behavior within this region is modeled with a *dependent* source of value av_D.

As v_D is increased, v_O increases in proportion until a point is reached where internal transistor saturation effects take place that cause the VTC to flatten out. This is the *positive saturation region,* where v_O no longer depends on v_D but remains fixed, making the op amp behave as an *independent* source of value V_{OH}. Similar considerations hold for the *negative saturation region,* where the op amp acts as an independent source of value V_{OL}. Note that in saturation v_D is no longer necessarily in the microvolt range!

For bipolar op amps, such as the 741, V_{OH} and V_{OL} are typically several *pn*-junction voltage drops (about 2 V) below V_{CC} and above V_{EE}. Thus, for symmetric ±15-V supplies we have $V_{OH} \cong 15 - 2 = 13$ V and $V_{OL} \cong -15 + 2 = -13$ V; that is, the saturation voltages are also approximately symmetric. In this case we simply

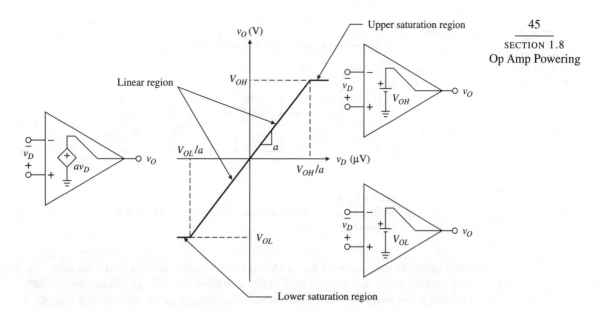

FIGURE 1.39
Regions of operation and approximate op amp models.

say that the 741 saturates at $\pm V_{\text{sat}} = \pm 13$ V. Moreover, since $13/200{,}000 = 65\ \mu$V, the input signal range corresponding to the linear region is $-65\ \mu$V $< v_D <$ $65\ \mu$V.

If the power supplies are other than ± 15 V, V_{OH} and V_{OL} will vary accordingly. For instance, in the case of a 741 op amp being powered from a single 9-V battery so that $V_{CC} = 9$ V and $V_{EE} = 0$ V, we have $V_{OH} \cong 9 - 2 = 7$ V and $V_{OL} \cong 0 + 2 = 2$ V, indicating a useful range, often called the *dynamic output range,* of about only $7 - 2 = 5$ V. In low power-supply systems the need arises for op amps with a maximized dynamic output range. Called *rail-to-rail* op amps, these devices are designed so that under moderate output loading they can swing v_O all the way up to V_{CC} and down to V_{EE}, so that $V_{OH} \cong V_{CC}$ and $V_{OL} \cong V_{EE}$. CMOS op amps are a familiar example. In general, V_{OH} and V_{OL} not only depend on the op amp type but also vary among different samples of the same type because of production variations, temperature drift, and output load variations. Consult the data sheets for more details.

In single-supply systems, such as mixed digital-analog systems with $V_{CC} = 5$ V and $V_{EE} = 0$ V, signals are usually constrained within the range of 0 V to 5 V. The need arises for a reference voltage at $(1/2)V_{CC} = 2.5$ V for termination of all analog sources and loads, and thus allow for symmetric voltage swings about this common reference. In Fig. 1.40 this voltage is synthesized by the *R-R* voltage divider, and is then buffered by OA_1 to provide a low-resistance drive. To maximize the dynamic range of signals, OA_2 is typically a device with rail-to-rail output capabilities, or $V_{OH} \cong 5$ V and $V_{OL} \cong 0$ V. The TLE2426 Rail Splitter (Texas Instruments) is a 3-terminal chip containing all circuitry needed for the synthesis of a precision 2.5-V common reference with a 7.5-mΩ output resistance.

When an op amp is used in the *negative-feedback mode,* its operation must be confined within the linear region because only there is the op amp capable of

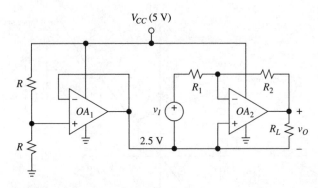

FIGURE 1.40
Synthesis of a 2.5-V common reference voltage in a 5-V single-supply system.

influencing its own input. If the device is inadvertently pushed into saturation, v_O will remain fixed and the op amp will no longer be able to influence v_D, thus yielding a completely different behavior. When analyzing op amp circuits it is often necessary to find the region of operation. To this end we start out assuming that the op amp is in the linear region and calculate v_O. If this falls within the range $V_{OL} < v_O < V_{OH}$, the assumption is correct. Otherwise, the op amp is saturated at either V_{OL} or V_{OH}, depending on whether the calculated value was less than V_{OL} or greater than V_{OH}.

Conversely, given a circuit in the lab, we may wish to find in which region the op amp is working at a given time. The answer lies in the value of v_O, which we can measure with a voltmeter or observe with an oscilloscope. If $V_{OL} < v_O < V_{OH}$, the device must be in the linear region, where, for instance, we can calculate v_D or v_I using $v_D = v_O/a$ or $v_I = v_O/A$. Otherwise, we must have either $v_O = V_{OL}$ or $v_O = V_{OH}$, and v_D will generally be significantly different from zero. The experimental determination of the operating region is very helpful in troubleshooting.

EXAMPLE 1.16. A 741 inverting amplifier with $A = -2$ V/V is driven by a ± 10-V peak-to-peak triangular wave. Sketch and label v_I, v_O, and v_N versus time.

Solution. With an input range of ± 10 V and a gain of 2, the output range would be ± 20 V, indicating that the op amp will saturate part of the time. The borderline between linear operation and saturation occurs when $v_I = \pm 13/2 = \pm 6.5$ V.

When -6.5 V $< v_I < 6.5$ V, the op amp is in the linear region, where $v_O = -2v_I$ and $v_N \cong 0$ V (virtual ground).

When $v_I > 6.5$ V, the op amp saturates at $v_O = -13$ V. By the superposition principle, $v_N = (R_2v_I + R_1v_O)/(R_1 + R_2) = (2/3)v_I + (1/3)(-13) = (2/3)v_I - 13/3$ V. For instance, when v_I peaks at 10 V, v_N will peak at $(2/3)10 - 13/3 = 2.333$ V. Clearly, the inverting input is no longer a virtual ground when $v_I > 6.5$ V.

When $v_I < -6.5$ V, circuit behavior is symmetric to the case in which $v_I > 6.5$ V. The circuit and its waveforms are shown in Fig. 1.41.

A common characteristic of saturating amplifiers is a *clipped* output waveform. Clipping is a form of distortion since the output of a linear amplifier should have

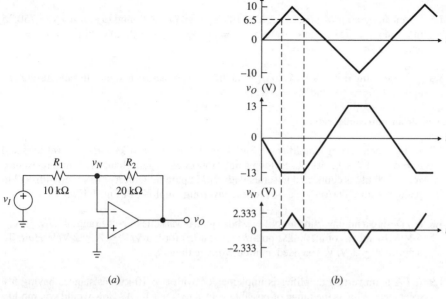

FIGURE 1.41
Waveforms of an inverting amplifier driven into saturation.

the same shape as the input. Clipping is generally undesirable, though there are situations in which it is exploited on purpose to achieve specific effects.

PROBLEMS

1.1 Amplifier fundamentals

1.1 In the voltage amplifier circuit of Fig. 1.1, let $v_S = 100\,\text{mV}$, $R_s = 100\,\text{k}\Omega$, $v_I = 75\,\text{mV}$, $R_L = 10\,\Omega$, and $v_O = 2\,\text{V}$. If connecting a 30-Ω resistance in parallel with R_L drops v_O to 1.8 V, find R_i, A_{oc}, and R_o.

1.2 Sketch the transconductance and transresistance amplifiers; derive expressions for their source-to-load gains.

1.3 (a) A transresistance amplifier with $R_i = 20\,\text{k}\Omega$, $A_{\text{oc}} = 1\,\text{V/ma}$, and $R_o = 300\,\Omega$ is driven by a source i_S with parallel resistance $R_s = 100\,\text{k}\Omega$ and drives a load $R_L = 600\,\Omega$. Find the transresistance gain v_L/i_S as well as the power gain p_L/p_S, where p_S is the power delivered by the source i_S and p_L that absorbed by the load R_L. (b) To what value must A_r be changed to achieve $v_L/i_S = 1\,\text{V/mA}$ exactly? What is the corresponding power gain?

1.4 A transconductance amplifier is driven by a source with $v_S = 30\,\text{mV}$ and $R_s = 100\,\text{k}\Omega$ and drives a load R_L. Digital multimeter (DMM) readings at the input and output ports yield $v_I = 25\,\text{mV}$, $i_L = 0.9\,\text{A}$ for $R_L = 20\,\Omega$, and $i_L = 0.8\,\text{A}$ for $R_L = 30\,\Omega$. Predict the DMM readings if the same amplifier is driven by a source with $v_S = 33\,\text{mV}$ and $R_s = 50\,\text{k}\Omega$ and drives a load $R_L = 40\,\Omega$.

1.2 The operational amplifier

1.5 Given an op amp with $r_d \cong \infty$, $a = 10^4$ V/V, and $r_o \cong 0$, find (*a*) v_O if $v_P = 750.25$ mV and $v_N = 751.50$ mV, (*b*) v_N if $v_O = -5$ V and $v_P = 0$, (*c*) v_P if $v_N = v_O = 5$ V, and (*d*) v_N if $v_P = -v_O = 1$ V.

1.6 A 741 op amp drives a 1-kΩ load. Find the voltages across and the currents through r_d and r_o if $v_P = 1$ V and $v_O = 5$ V.

1.3 Basic op amp configurations

1.7 In the noninverting amplifier of Fig. 1.6*a*, let $R_1 = 100$ kΩ, $R_2 = 200$ kΩ, and $a = \infty$. (*a*) What is its closed-loop gain? How does its gain change if a third resistance $R_3 = 100$ kΩ is connected in series with R_1? In parallel with R_1? In series with R_2? In parallel with R_2? (*b*) Repeat (*a*) for the inverting amplifier of Fig. 1.10*a*.

1.8 (*a*) Design a noninverting amplifier whose gain is variable over the range 1 V/V $\leq A \leq$ 5 V/V by means of a 100-kΩ pot. (*b*) Repeat (*a*) for 0.5 V/V $\leq A \leq$ 2 V/V. *Hint:* To achieve $A \leq 1$ V/V, you need an input voltage divider.

1.9 (*a*) A noninverting amplifier is implemented with two 10-kΩ resistances having 5% tolerance. What is the range of possible values for the gain A? How would you modify the circuit for the exact calibration of A? (*b*) Repeat, but for the inverting amplifier.

1.10 In the inverting amplifier of Fig. 1.10*a*, let $v_I = 0.1$ V, $R_1 = 10$ kΩ, and $R_2 = 100$ kΩ. Find v_O and v_N if (*a*) $a = 10^2$ V/V, (*b*) $a = 10^4$ V/V, (*c*) $a = 10^6$ V/V. Comment on your findings.

1.11 (*a*) Design an inverting amplifier whose gain is variable over the range -10 V/V $\leq A \leq 0$ by means of a 100-kΩ pot. (*b*) Repeat, but for -10 V/V $\leq A \leq -1$ V/V. *Hint:* To prevent A from reaching zero, you must use a suitable resistor in series with the pot.

1.12 (*a*) A source $v_S = 2$ V with $R_s = 10$ kΩ is to drive a gain-of-five inverting amplifier implemented with $R_1 = 20$ kΩ and $R_2 = 100$ kΩ. Find the amplifier output voltage and verify that because of loading its magnitude is *less* than $2 \times 5 = 10$ V. (*b*) Find the value to which R_2 must be changed if we want to compensate for loading and obtain a full output magnitude of 10 V.

1.13 (*a*) A source $v_S = 10$ V is fed to a voltage divider implemented with $R_A = 120$ kΩ and $R_B = 30$ kΩ, and the voltage across R_B is fed, in turn, to a gain-of-five noninverting amplifier having $R_1 = 30$ kΩ and $R_2 = 120$ kΩ. Sketch the circuit, and predict the amplifier output voltage v_O. (*b*) Repeat (*a*) for a gain-of-five inverting amplifier having $R_1 = 30$ kΩ and $R_2 = 150$ kΩ. Compare and comment on the differences.

1.14 An inverting amplifier is implemented with $R_1 = 10$ kΩ, $R_2 = 20$ kΩ and an op amp with $r_d \cong \infty$, $a = 1$ V/mV, and $r_o \cong 0$. Sketch and label v_I, v_O, and v_N versus time if v_I is a 1-kHz sine wave with ±5-V peak values.

1.4 Ideal op amp circuit analysis

1.15 Find v_N, v_P, and v_O in the circuit of Fig. P1.15, as well as the power released by the 4-V source; devise a method to check your results.

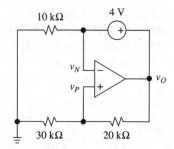

FIGURE P1.15

1.16 (a) Find v_N, v_P, and v_O in the circuit of Fig. P1.16. (b) Repeat (a) with a 5-kΩ resistance connected between A and B.

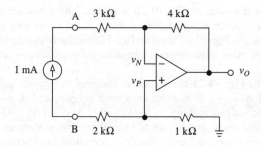

FIGURE P1.16

1.17 (a) Find v_N, v_P, and v_O in the circuit of Fig. P1.17 if $v_S = 9$ V. (b) Find the resistance R that, if connected between the inverting-input pin of the op amp and ground, causes v_O to double. Verify with PSpice.

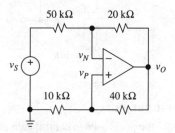

FIGURE P1.17

1.18 (a) Find v_N, v_P, and v_O in the circuit of Fig. P1.18. (b) Repeat (a) with a 40-kΩ resistance in parallel with the 0.3-mA source.

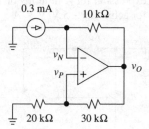

FIGURE P1.18

1.19 (a) Find v_N, v_P, and v_O in the circuit of Fig. P1.19 if $i_S = 1$ mA. (b) Find a resistance R that when connected in parallel with the 1-mA source will cause v_O to drop to half the value found in (a).

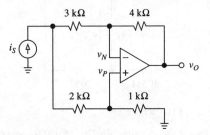

FIGURE P1.19

1.20 (a) If the current source of Fig. P1.16 is replaced by a voltage source v_S, find the magnitude and polarity of v_S so that $v_O = 10$ V. (b) If the wire connecting the 4-V source to node v_O in Fig. P1.15 is cut and a 5-kΩ resistance is inserted in series between the two, to what value must the source be changed to yield $v_O = 10$ V?

1.21 In the circuit of Fig. P1.21 the switch is designed to provide gain-polarity control. (a) Verify that $A = +1$ V/V when the switch is open, and $A = -R_2/R_1$ when the switch is closed, so that making $R_1 = R_2$ yields $A = \pm 1$ V/V. (b) To accommodate gains greater than unity, connect an additional resistance R_4 from the inverting-input pin of the op amp to ground. Derive separate expressions for A in terms of R_1 through R_4 with the switch open and with the switch closed. (c) Specify resistance values suitable for achieving $A = \pm 2$ V/V.

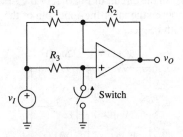

FIGURE P1.21

1.22 In the circuit of Fig. P1.22 the pot is used to control gain magnitude as well as polarity. (a) Letting k denote the fraction of R_3 between the wiper and ground, show that varying the wiper from bottom to top varies the gain over the range $-R_2/R_1 \le A \le 1$ V/V, so that making $R_1 = R_2$ yields -1 V/V $\le A \le +1$ V/V. (b) To accommodate gains greater than unity, connect an additional resistance R_4 from the op amp's inverting-input pin to ground. Derive an expression for A in terms of R_1, R_2, R_4, and k. (c) Specify resistance values suitable for achieving -5 V/V $\le A \le +5$ V/V.

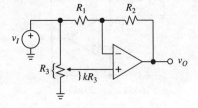

FIGURE P1.22

1.23 Consider the following statements about the input resistance R_i of the noninverting amplifier of Fig. 1.14a: (a) Since we are looking straight into the noninverting-input pin, which is an open circuit, we have $R_i = \infty$; (b) since the input pins are virtually shorted together, we have $R_i = 0 + (R_1 \parallel R_2) = R_1 \parallel R_2$; (c) since the noninverting-input pin is virtually shorted to the inverting-input pin, which is in turn a virtual-ground node, we have $R_i = 0 + 0 = 0$. Which statement is correct? How would you refute the other two?

1.24 (a) Show that the circuit of Fig. P1.24 has $R_i = \infty$ and $A = -(1 + R_3/R_4)R_1/R_2$. (b) Specify suitable components to make A variable over the range $-100\ \text{V/V} \le A \le 0$ by means of a 100-kΩ pot. Try minimizing the number of resistors you use.

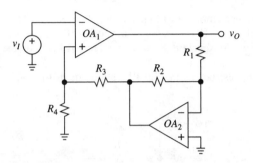

FIGURE P1.24

1.25 The audio panpot circuit of Fig. P1.25 is used to continuously vary the position of signal v_I between the left and the right stereo channels. (a) Discuss circuit operation. (b) Specify R_1 and R_2 so that $v_L/v_I = -1\ \text{V/V}$ when the wiper is fully down, $v_R/v_I = -1\ \text{V/V}$ when the wiper is fully up, and $v_L/v_I = v_R/v_I = -1/\sqrt{2}$ when the wiper is halfway.

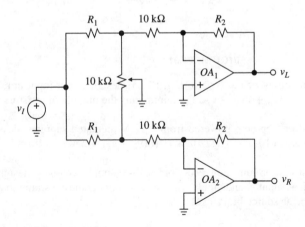

FIGURE P1.25

1.26 (a) Using standard 5% resistances in the kilohm range, design a circuit to yield $v_O = -100(4v_1 + 3v_2 + 2v_3 + v_4)$. (b) If $v_1 = 20$ mV, $v_2 = -50$ mV, and $v_4 = 100$ mV, find v_3 for $v_O = 0$ V.

1.27 (a) Using standard 5% resistances, design a circuit to give (a) $v_O = -10(v_I + 1\ \text{V})$; (b) $v_O = -v_I + V_O$, where V_O is variable over the range $-5\ \text{V} \le V_O \le +5\ \text{V}$ by means of a 100-kΩ pot. *Hint:* Connect the pot between the ±15-V supplies and use the wiper voltage as one of the inputs to your circuit.

1.28 In the circuit of Fig. 1.17 let $R_1 = R_3 = R_4 = 10 \text{ k}\Omega$ and $R_2 = 30 \text{ k}\Omega$. (*a*) If $v_1 = 3$ V, find v_2 for $v_O = 10$ V. (*b*) If $v_2 = 6$ V, find v_1 for $v_O = 0$ V. (*c*) If $v_1 = 1$ V, find the range of values for v_2 for which $-10 \text{ V} \leq v_O \leq +10$ V.

1.29 You can readily verify that if we put the output in the form $v_O = A_2 v_2 - A_1 v_1$ in the circuit of Fig. 1.17, then $A_2 \leq A_1 + 1$. Applications requiring $A_2 \geq A_1 + 1$ can be accommodated by connecting an additional resistance R_5 from the node common to R_1 and R_2 to ground. (*a*) Sketch the modified circuit and derive a relationship between its output and inputs. (*b*) Specify standard resistances to achieve $v_O = 5(2v_2 - v_1)$. Try minimizing the number of resistors you use.

1.30 (*a*) In the difference amplifier of Fig. 1.17 let $R_1 = R_3 = 10 \text{ k}\Omega$ and $R_2 = R_4 = 100$ kΩ. Find v_O if $v_1 = 10 \cos 2\pi 60t - 0.5 \cos 2\pi 10^3 t$ V, and $v_2 = 10 \cos 2\pi 60t + 0.5 \cos 2\pi 10^3 t$ V. (*b*) Repeat if R_4 is changed to 101 kΩ. Comment on your findings.

1.31 Show that if all resistances in Fig. P1.31 are equal, then $v_O = v_2 + v_4 + v_6 - v_1 - v_3 - v_5$.

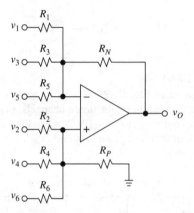

FIGURE P1.31

1.32 Using a topology of the type of Fig. P1.31, design a four-input amplifier such that $v_O = 4v_A - 3v_B + 2v_C - v_D$. Try minimizing the number of resistors you use.

1.33 Using just one op amp powered from ± 12-V regulated supplies, design a circuit to yield: (*a*) $v_O = 10v_I + 5$ V; (*b*) $v_O = 10(v_2 - v_1) - 5$ V.

1.34 Using just one op amp powered from ± 15-V supply voltages, design a circuit that accepts an ac input v_i and yields $v_O = v_i + 5$ V, under the constraint that the resistance seen by the ac source be 100 kΩ.

1.35 Design a two-input, two-output circuit that yields the sum and the difference of its inputs: $v_S = v_{I1} + v_{I2}$, and $v_D = v_{I1} - v_{I2}$. Try minimizing the component count.

1.36 Obtain a relationship between v_O and v_I if the differentiator of Fig. 1.18 includes also a resistance R_s in series with C. Discuss the extreme cases of v_I changing very slowly and very rapidly.

1.37 Obtain a relationship between v_O and v_I if the integrator of Fig. 1.19 includes also a resistance R_p in parallel with C. Discuss the extreme cases of v_I changing very rapidly and very slowly.

1.38 In the differentiator of Fig. 1.18 let $C = 10$ nF and $R = 100$ kΩ, and let v_I be a periodic signal alternating between 0 V and 2 V with a frequency of 100 Hz. Sketch and label v_I and v_O versus time if v_I is (a) a sine wave; (b) a triangular wave.

1.39 In the integrator of Fig. 1.19 let $R = 100$ kΩ and $C = 10$ nF. Sketch and label $v_I(t)$ and $v_O(t)$ if (a) $v_I = 5 \sin 2\pi 100t$ V and $v_O(0) = 0$; (b) $v_I = 5[u(t) - u(t - 2 \text{ ms})]$ V and $v_O(0) = 5$ V, where $u(t - t_0)$ is the unit step function defined as $u = 0$ for $t < t_0$, and $u = 1$ for $t > t_0$.

1.40 (a) In the integrator of Fig. 1.19 let $R = 10$ kΩ and $C = 0.1$ μF. Assuming that C is initially discharged, sketch and label $v_O(t)$ for $0 \le t \le 10$ ms if v_I is a 1-V step. (b) Repeat (a) with a 100-kΩ resistance connected in parallel with C.

1.41 If R_F in the summing amplifier of Fig. 1.15 is replaced by a capacitance C, the circuit becomes a *summing integrator*. (a) Derive a relationship between its output and its inputs. (b) Using a 10-nF capacitance, specify suitable resistances for $v_O(t) = v_O(0) - 10^3(\int_0^t v_1 \, d\xi + 2 \int_0^t v_2 \, d\xi + 0.5 \int_0^t v_3 \, d\xi)$.

1.42 Show that if the op amp of Fig. 1.20b has a finite gain a, then $R_{eq} = (-R_1 R / R_2) \times [1 + (1 + R_2/R_1)/a]/[1 - (1 + R_1/R_2)/a]$.

1.43 Find an expression for R_i in Fig. P1.43; discuss its behavior as R is varied over the range $0 \le R \le 2R_1$.

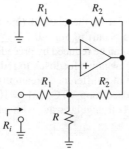

FIGURE P1.43

1.44 The circuit of Fig. P1.44 can be used to control the input resistance of the inverting amplifier based on OA_1. (a) Show that $R_i = R_1/(1 - R_1/R_3)$. (b) Specify resistances suitable for achieving $A = -10$ V/V with $R_i = \infty$.

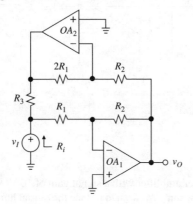

FIGURE P1.44

1.5 Negative feedback

1.45 A voltage amplifier has $a = 10^5$ V/V and $v_i = 10$ mV. Find v_d, v_f, v_o, A, T, and the percentage deviation of A from A_{ideal} for $\beta = 10^{-3}$ V/V, 10^{-2} V/V, 10^{-1} V/V, and 1 V/V. Compare the various cases and comment.

1.46 (a) Find the desensitivity factor of a negative-feedback system with $a = 10^3$ and $A = 10^2$. (b) Find A exactly via Eq. (1.40), and approximately via Eq. (1.49) if a drops by 10%. (c) Repeat (b) for a 50% drop in a; compare with (b) and comment.

1.47 You are asked to design an amplifier with a gain A of 10^2 V/V that is accurate to within $\pm 0.1\%$, or $A = 10^2$ V/V $\pm 0.1\%$. All you have available are amplifier stages with $a = 10^4$ V/V $\pm 25\%$ each. Your amplifier can be implemented using a cascade of basic stages, each employing a suitable amount of negative feedback. What is the minimum number of stages required? What is the β of each stage?

1.48 The open-loop VTC of a certain amplifier can be approximated piecewise by five segments with symmetric breakpoints at $(v_D, v_O) = \pm(80\ \mu\text{V}, 8\ \text{V})$, $\pm(280\ \mu\text{V}, 12\ \text{V})$, and $\pm(530\ \mu\text{V}, 13\ \text{V})$. (a) Sketch the above VTC; calculate and sketch the closed-loop VTC when the amplifier is placed in a feedback loop with $\beta = 0.5$ V/V. (b) Sketch v_I, v_O, and v_D versus time if v_I is a triangular wave with ± 5-V peak values; comment on the waveform of v_D. *Hint:* $v_D(t)$ can be derived point by point from $v_O(t)$ using the open-loop VTC of (a).

1.49 A crude BJT power amplifier of the class B (push-pull) type exhibits the VTC of Fig. P1.49b. The dead band occurring for $-0.7\ \text{V} \le v_1 \le +0.7\ \text{V}$ causes a crossover distortion at the output that can be reduced by preceding the power stage with a preamplifier stage and then using negative feedback to reduce the dead band. This is shown in Fig. P1.49a for the case of a difference preamplifier with gain a_1 and $\beta = 1$ V/V. (a) Sketch and label the closed-loop VTC if $a_1 = 10^2$ V/V. (b) Sketch v_I, v_1, and v_O versus time if v_I is a 100-Hz triangular wave with peak values of ± 1 V.

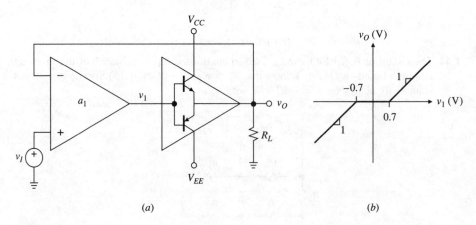

(a)   (b)

FIGURE P1.49

1.50 A certain audio power amplifier with a signal gain of 10 V/V is found to produce a 2-V peak-to-peak 120-Hz hum. We wish to reduce the output hum to less than 1 mV without changing the signal gain. To this end, we precede the power stage with a preamplifier stage with gain a_1 and then apply negative feedback around the composite amplifier. What are the required values of a_1 and β?

1.6 Feedback in op amp circuits

1.51 A voltage follower is implemented with an op amp having $r_d = 1$ MΩ, $a = 1$ V/mV, and $r_o = 1$ kΩ. (a) Find v_O if the follower is driven by a source $v_S = 10.000$ V with $R_s = 2$ MΩ. (b) Repeat (a) with a 1-kΩ output load.

1.52 An inverting amplifier is implemented with two precision resistors $R_1 = 100$ kΩ and $R_2 = 200$ kΩ and drives a 2-kΩ load. Assuming an op amp with $r_d = 1$ MΩ and $r_o = 100$ Ω, find the minimum gain a needed to contain the deviation of A from the ideal within (a) 1%, (b) 0.001%.

1.53 Let a voltage follower be implemented with an op amp having $r_d = 1$ kΩ, $r_o = 20$ kΩ, and $a = 10^6$ V/V (poor resistances, but excellent gain). Find A, R_i, and R_o, and comment on your findings.

1.7 The loop gain

1.54 (a) Find A_{ideal} in the circuit of Fig. P1.54 if all resistances are equal. (b) Assuming $r_d \cong \infty$ and $r_o \cong 0$, find a_{min} such that the deviation of A from A_{ideal} is less than 0.1%.

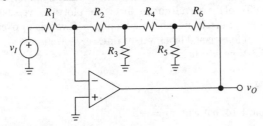

FIGURE P1.54

1.55 (a) Assuming that R_3 in Fig. 1.32a is a potentiometer connected as a variable resistance over the range $0 \leq R_3 \leq 1$ MΩ, specify suitable components for an input resistance of 500 kΩ and a continuously variable gain over the range -10^3 V/V $\leq A_{ideal} \leq -0.5$ V/V. (b) If $r_d = 1$ MΩ, $a = 10^5$ V/V, $r_o = 100$ Ω, and $R_L = 2$ kΩ, estimate the gain departure from the ideal at the two extremes of the range.

1.56 (a) Design a difference amplifier such that, ideally, $v_O = 100(v_2 - v_1)$. (b) Assuming an op amp with $r_d \cong \infty$ and $r_o \cong 0$, find the open-loop gain needed to approximate the ideal closed-loop gain within 0.1%.

1.57 Assuming that the op amp has $r_d \cong \infty$ and $r_o \cong 0$, find the feedback factor β in the circuits of Figs. P1.15 through P1.19.

1.58 For the dc-offsetting amplifier of Fig. 1.16 find the minimum open-loop gain needed to contain the deviation of its transfer characteristic from the ideal within 1%.

1.59 Using a single op amp, along with the ideas expressed in Problem 1.29, design a circuit that accepts two inputs v_1 and v_2 and yields $v_O = 100(3v_2 - 2v_1)$. Hence, assuming

$r_d = \infty$ and $r_o = 0$, find the minimum open-loop gain a needed to contain the transfer characteristic's deviation from ideality within 0.1%.

1.60 Assuming the op amp of Fig. P1.60 has $a = 3000$ V/V, $r_d = \infty$, and $r_o = 0$, find the loop gain T.

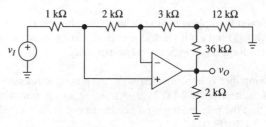

FIGURE P1.60

1.61 (*a*) Assuming the op amp of Fig. P1.16 has $r_d = \infty$ and $r_o = 0$, find β_N, β_P, and β. (*b*) Repeat, but with the current source replaced by a voltage source.

1.62 Repeat Problem 1.61, but for the circuit of Fig. P1.19.

1.63 In the circuit of Fig. P1.49*a* let $a_1 = 3000$ V/V and $R_L = 2$ kΩ, and suppose an additional 10-kΩ resistor is connected from node v_1 to node v_O. (*a*) Sketch and label the open-loop VTC of the overall circuit, that is, the plot of v_O versus the input difference $v_D = v_P - v_N$. (*b*) Sketch and label the loop gain T versus v_I over the range -0.3 V $\leq v_I \leq 0.3$ V. (*c*) Sketch and label, versus time, v_I, v_O, v_1, and v_D if v_I is a triangular wave with ± 0.3-V peak values.

1.8 Op amp powering

1.64 Repeat Example 1.14, but with $v_I = -5$ V.

1.65 Assuming that $I_Q = 1.5$ mA in the circuit of Fig. P1.65, calculate all currents and voltages, as well as the power dissipated inside the op amp, if (*a*) $v_I = +2$ V; (*b*) $v_I = -2$ V.

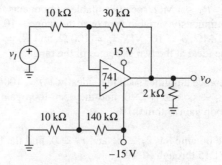

FIGURE P1.65

1.66 Using a 741 op amp powered from ± 12-V supplies, design a noninverting amplifier with a gain of 6 V/V. Sketch and label v_I, v_O, and v_N versus time if v_I is a sine wave with ± 3-V peak values.

1.67 (*a*) Assuming ±15-V power supplies, design a variable voltage source over the range $0 \text{ V} \leq v_S \leq 10 \text{ V}$. (*b*) Assuming a 1-k$\Omega$ grounded load and $I_Q = 1.5$ mA, find the maximum internal power dissipation of your op amp.

1.68 Assuming a 741 op amp in the dc-offsetting amplifier of Fig. 1.16, find: (*a*) v_I and v_N if $v_O = 5$ V; (*b*) find v_N and v_O if $v_I = 3$ V.

1.69 The noninverting amplifier of Fig. 1.14*a* is implemented with $R_1 = 10$ kΩ and $R_2 = 15$ kΩ, and a 741 op amp powered from ±12-V supplies. If the circuit includes also a third 30-kΩ resistor connected between the inverting input and the 12-V supply, find v_O and v_N if (*a*) $v_I = 4$ V, and (*b*) $v_I = -2$ V.

1.70 (*a*) Assuming $I_Q = 50$ μA and a grounded load of 100 kΩ at the output of the dc-offsetting amplifier of Fig. 1.16, find the values of v_I for which the op amp dissipates the maximum power. Show all corresponding voltages and currents. (*b*) Assuming $\pm V_{\text{sat}} = \pm 12$ V, find the range of values of v_I for which the op amp still operates within the linear region.

1.71 In the amplifier of Fig. 1.17 let $R_1 = 30$ kΩ, $R_2 = 120$ kΩ, $R_3 = 20$ kΩ, and $R_4 = 30$ kΩ, and let the op amp be a 741-type powered from ±15 V. (*a*) If $v_2 = 2 \sin \omega t$ V, find the range of values of v_1 for which the amplifier still operates in the linear region. (*b*) If $v_1 = V_m \sin \omega t$ and $v_2 = -1$ V, find the maximum value of V_m for which the op amp still operates in the linear region. (*c*) Repeat (*a*) and (*b*) for the case in which the power supplies are lowered to ±12 V.

1.72 Assuming that the op amps of Figs. P1.17 and P1.19 saturate at ±10 V, find the range of values of v_S and i_S for which the op amps still operate in the linear region.

1.73 In the inverting amplifier of Fig. 1.32*a* let v_I be a 1-kHz triangular wave with peak values $\pm V_{im}$, and let the op amp be ideal, except that its output saturates at ±10 V. Assuming that $R_1 = R_2 = 1$ MΩ, $R_3 = 18$ kΩ, and $R_4 = R_L = 2$ kΩ, sketch and label v_I, v_N, v_1, and v_O versus time if (*a*) $V_{im} = 0.5$ V; (*b*) $V_{im} = 2$ V.

1.74 The circuit of Fig. P1.74, called a *bridge amplifier,* allows one to double the linear output range as compared with a single op amp. (*a*) Show that if the resistances are in the ratios shown, then $v_O/v_I = 2A$. (*b*) If the individual op amps saturate at ±13 V, what is the maximum peak-to-peak output voltage that the circuit can provide without distortion?

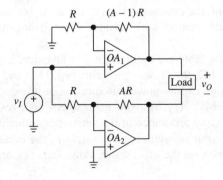

FIGURE P1.74

1.75 For the circuit of Fig. P1.65 sketch and label v_I, v_N, and v_O versus time if v_I is a triangular wave with ± 5-V peak values.

1.76 In the integrator of Fig. 1.19 let $R = 100$ kΩ and $C = 10$ nF, and let the op amp be ideal, except that its output saturates at ± 13 V. Assuming that $v_O(0) = 0$ V, sketch and label v_O and v_N versus time if (a) $v_I = 1$ V; (b) $v_I = 1$ mV; (c) $v_I = -1$ mV.

BIBLIOGRAPHY

Dostál, J.: *Operational Amplifiers,* 2d ed., Butterworth-Heinemann, Stoneham, MA, 1993.

Fredericksen, T. M.: *Intuitive IC Op Amps,* National Semiconductor Co., Santa Clara, CA, 1984.

Graeme, J. G.: *Optimizing Op Amp Performance,* McGraw-Hill, New York, 1997.

Graeme, J. G., G. E. Tobey, and L. P. Huelsman: *Operational Amplifiers: Design and Applications,* McGraw-Hill, New York, 1971.

Horowitz, P., and W. Hill: *The Art of Electronics,* 2d ed., Cambridge University Press, Cambridge, U.K., 1989.

Jung, W. G.: *IC Op Amp Cookbook,* 3d ed., Howard W. Sams, Carmel, IN, 1986.

Kennedy, E. J.: *Operational Amplifier Circuits: Theory and Applications,* Holt, Rinehart and Winston, Orlando, FL, 1988.

Pease, R. A.: *Troubleshooting Analog Circuits,* Butterworth-Heinemann, Stoneham, MA, 1991.

Roberge, J. K.: *Operational Amplifiers: Theory and Practice,* John Wiley & Sons, New York, 1975.

Rosenstark, S.: *Feedback Amplifier Principles,* Macmillan, New York, 1986.

Williams, J.: *Analog Circuit Design: Art, Science, and Personalities,* Butterworth-Heinemann, Stoneham, MA, 1991.

APPENDIX 1A
STANDARD RESISTOR VALUES

As a good work habit, always specify standard resistance values for the circuits you design (see Table 1A.1). In many applications 5% resistors are adequate; however, when higher precision is required, 1% resistors should be used. When even this tolerance is insufficient, the alternatives are either 0.1% (or better) resistors, or less precise resistors in conjunction with variable ones (trim pots) to allow for exact adjustments.

The numbers in the table are multipliers. For instance, if the calculations yield a resistance of 3.1415 kΩ, the closest 5% value is 3.0 kΩ and the closest 1% value is 3.16 kΩ. In the design of low-power circuits, the best resistance range is usually between 1 kΩ and 1 MΩ. Try to avoid excessively high resistances (e.g., above 10 MΩ), because the stray resistance of the surrounding medium tends to decrease the effective value of your resistance, particularly in the presence of moisture and salinity. Low resistances, on the other hand, cause unnecessarily high-power dissipation.

TABLE 1A.1
Standard resistance values

5% resistor values	1% resistor values			
10	100	178	316	562
11	102	182	324	576
12	105	187	332	590
13	107	191	340	604
15	110	196	348	619
16	113	200	357	634
18	115	205	365	649
20	118	210	374	665
22	121	215	383	681
24	124	221	392	698
27	127	226	402	715
30	130	232	412	732
33	133	237	422	750
36	137	243	432	768
39	140	249	442	787
43	143	255	453	806
47	147	261	464	825
51	150	267	475	845
56	154	274	487	866
62	158	280	499	887
68	162	287	511	909
75	165	294	523	931
82	169	301	536	953
91	174	309	549	976

2

CIRCUITS WITH RESISTIVE FEEDBACK

2.1 Current-to-Voltage Converters
2.2 Voltage-to-Current Converters
2.3 Current Amplifiers
2.4 Difference Amplifiers
2.5 Instrumentation Amplifiers
2.6 Instrumentation Applications
2.7 Transducer Bridge Amplifiers
    Problems
    References

In this chapter we investigate additional op amp circuits, this time with greater emphasis on practical applications. The circuits to be examined are designed to exhibit linear, frequency-independent transfer characteristics. Linear circuits that are deliberately intended for frequency-dependent behavior are more properly called *filters* and will be studied in Chapters 3 and 4. Finally, *nonlinear* op amp circuits will be studied in Chapters 9 and 13.

To get a feel for what a given circuit does, we first analyze it using the ideal op amp model. Then, in the spirit of Sections 1.6 and 1.7, we take a closer look at how op amp nonidealities, particularly the finite open-loop gain, affect its closed-loop parameters. A more systematic investigation of op amp nonidealities, such as static and dynamic errors, will be carried out in Chapters 5 and 6, after we have developed enough confidence with op amp circuits emphasizing the simpler op amp model. The circuits of the present and other chapters that are most directly affected by such limitations will be reexamined in greater detail then.

In the first half of the chapter we demonstrate how the op amp, which is basically a voltage-type amplifier, can be configured for other forms of amplification, such as current amplification and *V-I* and *I-V* conversion. This exceptional versatility stems from the negative-feedback ability to modify the closed-loop resistances as well as stabilize gain. The judicious application of this ability allows us to approach the ideal amplifier conditions of Table 1.1 to a highly satisfactory degree.

The second part of the chapter addresses instrumentation concepts and applications. The circuits examined include difference amplifiers, instrumentation amplifiers, and transducer-bridge amplifiers, which are the workhorses of today's automatic test, measurement, and control instrumentation.

2.1
CURRENT-TO-VOLTAGE CONVERTERS

A *current-to-voltage converter* (*I-V* converter), also called a *transresistance amplifier*, accepts an input current i_I and yields an output voltage of the type $v_O = Ai_I$, where A is the gain of the circuit in volts per ampere. Referring to Fig. 2.1, assume first that the op amp is ideal. Summing currents at the virtual-ground node gives $i_I + (v_O - 0)/R = 0$, or

$$v_O = -Ri_I \qquad (2.1)$$

The gain is $-R$ and is negative because of the choice of the reference direction of i_I; inverting this direction gives $v_O = Ri_I$. The magnitude of the gain is also called the *sensitivity* of the converter because it gives the amount of output voltage change for a given input current change. For instance, for a sensitivity of 1 V/mA we need $R = 1\ \text{k}\Omega$, for a sensitivity of 1 V/μA we need $R = 1\ \text{M}\Omega$, and so on. If desired, gain can be made variable by implementing R with a potentiometer. Note that the feedback element need not necessarily be limited to a resistance. In the more general case in which it is an impedance $Z(s)$, where s is the complex frequency, Eq. (2.1) takes on the Laplace-transform form $V_o(s) = -Z(s)I_i(s)$, and the circuit is called a *transimpedance amplifier*.

We observe that the op amp eliminates loading both at the input and at the output. In fact, should the input source exhibit some finite parallel resistance R_s, the op amp eliminates any current loss through it by forcing 0 V across it. Also, the op amp delivers v_O to an output load R_L with zero output resistance.

Closed-Loop Parameters

Let us now investigate the departure from ideal if a practical op amp is used. Comparing with Fig. 1.26*b* and *c*, we recognize the *shunt-shunt* topology. We can thus

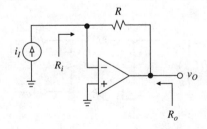

FIGURE 2.1
Basic *I-V* converter.

apply the techniques of Section 1.7 and write

$$T = \frac{ar_d}{r_d + R + r_o} \tag{2.2}$$

$$A = -R\frac{1}{1 + 1/T} \qquad R_i = \frac{r_d \parallel (R + r_o)}{1 + T} \qquad R_o \cong \frac{r_o}{1 + T} \tag{2.3}$$

EXAMPLE 2.1. Find the closed-loop parameters of the circuit of Fig. 2.1 if it is implemented with a 741 op amp and $R = 1$ MΩ.

Solution. Substituting the given component values, we get $T = 133,330$, $A = -0.999993$ V/μA, $R_i = 5$ Ω, and $R_o \cong 56$ mΩ.

High-Sensitivity *I-V* Converters

It is apparent that high-sensitivity applications may require unrealistically large resistances. Unless proper circuit fabrication measures are adopted, the resistance of the surrounding medium, being in parallel with R, will decrease the net feedback resistance and degrade the accuracy of the circuit. Figure 2.2 shows a widely used technique to avoid this drawback. The circuit utilizes a *T*-network to achieve high sensitivity without requiring unrealistically large resistances.

Summing currents at node v_1 yields $-v_1/R - v_1/R_1 + (v_O - v_1)/R_2 = 0$. But $v_1 = -Ri_I$, by Eq. (2.1). Eliminating v_1 yields

$$v_O = -kRi_I \tag{2.4a}$$

$$k = 1 + \frac{R_2}{R_1} + \frac{R_2}{R} \tag{2.4b}$$

The circuit in effect increases R by the multiplicative factor k. We can thus achieve a high sensitivity by starting out with a reasonable value of R and then multiplying it by the needed amount k.

EXAMPLE 2.2. In the circuit of Fig 2.2 specify suitable component values to achieve a sensitivity of 0.1 V/nA.

Solution. We have $kR = 0.1/10^{-9} = 100$ MΩ, a fairly large value. Start out with $R = 1$ MΩ and then multiply it by 100 to meet the specifications. Thus, $1 + R_2/R_1 + R_2/10^6 = 100$. Since we have one equation but two unknowns, fix one unknown; for example, let $R_1 = 1$ kΩ. Then, imposing $1 + R_2/10^3 + R_2/10^6 = 100$ yields $R_2 \cong 99$ kΩ (use 100 kΩ, the closest standard). If desired, R_2 can be made variable for the exact adjustment of kR.

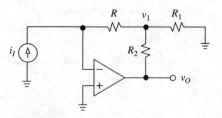

FIGURE 2.2
High-sensitivity *I-V* converter.

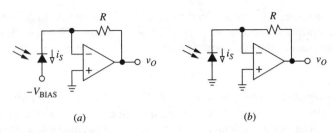

FIGURE 2.3
(*a*) Photoconductive and (*b*) photovoltaic detectors.

Real-life op amps do draw a small current at their input terminals. Called the *input bias current,* it may degrade the performance of high-sensitivity *I-V* converters, in which i_I itself is quite small. This drawback can be avoided by using op amps specifically rated for low input bias current, such as JFET-input and MOSFET-input op amps.

Photodetector Amplifiers

One of the most frequent *I-V* converter applications is in connection with current-type photodetectors such as photodiodes and photomultipliers.[1] Another common application, *I-V* conversion of current-output digital-to-analog converters, will be discussed in Chapter 12.

Photodetectors are transducers that produce electrical current in response to incident light or other forms of radiation, such as X rays. A transresistance amplifier is then used to convert this current to a voltage, as well as eliminate possible loading both at the input and at the output.

One of the most widely used photodetectors is the *silicon photodiode.* The reasons for its popularity are its solid-state reliability, low cost, small size, and low power dissipation.[1] The device can be used either with a reverse bias voltage, in the *photoconductive* mode, shown in Fig. 2.3*a,* or with zero bias, in the *photovoltaic* mode, shown in Fig. 2.3*b.* The photoconductive mode offers higher speed; it is therefore better suited to the detection of high-speed light pulses and to high-frequency light-beam modulation applications. The photovoltaic mode offers lower noise and is therefore better suited to measurement and instrumentation applications. The circuit of Fig. 2.3*b* can be used as a light meter by calibrating its output directly in units of light intensity.

2.2
VOLTAGE-TO-CURRENT CONVERTERS

A *voltage-to-current converter* (*V-I* converter), also called a *transconductance amplifier,* accepts an input voltage v_I and yields an output current of the type $i_O = Av_I$, where A is the *gain,* or *sensitivity,* of the circuit, in amperes per volt. For a practical converter, the characteristic takes on the more realistic form

$$i_O = Av_I - \frac{1}{R_o}v_L \tag{2.5a}$$

where v_L is the voltage developed by the output load in response to i_O, and R_o is the converter's output resistance as seen by the load. For true *V-I* conversion, i_O must be independent of v_L; that is, we must have

$$R_o = \infty \tag{2.5b}$$

Since it outputs a current, the circuit needs a load in order to work; leaving the output port open would result in circuit malfunction as i_O would have no path in which to flow. The *voltage compliance* is the range of permissible values of v_L for which the circuit still works properly, before the onset of any saturation effects on the part of the op amp.

If both terminals of the load are uncommitted, the load is said to be of the *floating* type. Frequently, however, one of the terminals is already committed to ground or to another potential. The load is then said to be of the *grounded* type, and the current from the converter must be fed to the uncommitted terminal.

Floating-Load Converters

Figure 2.4 shows two basic implementations, both of which use the load itself as the feedback element; if one of the load terminals were already committed, it would of course not be possible to use the load as the feedback element.

In the circuit of Fig. 2.4a the op amp outputs whatever current i_O it takes to make the inverting-input voltage follow v_I, or to make $Ri_O = v_I$. Solving for i_O yields

$$i_O = \frac{1}{R}v_I \tag{2.6}$$

This expression holds regardless of the type of load: it can be linear, as for a resistive transducer; it can be nonlinear, as for a diode; it can have time-dependent characteristics, as for a capacitor. No matter what the load, the op amp will force it to carry the current of Eq. (2.6), which depends on the control voltage v_I and the current-setting resistance R, but not on the load voltage v_L. To achieve this goal, the op amp must swing its output to the value $v_O = v_I + v_L$, something it will readily do as long as $V_{OL} < v_O < V_{OH}$. Consequently, the voltage compliance of the circuit is $(V_{OL} - v_I) < v_L < (V_{OH} - v_I)$.

In the circuit of Fig. 2.4b the op amp keeps its inverting input at 0 V. Consequently, its output terminal must draw the current $i_O = (v_I - 0)/R$, and it must

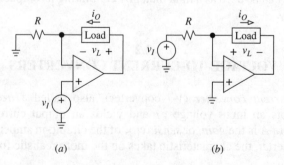

(a)   *(b)*

FIGURE 2.4
Floating-load *V-I* converters.

swing to the voltage $v_O = -v_L$. Apart from the polarity reversal, the current is the same as in Eq. (2.6); however, the voltage compliance is now $V_{OL} < v_L < V_{OH}$.

We observe that Eq. (2.6) holds for both circuits regardless of the polarity of v_I. The arrows of Fig. 2.4 show current direction for $v_I > 0$; making $v_I < 0$ will simply reverse the direction. The two converters are thus said to be *bidirectional*.

Of special importance is the case in which the load is a capacitor, so that the circuit is the familiar integrator. If v_I is kept constant, the circuit will force a constant current through the capacitor, causing it to charge or discharge, depending on the polarity of v_I, at a constant rate. This forms the basis of waveform generators such as sawtooth and triangular waveforms generators, *V-F* and *F-V* converters, and dual-ramp A-D converters.

A drawback of the converter of Fig. 2.4*b* is that i_O must come from the source v_I itself, whereas in Fig. 2.4*a* the source sees a virtually infinite input resistance. This advantage, however, is offset by a more restricted voltage compliance. The maximum current either circuit can deliver to the load depends on the op amp. For the 741, this is typically 25 mA. If larger currents are required, one can either use a power op amp or a low-power op amp with an output current booster.

> **EXAMPLE 2.3.** Let both circuits of Fig. 2.4 have $v_I = 5$ V, $R = 10$ kΩ, $\pm V_{\text{sat}} = \pm 13$ V, and a resistive load R_L. For both circuits find (*a*) i_O; (*b*) the voltage compliance; (*c*) the maximum permissible value of R_L.

Solution.

(*a*) $i_O = 5/10 = 0.5$ mA, flowing from right to left in the circuit of Fig. 2.4*a* and from left to right in that of Fig. 2.4*b*.

(*b*) For the circuit of Fig. 2.4*a*, -18 V $< v_L < 8$ V; for the circuit of Fig. 2.4*b*, -13 V $< v_L < 13$ V.

(*c*) With a purely resistive load, v_L will always be positive. For the circuit of Fig. 2.4*a*, $R_L < 8/0.5 = 16$ kΩ; for the circuit of Fig. 2.4*b*, $R_L < 13/0.5 = 26$ kΩ.

Practical Op Amp Limitations

We now wish to investigate the effect of using a practical op amp. After the op amp is replaced with its practical model, the circuit of Fig. 2.4*a* becomes as in Fig. 2.5. Summing voltages, we get $v_I - v_D + v_L + r_o i_O - a v_D = 0$. Summing currents,

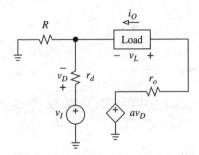

FIGURE 2.5
Investigating the effect of using a practical op amp.

$i_O + v_D/r_d - (v_I - v_D)/R = 0$. Eliminating v_D and rearranging, we can put i_O in the form of Eq. (2.5a) with

$$A = \frac{1}{R} \frac{a - R/r_d}{1 + a + r_o/R + r_o/r_d} \qquad R_o = (R \parallel r_d)(1 + a) + r_o \qquad (2.7)$$

It is apparent that as $a \to \infty$, we get the ideal results $A \to 1/R$ and $R_o \to \infty$. However, for a finite gain a, A will exhibit some error, and R_o, though large, will not be infinite, indicating a weak dependence of i_O on v_L. Similar considerations hold for the circuit of Fig. 2.4b.

Grounded-Load Converters

When one of its terminals is already committed, the load can no longer be placed within the feedback loop of the op amp. Figure 2.6a shows a converter suitable for grounded loads. Referred to as the *Howland current pump* after its inventor, the circuit consists of an input source v_I with series resistance R_1, and a negative-resistance converter synthesizing a grounded resistance of value $-R_2R_3/R_4$. The circuit seen by the load admits the Norton equivalent of Fig. 2.6b, whose *i-v* characteristic is given by Eq. (2.5a). We wish to find the overall output resistance R_o seen by the load.

To this end, we first perform a source transformation on the input source v_I and its resistance R_1, and then we connect the negative resistance in parallel, as depicted in Fig. 2.7. We have $1/R_o = 1/R_1 + 1/(-R_2R_3/R_4)$. Expanding and rearranging, we get

$$R_o = \frac{R_2}{R_2/R_1 - R_4/R_3} \qquad (2.8)$$

As we know, for true current-source behavior we must have $R_o = \infty$. To achieve this condition, the four resistances must form a *balanced bridge:*

$$\frac{R_4}{R_3} = \frac{R_2}{R_1} \qquad (2.9)$$

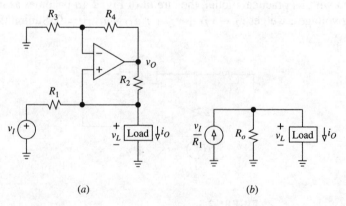

(a) (b)

FIGURE 2.6
Howland current pump and its Norton equivalent.

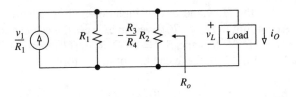

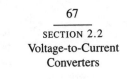

FIGURE 2.7
Using a negative resistance to control R_o.

When this condition is met, the output becomes independent of v_L:

$$i_O = \frac{1}{R_1} v_I \tag{2.10}$$

Clearly, the gain of the converter is $1/R_1$. For $v_I > 0$ the circuit will *source* current to the load, and for $v_I < 0$ it will *sink* current. Since $v_L = v_O R_3/(R_3 + R_4) = v_O R_1/(R_1 + R_2)$, the voltage compliance is, assuming symmetric output saturation,

$$|v_L| \le \frac{R_1}{R_1 + R_2} V_{\text{sat}} \tag{2.11}$$

For the purpose of extending the compliance it is thus desirable to keep R_2 sufficiently smaller than R_1 (e.g., $R_2 \cong 0.1R_1$).

If a fixed source or sink is needed, v_I can be obtained from one of the dc supply voltages, in the manner of the offsetting amplifier of Example 1.5.

EXAMPLE 2.4. Using a 741 op amp powered from ± 15-V regulated supplies, design a 1-mA dc source having a 10-V voltage compliance.

Solution. Letting $v_I = +15$ V, we obtain, from Eq. (2.10), $R_1 = 15/1 = 15$ kΩ. By Eq. (2.11), we want $10 \cong 13 \times R_1/(R_1 + R_2)$, that is, $R_2 = 0.3R_1$. Pick $R_1 = R_3 = 15.0$ kΩ (1%), and $R_2 = R_4 = 0.3 \times 15 = 4.5$ kΩ (use 4.42 kΩ, 1%). The circuit is shown in Fig. 2.8, along with its Norton equivalent.

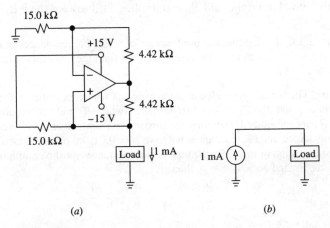

FIGURE 2.8
A 1-mA source and its Norton equivalent.

We observe that the Howland circuit includes both a *negative* and a *positive* feedback path. According to Eq. (1.76), we can express the corresponding feedback factors as $\beta_N = R_3/(R_3 + R_4)$ and $\beta_P = (R_1//R_L)/[(R_1//R_L) + R_2]$, where R_L denotes the load. By Eq. (2.9), these can be put in the form

$$\beta_N = \frac{1}{1 + R_2/R_1} \qquad \beta_P = \frac{1}{1 + R_2/R_1 + R_2/R_L}$$

It is apparent that as long as the circuit is terminated on some finite load $R_L < \infty$, we have $\beta_N > \beta_P$, indicating that negative feedback will prevail, thus resulting in a stable circuit.

Effect of Resistance Mismatches

In a practical circuit the resistive bridge is likely to be unbalanced because of resistance tolerances. This will inevitably degrade R_o, which should be infinite for true current-source behavior. It is therefore of interest to estimate the worst-case value of R_o for given resistance-tolerance specifications.

An unbalanced bridge implies unequal resistance ratios in Eq. (2.9), a condition that we can express in terms of the *imbalance factor* ϵ as

$$\frac{R_4}{R_3} = \frac{R_2}{R_1}(1 - \epsilon) \tag{2.12}$$

Substituting in Eq. (2.8) and simplifying, we obtain

$$R_o = \frac{R_1}{\epsilon} \tag{2.13}$$

As expected, the smaller the imbalance, the larger R_o. In the limit of perfect balance, or as $\epsilon \to 0$, we would of course have $R_o \to \infty$. We observe that ϵ and therefore R_o can be either positive or negative, depending on the direction in which the bridge is unbalanced. By Eq. (2.5a), $-1/R_o$ represents the slope of the i_O versus v_L characteristic. Consequently, $R_o = \infty$ implies a perfectly horizontal characteristic, $R_o > 0$ implies a tilt toward the right, and $R_o < 0$ implies a tilt toward the left.

EXAMPLE 2.5. (*a*) Discuss the implications of using 1% resistances in the circuit of Example 2.4. (*b*) Repeat for 0.1% resistances. (*c*) Find the resistance tolerance needed for $|R_o| \geq 50$ MΩ.

Solution. The worst-case bridge imbalance occurs when, for instance, the ratio R_2/R_1 is maximized and R_4/R_3 is minimized, that is, when R_2 and R_3 are maximized and R_1 and R_4 are minimized. Denoting the percentage tolerance of the resistances as p so that, for instance, for 1% resistances we have $p = 0.01$, we observe that to achieve the balanced condition of Eq. (2.9), the minimized resistances must be multiplied by $1 + p$, and the maximized ones by $1 - p$, thus giving

$$\frac{R_4(1 + p)}{R_3(1 - p)} = \frac{R_2(1 - p)}{R_1(1 + p)}$$

Rearranging, we get

$$\frac{R_4}{R_3} = \frac{R_2(1 - p)^2}{R_1(1 + p)^2} \cong \frac{R_2}{R_1}(1 - p)^2(1 - p)^2 \cong \frac{R_2}{R_1}(1 - 4p)$$

where we have exploited the fact that for $p \ll 1$ we can approximate $1/(1+p) \cong 1-p$ and we can ignore terms in p^n, $n \geq 2$. Comparison with Eq. (2.12) indicates that we can write

$$|\epsilon|_{max} \cong 4p$$

(a) For 1% resistances we have $|\epsilon|_{max} \cong 4 \times 0.01 = 0.04$, indicating a resistance ratio mismatch as large as 4%. Thus, $|R_o|_{min} = R_1/|\epsilon|_{max} \cong 15.0/0.04 = 375 \text{ k}\Omega$, indicating that with 1% resistances we can expect R_o to be anywhere in the range $|R_o| \geq 375 \text{ k}\Omega$.

(b) Improving the tolerance by an order of magnitude increases $|R_o|_{min}$ by the same amount, so $|R_o| \geq 3.75 \text{ M}\Omega$.

(c) For $|R_o|_{min} = 50 \text{ M}\Omega$, we need $|\epsilon|_{max} = R_1/|R_o|_{min} = (15 \times 10^3)/(50 \times 10^6) = 3 \times 10^{-4}$. Then $p \leq |\epsilon|_{max}/4 = 3 \times 10^{-4}/4 = 0.0075\%$, implying highly precise resistors!

An alternative to highly precise resistors is to make provision for resistance trimming. However, a good designer will strive to avoid trimmers whenever possible because they are mechanically and thermally unstable, they have finite resolution, and they are bulkier than ordinary resistors. Moreover, the calibration procedure increases production costs. There are, nonetheless, situations in which, after a careful analysis of cost, complexity, and other pertinent factors, trimming still proves preferable.

Figure 2.9 shows a setup for the calibration of the Howland circuit. The input is grounded, and the load is replaced by a sensitive ammeter initially connected to ground. In this state the ammeter reading should be zero; however, because of op amp nonidealities such as the input bias current and the input offset voltage, to be discussed in Chapter 5, the reading will generally be nonzero, albeit small. To calibrate the circuit for $R_O = \infty$, we flip the ammeter to some other voltage, such as 5 V, and we adjust the wiper for the same ammeter reading as when the ammeter is connected to ground.

EXAMPLE 2.6. In the circuit of Example 2.4 specify a suitable trimmer/resistor replacement for R_3 to allow bridge balancing in the case of 1% resistances.

Solution. Since $4pR_1 = 4 \times 0.01 \times 15.0 = 0.6 \text{ k}\Omega$, the series resistance R_s must be smaller than 15.0 kΩ by at least 0.6 kΩ, or $R_s \leq 15 - 0.6 = 14.4 \text{ k}\Omega$. To be on the safe side, let $R_s = 14.0 \text{ k}\Omega$ (1%). Then $R_{pot} = 2(15 - 14) = 2 \text{ k}\Omega$. Summarizing, $R_{pot} = 2 \text{ k}\Omega$, $R_s = 14.0 \text{ k}\Omega$ (1%), and all other resistors remain the same.

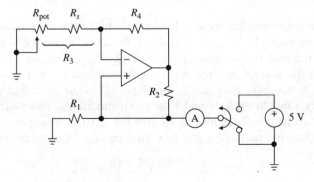

FIGURE 2.9
Howland circuit calibration.

Effect of Finite Open-Loop Gain

We now investigate the effect of a finite open-loop gain on the transfer characteristic of the Howland circuit. To evidence the effect of the op amp alone, we assume the resistances to form a perfectly balanced bridge. With reference to Fig. 2.6a, we have, by KCL, $i_O = (v_I - v_L)/R_1 + (v_O - v_L)/R_2$. The circuit can be viewed as a noninverting amplifier that amplifies v_L to yield $v_O = v_L a/[1 + a R_3/(R_3 + R_4)]$. Using Eq. (2.9), this can be written as $v_O = v_L a/[1 + a R_1/(R_1 + R_2)]$. Eliminating v_O and rearranging yields $i_O = (1/R_1)v_I - (1/R_o)v_L$, where

$$R_o = (R_1 \parallel R_2) \left(1 + \frac{a}{1 + R_2/R_1} \right) \tag{2.14}$$

[This expression could have been obtained also via Eq. (1.71).] A finite open-loop gain leaves the sensitivity $1/R_1$ unchanged; however, it decreases R_o from ∞ to the value given in Eq. (2.14).

> **EXAMPLE 2.7.** Find the output resistance of the 1-mA source of Example 2.4. Confirm your results using PSpice.
>
> **Solution.** $R_o = (15 \parallel 3) \times 10^3 [1 + 200 \times 10^3/(1 + 3/15)] = 417$ MΩ. Using the subcircuit OA discussed at the end of Section 1.2, we write the following circuit file:
>
> ```
> Finding Ro for the Howland circuit:
> R1 0 1 15k          ;bottom left resistance
> R2 1 3 3k           ;bottom right resistance
> R3 0 2 15k          ;top left resistance
> R4 2 3 3k           ;top right resistance
> X 1 2 3 OA          ;activates the op amp
> itest 0 1 1nA       ;applies a 1-nA test current
> .end
> ```
>
> The PSpice simulation gives a voltage of 0.4120 at node 1, so $R_o = 0.4120/10^{-9} = 412$ MΩ, which is close enough to the predicted value.

Improved Howland Current Pump

Depending on circuit conditions, the Howland circuit can be unnecessarily wasteful of power. As an example, let $v_I = 1$ V, $R_1 = R_3 = 1$ kΩ, and $R_2 = R_4 = 100$ Ω, and suppose the load is such that $v_L = 10$ V. By Eq. (2.10), $i_O = 1$ mA. Note, however, that the current through R_1 toward the left is $i_1 = (v_L - v_I)/R_1 = (10 - 1)/1 = 9$ mA, indicating that the op amp will have to waste 9 mA through R_1 to deliver only 1 mA to the load under the given conditions. This inefficient use of power can be avoided with the modification of Fig. 2.10, in which the resistance R_2 has been split into two parts, R_{2A} and R_{2B}, such that the balanced condition is now

$$\frac{R_4}{R_3} = \frac{R_{2A} + R_{2B}}{R_1} \tag{2.15a}$$

It is left as an exercise (see Problem 2.12) to prove that when this condition is met,

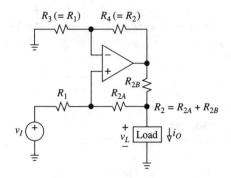

FIGURE 2.10
Improved Howland circuit.

the load still sees $R_o = \infty$, but the transfer characteristic is now

$$i_O = \frac{R_2/R_1}{R_{2B}} v_I \tag{2.15b}$$

Aside from the gain term R_2/R_1, the sensitivity is now set by R_{2B}, indicating that R_{2B} can be made as small as needed while the remaining resistances are kept high in order to conserve power. For instance, letting $R_{2B} = 1\text{ k}\Omega$, $R_1 = R_3 = R_4 = 100\text{ k}\Omega$, and $R_{2A} = 100 - 1 = 99\text{ k}\Omega$, we still get $i_O = 1$ mA with $v_I = 1$ V. However, even with $v_L = 10$ V, very little power is now wasted in the large 100-kΩ resistances. The voltage compliance is approximately $|v_L| \leq |V_{\text{sat}}| - R_{2B}|i_O|$. By Eq. (2.15b), this can be written as $|v_L| \leq |V_{\text{sat}}| - (R_2/R_1)|v_I|$.

Since Howland circuits employ both positive and negative feedback, they may become oscillatory under certain conditions.[2] Two small capacitors (typically on the order of 10 pF) in parallel with R_4 and R_1 are usually adequate to make negative feedback prevail over positive feedback at high frequencies and thus stabilize the circuit.

2.3
CURRENT AMPLIFIERS

Even though op amps are voltage amplifiers, they can also be configured for current amplification. The transfer characteristic of a practical current amplifier is of the type

$$i_O = A i_I - \frac{1}{R_o} v_L \tag{2.16a}$$

where A is the gain in amperes per ampere, v_L is the output load voltage, and R_o is the output resistance as seen by the load. To make i_O independent of v_L, a current amplifier must have

$$R_o = \infty \tag{2.16b}$$

Current-mode amplifiers are used in applications in which information is more conveniently represented in terms of current than in terms of voltage, for example, in two-wire remote sensing instrumentation, photodetector output conditioning, and V-F converter input conditioning.[3]

Figure 2.11 shows a current amplifier with a floating load. Assume first that the op amp is ideal. By KCL, i_O is the sum of the currents through R_1 and R_2, or

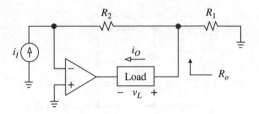

FIGURE 2.11
Floating-load current amplifier.

$i_O = i_I + (R_2 i_I)/R_1$, or $i_O = A i_I$, where

$$A = 1 + \frac{R_2}{R_1} \qquad (2.17)$$

This holds regardless of v_L, indicating that the circuit yields $R_o = \infty$. If the op amp has a finite gain a, one can prove (see Problem 2.20) that

$$A = 1 + \frac{R_2/R_1}{1 + 1/a} \qquad R_o = R_1(1 + a) \qquad (2.18)$$

indicating a gain error as well as a finite output resistance. One can readily verify that the voltage compliance is $-(V_{OH} + R_2 i_I) \leq v_L \leq -(V_{OL} + R_2 i_I)$.

Figure 2.12 shows a grounded-load current amplifier. Because of the virtual short, the voltage across the input source is v_L, so the current entering R_2 from the left is $i_S - v_L/R_s$. The op amp output is then $v_{OA} = v_L - R_2(i_S - v_L/R_s)$. By KCL and Ohm's law, $i_O = (v_{OA} - v_L)/R_1$. Eliminating v_{OA} gives $i_O = A i_S - (1/R_o)v_L$, where

$$A = -\frac{R_2}{R_1} \qquad R_o = -\frac{R_1}{R_2} R_s \qquad (2.19)$$

The negative gain indicates that the actual direction of i_O is opposite to that shown. Consequently, sourcing current to (or sinking current from) the circuit will cause it to sink current from (or source current to) the load. If $R_1 = R_2$, then $A = -1$ A/A and the circuit functions as a *current reverser,* or *current mirror.*

We observe that R_o is negative, something we could have anticipated by comparing our amplifier with the negative-resistance converter of Fig. 1.20*b*. The fact that R_o is finite indicates that i_O is not independent of v_L. To avoid this shortcoming, the circuit is used primarily in connection with loads of the virtual-ground type ($v_L = 0$), as in certain types of current-to-frequency converters and logarithmic amplifiers.

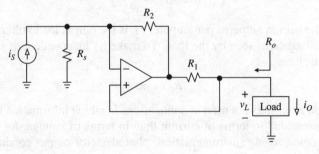

FIGURE 2.12
Grounded-load current amplifier.

2.4
DIFFERENCE AMPLIFIERS

The difference amplifier was introduced in Section 1.4, but since it forms the basis of other important circuits, such as instrumentation and bridge amplifiers, we now wish to analyze it in greater detail. Referring to Fig. 2.13a, we recall that as long as the resistances satisfy the balanced-bridge condition

$$\frac{R_4}{R_3} = \frac{R_2}{R_1} \tag{2.20a}$$

the circuit is a true difference amplifier, that is, its output is linearly proportional to the difference of its inputs,

$$v_O = \frac{R_2}{R_1}(v_2 - v_1) \tag{2.20b}$$

The unique characteristics of the difference amplifier are better appreciated if we introduce the *differential-mode* and the *common-mode* input components, defined as

$$v_{DM} = v_2 - v_1 \tag{2.21a}$$

$$v_{CM} = \frac{v_1 + v_2}{2} \tag{2.21b}$$

Inverting these equations, we can express the actual inputs in terms of the newly defined components:

$$v_1 = v_{CM} - \frac{v_{DM}}{2} \tag{2.22a}$$

$$v_2 = v_{CM} + \frac{v_{DM}}{2} \tag{2.22b}$$

This allows us to redraw the circuit in the form of Fig. 2.13b. We can now concisely define a true difference amplifier as a circuit that responds only to the differential-mode component v_{DM}, completely ignoring the common-mode component v_{CM}. In particular, if we tie the inputs together to make $v_{DM} = 0$, and we apply a common voltage $v_{CM} \neq 0$, a true difference amplifier will yield $v_O = 0$ regardless of the magnitude and polarity of v_{CM}. Conversely, this can serve as a test for finding how

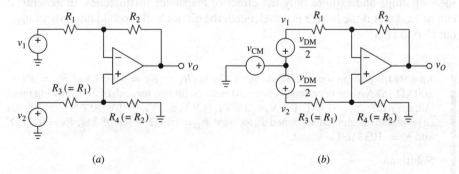

(a)   (b)

FIGURE 2.13
(a) Difference amplifier. (b) Expressing the inputs in terms of the common-mode and differential-mode components v_{CM} and v_{DM}.

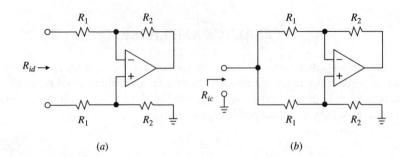

(a) $\qquad\qquad\qquad\qquad$ (b)

FIGURE 2.14
Differential-mode and common-mode input resistances.

close a practical difference amplifier is to ideal. The smaller the output variation due to a given variation of v_{CM} is, the closer the amplifier is to ideal.

The decomposition of v_1 and v_2 into the components v_{DM} and v_{CM} not only is a matter of mathematical convenience but also reflects a situation quite common in practice, that of a low-level differential signal riding on a high common-mode signal, as in the case of transducer signals. The useful signal is the differential one; extracting it from the high common-mode environment and then amplifying it can be a challenging task. Difference-type amplifiers are the natural candidates to meet this challenge.

Figure 2.14 illustrates the *differential-mode* and *common-mode input resistances*. It is readily seen (see Problem 2.26) that

$$R_{id} = 2R_1 \qquad R_{ic} = \frac{R_1 + R_2}{2} \qquad\qquad (2.23)$$

Effect of Resistance Mismatches

A difference amplifier will be insensitive to v_{CM} only as long as the op amp is ideal and the resistors satisfy the balanced-bridge condition of Eq. (2.20a). The effect of op amp nonidealities will be investigated in Chapters 5 and 6; here we shall assume ideal op amps and explore only the effect of resistance mismatches. In general, it can be said that if the bridge is unbalanced, the circuit will respond not only to v_{DM} but also to v_{CM}.

EXAMPLE 2.8. In the circuit of Fig. 2.13a let $R_1 = R_3 = 10\text{ k}\Omega$ and $R_2 = R_4 = 100\text{ k}\Omega$. (a) Assuming perfectly matched resistors, find v_O for each of the following input voltage pairs: $(v_1, v_2) = (-0.1\text{ V}, +0.1\text{ V})$, $(4.9\text{ V}, 5.1\text{ V})$, $(9.9\text{ V}, 10.1\text{ V})$. (b) Repeat (a) with the resistors mismatched as follows: $R_1 = 10\text{ k}\Omega$, $R_2 = 98\text{ k}\Omega$, $R_3 = 9.9\text{ k}\Omega$, and $R_4 = 103\text{ k}\Omega$. Comment.

Solution.

(a) $v_O = (100/10)(v_2 - v_1) = 10(v_2 - v_1)$. Since $v_2 - v_1 = 0.2$ V in each of the three cases, we get $v_O = 10 \times 0.2 = 2$ V regardless of the common-mode component, which is, in order, $v_{CM} = 0$ V, 5 V, and 10 V for the three input voltage pairs.

(b) By the superposition principle, $v_O = A_2 v_2 - A_1 v_1$, where $A_2 = (1 + R_2/R_1)/(1 + R_3/R_4) = (1 + 98/10)/(1 + 9.9/103) = 9.853$ V/V, and $A_1 = R_2/R_1 = 98/10 = 9.8$ V/V. Thus, for $(v_1, v_2) = (-0.1 \text{ V}, +0.1 \text{ V})$ we obtain $v_O = 9.853(0.1) - 9.8(-0.1) = 1.965$ V. Likewise, for $(v_1, v_2) = (4.9 \text{ V}, 5.1 \text{ V})$ we get $v_O = 2.230$ V, and for $(v_1, v_2) = (9.9 \text{ V}, 10.1 \text{ V})$ we get $v_O = 2.495$ V. As a consequence of mismatched resistors, not only do we have $v_O \neq 2$ V, but v_O also changes with the common-mode component. Clearly the circuit is no longer a true difference amplifier.

The effect of bridge imbalance can be investigated more systematically by introducing the *imbalance factor* ϵ, in the manner of the Howland circuit of Section 2.2. With reference to Fig. 2.15 we conveniently assume that three of the resistances possess their nominal values while the fourth is expressed as $R_2(1 - \epsilon)$ to account for the imbalance. Applying the superposition principle,

$$v_O = -\frac{R_2(1 - \epsilon)}{R_1}\left(v_{CM} - \frac{v_{DM}}{2}\right) + \frac{R_1 + R_2(1 - \epsilon)}{R_1} \times \frac{R_2}{R_1 + R_2}\left(v_{CM} + \frac{v_{DM}}{2}\right)$$

Multiplying out and collecting terms, we can put v_O in the insightful form

$$v_O = A_{dm}v_{DM} + A_{cm}v_{CM} \tag{2.24a}$$

$$A_{dm} = \frac{R_2}{R_1}\left(1 - \frac{R_1 + 2R_2}{R_1 + R_2}\frac{\epsilon}{2}\right) \tag{2.24b}$$

$$A_{cm} = \frac{R_2}{R_1 + R_2}\epsilon \tag{2.24c}$$

As expected, Eq. (2.24a) states that with an unbalanced bridge, the circuit responds not only to v_{DM} but also to v_{CM}. For obvious reasons A_{dm} and A_{cm} are called, respectively, the *differential-mode gain* and the *common-mode gain*. Only in the limit $\epsilon \to 0$ do we obtain the ideal results $A_{dm} = R_2/R_1$ and $A_{cm} = 0$.

The ratio A_{dm}/A_{cm} represents a figure of merit of the circuit and is called the *common-mode rejection ratio* (CMRR). Its value is expressed in decibels (dB) as

$$\text{CMRR}_{dB} = 20 \log_{10}\left|\frac{A_{dm}}{A_{cm}}\right| \tag{2.25}$$

For a true difference amplifier, $A_{cm} \to 0$ and thus $\text{CMRR}_{dB} \to \infty$. For a sufficiently small imbalance factor ϵ, the second term within parentheses in Eq. (2.24b) can be ignored in comparison with unity, and we can write $A_{dm}/A_{cm} \cong (R_2/R_1)/[R_2\epsilon/(R_1 +$

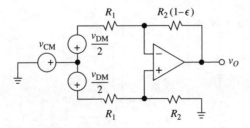

FIGURE 2.15
Investigating the effect of resistance mismatches.

$R_2)$], or

$$\text{CMRR}_{\text{dB}} \cong 20 \log_{10} \left| \frac{1 + R_2/R_1}{\epsilon} \right| \qquad (2.26)$$

The reason for using the absolute value is that ϵ can be positive or negative, depending on the direction of the imbalance. Note that for a given ϵ, the larger the differential gain R_2/R_1, the higher the CMRR of the circuit.

EXAMPLE 2.9. In Fig. 2.13a let $R_1 = R_3 = 10$ kΩ and $R_2 = R_4 = 100$ kΩ. (*a*) Discuss the implications of using 1% resistors. (*b*) Illustrate the case in which the inputs are tied together and are driven by a common 10-V source. (*c*) Estimate the resistance tolerance needed for a guaranteed CMRR of 80 dB.

Solution.

(*a*) Proceeding along lines similar to those in Example 2.5, we can write $|\epsilon|_{\text{max}} \cong 4p$, where p is the percentage tolerance. With $p = 1\% = 0.01$, we get $|\epsilon|_{\text{max}} \cong 0.04$. The worst-case scenario corresponds to $A_{\text{dm(min)}} \cong (100/10)[1-(210/110) \times 0.04/2] = 9.62$ V/V $\neq 10$ V/V, and $A_{\text{cm(max)}} \cong (100/110) \times 0.04 = 0.0364 \neq 0$. Thus, $\text{CMRR}_{\text{min}} = 20 \log_{10}(9.62/0.0364) = 48.4$ dB.

(*b*) With $v_{\text{DM}} = 0$ and $v_{\text{CM}} = 10$ V, the output error can be as large as $v_O = A_{\text{cm(max)}} \times v_{\text{CM}} = 0.0364 \times 10 = 0.364$ V $\neq 0$.

(*c*) To achieve a higher CMRR, we need to further decrease ϵ. By Eq. (2.26), $80 \cong 20 \log_{10}[(1 + 10)/|\epsilon|_{\text{max}}]$, or $|\epsilon|_{\text{max}} = 1.1 \times 10^{-3}$. Then $p = |\epsilon|_{\text{max}}/4 = 0.0275\%$.

It is apparent that for high CMRRs the resistors must be very tightly matched. The INA105 (Burr-Brown[4]) is a general-purpose monolithic difference amplifier with four identical resistors that are matched within 0.002%. In that case, Eq. (2.26) yields $\text{CMRR}_{\text{dB}} = 100$ dB.

The CMRR of a practical amplifier can be maximized by adjusting one of its resistors, usually R_4. This is shown in Fig. 2.16. The selection of the series resistance R_s and R_{pot} follows the lines of the Howland circuit of Example 2.6. Calibration is done with the inputs tied together to eliminate v_{DM} and evidence only v_{CM}. The latter is then flipped back and forth between two predetermined values, such as -5 V and $+5$ V, and the wiper is adjusted for a minimum variation at the output. To preserve bridge balance with temperature and aging, it is advisable to use a metal-film resistor array.

So far we have assumed ideal op amps. When studying their practical limitations in Chapter 5, we shall see that op amps are themselves sensitive to v_{CM}, so the

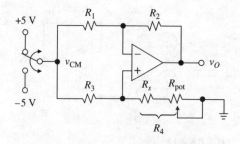

FIGURE 2.16
Difference-amplifier calibration.

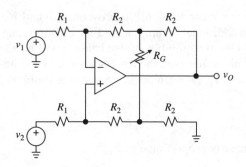

FIGURE 2.17
Difference amplifier with variable gain.

CMRR of a practical difference amplifier is actually the result of two effects: bridge imbalance and op amp nonideality. The two effects are interrelated so that it is possible to unbalance the bridge in such a way as to approximately cancel out the effect of the op amp. Indeed, this is what we do when we seek the minimum output variation during the calibration routine.

Variable Gain

Equation (2.20b) might leave the impression that gain can be varied by varying just one resistor, say, R_2. Since we must also satisfy Eq. (2.20a), two resistors rather than one would have to be varied, and in such a way as to maintain a very tight matching. This awkward task is avoided with the modification of Fig. 2.17, which makes it possible to vary the gain without disturbing bridge balance. It is left as an exercise (see Problem 2.27) to prove that if the various resistances are in the ratios shown, then

$$v_O = \frac{2R_2}{R_1}\left(1 + \frac{R_2}{R_G}\right)(v_2 - v_1) \tag{2.27}$$

so that gain can be varied by varying the single resistor R_G.

It is often desirable that gain vary linearly with the adjusting potentiometer to facilitate gain readings from potentiometer settings. Unfortunately, the circuit of

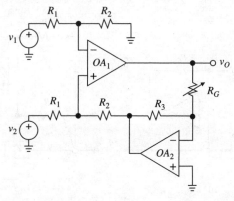

FIGURE 2.18
Difference amplifier with linear gain control.

Fig. 2.17 exhibits a nonlinear relationship between gain and R_G. This drawback is avoided by using an additional op amp, as in Fig. 2.18. As long as the closed-loop output resistance of OA_2 is negligible, bridge balance will be unaffected. Moreover, since OA_2 provides phase reversal, the feedback signal must now be applied to the noninverting input of OA_1. One can readily prove (see Problem 2.28) that

$$v_O = \frac{R_2 R_G}{R_1 R_3}(v_2 - v_1) \tag{2.28}$$

so that the gain is linearly proportional to R_G.

Ground-Loop Interference Elimination

In practical installations source and amplifier are often far apart and share the common ground bus with a variety of other circuits. Far from being a perfect conductor, the ground bus has a small distributed resistance, inductance, and capacitance and thus behaves as a distributed impedance. Under the effect of the various currents flowing on the bus, this impedance will develop a small voltage drop, causing different points on the bus to be at slightly different potentials. In Fig. 2.19, Z_g denotes the ground-bus impedance between the input signal common N_i and the output signal common N_o, and v_g is the corresponding voltage drop. Ideally, v_g should have no effect on circuit performance.

Consider the arrangement of Fig. 2.19a, where v_i is to be amplified by an ordinary inverting amplifier. Unfortunately, the amplifier sees v_i and v_g in series, so

$$v_o = -\frac{R_2}{R_1}(v_i + v_g) \tag{2.29}$$

The presence of the v_g term, generally referred to as *ground-loop interference* or also *cross-talk for common return impedance,* may degrade the quality of the output signal appreciably, especially if v_i happens to be a low-level signal of magnitude comparable to v_g, as is often the case with transducer signals in industrial environments.

We can get rid of the v_g term by regarding v_i as a differential signal and v_g as a common-mode signal. Doing so requires changing the original amplifier to a

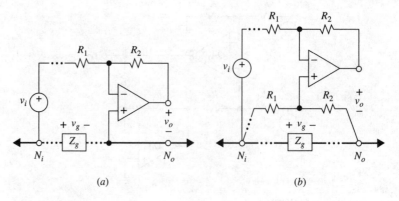

FIGURE 2.19
Using a difference amplifier to eliminate ground-loop interference.

difference-type amplifier and using an additional wire for direct access to the input signal common, in the manner shown in Fig. 2.19b. By inspection, we now have

$$v_O = -\frac{R_2}{R_1}v_i \qquad (2.30)$$

The price we are paying in increased circuit complexity and wiring is certainly worth the benefits derived from the elimination of the v_g term.

2.5
INSTRUMENTATION AMPLIFIERS

An *instrumentation amplifier* (IA) is a difference amplifier meeting the following specifications: (*a*) extremely high (ideally infinite) common-mode and differential-mode input impedances; (*b*) very low (ideally zero) output impedance; (*c*) accurate and stable gain, typically in the range of 1 V/V to 10^3 V/V; and (*d*) extremely high common-mode rejection ratio. The IA is used to accurately amplify a low-level signal in the presence of a large common-mode component, such as a transducer output in process control and biomedicine. For this reason, IAs find widespread application in test and measurement instrumentation—hence the name.

With proper trimming, the difference amplifier of Fig. 2.13 can be made to meet the last three specifications satisfactorily. However, by Eq. (2.23), it fails to meet the first specification because both its differential-mode and its common-mode input resistances are finite; consequently, it will generally load down the circuit supplying the voltages v_1 and v_2, not to mention the ensuing degradation in the CMRR. These drawbacks are eliminated by preceding it with two high-input-impedance buffers. The result is a classic circuit known as the *triple-op-amp IA*.

Triple-Op-Amp IAs

In Fig. 2.20 OA_1 and OA_2 form what is often referred to as the *input* or *first* stage, and OA_3 forms the *output* or *second* stage. By the input voltage constraint, the voltage across R_G is $v_1 - v_2$. By the input current constraint, the resistances denoted R_3 carry the same current as R_G. Applying Ohm's law yields $v_{O1} - v_{O2} = (R_3 + R_G + R_3)(v_1 - v_2)/R_G$, or

$$v_{O1} - v_{O2} = \left(1 + \frac{2R_3}{R_G}\right)(v_1 - v_2)$$

For obvious reasons the input stage is also referred to as a *difference-input, difference-output amplifier*. Next, we observe that OA_3 is a difference amplifier, and thus

$$v_O = \frac{R_2}{R_1}(v_{O2} - v_{O1})$$

Combining the last two equations gives

$$v_O = A(v_2 - v_1) \qquad (2.31a)$$

$$A = A_I \times A_{II} = \left(1 + 2\frac{R_3}{R_G}\right) \times \left(\frac{R_2}{R_1}\right) \qquad (2.31b)$$

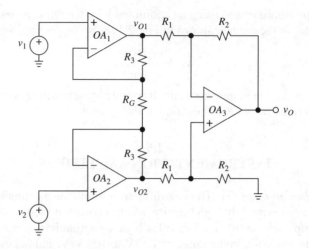

FIGURE 2.20
Triple-op-amp instrumentation amplifier.

indicating that the overall gain A is the product of the first- and second-stage gains A_I and A_{II}.

The gain depends on external resistance ratios, so it can be made quite accurate and stable by using resistors of suitable quality. Since OA_1 and OA_2 are operated in the noninverting configuration, their closed-loop input resistances are extremely high. Likewise, the closed-loop output resistance of OA_3 is quite low. Finally, the CMRR can be maximized by proper trimming of one of the second-stage resistances. We conclude that the circuit meets all the IA requisites listed earlier.

Equation (2.31b) points the way to go if variable gain is desired. To avoid perturbing bridge balance, we leave the second stage undisturbed and we vary gain by varying the single resistance R_G. If linear gain control is desired, we can use an arrangement of the type of Fig. 2.18.

EXAMPLE 2.10. (a) Design an IA whose gain can be varied over the range 1 V/V $\leq A \leq 10^3$ V/V by means of a 100-kΩ pot. (b) Make provisions for a trimmer to optimize its CMRR. (c) Outline a procedure for calibrating the trimmer.

Solution.

(a) Connect the 100-kΩ pot as a variable resistor, and use a series resistance R_4 to prevent R_G from going to zero. Since $A_I > 1$ V/V, we require $A_{II} < 1$ V/V in order to allow A to go all the way down to 1 V/V. Arbitrarily impose $A_{II} = R_2/R_1 = 0.5$ V/V, and use $R_1 = 100$ kΩ and $R_2 = 49.9$ kΩ, both 1%. By Eq. (2.31b), A_I must be variable from 2 V/V to 2000 V/V. At these extremes we have $2 = 1 + 2R_3/(R_4 + 100$ k$\Omega)$ and $2000 = 1 + 2R_3/(R_4 + 0)$. Solving, we obtain $R_4 = 50$ Ω and $R_3 = 50$ kΩ. Use $R_4 = 49.9$ Ω and $R_3 = 49.9$ kΩ, both 1%.

(b) Following Example 2.6, $4pR_2 = 4 \times 0.01 \times 49.9$ k$\Omega = 2$ kΩ. To be on the safe side, use a 47.5-kΩ, 1% resistor in series with a 5-kΩ pot. A suitable op amp is the OP-27 precision op amp (Analog Devices). The circuit is shown in Fig. 2.21.

(c) To calibrate the circuit, tie the inputs together and set the 100-kΩ pot for the maximum gain (wiper all the way up). Then, while switching the common inputs back and forth between -5 V and $+5$ V, adjust the 5-kΩ pot for the minimum change at the output.

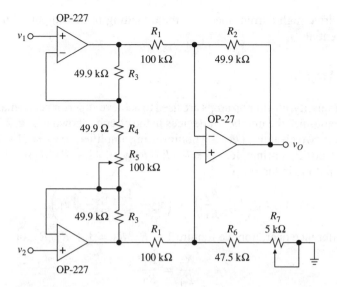

FIGURE 2.21
IA of Example 2.10.

The triple-op-amp IA configuration is available in IC form from various manu-facturers. Familiar examples are the AD522 (Analog Devices) and INA101 (Burr-Brown). These devices contain all components except for R_G, which is supplied externally by the user to set the gain, usually from 1 V/V to 10^3 V/V. Figure 2.22 shows a frequently used circuit symbol for the IA, along with its interconnection for remote sensing. In this arrangement, the *sense* and *reference* voltages are sensed right at the load terminals, so the effect of any signal losses in the long wires is eliminated by including these losses within the feedback loop. The accessibility to these terminals affords additional flexibility, such as the inclusion of an output power

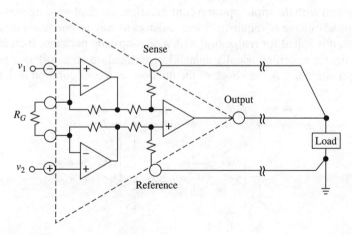

FIGURE 2.22
Standard IA symbol and connection for remote sensing.

booter to drive high-current loads, or the offsetting of the output with respect to ground potential.

Dual-Op-Amp IAs

When high-quality, costlier op amps are used to achieve superior performance, it is of interest to minimize the number of devices in the circuit. Shown in Fig. 2.23 is an IA that uses only two op amps. OA_1 is a noninverting amplifier, so $v_3 = (1 + R_3/R_4)v_1$. By the superposition principle, $v_O = -(R_2/R_1)v_3 + (1 + R_2/R_1)v_2$. Eliminating v_3, we can put v_O in the form

$$v_O = \left(1 + \frac{R_2}{R_1}\right) \times \left(v_2 - \frac{1 + R_3/R_4}{1 + R_1/R_2}v_1\right) \tag{2.32}$$

For true difference operation we require $1 + R_3/R_4 = 1 + R_1/R_2$, or

$$\frac{R_3}{R_4} = \frac{R_1}{R_2} \tag{2.33}$$

When this condition is met, we have

$$v_O = \left(1 + \frac{R_2}{R_1}\right)(v_2 - v_1) \tag{2.34}$$

Moreover, the circuit enjoys high input resistances and low output resistance. To maximize the CMRR, one of the resistors, say, R_4, should be trimmed. The adjustment of the trimmer proceeds as in the triple-op-amp case.

Adding a variable resistance between the inverting inputs of the two op amps as in Fig. 2.24 makes the gain adjustable. It can be shown (see Problem 2.38) that $v_O = A(v_2 - v_1)$, where

$$A = 1 + \frac{R_2}{R_1} + \frac{2R_2}{R_G} \tag{2.35}$$

Compared with the triple-op-amp configuration, the dual-op-amp version offers the obvious advantage of requiring fewer resistors as well as one fewer op amp. The configuration is suited for realization with a dual-op-amp package, such as the OP-227. The tighter matching usually available with dual op amps offers a significant boost in performance. A drawback of the dual-op-amp configuration is that it treats

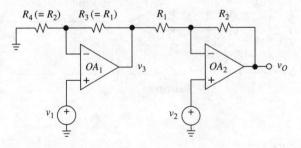

FIGURE 2.23
Dual-op-amp instrumentation amplifier.

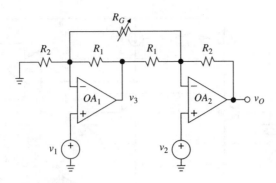

FIGURE 2.24
Dual-op-amp IA with variable gain.

the inputs asymmetrically because v_1 has to propagate through OA_1 before catching up with v_2. Because of this additional delay, the common-mode components of the two signals will no longer cancel each other out as frequency is increased, leading to a premature degradation of the CMRR with frequency. Conversely, the triple-op-amp configuration enjoys a higher degree of symmetry and usually maintains high CMRR performance over a broader frequency range. The factors limiting the CMRR here are mismatches in the delays through the first-stage op amps, as well as bridge imbalance and common-mode limitations of the second-stage op amp.

Monolithic IAs

The need for instrumentation amplification arises so often that it justifies the manufacture of special ICs to perform just this function.[5] Compared with realizations built using general-purpose op amps, this approach allows better optimization of the parameters that are critical to this application, particularly the CMRR, gain linearity, and noise.

The task of first-stage difference amplification as well as common-mode rejection is delegated to highly matched transistor pairs. A transistor pair is faster than a pair of full-fledged op amps and can be made to be less sensitive to common-mode signals, thus relaxing the need for very tightly matched resistances. Examples of dedicated IC IAs are the AD521/524/624/625 and the AMP-01 and AMP-05 (Analog Devices).

Figure 2.25 shows a simplified circuit diagram of the AMP-01, and Fig. 2.26 shows the basic interconnection to make it work with gains ranging from 0.1 V/V to 10^4 V/V. As shown, the gain is set by the ratio of two user-supplied resistors R_S and R_G as

$$A = 20 \frac{R_S}{R_G} \tag{2.36}$$

With this arrangement one can achieve highly stable gains by using a pair of temperature-tracking resistors.

Referring to Fig. 2.25 and the connection of Fig. 2.26, we can describe circuit operation as follows. Applying a differential signal between the inputs unbalances

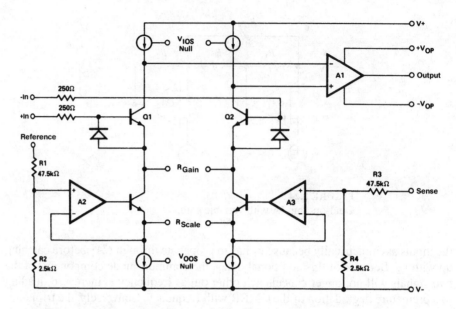

FIGURE 2.25
Simplified circuit diagram of the AMP-01 low-noise precision IA. (Courtesy of Analog
Devices.)

the currents through Q_1 and Q_2. A_1 reacts to this by unbalancing Q_1 and Q_2 in
the opposite direction in order to restore the balanced condition $v_N = v_P$ at its own
inputs. A_1 achieves this by applying a suitable drive to the bottom transistor pair
via A_3. The amount of drive needed depends on the ratio R_S/R_G as well as on the
magnitude of the input difference. This drive forms the output of the IA. Table 2.1
summarizes the salient features of the device.

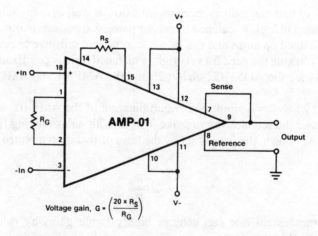

FIGURE 2.26
Basic AMP-01 connection for gains from 0.1 V/V to
10 V/mV. (Courtesy of Analog Devices.)

TABLE 2.1
Summary of AMP-01 characteristics

Offset voltage	$15\ \mu V$
Offset voltage drift	$0.1\ \mu V/^\circ C$
Noise	$0.2\ \mu V_{\text{p-p}}$ (0.1 Hz to 10 Hz)
Output drive	± 10 V @ ± 50 mA
Capacitive load stability	To 1 μF
Gain range	0.1 to 10,000
Linearity	16 bit at $G = 1000$ V/V
CMRR_{dB}	140 dB at $G = 1000$ V/V
Bias current	1 nA
Output stage thermal shutdown	

Courtesy of Analog Devices.

Flying-Capacitor Techniques

A popular alternative for achieving high CMRRs is the *flying-capacitor technique,*
so called because it flips a capacitor back and forth between source and amplifier. As
exemplified[6] in Fig. 2.27, flipping the switches to the left charges C_1 to the voltage
difference $v_2 - v_1$, and flipping the switches to the right transfers charge from C_1
to C_2. Continuous switch clocking causes C_2 to charge up until the equilibrium
condition is reached in which the voltage across C_2 becomes equal to that across
C_1. This voltage is magnified by the noninverting amplifier to give

$$v_O = \left(1 + \frac{R_2}{R_1}\right)(v_2 - v_1) \tag{2.37}$$

To achieve high performance, the circuit shown uses the LTC1043 precision
instrumentation switched-capacitor building block and the LT1013 precision op

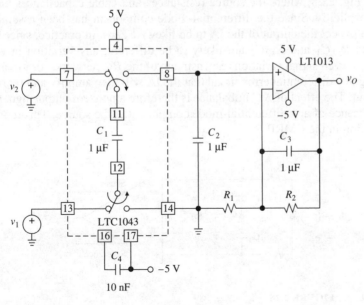

FIGURE 2.27
Flying-capacitor IA. (Courtesy of Linear Technology.)

amp. The former includes an on-chip clock generator to operate the switches at a frequency set by C_4. With $C_4 = 10$ nF, this frequency is 500 Hz. The function of C_3 is to provide low-pass filtering to ensure a clean output. Thanks to the flying-capacitor technique, the circuit completely ignores common-mode input signals to achieve a high CMRR, typically[6] in excess of 120 dB at 60 Hz.

2.6
INSTRUMENTATION APPLICATIONS

In this section we examine some issues arising in the application of instrumentational amplifiers.[7] Additional applications will be discussed in the next section.

Active Guard Drive

In applications such as the monitoring of hazardous industrial conditions, source and amplifier may be located far apart from each other. To help reduce the effect of noise pickup as well as ground-loop interference, the input signal is transmitted in double-ended form over a pair of shielded wires and then processed with a difference amplifier, such as an IA. The advantage of double-ended over single-ended transmission is that since the two wires tend to pick up identical noise, this noise will appear as a common-mode component and will thus be rejected by the IA. For this reason, double-ended transmission is also referred to as *balanced transmission*. The purpose of shielding is to help reduce differential-mode noise pickup.

Unfortunately, because of the distributed capacitance of the cable, another problem arises, namely, CMRR degradation with frequency. To investigate this aspect, refer to Fig. 2.28, where the source resistances and cable capacitances have been shown explicitly. Since the differential-mode component has been assumed to be zero, we expect the output of the IA to be likewise zero. In practice, since the time constants $R_{s1}C_1$ and $R_{s2}C_2$ are likely to be different, any variation in v_{CM} will produce uneven signal variations downstream of the RC networks, or $v_1 \neq v_2$, thus resulting in a differential error signal that the IA will then amplify and reproduce at the output. The effect of RC imbalance is therefore a nonzero output signal in spite of the absence of any differential-mode component at the source. This represents a degradation in the CMRR.

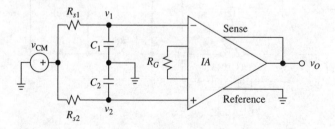

FIGURE 2.28
Model of nonzero source resistance and distributed cable capacitance.

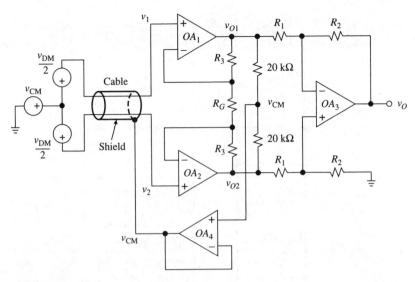

FIGURE 2.29
IA with active guard drive.

The CMRR due to RC imbalance is[7]

$$\mathrm{CMRR_{dB}} \cong 20 \log_{10} \frac{1}{2\pi f R_{\mathrm{dm}} C_{\mathrm{cm}}} \tag{2.38}$$

where $R_{\mathrm{dm}} = |R_{s1} - R_{s2}|$ is the source resistance imbalance, $C_{\mathrm{cm}} = (C_1 + C_2)/2$ is the common-mode capacitance between each wire and the grounded shield, and f is the frequency of the common-mode input component. For instance, at 60 Hz, a source resistance imbalance of 1 kΩ in conjunction with a 100-foot cable having a distributed capacitance of 1 nF would degrade the CMRR to $20 \log_{10}[1/(2\pi 60 \times 10^3 \times 10^{-9})] = 68.5$ dB, even with an IA having infinite CMRR.

The effect of C_{cm} can, to a first approximation, be neutralized by driving the shield with the common-mode voltage itself so as to reduce the common-mode swing across C_{cm} to zero. Figure 2.29 shows a popular way of achieving this goal. By op amp action, the voltages at the top and bottom nodes of R_G are v_1 and v_2. Denoting the voltage across R_3 as v_3, we can write $v_{\mathrm{CM}} = (v_1 + v_2)/2 = (v_1 + v_3 + v_2 - v_3)/2 = (v_{O1} + v_{O2})/2$, indicating that v_{CM} can be extracted by computing the mean of v_{O1} and v_{O2}. This mean is found via the two 20-kΩ resistors and is then buffered to the shield by OA_4.

Digitally Programmable Gain

In automatic instrumentation, such as data acquisition systems, it is often desirable to program the gain of the IA electronically, usually by means of JFET or MOSFET switches. The method depicted in Fig. 2.30 programs the first-stage gain A_I by using a string of symmetrically valued resistors, and a string of simultaneously activated switch pairs to select the tap pair corresponding to a given gain. At any given time, only one switch pair is closed and all others are open. By Eq. (2.31b), A_I can be put

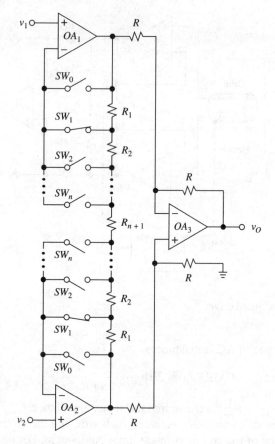

FIGURE 2.30
Digitally programmable IA.

in the form

$$A_I = 1 + \frac{R_{\text{outside}}}{R_{\text{inside}}} \tag{2.39}$$

where R_{inside} is the sum of the resistances located between the two selected switches and R_{outside} is the sum of all remaining resistances. For the case shown, the selected switch pair is SW_1, so $R_{\text{outside}} = 2R_1$ and $R_{\text{inside}} = 2(R_2 + R_3 + \cdots + R_n) + R_{n+1}$. Selecting SW_2 gives $R_{\text{outside}} = 2(R_1 + R_2)$ and $R_{\text{inside}} = 2(R_3 + \cdots + R_n) + R_{n+1}$. It is apparent that changing to a different switch pair increases (or decreases) R_{outside} at the expense of an equal decrease (or increase) in R_{inside}, thus yielding a different resistance ratio and, hence, a different gain.

The advantage of this topology is that the current flowing through any closed switch is the negligible input current of the corresponding op amp. This is particularly important when the switches are implemented with FETs because FETs have a nonzero on-resistance and the ensuing voltage drop could degrade the accuracy of the IA. With zero current this drop is also zero, in spite of the nonideality of the switch.

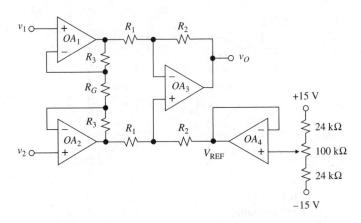

FIGURE 2.31
IA with output offset control.

The two groups of switches of Fig. 2.30 can easily be implemented with CMOS analog multiplexers/demultiplexers, such as the CD4051 or CD4052. Digitally programmable IAs, containing all the necessary resistors, analog switches, and TTL-compatible decoder and switch-driver circuitry, are also available in IC form. Consult the manufacturer catalogs for more information.

Output-Offsetting

There are applications that call for a prescribed amount of offset at the output of an IA, as when an IA is fed to a voltage-to-frequency converter, which requires that its input range be of only one polarity. Since the IA output is usually bipolar, it must be suitably offset to ensure a unipolar range. In the circuit of Fig. 2.31 the reference node is driven by voltage V_{REF}. This voltage, in turn, is obtained from the wiper of a pot and is buffered by OA_4, whose low output resistance prevents disturbance of the bridge balance. Applying the superposition principle, we obtain $v_O = A(v_2 - v_1) + (1 + R_2/R_1) \times [R_1/(R_1 + R_2)]V_{\text{REF}}$, or

$$v_O = A(v_2 - v_1) + V_{\text{REF}} \tag{2.40}$$

where A is given by Eq. (2.31b). With the component values shown, V_{REF} is variable from -10 V to $+10$ V.

Current-Output IAs

By turning the second stage into a Howland circuit, in the manner depicted in Fig. 2.32, we can configure the triple-op-amp IA for current-output operation. This type of operation is desirable when transmitting signals over long wires since the stray wire resistance does not degrade current signals. Combining the results of Problem 2.9 with Eq. (2.31b), we readily obtain

$$i_O = \frac{1 + 2R_3/R_G}{R_1}(v_2 - v_1) \tag{2.41}$$

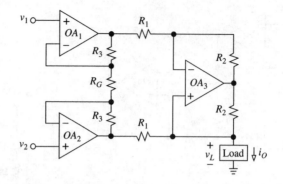

FIGURE 2.32
Current-output IA.

The gain can be adjusted via R_G, as usual. For efficient operation, the Howland stage can be improved with the modification of Fig. 2.10. For high CMRR, the top left resistance should be trimmed.

The dual-op-amp IA is configured for current-output operation by the bootstrapping technique[8] of Fig. 2.33. It is left as an exercise (see Problem 2.44) to prove that the transfer characteristic of the circuit is of the type

$$i_O = \frac{1}{R}(v_2 - v_1) - \frac{1}{R_o}v_L \tag{2.42a}$$

$$R_o = \frac{R_2/R_1}{R_5/R_4 - (R_2 + R_3)/R_1}R_3 \tag{2.42b}$$

so that imposing $R_2 + R_3 = R_1 R_5/R_4$ yields $R_o = \infty$. If adjustable gain is desired, it is readily obtained by connecting a variable resistance R_G between the inverting input pins of the two op amps, in the manner of Fig. 2.24.

Besides offering difference-input operation with high input resistances, the circuit enjoys the efficiency advantages of the improved Howland circuit because R_2 can be kept as small as needed while all remaining resistances can be made relatively

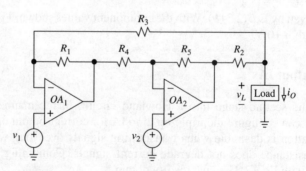

FIGURE 2.33
Dual-op-amp IA with current output.

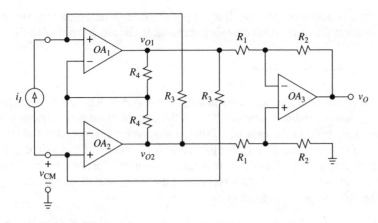

FIGURE 2.34
Current-input IA.

large to conserve power. When this constraint is imposed, the voltage compliance is approximately $|v_L| \le V_{\text{sat}} - R_2|i_O| = V_{\text{sat}} - 2|v_2 - v_1|$.

Current-Input IA

In current-loop instrumentation the need arises for sensing a floating current and converting it to a voltage. To avoid perturbing the characteristics of the loop, it is desirable that the circuit downstream appear as a virtual short. An IA can once again be suitably modified to meet this requirement. In Fig. 2.34 we observe that OA_1 and OA_2 force the voltages at their input pins to track v_{CM}, thus ensuring 0 V across the input source. By KVL and Ohm's law, $v_{O2} = v_{\text{CM}} - R_3 i_I$ and $v_{O1} = v_{\text{CM}} + R_3 i_I$. But $v_O = (R_2/R_1) \times (v_{O2} - v_{O1})$. Combining, we get

$$v_O = -\frac{2R_2}{R_1} R_3 i_I \tag{2.43}$$

If variable gain is desired, this can be obtained by modifying the difference stage as in Fig. 2.17 or 2.18. If, on the other hand, the difference stage is modified as in Fig. 2.32, the circuit becomes a floating-input current amplifier.

2.7
TRANSDUCER BRIDGE AMPLIFIERS

Resistive transducers are devices whose resistance varies as a consequence of some environmental condition, such as temperature (thermistors; resistance temperature detectors, or RTDs), light (photoresistors), strain (strain gauges), and pressure (piezoresistive transducers). By making these devices part of a circuit, it is possible to produce an electric signal that, after suitable conditioning, can be used to monitor as well as control the physical process affecting the transducer.[9] In general it is desirable that the relationship between the final signal and the original physical variable be linear, so that the former can directly be calibrated in the physical units of the latter. Transducers play such an important role in measurement and control instrumentation that it is worth studying transducer circuits in some detail.

Transducer Resistance Deviation

Transducer resistances are expressed in the form $R + \Delta R$, where R is the resistance at some *reference* condition, such as 0 °C in the case of temperature transducers, or the absence of strain in the case of strain gauges, and ΔR represents the *deviation* from the reference value as a consequence of a change in the physical condition affecting the transducer. Transducer resistances are also expressed in the alternative form $R(1 + \delta)$, where $\delta = \Delta R/R$ represents the *fractional deviation*. Multiplying δ by 100 yields the *percentage deviation*.

> **EXAMPLE 2.11.** Platinum resistance temperature detectors (Pt RTDs) have a temperature coefficient[10] $\alpha = 0.00392/°C$. A popular Pt RTD reference value at $T = 0 °C$ is 100 Ω. (*a*) Write an expression for the resistance as a function of T. (*b*) Compute $R(T)$ for $T = 25 °C, 100 °C, -15 °C$. (*c*) Calculate ΔR and δ for a temperature change $\Delta T = 10 °C$.
>
> **Solution.**
>
> (*a*) $R(T) = R(0 °C)(1 + \alpha T) = 100(1 + 0.00392T)$ Ω.
> (*b*) $R(25 °C) = 100(1 + 0.00392 \times 25) = 109.8$ Ω. Likewise, $R(100 °C) = 139.2$ Ω and $R(-15 °C) = 94.12$ Ω.
> (*c*) $R + \Delta R = 100 + 100\alpha T = 100 + 100 \times 0.00392 \times 10 = 100$ $\Omega + 3.92$ Ω; $\delta = \alpha \Delta T = 0.00392 \times 10 = 0.0392$. This corresponds to a change of $0.0392 \times 100 = 3.92\%$.

The Transducer Bridge

To measure resistance deviation, we must find a method to convert ΔR to a voltage variation ΔV. The simplest technique is to make the transducer part of a voltage divider, as shown in Fig. 2.35. The transducer voltage is $v_1 = V_{REF}R(1 + \delta)/[R_1 + R(1 + \delta)]$, which can be put in the insightful form

$$v_1 = \frac{R}{R_1 + R}V_{REF} + \frac{\delta V_{REF}}{2 + R_1/R + R/R_1 + (1 + R/R_1)\delta} \tag{2.44}$$

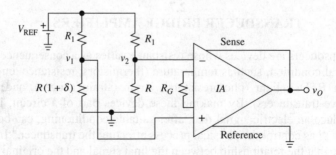

FIGURE 2.35
Transducer bridge and IA.

where $\delta = \Delta R/R$. We observe that v_1 consists of a fixed term plus a term controlled by $\delta = \Delta R/R$. It is precisely the latter that interests us, so we must find a means for amplifying it while ignoring the former. This is achieved by using a second voltage divider to synthesize the term

$$v_2 = \frac{R}{R_1 + R} V_{REF} \tag{2.45}$$

and then using an IA to take the difference $v_1 - v_2$. Denoting the IA gain as A, we get $v_O = A(v_1 - v_2)$, or

$$v_O = A V_{REF} \frac{\delta}{1 + R_1/R + (1 + R/R_1)(1 + \delta)} \tag{2.46}$$

The four-resistor structure is the familiar resistive bridge, and the two voltage dividers are referred to as the *bridge legs*.

It is apparent that v_O is a nonlinear function of δ. In microprocessor-based systems, a nonlinear function can easily be linearized in the software. Quite often, however, we have $\delta \ll 1$, so

$$v_O \cong \frac{A V_{REF}}{2 + R_1/R + R/R_1} \delta \tag{2.47}$$

indicating a linear dependence of v_O on δ. Many bridges are designed with $R_1 = R$, in which case Eqs. (2.46) and (2.47) become

$$v_O = \frac{A V_{REF}}{4} \frac{\delta}{1 + \delta/2} \tag{2.48}$$

$$v_O \cong \frac{A V_{REF}}{4} \delta \tag{2.49}$$

EXAMPLE 2.12. Let the tranducer of Fig. 2.35 be the Pt RTD of Example 2.11, and let $V_{REF} = 15$ V. (a) Specify values for R_1 and A suitable for achieving an output sensitivity of 0.1 V/°C near 0 °C. To avoid self-heating in the RTD, limit its power dissipation to less than 0.2 mW. (b) Compute $v_O(100 °C)$ and estimate the equivalent error, in degrees Celsius, in making the approximation of Eq. (2.47).

Solution.

(a) Denoting the transducer current as i, we have $P_{RTD} = Ri^2$. Thus, $i^2 \leq P_{RTD(max)}/R = 0.2 \times 10^{-3}/100$, or $i = 1.41$ mA. To be on the safe side, impose $i \cong 1$ mA, or $R_1 = 15$ kΩ. For $\Delta T = 1$ °C we have $\delta = \alpha \times 1 = 0.00392$, and we want $\Delta v_O = 0.1$ V. By Eq. (2.47) we need $0.1 = A \times 15 \times 0.00392/(2+15/0.1+0.1/15)$, or $A = 258.5$ V/V.

(b) For $\Delta T = 100$ °C we have $\delta = \alpha \Delta T = 0.392$. Inserting into Eq. (2.46), we get $v_O(100 °C) = 9.974$ V. Equation (2.47) predicts that $v_O(100 °C) = 10.0$ V, which exceeds the actual value by $10 - 9.974 = 0.026$ V. Since 0.1 V corresponds to 1 °C, 0.026 V corresponds to $0.026/0.1 = 0.26$ °C. Therefore, in using the approximated expression, we cause, at 100 °C, an error of about one-quarter of a degree Celsius.

Bridge Calibration

With $\Delta R = 0$, a transducer bridge should be balanced and yield a zero voltage difference between its taps. In practice, because of resistance tolerances, including

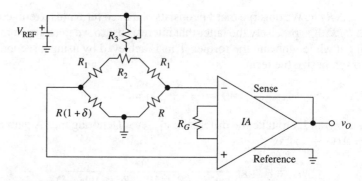

FIGURE 2.36
Bridge calibration.

the tolerance of the transducer's reference value, the bridge is likely to be unbalanced and a trimmer should be included to balance it. Moreover, the tolerances in the values of the resistances and of V_{REF} will affect the bridge sensitivity $(v_1 - v_2)/\delta$, thus creating the need for adjustment of this parameter as well.

Figure 2.36 shows a circuit that allows for both adjustments. Varying R_2's wiper from its midway position will assign more resistance to one leg and less to the other, thus allowing the compensation of their inherent mismatches. Varying R_3 changes the bridge current and hence the magnitude of the voltage variation produced by the transducer, thus allowing the adjustment of the sensitivity.

EXAMPLE 2.13. Let all resistors in Example 2.12 have a 1% tolerance, and let V_{REF} have a 5% tolerance. (a) Design a circuit to calibrate the bridge. (b) Outline the calibration procedure.

Solution.

(a) A 5% tolerance in V_{REF} means that its actual value can deviate from its nominal value by as much as $\pm 0.05 \times 15 = \pm 0.75$ V. To be on the safe side and to also include the effect of 1% resistance tolerance, assume a maximum deviation of ± 1 V, and thus design for $14 \text{ V} \pm 1$ V at R_2's wiper. To ensure a current of 1 mA at each leg, we need $R_3 = 2/(1 + 1) = 1$ kΩ and $R + R_1 + R_2/2 = 14/1 = 14$ kΩ. Since R_2 must compensate for up to a 1% variation on each leg, we need $R_2 = 2 \times 0.01 \times 14$ k$\Omega = 280$ Ω. To be on the safe side, pick $R_2 = 500$ Ω. Then $R_1 = 14$ k$\Omega - 100$ $\Omega - 500/2$ $\Omega = 13.65$ kΩ (use 13.7 kΩ, 1%). The IA gain A must be recomputed via Eq. (2.47), but with $V_{REF} = 14$ V and with 13.7 k$\Omega + 500/2$ $\Omega = 13.95$ kΩ in place of R_1. This yields $A = 257.8$ V/V. Summarizing, we need $R_1 = 13.7$ kΩ, 1%; $R_2 = 500$ Ω; $R_3 = 1$ kΩ; and $A = 257.8$ V/V.

(b) To calibrate, first set $T = 0$ °C and adjust R_2 for $v_O = 0$ V. Then set $T = 100$ °C and adjust R_3 for $v_O = 10.0$ V.

Strain-Gauge Bridges

The resistance of a wire having resistivity ρ, cross-sectional area S, and length ℓ is $R = \rho\ell/S$. Straining the wire changes its length to $\ell + \Delta\ell$, its area to $S - \Delta S$, and its resistance to $R + \Delta R = \rho(\ell + \Delta\ell)/(S - \Delta S)$. Since its volume must remain constant, we have $(\ell + \Delta\ell) \times (S - \Delta S) = S\ell$. Eliminating $S - \Delta S$, we get $\Delta R = R(\Delta\ell/\ell)(2 + \Delta\ell/\ell)$. But $\Delta\ell/\ell \ll 2$, so

$$\Delta R = 2R\frac{\Delta\ell}{\ell} \qquad (2.50)$$

where R is the *unstrained* resistance and $\Delta\ell/\ell$ is the *fractional elongation*. A strain gauge is fabricated by depositing resistive material on a flexible backing according to a pattern designed to maximize its fractional elongation for a given strain. Since strain gauges are sensitive also to temperature, special precautions must be taken to mask out temperature-induced variations. A common solution is to work with gauge pairs designed to compensate for each other's temperature variations.

The strain-gauge arrangement of Fig. 2.37 is referred to as a *load cell*. Denoting the bridge voltage as V_B and ignoring R_1 for a moment, the voltage divider formula yields $v_1 = V_B(R + \Delta R)/(R + \Delta R + R - \Delta R) = V_B(R + \Delta R)/2R$, $v_2 = V_B(R - \Delta R)/2R$, and $v_1 - v_2 = V_B\Delta R/R = V_B\delta$, so

$$v_O = AV_{REF}\delta \qquad (2.51)$$

The sensitivity is now four times as large as that given in Eq. (2.49), thus relaxing the demands upon the IA. Furthermore, the dependence of v_O on δ is now perfectly linear—another advantage of working with gauge pairs. To achieve the $+\Delta R$ and $-\Delta R$ variations, two of the gauges will be bonded to one side of the structure under strain, and the other two to the opposite side. Even in installations in which only one side is accessible, it pays to work with four gauges because two can be used as dummy gauges to provide temperature compensation for the active ones. Piezoresistive pressure sensors also use this arrangement.

Figure 2.37 also illustrates an alternative technique for balancing the bridge. In the absence of strain, each tap voltage should be $V_B/2$. In practice there will be deviations due to the initial tolerances of the four gauges. By varying R_2's wiper, we can force an adjustable amount of current through R_1 that will increase or decrease the corresponding tap voltage until the bridge is nulled. Resistors R_3 and R_4 drop V_{REF} to V_B, and R_3 adjusts the sensitivity.

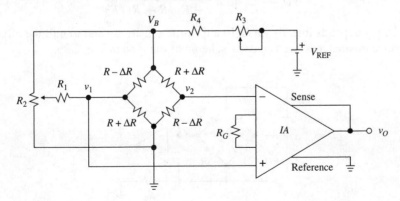

FIGURE 2.37
Strain-gauge bridge and IA.

EXAMPLE 2.14. Let the strain gauges of Fig. 2.37 be 120-Ω, $\pm 1\%$ types, and let their maximum current be limited to 20 mA to avoid excessive self-heating. (*a*) Assuming that $V_{REF} = 15$ V $\pm 5\%$, specify suitable values for R_1 through R_4. (*b*) Outline the calibration procedure.

Solution.

(*a*) By Ohm's law, $V_B = 2 \times 120 \times 20 \times 10^{-3} = 4.8$ V. In the absence of strain, the tap voltages are nominally $V_B/2 = 2.4$ V. Their actual values may deviate from $V_B/2$ by as much as $\pm 1\%$ of 2.4 V, that is, by as much as ± 0.024 V. Consider the case in which $v_1 = 2.424$ V and $v_2 = 2.376$ V. By moving R_2's wiper to ground, we must be able to lower v_1 to 2.376 V, that is, to change v_1 by 0.048 V. To achieve this, R_1 must sink a current $i = 0.048/(120 \| 120) = 0.8$ mA, so $R_1 \cong 2.4/0.8 = 3$ kΩ (to be on the safe side, use $R_1 = 2.37$ kΩ, 1%). To prevent excessive loading of R_2's wiper by R_1, use $R_2 = 1$ kΩ. Under nominal conditions we have $i_{R_3} = i_{R_4} = 2 \times 20 \times 10^{-3} + 4.8/10^3 \cong 45$ mA. Following Example 2.13, we wish R_3 to drop a maximum of 2 V. So $R_3 = 2/45 = 44$ Ω (use $R_3 = 50$ Ω). With R_3's wiper halfway we have $R_4 = (15 - 25 \times 45 \times 10^{-3} - 4.8)/(45 \times 10^{-3}) = 202$ Ω (use 200 Ω). Summarizing, $R_1 = 2.37$ kΩ, $R_2 = 1$ kΩ, $R_3 = 50$ Ω, and $R_4 = 200$ Ω.

(*b*) To calibrate, first adjust R_2 so that with no strain we get $v_O = 0$ V. Then apply a known strain, preferably near the full scale, and adjust R_3 for the desired value of v_O.

Single-Op-Amp Amplifier

For reasons of cost it is sometimes desirable to use a simpler amplifier than the full-fledged IA. Figure 2.38 shows a bridge amplifier implemented with a single op amp. After applying Thévenin's theorem to the two legs of the bridge, we end up with the familiar difference amplifier. One can then show (see Problem 2.49) that

$$v_O = \frac{R_2}{R} V_{REF} \frac{\delta}{R_1/R + (1 + R_1/R_2)(1 + \delta)} \tag{2.52}$$

For $\delta \ll 1$ this simplifies to

$$v_O \cong \frac{R_2}{R} V_{REF} \frac{\delta}{1 + R_1/R + R_1/R_2} \tag{2.53}$$

That is, v_O depends linearly on δ. To adjust the sensitivity and to null the effect of resistance mismatches, we can use a scheme of the type of Fig. 2.36.

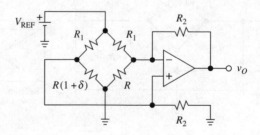

FIGURE 2.38
Single-op-amp bridge amplifier.

Bridge Linearization

With the exception of the strain-gauge circuit of Fig. 2.37, all bridge circuits discussed so far suffer from the fact that the response is reasonably linear only as long as $\delta \ll 1$. It is therefore of interest to seek circuit solutions capable of a linear response regardless of the magnitude of δ.

The design of Fig. 2.39 linearizes the bridge by driving it with a constant current.[11] This is achieved by placing the entire bridge within the feedback loop of a floating-load V-I converter. The bridge current is $I_B = V_{REF}/R_1$. By using a transducer pair as shown, I_B will split equally between the two legs. Since OA keeps the bottom node of the bridge at V_{REF}, we have $v_1 = V_{REF} + R(1+\delta)I_B/2$, $v_2 = V_{REF} + RI_B/2$, and $v_1 - v_2 = R\delta I_B/2$, so

$$v_O = \frac{AR V_{REF}}{2R_1}\delta \qquad (2.54)$$

The alternative design of Fig. 2.40 uses a single-transducer element and a pair of inverting-type op amps.[11] The response is again linearized by placing the bridge within the feedback loop of the V-I converter OA_1. It is left as an exercise (see Problem 2.49) to show that

$$v_O = \frac{R_2 V_{REF}}{R_1}\delta \qquad (2.55)$$

For additional bridge circuit examples, see references 9, 11, 12, and 13 and the end-of-chapter problems.

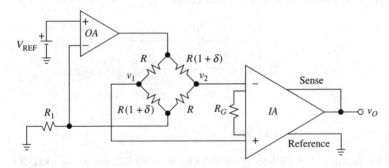

FIGURE 2.39
Bridge linearization by constant-current drive.

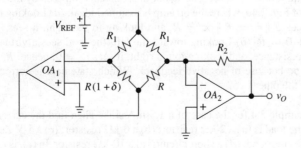

FIGURE 2.40
Single-transducer circuit with linear response.

PROBLEMS

2.1 Current-to-voltage converters

2.1 Using two op amps, design a circuit that accepts two current *sources* i_1 and i_2 having parallel resistances R_1 and R_2 and yields $v_O = (0.1 \text{ V}/\mu\text{A}) \times (i_1 - i_2)$ to a load R_L regardless of R_1, R_2, and R_L. The reference directions of both sources are from ground into your circuit. Try minimizing the number of resistors you use.

2.2 Design a circuit to convert a 4-mA-to-20-mA input current to a 0-V-to-10-V output voltage. The reference direction of the input source is from ground into your circuit, and the circuit is powered from ±15-V regulated supplies.

2.3 Estimate the closed-loop parameters if the circuit of Example 2.2 is implemented with a 741 op amp.

2.4 (*a*) Using an op amp powered from ±15-V regulated supplies, design a photodetector amplifier such that as i_I changes from 0 to 1 μA, v_O changes from -5 V to $+5$ V.
(*b*) What is the minimum open-loop gain for a deviation of the transfer characteristic from the ideal of less than 1%?

2.2 Voltage-to-current converters

2.5 (*a*) Show that the floating-load *V-I* converter of Fig. P2.5 yields $i_O = v_I/(R_1/k)$, $k = 1 + R_2/R_3$. (*b*) Specify standard 5% resistances for a sensitivity of 1 mA/V and $R_i = 1 \text{ M}\Omega$, where R_i is the resistance seen by the input source. (*c*) If $\pm V_{\text{sat}} = \pm13$ V, what is the voltage compliance of your circuit?

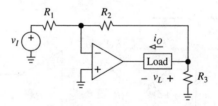

FIGURE P2.5

2.6 In the circuit of Fig. P2.5 let $R_1 = 100 \text{ k}\Omega$, $R_2 = 99 \text{ k}\Omega$, and $R_3 = 1 \text{ k}\Omega$. If $r_d \cong \infty$, $a = 10^3$ V/V, and $r_o \cong 0$, estimate the resistance R_o seen by the load.

2.7 Consider the following statements about the resistance R_o seen by the load in the *V-I* converter of Fig. 2.4*b*, where the op amp is assumed ideal: (*a*) Looking toward the left, the load sees $R \parallel r_d = R \parallel \infty = R$, and looking to the right, it sees $r_o = 0$; hence, $R_o = R + 0 = R$. (*b*) Looking toward the left, the load sees a virtual-ground node with zero resistance, and looking to the right, it sees $r_o = 0$; hence, $R_o = 0 + 0 = 0$. (*c*) $R_o = \infty$ because of negative feedback. Which statement is correct? How would you refute the other two?

2.8 Repeat Example 2.4 for the case of a 1.5-mA sink. Then find the currents through R_1 and R_2 if the load is (*a*) a 2-kΩ resistor; (*b*) a 6-kΩ resistor; (*c*) a 5-V Zener diode with the cathode at ground; (*d*) a short circuit; (*e*) a 10-kΩ resistor. In (*e*), is i_O still 1.5 mA? Explain.

2.9 Suppose in the Howland circuit of Fig. 2.6a we lift the left terminal of R_3 off ground and simultaneously apply an input v_1 via R_3 and an input v_2 via R_1. Show that the circuit is a *difference V-I converter* with $i_O = (1/R_1)(v_2 - v_1) - (1/R_o)v_L$, where R_o is given by Eq. (2.8).

2.10 Design a grounded-load V-I converter that converts a 0-V to 10-V input to a 4-mA to 20-mA output. The circuit is to be powered from ±15-V regulated supplies.

2.11 Design a grounded-load current generator meeting the following specifications: i_O is to be variable over the range $-2\,\text{mA} \le i_O \le +2\,\text{mA}$ by means of a 100-kΩ pot; the voltage compliance must be 10 V; the circuit is to be powered from ±15-V regulated supplies.

2.12 (a) Prove Eq. (2.15). (b) Using a 741 op amp powered from ±15-V supplies, design an improved Howland circuit with a sensitivity of 1 mA/V for $-10\,\text{V} \le v_I \le 10\,\text{V}$. The voltage compliance of the circuit must also be 10 V.

2.13 Design an improved Howland circuit whose sensitivity is variable from 0.1 mA/V to 1 mA/V by means of a 10-kΩ pot.

2.14 (a) Given that the circuit of Fig. P2.14 yields $i_O = A(v_2 - v_1) - (1/R_o)v_L$, find expressions for A and R_o, as well as the condition among the resistances that yields $R_o = \infty$. (b) Discuss the effect of using 1% resistances.

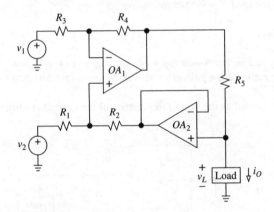

FIGURE P2.14

2.15 (a) Given that the circuit of Fig. P2.15 yields $i_O = Av_I - (1/R_o)v_L$, find expressions for A and R_o, as well as the condition among its resistances that yields $R_o = \infty$. (b) Discuss the effect of using 1 % resistances.

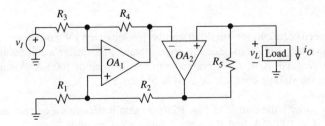

FIGURE P2.15

2.16 Repeat Problem 2.15 for the circuit of Fig. P2.16.

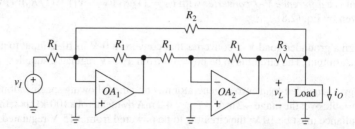

FIGURE P2.16

2.17 The current source of Example 2.4 drives a $0.1\text{-}\mu\text{F}$ load. (*a*) Assuming that the capacitance is initially discharged, sketch and label $v_O(t \geq 0)$. (*b*) Find the time it takes for the op amp to enter the saturation region.

2.18 Repeat Problem 2.17 with R_4 (*a*) decreased by 10%, and (*b*) increased by 10%.

2.19 Assuming an ideal op amp, find the input resistance R_i of a Howland current pump as a function of the load R_L. Comment.

2.3 Current amplifiers

2.20 (*a*) Prove Eq. (2.18). (*b*) Assuming a 741 op amp in Fig. 2.11, specify resistances for $A = 10$ A/A; estimate the gain error as well as the output resistance of the circuit.

2.21 Find the gain as well as the output impedance of the current amplifier of Fig. P2.21.

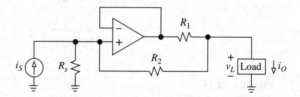

FIGURE P2.21

2.22 Show that if $R_s = \infty$ and $a \neq \infty$ in the current amplifier of Fig. 2.12, then Eq. (2.18) holds.

2.23 A grounded-load current amplifier can be implemented by cascading an I-V and a V-I converter. Using resistances no greater than 1 MΩ, design a current amplifier with $R_i = 0$, $A = 10^5$ A/A, $R_o = \infty$, and a full-scale input of 100 nA. Assuming ± 15-V supplies, the voltage compliance must be at least 5 V.

2.24 Suitably modify the circuit of Fig. P2.16 so that it becomes a current amplifier with $R_i = 0$, $A = 100$ A/A, and $R_o = \infty$. Assume ideal op amps.

2.25 In Fig. P2.25 the odd-numbered inputs are fed to OA_2's summing junction directly, and the even-numbered inputs are fed via a current reverser. Obtain a relationship between v_O and the various inputs. What happens if any of the inputs is left floating? Will it affect the contribution from the other inputs? What is an important advantage of this circuit compared to that of Problem 1.31?

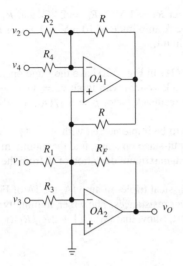

FIGURE P2.25

2.4 Difference amplifiers

2.26 Derive Eq. (2.23).

2.27 (*a*) Derive Eq. (2.27). (*b*) Using a 100-kΩ pot, specify suitable resistances such that varying the wiper from end to end varies the gain from 10 V/V to 100 V/V.

2.28 (*a*) Derive Eq. (2.28). (*b*) Specify suitable component values such that gain can be varied from 1 V/V to 100 V/V.

2.29 (*a*) A difference amplifier has $v_1 = 10\cos 2\pi 60t$ V $- 5\cos 2\pi 10^3 t$ mV, and $v_2 = 10\cos 2\pi 60t$ V $+ 5\cos 2\pi 10^3 t$ mV. If $v_O = 100\cos 2\pi 60t$ mV $+ 2\cos 2\pi 10^3 t$ V, find A_{dm}, A_{cm}, and CMRR$_{dB}$. (*b*) Repeat (*a*) with $v_1 = 10.01\cos 2\pi 60t$ V $- 5\cos 2\pi 10^3 t$ mV, $v_2 = 10.00\cos 2\pi 60t$ V $+ 5\cos 2\pi 10^3 t$ mV, and $v_O = 0.5\cos 2\pi 60t$ V $+ 2.5\cos 2\pi 10^3 t$ V.

2.30 If the actual resistance values in Fig. 2.13a are found to be $R_1 = 1.01$ kΩ, $R_2 = 99.7$ kΩ, $R_3 = 0.995$ kΩ, and $R_4 = 102$ kΩ, estimate A_{dm}, A_{cm}, and CMRR$_{dB}$.

2.31 If the difference amplifier of Fig. 2.13a has a differential-mode gain of 60 dB and CMRR$_{dB} = 100$ dB, find v_O if $v_1 = 4.001$ V and $v_2 = 3.999$ V. What is the percentage error of the output due to finite CMRR?

2.32 If the resistance pairs are perfectly balanced and the op amp is ideal in the difference amplifier of Fig. 2.13a, then we have CMRR$_{dB} = \infty$. But what if the open-loop gain a is finite, everything else being ideal? Is the CMRR still infinite? Justify your finding intuitively.

2.5 Instrumentation amplifiers

2.33 In the IA of Fig. 2.20 let $R_3 = 1$ MΩ, $R_G = 2$ kΩ, and $R_1 = R_2 = 100$ kΩ. If v_{DM} is an ac voltage with a peak amplitude of 10 mV and v_{CM} is a dc voltage of 5 V, find all node voltages in the circuit.

2.34 Show that if OA_1 and OA_2 in Fig. 2.20 have the same open-loop gain a, together they form a negative-feedback system with input $v_I = v_1 - v_2$, output $v_O = v_{O1} - v_{O2}$, open-loop gain a, and feedback factor $\beta = R_G/(R_G + 2R_3)$.

2.35 A triple-op-amp IA is to be implemented with $A = A_I \times A_{II} = 50 \times 20 = 10^3$ V/V. Assuming matched input-stage op amps, find the minimum open-loop gain required of each op amp for a 0.1% maximum deviation of A from the ideal.

2.36 Compared with the classical triple-op-amp IA, the IA of Fig. P2.36 (see *EDN*, Oct. 1, 1992, p. 115) uses fewer resistances. The wiper, nominally positioned halfway, is used to maximize the CMRR. Show that $v_O = (1 + 2R_2/R_1)(v_2 - v_1)$.

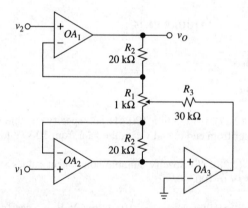

FIGURE P2.36

2.37 (*a*) To investigate the effect of mismatched resistances in the IA of Fig. 2.23, assume that $R_3/R_4 = (R_1/R_2)(1 - \epsilon)$. Show that $v_O = A_{\mathrm{dm}}v_{\mathrm{DM}} + A_{\mathrm{cm}}v_{\mathrm{CM}}$, where $A_{\mathrm{dm}} = 1 + R_2/R_1 - \epsilon/2$ and $A_{\mathrm{cm}} = \epsilon$. (*b*) Discuss the implications of using 1% resistors without trimming for the case $A = 10^2$ V/V.

2.38 (*a*) Derive Eq. (2.35). (*b*) Specify suitable components such that A can be varied over the range 10 V/V $\leq A \leq$ 100 V/V by means of a 10-kΩ pot.

2.39 The gain of the dual-op-amp IA of Fig. P2.39 (see *EDN*, Feb. 20, 1986, pp. 241–242) is adjustable by means of a single resistor R_G. (*a*) Show that $v_O = 2(1 + R/R_G)(v_2 - v_1)$. (*b*) Specify suitable components to make A variable from 10 V/V to 100 V/V by means of a 10-kΩ pot.

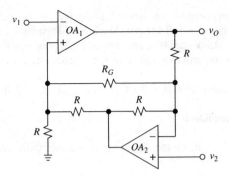

FIGURE P2.39

2.40 The dual-op-amp IA of Fig. P2.40 (see *Signals and Noise, EDN,* May 29, 1986) offers the advantage that by proper adjustment of the pot, a fairly high CMRR can be achieved and maintained well into the kilohertz range. Show that $v_O = (1 + R_2/R_1)(v_2 - v_1)$.

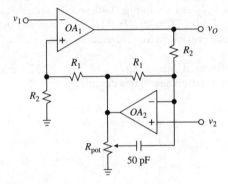

FIGURE P2.40

2.41 Assuming perfectly matched resistances as well as perfectly matched op amps in the dual-op-amp IA of Fig. 2.23, investigate the effect of finite open-loop op amp gain a upon the CMRR of the circuit (except for their finite gain, both op amps are ideal). Assuming $a = 10^5$ V/V, find CMRR$_{dB}$ if $A = 10^3$ V/V. Repeat, but if $A = 10$ V/V, and comment on your findings.

2.6 Instrumentation applications

2.42 Design a digitally programmable IA having an overall gain of 1 V/V, 10 V/V, 100 V/V, and 1000 V/V. Show the final design.

2.43 Assuming ± 15-V regulated power supplies, design a programmable IA with two operating modes: in the first mode the gain is 100 V/V and the output offset is 0 V; in the second mode the gain is 200 V/V and the output offset is -5 V.

2.44 (*a*) Derive Eq. (2.42). (*b*) In the current-output IA of Fig. 2.33 specify suitable components for a sensitivity of 1 mA/V. (*c*) Investigate the effect of using 0.1% resistances.

2.45 In the circuit of Fig. 2.33 let $R_1 = R_4 = R_5 = 10$ kΩ, $R_2 = 1$ kΩ, and $R_3 = 9$ kΩ. If an additional resistance R_G is connected between the inverting input nodes of the two op amps, find the gain as a function of R_G.

2.46 (*a*) Design a current-output IA whose sensitivity can be varied from 1 mA/V to 100 mA/V by means of a 100-kΩ pot. The circuit must have a voltage compliance of at least 5 V with ±15-V supplies, and it must have provision for CMRR optimization by means of a suitable trimmer. (*b*) Outline the procedure for calibrating the trimmer.

2.47 Design a current-input, voltage-output IA with a gain of 10 V/mA.

2.7 Transducer bridge amplifiers

2.48 Repeat Example 2.12 using the single-op-amp configuration of Fig. 2.38. Show the final circuit.

2.49 (*a*) Derive Eqs. (2.52) and (2.53). (*b*) Derive Eq. (2.55).

2.50 Assuming that $V_{REF} = 2.5$ V in Fig. 2.39, specify suitable component values for an output sensitivity of 0.1 V/°C with a Pt RTD.

2.51 (*a*) Assuming that $V_{REF} = 15$ V in Fig. 2.40, specify suitable component values for an output sensitivity of 0.1 V/°C with a Pt RTD. (*b*) Assuming the same tolerances as in Example 2.13, make provisions for bridge calibration.

2.52 Show that the linearized bridge circuit of Fig. P2.52 yields $v_O = -RV_{REF}\delta/(R_1 + R)$. Name a disadvantage of this circuit.

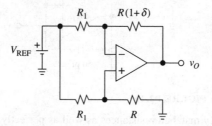

FIGURE P2.52

2.53 Using the circuit of Fig. P2.52 with $V_{REF} = 2.5$ V and an additional gain stage, design an RTD amplifier circuit with a sensitivity of 0.1 V/°C. The circuit is to have provisions for bridge calibration. Outline the calibration procedure.

2.54 Show that the linearized bridge circuit[11] of Fig. P2.54 (U.S. Patent 4,229,692) yields $v_O = R_2V_{REF}\delta/R_1$. Discuss how you would make provisions for calibrating the circuit.

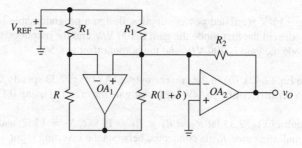

FIGURE P2.54

REFERENCES

1. "Silicon Photovoltaic Detectors and Detector/Amplifier Combinations," Application Note D3011C-8, EG&G Electro-Optics, Salem, MA, 1984.

2. J. Steele and T. Green, "Tame Those Versatile Current-Source Circuits," *Electronic Design,* Oct. 15, 1992, pp. 61–72.

3. J. Graeme, "Manipulate Current Signals with Op Amps," *EDN,* Aug. 8, 1985, pp. 147–158.

4. R. M. Stitt, "Monolithic Difference Amp Eases the Design of a Variety of Circuits," *EDN,* March 20, 1986, pp. 181–188.

5. J. R. Riskin, "A User's Guide to IC Instrumentation Amplifiers," Application Note AN-244, *Applications Reference Manual,* Analog Devices, Norwood, MA, 1993.

6. J. Williams, "Applications for a Switched-Capacitor Instrumentation Building Block," Application Note AN-3, *Linear Applications Handbook,* vol. 1, Linear Technology, Milpitas, CA, 1990.

7. "Instrumentation Amplifiers: Versatile Differential Input Gain Blocks," Application Note AN-75, *Burr-Brown Handbook of Linear IC Applications,* Burr-Brown, Tucson, AZ, 1987.

8. J. Graeme, "Bootstrapped Amp Makes Current Source," *EDN,* Jan. 21, 1991, pp. 152–154.

9. Analog Devices Engineering Staff, *Practical Design Techniques for Sensor Signal Conditioning,* Analog Devices, Norwood, MA, 1999.

10. "Practical Temperature Measurements," Application Note 290, Hewlett-Packard, Palo Alto, CA, 1980.

11. J. Graeme, "Tame Transducer Bridge Errors with Op Amp Feedback Control," *EDN,* May 26, 1982, pp. 173–176.

12. J. Williams, "Good Bridge-Circuit Design Satisfies Gain and Balance Criteria," *EDN,* Oct. 25, 1990, pp. 161–174.

13. J. Wong and A. Garcia, "Precision Transducer Interfaces," *Amplifier Applications Guide,* Analog Devices, Norwood, MA, 1992.

3

ACTIVE FILTERS: PART I

3.1 The Transfer Function
3.2 First-Order Active Filters
3.3 Audio Filter Applications
3.4 Standard Second-Order Responses
3.5 *KRC* Filters
3.6 Multiple-Feedback Filters
3.7 State-Variable and Biquad Filters
3.8 Sensitivity
   Problems
   References

A filter is a circuit that processes signals on a frequency-dependent basis. The manner in which its behavior varies with frequency is called the *frequency response* and is expressed in terms of the *transfer function* $H(j\omega)$, where $\omega = 2\pi f$ is the *angular frequency*, in radians per second (rad/s), and j is the *imaginary unit* ($j^2 = -1$). This response is further specialized as the *magnitude* response $|H(j\omega)|$ and the *phase* response $\sphericalangle H(j\omega)$, giving, respectively, the *gain* and *phase shift* experienced by an ac signal in going through the filter.

Common Frequency Responses

On the basis of magnitude response, filters are classified as *low-pass, high-pass, band-pass,* and *band-reject* (or *notch*) filters. A fifth category is provided by *all-pass* filters, which process phase but leave magnitude constant. With reference to Fig. 3.1, we ideally define these responses as follows.

The low-pass response is characterized by a frequency ω_c, called the *cutoff frequency,* such that $|H| = 1$ for $\omega < \omega_c$ and $|H| = 0$ for $\omega > \omega_c$, indicating that input

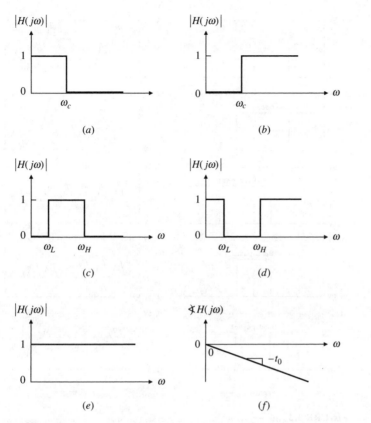

FIGURE 3.1
Idealized filter responses: (*a*) low-pass, (*b*) high-pass, (*c*) band-pass,
(*d*) band-reject, and (*e*), (*f*) all-pass.

signals with frequency less than ω_c go through the filter with unchanged amplitude,
while signals with $\omega > \omega_c$ undergo complete attentuation. A common low-pass filter
application is the removal of high-frequency noise from a signal.

The high-pass response is complementary to the low-pass response. Signals with
frequency greater than the cutoff frequency ω_c emerge from the filter unattenuated,
and signals with $\omega < \omega_c$ are completely blocked out.

The band-pass response is characterized by a *frequency band* $\omega_L < \omega < \omega_H$,
called the *passband,* such that input signals within this band emerge unattenuated,
while signals with $\omega < \omega_L$ or $\omega > \omega_H$ are cut off. A familiar band-pass filter is the
tuning circuitry of a radio, which allows the user to select a particular station and
block out all others.

The band-reject response is complementary to the band-pass response because it
blocks out frequency components within the *stopband* $\omega_L < \omega < \omega_H$, while passing
all the others. When the stopband is sufficiently narrow, the response is called a
notch response. An application of notch filters is the elimination of unwanted 60-Hz
pickup in medical equipment.

The all-pass response is characterized by $|H| = 1$ regardless of frequency, and
$\sphericalangle H = -t_0\omega$, where t_0 is a suitable proportionality constant, in seconds. This filter

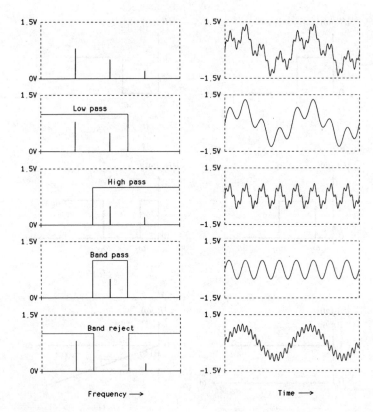

FIGURE 3.2
Effect of filtering in the frequency domain (left) and in the time domain
(right).

passes an ac signal without affecting its amplitude, but it delays it in proportion to its
frequency ω. For obvious reasons, all-pass filters are also called *delay filters*. Delay
equalizers and wideband 90° phase-shift networks are examples of all-pass filters.

Figure 3.2 illustrates the effects of the first four ideal filter types using the input
voltage

$$v_I(t) = 0.8 \sin \omega_0 t + 0.5 \sin 4\omega_0 t + 0.2 \sin 16\omega_0 t \text{ V}$$

as an example. Shown at the left are the spectra that we would observe with a
spectrum analyzer; shown at the right are the waveforms that we would observe
with an oscilloscope. The spectrum and waveform at the top pertain to the input
signal, and those below pertain, respectively, to the low-pass, high-pass, band-pass,
and band-reject outputs. For instance, if we send $v_I(t)$ through a low-pass filter with
ω_c somewhere between $4\omega_0$ and $16\omega_0$, the first two components are multiplied by
1 and thus passed, but the third component is multiplied by 0 and is thus blocked:
the result is $v_O(t) = 0.8 \sin \omega_0 t + 0.5 \sin 4\omega_0 t$ V.

As we proceed we shall see that the practical filters provide only approximations
to the idealized brick-wall magnitudes shown in the figure and also that they affect
phase.

Filter theory is a vast discipline, and it is documented in a number of textbooks dedicated only to it.[1-4] Filters can be built solely from resistors, inductors and capacitors (*RLC* filters), which are passive components. However, after the emergence of the feedback concept, it was realized that incorporating an amplifier in a filter circuit made it possible to achieve virtually any response, but without the use of inductors. This is a great advantage because inductors are the least ideal among the basic circuit elements, and are also bulky, heavy, and expensive—they do not lend themselves to IC-type mass production.

How amplifiers manage to displace inductors is an intriguing issue that we shall address. Here, we intuitively justify how by noting that an amplifier can take energy from its power supplies and inject it into the surrounding circuitry to make up for energy losses in the resistors. Inductors and capacitors are nondissipative elements that can store energy during part of a cycle and release it during the rest of the cycle. An amplifier, backed by its power supply, can do the same and more because, unlike inductors and capacitors, it can be made to release more energy than is actually absorbed by the resistors. Amplifiers are said to be active elements because of this, and filters incorporating amplifiers are called *active filters*. These filters provide one of the most fertile areas of application for op amps.

An active filter will work properly only to the extent that the op amp will. The most serious op amp limitation is the open-loop gain rolloff with frequency, an issue addressed at length in Chapter 6. This limitation generally restricts active-filter applications below the megahertz range. This includes the audio and instrumentation ranges, where op amp filters find their widest application and where inductors would be too bulky to compete with the miniaturization available with ICs. Beyond the frequency reach of op amps, inductors take over again, so high-frequency filters are still implemented with passive *RLC* components. In these filters, inductor sizes and weights are more manageable as inductance and capacitance values decrease with the operating frequency range.

In the present chapter we study first-order and second-order active filters. Higher-order filters are covered in Chapter 4, along with switched-capacitor filters.

3.1
THE TRANSFER FUNCTION

Filters are implemented with devices exhibiting frequency-dependent characteristics, such as capacitors and inductors. When subjected to ac signals, these elements oppose current flow in a frequency-dependent manner and also introduce a 90° phase shift between voltage and current. To account for this behavior, we use the *complex impedances* $Z_L = sL$ and $Z_C = 1/sC$, where $s = \sigma + j\omega$ is the *complex frequency*, in complex nepers per second (complex Np/s). Here, σ is the *Neper frequency*, in nepers per second (Np/s) and ω is the *angular frequency*, in radians per second (rad/s).

The behavior of a circuit is uniquely characterized by its transfer function $H(s)$. To find this function, we first derive an expression for the output X_o in terms of the input X_i (X_o and X_i can be voltages or currents) using familiar tools such as

Ohm's law $V = Z(s)I$, KVL, KCL, the voltage and current divider formulas, and the superposition principle. Then, we solve for the ratio

$$H(s) = \frac{X_o}{X_i} \tag{3.1}$$

Once $H(s)$ is known, the response $x_o(t)$ to a given input $x_i(t)$ can be found as

$$x_o(t) = \mathcal{L}^{-1}\{H(s)X_i(s)\} \tag{3.2}$$

where $\mathcal{L}^{-1}$ denotes the inverse Laplace transform, and $X_i(s)$ is the Laplace transform of $x_i(t)$.

Transfer functions turn out to be *rational functions* of s,

$$H(s) = \frac{N(s)}{D(s)} = \frac{a_m s^m + a_{m-1} s^{m-1} + \cdots + a_1 s + a_0}{b_n s^n + b_{n-1} s^{n-1} + \cdots + b_1 s + b_0} \tag{3.3}$$

where $N(s)$ and $D(s)$ are suitable polynomials of s with real coefficients and with degrees m and n. The degree of the denominator determines the *order* of the filter (first-order, second-order, etc.). The roots of the equations $N(s) = 0$ and $D(s) = 0$ are called, respectively, the *zeros* and the *poles* of $H(s)$, and are denoted as $z_1, z_2, \ldots, z_m$, and $p_1, p_2, \ldots, p_n$. Factoring out $N(s)$ and $D(s)$ in terms of their respective roots, we can write

$$H(s) = H_0 \frac{(s - z_1)(s - z_2) \cdots (s - z_m)}{(s - p_1)(s - p_2) \cdots (s - p_n)} \tag{3.4}$$

where $H_0 = a_m/b_n$ is called the *scaling factor*. Aside from H_0, $H(s)$ is uniquely determined once its zeros and poles are known. Roots are also referred to as *critical* or *characteristic frequencies* because they depend solely on the circuit, that is, on its elements and the way they are interconnected, irrespective of its signals or the energy stored in its reactive elements. In fact, essential circuit specifications are often given in terms of the roots.

Roots can be real or complex. When zeros or poles are complex, they occur in conjugate pairs. For instance, if $p_k = \sigma_k + j\omega_k$ is a pole, then $p_k^* = \sigma_k - j\omega_k$ is also a pole. Roots are conveniently visualized as points in the *complex plane*, or *s plane*: σ_k is plotted against the horizontal, or *real*, axis, which is calibrated in nepers per second (Np/s); ω_k is plotted against the vertical, or *imaginary*, axis, which is calibrated in radians per second (rad/s). In these plots a zero is represented as "o" and a pole as "×". Just by looking at the pole-zero pattern of a circuit, a designer can predict important characteristics, such as stability and frequency response. Because these characteristics will arise frequently as we proceed, we wish to give them a definitive review.

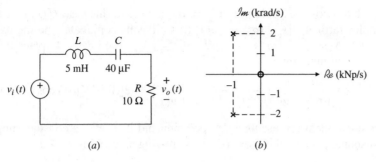

FIGURE 3.3
Circuit of Example 3.1 and its pole-zero plot.

EXAMPLE 3.1. Find the pole-zero plot of the circuit of Fig. 3.3a.

Solution. Using the generalized voltage divider formula, $V_o = [R/(sL+1/sC+R)]V_i$. Rearranging,

$$H(s) = \frac{V_o}{V_i} = \frac{RCs}{LCs^2 + RCs + 1} = \frac{R}{L} \times \frac{s}{s^2 + (R/L)s + 1/LC}$$

Substituting the given component values and factoring out,

$$H(s) = 2 \times 10^3 \times \frac{s}{[s - (-1 + j2)10^3] \times [s - (-1 - j2)10^3]}$$

This function has $H_0 = 2 \times 10^3$ V/V, a zero at the origin, and a conjugate pole pair at $-1 \pm j2$ complex kNp/s. Its pole-zero plot is shown in Fig 3.3b.

$H(s)$ and Stability

A circuit is said to be stable if it produces a bounded output in response to any bounded input. One way to assess whether a circuit is stable or not is to inject some energy into one or more of its reactive elements and then observe how the circuit does on its own, in the absence of any applied sources. The circuit response is in this case called the *source-free,* or *natural, response.* A convenient method of injecting energy is to apply an impulsive input, whose Laplace transform is unity. By Eq. (3.2), the ensuing response, or impulse response, is then $h(t) = \mathcal{L}^{-1}\{H(s)\}$. Interestingly enough, this response is determined by the poles. We identify two representative cases:

1. $H(s)$ has a real pole at $s = \sigma_k \pm j0 = \sigma_k$. Using well-known Laplace-transform techniques,[5] one can prove that $H(s)$ contains the term $A_k/(s - \sigma_k)$, where A_k is called the *residue* of $H(s)$ at that pole, and is found as $A_k = (s - \sigma_k)H(s)|_{s=\sigma_k}$. From the Laplace-transform tables we find

$$\mathcal{L}^{-1}\left\{\frac{A_k}{s - \sigma_k}\right\} = A_k e^{\sigma_k t} u(t) \tag{3.5}$$

where $u(t)$ is the unit step function ($u = 0$ for $t < 0$, $u = 1$ for $t > 0$). A real pole contributes an exponential component to the response $x_o(t)$, and this component decays if $\sigma_k < 0$, remains constant if $\sigma_k = 0$, and diverges if $\sigma_k > 0$.

2. $H(s)$ has a complex pole pair at $s = \sigma_k \pm j\omega_k$. In this case $H(s)$ contains the complex term $A_k/[s - (\sigma_k + j\omega_k)]$ as well as its conjugate, and the residue is found as $A_k = [s - (\sigma_k + j\omega_k)]H(s)|_{s=\sigma_k+j\omega_k}$. The inverse Laplace transform of their combination is

$$\mathscr{L}^{-1}\left\{\frac{A_k}{s - (\sigma_k + j\omega_k)} + \frac{A_k^*}{s - (\sigma_k - j\omega_k)}\right\} = 2|A_k|e^{\sigma_k t}u(t)\cos(\omega_k t + \sphericalangle A_k)$$

(3.6)

This component represents a damped sinusoid if $\sigma_k < 0$, a constant-amplitude, or sustained, sinusoid if $\sigma_k = 0$, and a growing sinusoid if $\sigma_k > 0$.

It is apparent that for a circuit to be stable, *all poles must lie in the left half of the s plane* (LHP), where $\sigma < 0$. Passive *RLC* circuits, such as that of Example 3.1, meet this constraint and are thus stable. However, if a circuit contains dependent sources such as op amps, its poles may spill into the right half-plane and thus lead to instability. Its output will grow until the saturation limits of the op amp are reached. If the circuit has a complex pole pair, the outcome of this is a sustained oscillation. Instability is generally undesirable, and stabilization techniques are covered in Chapter 8. There are nevertheless situations in which instability is exploited on purpose. A common example is the design of sine wave oscillators, to be addressed in Chapter 10.

EXAMPLE 3.2. Find the impulse response of the circuit of Example 3.1.

Solution. We have $A_1 = [s - (-1 + j2)10^3]H(s)|_{s=(-1+j2)10^3} = 1000 + j500 = 500\sqrt{5}\underline{/26.57°}$. So, $v_o(t) = 10^3\sqrt{5}e^{-10^3 t}u(t)\cos(2 \times 10^3 t + 26.57°)$ V.

$H(s)$ and the Frequency Response

In the study of filters we are interested in the response to an ac input of the type

$$x_i(t) = X_{im}\cos(\omega t + \theta_i)$$

where X_{im} is the amplitude, ω the angular frequency, and θ_i the phase angle. In general, the complete response $x_o(t)$ of Eq. (3.2) consists of two components,[5] namely, a *transient* component functionally similar to the natural response, and a *steady-state* component having the same frequency as the input, but differing in amplitude and phase. If all poles are in the LHP, the transient component will die out, leaving only the steady-state component,

$$x_o(t) = X_{om}\cos(\omega t + \theta_o)$$

This is illustrated in Fig. 3.4. Since we are narrowing our scope to this component alone, we wonder whether we can simplify our math, bypassing the general Laplace approach of Eq. (3.2). Such a simplification is possible, and it merely requires that we compute $H(s)$ on the imaginary axis. We do this by letting $s \to j\omega$ (or $s \to j2\pi f$ when working with the cyclical frequency f, in hertz.) Then, the output parameters are found as

$$X_{om} = |H(j\omega)| \times X_{im}$$

(3.7a)

$$\theta_o = \sphericalangle H(j\omega) + \theta_i$$

(3.7b)

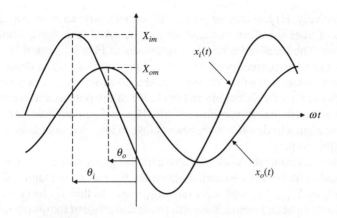

FIGURE 3.4
In general, a filter affects both amplitude and phase.

In the course of complex-number manipulations, we shall often use the following important properties: Let

$$H = |H|\underline{/\sphericalangle H} = H_r + jH_i \tag{3.8}$$

where $|H|$ is the *modulus* or *magnitude of H, $\sphericalangle H$* its *argument* or *phase angle,* and H_r and H_i the *real* and *imaginary* parts. Then,

$$|H| = \sqrt{H_r^2 + H_i^2} \tag{3.9a}$$

$$\sphericalangle H = \tan^{-1}(H_i/H_r) \qquad \text{if } H_r > 0 \tag{3.9b}$$

$$\sphericalangle H = 180° - \tan^{-1}(H_i/H_r) \qquad \text{if } H_r < 0 \tag{3.9c}$$

$$|H_1 \times H_2| = |H_1| \times |H_2| \tag{3.10a}$$

$$\sphericalangle(H_1 \times H_2) = \sphericalangle H_1 + \sphericalangle H_2 \tag{3.10b}$$

$$|H_1/H_2| = |H_1| - |H_2| \tag{3.11a}$$

$$\sphericalangle(H_1/H_2) = \sphericalangle H_1 - \sphericalangle H_2 \tag{3.11b}$$

EXAMPLE 3.3. Find the steady-state response of the circuit of Example 3.1 to the signal $v_i(t) = 10\cos(10^3 t + 45°)$ V.

Solution. Letting $s \to j10^3$ rad/s in Example 3.1 we get $H(j10^3) = j1/(2 + j1) = (1/\sqrt{5})\underline{/63.43°}$ V/V. So $V_{om} = 10/\sqrt{5}$ V, $\theta_o = 63.43° + 45° = 108.43°$, and $v_o(t) = \sqrt{20}\cos(10^3 t + 108.43°)$ V.

There are various viewpoints we can take in regard to $H(j\omega)$. Presented with the circuit diagram of a filter, we may wish to find $H(s)$ analytically, and then plot $|H(j\omega)|$ and $\sphericalangle H(j\omega)$ versus ω (or f) for a visual display of the frequency response. These plots, referred to as *Bode plots,* can be generated by hand or via PSpice.

Conversely, given $H(j\omega)$, we may want to let $j\omega \to s$ to obtain $H(s)$, find its roots, and construct the pole-zero plot.

Alternatively, $H(j\omega)$ may be given to us, either analytically or in graphical form or in terms of filter specifications, and we may be asked to design a circuit realizing this function. The idealized brick-wall responses of Fig. 3.1 cannot be achieved in practice but can be approximated via rational functions of s. The degree n of $D(s)$ determines the order of the filter (first-order, second-order, etc.). As a general rule, the higher n, the greater the flexibility in the choice of the polynomial coefficients best suited to a given frequency-response profile. However, circuit complexity increases with n, indicating a trade-off between how close to ideal we want to be and the price we are willing to pay.

Another viewpoint yet is one in which a filter is given to us in black-box form and we are asked to find $H(j\omega)$ experimentally. By Eq. (3.7), the magnitude and phase are $|H(j\omega)| = X_{om}/X_{im}$ and $\sphericalangle H(j\omega) = \theta_o - \theta_i$. To find $H(j\omega)$ experimentally, we apply an ac input and measure the amplitude and phase of the output relative to the input at different frequencies. We then plot measured data versus frequency point-by-point and obtain the experimental profiles of $|H(j\omega)|$ and $\sphericalangle H(j\omega)$. If desired, measured data can be processed with suitable curve-fitting algorithms to obtain an analytical expression for $H(j\omega)$ in terms of its critical frequencies. In the case of voltage signals, the measurements are easily done with a dual-trace oscilloscope. To simplify the calculations, it is convenient to set $V_{im} = 1$ V, and to adjust the trigger so that $\theta_i = 0$. Then we have $|H(j\omega)| = V_{om}$ and $\sphericalangle H(j\omega) = \theta_o$.

Bode Plots

The magnitude and frequency range of a filter can be quite wide. For instance, in audio filters the frequency range is typically from 20 Hz to 20 kHz, which represents a 1000:1 range. In order to visualize small as well as large details with the same degree of clarity, $|H|$ and $\sphericalangle H$ are plotted on *logarithmic* and *semilogarithmic* scales, respectively. That is, frequency intervals are expressed in *decades* ($\ldots, 0.01, 0.1, 1, 10, 100, \ldots$) or in *octaves* ($\ldots, \frac{1}{8}, \frac{1}{4}, \frac{1}{2}, 1, 2, 4, 8, \ldots$), and $|H|$ is expressed in *decibels* (dB) as

$$|H|_{dB} = 20\log_{10}|H| \tag{3.12}$$

The Bode plots are plots of decibels and degrees versus decades (or octaves). Another advantage of these plots is that the following useful properties hold:

$$|H_1 \times H_2|_{dB} = |H_1|_{dB} + |H_2|_{dB} \tag{3.13a}$$

$$|H_1/H_2|_{dB} = |H_1|_{dB} - |H_2|_{dB} \tag{3.13b}$$

$$|1/H_1|_{dB} = -|H|_{dB} \tag{3.13c}$$

To speed up the hand generation of these plots, it is often convenient to effect asymptotic approximations. To this end, the following properties are useful:

$$H \cong H_r \qquad \text{if } |H_r| \gg |H_i| \tag{3.14a}$$

$$H \cong jH_i \qquad \text{if } |H_i| \gg |H_r| \tag{3.14b}$$

Keep Eqs. (3.13) and (3.14) in mind because we shall use them frequently.

3.2
FIRST-ORDER ACTIVE FILTERS

The simplest active filters are obtained from the basic op amp configurations by using a capacitance as one of its external components. Since $Z_C = 1/sC = 1/j\omega C$, the result is a gain with frequency-dependent magnitude and phase. As you study filters, it is important that you try justifying your mathematical findings using physical insight. In this respect, a most valuable tool is asymptotic verification, which is based on the following properties:

$$\lim_{\omega \to 0} Z_C = \infty \tag{3.15a}$$

$$\lim_{\omega \to \infty} Z_C = 0 \tag{3.15b}$$

In words, at low frequencies a capacitance tends to behave as an open circuit compared with the surrounding elements, and at high frequencies it tends to behave as a short circuit.

The Differentiator

In the inverting configuration of Fig. 3.5a we have $V_o = (-R/Z_C)V_i = -RCsV_i$. By a well-known Laplace-transform property, multiplication by s in the frequency domain is equivalent to differentiation in the time domain. This confirms the designation *differentiator* for the circuit. Solving for the ratio V_o/V_i gives

$$H(s) = -RCs \tag{3.16}$$

indicating a zero at the origin.

Letting $s \to j\omega$ and introducing the scaling frequency

$$\omega_0 = \frac{1}{RC} \tag{3.17}$$

we can express $H(j\omega)$ in the normalized form

$$H(j\omega) = -j\omega/\omega_0 = (\omega/\omega_0) \underline{/-90^\circ} \tag{3.18}$$

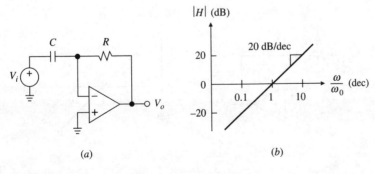

(a)　　　　　　　　　(b)

FIGURE 3.5
The differentiator and its magnitude Bode plot.

Considering that $|H|_{dB} = 20\log_{10}(\omega/\omega_0)$, the plot of $|H|_{dB}$ versus $\log_{10}(\omega/\omega_0)$ is a straight line of the type $y = 20x$. As shown in Fig. 3.5*b*, its slope is 20 dB/dec, indicating that for every decade increase (or decrease) in frequency, magnitude increases (or decreases) by 20 dB. Equation (3.18) indicates that the circuit introduces a 90° phase lag, and amplifies in proportion to frequency. Physically, we observe that at low frequencies, where $|Z_C| > R$, the circuit provides attenuation (negative decibels); at high frequencies, where $|Z_C| < R$, it provides magnification (positive decibels); at $\omega = \omega_0$, where $|Z_C| = R$, it provides unity gain (0 db). Consequently, ω_0 is called the *unity-gain frequency*.

Integrators

Also called *Miller integrator* because the capacitor is in the feedback path, the circuit of Fig. 3.6*a* gives $V_o = (-Z_C/R)V_i = -(1/RCs)V_i$. The fact that division by s in the frequency domain corresponds to integration in the time domain confirms the designation *integrator*. Its transfer function

$$H(s) = -\frac{1}{RCs} \qquad (3.19)$$

has a pole at the origin. Letting $s \to j\omega$, we can write

$$H(j\omega) = -\frac{1}{j\omega/\omega_0} = \frac{1}{\omega/\omega_0}\underline{/+90°} \qquad (3.20)$$

where $\omega_0 = 1/RC$, as in Eq. (3.17). Observing that the transfer function is the reciprocal of that of the differentiator, we can apply Eq. (3.13*c*) and construct the integrator magnitude plot simply by reflecting that of the differentiator about the 0-dB axis. The result, shown in Fig. 3.6*b*, is a straight line with a slope of -20 dB/dec and with ω_0 as the *unity-gain frequency*. Moreover, the circuit introduces a 90° phase lead.

Because of the extremely high gain at low frequencies, where $|Z_C| \gg R$, a practical integrator circuit is seldom used alone as it tends to saturate. As mentioned in Chapter 1, an integrator is usually placed inside a control loop designed to keep the op amp within the linear region. We shall see examples when studying state-variable and biquad filters in Section 3.7, and sine wave oscillators in Section 10.1.

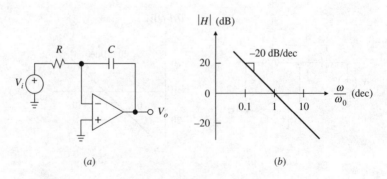

(a)   (b)

FIGURE 3.6
The integrator and its magnitude Bode plot.

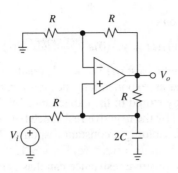

FIGURE 3.7
Noninverting, or Deboo, integrator.

Due to the negative sign in Eq. (3.19), the Miller integrator is also said to be an *inverting integrator*. The circuit of Fig. 3.7, called the *Deboo integrator*, for its inventor, uses a Howland current pump with a capacitance as load to achieve noninverting integration. As we know, the pump forces a current $I = V_i/R$ into the capacitance, resulting in a noninverting-input voltage $V_p = (1/s2C)I = V_i/2sRC$. The op amp then amplifies this voltage to give $V_o = (1 + R/R)V_p = V_i/sRC$, so

$$H(s) = \frac{1}{RCs} \tag{3.21}$$

The magnitude plot is the same as for the inverting integrator. However, the phase angle is now $-90°$, rather than $+90°$.

It is instructive to investigate the circuit from the more general viewpoint of Fig. 3.8a, where we identify two blocks: the RC network shown at the bottom, and the rest of the circuit forming a negative resistance converter. The converter provides a variable resistance $-R(R/kR) = -R/k, k \geq 0$, so the net resistance seen by C is $R \parallel (-R/k) = R/(1-k)$, indicating the pole

$$p = -\frac{1-k}{RC} \tag{3.22}$$

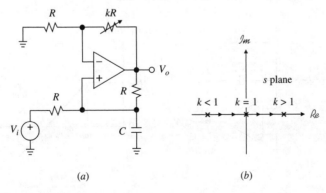

(a)   (b)

FIGURE 3.8
Varying k varies pole location.

The natural response is then

$$v_O(t) = v_O(0)e^{-t(1-k)/RC}u(t) \qquad (3.23)$$

We identify three important cases: (a) For $k < 1$, positive resistance prevails, indicating a negative pole and an exponentially decaying response. The decay is due to dissipation of the energy stored in the capacitance by the net resistance. (b) For $k = 1$, the energy supplied by the negative resistance balances the energy dissipated by the positive resistance, yielding a constant response. The net resistance is now infinite, and the pole is right at the origin. (c) For $k > 1$, the negative resistance supplies more energy than the positive resistance can dissipate, causing an exponential buildup. Negative resistance prevails, the pole is now in the right half plane, and the response diverges. Figure 3.8b shows the root locus as k is increased.

Low-Pass Filter with Gain

Placing a resistor in parallel with the feedback capacitor, as in Fig. 3.9a, turns the integrator into a low-pass filter with gain. Letting $1/Z_2 = 1/R_2 + 1/(1/sC) = (R_2Cs + 1)/R_2$ gives $H(s) = -Z_2/R_1$, or

$$H(s) = -\frac{R_2}{R_1}\frac{1}{R_2Cs + 1} \qquad (3.24)$$

indicating a real pole at $s = -1/R_2C$. Letting $s \to j\omega$, we can express $H(s)$ in the normalized form

$$H(j\omega) = H_0\frac{1}{1 + j\omega/\omega_0} \qquad (3.25a)$$

$$H_0 = -\frac{R_2}{R_1} \qquad \omega_0 = \frac{1}{R_2C} \qquad (3.25b)$$

Physically, the circuit works as follows. At sufficiently low frequencies, where $|Z_C| \gg R_2$, we can ignore Z_C compared with R_2 and thus regard the circuit as an inverting amplifier with gain $H \cong -R_2/R_1 = H_0$. For obvious reasons, H_0 is called the *dc gain*. As shown in Fig. 3.9b, the low-frequency asymptote of the magnitude Bode plot is a horizontal line positioned at $|H_0|_{dB}$.

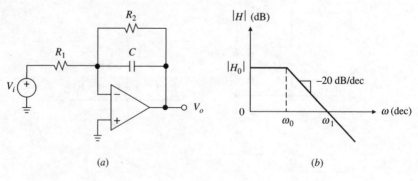

(a)   (b)

FIGURE 3.9
Low-pass filter with gain.

At sufficiently high frequencies, where $|Z_C| \ll R_2$, we can ignore R_2 compared with Z_C and thus regard the circuit as an integrator. As we know, its high-frequency asymptote is a line with a slope of -20 dB/dec and passing through the unity-gain frequency $\omega_1 = 1/R_1 C$. Since the circuit approximates integrator behavior over only a limited frequency range, it is also called a *lossy integrator.*

The borderline between amplifier and integrator behavior occurs at the frequency that makes $|Z_C| = R_2$, or $1/\omega C = R_2$. Clearly, this is the frequency ω_0 of Eq. (3.25b). For $\omega/\omega_0 = 1$, Eq. (3.25a) predicts $|H| = |H_0/(1 + j1)| = |H_0|/\sqrt{2}$, or, equivalently, $|H|_{dB} = |H_0|_{dB} - 3$ dB. Hence, ω_0 is called the *−3-dB frequency.*

The magnitude profile indicates that this is a low-pass filter with H_0 as dc gain and with ω_0 as cutoff frequency. Signals with $\omega < \omega_0$ are passed with gain close to H_0, but signals with $\omega > \omega_0$ are progressively attenuated, or cut. For every decade increase in ω, $|H|$ decreases by 20 dB. Clearly, this is only a crude approximation to the brick-wall profile of Fig. 3.1b.

> **EXAMPLE 3.4.** (a) In the circuit of Fig. 3.9a, specify suitable components to achieve a -3-dB frequency of 1 kHz with a dc gain of 20 dB and an input resistance of at least 10 kΩ. (b) At what frequency does gain drop to 0 dB? What is the phase there?

Solution.

(a) Since 20 dB corresponds to $10^{20/20} = 10$ V/V, we need $R_2 = 10R_1$. To ensure $R_i > 10$ kΩ, try $R_1 = 20$ kΩ. Then, $R_2 = 200$ kΩ, and $C = 1/\omega_0 R_2 = 1/(2\pi \times 10^3 \times 200 \times 10^3) = 0.796$ nF. Use $C = 1$ nF, which is a more readily available value. Then, scale the resistances as $R_2 = 200 \times 0.796 = 158$ kΩ and $R_1 = 15.8$ kΩ, both 1%.

(b) Imposing $|H| = 10/\sqrt{1^2 + (f/10^3)^2} = 1$ and solving yields $f = 10^3\sqrt{10^2 - 1} = 9.950$ kHz. Moreover, $\sphericalangle H = 180° - \tan^{-1} 9950/10^3 = 95.7°$.

High-Pass Filter with Gain

Placing a capacitor in series with the input resistor as in Fig. 3.10a turns the differentiator into a high-pass filter with gain. Letting $Z_1 = R_1 + 1/sC = (R_1 Cs + 1)/sC$ and $H(s) = -R_2/Z_1$ gives

$$H(s) = -\frac{R_2}{R_1}\frac{R_1 Cs}{R_1 Cs + 1} \tag{3.26}$$

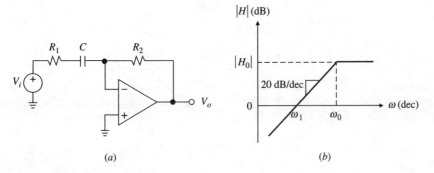

(a)   (b)

FIGURE 3.10
High-pass filter with gain.

indicating a zero at the origin and a real pole at $s = -1/R_1C$. Letting $s \to j\omega$, we can express $H(s)$ in the normalized form

$$H(j\omega) = H_0 \frac{j\omega/\omega_0}{1 + j\omega/\omega_0} \tag{3.27a}$$

$$H_0 = -\frac{R_2}{R_1} \qquad \omega_0 = \frac{1}{R_1C} \tag{3.27b}$$

where H_0 is called the *high-frequency gain* and ω_0 is again the -3-dB frequency. As shown in Fig. 3.10b, which you are encouraged to justify asymptotically, the circuit is a high-pass filter.

Wideband Band-Pass Filter

The last two circuits can be merged as in Fig. 3.11a to give a *band-pass* response. Letting $Z_1 = (R_1C_1s + 1)/C_1s$ and $Z_2 = R_2/(R_2C_2s + 1)$, we get $H(s) = -Z_2/Z_1$, or

$$H(s) = -\frac{R_2}{R_1} \frac{R_1C_1s}{R_1C_1s + 1} \frac{1}{R_2C_2s + 1} \tag{3.28}$$

indicating a zero at the origin and two real poles at $-1/R_1C_1$ and $-1/R_2C_2$. Though this is a second-order filter, we have chosen to discuss it here to demonstrate the use of lower-order building blocks to synthetize higher-order filters. Letting $s \to j\omega$ yields

$$H(j\omega) = H_0 \frac{j\omega/\omega_L}{(1 + j\omega/\omega_L)(1 + j\omega/\omega_H)} \tag{3.29a}$$

$$H_0 = -\frac{R_2}{R_1} \qquad \omega_L = \frac{1}{R_1C_1} \qquad \omega_H = \frac{1}{R_2C_2} \tag{3.29b}$$

where H_0 is called the *midfrequency gain*. The filter is useful with $\omega_L \ll \omega_H$, in which case ω_L and ω_H are called the *low* and *high* -3-dB frequencies. This circuit is used especially in audio applications, where it is desired to amplify signals within the audio range while blocking out subaudio components, such as dc, as well as noise above the audio range.

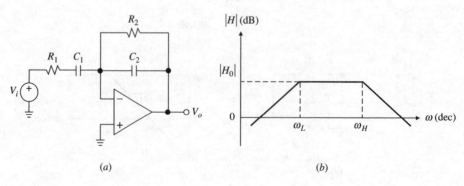

(a)   (b)

FIGURE 3.11
Wideband band-pass filter.

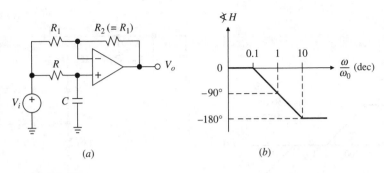

FIGURE 3.12
Phase shifter.

EXAMPLE 3.5. In the circuit of Fig. 3.11a specify suitable component values for a band-pass response with a gain of 20 dB over the audio range.

Solution. For a gain of 20 dB we need $R_2/R_1 = 10$. Try $R_1 = 10$ kΩ and $R_2 = 100$ kΩ. Then, for $\omega_L = 2\pi \times 20$ rad/s we need $C_1 = 1/(2\pi \times 20 \times 10 \times 10^3) = 0.7958$ μF. Use 1 μF, and rescale the resistances as $R_1 = 10^4 \times 0.7958 \cong 7.87$ kΩ and $R_2 = 78.7$ kΩ. For $\omega_H = 2\pi \times 20$ krad/s, use $C_2 = 1/(2\pi \times 20 \times 10^3 \times 78.7 \times 10^3) \cong 100$ pF.

Phase Shifters

In Fig. 3.12a the noninverting-input voltage V_p is related to V_i by the low-pass function as $V_p = V_i/(RCs+1)$. Moreover, $V_o = -(R_2/R_1)V_i + (1+R_2/R_1)V_p = 2V_p - V_i$. Eliminating V_p yields

$$H(s) = \frac{-RCs + 1}{RCs + 1} \tag{3.30}$$

indicating a zero at $s = 1/RC$ and a pole at $s = -1/RC$. Letting $s \to j\omega$ yields

$$H(j\omega) = \frac{1 - j\omega/\omega_0}{1 + j\omega/\omega_0} = 1\underline{/-2\tan^{-1}(\omega/\omega_0)} \tag{3.31}$$

With a gain of 1 V/V, this circuit passes all signals without altering their amplitude. However, as shown in Fig. 3.12b, it introduces a variable phase lag from from 0° to $-180°$, with a value of $-90°$ at $\omega = \omega_0$. Can you justify using physical insight?

3.3
AUDIO FILTER APPLICATIONS

Audio signal processing provides a multitude of uses for active filters. Common functions required in high-quality audio systems are equalized preamplifiers, active tone control, and graphic equalizers.[6] Equalized preamplifiers are used to compensate for the varying levels at which different parts of the audio spectrum are recorded commercially. Tone control and graphic equalization refer to response adjustments

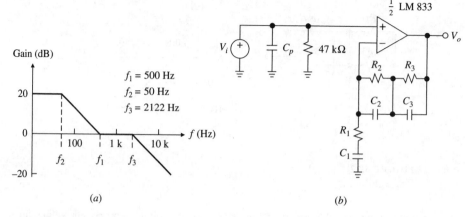

FIGURE 3.13
RIAA playback equalization curve and phono preamplifier.

that the listener can effect to compensate for nonideal loudspeaker response, to match apparent room acoustics, or simply to suit one's taste.

Phono Preamplifier

The function of a phono preamplifier is to provide amplification as well as amplitude equalization for the signal from a moving-magnet or a moving-coil cartridge. The response must conform to the standard RIAA (Record Industry Association of America) curve of Fig. 3.13a.

Preamplifier gains are usually specified at 1 kHz. The required amount of gain is typically 30 to 40 dB for moving-magnet cartridges, and 50 to 60 dB for moving-coil types. Since the RIAA curve is normalized for unity gain, the actual preamp response will be shifted upward by an amount equal to its gain.

Figure 3.13b shows one[7] of several topologies commonly used to approximate the RIAA response. The input-shunting network provides impedance matching for the source, while C_1 provides a low-frequency breakpoint (usually below 20 Hz) to block out dc and any subaudio frequency components. Since $|Z_{C_1}| \ll R_1$ over the frequency range of interest, the transfer function can be found as $H \cong 1 + Z_f/R_1$, where Z_f is the impedance of the feedback network. The result is (see Problem 3.17)

$$H(jf) \cong 1 + \frac{R_2 + R_3}{R_1} \frac{1 + jf/f_1}{(1 + jf/f_2)(1 + jf/f_3)} \tag{3.32}$$

$$f_1 = \frac{1}{2\pi(R_2 \parallel R_3)(C_2 + C_3)} \qquad f_2 = \frac{1}{2\pi R_2 C_2} \qquad f_3 = \frac{1}{2\pi R_3 C_3} \tag{3.33}$$

As long as the circuit is configured for substantially high gain, the unity term in Eq. (3.32) can be ignored, indicating that $H(jf)$ approximates the standard RIAA curve over the audio range.

EXAMPLE 3.6. Design a 40-dB gain, RIAA phono amplifier.

Solution. The RIAA curve must be shifted upward by 40 dB, so the gain below f_2 must be $40 + 20 = 60$ dB $= 10^3$ V/V. Thus, $(R_2 + R_3)/R_1 \cong 10^3$. The expressions for f_1 through f_3 provide three equations in four unknowns. Fix one, say, let $C_2 = 10$ nF. Then, Eq. (3.33) gives $R_2 = 1/(2\pi \times 50 \times 10 \times 10^{-9}) = 318$ kΩ (use 316 kΩ). We also have $1/R_2 + 1/R_3 = 2\pi f_1(C_2 + C_3)$ and $1/R_3 = 2\pi f_3 C_3$. Eliminating $1/R_3$ gives $C_3 = 2.77$ nF (use 2.7 nF). Back substituting gives $R_3 = 27.7$ kΩ (use 28.0 kΩ). Finally, $R_1 = (316 + 28)/10^3 = 344$ Ω (use 340 Ω) and $C_1 = 1/(2\pi \times 340 \times 20) = 23$ μF (use 33 μF). Summarizing, $R_1 = 340$ Ω, $R_2 = 316$ kΩ, $R_3 = 28.0$ kΩ, $C_1 = 33$ μF, $C_2 = 10$ nF, and $C_3 = 2.7$ nF.

Tape Preamplifier

A tape preamplifier must provide gain as well as amplitude and phase equalizations, for the signal from a tape head. The response is governed by the standard NAB (National Association of Broadcasters) curve of Fig. 3.14a. A circuit[7] to approximate this response is shown in Fig. 3.14b. As long as $|Z_{C_1}| \ll R_1$, we have (see Problem 3.18)

$$H(jf) \cong 1 + \frac{R_3}{R_1} \frac{1 + jf/f_1}{1 + jf/f_2} \tag{3.34}$$

$$f_1 = \frac{1}{2\pi R_2 C_2} \qquad f_2 = \frac{1}{2\pi (R_2 + R_3) C_2} \tag{3.35}$$

Active Tone Control

The most common form of tone control is *bass* and *treble* control, which allows the independent adjustment of gain over the lower (bass) and higher (treble) portions

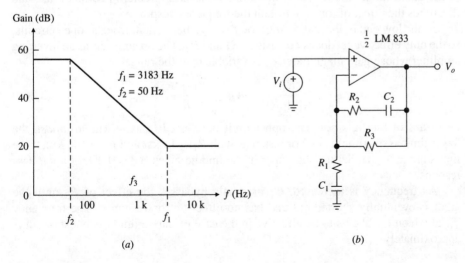

(a)

(b)

FIGURE 3.14
NAB equalization curve and tape preamplifier.

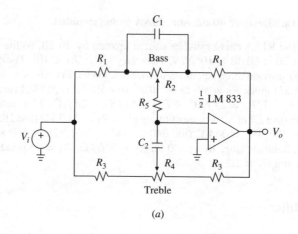

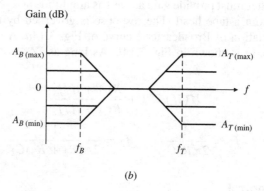

(b)

FIGURE 3.15
Bass and treble control.

of the audio range. Figure 3.15 shows one of several circuits in common use and illustrates the effect of tone control on the frequency response.

At the low end of the audio range, or $f < f_B$, the capacitors act as open circuits, so the only effective feedback consists of R_1 and R_2. The op amp acts as an inverting amplifier whose gain magnitude A_B is variable over the range

$$\frac{R_1}{R_1 + R_2} \le A_B \le \frac{R_1 + R_2}{R_1} \qquad (3.36a)$$

by means of the bass pot. The upper limit is referred to as maximum *boost,* the lower limit as maximum *cut.* For instance, with $R_1 = 11$ kΩ and $R_2 = 100$ kΩ, these limits are ± 20 dB. Setting the wiper in the middle gives $A_B = 0$ dB, or a *flat* bass response.

As frequency is increased, C_1 gradually bypasses the effect of R_2 until the latter is eventually shorted out and has no effect on the response. The frequency f_B at which C_1 begins to be effective in the case of maximum bass boost or cut is approximately

$$f_B = \frac{1}{2\pi R_2 C_1} \qquad (3.36b)$$

Above this frequency the response approaches the flat curve with a slope of about ±6 dB/oct, depending on whether the pot is set for maximum cut or boost.

At the high end of the audio range, or $f > f_T$, the capacitors act as short circuits, so the gain is now controlled by the treble pot. (The bass pot is ineffectual since it is being shorted out by C_1.) It can be proven that if the condition $R_4 \gg (R_1 + R_3 + 2R_5)$ is met, the range of variability of the treble gain A_T is

$$\frac{R_3}{R_1 + R_3 + 2R_5} \leq A_T \leq \frac{R_1 + R_3 + 2R_5}{R_3} \qquad (3.37a)$$

and the frequency f_T below which the treble control gradually ceases to affect the response is approximately

$$f_T = \frac{1}{2\pi R_3 C_2} \qquad (3.37b)$$

EXAMPLE 3.7. Design a bass/treble control with $f_B = 30$ Hz, $f_T = 10$ kHz, and ±20 dB maximum boost/cut at both ends.

Solution. Since 20 dB corresponds to 10 V/V, we must have $(R_1 + R_2)/R_1 = 10$ and $(R_1 + R_3 + 2R_5)/R_3 = 10$. Let R_2 be a 100-kΩ pot so that $R_1 = 11$ kΩ. Arbitrarily impose $R_5 = R_1 = 11$ kΩ. Then $R_3 = 3.67$ kΩ (use 3.6 kΩ). To meet the condition $R_4 \gg (R_1 + R_3 + 2R_5) \cong 37$ kΩ, let R_4 be a 500-kΩ pot. Then $C_1 = 1/2\pi R_2 f_B = 53$ nF (use 51 nF), and $C_2 = 1/2\pi R_3 f_T = 4.4$ nF (use 5.1 nF). Summarizing, $R_1 = 11$ kΩ, $R_2 = 100$ kΩ, $R_3 = 3.6$ kΩ, $R_4 = 500$ kΩ, $R_5 = 11$ kΩ, $C_1 = 51$ nF, and $C_2 = 5.1$ nF.

Graphic Equalizers

The function of a graphic equalizer is to provide boost and cut control not just at the bass and treble extremes, but also within intermediate frequency bands. Equalizers are implemented with arrays of narrow-band filters whose individual responses are adjusted by vertical slide pots arranged side by side to provide a graphic visualization of the equalized response (hence the name).

Figure 3.16 shows a familiar realization of one of the equalizer sections. The circuit is designed so that over a specified frequency band, C_1 acts as an open circuit

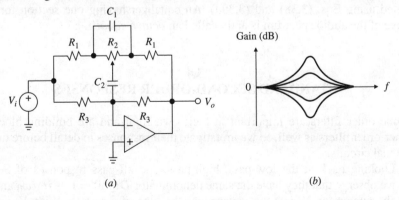

(a)   (b)

FIGURE 3.16
Section of a graphic equalizer.

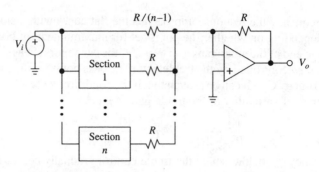

FIGURE 3.17
Graphic equalizer with n bands.

while C_2 acts as a short, thus allowing for boost or cut control, depending on whether the wiper position is to the left or to the right, respectively. Outside the band the circuit provides unity gain, regardless of the wiper position. This stems from the fact that C_2 acts as an open circuit at low frequencies, and C_1 acts as a short at high frequencies. The result is a flat response, but with a peak or a dip over the specified band.

It can be proven[8] that if the component values are chosen so that

$$R_3 \gg R_1 \qquad R_3 = 10R_2 \qquad C_1 = 10C_2 \qquad (3.38)$$

then the center of the band is

$$f_0 = \frac{\sqrt{2 + R_2/R_1}}{20\pi R_2 C_2} \qquad (3.39a)$$

and the gain magnitude A_0 at this frequency is variable over the range

$$\frac{3R_1}{3R_1 + R_2} \leq A_0 \leq \frac{3R_1 + R_2}{3R_1} \qquad (3.39b)$$

An n-band equalizer is implemented by paralleling n sections and summing the individual outputs with the input in a $1:(n-1)$ ratio.[8] This is done with an ordinary summing amplifier, as in Fig. 3.17. Common choices for the resistances of each section are $R_1 = 10\text{ k}\Omega$, $R_2 = 100\text{ k}\Omega$, and $R_3 = 1\text{ M}\Omega$. The capacitances are calculated using Eqs. (3.38) and (3.39a). An equalizer having one section for each octave of the audio spectrum is aptly called an *octave equalizer.*

3.4
STANDARD SECOND-ORDER RESPONSES

Second-order filters are important in their own right and are building blocks of higher-order filters as well, so we investigate their responses in detail before turning to actual circuits.

Looking back at the low-pass, high-pass, and all-pass responses of Section 3.2, we observe that they have the same denominator $D(j\omega) = 1 + j\omega/\omega_0$ and that it is the numerator $N(j\omega)$ that determines the type of response. With $N(j\omega) = 1$ we get the low-pass, with $N(j\omega) = j\omega/\omega_0$ the high-pass, and with $N(j\omega) = 1 -$

$j\omega/\omega_0 = D(j\omega)$ the all-pass response. Moreover, the presence of a scaling factor H_0 does not change the response type; it only shifts its magnitude plot up or down, depending on whether $|H_0| > 1$ or $|H_0| < 1$.

Similar considerations hold for second-order responses. However, since the degree of the denominator is now 2, we have an additional filter parameter besides ω_0. All second-order functions can be put in the standard form

$$H(s) = \frac{N(s)}{(s/\omega_0)^2 + 2\zeta(s/\omega_0) + 1} \tag{3.40}$$

where $N(s)$ is a polynomial in s of degree $m \le 2$; ω_0 is called the *undamped natural frequency,* in radians per second; and ζ (zeta) is a dimensionless parameter called the *damping ratio.* This function has two poles, $p_{1,2} = (-\zeta \pm \sqrt{\zeta^2 - 1})\omega_0$, whose location in the s plane is controlled by ζ as follows:

1. For $\zeta > 1$, the poles are real and negative. The natural response consists of two decaying exponentials and is said to be *overdamped.*
2. For $0 < \zeta < 1$, the poles are complex conjugate and can be expressed as

$$p_{1,2} = -\zeta\omega_0 \pm j\omega_0\sqrt{1 - \zeta^2} \tag{3.41}$$

These poles lie in the left half plane, and the natural response, now called *underdamped,* is the damped sinusoid $x_o(t) = 2|A|e^{-\zeta\omega_0 t}\cos(\omega_0\sqrt{1-\zeta^2}t + \angle A)$, where A is the residue at the upper pole.
3. For $\zeta = 0$, Eq. (3.41) yields $p_{1,2} = \pm j\omega_0$, indicating that the poles lie right on the imaginary axis. The natural response is a sustained, or *undamped,* sinusoid with frequency ω_0; hence the name for ω_0.
4. For $\zeta < 0$, the poles lie in the right half plane, thus causing a *diverging* response because the exponent in the term $e^{-\zeta\omega_0 t}$ is now positive. Filters must have $\zeta > 0$ in order to be stable.

The system of trajectories described by the roots as a function of ζ is the root locus depicted in Fig. 3.18. Note that for $\zeta = 1$ the poles are real and coincident.

Letting $s \to j\omega$ yields the frequency response, which we shall express in terms of the alternative dimensionless parameter Q as

$$H(j\omega) = \frac{N(j\omega)}{1 - (\omega/\omega_0)^2 + (j\omega/\omega_0)/Q} \tag{3.42}$$

$$Q = \frac{1}{2\zeta} \tag{3.43}$$

The meaning of Q will become clear as we proceed.

The Low-Pass Response H_{LP}

All second-order low-pass functions can be put in the standard form $H(j\omega) = H_{0\text{LP}}H_{\text{LP}}(j\omega)$, where $H_{0\text{LP}}$ is a suitable constant referred to as the *dc gain,* and

$$H_{\text{LP}}(j\omega) = \frac{1}{1 - (\omega/\omega_0)^2 + (j\omega/\omega_0)/Q} \tag{3.44}$$

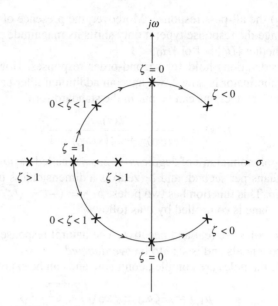

FIGURE 3.18
Root locus for a second-order transfer function.

To construct the magnitude plot we use asymptotic approximations.

1. For $\omega/\omega_0 \ll 1$, the second and third denominator terms can be ignored in comparison with unity, so $H_{LP} \to 1$. The low-frequency asymptote is thus

$$|H_{LP}|_{dB} = 0 \qquad (\omega/\omega_0 \ll 1) \tag{3.45a}$$

2. For $\omega/\omega_0 \gg 1$, the second denominator term dominates over the other two, so $H_{LP} \to -1/(\omega/\omega_0)^2$. The high-frequency asymptote is $|H_{LP}|_{dB} = 20\log_{10}[1/(\omega/\omega_0)^2]$, or

$$|H_{LP}|_{dB} = -40\log_{10}(\omega/\omega_0) \qquad (\omega/\omega_0 \gg 1) \tag{3.45b}$$

This equation is of the type $y = -40x$, or a straight line with a slope of -40 dB/dec. Compared to the first-order response, which has a slope of only -20 dB/dec, the second-order response is closer to the idealized brick-wall profile.

3. For $\omega/\omega_0 = 1$, the two asymptotes meet since letting $\omega/\omega_0 = 1$ in Eq. (3.45b) gives Eq. (3.45a). Moreover, the first and second denominator terms cancel each other out to give $H_{LP} = -jQ$, or

$$|H_{LP}|_{dB} = Q_{dB} \qquad (\omega/\omega_0 = 1) \tag{3.45c}$$

In the frequency region near $\omega/\omega_0 = 1$ we now have a family of curves, depending on the value of Q. Contrast this with the first-order case, where only one curve was possible.

The second-order response, besides providing a high-frequency asymptotic slope twice as steep, offers an additional degree of freedom in specifying the magnitude profile in the vicinity of $\omega/\omega_0 = 1$. In actual applications, Q may range from as low as 0.5 to as high as 100, with values near unity being by far the most common. The magnitude plot is shown in Fig. 3.19a for different values of Q. For low Qs the

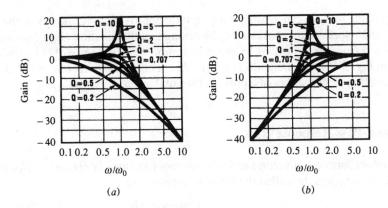

FIGURE 3.19
Standard second-order responses for different values of Q: (a) low-pass and (b) high-pass.

transition from one asymptote to the other is very gradual, while for high Qs there is a range of frequencies in the vicinity of $\omega/\omega_0 = 1$ where $|H_{LP}| > 1$, a phenomenon referred to as *peaking*.

One can prove that the largest Q before the onset of peaking is $Q = 1/\sqrt{2} = 0.707$. The corresponding curve is said to be *maximally flat* and is also referred to as the *Butterworth response*. This curve is the closest to the brick-wall model, hence its widespread use. By Eq. (3.45c), $|H_{LP}|_{dB} = (1/\sqrt{2})_{dB} = -3$ dB. The meaning of ω_0 for the Butterworth response is the same as for the first-order case, that is, ω_0 represents the *−3-dB frequency,* also called the *cutoff frequency.*

It can be proven[5] that in the case of peaked responses, or $Q > 1/\sqrt{2}$, the frequency at which $|H_{LP}|$ is maximized and the corresponding maximum are

$$\omega/\omega_0 = \sqrt{1 - 1/2Q^2} \tag{3.46a}$$

$$|H_{LP}|_{max} = \frac{Q}{\sqrt{1 - 1/4Q^2}} \tag{3.46b}$$

For sufficiently large Qs, say, $Q > 5$, we have $\omega/\omega_0 \cong 1$ and $|H_{LP}|_{max} \cong Q$. Of course, in the absence of peaking, or $Q < 1/\sqrt{2}$, the maximum is reached at $\omega/\omega_0 = 0$, that is, at dc. Peaked responses are useful in the cascade synthesis of higher-order filters, to be covered in Chapter 4.

The High-Pass Response H_{HP}

The standard form of all second-order high-pass functions is $H(j\omega) = H_{0HP} H_{HP}(j\omega)$, where H_{0HP} is called the *high-frequency gain,* and

$$H_{HP}(j\omega) = \frac{-(\omega/\omega_0)^2}{1 - (\omega/\omega_0)^2 + (j\omega/\omega_0)/Q} \tag{3.47}$$

(Note that the negative sign in the numerator is part of the definition.) Letting $j\omega \to s$ reveals that $H(s)$, besides the pole pair, has a double zero at the origin. To construct the magnitude plot we can again use asymptotic approximations; however, the procedure can be speeded up considerably by noting that the function $H_{HP}(j\omega/\omega_0)$ can be obtained from $H_{LP}(j\omega/\omega_0)$ by the substitution of $(j\omega/\omega_0) \to 1/(j\omega/\omega_0)$. As shown in Fig. 3.19b, the magnitude plot of H_{HP} is thus the mirror image of that of H_{LP}. Equation (3.46) still holds, provided we replace ω/ω_0 with ω_0/ω.

The Band-Pass Response H_{BP}

The standard form of all second-order band-pass functions is $H(j\omega) = H_{0BP}H_{BP}(j\omega)$, where H_{0BP} is the called the *resonance gain,* and

$$H_{BP}(j\omega) = \frac{(j\omega/\omega_0)/Q}{1 - (\omega/\omega_0)^2 + (j\omega/\omega_0)/Q} \tag{3.48}$$

(Note that Q in the numerator is part of the definition.) Besides the pole pair, this function has a zero at the origin. To construct the magnitude plot we use asymptotic approximations.

1. For $\omega/\omega_0 \ll 1$, we can ignore the second and third denominator terms and write $H_{BP} \to (j\omega/\omega_0)/Q$. The low-frequency asymptote is thus $|H_{BP}|_{dB} = 20\log_{10}[(\omega/\omega_0)/Q]$, or

$$|H_{BP}|_{dB} = 20\log_{10}(\omega/\omega_0) - Q_{dB} \qquad (\omega/\omega_0 \ll 1) \tag{3.49a}$$

This equation is of the type $y = 20x - Q_{dB}$, indicating a straight line with a slope of $+20$ dB/dec, but shifted by $-Q_{dB}$ with respect to the 0-dB axis at $\omega/\omega_0 = 1$.

2. For $\omega/\omega_0 \gg 1$, the second term dominates in the denominator, so $H_{BP} \to -j1/(\omega/\omega_0)Q$. The high-frequency asymptote is thus

$$|H_{BP}|_{dB} = -20\log_{10}(\omega/\omega_0) - Q_{dB} \qquad (\omega/\omega_0 \gg 1) \tag{3.49b}$$

This is a straight line with the same amount of downshift as before, but with a slope of -20 dB/dec.

3. For $\omega/\omega_0 = 1$, we get $H_{BP} = 1$, or

$$|H_{BP}|_{dB} = 0 \qquad (\omega/\omega_0 = 1) \tag{3.49c}$$

One can prove that $|H_{BP}|$ peaks at $\omega/\omega_0 = 1$ regardless of Q, this being the reason why ω_0 is called the *peak,* or *resonance, frequency.*

Magnitude is plotted in Fig. 3.20a for different Qs. All curves peak at 0 dB. Those corresponding to low Qs are broad, but those corresponding to high Qs are narrow, indicating a higher degree of selectivity. In the vicinity of $\omega/\omega_0 = 1$ the high-selectivity curves are much steeper than ± 20 dB/dec, though away from resonance they roll off at the same ultimate rate of ± 20 dB/dec.

To express selectivity quantitatively, we introduce the *bandwidth*

$$BW = \omega_H - \omega_L \tag{3.50}$$

where ω_L and ω_H are the -3-dB frequencies, that is, the frequencies at which the response is 3 dB below its maximum, as depicted in Fig. 3.20b. One can prove[5]

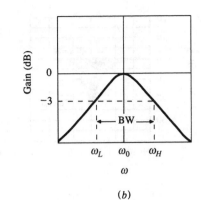

FIGURE 3.20
(a) Standard second-order band-pass response as a function of Q, and
(b) its bandwidth BW.

that

$$\omega_L = \omega_0(\sqrt{1 + 1/4Q^2} - 1/2Q) \tag{3.51a}$$

$$\omega_H = \omega_0(\sqrt{1 + 1/4Q^2} + 1/2Q) \tag{3.51b}$$

$$\omega_0 = \sqrt{\omega_L \omega_H} \tag{3.52}$$

The resonance frequency ω_0 is the *geometric mean* of ω_L and ω_H, indicating that on a logarithmic scale ω_0 appears halfway between ω_L and ω_H. It is apparent that the narrower the bandwidth, the more selective the filter. However, selectivity depends also on ω_0, since a filter with BW $= 10$ rad/s and $\omega_0 = 1$ krad/s is certainly more selective than one with BW $= 10$ rad/s but $\omega_0 = 100$ rad/s. A proper measure of selectivity is the ratio $\omega_0/$BW. Subtracting Eq. (3.51a) from Eq. (3.51b) and taking the reciprocal, we get

$$Q = \frac{\omega_0}{\text{BW}} \tag{3.53}$$

that is, Q *is the selectivity*. We now have a more concrete interpretation for this parameter.

The Notch Response H_N

The most common form for the notch function is $H(j\omega) = H_{0N} H_N(j\omega)$, where H_{0N} is an appropriate gain constant, and

$$H_N(j\omega) = \frac{1 - (\omega/\omega_0)^2}{1 - (\omega/\omega_0)^2 + (j\omega/\omega_0)/Q} \tag{3.54}$$

(In Section 3.7 we shall see that other notch functions are possible, in which ω_0 in the numerator has not necessarily the same value as ω_0 in the denominator.) Letting $j\omega \rightarrow s$ reveals that $H(s)$, besides the pole pair, has a zero pair on the imaginary

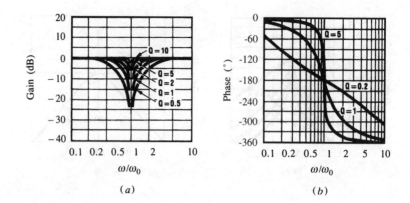

FIGURE 3.21
Standard second-order responses for different values of Q: (a) notch and
(b) all-pass.

axis, or $z_{1,2} = \pm j\omega_0$. We observe that at sufficiently low and high frequencies, $H_N \to 1$. However, for $\omega/\omega_0 = 1$ we get $H_N \to 0$, or $|H_N|_{dB} \to -\infty$. The notch response is shown in Fig. 3.21a, where we note that the higher the Q, the narrower the notch. For obvious reasons, ω_0 is called the *notch frequency*. In a practical circuit, due to component nonidealities, an infinitely deep notch is unrealizable.

It is interesting to note that

$$H_N = H_{LP} + H_{HP} = 1 - H_{BP} \tag{3.55}$$

indicating alternative ways of synthesizing the notch response once the other responses are available.

The All-Pass Response H_{AP}

Its general form is $H(j\omega) = H_{0AP} H_{AP}(j\omega)$, where H_{0AP} is the usual gain term, and

$$H_{AP}(j\omega) = \frac{1 - (\omega/\omega_0)^2 - (j\omega/\omega_0)/Q}{1 - (\omega/\omega_0)^2 + (j\omega/\omega_0)/Q} \tag{3.56}$$

This function has two poles and two zeros. For $Q > 0.5$, the zeros and poles are complex and are symmetrical about the $j\omega$ axis. Since $N(j\omega) = D(j\omega)$, we have $|H_{AP}| = 1$, or $|H_{AP}|_{dB} = 0$ dB, regardless of frequency. The argument is

$$\sphericalangle H_{AP} = -2\tan^{-1}\frac{(\omega/\omega_0)/Q}{1 - (\omega/\omega_0)^2} \qquad \text{for } \omega/\omega_0 < 1 \tag{3.57a}$$

$$\sphericalangle H_{AP} = -360° - 2\tan^{-1}\frac{(\omega/\omega_0)/Q}{1 - (\omega/\omega_0)^2} \qquad \text{for } \omega/\omega_0 > 1 \tag{3.57b}$$

indicating that as ω/ω_0 is swept from 0 to ∞, the argument changes from $0°$, through $-180°$, to $-360°$. This is shown in Fig. 3.21b. The all-pass function can also be synthesized as

$$H_{AP} = H_{LP} - H_{BP} + H_{HP} = 1 - 2H_{BP} \qquad (3.58)$$

Filter Measurements

Because of component tolerances and other nonidealities, the parameters of a practical filter are likely to deviate from their design values. We thus need to measure them and, if necessary, to tune them via suitable potentiometers.

For a low-pass filter we have $H_{LP}(j0) = H_{0LP}$ and $H_{LP}(j\omega_0) = -jH_{0LP}Q$. To measure ω_0 we look for the frequency at which the output is shifted by $90°$ with respect to the input, and to measure Q we take the ratio $Q = |H_{LP}(j\omega_0)|/|H_{0LP}|$.

For a band-pass filter we have $H_{BP}(j\omega_0) = H_{0BP}$, $\sphericalangle H_{BP}(j\omega_L) = \sphericalangle H_{0BP} - 45°$, and $\sphericalangle H_{BP}(j\omega_H) = \sphericalangle H_{0BP} - 135°$. Thus, ω_0 is measured as the frequency at which the output is in phase with the input if $H_{0BP} > 0$, or $180°$ out of phase if $H_{0BP} < 0$. To find Q, we measure the frequencies ω_L and ω_H at which the output is shifted by $\pm 45°$ with respect to the input. Then, $Q = \omega_0/(\omega_H - \omega_L)$. The reader can apply similar considerations to measure the parameters of the other responses.

3.5
KRC FILTERS

Since an R-C stage provides a first-order low-pass response, cascading two such stages as in Fig. 3.22a ought to provide a second-order response, and without using any inductances. Indeed, at low frequencies the capacitors act as open circuits, thus letting the input signal pass through with $H \to 1$ V/V. At high frequencies the incoming signal will be shunted to ground first by C_1 and then by C_2, thus providing a two-step attenuation; hence the designation *second-order*. Since at high frequencies a single R-C stage gives $H \to 1/(j\omega/\omega_0)$, the cascade combination of two stages gives $H \to [1/(j\omega/\omega_1)] \times [1/(j\omega/\omega_2)] = -1/(\omega/\omega_0)^2$, $\omega_0 = \sqrt{\omega_1\omega_2}$, indicating an asymptotic slope of -40 dB/dec. The filter of Fig. 3.22a does meet the asymptotic criteria for a second-order low-pass response; however, it does not offer sufficient flexibility for controlling the magnitude profile in the vicinity of $\omega/\omega_0 = 1$. In fact, one can prove[5] that this all-passive filter yields $Q < 0.5$.

If we wish to increase Q above 0.5, we must bolster the magnitude response near $\omega = \omega_0$. One way to achieve this is by providing a controlled amount of *positive*

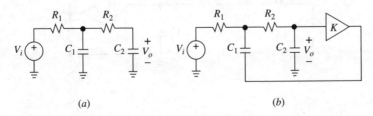

(a)   (b)

FIGURE 3.22
(a) Passive and (b) active realization of a second-order low-pass filter.

feedback. In Fig. 3.22*b* the output of the R_2-C_2 stage is magnified by an amplifier with gain K, and then is fed back to the interstage node via C_1, whose bottom terminal has been lifted off ground to create the positive feedback path. This feedback must be effective only in the vicinity of $\omega = \omega_0$, where bolstering is specifically needed. We can use physical insight to verify the band-pass nature of the feedback: for $\omega/\omega_0 \ll 1$ the impedance of C_1 is simply too large to feed back much signal, whereas for $\omega/\omega_0 \gg 1$ the shunting action by C_2 makes V_o too small to do much good; however, near $\omega/\omega_0 = 1$ there will be feedback, which we can adjust for the desired amount of peaking by acting on K. Filters of the type of Fig. 3.22*b* are aptly called *KRC* filters—or also *Sallen-Key filters,* for their inventors.

Low-Pass *KRC* Filters

In Fig. 3.23 the gain block is implemented with an op amp operating as a noninverting amplifier, and

$$K = 1 + \frac{R_B}{R_A} \tag{3.59}$$

Note that V_o is obtained from the output node of the op amp to take advantage of its low impedance. By inspection,

$$V_o = K \frac{1}{R_2 C_2 s + 1} V_1$$

Summing currents at node V_1,

$$\frac{V_i - V_1}{R_1} + \frac{V_o/K - V_1}{R_2} + \frac{V_o - V_1}{1/C_1 s} = 0$$

Eliminating V_1 and collecting, we get

$$H(s) = \frac{V_o}{V_i} = \frac{K}{R_1 C_1 R_2 C_2 s^2 + [(1 - K)R_1 C_1 + R_1 C_2 + R_2 C_2]s + 1}$$

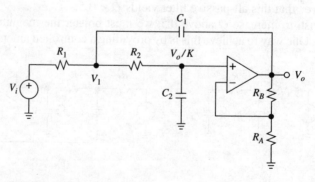

FIGURE 3.23
Low-pass *KRC* filter.

Letting $s \to j\omega$ yields

$$H(j\omega) = K \frac{1}{1 - \omega^2 R_1 C_1 R_2 C_2 + j\omega[(1 - K)R_1 C_1 + R_1 C_2 + R_2 C_2]}$$

Next, we put this function in the standard form $H(j\omega) = H_{0LP} H_{LP}(j\omega)$, with $H_{LP}(j\omega)$ as in Eq. (3.44). To do so, we equate the coefficients pairwise. By inspection,

$$H_{0LP} = K \qquad (3.60a)$$

Letting $\omega^2 R_1 C_1 R_2 C_2 = (\omega/\omega_0)^2$ gives

$$\omega_0 = \frac{1}{\sqrt{R_1 C_1 R_2 C_2}} \qquad (3.60b)$$

indicating that ω_0 is the geometric mean of the individual-stage frequencies $\omega_1 = 1/R_1 C_1$ and $\omega_2 = 1/R_2 C_2$. Finally, letting $j\omega[(1 - K)R_1 C_1 + R_1 C_2 + R_2 C_2] = (j\omega/\omega_0)/Q$ gives

$$Q = \frac{1}{(1 - K)\sqrt{R_1 C_1/R_2 C_2} + \sqrt{R_1 C_2/R_2 C_1} + \sqrt{R_2 C_2/R_1 C_1}} \qquad (3.60c)$$

We observe that K and Q depend on component *ratios*, while ω_0 depends on component *products*. Because of component tolerances and op amp nonidealities, the parameters of an actual filter are likely to depart from their intended values. Our filter can be tuned as follows: (*a*) adjust R_1 for the desired ω_0 (this adjustment varies also Q); (*b*) once ω_0 has been tuned, adjust R_B for the desired Q (this leaves ω_0 unchanged; however, it varies K, but this is of little concern because it does not affect the frequency behavior).

Since we have five parameters (K, R_1, C_1, R_2, and C_2) but only three equations, we have the choice of fixing two so we can specify design equations for the remaining three. Two common designs are the *equal-component* and the *unity-gain* designs (other designs are discussed in the end-of-chapter problems).

Equal-Component *KRC* Circuit

Imposing $R_1 = R_2 = R$ and $C_1 = C_2 = C$ simplifies inventory and reduces Eq. (3.60) to

$$H_{0LP} = K \qquad \omega_0 = \frac{1}{RC} \qquad Q = \frac{1}{3 - K} \qquad (3.61)$$

The design equations are then

$$RC = 1/\omega_0 \qquad K = 3 - 1/Q \qquad R_B = (K - 1)R_A \qquad (3.62)$$

EXAMPLE 3.8. Using the equal-component design, specify elements for a second-order low-pass filter with $f_0 = 1$ kHz and $Q = 5$. What is its dc gain?

Solution. Arbitrarily select $C = 10$ nF, which is an easily available value. Then, $R = 1/(\omega_0 C) = 1/(2\pi 10^3 \times 10 \times 10^{-9}) = 15.92$ kΩ (use 15.8 kΩ, 1%). Moreover, $K = 3 - 1/5 = 2.80$, and $R_B/R_A = 2.80 - 1 = 1.80$. Let $R_A = 10.0$ kΩ, 1%; then, $R_B = 17.8$ kΩ, 1%. The circuit, shown in Fig. 3.24a, has a dc gain of 2.78 V/V.

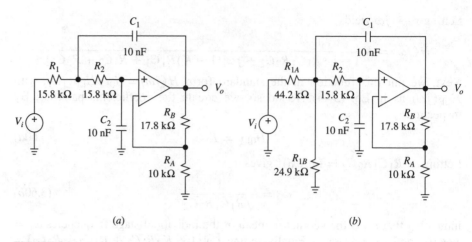

(a)   (b)

FIGURE 3.24
Filter realizations of Examples 3.8 and 3.9.

> **EXAMPLE 3.9.** Modify the circuit of Example 3.8 for a dc gain of 0 dB.
>
> **Solution.** This situation arises often enough to merit a detailed treatment. To reduce gain from an existing value A_{old} to a different value A_{new}, apply Thévenin's theorem and replace R_1 with a voltage divider R_{1A} and R_{1B} such that
>
> $$A_{new} = \frac{R_{1B}}{R_{1A} + R_{1B}} A_{old} \qquad R_{1A} \parallel R_{1B} = R_1$$
>
> where the second constraint ensures that ω_0 is unaffected by the replacement. Solving, we get
>
> $$R_{1A} = R_1 \frac{A_{old}}{A_{new}} \qquad R_{1B} = \frac{R_1}{1 - A_{new}/A_{old}} \qquad (3.63)$$
>
> In our case, $A_{old} = 2.8$ V/V and $A_{new} = 1$ V/V. So, $R_{1A} = 15.92 \times 2.8/1 = 44.56$ kΩ (use 44.2 kΩ, 1%) and $R_{1B} = 15.92/(1 - 1/2.8) = 24.76$ kΩ (use 24.9 kΩ, 1%). The circuit is shown in Fig. 3.24b.

Unity-Gain *KRC* Circuit

Imposing $K = 1$ minimizes the number of components and also maximizes the bandwidth of the op amp, an issue that will be studied in Chapter 6. To simplify the math, we relabel the components as $R_2 = R$, $C_2 = C$, $R_1 = mR$, and $C_1 = nC$. Then, Eq. (3.60) reduces to

$$H_{0LP} = 1 \text{ V/V} \qquad \omega_0 = \frac{1}{\sqrt{mn}RC} \qquad Q = \frac{\sqrt{mn}}{m + 1} \qquad (3.64)$$

You can verify that for a given n, Q is maximized when $m = 1$, that is, when the resistances are equal. With $m = 1$, Eq. (3.64) gives $n = 4Q^2$. In practice, one starts out with two easily available capacitances in a ratio $n \geq 4Q^2$; then m is found as $m = k + \sqrt{k^2 - 1}$, where $k = n/2Q^2 - 1$.

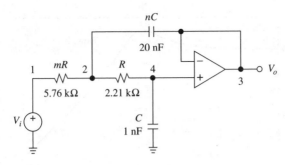

FIGURE 3.25
Filter of Example 3.10.

EXAMPLE 3.10. (*a*) Using the unity-gain option, design a low-pass filter with $f_0 = 10$ kHz and $Q = 2$. (*b*) Use PSpice to visualize its frequency response.

Solution.

(*a*) Arbitrarily pick $C = 1$ nF. Since $4Q^2 = 4 \times 2^2 = 16$, let $n = 20$. Then, $nC = 20$ nF, $k = 20/(2 \times 2^2) - 1 = 1.5$, $m = 1.5 + \sqrt{1.5^2 - 1} = 2.618$, $R = 1/(\sqrt{mn}\omega_0 C) = 1/(\sqrt{2.618 \times 20} \times 2\pi 10^4 \times 10^{-9}) = 2.199$ kΩ (use 2.21 kΩ, 1%), and $mR = 5.758$ kΩ (use 5.76 kΩ, 1%). The filter is shown in Fig. 3.25.

(*b*) Using the node numbering shown, we write the PSpice file:

```
KRC low-pass filter: f0 = 10 kHz, Q = 2.
Vi 1 0 ac 1V
Rm 1 2 5.76k
R 2 4 2.21k
Cn 2 3 20nF
C 4 0 1nF
eOA 3 0 4 3 1G
.ac dec 100 1kHz 100kHz
.probe
.end
```

The frequency response is shown in Fig. 3.26.

EXAMPLE 3.11. (*a*) Design a second-order low-pass Butterworth filter with a -3-dB frequency of 10 kHz. (*b*) If $v_i(t) = 10 \cos (4\pi 10^4 t - 90°)$ V, find $v_o(t)$.

Solution.

(*a*) The Butterworth response, for which $Q = 1/\sqrt{2}$, is implemented with $m = 1$ and $n = 2$. Letting $C = 1$ nF, we get $nC = 2$ nF and $mR = R = 11.25$ kΩ (use 11.3 kΩ, 1%).

(*b*) Since $\omega/\omega_0 = 2$, we have $H(j4\pi 10^4) = 1/[1 - 2^2 + j2/(1/\sqrt{2})] = (1/\sqrt{17})$ $\underline{/136.69°}$ V/V. So, $V_{om} = 10/\sqrt{17} = 2.426$ V, $\theta_o = 136.69° - 90° = 46.69°$, and $v_o(t) = 2.426 \cos (4\pi 10^4 t + 46.69°)$ V.

The advantages of the unity-gain design are offset by a quadratic increase of the capacitance spread n with Q. Moreover, the circuit does not enjoy the tuning advantages of the equal-component design because the adjustments of ω_0 and Q interfere with each other, as revealed by Eq. (3.64). On the other hand, at high Qs the equal-component design becomes too sensitive to the tolerances of R_B and R_A,

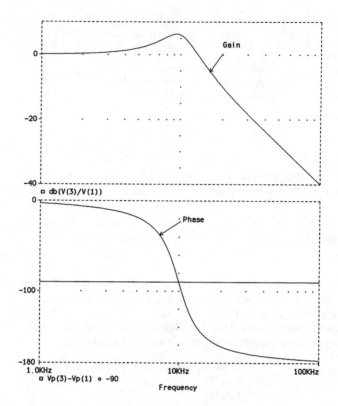

FIGURE 3.26
Frequency response of the filter of Fig. 3.25.

when their ratio is very close to 2. A slight mismatch may cause an intolerable departure of Q from the desired value. Should this ratio reach (or even surpass) 2, Q will become infinite (or even negative), causing the filter to oscillate. For these reasons, *KRC* filters are used for Qs below 10. Section 3.7 presents filter topologies suited to high Qs.

High-Pass *KRC* Filters

Interchanging the components of a low-pass *R-C* stage with each other turns it into a high-pass *C-R* stage. Interchanging resistances and capacitances in the low-pass filter of Fig. 3.23 leads to the filter of Fig. 3.27, which you can readily classify as a high-pass type using physical insight. By similar analysis, we find that $V_o/V_i = H_{0HP}H_{HP}$, where H_{HP} is given in Eq. (3.47), and

$$H_{0HP} = K \qquad \omega_0 = \frac{1}{\sqrt{R_1C_1R_2C_2}} \qquad (3.65a)$$

$$Q = \frac{1}{(1-K)\sqrt{R_2C_2/R_1C_1} + \sqrt{R_1C_2/R_2C_1} + \sqrt{R_1C_1/R_2C_2}} \qquad (3.65b)$$

As in the low-pass case, two interesting options available to the designer are the *equal-component* and the *unity-gain* designs.

EXERCISE 3.1. Derive Eq. (3.65).

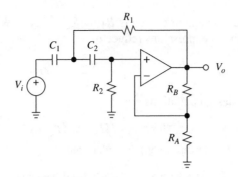

FIGURE 3.27
High-pass *KRC* filter.

EXAMPLE 3.12. Design a second-order high-pass filter with $f_0 = 200$ Hz and $Q = 1.5$.

Solution. To minimize the component count, choose the unity-gain option, for which $R_A = \infty$ and $R_B = 0$. Letting $C_1 = nC_2$ and $R_1 = mR_2$ in Eq. (3.65) gives $\omega_0 = 1/\sqrt{mn}RC$ and $Q = (\sqrt{n/m})/(n+1)$. Let $C_1 = C_2 = 0.1\,\mu$F, so that $n = 1$. Imposing $1.5 = (\sqrt{1/m})/2$ gives $m = 1/9$, and imposing $2\pi 200 = 1/(\sqrt{1/9}R_2 \times 10^7)$ gives $R_2 = 23.87$ kΩ and $R_1 = mR_2 = 2.653$ kΩ.

Band-Pass *KRC* Filters

The circuit of Fig. 3.28 consists of an *R-C* stage followed by a *C-R* stage to synthesize a band-pass block, and a gain block to provide positive feedback via R_3. This feedback is designed to bolster the response near $\omega/\omega_0 = 1$. The ac analysis of the filter yields $V_o/V_i = H_{0BP}H_{BP}$, where H_{BP} is given in Eq. (3.48), and

$$H_{0BP} = \frac{K}{1 + (1-K)R_1/R_3 + (1 + C_1/C_2)R_1/R_2} \qquad \omega_0 = \frac{\sqrt{1 + R_1/R_3}}{\sqrt{R_1C_1R_2C_2}}$$

$$(3.66a)$$

$$Q = \frac{\sqrt{1 + R_1/R_3}}{[1 + (1-K)R_1/R_3]\sqrt{R_2C_2/R_1C_1} + \sqrt{R_1C_2/R_2C_1} + \sqrt{R_1C_1/R_2C_2}}$$

$$(3.66b)$$

We again note that one can vary R_1 to tune ω_0 and R_B to adjust Q.

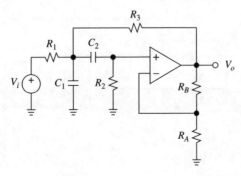

FIGURE 3.28
Band-pass *KRC* filter.

If $Q > \sqrt{2}/3$, a convenient choice is $R_1 = R_2 = R_3 = R$ and $C_1 = C_2 = C$, in which case the above expressions reduce to

$$H_{0BP} = \frac{K}{4-K} \qquad \omega_0 = \frac{\sqrt{2}}{RC} \qquad Q = \frac{\sqrt{2}}{4-K} \qquad (3.67)$$

The corresponding design equations are

$$RC = \sqrt{2}/\omega_0 \qquad K = 4 - \sqrt{2}/Q \qquad R_B = (K-1)R_A \qquad (3.68)$$

EXERCISE 3.2. Derive Eqs. (3.66) through (3.68).

EXAMPLE 3.13. (*a*) Design a second-order band-pass filter with $f_0 = 1$ kHz and $BW = 100$ Hz. What is its resonance gain? (*b*) Modify the circuit for a resonance gain of 20 dB.

Solution.

(*a*) Use the equal-component option with $C_1 = C_2 = 10$ nF and $R_1 = R_2 = R_3 = \sqrt{2}/(2\pi 10^3 \times 10^{-8}) = 22.5$ kΩ (use 22.6 kΩ, 1%). We need $Q = f_0/BW = 10$, so $K = 4 - \sqrt{2}/10 = 3.858$. Pick $R_A = 10.0$ kΩ, 1%. Then, $R_B = (K-1)R_A = 28.58$ kΩ (use 28.7 kΩ, 1%). The resonance gain is $K/(4-K) = 27.28$ V/V.

(*b*) Replace R_1 with two resistances R_{1A} and R_{1B}, in the manner of Example 3.9, whose values are found via Eq. (3.63) with $A_{old} = 27.28$ V/V and $A_{new} = 10^{20/20} = 10$ V/V. This gives $R_{1A} = 61.9$ kΩ, 1%, and $R_{1B} = 35.7$ kΩ, 1%.

Band-Reject *KRC* Filters

The circuit of Fig. 3.29 consists of a *twin-T*-network and a gain block to provide positive feedback via the top capacitance. The *T*-networks provide alternative forward paths through which V_i can reach the amplifier's input: the low-frequency path *R-R*, and the high-frequency path *C-C*, indicating $H \rightarrow K$ at the frequency extremes. At intermediate frequencies, however, the two paths provide opposing phase angles, indicating a tendency of the two forward signals to cancel each other out at the amplifier's input. We thus anticipate a notch response. The ac analysis of the circuit gives $V_o/V_i = H_{0N}H_N$, where H_N is given in Eq. (3.54), and

$$H_{0N} = K \qquad \omega_0 = \frac{1}{RC} \qquad Q = \frac{1}{4-2K} \qquad (3.69)$$

EXERCISE 3.3. Derive Eq. (3.69).

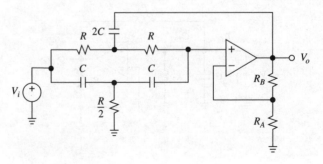

FIGURE 3.29
Band-reject *KRC* filter.

EXAMPLE 3.14. Design a second-order notch filter with $f_0 = 60$ Hz and BW = 5 Hz. What is its low- and high-frequency gain?

Solution. Let $C = 100$ nF and $2C = 200$ nF. Then, $R = 1/(2\pi 60 \times 10^{-7}) = 26.53$ kΩ, and $R/2 = 13.26$ kΩ. Since $Q = 60/5 = 12$, we get $2K = 4 - 1/12$, or $K = 47/24$, which represents the low- as well as the high-frequency gain of the filter. Use $R_A = 10.0$ kΩ and $R_B = 9.53$ kΩ.

3.6
MULTIPLE-FEEDBACK FILTERS

Multiple-feedback filters utilize more than one feedback path. Unlike their *KRC* counterparts, which configure the op amp for a *finite* gain K, multiple-feedback filters exploit the full open-loop gain and are also referred to as *infinite-gain filters*. Together with *KRC* filters, they are the most popular single-op-amp realizations of the second-order responses.

Band-Pass Filters

In the circuit of Fig. 3.30, also called the *Delyiannis-Friend* filter, named after its inventors, the op amp acts as a differentiator with respect to V_1, so we write

$$V_o = -sR_2C_2V_1$$

Summing currents at node V_1,

$$\frac{V_i - V_1}{R_1} + \frac{V_o - V_1}{1/sC_1} + \frac{0 - V_1}{1/sC_2} = 0$$

Eliminating V_1, letting $s \to j\omega$, and rearranging,

$$H(j\omega) = \frac{V_o}{V_i} = \frac{-j\omega R_2 C_2}{1 - \omega^2 R_1 R_2 C_1 C_2 + j\omega R_1(C_1 + C_2)}$$

To put this function in the standard form $H(j\omega) = H_{0BP}H_{BP}(j\omega)$, we impose $\omega^2 R_1 R_2 C_1 C_2 = (\omega/\omega_0)^2$ to get

$$\omega_0 = \frac{1}{\sqrt{R_1 R_2 C_1 C_2}} \tag{3.70a}$$

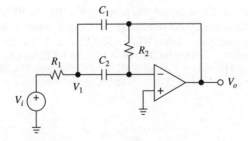

FIGURE 3.30
Multiple-feedback band-pass filter.

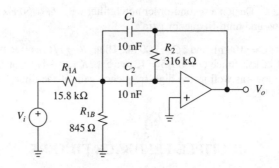

FIGURE 3.31
Band-pass filter of Example 3.15.

and $j\omega R_1(C_1 + C_2) = (j\omega/\omega_0)/Q$ to get

$$Q = \frac{\sqrt{R_2/R_1}}{\sqrt{C_1/C_2} + \sqrt{C_2/C_1}} \tag{3.70b}$$

Finally, we impose $-j\omega R_2 C_2 = H_{0BP} \times (j\omega/\omega_0)/Q$ to get

$$H_{0BP} = \frac{-R_2/R_1}{1 + C_1/C_2} \tag{3.70c}$$

Clearly, this filter is of the inverting type. It is customary to impose $C_1 = C_2 = C$, after which the above expressions simplify to

$$\omega_0 = \frac{1}{\sqrt{R_1 R_2} C} \qquad Q = 0.5\sqrt{R_2/R_1} \qquad H_{0BP} = -2Q^2 \tag{3.71}$$

The corresponding design equations are

$$R_1 = 1/2\omega_0 QC \qquad R_2 = 2Q/\omega_0 C \tag{3.72}$$

Denoting resonance-gain magnitude as $H_0 = |H_{0BP}|$ for simplicity, we observe that it increases quadratically with Q. If we want $H_0 < 2Q^2$, we must replace R_1 with a voltage divider in the manner of Example 3.9. The design equations are then

$$R_{1A} = Q/H_0\omega_0 C \qquad R_{1B} = R_{1A}/(2Q^2/H_0 - 1) \tag{3.73}$$

EXAMPLE 3.15. Design a multiple-feedback band-pass filter with $f_0 = 1$ kHz, $Q = 10$, and $H_0 = 20$ dB. Show the final circuit.

Solution. Let $C_1 = C_2 = 10$ nF. Then, $R_2 = 2 \times 10/(2\pi 10^3 \times 10^{-8}) = 318.3$ kΩ (use 316 kΩ, 1%). Since 20 dB implies $H_0 = 10$ V/V, which is less than $2Q^2 = 200$, we need an input attenuator. Thus, $R_{1A} = 10/(10 \times 2\pi 10^3 \times 10^{-8}) = 15.92$ kΩ (use 15.8 kΩ, 1%), and $R_{1B} = 15.92/(200/10 - 1) = 837.7$ Ω (use 845 Ω, 1%). The circuit is shown in Fig. 3.31.

Low-Pass Filters

The circuit of Fig. 3.32 consists of the low-pass stage R_1-C_1 followed by the integrator stage made up of R_2, C_2, and the op amp, so we anticipate a low-pass response.

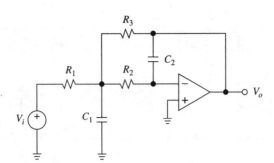

FIGURE 3.32
Multiple-feedback low-pass filter.

Moreover, the presence of positive feedback via R_3 should allow for Q control. The ac analysis of the circuit gives $V_o/V_i = H_{0LP}H_{LP}$, where

$$H_{0LP} = -\frac{R_3}{R_1} \qquad \omega_0 = \frac{1}{\sqrt{R_2R_3C_1C_2}}$$

$$Q = \frac{\sqrt{C_1/C_2}}{\sqrt{R_2R_3/R_1^2} + \sqrt{R_3/R_2} + \sqrt{R_2/R_3}} \qquad (3.74)$$

These expressions indicate that we can vary R_3 to adjust ω_0, and R_1 to adjust Q.

EXERCISE 3.4. Derive Eq. (3.74).

A possible design procedure[2] is to choose a convenient value for C_2 and calculate $C_1 = nC_2$, where n is the capacitance spread,

$$n \geq 4Q^2(1 + H_0) \qquad (3.75)$$

H_0 being the desired dc-gain magnitude. The resistances are then found as

$$R_3 = \frac{1 + \sqrt{1 - 4Q^2(1 + H_0)/n}}{2\omega_0 QC_2} \qquad R_1 = \frac{R_3}{H_0} \qquad R_2 = \frac{1}{\omega_0^2 R_3 C_1 C_2} \qquad (3.76)$$

A disadvantage of this filter is that the higher the Q and H_0, the greater the capacitance spread.

EXAMPLE 3.16. Design a multiple-feedback low-pass filter with $H_0 = 2$ V/V, $f_0 = 10$ kHz, and $Q = 4$.

Solution. Substituting the given values yields $n \geq 192$. Let $n = 200$. Start out with $C_2 = 1$ nF. Then, $C_1 = 0.2$ μF, $R_3 = 2.387$ kΩ (use 2.37 kΩ, 1%), $R_1 = 1.194$ kΩ (use 1.18 kΩ, 1%), and $R_3 = 530.5$ Ω (use 536 Ω, 1%).

Notch Filters

The circuit of Fig. 3.33 exploits Eq. (3.55) to synthesize the notch response using the band-pass response. By inspection, $V_o = -(R_5/R_3)(-H_0H_{BP})V_i - (R_5/R_4)V_i = -(R_5/R_4)[1 - (H_0R_4/R_3)H_{BP}]V_i$. It is apparent that imposing $H_0R_4/R_3 = 1$

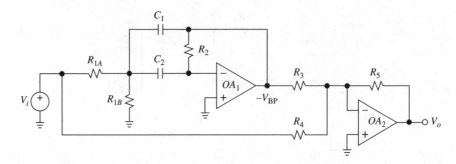

FIGURE 3.33
Synthesis of H_N using H_{BP}.

leads to a mutual cancellation of the $(j\omega/\omega_0)/Q$ terms in the numerator, giving $V_o/V_i = H_{0N}H_N$, $H_{0N} = -R_5/R_4$.

EXAMPLE 3.17. Design a notch filter with $f_0 = 1$ kHz, $Q = 10$, and $H_{0N} = 0$ dB.

Solution. First, implement a band-pass stage with $f_0 = 1$ kHz, $Q = 10$, and $H_0 = 1$ V/ V. Using $C_1 = C_2 = 10$ nF, this requires $R_2 = 318.3$ kΩ, $R_{1A} = 159.2$ kΩ, and $R_{1B} = 799.8$ Ω. Then, use $R_3 = R_4 = R_5 = 10.00$ kΩ.

3.7
STATE-VARIABLE AND BIQUAD FILTERS

The second-order filters investigated so far use a single op amp with a minimum or near-minimum number of external components. Simplicity, however, does not come without a price. Drawbacks such as wide component spreads; awkward tuning capabilities; and high sensitivity to component variations, particularly to the gain of the amplifier, generally limit these filters to $Q \leq 10$.

Component minimization, especially minimization of the number of op amps, was of concern when these devices were expensive. Nowadays, multiple-op-amp packages such as duals and quads are cost-competitive with precision passive components. The question then arises whether filter performance and versatility can be improved by shifting the burden from passive to active devices. The answer is provided by multiple-op-amp filters, such as the *state-variable* and *biquad types,* which, though using more components, are generally easier to tune, are less sensitive to passive component variations, and do not require extravagant component spreads. Since they provide more than one response simultaneously, they are also referred to as *universal filters.*

State-Variable (SV) Filters

The SV filter—also known as the *KHN filter* for inventors W. J. Kerwin, L. P. Huelsman, and R. W. Newcomb, who first reported it in 1967—uses two integrators and a summing amplifier to provide the second-order low-pass, band-pass, and

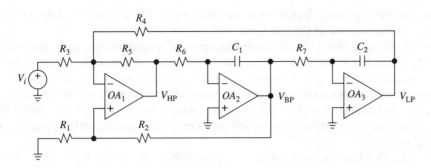

FIGURE 3.34
State-variable filter (inverting).

high-pass responses. A fourth op amp can be used to combine the existing responses and synthesize the notch or the all-pass responses. The circuit realizes a second-order differential equation, hence its name.

In the SV version of Fig. 3.34, OA_1 forms a linear combination of the input and the outputs of the remaining op amps. Using the superposition principle, we write

$$V_{HP} = -\frac{R_5}{R_3}V_i - \frac{R_5}{R_4}V_{LP} + \left(1 + \frac{R_5}{R_3 \parallel R_4}\right)\frac{R_1}{R_1 + R_2}V_{BP}$$

$$= -\frac{R_5}{R_3}V_i - \frac{R_5}{R_4}V_{LP} + \frac{1 + R_5/R_3 + R_5/R_4}{1 + R_2/R_1}V_{BP} \qquad (3.77)$$

Since OA_2 and OA_3 are integrators, we have

$$V_{BP} = \frac{-1}{R_6C_1s}V_{HP} \qquad V_{LP} = \frac{-1}{R_7C_2s}V_{BP} \qquad (3.78)$$

or $V_{LP} = (1/R_6C_1R_7C_2s^2)V_{HP}$. Substituting V_{BP} and V_{LP} into Eq. (3.77) and collecting, we get

$$\frac{V_{HP}}{V_i} = -\frac{R_5}{R_3}\frac{R_4R_6C_1R_7C_2s^2/R_5}{R_4R_6C_1R_7C_2s^2/R_5 + R_4(1 + R_5/R_3 + R_5/R_4)s/(1 + R_2/R_1)R_5 + 1}$$

Putting this expression in the standard form $V_{HP}/V_i = H_{0HP}H_{HP}$ allows us to find $H_{0HP} = -R_5/R_3$ and

$$\omega_0 = \frac{\sqrt{R_5/R_4}}{\sqrt{R_6C_1R_7C_2}} \qquad Q = \frac{(1 + R_2/R_1)\sqrt{R_5R_6C_1/R_4R_7C_2}}{1 + R_5/R_3 + R_5/R_4} \qquad (3.79)$$

Using $V_{BP}/V_i = (-1/R_6C_1s)V_{HP}/V_i$ indicates that $V_{BP}/V_i = H_{0BP}H_{BP}$, and also allows us to find H_{0BP}. We similarly find $V_{LP}/V_i = (-1/R_7C_2s)V_{BP}/V_i = H_{0LP}H_{LP}$. The results are

$$H_{0HP} = -\frac{R_5}{R_3} \qquad H_{0BP} = \frac{1 + R_2/R_1}{1 + R_3/R_4 + R_3/R_5} \qquad H_{0LP} = -\frac{R_4}{R_3} \qquad (3.80)$$

The above derivations reveal some interesting properties: first, the band-pass response is generated by integrating the high-pass response, and the low-pass is in turn generated by integrating the band-pass; second, since the product of two transfer functions corresponds to the addition of their decibel plots, and since the integrator plot has a constant slope of -20 dB/dec, the band-pass decibel plot is obtained by rotating the high-pass decibel plot clockwise by 20 dB/dec, and the low-pass plot by a similar rotation of the band-pass plot.

We observe that Q is no longer the result of a cancellation, as in the case of *KRC* filters, but depends on the resistor ratio R_2/R_1 in a straightforward manner. We therefore expect Q to be much less sensitive to resistance tolerances and drift. Indeed, with proper component selection and circuit construction, the SV filter can easily yield dependable Qs in the range of hundreds. For best results, use metal-film resistors and polystyrene or polycarbonate capacitors, and properly bypass the op amp supplies.

The SV filter is usually implemented with $R_5 = R_4 = R_3$, $R_6 = R_7 = R$, and $C_1 = C_2 = C$, so the earlier expressions simplify to

$$\omega_0 = 1/RC \qquad Q = \frac{1}{3}(1 + R_2/R_1) \qquad (3.81a)$$

$$H_{0HP} = -1 \qquad H_{0BP} = Q \qquad H_{0LP} = -1 \qquad (3.81b)$$

The filter is tuned as follows: (*a*) adjust R_3 for the desired magnitude of the response of interest; (*b*) adjust R_6 (or R_7) to tune ω_0; (*c*) adjust the ratio R_2/R_1 to tune Q.

EXAMPLE 3.18. In the circuit of Fig 3.34 specify component values for a band-pass response with a bandwidth of 10 Hz centered at 1 kHz. What is the resonance gain?

Solution. Pick the convenient values $C_1 = C_2 = 10$ nF. Then, $R = 1/(2\pi 10^3 \times 10^{-8}) = 15.92$ kΩ (use 15.8 kΩ, 1%). By definition, $Q = f_0/BW = 10^3/10 = 100$. Imposing $(1 + R_2/R_1)/3 = 100$ gives $R_2/R_1 = 299$. Pick $R_1 = 1.00$ kΩ, 1%, and $R_2 = 301$ kΩ, 1%. To simplify inventory, let also $R_3 = R_4 = R_5 = 15.8$ kΩ, 1%. The gain at resonance is $H_{0BP} = 100$ V/V.

Equation (3.81*b*) indicates that at $\omega = \omega_0$ all three responses exhibit a magnitude of Q V/V. In high-Q situations this may cause the op amps to saturate, unless the input signal level is kept suitably low. Low-input levels can be obtained by replacing R_3 with a suitable voltage divider, in the manner of Example 3.9 (see Problem 3.35).

Moving the input signal from the inverting to the noninverting side of OA_1 results in the circuit of Fig. 3.35, which represents another popular form of the SV filter. It

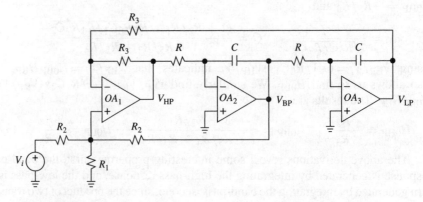

FIGURE 3.35
State-variable filter (noninverting).

can be shown (see Problem 3.36) that with the components shown, we now have

$$\omega_0 = 1/RC \qquad Q = 1 + R_2/2R_1 \qquad (3.82a)$$

$$H_{0HP} = 1/Q \qquad H_{0BP} = -1 \qquad H_{0LP} = 1/Q \qquad (3.82b)$$

indicating that for $\omega = \omega_0$ all three responses now exhibit 0-dB magnitudes. The band-pass plot is as in Fig. 3.20a; the low- and high-pass plots are as in Fig. 3.19, but shifted downward by Q_{dB}.

The Biquad Filter

Also known as the *Tow-Thomas filter*, for its inventors, the circuit of Fig. 3.36 consists of two integrators, one of which is of the lossy type. The third op amp is a unity-gain inverting amplifier whose sole purpose is to provide polarity reversal. If one of the integrators is allowed to be of the noninverting type, the inverting amplifier is omitted and only two op amps are required.

To analyze the circuit, we sum currents at the inverting-input node of OA_1,

$$\frac{V_i}{R_1} + \frac{-V_{LP}}{R_5} + \frac{V_{BP}}{R_2} + \frac{V_{BP}}{1/sC_1} = 0$$

Letting $V_{LP} = (-1/R_4C_2s)V_{BP}$ and collecting gives $V_{BP}/V_i = H_{0BP}H_{BP}$ and $V_{LP}/V_i = (-1/R_4C_2s)V_{BP}/V_i = H_{0LP}H_{LP}$, with

$$H_{0BP} = -\frac{R_2}{R_1} \qquad H_{0LP} = \frac{R_5}{R_1} \qquad \omega_0 = \frac{1}{\sqrt{R_4R_5C_1C_2}} \qquad Q = \frac{R_2\sqrt{C_1}}{\sqrt{R_4R_5C_2}}$$
$$(3.83)$$

We observe that unlike the SV filter, the biquad yields only two significant responses. However, since all its op amps are operated in the inverting mode, the circuit is immune from common-mode limitations, an issue to be studied in Chapter 5.

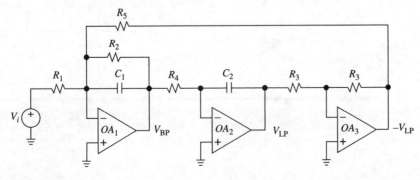

FIGURE 3.36
Biquad filter.

The biquad filter is usually implemented with $R_4 = R_5 = R$ and $C_1 = C_2 = C$, after which the above expressions simplify as

$$H_{0BP} = -\frac{R_2}{R_1} \qquad H_{0LP} = \frac{R}{R_1} \qquad \omega_0 = \frac{1}{RC} \qquad Q = \frac{R_2}{R} \qquad (3.84)$$

The filter is tuned as follows: (a) adjust R_4 (or R_5) to tune ω_0; (b) adjust R_2 to tune Q; (c) adjust R_1 for the desired value of H_{0BP} or of H_{0LP}.

EXAMPLE 3.19. Design a biquad filter with $f_0 = 8$ kHz, BW $= 200$ Hz, and a 20-dB resonance gain. What is the value of H_{0LP}?

Solution. Let $C_1 = C_2 = 1$ nF. Then, $R_4 = R_5 = 1/(2\pi \times 8 \times 10^3 \times 10^{-9}) = 19.89$ kΩ (use 20.0 kΩ, 1%); $Q = 8 \times 10^3/200 = 40$; $R_2 = 40 \times 19.89 = 795.8$ kΩ (use 787 kΩ, 1%); $R_1 = R_2/10^{20/20} = 78.7$ kΩ, 1%; $H_{0LP} = 20.0/78.7 = 0.254$ V/V, or -11.9 dB.

The Notch Response

With the help of a fourth op amp and a few resistors, both the biquad and the SV circuits can be configured for the notch response, which explains why these filters are also called *universal*. With a quad package, the fourth op amp is already available, so it only takes a few resistors to synthesize a notch.

The filter of Fig. 3.37 uses the biquad circuit to generate the notch response as $V_N = -[(R_5/R_2)(V_i - V_{BP}) \pm (R_5/R_4)V_{LP}]$, where the $\pm$ sign depends on the

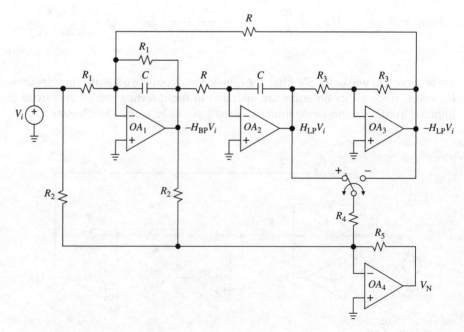

FIGURE 3.37
Synthesizing notch responses.

switch position, as indicated. It can be shown (see Exercise 3.5) that

$$\frac{V_N}{V_i} = -\frac{R_5\omega_z^2}{R_2\omega_0^2} \times \frac{1 - (\omega/\omega_z)^2}{1 - (\omega/\omega_0)^2 + (j\omega/\omega_0)/Q} \tag{3.85a}$$

$$\omega_0 = \frac{1}{RC} \qquad Q = \frac{R_1}{R} \qquad \omega_z = \omega_0\sqrt{1 \pm R_2/R_4 Q} \tag{3.85b}$$

This response presents a notch at $\omega = \omega_z$. We identify three cases:

1. R_4 is absent, or $R_4 = \infty$. By Eq. (3.85), we have

$$\omega_z = \omega_0 \qquad H_{0N} = -\frac{R_5}{R_2} \tag{3.86}$$

This is the familiar *symmetric notch* shown in Fig. 3.38b for the case $|H_{0N}| = 0$ dB. It is obtained by subtracting V_{BP} from V_i, in the manner depicted in Fig. 3.33.

2. The switch is in the left position, so also a low-pass term is now being added to the existing combination of V_i and $-V_{BP}$. The result is a *low-pass notch*. By Eq. (3.85), we now have

$$\omega_z = \omega_0\sqrt{1 + R_2/R_4 Q} \qquad H_{0LP} = -\frac{R_5\omega_z^2}{R_2\omega_0^2} \tag{3.87}$$

indicating $\omega_z > \omega_0$. The scaling term is called the *dc gain* H_{0LP}. The low-pass notch is shown in Fig. 3.38a for the case $|H_{0LP}| = 0$ dB. By Eq. (3.85a), the high-frequency gain is $H_{0HP} = H_{0LP}(1/\omega_z^2)/(1/\omega_0^2) = -R_5/R_2$.

3. The switch is in the right position, so the low-pass term is now being subtracted. The result is a *high-pass notch* with

$$\omega_z = \omega_0\sqrt{1 - R_2/R_4 Q} \qquad H_{0HP} = -\frac{R_5}{R_2} \tag{3.88}$$

We now have $\omega_z < \omega_0$, and the scaling factor is called the *high-frequency gain* H_{0HP}. This notch is shown in Fig. 3.38c for the case $|H_{0HP}| = 0$ dB. The dc gain is $H_{0LP} = -R_5\omega_z^2/R_2\omega_0^2$.

EXERCISE 3.5. Derive Eq. (3.85).

In Chapter 4 we shall use low- and high-pass notches to synthesize a class of higher-order filters known as *elliptic filters*. The above expressions can be turned

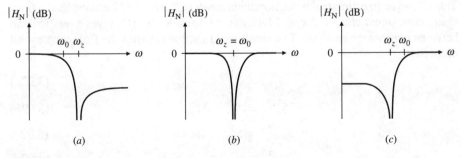

FIGURE 3.38
Notch responses: (a) low-pass notch, (b) symmetric notch, and (c) high-pass notch.

around to yield the design equations:

$$R = \frac{1}{\omega_0 C} \qquad R_1 = QR \qquad R_4 = \frac{R_2}{Q} \frac{\omega_0^2}{|\omega_0^2 - \omega_z^2|} \qquad (3.89a)$$

$$R_5 = R_2 \left(\frac{\omega_0}{\omega_z}\right)^2 \quad \text{for } \omega_z > \omega_0 \qquad R_5 = R_2 \quad \text{for } \omega_z < \omega_0 \qquad (3.89b)$$

where R_2 and R_3 are arbitrary and R_5 has been specified for $H_{0\mathrm{LP}}$ and $H_{0\mathrm{HP}}$ of 0 dB. These gains can be raised or lowered by changing R_5 in proportion.

> **EXAMPLE 3.20.** Specify the components of Fig. 3.37 for a low-pass notch with $f_0 = 1$ kHz, $f_z = 2$ kHz, $Q = 10$, and 0-dB dc gain. What is the high-frequency gain?
>
> **Solution.** Let $C = 10$ nF; then $R = 1/\omega_0 C = 15.9\,\mathrm{k\Omega}$ (use $15.8\,\mathrm{k\Omega}$); $R_1 = QR = 158$ $\mathrm{k\Omega}$; let $R_2 = 100\,\mathrm{k\Omega}$; then $R_4 = (100/10) \times 1^2/|1^2 - 2^2| = 3.333\,\mathrm{k\Omega}$ (use $3.32\,\mathrm{k\Omega}$, 1%); $R_5 = 100 \times (1/2)^2 = 25\,\mathrm{k\Omega}$ (use $24.9\,\mathrm{k\Omega}$, 1%); $H_{0\mathrm{HP}} = (1/2)^2 = 0.25$ V/V $\cong -12$ dB.

3.8
SENSITIVITY

Because of component tolerances and op amp nonidealities, the response of a practical filter is likely to deviate from that predicted by theory. Even if some of the components are made adjustable to allow for fine tuning, deviations will still arise because of component aging and thermal drift. It is therefore of interest to know how sensitive a given filter is to component variations. For instance, the designer of a second-order band-pass filter may want to know the extent to which a 1% variation in a given resistance or capacitance affects ω_0 and BW.

Given a filter parameter y such as ω_0 and Q, and given a filter component x such as a resistance R or a capacitance C, the *classical sensitivity function* S_x^y is defined as

$$S_x^y = \frac{\partial y/y}{\partial x/x} = \frac{x}{y} \frac{\partial y}{\partial x} \qquad (3.90)$$

where we use partial derivatives to account for the fact that filter parameters usually depend on more than just one component. For small changes, we can approximate

$$\frac{\Delta y}{y} \cong S_x^y \frac{\Delta x}{x} \qquad (3.91)$$

This allows us to estimate the *fractional parameter change* $\Delta y/y$ caused by the *fractional component change* $\Delta x/x$. Multiplying both sides by 100 gives a relationship between *percentage changes*. The sensitivity function satisfies the following useful properties:

$$S_{1/x}^y = S_x^{1/y} = -S_x^y \qquad (3.92a)$$

$$S_x^{y_1 y_2} = S_x^{y_1} + S_x^{y_2} \qquad (3.92b)$$

$$S_x^{y_1/y_2} = S_x^{y_1} - S_x^{y_2} \qquad (3.92c)$$

$$S_x^{x^n} = n \qquad (3.92d)$$

$$S_{x_1}^y = S_{x_2}^y S_{x_1}^{x_2} \qquad (3.92e)$$

(See Problem 3.41 for the derivations.) To gain an understanding of sensitivity, we examine some popular filter configurations.

151

SECTION 3.8
Sensitivity

KRC Filter Sensitivities

With reference to the *low-pass KRC filter* of Fig. 3.23, we have, by Eq. (3.60b), $\omega_0 = R_1^{-1/2} C_1^{-1/2} R_2^{-1/2} C_2^{-1/2}$. Consequently, Eq. (3.92d) gives

$$S_{R_1}^{\omega_0} = S_{C_1}^{\omega_0} = S_{R_2}^{\omega_0} = S_{C_2}^{\omega_0} = -\frac{1}{2} \qquad (3.93)$$

Applying Eqs. (3.90) and (3.92) to the expression for Q given in Eq. (3.60c), we obtain

$$S_{R_1}^{Q} = -S_{R_2}^{Q} = Q\sqrt{R_2 C_2 / R_1 C_1} - \frac{1}{2} \qquad (3.94a)$$

$$S_{C_1}^{Q} = -S_{C_2}^{Q} = Q(\sqrt{R_2 C_2 / R_1 C_1}) + \sqrt{R_1 C_2 / R_2 C_1}) - \frac{1}{2} \qquad (3.94b)$$

$$S_{K}^{Q} = QK\sqrt{R_1 C_1 / R_2 C_2} \qquad (3.94c)$$

$$S_{R_A}^{Q} = -S_{R_B}^{Q} = Q(1 - K)\sqrt{R_1 C_1 / R_2 C_2} \qquad (3.94d)$$

For the *equal-component* design, the Q sensitivities simplify to

$$S_{R_1}^{Q} = -S_{R_2}^{Q} = Q - \frac{1}{2} \qquad S_{C_1}^{Q} = -S_{C_2}^{Q} = 2Q - \frac{1}{2} \qquad (3.95a)$$

$$S_{K}^{Q} = 3Q - 1 \qquad S_{R_A}^{Q} = -S_{R_B}^{Q} = 1 - 2Q \qquad (3.95b)$$

and for the *unity-gain* design they simplify to

$$S_{R_1}^{Q} = -S_{R_2}^{Q} = \frac{1 - R_1 / R_2}{2(1 + R_1 / R_2)} \qquad S_{C_1}^{Q} = -S_{C_2}^{Q} = \frac{1}{2} \qquad (3.96)$$

Since the Q sensitivities of the equal-component design increase with Q, they may become unacceptable at high Qs. As we already know, S_{K}^{Q} is of particular concern at high Qs because a slight mismatch in the R_B / R_A ratio may drive Q to infinity or even make it negative, thus leading to oscillatory behavior. By contrast, the unity-gain design offers much lower sensitivities. It is apparent that the designer must carefully weigh a number of conflicting factors before choosing a particular filter design for the given application. These include circuit simplicity, cost, component spread, tunability, and sensitivity.

EXAMPLE 3.21. Investigate the effect of a 1% variation of each component in the low-pass filter of (*a*) Example 3.8 and (*b*) Example 3.10.

Solution. By Eq. (3.93), a 1% increase (decrease) in any of R_1, C_1, R_2, and C_2 causes a 0.5% decrease (increase) in ω_0 in either circuit.

(a) By Eq. (3.95), a 1% increase (decrease) in R_1 increases (decreases) Q by approximately $5 - 0.5 = 4.5\%$ (the opposite holds for R_2). Similarly, 1% capacitance variations result in Q variations of about 9.5%. Finally, since $1 - 2Q = 1 - 2 \times 5 = -9$, it follows that 1% variations in R_A or in R_B result in Q variations of about 9%.

(b) With $R_1/R_2 = 5.76/2.21$, Eq. (3.96) gives $S_{R_1}^Q = -S_{R_2}^Q \cong -0.22$. Thus, 1% resistance and 1% capacitance variations result in Q variations of 0.22% and 0.5%, respectively.

Multiple-Feedback Filter Sensitivities

The sensitivities of the *multiple-feedback band-pass filter* of Fig. 3.30 are found from Eq. (3.70), and they are

$$S_{R_1}^{\omega_0} = S_{C_1}^{\omega_0} = S_{R_2}^{\omega_0} = S_{C_2}^{\omega_0} = -\frac{1}{2} \tag{3.97a}$$

$$S_{R_1}^Q = -S_{R_2}^Q = -\frac{1}{2} \qquad S_{C_1}^Q = -S_{C_2}^Q = \frac{1}{2}\frac{C_2 - C_1}{C_2 + C_1} \tag{3.97b}$$

Note that the equal-capacitance design results in $S_{C_1}^Q = S_{C_2}^Q = 0$. The sensitivities of the *multiple-feedback low-pass filter* of Fig. 3.32 can be computed likewise, and they are found to be[2]

$$S_{R_2}^{\omega_0} = S_{C_1}^{\omega_0} = S_{R_3}^{\omega_0} = S_{C_2}^{\omega_0} = -\frac{1}{2} \tag{3.98a}$$

$$|S_{R_1}^Q| < 1 \qquad |S_{R_2}^Q| < \frac{1}{2} \qquad |S_{R_3}^Q| < \frac{1}{2} \qquad S_{C_1}^Q = -S_{C_2}^Q = \frac{1}{2} \tag{3.98b}$$

It is apparent that multiple-feedback configurations enjoy low sensitivities and are therefore popular.

Multiple-Op-Amp Filter Sensitivities

The sensitivities of the *biquad filter* of Fig. 3.36 are found from Eq. (3.83), and the results are

$$S_{R_4}^{\omega_0} = S_{R_5}^{\omega_0} = S_{C_1}^{\omega_0} = S_{C_2}^{\omega_0} = -\frac{1}{2} \tag{3.99a}$$

$$S_{R_2}^Q = 1 \qquad S_{R_4}^Q = S_{R_5}^Q = -S_{C_1}^Q = S_{C_2}^Q = \frac{1}{2} \tag{3.99b}$$

These sensitivities are fairly low and are similar to those of a passive *RLC* filter yielding the same responses. The sensitivities of *state-variable filters* are similarly low (see Problem 3.44). Considering also the advantages of tuning, low parameter spread, and multiple simultaneous responses, we now appreciate why these filters are widely used.

3.1 The transfer function

3.1 A transfer function with $H_0 = 1$ has a zero at $s = +1$ kNp/s and a pole pair at $-1 \pm j1$ complex kNp/s. (a) Find its impulse response. (b) Find its steady-state response to an ac input with unity amplitude, zero phase, and $\omega = 1$ krad/s.

3.2 First-order active filters

3.2 The circuit of Fig. P3.2 is a noninverting differentiator. (a) Derive its transfer function. (b) Specify component values for a unity-gain frequency of 100 Hz.

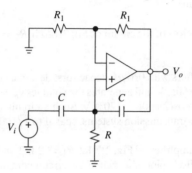

FIGURE P3.2

3.3 If $R_1C_1 = R_2C_2$, the circuit of Fig. P3.3 is a noninverting integrator. (a) Find its transfer function. (b) Specify component values for a gain of 20 dB at 100 Hz.

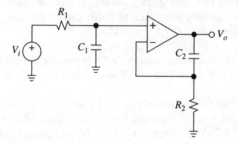

FIGURE P3.3

3.4 (a) Specify suitable component values for a unity-gain frequency of 1 kHz in the Deboo integrator of Fig. 3.7. (b) What happens if the upper-right resistance is 1% less than its nominal value? Illustrate via the magnitude plot. *Hint:* Replace the Howland current pump with its Norton equivalent.

3.5 Suppose the time constants in the circuit of Fig. P3.3 are mismatched, say, $R_1C_1 = R_2C_2$ $(1-\epsilon)$. (a) Investigate the effect of the mismatch and illustrate via the magnitude plot. (b) Devise a method for balancing out the mismatch, and outline the calibration procedure.

3.6 Inserting a resistance R_3 in series with C in the low-pass filter of Fig. 3.9a turns it into a circuit known as a *pole-zero* circuit, which finds application in control. (a) Sketch the modified circuit, and find its transfer function to justify its name. (b) Specify component values for a pole frequency of 1 kHz, a zero frequency of 10 kHz, and a dc gain of 0 dB; sketch its magnitude plot.

3.7 Inserting a resistance R_3 in parallel with C in the high-pass filter of Fig 3.10a turns it into a circuit known as a *zero-pole* circuit, which finds application in control. (*a*) Sketch the modified circuit, and find its transfer function to justify its name. (*b*) Specify component values for a zero frequency of 100 Hz, a pole frequency of 1 kHz, and a high-frequency gain of 0 dB; sketch its magnitude plot.

3.8 Redraw the phase shifter of Fig 3.12a, but with R and C interchanged with each other; derive its transfer function and sketch its Bode plots. What is the main difference between the responses of the original and the modified circuit? Name a possible disadvantage of the modified circuit.

3.9 (*a*) Sketch the Bode plots of the circuit of Fig 3.12a if $R_2 = 10R_1$. (*b*) Repeat, but with $R_1 = 10R_2$.

3.10 Using two phase shifters with 0.1-μF capacitors, design a circuit that accepts a voltage $v_a = 1.20\sqrt{2}\cos(2\pi 60t)$ V, and generates the voltages $v_b = 1.20\sqrt{2}\cos(2\pi 60t-120°)$ V and $v_c = 1.20\sqrt{2}\cos(2\pi 60t + 120°)$ V. Such a circuit simulates the voltages used in three-phase power transmission systems, scaled to $1/100$ of their actual values.

3.11 In the noninverting amplifier of Fig. 1.7 let $R_1 = 2$ kΩ and $R_2 = 18$ kΩ. Sketch and label the magnitude Bode plot of its gain if the circuit contains also a 10-nF capacitance in parallel with R_2.

3.12 Suppose the inverting amplifier of Fig. 1.11 has also a capacitance C_1 in parallel with R_1 and a capacitance C_2 in parallel with R_2. Derive its transfer function, sketch and label the magnitude Bode plot, and specify suitable component values for a low-frequency gain of 40 dB, a high-frequency gain of 0 dB, and so that the geometric mean of its pole and zero frequencies $(f_p f_z)^{1/2}$ is 1 kHz.

3.13 Sketch and label the linearized magnitude Bode plot for the circuit of Fig. P3.3 if: (*a*) $R_2C_2 = 1$ ms and $R_1C_1 = 0.1$ ms. (*b*) Repeat, but with $R_1C_1 = 10$ ms.

3.14 In the wideband band-pass filter of Fig. 3.11a let $R_1 = R_2 = R$ and $C_1 = C_2 = C$. (*a*) Find the output $v_o(t)$ if the input is $v_i(t) = 1\cos(t/RC)$ V. (*b*) Repeat, but for $v_i(t) = 1\cos(t/2RC)$ V. (*c*) Repeat, but for $v_i(t) = 1\cos(t/0.5RC)$ V.

3.15 The circuit of Fig. P3.15 is a capacitance multiplier. (*a*) Show that $C_{eq} = (1+R_2/R_1)C$. (*b*) Using a 0.1-μF capacitance, specify component values to simulate a variable capacitance from 0.1 μF to 100 μF by means of a 1-MΩ pot. *Hint:* In part (*a*), apply a test voltage V, find the resulting current I, and obtain C_{eq} as $1/sC_{eq} = V/I$.

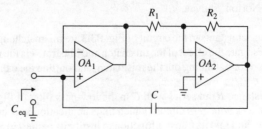

FIGURE P3.15

3.16 The circuit of Fig. P3.16 is a capacitance simulator. (*a*) Show that $C_{eq} = (R_2 R_3 / R_1 R_4)C$. (*b*) Using a 1-nF capacitance, specify component values to simulate a 1-mF capacitance. List a possible application of such a large capacitance. *Hint:* See Problem 3.15.

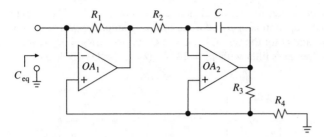

FIGURE P3.16

3.3 Audio filter applications

3.17 Derive Eqs. (3.32) and (3.33).

3.18 (*a*) Derive Eqs. (3.34) and (3.35). (*b*) Specify component values to approximate the NAB curve with a 30-dB gain at 1 kHz. Show the final circuit.

3.19 Using standard component values, design an octave equalizer with center frequencies at approximately $f_0 = 32$ Hz, 64 Hz, 125 Hz, 250 Hz, 500 Hz, 1 kHz, 2 kHz, 4 kHz, 8 kHz, and 16 kHz. Show the final circuit.

3.4 Standard second-order responses

3.20 (*a*) By proper manipulation, put the wideband band-pass function of Eq. (3.29a) in the standard form $H(j\omega) = H_{0BP} H_{BP}$. (*b*) Show that no matter how you select ω_L and ω_H, the Q of that filter can never exceed $\frac{1}{2}$. This is why the filter is called *wideband*.

3.21 Construct the phase plots of H_{LP}, H_{HP}, H_{BP}, and H_N for $Q = 0.2$, 1, and 10.

3.5 *KRC* filters

3.22 An alternative design procedure for the low-pass *KRC* filter of Fig. 3.23 is $R_A = R_B$ and $R_2/R_1 = C_1/C_2 = Q$. (*a*) Develop design equations for this option. (*b*) Hence, use it to redesign the filter of Example 3.8.

3.23 An alternative design procedure for the low-pass *KRC* filter of Fig. 3.23 that allows us to specify also H_{0LP}, $H_{0LP} > 2$ V/V, is $C_1 = C_2 = C$. (*a*) Show that the design equations for this option are $R_2 = [1 + \sqrt{1 + 4Q^2(H_{0LP} - 2)}]/2\omega_0 QC$ and $R_1 = 1/\omega_0^2 R_2 C^2$. (*b*) Use this option to redesign the filter of Example 3.8, but with $H_{0LP} = 10$ V/V.

3.24 (*a*) Design a high-pass *KRC* filter with $f_0 = 100$ Hz and Q variable from 0.5 to 5 by means of a 100-kΩ potentiometer. (*b*) If the input is a 60-Hz, 5-V (rms) ac wave with a dc component of 3 V, what comes out of the filter with the wiper at either extreme?

3.25 An alternative design procedure for the high-pass *KRC* filter of Fig. 3.27 that allows us to specify also H_{0HP}, $H_{0HP} > 1$, is $C_1 = C_2 = C$. (*a*) Show that the design equations are then $R_1 = [1 + \sqrt{1 + 8Q^2(H_{0HP} - 1)}]/4\omega_0 QC$ and $R_2 = 1/\omega_0^2 R_1 C^2$. (*b*) Use this option to implement a high-pass Butterworth response with $H_{0HP} = 10$ V/V and $f_0 = 1$ kHz.

3.26 An alternative design procedure for the band-pass *KRC* filter of Fig. 3.28 is $R_A = R_B$ and $C_1 = C_2 = C$. Develop design equations for this option. Hence, use it to design a band-pass filter with $H_{0BP} = 0$ dB, $f_0 = 1$ kHz, and $Q = 5$.

3.27 The low-pass filter of Fig. P3.27 is referred to as a $-KRC$ *filter* ("minus" *KRC* filter) because the op amp is operated as an inverting amplifier with a gain of $-K$. (*a*) Find H_{0LP}, ω_0, and Q for the case $C_1 = C_2 = C$ and $R_1 = R_2 = R_3 = R_4 = R$. (*b*) Design a $-KRC$ low-pass filter with $f_0 = 2$ kHz, $Q = 5$, and 0-dB dc gain.

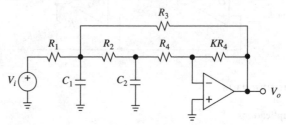

FIGURE P3.27

3.28 The band-pass filter of Fig. P3.28 is referred to as a $-KRC$ *filter* ("minus" *KRC* filter) because the op amp is operated as an inverting amplifier with a gain of $-K$. (*a*) Find H_{0BP}, ω_0, and Q for the case $C_1 = C_2 = C$ and $R_1 = R_2 = R$. (*b*) Design a $-KRC$ band-pass filter with $f_0 = 1$ kHz, $Q = 10$, and unity-resonance gain.

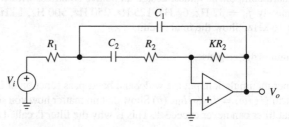

FIGURE P3.28

3.29 The notch filter of Fig. P3.29 allows Q tuning via the ratio R_2/R_1. (*a*) Show that $V_o/V_i = H_N$ with $\omega_0 = 1/RC$ and $Q = (1 + R_1/R_2)/4$. (*b*) Specify component values for $f_0 = 60$ Hz and $Q = 25$.

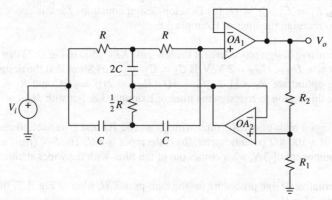

FIGURE P3.29

3.30 An alternative design procedure for the multiple-feedback low-pass filter of Fig. 3.32 is $R_1 = R_2 = R_3 = R$. Find expressions for H_{0LP}, ω_0, and Q. Hence, develop the design equations.

3.31 In the circuit of Fig. 3.33 let $R_3 = R_4 = R$, and $R_5 = KR$. (a) Show that if $H_{0BP} = -2$ V/V, the circuit gives the all-pass response with gain $-K$. (b) Specify component values for $f_0 = 1$ kHz, $Q = 5$, and a gain of 20 dB.

3.32 Show that the circuit of Fig. P3.32 realizes the all-pass function with $H_{0AP} = 1/3$, $\omega_0 = \sqrt{2}/RC$, and $Q = 1/\sqrt{2}$.

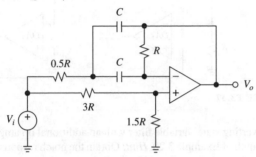

FIGURE P3.32

3.33 The circuit of Fig P3.33, known as a *Q multiplier*, uses a summing amplifier OA_1 and a band-pass stage OA_2 to increase the Q of the band-pass stage without changing ω_0. This allows for high Qs without unduly taxing OA_2. (a) Show that the gain and Q of the composite circuit are related to those of the basic band-pass stage as $Q_{comp} = Q/[1 - (R_5/R_4)|H_{0BP}|]$, and $H_{0BP(comp)} = (R_5/R_3)(Q_{comp}/Q)H_{0BP}$. (b) Specify component values for $f_0 = 3600$ Hz, $Q_{comp} = 60$, and $H_{0BP(comp)} = 2$ V/V, starting with $Q = 10$.

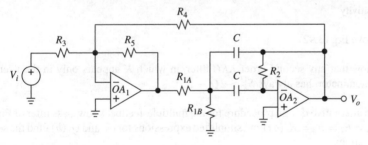

FIGURE P3.33

3.34 With reference to the multiple-feedback low-pass filter of Fig. 3.32, show that the circuit consisting of R_2, R_3, C_2, and the op amp acts as a resistance $R_{eq} = R_2 \| R_3$ and an inductance $L_{eq} = R_2 R_3 C_2$, both in parallel with C_1. Hence, explain circuit operation in terms of the above equivalence.

3.7 State-variable and biquad filters

3.35 Suitably modify the filter of Example 3.18 so that $H_{0BP} = 1$ V/V. Show your final design.

3.36 (a) Derive Eqs. (3.82a) and (3.82b). (b) Specify suitable component values to achieve a band-pass response with $f_L = 594$ Hz and $f_H = 606$ Hz. (c) What is the dc gain of the low-pass response?

3.37 The simplified state-variable filter of Fig. P3.37 provides the low-pass and band-pass responses using only two op amps. (a) Show that $H_{0BP} = -n$, $H_{0LP} = m/(m + 1)$, $Q = \sqrt{n(1 + 1/m)}$, and $\omega_0 = Q/nRC$. (b) Specify component values for a band-pass response with $f_0 = 2$ kHz and $Q = 10$. (c) What is the resonance gain of your circuit? What is the most serious drawback of this circuit?

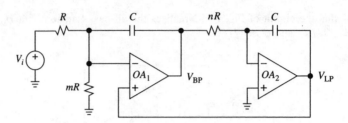

FIGURE P3.37

3.38 Use the noninverting state-variable filter with an additional op amp adder to synthesize the low-pass notch of Example 3.20. *Hint:* Obtain the notch response as $V_N = A_L V_{LP} + A_H V_{HP}$, where A_L and A_H are suitable coefficients.

3.39 Consider the dual-op-amp biquad obtained from the standard biquad of Fig. 3.36 by replacing OA_2 and OA_3 with the Deboo integrator of Fig. 3.7. Find its responses; specify component values for a low-pass response with $f_0 = 10$ kHz, $Q = 5$, and $H_{0LP} = 0$ dB.

3.40 Using the state-variable filter, along with a fourth op amp adder, design an all-pass circuit with $f_0 = 1$ kHz and $Q = 1$. *Hint:* Apply Eq. (3.58).

3.8 Sensitivity

3.41 Prove Eq. (3.92).

3.42 Show that any second-order *KRC* filter, in which K appears only in the s-term in the denominator, has always $S_K^Q > 2Q - 1$.

3.43 An alternative design procedure for the multiple-feedback low-pass filter of Fig. 3.32 is $R_1 = R_2 = R_3 = R$. (a) Find simplified expressions for ω_0 and Q. (b) Find the sensitivity functions.

3.44 Calculate the sensitivities of the state-variable filter of Example 3.18.

REFERENCES

1. M. E. Van Valkenburg, *Analog Filter Design,* Holt, Rinehart and Winston, Orlando, FL, 1982.
2. L. P. Huelsman, *Active and Passive Analog Filter Design: An Introduction,* McGraw-Hill, New York, 1993.

3. F. W. Stephenson, *RC Active Filter Design Handbook,* John Wiley and Sons, New York, 1985.

4. A. B. Williams and F. J. Taylor, *Electronic Filter Design Handbook: LC, Active, and Digital Filters,* 2d ed., McGraw-Hill, New York, 1988.

5. S. Franco, *Electric Circuits Fundamentals,* Oxford University Press, New York, 1995.

6. W. G. Jung, *Audio IC Op Amp Applications,* 3d ed., Howard W. Sams, Carmel, IN, 1987.

7. K. Lacanette, "High Performance Audio Applications of the LM833," Application Note AN-346, *Linear Applications Handbook,* National Semiconductor, Santa Clara, CA, 1994.

8. R. A. Greiner and M. Schoessow, "Design Aspects of Graphic Equalizers," *J. Audio Eng. Soc.,* Vol. 31, No. 6, June 1983, pp. 394–407.

4

ACTIVE FILTERS: PART II

4.1 Filter Approximations
4.2 Cascade Design
4.3 Generalized Impedance Converters
4.4 Direct Design
4.5 The Switched Capacitor
4.6 Switched-Capacitor Filters
4.7 Universal SC Filters
   Problems
   References

Having studied first-order and second-order filters, we now turn to higher-order filters, which are required when the cutoff characteristics of the lower-order types are not sufficiently sharp to meet the demands of the given application. Among the various methods of realizing higher-order active filters, the ones that have gained prominence are the *cascade design* approach and the *direct synthesis* approach. The cascade approach realizes the desired response by cascading second-order filter stages (and possibly a first-order stage) of the types studied in Chapter 3. The direct approach uses active impedance converters, such as gyrators and frequency-dependent negative resistances, to simulate a passive *RLC* filter prototype meeting the given specifications.

Regardless of the complexity of their responses, the above filters, also known as *continuous-time filters,* do not lend themselves to monolithic fabrication due to the large sizes of the capacitances involved, and the stringent requirements on the accuracy and stability of the *RC* products controlling characteristic frequencies. On the other hand, today's very large scale integration (VLSI) applications often call for digital as well as analog functions on the same chip. To meet this requirement in the area of filtering and other traditional analog areas, switched-capacitor techniques have been developed, which use MOS op amps, capacitors, and switches, but no resistors, to realize fairly stable filter functions—if over comparatively limited frequency ranges.

Switched capacitor (SC) circuits belong to the category of sampled-data systems, where information is processed at discrete time intervals rather than continuously. This generally limits their usage to voice-band applications, such as tone coding/decoding (Codecs), speech processing, and audio spectrum analysis.

4.1
FILTER APPROXIMATIONS

If the signals to be rejected are very close in frequency to those that must be passed, the cutoff characteristics of a second-order filter may not prove sufficiently sharp, so a higher-order filter may be needed. Actual filters can only approximate the brick-wall responses of Fig. 3.1. In general, the closer the desired approximation, the higher the order of the filter.

The departure of a practical filter from its brick-wall model is visualized in terms of a shaded area,[1] as shown in Fig. 4.1a for the low-pass case. Introducing the attenuation $A(\omega)$ as

$$A(\omega) = -20 \log_{10} |H(j\omega)| \qquad (4.1)$$

we observe that the range of frequencies that are passed with little or no attenuation defines the *passband*. For a low-pass filter, this band extends from dc to some frequency ω_c, called the *cutoff frequency*. Gain is not necessarily constant within the passband but is allowed a maximum variation A_{max}, such as $A_{max} = 1$ dB. Gain may exhibit ripple within the passband, in which case A_{max} is called the *maximum passband ripple* and the passband is called the *ripple band*. In this case ω_c represents the frequency at which the response departs from the ripple band.

Past ω_c the magnitude drops off to the *stopband,* or the frequency region of substantial attenuation. This band is specified in terms of some minimum allowable attenuation, such as $A_{min} = 60$ dB. The frequency at which the stopband begins is denoted as ω_s. The ratio ω_s/ω_c is called the *selectivity factor* because it gives a measure of the sharpness of the response. The frequency region between ω_c and ω_s is called the *transition band,* or *skirt.* Certain filter approximations maximize the rate of descent within this band at the expense of ripples within the other bands.

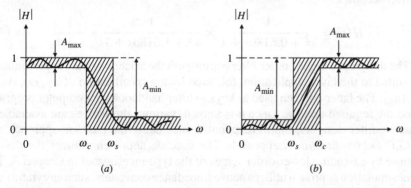

FIGURE 4.1
Magnitude limits for (*a*) the low-pass and (*b*) the high-pass responses.

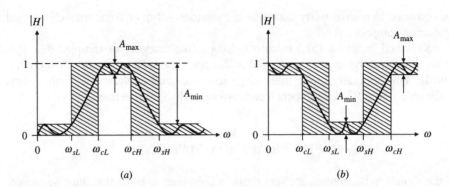

FIGURE 4.2
Magnitude limits for (*a*) the band-pass and (*b*) the band-reject responses.

The terminology developed for the low-pass case is readily extended to the high-pass case depicted in Fig. 4.1*b*, and to the band-pass and band-reject cases depicted in Fig. 4.2.

As the order n of a transfer function is increased, additional parameters are brought into play in the form of the higher-order polynomial coefficients. These coefficients provide the designer with additional freedom in specifying the frequency profiles of magnitude or phase, thus allowing for an increased degree of optimization. Among the various approximations, some have been found to be consistently satisfactory to justify the tabulation of their coefficients in filter handbooks. These include the *Butterworth, Chebyshev, Cauer,* and *Bessel* approximations.

Filter tables list the denominator polynomial coefficients of the various approximations for a cutoff frequency of 1 rad/s. As an example, the coefficients of the fifth-order Butterworth response are[2] $b_0 = b_5 = 1$, $b_2 = b_4 = 3.236$, and $b_3 = 5.236$, so

$$H(s) = \frac{1}{s^5 + 3.236s^4 + 5.236s^3 + 5.236s^2 + 3.236s + 1} \tag{4.2}$$

An alternative approach is to factor out $H(s)$ into the product of terms of order ≤ 2 and tabulate the coefficients of these terms instead. Expressed in this form, the above function becomes

$$H(s) = \frac{1}{s^2 + 0.6180s + 1} \times \frac{1}{s^2 + 1.6180s + 1} \times \frac{1}{s + 1} \tag{4.3}$$

The design of a higher-order filter begins with the selection of the approximation best suited to the given application, followed by the specification of ω_c, ω_s, A_{max}, and A_{min}. The latter are then used as keys to filter handbooks or computer programs to find the required order n. Once n is known, various alternatives are available to the active-filter designer, the most popular ones being the *cascade* approach and the *RLC ladder simulation* approach. The cascade approach realizes the desired response by cascading lower-order stages of the type investigated in Chapter 3. The ladder simulation approach utilizes active impedance converters, such as gyrators and frequency-dependent negative resistors, to simulate a passive *RLC* filter prototype meeting the desired specifications.

Once an approach has been chosen, one must find the individual-stage values of ω_0 and Q (and possibly ω_z) in the case of cascade design, or the individual values of R, L, and C in the case of ladder simulation. These data are again found with the help of filter tables or computer programs, the latter being provided by op amp manufacturers to promote the application of their products. One such program is the FILDES program, written by National Semiconductor, which we shall use extensively in our cascade design examples. This program can be downloaded from the World Wide Web; please check our Web site at http://www.mhhe.com/franco, as described in the preface.

Plotting $H(j\omega)$ Using PSpice

The frequency behavior of a function $H(s)$ can be visualized with PSpice using voltage-controlled sources with values that are functions of s. Using the factored form of $H(s)$, we create a cascade of VCVSs whose values are given by the individual terms of $H(s)$. Figure 4.3 shows the cascade for the function of Eq. (4.3). Scaling s by 2π to obtain $f_c = 1$ Hz, we write the file:

```
5th-Order Butterworth low-pass response:
Vi 1 0 ac 1V
Ri 1 0 1
E1 2 0 Laplace {V(1)} = {1/(1+(s/6.283)*(s/6.283+0.6180))}
R1 2 0 1
E2 3 0 Laplace {V(2)} = {1/(1+(s/6.283)*(s/6.283+1.618))}
R2 3 0 1
E3 4 0 Laplace {V(3)} = {1/(1+s/6.283)}
R3 4 0 1
.ac dec 100 0.01Hz 100Hz
.probe ;Vdb(4)
.end
```

The magnitude plot is shown in Fig. 4.5 (page 166) along with the plots of the other three response types, which are obtained by a similar procedure.

Butterworth Approximation

The gain of the Butterworth approximation is[3]

$$|H(j\omega)| = \frac{1}{\sqrt{1 + \epsilon^2 (\omega/\omega_c)^{2n}}} \tag{4.4}$$

where n is the order of the filter, ω_c is the cutoff frequency, and ϵ is a constant

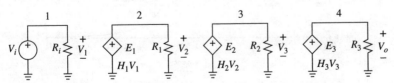

FIGURE 4.3
PSpice circuit to find the frequency behavior of the function $H(s) = H_1(s) \times H_2(s) \times H_3(s)$.

that determines the maximum passband variation as $A_{max} = A(\omega_c) = 20 \times \log_{10} \sqrt{1 + \epsilon^2} = 10 \log_{10}(1 + \epsilon^2)$. The first $2n - 1$ derivatives of $|H(j\omega)|$ are zero at $\omega = 0$, indicating a curve as flat as possible at $\omega = 0$. Aptly referred to as *maximally flat*, a Butterworth curve becomes somewhat rounded near ω_c and rolls off at an ultimate rate of $-20n$ dB/dec in the stopband. As shown in Fig. 4.4a for $\epsilon = 1$, the higher the order n, then the closer the response is to the brick-wall model.

EXAMPLE 4.1. Find n for a low-pass Butterworth response with $f_c = 1$ kHz, $f_s = 2$ kHz, $A_{max} = 1$ dB, and $A_{min} = 40$ dB.

Solution. Letting $A_{max} = A(\omega_c) = 20 \log_{10} \sqrt{1 + \epsilon^2} = 1$ dB gives $\epsilon = 0.5088$. Letting $A(\omega_s) = 10 \log_{10}[1 + \epsilon^2(2/1)^{2n}] = 40$ dB, we find that $n = 7$ gives $A(\omega_s) = 36.3$ dB and $n = 8$ gives $A(\omega_s) = 42.2$ dB. For $A_{min} = 40$ dB we thus select $n = 8$.

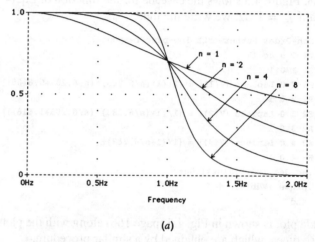

(a)

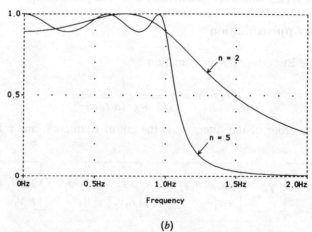

(b)

FIGURE 4.4
(a) Butterworth and (b) 1-dB Chebyshev responses.

Chebyshev Approximation

There are applications where sharp cutoff is more important than maximal flatness. Chebyshev filters maximize the transition-band cutoff rate at the price of introducing passband ripples, as shown in Fig. 4.4b. As a general rule, the higher A_{max}, the narrower the transition band for a given A_{min}. The gain of an nth-order Chebyshev approximation with cutoff frequency ω_c and $A_{max} = 10 \log_{10}(1 + \epsilon^2)$ is[3]

$$|H(j\omega)| = \frac{1}{\sqrt{1 + \epsilon^2 C_n^2(\omega/\omega_c)}} \tag{4.5}$$

where $C_n(\omega/\omega_c)$ is the Chebyshev polynomial of order n, defined as

$$C_n(\omega/\omega_c \leq 1) = \cos[n \cos^{-1}(\omega/\omega_c)] \tag{4.6a}$$

$$C_n(\omega/\omega_c \geq 1) = \cosh[n \cosh^{-1}(\omega/\omega_c)] \tag{4.6b}$$

We observe that $C_n^2(\omega/\omega_c \leq 1) \leq 1$, and $C_n^2(\omega/\omega_c \geq 1) \geq 1$. Moreover, within the passband $|H(j\omega)|$ exhibits peak values of 1 and valley values of $1/\sqrt{1 + \epsilon^2}$ at the frequencies that make the cosine term zero and unity, respectively. The number of these peaks or valleys, including the one at the origin, is n.

Compared to the Butterworth approximation, which exhibits appreciable departure from its dc value only at the upper end of the passband, the Chebyshev approximation improves the transition-band characteristic by spreading its equal-sized ripples throughout the passband. At dc, the decibel value of a Chebyshev response is 0 if n is odd, and $0 - A_{max}$ if n is even. A Chebyshev filter can achieve a given transition-band cutoff rate with a lower order than a Butterworth filter, thus reducing circuit complexity and cost. Past the transition band, however, the Chebyshev response rolls off at an ultimate rate of $-20n$ dB/dec, just like a Butterworth response of the same order.

Cauer Approximation

Cauer filters, also called *elliptic filters,* carry the Chebyshev approach one step further by trading ripples in both the passband and the stopband for an even sharper characteristic in the transition band. Consequently, they can provide a given transition-band cutoff rate with an even lower order n than Chebyshev filters. The idea is to follow an existing low-pass response with a notch just above ω_c to further sharpen the response. To be effective, the notch must be narrow, indicating that the curve will come back up just past this notch. At this point another notch is created to press the curve back down, and the process is repeated until the overall profile within the stopband is pushed below the level specified by A_{min}. The various approximations are compared in Fig. 4.5 for $n = 5$ and $A_{max} = 3$ dB. Shown at the top is an expanded view of the passband.

Bessel Approximation

In general, filters introduce a frequency-dependent phase shift. If this shift varies linearly with frequency, its effect is simply to delay the signal by a constant amount.

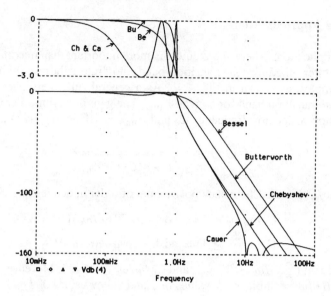

FIGURE 4.5
Comparison of fifth-order 3-dB responses.

However, if phase varies nonlinearly, different input frequency components will experience different delays, so nonsinusoidal signals may experience significant phase distortion in propagating through the filter. In general, the steeper the transition-band magnitude characteristic, the higher the distortion.[1]

Bessel filters, also called *Thomson filters,* maximize the passband delay just as Butterworth filters maximize the passband magnitude. The result is a nearly linear phase characteristic within the passband, if at the price of a less sharp magnitude characteristic in the transition band. Figure 4.6 shows that a pulse emerges fairly undistorted from a Bessel filter, but exhibits appreciable overshoot and ringing when processed with a Chebyshev filter, whose phase response is less linear than Bessel's.

4.2
CASCADE DESIGN

This approach is based on the factorization of a transfer function $H(s)$ into the product of lower-order terms. If the order n is even, the decomposition consists of $n/2$ second-order terms,

$$H(s) = H_1(s) \times H_2(s) \times \cdots \times H_{n/2}(s) \qquad (4.7)$$

If n is odd, the factorization includes also a first-order term. Sometimes this term is combined with one of the second-order terms to create a third-order filter stage. The first-order term, if any, can be implemented with a plain RC or CR network, so all we need to know is its required frequency ω_0. The second-order terms can be implemented with any of the filters of Sections 3.5 through 3.7. For each of these stages we need to know ω_0 and Q, and ω_z if the stage is a notch stage. As mentioned, these data are tabulated in filter handbooks[3] or can be calculated by computer.[4]

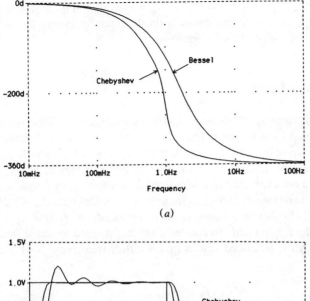

(a)

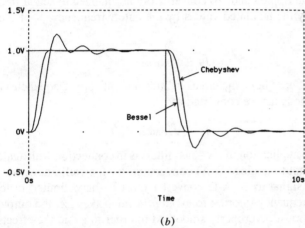

(b)

FIGURE 4.6
Phase and pulse responses of the fourth-order Bessel and 1-dB Chebyshev
filters.

The cascade approach offers a number of advantages. The design of each section is relatively simple, and the component count is usually low. The low-output
impedance of the individual sections eliminates interstage loading, so each section
can be regarded as isolated from the others and can be tuned independently, if
needed. The inherent modularity of this approach is also attractive from the eco-
nomic standpoint, since one can use a few standardized blocks to design a variety
of more complex filters.

Mathematically, the order in which the various sections are cascaded is irrelevant. In practice, to avoid loss of dynamic range and filter accuracy due to possible
signal clipping in the high-Q sections, the sections are cascaded in order of ascend-
ing Qs, with the low-Q stages first in the signal path. This ordering, however, does
not take into account internal noise, which may be of concern in the high-Q stages,
where any noise component falling under the resonance peak may be amplified

significantly. So, to minimize noise, high-Q stages should go first in the cascade. In general, the optimum ordering depends on the input spectrum, the filter type, and the noise characteristics of its components.

Low-Pass Filter Design

Table 4.1 gives examples of tabulated data for cascade design. Butterworth and Bessel data are tabulated for different values of n, Chebyshev data for different values of n and A_{max} (shown in the table are the data for $A_{max} = 0.1$ dB and $A_{max} = 1.0$ dB), and Cauer data (not shown in the table) for different values of n, A_{max}, and A_{min}. Frequency data are expressed in normalized form for a cutoff frequency of 1 Hz. In the Butterworth and Bessel cases this frequency coincides with the -3-dB frequency, while in the Chebyshev and Cauer cases it represents the frequency at which gain departs from the ripple band. To convert from normalized to actual frequencies, we simply multiply the tabulated values by the cutoff frequency f_c of the filter being designed, or

$$f_0 = f_{0(table)} \times f_c \qquad (4.8a)$$

In the case of Cauer filters, the tables include not only pole frequencies but also zero frequencies. The latter are converted as

$$f_z = f_{z(table)} \times f_c \qquad (4.8b)$$

A common application of low-pass filters is in connection with analog-to-digital (A-D) and digital-to-analog (D-A) conversion. By the well-known sampling theorem, the input signal to an A-D converter must be band limited to less than half the sampling frequency in order to avoid *aliasing*. Likewise, the output signal of a D-A converter must be properly smoothed in order to avoid the effects of discrete quantization and time sampling. Both tasks are accomplished with sharp low-pass filters designed to provide adequate attenuation at half the sampling frequency.

> **EXAMPLE 4.2.** The output of a D-A converter with a sampling rate of 40 kHz is to be smoothed with a sixth-order 1.0-dB Chebyshev low-pass filter providing an attenuation of 40 dB at half the sampling frequency, or 20 kHz. This attenuation requirement is met by letting $f_c = 13.0$ kHz. (a) Design such a filter. (b) Verify with PSpice.
>
> **Solution.**
>
> (a) From Table 4.1 we find that a 1.0-dB Chebyshev filter with $n = 6$ requires three second-order stages with
>
> $$f_{01} = 0.995 f_c = 12.9 \text{ kHz} \qquad Q_1 = 8.00$$
>
> $$f_{02} = 0.747 f_c = 9.71 \text{ kHz} \qquad Q_2 = 2.20$$
>
> $$f_{03} = 0.353 f_c = 4.59 \text{ kHz} \qquad Q_3 = 0.761$$
>
> Use three unity-gain Sallen-Key sections and cascade them in order of ascending Qs. Retracing the design steps of Example 3.10, we find the component values shown in Fig. 4.7, where the resistances have been rounded off to the nearest 1% standard values.

TABLE 4.1
Examples of normalized (1 Hz) low-pass filter tables

Butterworth low-pass filter

n	f_{01}	Q_1	f_{02}	Q_2	f_{03}	Q_3	f_{04}	Q_4	f_{05}	Q_5	Att (dB) at $2f_c$
2	1	0.707	1								12.30
3	1	1.000	1								18.13
4	1	0.541	1	1.306							24.10
5	1	0.618	1	1.620	1						30.11
6	1	0.518	1	0.707	1	1.932					36.12
7	1	0.555	1	0.802	1	2.247	1				42.14
8	1	0.510	1	0.601	1	0.900	1	2.563			48.16
9	1	0.532	1	0.653	1	1.000	1	2.879	1		54.19
10	1	0.506	1	0.561	1	0.707	1	1.101	1	3.196	60.21

Bessel low-pass filter

n	f_{01}	Q_1	f_{02}	Q_2	f_{03}	Q_3	f_{04}	Q_4	f_{05}	Q_5
2	1.274	0.577								
3	1.453	0.691	1.327							
4	1.419	0.522	1.591	0.806						
5	1.561	0.564	1.760	0.917	1.507					
6	1.606	0.510	1.691	0.611	1.907	1.023				
7	1.719	0.533	1.824	0.661	2.051	1.127	1.685			
8	1.784	0.506	1.838	0.560	1.958	0.711	2.196	1.226		
9	1.880	0.520	1.949	0.589	2.081	0.760	2.324	1.322	1.858	
10	1.949	0.504	1.987	0.538	2.068	0.620	2.211	0.810	2.485	1.415

0.10-dB ripple Chebyshev low-pass filter

n	f_{01}	Q_1	f_{02}	Q_2	f_{03}	Q_3	f_{04}	Q_4	f_{05}	Q_5	Att (dB) at $2f_c$
2	1.820	0.767									3.31
3	1.300	1.341	0.969								12.24
4	1.153	2.183	0.789	0.619							23.43
5	1.093	3.282	0.797	0.915	0.539						34.85
6	1.063	4.633	0.834	1.332	0.513	0.599					46.29
7	1.045	6.233	0.868	1.847	0.575	0.846	0.377				57.72
8	1.034	8.082	0.894	2.453	0.645	1.183	0.382	0.593			69.16
9	1.027	10.178	0.913	3.145	0.705	1.585	0.449	0.822	0.290		80.60
10	1.022	12.522	0.928	3.921	0.754	2.044	0.524	1.127	0.304	0.590	92.04

1.00-dB ripple Chebyshev low-pass filter

n	f_{01}	Q_1	f_{02}	Q_2	f_{03}	Q_3	f_{04}	Q_4	f_{05}	Q_5	Att (dB) at $2f_c$
2	1.050	0.957									11.36
3	0.997	2.018	0.494								22.46
4	0.993	3.559	0.529	0.785							33.87
5	0.994	5.556	0.655	1.399	0.289						45.31
6	0.995	8.004	0.747	2.198	0.353	0.761					56.74
7	0.996	10.899	0.808	3.156	0.480	1.297	0.205				68.18
8	0.997	14.240	0.851	4.266	0.584	1.956	0.265	0.753			79.62
9	0.998	18.029	0.881	5.527	0.662	2.713	0.377	1.260	0.159		91.06
10	0.998	22.263	0.902	6.937	0.721	3.561	0.476	1.864	0.212	0.749	102.50

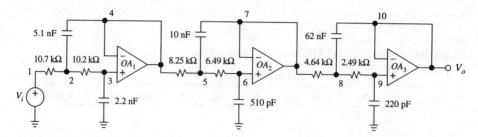

FIGURE 4.7
Sixth-order 1-dB Chebyshev low-pass filter.

(*b*) Using the node numbering shown, we write the circuit file:

```
Cascade Design:
Vi 1 0 ac 1V
R1 1 2 10.69k
R2 2 3 10.02k
C1 2 4 5.1nF
C2 3 0 2.2nF
EOA1 4 0 3 4 1G
R3 4 5 8.191k
R4 5 6 6.434k
C3 5 7 10nF
C4 6 0 510pF
EOA2 7 0 6 7 1G
R5 7 8 4.554k
R6 8 9 2.438k
C5 8 10 62nF
C6 9 0 220pF
EOA3 10 0 9 10 1G
.ac dec 100 1kHz 100kHz
.probe
.end
```

Figure 4.8 shows the overall response as well as the individual-stage responses. It is interesting to observe how the latter combine to create the ripple and cutoff characteristics of the overall response.

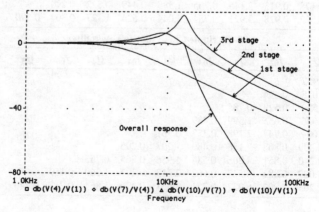

FIGURE 4.8
Overall as well as individual-stage responses of the filter of Fig. 4.7.

EXAMPLE 4.3. Design a Cauer low-pass filter with $f_c = 1$ kHz, $f_s = 1.3$ kHz, $A_{max} = 0.1$ dB, $A_{min} = 40$ dB, and dc gain $H_0 = 0$ dB.

Solution. Using the aforementioned filter design program FILDES (check our Web site for information on how to download this program), we find that a sixth-order implementation is required, with the following individual-stage parameters:

$$f_{01} = 648.8 \text{ Hz} \qquad f_{z1} = 4130.2 \text{ Hz} \qquad Q_1 = 0.625$$

$$f_{02} = 916.5 \text{ Hz} \qquad f_{z2} = 1664.3 \text{ Hz} \qquad Q_2 = 1.789$$

$$f_{03} = 1041.3 \text{ Hz} \qquad f_{z3} = 1329.0 \text{ Hz} \qquad Q_3 = 7.880$$

Moreover, the program indicates that the actual attenuation at 1.3 kHz is 47 dB, and the -3-dB frequency is 1.055 kHz.

We shall implement the filter with three low-pass notch sections of the biquad type of Fig. 3.37. Using Eq. (3.89) and retracing the steps of Example 3.20, we find the component values shown in Fig. 4.9, where the resistances have been rounded off to the nearest 1% standard values. The entire filter can be built with three quad-op-amp packages.

High-Pass Filter Design

Owing to the fact that a high-pass transfer function can be obtained from a low-pass function via the substitution $s/\omega_0 \rightarrow 1/(s/\omega_0)$, the normalized frequency data of Table 4.1 can also be used in the cascade design of high-pass filters, provided actual frequencies are obtained from tabulated frequencies as

$$f_0 = f_c/f_{0(\text{table})} \qquad (4.9a)$$

$$f_z = f_c/f_{z(\text{table})} \qquad (4.9b)$$

where f_c is the cutoff frequency of the filter being designed.

EXAMPLE 4.4. Design a third-order, 0.1-dB Chebyshev high-pass filter with $f_c = 100$ Hz and high-frequency gain $H_0 = 20$ dB.

Solution. Table 4.1 indicates that we need a second-order high-pass section with $f_{01} = 100/1.300 = 76.92$ Hz and $Q_1 = 1.341$, and a first-order high-pass section with $f_{02} = 100/0.969 = 103.2$ Hz. As shown in Fig. 4.10, we implement the filter with a second-order unity-gain Sallen-Key high-pass stage, followed by a first-order high-pass stage with a high-frequency gain of 10 V/V.

Band-Pass Filter Design

EXAMPLE 4.5. Design a Butterworth band-pass filter with center frequency $f_0 = 1$ kHz, BW $= 100$ Hz, $A(f_0/2) = A(2f_0) \geq 60$ dB, and resonance gain $H_0 = 0$ dB.

Solution. Using the aforementioned FILDES program, we find that the given specifications can be met with a sixth-order filter having the following individual-stage

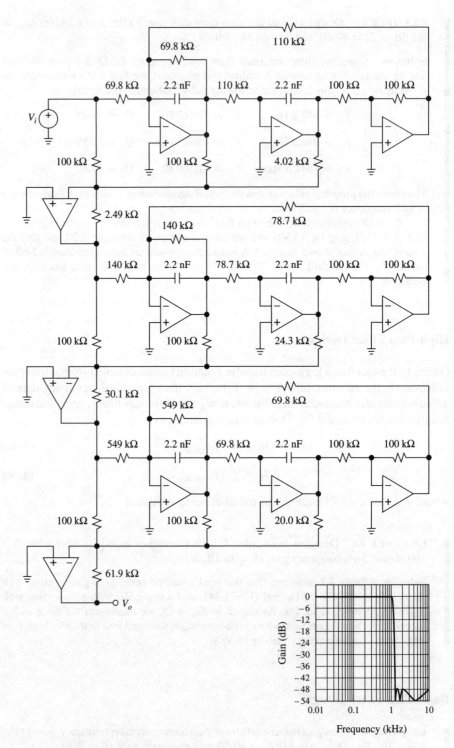

FIGURE 4.9
Sixth-order 0.1/40-dB elliptic low-pass filter.

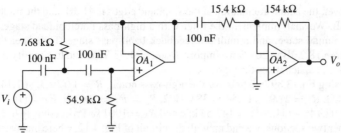

FIGURE 4.10
Third-order 0.1-dB Chebyshev high-pass filter of Example 4.4.

parameters:

$$f_{01} = 957.6 \text{ Hz} \qquad Q_1 = 20.02$$
$$f_{02} = 1044.3 \text{ Hz} \qquad Q_2 = 20.02$$
$$f_{03} = 1000.0 \text{ Hz} \qquad Q_3 = 10.0$$

Furthermore, the actual attenuation at 500 Hz and 2 kHz is 70.5 dB, and the midband gain is −12 dB, that is, 0.25 V/V. To raise it to 0 dB we shall impose $H_{0BP1} = H_{0BP2} = 2$ V/V, and $H_{0BP3} = 1$ V/V.

We shall implement the filter with three multiple-feedback band-pass sections equipped with input resistance attenuators. Retracing the steps of Example 3.15 we find the components of Fig. 4.11, where the resistances have been rounded off to 1% standard values, and the second leg of each attenuator has been made variable for tuning purposes. To tune a given section, apply an ac input at the desired resonance frequency of that section, and adjust its pot until the Lissajous figure changes from an ellipse to a straight segment.

EXAMPLE 4.6. Design an elliptic band-pass filter with $f_0 = 1$ kHz, passband = 200 Hz, stopband = 500 Hz, $A_{max} = 1$ dB, $A_{min} = 40$ dB, and $H_0 = 20$ dB.

Solution. The abovementioned FILDES program indicates that we need a sixth-order filter with the following individual-stage parameters:

$$f_{01} = 907.14 \text{ Hz} \qquad f_{z1} = 754.36 \text{ Hz} \qquad Q_1 = 21.97$$
$$f_{02} = 1102.36 \text{ Hz} \qquad f_{z2} = 1325.6 \text{ Hz} \qquad Q_2 = 21.97$$
$$f_{03} = 1000.0 \text{ Hz} \qquad \qquad \qquad \qquad \qquad Q_3 = 9.587$$

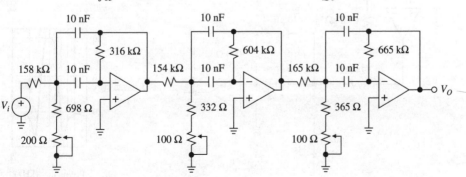

FIGURE 4.11
Sixth-order Butterworth band-pass filter.

Moreover, the actual attenuation at the stopband edges is 41 dB, and the midband gain is 18.2 dB. We shall implement the filter with a high-pass notch biquad stage, a low-pass notch biquad stage, and a multiple-feedback band-pass stage. To bolster the midband gain from 18.2 dB to 20 dB we impose $H_{0BP3} = 1.23$ V/V, and to simplify inventory we use 10-nF capacitances throughout.

Using Eq. (3.89) we find, for the high-pass notch, $R = 1/(2\pi \times 907.14 \times 10^{-8}) = 17.54$ kΩ, $R_1 = 21.97 \times 17.54 = 385.4$ kΩ, $R_2 = R_3 = 100$ kΩ, $R_4 = (100/21.97)907.14^2/(907.14^2 - 754.36^2) = 14.755$ kΩ, and $R_5 = 100$ kΩ. Proceeding in like manner for the other two sections, we end up with the circuit of Fig. 4.12, where the resistances have been rounded off to 1% standard values, and provisions have been made for frequency and Q tuning.

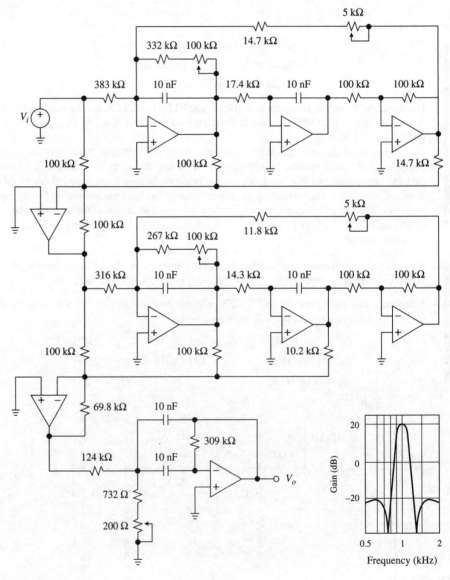

FIGURE 4.12
Sixth-order 1.0/40-dB elliptic band-pass filter.

Band-Reject Filter Design

EXAMPLE 4.7. A 0.1-dB Chebyshev band-reject filter is to be designed with notch frequency $f_z = 3600$ Hz, passband = 400 Hz, stopband = 60 Hz, $A_{max} = 0.1$ dB, and $A_{min} = 40$ dB. The circuit must have provision for frequency tuning of its individual stages.

Solution. The aforementioned FILDES program indicates that we need a sixth-order filter with the following individual-stage parameters:

$$f_{01} = 3460.05 \text{ Hz} \qquad f_{z1} = 3600 \text{ Hz} \qquad Q_1 = 31.4$$

$$f_{02} = 3745.0 \text{ Hz} \qquad f_{z2} = 3600 \text{ Hz} \qquad Q_2 = 31.4$$

$$f_{03} = 3600.0 \text{ Hz} \qquad f_{z3} = 3600 \text{ Hz} \qquad Q_3 = 8.72$$

Moreover, the actual stopband attenuation is 45 dB. This filter is readily designed using three biquad sections, namely, a high-pass notch, followed by a low-pass notch, followed by a symmetric notch (see Problem 4.13).

4.3
GENERALIZED IMPEDANCE CONVERTERS

Impedance converters are active RC circuits designed to simulate frequency-dependent elements such as inductances for use in active filter synthesis. Among the various configurations, one that has gained prominence is the *generalized impedance converter* (GIC) of Fig. 4.13, which can be used not only to simulate inductances, but also to synthesize frequency-dependent resistances.

To find the equivalent impedance Z seen looking into node A, we apply a test voltage V as in Fig. 4.14, we find the resulting current I, and then let $Z = V/I$.

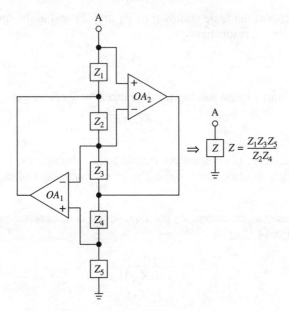

FIGURE 4.13
Generalized impedance converter (GIC).

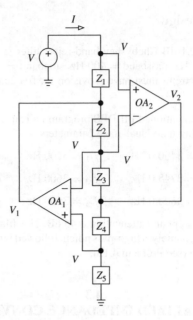

FIGURE 4.14
Finding the equivalent impedance
of a GIC toward ground.

Exploiting the fact that each op amp keeps $V_n = V_p$, we have labeled the voltages
at the input nodes of both op amps as V. By Ohm's law, we have

$$I = \frac{V - V_1}{Z_1}$$

Summing currents at the node common to Z_2 and Z_3 and at the node common to
Z_4 and Z_5 we obtain, respectively,

$$\frac{V_1 - V}{Z_2} + \frac{V_2 - V}{Z_3} = 0 \qquad \frac{V_2 - V}{Z_4} + \frac{0 - V}{Z_5} = 0$$

Eliminating V_1 and V_2, and solving for the ratio $Z = V/I$, we get

$$Z = \frac{Z_1 Z_3 Z_5}{Z_2 Z_4} \qquad (4.10)$$

Depending on the type of components we use for Z_1 through Z_5, we can configure
the circuit for various impedance types. The most interesting and useful ones are as
follows:

1. All Zs are resistances, except Z_2 (or Z_4), which is a capacitance. Letting $Z_2 =
   1/j\omega C_2$ in Eq. (4.10) gives

$$Z = \frac{R_1 R_3 R_5}{(1/j\omega C_2) R_4} = j\omega L \qquad (4.11a)$$

$$L = \frac{R_1 R_3 R_5 C_2}{R_4} \qquad (4.11b)$$

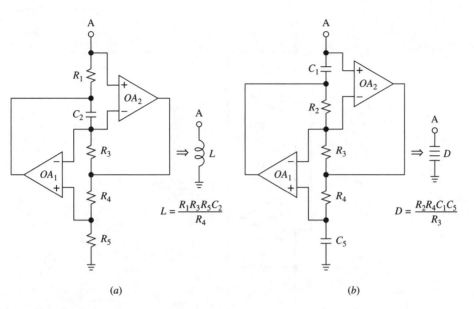

FIGURE 4.15
(*a*) Inductance simulator and (*b*) *D*-element realization.

indicating that the circuit simulates a *grounded inductance*. This is depicted in Fig. 4.15*a*. If desired, this inductance can be adjusted by varying one of the resistances, say, R_5.

2. All Zs are resistances, except for Z_1 and Z_5, which are capacitances. Letting $Z_1 = 1/j\omega C_1$ and $Z_5 = 1/j\omega C_5$ in Eq. (4.10) gives

$$Z = \frac{(1/j\omega C_1)R_3(1/j\omega C_5)}{R_2 R_4} = -\frac{1}{\omega^2 D} \qquad (4.12a)$$

$$D = \frac{R_2 R_4 C_1 C_5}{R_3} \qquad (4.12b)$$

The circuit now simulates a *grounded frequency-dependent negative resistance* (grounded FDNR). Since a capacitance produces a voltage proportional to the integral of the current, the FDNR (or *D element,* as it is often called) can be viewed as an element that integrates current twice. Its GIC realization and circuit symbol are shown in Fig. 4.15*b*, and its application will be illustrated shortly. The *D* element can be adjusted by varying one of the resistances.

Figure 4.16 shows another popular realization of the *D* element (see Problem 4.17). Needless to say, the simulated impedances can be no better than the resistances, capacitances, and op amps utilized in their simulation. For good results, use metal-film resistors and NPO ceramic capacitors for temperature stability and polypropylene capacitors for high-*Q* performance. And use a dual op amp with sufficiently fast dynamics (see Section 6.5).

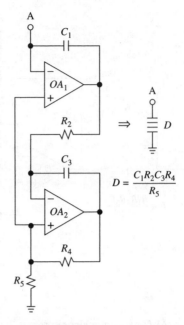

FIGURE 4.16
Alternative D-element realization.

Synthesis Using Grounded Inductances

A popular GIC application is the realization of inductorless filters starting from passive *RLC* filter prototypes. To this end we first design an *RLC* filter meeting the given specifications, then we replace its inductances with synthetic inductances realized with GICs. Note, however, that this direct one-to-one replacement is applicable only if the inductances in the prototype are of the grounded type.

A classic example is offered by the band-pass prototype of Fig. 4.17*a*. This is a band-pass filter because low-frequency signals are shunted by L, high-frequency signals are shunted by C, and intermediate-frequency signals are passed because of resonance. Once the filter specifications are known, we first find a set of *RLC* values meeting the specification, then we replace the original inductance with a GIC inductance simulator to end up with a circuit containing only resistances and capacitances. The result is the *dual-amplifier band-pass* (DABP) filter of Fig. 4.17*b*.

EXAMPLE 4.8. In the circuits of Fig. 4.17 specify component values for a band-pass response with $f_0 = 2$ kHz and $Q = 25$.

Solution. The *RLC* prototype gives $V_o/V_i = (Z_C \parallel Z_L)/(R + Z_C \parallel Z_L)$, $Z_C = 1/(j\omega C)$, $Z_L = j\omega L$. Expanding and collecting gives $V_o/V_i = H_{\text{BP}}$, with

$$\omega_0 = 1/\sqrt{LC} \qquad Q = R\sqrt{C/L}$$

Choose $C = 10$ nF so that $L = 1/(2\pi f_0)^2 C = 1/[(2\pi \times 2 \times 10^3)^2 \times 10^{-8}] = 0.633$ H, and $R = Q/\sqrt{C/L} = 199$ kΩ (use 200 kΩ, 1%).

Next, specify the components for the GIC. To simplify inventory, use equal capacitances and equal resistances. Thus, $C_2 = C = 10$ nF. By Eq. (4.11*b*), $R_1 = R_3 = R_4 = R_5 = \sqrt{L/C_2} = \sqrt{0.633/10^{-8}} = 7.96$ kΩ (use 7.87 kΩ, 1%).

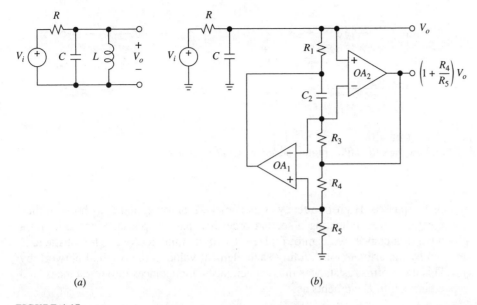

(a) (b)

FIGURE 4.17
(a) Passive band-pass filter prototype and (b) active realization using an inductance simulator.

We observe that the node designated as V_o in Fig. 4.17b is prone to external loading. This can be avoided by using the response from the low-impedance output of OA_2, where it is available with a gain of $1 + R_4/R_5$. With equal resistances, this gain is 2 V/V. If unity gain is desired, replace R with a voltage divider, in the manner of Example 3.9.

Using L as given in Eq. (4.11b), we have $\omega_0 = \sqrt{R_4/R_1 R_3 R_5 C_2 C}$ and $Q = R\sqrt{R_4 C/R_1 R_3 R_5 C_2}$, so the sensitivities are

$$S_{R_1}^{\omega_0} = S_{C_2}^{\omega_0} = S_{R_3}^{\omega_0} = -S_{R_4}^{\omega_0} = S_{R_5}^{\omega_0} = S_C^{\omega_0} = -1/2$$

$$S_R^Q = 1 \qquad S_{R_1}^Q = S_{C_2}^Q = S_{R_3}^Q = -S_{R_4}^Q = S_{R_5}^Q = -S_C^Q = -1/2$$

These fairly low values are typical of filters based on the ladder simulation approach. If the circuit is implemented with $C_2 = C$ and $R_5 = R_4 = R_3 = R_1$, then $\omega_0 = 1/RC$ and $Q = R/R_1$. This resistance spread compares quite favorably with that of the multiple-feedback band-pass filter, which is $4Q^2$. Moreover, the DABP filter is easily tuned since R_1 (or R_3) adjusts ω_0, and R adjusts Q. Even though the circuit uses two op amps instead of one, it has been proved[5] that if their open-loop frequency characteristics are matched, as is usually the case with dual packages, the op amps tend to compensate for each other's deficiencies, resulting in fairly small deviations of Q and ω_0 from their design values. Owing to these advantages, the DABP filter is a highly recommended configuration.

Synthesis Using FDNRs

As an example of active filter synthesis using FDNRs, consider the RLC filter of Fig. 4.18a. Low-frequency signals make L a short circuit and C an open, so these

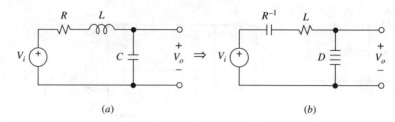

(a)   (b)

FIGURE 4.18
Low-pass RLC filter prototype and its CRD equivalent.

signals are passed. High-frequency signals make L an open and C a short, so they are rejected twice, indicating a second-order low-pass response. Since L is not a grounded inductance, we cannot replace it with a simulated one. This obstacle is avoided by the artifice[6] of dividing each element value in the original network by $j\omega$. This transforms resistances into capacitances, inductances into resistances, and capacitances into D elements as

$$\frac{R}{j\omega} \rightarrow \frac{1}{j\omega R^{-1}} \qquad \text{(capacitance of value } R^{-1}\text{)} \qquad (4.13a)$$

$$\frac{j\omega L}{j\omega} \rightarrow L \qquad \text{(resistance of value } L\text{)} \qquad (4.13b)$$

$$\frac{1/j\omega C}{j\omega} \rightarrow -\frac{1}{\omega^2 C} \qquad \text{(} D \text{ element of value } C\text{)} \qquad (4.13c)$$

The transformed network is shown in Fig. 4.18b. It can be proven[3] that dividing all the impedances of a network by the same factor yields a modified network with the same transfer function as the original one. Consequently, the modified circuit of Fig. 4.18b not only retains the original response, but is also realizable with a GIC since the transformation has eliminated the floating inductance while creating a grounded D element, which is amenable to GIC simulation.

EXAMPLE 4.9. Using the RLC circuit of Fig. 4.18a as a prototype, design a GIC low-pass filter with $f_0 = 1$ kHz and $Q = 5$.

Solution. The transformed circuit of Fig 4.18b gives, by the voltage divider formula, $V_o/V_i = (-1/\omega^2 C)/(1/j\omega R^{-1} + L - 1/\omega^2 C) = 1/(1 - \omega^2 LC + j\omega RC) = H_{\text{LP}}$, where

$$\omega_0 = 1/\sqrt{LC} \qquad Q = \sqrt{L/C}/R$$

Let the capacitance denoted as R^{-1} be 100 nF. Since $Q\omega_0 = 1/RC$, the value of the D element is $R^{-1}/Q\omega_0 = (100 \times 10^{-9})/(5 \times 2\pi \times 10^3) = 10^{-11}/\pi \ \text{s}^2/\Omega$. Finally, the resistance denoted as L is $1/\omega_0^2 C = 1/[(2\pi \times 10^3)^2 \times 10^{-11}/\pi] = 7.958$ kΩ (use 8.06 kΩ, 1%).

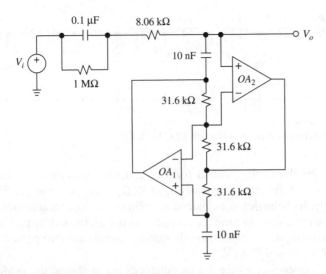

FIGURE 4.19
Low-pass filter using an FDNR.

Next, specify the components of the GIC, using equal components to simplify inventory. Let $C_1 = C_2 = 10\,\text{nF}$. By Eq. (4.12b), $R_2 = R_3 = R_4 = D/C_2 C_5 = (10^{-11}/\pi)/(10^{-8})^2 = 31.83\,\text{k}\Omega$ (use $31.6\,\text{k}\Omega$, 1%). The circuit is shown in Fig. 4.19.

Remark. In order to provide a dc path for the tiny inverting-input bias current of OA_2, a resistive termination is required. This is performed by the 1-MΩ resistance, whose large value will have little effect on filter performance over the frequency range of interest. A good choice for the op amps is a FET-input dual op amp. To avoid output loading, a buffer can be used.

4.4
DIRECT DESIGN

The interstage isolation properties of cascaded filters, while desirable from the viewpoint of modularity, render the overall response particularly sensitive to individual-stage parameter variations stemming from tolerance, thermal drift, and aging. Of special concern are the high-Q stages, where even a small component variation in a single stage may drastically alter the response of the entire cascade. On the other hand, it has long been recognized that RLC filters of the doubly terminated ladder type enjoy the lowest sensitivities to component variations. The ladder structure is a tightly coupled system in which sensitivity is spread out over its elements as a group rather than being confined to specific ones. Sensitivity considerations, together with the wealth of knowledge available in the area of passive RLC network synthesis, provide the motivation for the ladder simulation approach.

The starting point is a passive RLC ladder prototype, which is designed using suitable filter tables or computer programs. The filter is then realized in active form by replacing its inductors with simulated ones, that is, with active circuits specifically designed to simulate inductance behavior. The resulting active network retains the low-sensitivity advantages of its RLC prototype, a feature that makes it suited to applications with stringent specifications.

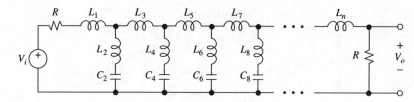

FIGURE 4.20
Doubly terminated series-resonant *RLC* ladder.

Figure 4.20 shows the general form of a doubly terminated, series-resonant *RLC* ladder, one of the most frequently used *RLC* prototypes in active filter synthesis. Physically, its behavior is explained as follows. At low frequencies, where the inductances act as shorts and the capacitances as opens, the ladder provides a direct signal path from input to output. Low-frequency signals are thus passed, and the dc gain is $R/(R + R) = 1/2$ V/V.

At high frequencies, where the capacitances act as shorts, the ladder becomes predominantly inductive and, as such, it presents considerable impedance to signal propagation. Thus, high-frequency signals are attenuated.

At intermediate frequencies, due to the series resonance of the *LC* elements in each leg, the response exhibits a series of notches, one for each leg. Consequently, the ladder provides a low-pass response with notches, or an *elliptic low-pass response*. The order n of the response is twice the number of legs plus 1, that is, n is odd. If the rightmost inductance is eliminated, then n is decreased by 1 and becomes even. Suppressing the inductances in the ladder legs eliminates the resonances and therefore the stopband notches. This reduced ladder version, referred to as *all-pole ladder,* can then be used to synthesize the Butterworth, Chebyshev, or Bessel responses.

The individual element values are tabulated in filter handbooks[7] or can be calculated by computer.[8] Table 4.2 shows an example of tabulated data. Element values are normalized for a cutoff frequency of 1 rad/s and 1 Ω; however, they are readily adapted to actual frequencies by dividing all reactive elements by the desired cutoff frequency ω_c of the filter.

Low-Pass Filter Design

As is, the ladder of Fig. 4.20 is not amenable to GIC simulation because it contains floating inductances. This obstacle is overcome by applying the $1/j\omega$ transformation discussed in Section 4.3, after which the resistances are changed to capacitances, the inductances to resistances, and the capacitances to D elements. The resulting *CRD* structure is then simulated with grounded FDNRs.

In addition to applying the $1/j\omega$ transformation, we must also frequency-scale the normalized ladder elements to achieve the desired cutoff frequency, and impedance-scale the resulting elements to obtain practical values in the final circuit. The three steps can be carried out at once via the following transformations:[3]

$$C_{\text{new}} = 1/k_z R_{\text{old}} \tag{4.14a}$$

$$R_{j(\text{new})} = (k_z/\omega_c)L_{j(\text{old})} \tag{4.14b}$$

$$D_{j(\text{new})} = (1/k_z\omega_c)C_{j(\text{old})} \tag{4.14c}$$

TABLE 4.2

183

SECTION 4.4
Direct Design

Element values for doubly terminated Butterworth and Chebyshev low-pass filters

	Butterworth low-pass element values (1-rad/s bandwidth)									
n	L_1	C_2	L_3	C_4	L_5	C_6	L_7	C_8	L_9	C_{10}
2	1.414	1.414								
3	1.000	2.000	1.000							
4	0.7654	1.848	1.848	0.7654						
5	0.6180	1.618	2.000	1.618	0.6180					
6	0.5176	1.414	1.932	1.932	1.414	0.5176				
7	0.4450	1.247	1.802	2.000	1.802	1.247	0.4450			
8	0.3902	1.111	1.663	1.962	1.962	1.663	1.111	0.3902		
9	0.3473	1.000	1.532	1.879	2.000	1.879	1.532	1.000	0.3473	
10	0.3129	0.9080	1.414	1.782	1.975	1.975	1.782	1.414	0.9080	0.3129

	Chebyshev low-pass element values (1-rad/s bandwidth)								
n	L_1	C_2	L_3	C_4	L_5	C_6	L_7	C_8	R_2
				0.1-dB ripple					
2	0.84304	0.62201							0.73781
3	1.03156	1.14740	1.03156						1.00000
4	1.10879	1.30618	1.77035	0.81807					0.73781
5	1.14681	1.37121	1.97500	1.37121	1.14681				1.00000
6	1.16811	1.40397	2.05621	1.51709	1.90280	0.86184			0.73781
7	1.18118	1.42281	2.09667	1.57340	2.09667	1.42281	1.18118		1.00000
8	1.18975	1.43465	2.11990	1.60101	2.16995	1.58408	1.94447	0.87781	0.73781
				0.5-dB ripple					
3	1.5963	1.0967	1.5963						1.0000
5	1.7058	1.2296	2.5408	1.2296	1.7058				1.0000
7	1.7373	1.2582	2.6383	1.3443	2.6383	1.2582	1.7373		1.0000
				1.0-dB ripple					
3	2.0236	0.9941	2.0236						1.0000
5	2.1349	1.0911	3.0009	1.0911	2.1349				1.0000
7	2.1666	1.1115	3.0936	1.1735	3.0936	1.1115	2.1666		1.0000

where $j = 1, 2, \ldots, n$. Here the element values of the *RLC* prototype are referred to as *old,* those of the transformed *RCD* network as *new,* ω_c is the desired cutoff frequency, and k_z is an appropriate impedance-scaling factor to be chosen on the basis of the desired impedance levels in the final circuit.

EXAMPLE 4.10. Figure 4.21 (top) shows a ladder prototype suitable for the GIC realization of a sharp-cutoff smoothing filter for audio D/A converters.[9] The ladder provides a seventh-order Cauer low-pass response with $A_{max} = 0.28$ dB and $A_{min} = 60$ dB at $f_s = 1.252 f_c$. Design an FDNR implementation with $f_c = 15$ kHz.

Solution. First convert the normalized *RLC* prototype to a *CRD* network. Let us arbitrarily decide to use 1-nF capacitances throughout. Since the 1-Ω resistances must change to 1-nF capacitances, Eq. (4.14a) gives $k_z = 1/10^{-9} = 10^9$.

By Eq. (4.14b), $R_{1(new)} = L_{1(old)} \times 10^9/(2\pi \times 15 \times 10^3) = 1.367 \times 10{,}610 = 14.5$ kΩ, and $R_{2(new)} = 0.1449 \times 10{,}610 = 1.54$ kΩ; by Eq. (4.14c), $D_{2(new)} = C_{2(old)}/(10^9 \times 2\pi \times 15 \times 10^3) = 1.207 \times 1.061 \times 10^{-14} = 1.281 \times 10^{-14}$ s^2/Ω. Applying similar transformations to the other elements, we end up with the *CRD* network of Fig. 4.21 (center).

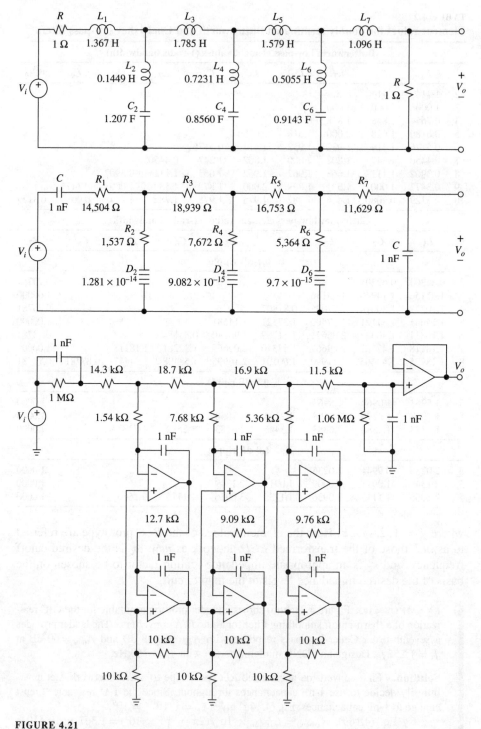

FIGURE 4.21
Seventh-order 0.28/60-dB elliptic low-pass filter. Top: normalized *RLC* prototype; center: *CRD* equivalent, with *D*-element values in square seconds per ohm; and bottom: active realization using FDNRs.

Finally, we find the elements in the FDNRs. Let us use the FDNRs of Fig. 4.16 with $R_4 = R_5 = 10$ kΩ. Then, Eq. (4.12b) gives, for the leftmost FDNR, $R_2 = D/C^2 = 1.281 \times 10^{-14}/(10^{-9})^2 = 12.81$ kΩ (use 12.7 kΩ, 1%). We similarly calculate the remaining FDNRs and end up with the realization of Fig. 4.21 (bottom), where the resistances have been rounded off to 1% standard values.

Note again the use of the 1-MΩ resistance at the input end to provide a dc path for the op amps. To ensure a dc gain of $\frac{1}{2}$ V/V, this resistance must be counterbalanced by a 1.061-MΩ resistance at the output. To avoid loading problems, an output buffer is used. The FDNRs can be implemented with dual FET-input op amps. If desired, each FDNR can be tuned by adjusting one of its resistances.

High-Pass Filter Design

The ladder network of Fig. 4.20, though of the low-pass type, can also serve as prototype for high-pass filters provided we replace the inductances with capacitances, the capacitances with inductances, and use reciprocal element values to maintain frequency normalization at 1 rad/s. The transformed network provides a response with characteristics reciprocal to the original one, that is, a Cauer high-pass response with a cutoff frequency of 1 rad/s and with notches located at reciprocal positions of the low-pass prototype. Suppressing the capacitances in the legs of the transformed ladder eliminates the stopband notches. This reduced ladder can then be used to synthesize the Butterworth, Chebyshev, or Bessel responses.

In either case, the inductances of the transformed ladder are of the grounded type and as such can be simulated with GICs. After the low-pass to high-pass transformation, the elements must be frequency-scaled to the desired cutoff frequency and impedance-scaled to practical impedance levels. The three steps can be carried out at once via the following transformations:[3]

$$R_{\text{new}} = k_z / R_{\text{old}} \tag{4.15a}$$

$$C_{j(\text{new})} = 1/(k_z \omega_c L_{j(\text{old})}) \tag{4.15b}$$

$$L_{j(\text{new})} = k_z/(\omega_c C_{j(\text{old})}) \tag{4.15c}$$

where the meaning of the notation is similar to Eq. (4.14).

EXAMPLE 4.11. Design an elliptic high-pass filter with $f_c = 300$ Hz, $f_s = 150$ Hz, $A_{\max} = 0.1$ dB, and $A_{\min} = 40$ dB.

Solution. Using standard filter tables[7] or filter-design computer programs,[8] it is found that the specifications can be met with a fifth-order filter whose low-pass prototype has the element values of Fig. 4.22 (top). The actual attenuation at the edge of the stopband is $A(f_s) = 43.4$ dB.

Let us arbitrarily impose $R_{\text{new}} = 100$ kΩ, so $k_z = 10^5$, by Eq. (4.15a). Using Eq. (4.15b), $C_{1(\text{new})} = 1/(10^5 \times 2\pi \times 300 \times 1.02789) = 5.161$ nF. Using Eq. (4.15c), $L_{2(\text{new})} = 10^5/(2\pi \times 300 \times 1.21517) = 43.658$ H. Applying similar transformations to the other elements we end up with the high-pass ladder of Fig. 4.22 (center).

Finally, we find the elements in the GICs. Let $C = 10$ nF and impose equal resistances. Then, Eq. (4.11) requires, for the leftmost GIC, $R_1 = R_3 = R_4 = R_5 = \sqrt{L/C} = \sqrt{43.658/10^{-8}} = 66.07$ kΩ. Likewise, the resistances for the other GIC are found to be 75.32 kΩ. The final circuit is shown in Fig. 4.22 (bottom), where the resistances have been rounded off to 1% standard values. To avoid output loading, a voltage buffer can be used.

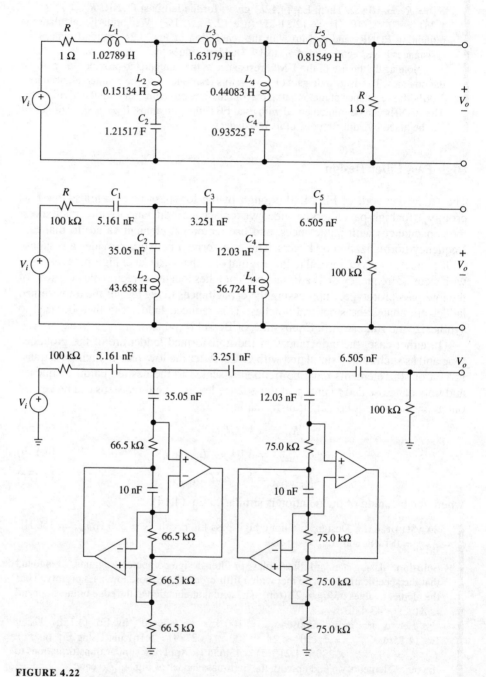

FIGURE 4.22
Fifth-order 0.1/40-dB elliptic high-pass filter. Top: normalized *RLC* prototype; center: high-pass equivalent; and bottom: active realization using simulated inductances.

THE SWITCHED CAPACITOR

The filters investigated so far, known as *continuous-time filters,* are characterized by the fact that H_0 and Q are usually controlled by component ratios and ω_0 is controlled by component products. Though ratios can easily be maintained with temperature and time by using devices with adequate tracking capabilities, products are inherently more difficult to control. Moreover, IC processes do not lend themselves to the fabrication of resistances and capacitances with the magnitudes (10^3 to 10^6 Ω and 10^{-9} to 10^{-6} F) and accuracies (1% or better) typically required in audio and instrumentation applications.

If filter functions are to coexist with digital functions on the same chip, filters must be realized with the components that are most natural to VLSI processes, namely, MOS transistors and small MOS capacitors. This constraint has led to the development of switched-capacitor (SC) filters,[10-12] which simulate resistors by periodically operating MOS capacitors with MOSFET switches, and produce time constants that depend on capacitance ratios rather than *R-C* products.

To illustrate, let us start with the basic MOSFET-capacitor arrangement of Fig. 4.23*a*. The transistors are *n*-channel enhancement types, characterized by a low channel resistance (typically $<10^3$ Ω) when the gate voltage is high, and a high resistance (typically $>10^{12}$ Ω) when the gate voltage is low. With an off/on ratio this high, a MOSFET can be regarded for all practical purposes as a switch. If the gates are driven with nonoverlapping out-of-phase clock signals of the type in Fig. 4.23*b*, the transistors will conduct on alternate half cycles, thus providing a single-pole double-throw (SPDT) switch function with break-before-make characteristics.

Referring to the symbolic switch representation of Fig. 4.24*a* and assuming $V_1 > V_2$, we observe that flipping the switch to the left charges C to V_1, and flipping it to the right discharges C to V_2. The net charge transfer from V_1 to V_2 is $\Delta Q = C(V_1 - V_2)$. If the switch is flipped back and forth at a rate of f_{CK} cycles per second, the charge transferred in 1 second from V_1 to V_2 defines an average current

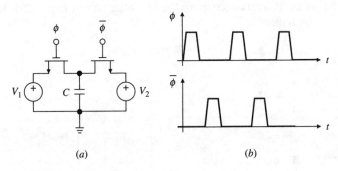

(a)   (b)

FIGURE 4.23
Switched capacitor using a MOSFET SPDT switch, and clock drive for the MOSFETs.

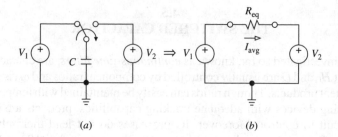

FIGURE 4.24
Resistance simulation using a switched capacitor.

$$I_{avg} = f_{CK} \times \Delta Q, \text{ or}$$

$$I_{avg} = C f_{CK}(V_1 - V_2) \qquad (4.16)$$

Note that charge is flowing in packets rather than continuously. However, if f_{CK} is made sufficiently higher than the highest-frequency components of V_1 and V_2, the process can be regarded as continuous, and the switch-capacitor combination can be modeled with an equivalent resistance

$$R_{eq} = \frac{V_1 - V_2}{I_{avg}} = \frac{1}{C f_{CK}} \qquad (4.17)$$

The model is depicted in Fig. 4.24b. Let us investigate how such a resistance can be used to implement, what by now has proved to be the workhorse of active filters, namely, the integrator.

SC Integrators

As we know, the *RC* integrator of Fig. 4.25a yields $H(j\omega) = -1/(j\omega/\omega_0)$, where the unity-gain frequency is given by

$$\omega_0 = \frac{1}{R_1 C_2} \qquad (4.18)$$

Replacing R_1 by an SC resistance gives the SC integrator of Fig. 4.25b. If the input frequency ω is such that

$$\omega \ll \omega_{CK} \qquad (4.19)$$

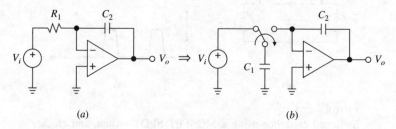

FIGURE 4.25
Converting an *RC* integrator to an SC integrator.

where $\omega_{CK} = 2\pi f_{CK}$, then current flow from V_i to the summing node can be regarded as continuous, and ω_0 is found by substituting R_{eq} into Eq. (4.18),

$$\omega_0 = \frac{C_1}{C_2} f_{CK} \qquad (4.20)$$

This expression reveals three important features that hold for SC filters in general, not just for SC integrators:

1. There are no resistors. This is highly desirable from the viewpoint of IC fabrication, since monolithic resistors are plagued by large tolerances and thermal drift, and also take up precious chip area. Switches, on the other hand, are implemented with MOSFETs, which are the basic ingredients of VLSI technology and occupy very little chip area.
2. The characteristic frequency ω_0 depends on capacitance ratios, which are much easier to control and maintain with temperature and time than $R\text{-}C$ products. With present technology, ratio tolerances as low as 0.1% are readily achievable.
3. The characteristic frequency ω_0 is proportional to the clock frequency f_{CK}, indicating that SC filters are inherently of the programmable type. Varying f_{CK} will shift the response up or down the frequency spectrum. If, on the other hand, a fixed and stable characteristic frequency is desired, f_{CK} can be generated with a quartz crystal oscillator.

Equation (4.20) also shows that by judicious choice of the values of f_{CK} and the C_1/C_2 ratio, it is possible to avoid undesirably large capacitances even when low values of ω_0 are desired. For instance, with $f_{CK} = 1$ kHz, $C_1 = 1$ pF, and $C_2 = 15.9$ pF, the SC integrator gives $f_0 = (1/2\pi)(1/15.9)10^3 = 10$ Hz. An RC integrator with the same f_0 could be implemented, for instance, with $R_1 = 1.59$ MΩ and $C_2 = 10$ nF. Fabricating these components monolithically and maintaining the value of their product within 0.1% would be unrealistic. Current SC filters use capacitances in the range of 0.1 pF to 100 pF, with the 1-pF to 10-pF range being the most common. The upper limit is dictated by die area considerations, and the lower limit by parasitic capacitances of the SC structure.

To minimize the effect of parasitic capacitances and also increase circuit versatility, practical SC integrators are implemented with SPDT switch pairs, in the manner of Fig. 4.26. In Fig. 4.26a, flipping the switches down discharges C_1 to zero, and flipping the switches up charges C_1 to V_i. Current will thus flow into the summing junction of the op amp if $V_i > 0$, and out if $V_i < 0$, indicating that the integrator is of the inverting type.

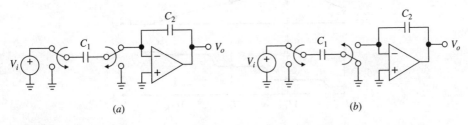

(a)   (b)

FIGURE 4.26
Inverting and noninverting SC integrators.

Changing the phase of one of the switches yields the circuit of Fig. 4.26b. With the switches in the positions shown, the left plate of C_1 is at V_i and the right plate at 0 V. Commutating both switches will discharge C_1 to 0 V, thus pulling charge out of the summing junction if $V_i > 0$, and pushing current into the junction if $V_i < 0$. A simple phase rewiring of the two MOSFETs making up one of the switches inverts the direction of I_{avg}, resulting in an SC integrator of the noninverting type. We shall exploit the availability of this type of integrator in the next sections.

Practical Limitations of SC Filters

There are some important limitations that we need be aware of when applying SC filters.[10] First, there are limits on the permissible range of f_{CK}. The upper limit is dictated by the quality of the MOS switches and the speed of the op amps. Taking 10 pF as a typical switched capacitance and 1 kΩ as a typical resistance of a closed MOS switch, we observe that the time constant is on the order of $10^3 \times 10^{-11} = 10$ ns. Considering that to charge a capacitance to within 0.1% of its final voltage takes about seven time constants ($e^{-7} \cong 10^{-3}$), it follows that the minimum time interval between consecutive switch commutations is on the order of 10^2 ns. This also happens to be the typical time it takes for the step response of a MOS op amp to settle within 0.1% of its final value. Consequently, the upper limit for f_{CK} is in the megahertz range.

The lower practical limit for f_{CK} is dictated by the leakage of open MOS switches and the input bias currents of op amps, both of which tend to discharge the capacitors and, hence, to destroy the accumulated information. At room temperature these currents are in the picoampere range. Assuming a maximum acceptable droop of 1 mV across a capacitor of 10 pF, we have $f_{CK} \geq (1 \text{ pA})/[(10 \text{ pF}) \times (1 \text{ mV})] = 10^2$ Hz. In summary, the permissible clock range is typically 10^2 Hz $< f_{CK} < 10^6$ Hz.

The other important limitation of SC filters stems from their discrete-time rather than continuous-time operation. This is evidenced in Fig. 4.27, which shows the

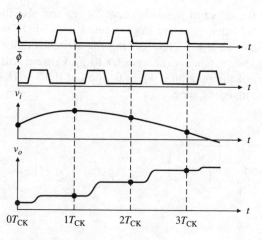

FIGURE 4.27
Noninverting SC integrator waveforms.

input and output waveforms for the noninverting integrator of Fig. 4.26a. Time has been divided into equal intervals according to the clock period T_{CK}. Referring to the actual circuit, we observe that ϕ pulses charge C_1 to v_i, while $\bar{\phi}$ pulses pull the charge accumulated in C_1 out of C_2, causing a step increase in v_o. Because of nonzero switch resistance, this step is gradual.

Letting n denote an arbitrary clock period, we have $v_o[nT_{CK}] = v_o[(n-1)T_{CK}] + \Delta Q[(n-1)T_{CK}]/C_2$, or

$$v_o[nT_{CK}] = v_o[(n-1)T_{CK}] + \frac{C_1}{C_2}v_i[(n-1)T_{CK}] \tag{4.21}$$

where $\Delta Q[(n-1)T_{CK}] = C_1 v_i[(n-1)T_{CK}]$ denotes the charge accumulated by C_1 during the previous ϕ pulse. Equation (4.21) represents a discrete time sequence relating input and output values, which have been emphasized with dots. A well-known Fourier transform property states that delaying a signal by one clock period T_{CK} is equivalent to multiplying its Fourier transform by $\exp(-j\omega T_{CK})$. Taking the Fourier transforms of both sides of Eq. (4.21) gives

$$V_o(j\omega) = V_o(j\omega)e^{-j\omega T_{CK}} + \frac{C_1}{C_2}V_i(j\omega)e^{-j\omega T_{CK}} \tag{4.22}$$

Collecting, solving for the ratio $H(j\omega) = V_o(j\omega)/V_i(j\omega)$, and using Euler's identity $\sin\alpha = (e^\alpha - e^{-\alpha})/2j$, we finally obtain the *exact* transfer function of the SC noninverting integrator,

$$H(j\omega) = \frac{1}{j\omega/\omega_0} \times \frac{\pi\omega/\omega_{CK}}{\sin(\pi\omega/\omega_{CK})} \times e^{-j\pi\omega/\omega_{CK}} \tag{4.23}$$

where $\omega_0 = (C_1/C_2)f_{CK}$ and $\omega_{CK} = 2\pi/T_{CK} = 2\pi f_{CK}$.

We observe that in the limit $\omega/\omega_{CK} \to 0$ we obtain the familiar integrator function $H(j\omega) = 1/(j\omega/\omega_0)$, confirming that as long as $\omega_{CK} \gg \omega$, the SC process can be regarded as a continuous-time process. Writing $H(j\omega) = [1/(j\omega/\omega)] \times \epsilon_m \times \exp(-j\epsilon_\phi)$ indicates that in general the SC process introduces a *magnitude error* $\epsilon_m = (\pi\omega/\omega_{CK})/[\sin(\pi\omega/\omega_{CK})]$ and a *phase error* $\epsilon_\phi = -\pi\omega/\omega_{CK}$. The effect of these errors is illustrated in the linear plots of Fig. 4.28 for a noninverting integrator with $\omega_0 = \omega_{CK}/10$.

The ideal magnitude and phase responses are $|H| = 1/(\omega/\omega_0)$ and $\sphericalangle H = -90°$. The SC integrator deviation increases with ω until, for $\omega = \omega_{CK}$, the magnitude error becomes infinite and phase undergoes polarity reversal. These results are consistent with well-known sampled-data principles, stating that the effect of sampling a function of time at the rate of f_{CK} samples per second is a replication of its frequency spectrum at integral multiples of f_{CK}.

For $\omega \ll \omega_{CK}$, the effect of the magnitude error is similar to the effect of component tolerance or drift in ordinary *RC* integrators. As such, it may not be detrimental, especially if the performance requirements are not stringent. To contain this error within tolerable limits, the useful frequency range is limited to a couple of decades below ω_{CK}.

The effect of the phase error, however, is critical since it may cause Q enhancement or even instability. One method of compensating for this error is by alternating the clock phasing of consecutive integrators,[10] as we shall see in Section 4.6.

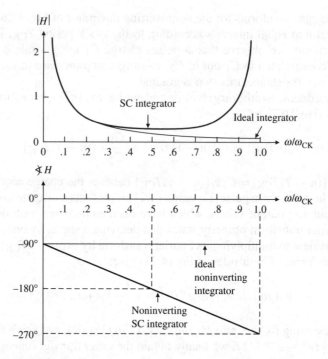

FIGURE 4.28
Magnitude and phase responses of a noninverting SC integrator
for the case $\omega_0 = \omega_{CK}/10$.

4.6
SWITCHED-CAPACITOR FILTERS

Switched-capacitor filters are based on the integrator configurations of the previous section. As in the case of continuous-time filters, two popular approaches to SC filter synthesis are the cascade approach and the ladder simulation approach.

Dual-Integrator-Loop Filters

A dual-integrator-loop SC filter can be synthesized by replacing the resistors of a continuous-time prototype with SC equivalents. Figure 4.29 shows the SC implementation of the popular biquad topology of Fig. 3.36. Here OA_2 is a lossless noninverting integrator, a function that requires only one op amp when implemented in SC form. We thus have, for $\omega \ll \omega_{CK}$,

$$V_{LP} = \frac{1}{j\omega/\omega_0} V_{BP}$$

where $\omega_0 = (C_1/C_2) f_{CK}$, by Eq. (4.20). The op amp OA_1 forms a lossy inverting integrator, whose equivalent feedback resistance, simulated by C_3 and the associated switch, sets the value of Q. By Eq. (4.17), this resistance is $R_Q = 1/C_3 f_{CK}$. With the input switches in the position shown, the leftmost capacitance C_1 is charged to $V_{LP} - V_i$. Flipping the switches down transfers the charge $\Delta Q = C_1(V_{LP} - V_i)$

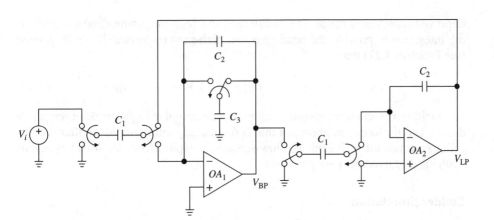

FIGURE 4.29
SC biquad filter.

into the summing junction of OA_1, so the corresponding average current is $I_1 = C_1 f_{CK}(V_{LP} - V_i)$. Summing currents at this junction gives, for $\omega \ll \omega_{CK}$,

$$C_1 f_{CK}(V_{LP} - V_i) + C_3 f_{CK} V_{BP} + j\omega C_2 V_{BP} = 0$$

Substituting $V_{LP} = V_{BP}/(j\omega/\omega_0)$ and collecting gives $V_{BP}/V_i = H_{0BP}H_{BP}$ and $V_{LP}/V_i = H_{0LP}H_{LP}$, where H_{LP} and H_{BP} are the standard second-order low-pass and band-pass responses, and

$$\omega_0 = \frac{C_1}{C_2} f_{CK} \qquad Q = \frac{C_1}{C_3} \qquad H_{0BP} = Q \qquad H_{0LP} = 1 \text{ V/V} \qquad (4.24)$$

EXAMPLE 4.12. Assuming $f_{CK} = 100$ kHz in the circuit of Fig. 4.29, specify suitable capacitances for a Butterworth low-pass response with $f_0 = 1$ kHz and a total capacitance of 100 pF or less.

Solution. We have $C_2/C_1 = f_{CK}/(2\pi f_0) = 15.9$ and $C_3/C_1 = 1/Q = \sqrt{2}$. Choose $C_1 = 1$ pF, $C_2 = 15.9$ pF, and $C_3 = 1.41$ pF.

The realization of Fig. 4.29 is by no means unique, nor is it necessarily the best. In fact (see Problem 4.26), its capacitance spread increases with Q to the point of making this arrangement unfeasible. Figure 4.30 shows an SC realization with

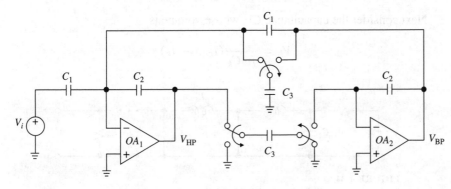

FIGURE 4.30
SC biquad filter with improved capacitance spread.

improved capacitance ratios. The circuit uses an integrator/summer and a noninverting integrator to provide the band-pass and high-pass responses. It can be proved (see Problem 4.27) that

$$\omega_0 = \frac{C_3}{C_2} f_{CK} \qquad Q = \frac{C_2}{C_1} \qquad H_{0BP} = -1 \text{ V/V} \qquad H_{0HP} = \frac{-1}{Q} \qquad (4.25)$$

In the next section we investigate the cascade design of higher-order filters using dual-integrator loops, an approach that is particularly attractive when filter specifications are not too stringent. For low-sensitivity applications, the direct synthesis methods discussed next are preferable.

Ladder Simulation

Direct SC filter synthesis uses SC integrators to simulate passive *RLC* ladders. Since it retains the low-sensitivity advantages of ladders, this approach is preferable when filter specifications are more stringent. One of the most frequently used structures is the doubly terminated all-pole ladder of Fig. 4.31, which can be configured for Butterworth, Chebyshev, or Bessel responses, the order *n* coinciding with the number of reactive elements present. As we know, the required component values are tabulated in filter handbooks or can be calculated by computer.

We observe that the ladder is a repetitive structure of *LC* pairs of the type of Fig. 4.32*a*. The inductance current is

$$I_{k-1} = \frac{V_{k-1} - V_k}{j\omega L_{k-1}}$$

SC integrators are inherently voltage-processing blocks, so to make the above function amenable to SC implementation, we use the artifice of multiplying both sides by a scaling resistance R_s, which converts the current I_{k-1} to a voltage $V'_{k-1} = R_s I_{k-1}$, or

$$V'_{k-1} = \frac{1}{j\omega/\omega_{L_{k-1}}}(V_{k-1} - V_k) \qquad \omega_{L_{k-1}} = \frac{1}{L_{k-1}/R_s}$$

This integration is implemented with an *L*-integrator of the type also shown in Fig. 4.32*b*. By Eq. (4.20), its capacitances must satisfy $C_0/C_{L_{k-1}} = \omega_{L_{k-1}}$, or

$$C_{L_{k-1}}/C_0 = (L_{k-1}/R_s) f_{CK} \qquad (4.26)$$

Next consider the capacitance C_k, whose voltage is

$$V_k = \frac{1}{j\omega C_k}(I_{k-1} - I_k)$$

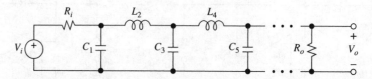

FIGURE 4.31
Doubly terminated all-pole *RLC* ladder.

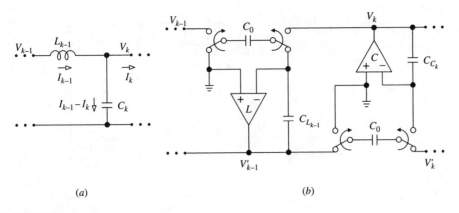

(a)   (b)

FIGURE 4.32
LC ladder section and its realization in SC form.

Multiplying numerator and denominator by R_s to convert the currents I_{k-1} and I_k to the voltages $V'_{k-1} = R_s I_{k-1}$ and $V'_k = R_s I_k$, we obtain

$$V_k = \frac{1}{j\omega/\omega_{C_k}}(V'_{k-1} - V'_k) \qquad \omega_{C_k} = \frac{1}{R_s C_k}$$

This integration is implemented with a *C*-integrator of the type also shown in Fig. 4.32*b*. By Eq. (4.20), its capacitances must satisfy $C_0/C_{C_k} = \omega_{C_k}$, or

$$C_{C_k}/C_0 = R_s C_k f_{CK} \qquad (4.27)$$

We thus conclude that if the conditions of Eqs. (4.26) and (4.27) are met, the SC integrators of Fig. 4.32*b* will simulate the *LC* pair of Fig. 4.32*a*. The by-product variables V'_{k-1} and V'_k need not concern us as they are internal to the circuit.

To complete the ladder simulation, we also need SC equivalents of the terminating resistors. This is readily achieved by making the first and last SC integrators of the lossy type. Denoting the capacitances simulating these resistances as C_{R_i} and C_{R_o}, we have

$$C_{R_i}/C_0 = R_i/R_s \qquad C_{R_o}/C_0 = R_o/R_s \qquad (4.28)$$

For simplicity we can let $R_i = R_o = R_s = 1\,\Omega$, after which we get $C_{R_i} = C_{R_o} = C_0$.

As an example, Fig. 4.33 shows a fifth-order low-pass SC filter. Since the leftmost reactive element in the ladder prototype is a capacitance, the leftmost integrator is a *C*-integrator. The rightmost integrator is either a *C*-integrator or an *L*-integrator, depending on whether the order *n* of the filter is odd (as in the example) or even. Moreover, the leftmost and rightmost integrators must be of the lossy type to simulate the terminating resistances. Note also the alternation in the switch phases of adjacent integrators in order to minimize the effects of sampling delays, as mentioned at the end of the previous section.

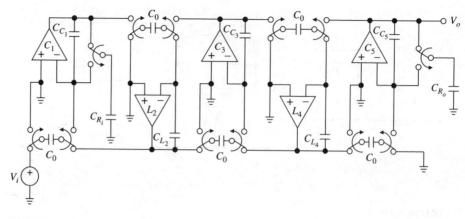

FIGURE 4.33
Fifth-order SC low-pass filter.

Direct Synthesis of Low-Pass Filters

Although the element values of Table 4.2 refer to all-pole ladders with an inductance as the leftmost reactive element, they are readily adapted to ladders with a capacitance as the leftmost reactive element, provided we change the column headings from L_1, $C_2, L_3, C_4, \ldots$ to $C_1, L_2, C_3, L_4, \ldots$. Since the tabulated RLC values are normalized for a cutoff frequency of 1 rad/s, they must be frequency-scaled before Eqs. (4.26) and (4.27) can be applied. As discussed in Section 4.4, this requires dividing all reactive values by the cutoff frequency ω_c. Assuming $R_s = 1 \Omega$, the above equations become

$$C_{C_k}/C_0 = (C_k/\omega_c)f_{CK} \qquad C_{L_k}/C_0 = (L_k/\omega_c)f_{CK} \qquad (4.29)$$

where C_k and L_k represent the kth normalized reactive element values of the filter prototype.

EXAMPLE 4.13. In the circuit of Fig. 4.33, specify capacitances for a fifth-order Butterworth low-pass response with $f_c = 1$ kHz and $f_{CK} = 100$ kHz.

Solution. From Table 4.2 we find the following normalized element values: $C_1 = C_5 = 0.618$, $C_3 = 2.000$, and $L_2 = L_4 = 1.618$. Using Eq. (4.29), we obtain $C_{C_1}/C_0 = 0.618 \times 10^5/2\pi 10^3 = 9.836$, $C_{L_2}/C_0 = 1.618 \times 10^5/2\pi 10^3 = 25.75$, etc., and $C_{R_i}/C_0 = C_{R_o}/C_0 = 1$. A set of capacitances meeting the above constraints is $C_{R_i} = C_{R_o} = C_0 = 1$ pF, $C_{C_1} = C_{C_5} = 9.84$ pF, $C_{L_2} = C_{L_4} = 25.75$ pF, and $C_{C_3} = 31.83$ pF.

Direct Synthesis of Band-Pass Filters

The low-pass ladder of Fig. 4.31 can also serve as the prototype for other responses. For example, replacing each capacitance by an inductance and vice versa, and using reciprocal element values, the ladder becomes of the high-pass type. Replacing each inductance in the original ladder by a parallel LC pair yields a low-pass response with notches, that is, an elliptic low-pass response. Replacing each capacitance in the original ladder by a parallel LC pair and each inductance by a series LC pair yields a band-pass response. Replacing each capacitance in the original ladder by a series LC pair and each inductance by a parallel LC pair yields a band-reject response.

Once the ladder has been transformed, we write circuit equations for each node and branch, and use resistance scaling to convert currents to voltages to render the equations amenable to SC simulation. We shall illustrate the procedure for the band-pass case.

The ladder of Fig. 4.34 (top) is a second-order low-pass prototype. If we replace its capacitance by a parallel LC pair and its inductance by a series LC pair, we end up with the fourth-order band-pass ladder of Fig. 4.34 (center). RLC filter theory states[3] that to achieve a center frequency of 1 rad/s with a normalized bandwidth BW, the element values of the transformed ladder must be related to those of the low-pass prototype as

$$C_{1(\text{new})} = C_{1(\text{old})}/\text{BW} \qquad L_{1(\text{new})} = \text{BW}/C_{1(\text{old})} \qquad (4.30a)$$

$$C_{2(\text{new})} = \text{BW}/L_{2(\text{old})} \qquad L_{2(\text{new})} = L_{2(\text{old})}/\text{BW} \qquad (4.30b)$$

where the low-pass elements are referred to as *old*, and the band-pass ones as *new*. The former are tabulated in filter handbooks.

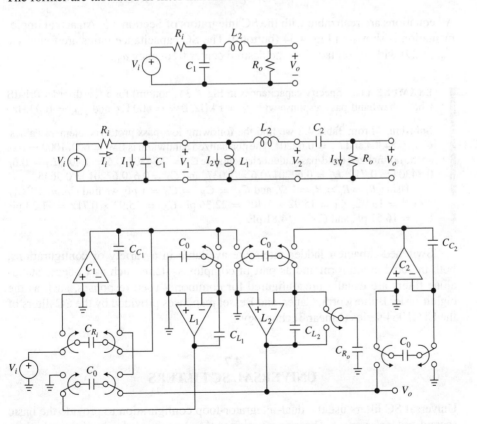

FIGURE 4.34
Fourth-order band-pass filter. Top: second-order RLC prototype; center: fourth-order RLC equivalent; and bottom: SC realization.

Let us now develop the necessary circuit equations. By KCL, $V_1 = (1/j\omega C_1) \times (I_i - I_2 - I_3)$. Multiplying numerator and denominator by the scaling resistance R_s to convert currents to voltages as $V_i' = R_s I_i$, $V_2' = R_s I_2$, and $V_3' = R_s I_3$, we obtain

$$V_1 = \frac{1}{j\omega/\omega_{C_1}}(V_i' - V_2' - V_3') \qquad \omega_{C_1} = \frac{1}{R_s C_1}$$

By Ohm's law, $I_2 = V_1/j\omega L_1$. Multiplying both sides by R_s gives

$$V_2' = \frac{1}{j\omega/\omega_{L_1}}V_1 \qquad \omega_{L_1} = \frac{1}{L_1/R_s}$$

By Ohm's law, $I_3 = (V_1 - V_2)/j\omega L_2$, or

$$V_3' = \frac{1}{j\omega/\omega_{L_2}}(V_1 - V_2) \qquad \omega_{L_2} = \frac{1}{L_2/R_s}$$

By KVL, $V_2 = V_o + I_3/j\omega C_2$, or

$$V_2 = V_o + \frac{1}{j\omega/\omega_{C_2}}V_3' \qquad \omega_{C_2} = \frac{1}{R_s C_2}$$

All equations are realizable with the SC integrators of Section 4.5. An actual implementation is shown in Fig. 4.34 (bottom). The SC capacitance ratios are found via Eq. (4.29) with ω_c replaced by the desired center frequency ω_0.

EXAMPLE 4.14. Specify capacitances in Fig. 4.34 (bottom) for a fourth-order 0.1-dB Chebyshev band-pass response with $f_0 = 1$ kHz, BW = 600 Hz, and $f_{CK} = 100$ kHz.

Solution. From Table 4.2 we find the following low-pass prototype element values: $C_1 = 0.84304$ and $L_2 = 0.62201$. The normalized bandwidth is BW = 600/1000 = 0.6, so the normalized band-pass ladder elements are $C_1 = 0.84304/0.6 = 1.405$, $L_1 = 0.6/0.84304 = 0.712$, $L_2 = 0.62201/0.6 = 1.037$, and $C_2 = 0.6/0.62201 = 0.9646$.

    Using $R_i = R_o = R_s = 1\ \Omega$, and $C_{R_i} = C_{R_o} = C_0 = 1$ pF, we find $C_{C_1} = 10^5 C_1/2\pi 10^3 = 15.92$, $C_1 = 15.92 \times 1.405 = 22.36$ pF, $C_{L_1} = 15.92 \times 0.712 = 11.33$ pF, $C_{L_2} = 16.51$ pF, and $C_{C_2} = 14.81$ pF.

Switched-capacitor ladder filters are available in a variety of configurations, both in stand-alone form and as part of complex systems such as Codecs. Stand-alone filters are usually preconfigured for commonly used responses, such as the eighth-order Butterworth, Cauer, and Bessel responses provided by the SC filters of the LTC1064 series (Linear Technology).

4.7
UNIVERSAL SC FILTERS

Universal SC filters use the dual-integrator-loop configuration to provide the basic second-order responses. These responses can then be cascaded to implement higher-order filters. Two popular and well-documented examples are the LTC1060 (Linear Technology) and the MF10 (National Semiconductor).

The MF10 Universal SC Filter

The MF10 filter, whose block diagram is shown in Fig. 4.35, consists of two dual-integrator-loop sections, each equipped with an uncommitted op amp to add versatility and facilitate cascading. Each section can independently be configured for the low-pass, band-pass, high-pass, notch, and all-pass responses by means of external resistances. Though these resistances could have been synthesized on-chip using SC techniques, placing them under the control of the user increases the versatility of the circuit. Furthermore, filter parameters are made to depend on resistance ratios, rather than on absolute values, to take advantage of component tracking.

The integrators are of the noninverting type, with the transfer function

$$H(jf) = \frac{1}{jf/f_1} \tag{4.31}$$

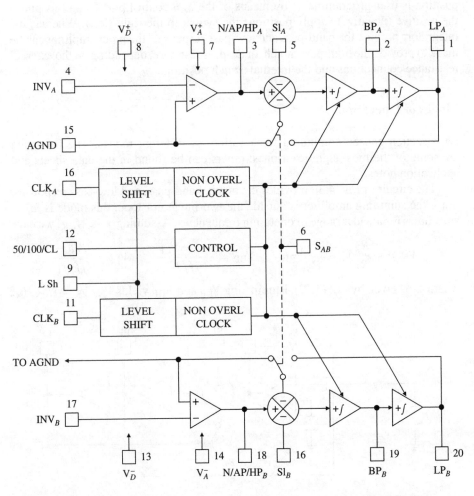

FIGURE 4.35
Block diagram of the MF10 universal monolithic dual SC filter. (Courtesy of National Semiconductor.)

where f_1 is the *integration unity-gain frequency*, and

$$f_1 = \frac{f_{CK}}{100} \quad \text{or} \quad \frac{f_{CK}}{50} \tag{4.32}$$

depending on the voltage level applied at the 50/100/CL frequency-ratio programming pin: tying it to ground enables the 100 ratio, and tying it to the positive supply enables the 50 ratio.

In general, the characteristic frequency f_0 of a section coincides with the unity-gain frequency f_1 of its integrators; however, connecting an external resistance between the LP and INV pins shifts f_0 away from f_1 by an amount controlled by an external resistance ratio. This feature is useful in cascade design, where the resonance frequency of each stage must be set independently while all sections are controlled by the same clock frequency f_{CK}.

For additional flexibility, an internal programming switch is provided, whose position is user-programmable by means of the S_{AB} control pin. Tying this pin to the positive (negative) supply positions the switch to the right (left). Whereas the integrators provide the band-pass and low-pass responses, the input amplifier can be made to provide the high-pass, notch, or all-pass response, depending on the external resistance connections and the internal switch position.

Modes of Operation

Each section can be configured for a variety of different modes. The following are some of the most significant ones; others can be found in the data sheets and application notes.[4]

The circuit of Fig. 4.36 provides the notch, band-pass, and low-pass responses. Since the summing amplifier is outside the two-integrator loop, this mode is faster and allows for a wider range of operating frequencies. Assuming $f \ll f_{CK}$, we have

$$V_N = -\frac{R_2}{R_1}V_i - \frac{R_2}{R_3}V_{BP} \qquad V_{BP} = \frac{V_N - V_{LP}}{jf/f_1} \qquad V_{LP} = \frac{V_{BP}}{jf/f_1}$$

where f_1 is given by Eq. (4.32). Eliminating V_{LP} and V_{BP} yields $V_N/V_i = H_{0N}H_N$,

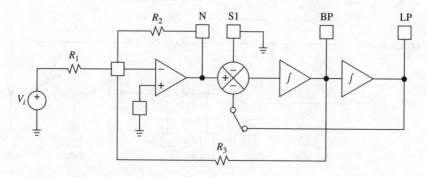

FIGURE 4.36
Basic MF10 connection for the notch, band-pass, and low-pass responses. (Courtesy of National Semiconductor.)

$V_{\text{BP}}/V_i = H_{0\text{BP}}H_{\text{BP}}$, and $V_{\text{LP}}/V_i = H_{0\text{LP}}H_{\text{LP}}$, where

$$f_z = f_0 = f_1 \qquad Q = R_3/R_2 \qquad (4.33a)$$

$$H_{0\text{N}} = H_{0\text{LP}} = -R_2/R_1 \qquad H_{0\text{BP}} = -R_3/R_1 \qquad (4.33b)$$

Note that in this mode both f_z and f_0 coincide with the integration unity-gain frequency $f_1 = f_{\text{CK}}/100(50)$.

EXAMPLE 4.15. In the circuit of Fig. 4.36, specify suitable resistances for a band-pass response with $f_0 = 1$ kHz, BW = 50 Hz, and $H_{0\text{BP}} = 20$ dB.

Solution. Impose $R_3/R_2 = Q = f_0/\text{BW} = 10^3/50 = 20$, and $R_3/R_1 = |H_{0\text{BP}}| = 10^{20/20} = 10$. Pick $R_1 = 20$ kΩ, $R_2 = 10$ kΩ, $R_3 = 200$ kΩ, $f_{\text{CK}} = 100$ kHz, and tie the 50/100/CL pin to ground to make $f_1 = f_{\text{CK}}/100$.

The mode of Fig. 4.37 is referred to as the *state-variable* mode because it provides the high-pass, band-pass, and low-pass responses by direct consecutive integrations. One can readily show (see Problem 4.29) that, if $f \ll f_{\text{CK}}$, the circuit gives $V_{\text{HP}}/V_i = H_{0\text{HP}}H_{\text{HP}}$, $V_{\text{BP}}/V_i = H_{0\text{BP}}H_{\text{BP}}$, and $V_{\text{LP}}/V_i = H_{0\text{LP}}H_{\text{LP}}$, where

$$f_0 = f_1\sqrt{R_2/R_4} \qquad Q = (R_3/R_2)\sqrt{R_2/R_4} \qquad (4.34a)$$

$$H_{0\text{HP}} = -R_2/R_1 \qquad H_{0\text{BP}} = -R_3/R_1 \qquad H_{0\text{LP}} = -R_4/R_1 \quad (4.34b)$$

A distinctive feature of this mode is that f_0 can be tuned independently of the integration unity-gain frequency $f_1 = f_{\text{CK}}/100(50)$ by means of the ratio R_2/R_4, a feature we shall exploit in cascade design. Since the summing amplifier is now inside the integrator loop, the frequency limitations of its open-loop gain are likely to cause Q enhancement, a subject that will be addressed in Chapter 6. Suffice it to say here that this enhancement can be compensated by placing a phase-lead capacitance on the order of 10 pF to 100 pF in parallel with R_4.

By combining the high-pass and low-pass responses with an external summing amplifier, in the familiar manner of Fig. 4.38, the notch response is synthesized. One

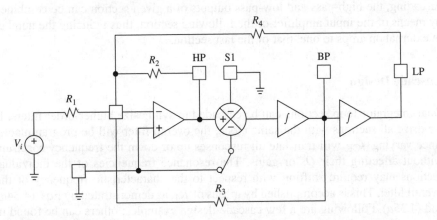

FIGURE 4.37
State-variable configuration using the MF10. (Courtesy of National Semiconductor.)

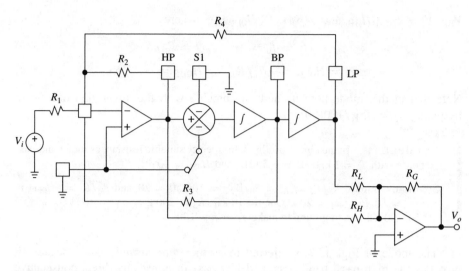

FIGURE 4.38
MF10 with an external op amp to provide the notch response. (Courtesy of National Semi-conductor.)

can readily show (see Problem 4.29) that, if $f \ll f_{CK}$, the circuit gives

$$\frac{V_o}{V_i} = H_{0N} \frac{1 - (f/f_z)^2}{1 - (f/f_0)^2 + (jf/f_0)/Q}$$

$$f_0 = f_1\sqrt{R_2/R_4} \qquad f_z = f_1\sqrt{R_H/R_L} \qquad Q = R_3/R_2\sqrt{R_2/R_4} \qquad (4.35a)$$

$$H_{0N} = \frac{R_G R_4}{R_L R_1} \qquad H_{0HP} = -\frac{R_2}{R_1} \qquad H_{0BP} = -\frac{R_3}{R_1} \qquad H_{0LP} = -\frac{R_4}{R_1} \qquad (4.35b)$$

Depending on how one specifies the various resistances, the notch can be of the high-pass or low-pass type, and it can be utilized in the synthesis of Cauer filters. When cascading, the high-pass and low-pass outputs of a given section can be combined by means of the input amplifier of the following section, thus reducing the number of external op amps to one, that of the last section.

Cascade Design

Dual-integrator-loop sections can be cascaded to synthesize higher-order filters. If we drive all sections with the same clock, the overall filter will be programmable, since varying f_{CK} will translate all responses up or down the frequency spectrum without affecting their Qs or gains. The resonance frequencies of the individual sections may require shifting with respect to the characteristic frequency of the overall filter. This is accomplished by means of R_4, as demonstrated by Eqs. (4.34a) and (4.35a). Following are a few cascade-design examples; others can be found in the manufacturer's literature.[4]

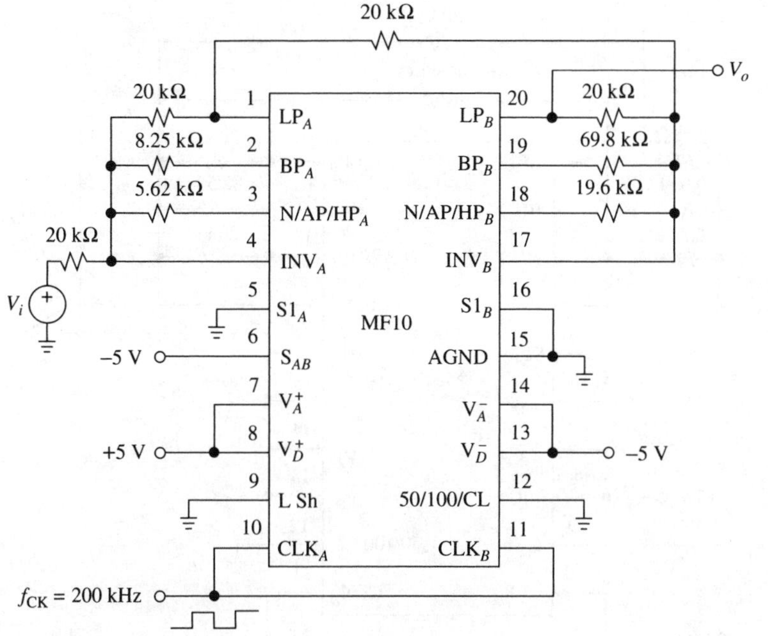

The circuit diagram shows the MF10 filter with the following connections:

- 20 kΩ (top, feedback)
- 20 kΩ, pin 1 LP$_A$; pin 20 LP$_B$, 20 kΩ → V_o
- 8.25 kΩ, pin 2 BP$_A$; pin 19 BP$_B$, 69.8 kΩ
- 5.62 kΩ, pin 3 N/AP/HP$_A$; pin 18 N/AP/HP$_B$, 19.6 kΩ
- 20 kΩ, pin 4 INV$_A$; pin 17 INV$_B$
- V_i source; pin 5 S1$_A$; pin 16 S1$_B$
- pin 6 S$_{AB}$; MF10 ; pin 15 AGND
- −5 V ; pin 7 V$_A^+$; pin 14 V$_A^-$
- +5 V ; pin 8 V$_D^+$; pin 13 V$_D^-$ → −5 V
- pin 9 L Sh ; pin 12 50/100/CL
- pin 10 CLK$_A$; pin 11 CLK$_B$
- f_{CK} = 200 kHz → CLK

FIGURE 4.39
Fourth-order, 1-dB, 2-kHz Chebyshev low-pass filter.

EXAMPLE 4.16. Using the MF10 filter, design a fourth-order 1.0-dB Chebyshev low-pass filter with $f_c = 2$ kHz and 0-dB dc gain.

Solution. Let $f_{CK} = 100 f_c = 200$ kHz. From Table 4.1 we find that the following individual-stage parameters are needed: $f_{01} = 0.993 f_c$, $Q_1 = 3.559$, $f_{02} = 0.529 f_c$, and $Q_2 = 0.785$. Let section A be the low-Q stage, and section B the high-Q stage, and let us cascade them in this order to maximize filter dynamics. Since both sections require frequency shifting with respect to f_c, we use the configuration of Fig. 4.37.

By Eq. (4.34), $\sqrt{R_{2A}/R_{4A}} = 0.529$, or $R_{2A}/R_{4A} = 0.2798$; $R_{3A}/R_{2A} = Q_A/\sqrt{R_{2A}/R_{4A}} = 0.785/0.529 = 1.484$; $R_{4A}/R_{1A} = |H_{0LPA}| = 1$. Let $R_{1A} = R_{4A} = 20$ kΩ. Then, $R_{2A} = 5.60$ kΩ and $R_{3A} = 8.30$ kΩ. Likewise, we find $R_{1B} = R_{4B} = 20$ kΩ, $R_{2B} = 19.7$ kΩ, and $R_{3B} = 70.7$ kΩ. The final circuit is shown in Fig. 4.39, where the resistances have been rounded off to 1% standard values. For optimum performance, bypass the power supplies with 0.1-μF disk capacitors right at the supply pins.

EXAMPLE 4.17. Design an elliptic low-pass filter meeting the following specifications: $f_c = 1$ kHz, $f_s = 2$ kHz, $A_{max} = 1.0$ dB, $A_{min} = 50$ dB, and 0-dB dc gain.

Solution. The aforementioned FILDES program indicates that we need a fourth-order filter with the following individual-stage parameters:

$$f_{01} = 0.5650 \text{ kHz} \qquad f_{z1} = 2.1432 \text{ kHz} \qquad Q_1 = 0.8042$$

$$f_{02} = 0.9966 \text{ kHz} \qquad f_{z2} = 4.9221 \text{ kHz} \qquad Q_2 = 4.1020$$

Moreover, the actual attenuation at 2 kHz is 51.9 dB.

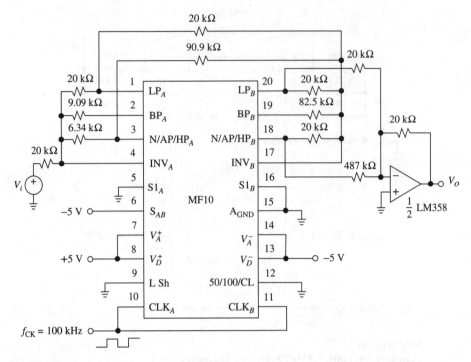

FIGURE 4.40
Fourth-order, 1-dB, 1-kHz elliptic low-pass filter.

Use the notch configuration of Fig. 4.38 with $f_{CK} = 100 f_c = 100$ kHz. Design section A first. Let $R_{1A} = 20$ kΩ. Imposing $|H_{0LPA}| = 1$ V/V gives $R_{4A} = R_{1A} = 20$ kΩ. To achieve the given f_{01} we need $R_{2A}/R_{4A} = 0.5650^2$, or $R_{2A} = 6.384$ kΩ. To achieve the given Q_1 we need $R_{3A} = R_{2A}Q_1/\sqrt{R_{2A}/R_{4A}} = 6.384 \times 0.8042/0.5650 = 9.087$ kΩ. Let $R_{LA} = 20$ kΩ, so that to achieve the given f_{z1} we need $R_{HA}/R_{LA} = 2.1432^2$, or $R_{HA} = 91.87$ kΩ.

Now design section B using the input amplifier of section B to combine the high-pass and low-pass responses of section A. Imposing $|H_{0LPB}| = 1$ V/V gives $R_{4B} = R_{LA} = 20$ kΩ. Repeating similar calculations we obtain $R_{2B} = 19.86$ kΩ, $R_{3B} = 81.76$ kΩ, $R_{LB} = 20$ kΩ, and $R_{HB} = 484.5$ kΩ. The last notch requires an external op amp with $R_G = R_{LB} = 20$ kΩ to ensure a 0-dB dc gain. The final circuit is shown in Fig. 4.40, where the resistances have been rounded off to 1% standard values.

PROBLEMS

4.1 Filter approximations

4.1 (*a*) Find n for a low-pass Butterworth filter with $A_{max} = 1$ dB, $A_{min} = 20$ dB, and $\omega_s/\omega_c = 1.2$. (*b*) Find the actual value of $A(\omega_s)$. (*c*) Find A_{max} so that $A(\omega_s) = 20$ dB exactly.

4.2 Using Eq. (4.5), find n for a low-pass Chebyshev response with the same specifications as the Butterworth response of Example 4.1.

4.3 Using Eq. (4.6), find the passband frequencies at which the gain of a seventh-order 0.5-dB Chebyshev filter exhibits its peaks and valleys, as well as the gain at $2\omega_c$, $10\omega_c$.

4.4 (*a*) Sketch the magnitude plots of the Butterworth and Chebyshev responses for $n = 5$ and $A_{\max} = 1$ dB. (*b*) Compare the attenuations provided at $\omega = 2\omega_c$.

4.5 The normalized third-order Butterworth low-pass response is $H(s) = 1/(s^3 + 2s^2 + 2s + 1)$. (*a*) Verify that it satisfies Eq. (4.4) with $\epsilon = 1$. (*b*) Show that if $k_1 = 0.14537$ and $k_2 = 2.5468$, the single-op-amp filter of Fig. P4.5 implements the third-order Butterworth response with $\omega_c = 1/RC(k_1 k_2)^{1/3}$. (*c*) Specify components for $f_c = 1$ kHz.

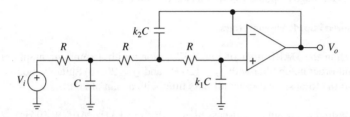

FIGURE P4.5

4.2 Cascade design

4.6 The normalized fourth-order Butterworth low-pass response can be factored as $H(s) = [s^2 + s(2 - 2^{1/2})^{1/2} + 1]^{-1} \times [s^2 + s(2 + 2^{1/2})^{1/2} + 1]^{-1}$. (*a*) Verify that it satisfies the condition of Eq. (4.4) with $\epsilon = 1$. (*b*) Design a fourth-order Butterworth low-pass filter with $f_c = 880$ Hz and $H_0 = 0$ dB.

4.7 A drawback of the implementation of Fig. 4.7 is its high capacitance spread, especially in the high-Q stage. This can be avoided by using $K > 1$. Redesign the filter so that the capacitance spread is kept below 10 while still ensuring 0-dB dc gain.

4.8 The smoothing filter of Fig. 4.7 is adequate for moderate performance requirements. Ultra-high fidelity audio applications require lower passband ripple and even sharper cutoff characteristics. For a 40-kHz sampling rate, these demands can be met[9] with a tenth-order 0.25-dB Chebyshev low-pass filter having $f_c = 15$ kHz. Such a filter provides $A(20\text{ kHz}) = 50.5$ dB with a -3-dB frequency of 15.35 kHz. The individual-stage parameters are: $f_{01} = 3.972$ kHz, $Q_1 = 0.627$, $f_{02} = 7.526$ kHz, $Q_2 = 1.318$, $f_{03} = 11.080$ Hz, $Q_3 = 2.444$, $f_{04} = 13.744$ kHz, $Q_4 = 4.723$, $f_{05} = 15.158$ kHz, and $Q_5 = 15.120$. Design such a filter and show your final circuit.

4.9 Using equal-component *KRC* sections, design a fifth-order Bessel low-pass filter with $f_c = 1$ kHz and $H_0 = 0$ dB.

4.10 Using *KRC* sections with $C_1 = C_2$ and $R_A = R_B$, design a seventh-order Butterworth low-pass filter with $f_c = 1$ kHz and $H_0 = 20$ dB.

4.11 Design a fifth-order 1.0-dB Chebyshev high-pass filter with $f_c = 360$ Hz and high-frequency gain H_0 adjustable from 0 to 20 dB. Use equal capacitances throughout.

4.12 A band-pass filter is to be designed with center frequency $f_0 = 300$ Hz, $A(300 \pm 10\text{ Hz}) = 3$ dB, $A(300 \pm 40\text{ Hz}) \geq 25$ dB, and resonance gain $H_0 = 12$ dB. These specifications[3] can be met with a sixth-order cascade filter having the following individual-stage parameters: $f_{01} = 288.0$ Hz, $Q_1 = 15.60$, $H_{0BP1} = 2.567$ V/V; $f_{02} = 312.5$ Hz, $Q_2 = 15.60$, $H_{0BP2} = 2.567$ V/V; $f_{03} = 300.0$ Hz, $Q_3 = 15.34$, $H_{0BP3} = 1.585$ V/V. Design such a filter using three individually tunable multiple-feedback stages.

4.13 Complete the design of Example 4.7, and show the final circuit.

4.14 Using the cascade-design approach, along with the FILDES program, design a 0.5-dB Chebyshev low-pass filter with a cutoff frequency of 10 kHz, a stopband frequency of 20 kHz, a minimum stopband attenuation of 60 dB, and a dc gain of 12 dB. Then, run a PSpice simulation of your circuit, showing the magnitude Bode plots of the individual-stage responses as well as the overall response.

4.3 Generalized impedance converters

4.15 (a) Using the DABP filter of Fig. 4.17b, along with a summing amplifier, design a second-order notch filter with $f_z = 120$ Hz and $Q = 20$. (b) Suitably modify the circuit of part (a) for a second-order all-pass filter with a gain of 20 dB.

4.16 It is desired to design a band-pass filter with $f_0 = 1$ kHz, $A(f_0 \pm 10$ Hz$) = 3$ dB, and $A(f_0 \pm 40$ Hz$) \geq 20$ dB. Such a filter[3] can be implemented by cascading two second-order band-pass stages with $f_{01} = 993.0$ Hz, $f_{02} = 1007$ Hz, and $Q_1 = Q_2 = 70.7$. Design an implementation using the DABP filter of Fig. 4.17b. Make provision for frequency tuning of the individual stages.

4.17 (a) Show that Eq. (4.12) holds also for the D element of Fig. 4.16. (b) Using this element, along with the RLC prototype of Fig. 4.18a, design a low-pass filter with $f_0 = 800$ Hz and $Q = 4$.

4.18 Provided $R = \sqrt{2L/C}$, the circuit of Fig. P4.18 yields a third-order, high-pass Butterworth response with -3-dB frequency $\omega_c = 1/\sqrt{2LC}$. (a) Specify suitable components for $f_c = 1$ kHz. (b) Convert the circuit to a GIC realization.

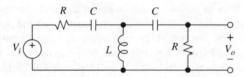

FIGURE P4.18

4.19 Show that the circuit of Fig. P4.19 simulates a grounded inductance $L = R_1 R_3 R_4 C / R_2$.

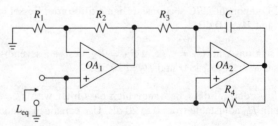

FIGURE P4.19

4.20 The circuit of Fig. P4.20 simulates an impedance Z_1 proportional to the reciprocal of Z_2. Called a *gyrator*, it finds application as an inductance by letting Z_2 be a capacitance.

(a) Show that $Z_1 = R^2/Z_2$. (b) Using this circuit, design a second-order band-pass filter with $f_0 = 1$ Hz, $Q = 10$, and zero output impedance. What is the resonance gain of your circuit?

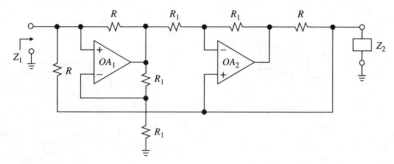

FIGURE P4.20

4.4 Direct design

4.21 It is desired to design a seventh-order 0.5-dB Chebyshev low-pass filter with a -3-dB frequency of 10 kHz. From Table 4.2 we find the *RLC* element values shown in Fig. P4.21. Using this ladder as a prototype, design an FDNR implementation.

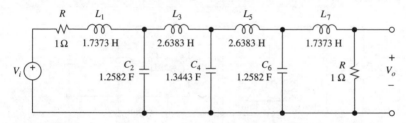

FIGURE P4.21

4.22 Using GICs and the information of Table 4.2, design a seventh-order 1-dB Chebyshev high-pass filter with $f_c = 500$ Hz.

4.5 The switched capacitor

4.23 Find a relationship beween V_o and V_1 and V_2 in the circuits of Fig. P4.23 for $f \ll f_{CK}$, and give the circuits' descriptive names.

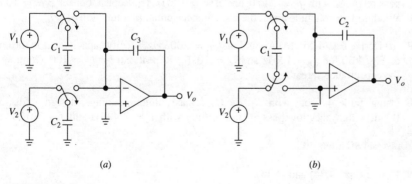

(a)   (b)

FIGURE P4.23

4.24 Find the transfer function of the circuits of Fig. P4.24 for $f \ll f_{CK}$, and give the circuits' descriptive names.

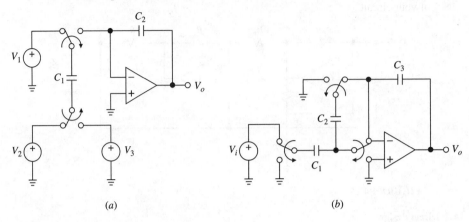

(a) (b)

FIGURE P4.24

4.25 (a) Assuming $f \ll f_{CK}$, show that the circuit of Fig. P4.25 gives the notch response. (b) Assuming $f_{CK} = 100$ kHz, specify capacitances for a 1-kHz notch with $Q = 10$.

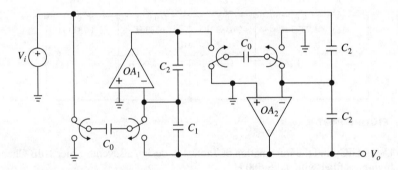

FIGURE P4.25

4.6 Switched-capacitor filters

4.26 (a) Assuming $f_{CK} = 250$ kHz in the circuit of Fig. 4.29, specify capacitances for a band-pass response with $f_0 = 2$ kHz and BW $= 1$ kHz. (b) Repeat, but for BW $= 100$ Hz. What do you conclude about the capacitance spread as a function of Q?

4.27 (a) Derive Eq. (4.25). (b) Assuming $f_{CK} = 200$ kHz, specify capacitances in the circuit of Fig. 4.30 for $f_0 = 1$ kHz and $Q = 10$. (c) Repeat, but for $Q = 100$. Comment on the capacitance spread.

4.28 Using Table 4.2, but with $C_1, L_2, C_3, \ldots$ as column headings, design a fifth-order 0.1-dB Chebyshev low-pass SC ladder filter with $f_c = 3.4$ kHz and $f_{CK} = 128$ kHz.

4.7 Universal SC filters

4.29 Derive Eqs. (4.34) and (4.35).

4.30 Consider the circuit obtained from that of Fig. 4.36 by removing R_1, lifting the S1 pin off ground, and applying V_i to S1, so that only two resistances are used. (*a*) Sketch the modified circuit and show that $V_{BP}/V_i = -QH_{BP}$ and $V_{LP}/V_i = -H_{LP}$, with f_0 and Q given by Eq. (4.33*a*). (*b*) Specify resistances for $f_0 = 500$ Hz and $Q = 10$.

4.31 The MF10 configuration of Fig. P4.31 provides the notch, band-pass, and low-pass responses, with the notch frequency f_z and the resonance frequency f_0 independently tunable by means of the resistance ratio R_2/R_4. Find expressions for f_0, f_z, Q, and the low-frequency gain.

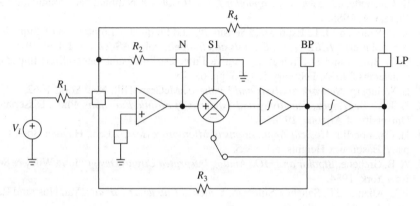

FIGURE P4.31

4.32 If in the circuit of Fig. P4.31 we lift the S1 input off ground and we connect it to V_i, with everything else remaining the same, then the output of the leftmost op amp changes from the notch to the all-pass response, with the numerator and denominator Qs separately adjustable. Assuming $f \ll f_{CK}$, find f_0, the numerator and denominator Qs, and the gain.

4.33 Using the MF10 in the configuration of Problem 4.30, design a minimum-component fourth-order Butterworth low-pass filter with $f_c = 1$ kHz and 20-dB dc gain.

4.34 A fourth-order 0.5-dB Chebyshev band-pass filter is to be designed with $f_0 = 2$ kHz and BW $= 1$ kHz. Using the FILDES program, it is found that the cascade realization requires the following individual-stage parameters: $f_{01} = 1554.2$ Hz, $f_{02} = 2473.6$ Hz, and $Q_1 = Q_2 = 2.8955$. Design such a filter using the MF10.

4.35 A fourth-order 1.0-dB Chebyshev notch filter with $f_0 = 1$ kHz is to be implemented by cascading two second-order sections with $f_{01} = 1.0414 f_0$, $f_{02} = 0.9602 f_0$, $f_{z1} = f_{z2} = f_0$, and $Q_1 = Q_2 = 20.1$. Design such a filter using the MF10.

4.36 It is desired to design a 0.5-dB elliptic band-pass filter with a center frequency $f_0 = 2$ kHz, a passband of 100 Hz, a stopband of 300 Hz, and a minimum stopband attenuation of 20 dB. Using the FILDES program, it is found that this filter requires a fourth-order implementation with the following individual-stage parameters: $f_{01} = 1.948$ kHz, $f_{z1} = 1.802$ kHz, $f_{02} = 2.053$ kHz, $f_{z2} = 2.220$ kHz, and $Q_1 = Q_2 = 29.48$. Moreover, the actual attenuation at the stopband edges is 21.5 dB. Design such a filter using the MF10 and an external op amp.

4.37 Using two MF10s, design an eighth-order 0.1-dB Chebyshev high-pass filter with $f_c = 500$ Hz and 0-dB high-frequency gain.

REFERENCES

1. L. P. Huelsman, *Active and Passive Analog Filter Design: An Introduction,* McGraw-Hill, New York, 1993.

2. K. Lacanette, "A Basic Introduction to Filters: Active, Passive, and Switched-Capacitor," Application Note AN-779, *Linear Applications Handbook,* National Semiconductor, Santa Clara, CA, 1994.

3. A. B. Williams and F. J. Taylor, *Electronic Filter Design Handbook: LC, Active, and Digital Filters,* 2d ed., McGraw-Hill, New York, 1988.

4. K. Lacanette, ed., *Switched Capacitor Filter Handbook,* National Semiconductor, Santa Clara, CA, 1985.

5. A. S. Sedra and J. L. Espinoza, "Sensitivity and Frequency Limitations of Biquadratic Active Filters," *IEEE Trans. Circuits and Systems,* vol. CAS-22, no. 2, Feb. 1975.

6. L. T. Bruton and D. Treleaven, "Active Filter Design Using Generalized Impedance Converters," *EDN,* February 5, 1973, pp. 68–75.

7. L. Weinberg, *Network Analysis and Synthesis,* McGraw-Hill, New York, 1962.

8. D. J. M. Baezlopez, *Sensitivity and Synthesis of Elliptic Functions,* Ph.D. Dissertation, University of Arizona, 1978.

9. H. Chamberlin, *Musical Applications of Microprocessors,* 2d ed., Hayden Book Company, Hasbrouck Heights, NJ, 1985.

10. A. B. Grebene, *Bipolar and MOS Analog Integrated Circuit Design,* John Wiley & Sons, New York, 1984.

11. P. E. Allen and E. Sanchez-Sinencio, *Switched Capacitor Circuits,* Van Nostrand Reinhold, New York, 1984.

12. R. Gregorian and G. C. Temes, *Analog MOS Integrated Circuits for Signal Processing,* John Wiley & Sons, New York, 1986.

5

STATIC OP AMP LIMITATIONS

5.1 Simplified Op Amp Circuit Diagram
5.2 Input Bias and Offset Currents
5.3 Low-Input-Bias-Current Op Amps
5.4 Input Offset Voltage
5.5 Low-Input-Offset-Voltage Op Amps
5.6 Input Offset-Error Compensation
5.7 Maximum Ratings
      Problems
      References
      Appendix 5A

If you have had the opportunity to experiment with the op amp circuits covered so far, you may have noted that as long as the op amps are operated at moderate frequencies and moderate dc gains there is generally a remarkable agreement between actual behavior and behavior predicted by the ideal op amp model. Increasing frequency or gain, however, is accompanied by a progressive degradation in performance because various limitations come into play. The objectives of the present and following chapters are to study these limitations systematically, to predict their effect on circuit performance, and to find possible cures.

One of the most serious limitations is the fact that the open-loop gain is high only from dc up to a few hertz, and it decreases with frequency thereafter, causing a progressive degradation in closed-loop performance. A related drawback is the fact that there is a limit to how fast an op amp can respond to sudden changes at the input. Frequency- and time-related limitations will be covered in Chapter 6.

Even if the operating frequencies are kept suitably low, other limitations come into play. Generally designated as *input-referred errors,* they are particularly notice-able in high-dc-gain applications. The most common ones are the *input bias current*

I_B, the *input offset current* I_{OS}, the *input offset voltage* V_{OS}, and the *ac noise densities* e_n and i_n. Related topics are the thermal drift $TC(V_{OS})$, the *common-mode* and the *power-supply rejection ratios* CMRR and PSRR, and *gain nonlinearity*. These non-idealities are generally impervious to the curative properties of negative feedback, and their effects must be alleviated on a one-to-one basis by other means. Finally, in order for an op amp to function properly, certain operating limits must be respected. These include the maximum operating temperature, supply voltage, and power dissipation, the input common-mode voltage range, and the output short-circuit current. Except for ac noise, which will be covered in Chapter 7, all these limitations are addressed in the present chapter.

However discouraging all this may sound, you should by no means relinquish your confidence in the ideal op amp model, for it still is a powerful tool for a preliminary understanding of most circuits. Only in the course of a second, more refined analysis does the user examine the impact of practical limitations in order to identify the offenders and apply corrective measures, if needed.

To facilitate our study, we shall concentrate on one limitation at a time, assuming the op amp to be otherwise ideal. In practice, all limitations are present simultaneously; however, assessing their effects individually will allow us to better weigh their relative importance and identify the most critical ones for the application at hand.

In principle, each limitation can be estimated either by calculation or by computer simulation once the op amp's internal circuit schematic and process parameters are known. An alternative approach is to regard the device as a black box and utilize the information available in the data sheets to model it and then predict its behavior. If the actual performance does not meet the objectives, the designer will either change the circuit approach or select a different device, or a combination of both, until a satisfactory solution is found.

Proper interpretation of data-sheet information is, therefore, an integral part of the design process. In the following sections, this procedure will be illustrated using the 741 data sheets of Appendix 5A as a vehicle. Since space does not permit the inclusion of data sheets for other devices, you are encouraged to build your own library of linear products catalogs. Once you have learned to interpret the data sheets of the 741, you can readily extend your skills to the interpretation of other devices.

5.1
SIMPLIFIED OP AMP CIRCUIT DIAGRAM

Even though the data sheets provide all the information the user needs to know, it is instructive to examine the simplified diagram[1] of Fig. 5.1 for an intuitive understanding of how the various op amp limitations originate. This diagram contains the building blocks found in a wide variety of IC op amps, including the popular 741. They are the *input stage,* the *second,* or *intermediate, stage,* and the *output stage.* The following discussion is based on simple transistor theory, but the unfamiliar reader may skip the rest of this section without serious loss of continuity.

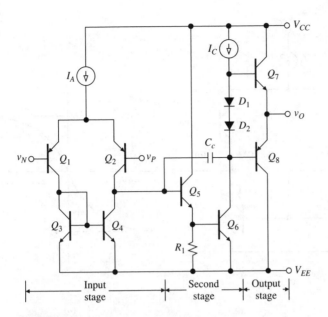

FIGURE 5.1
Simplified op amp circuit diagram.

The Input Stage

This stage senses any imbalance between the inverting and noninverting input voltages v_N and v_P, and converts it to a single-ended output current i_{O1} according to

$$i_{O1} = g_{m1}(v_P - v_N) \tag{5.1}$$

where g_{m1} is the input-stage *transconductance*. This stage is designed to also provide high input impedance and draw negligible input currents. As shown again in Fig. 5.2a, the input stage consists of two matched transistor pairs, namely, the *differential pair* Q_1 and Q_2, and the *current mirror* Q_3 and Q_4.

The input-stage bias current I_A splits between Q_1 and Q_2. Ignoring transistor base currents and applying KCL, we have

$$i_{C1} + i_{C2} = I_A \tag{5.2}$$

For a *pnp* transistor, the collector current i_C is related to its emitter-base voltage drop v_{EB} by the well-known exponential law,

$$i_C = I_s \exp(v_{EB}/V_T) \tag{5.3}$$

where I_s is the *collector saturation current* and V_T the *thermal voltage* ($V_T \cong 26\,\text{mV}$ at room temperature). Assuming matched BJTs ($I_{s1} = I_{s2}$), we can write

$$\frac{i_{C1}}{i_{C2}} = \exp\left(\frac{v_{EB1} - v_{EB2}}{V_T}\right) = \exp\left(\frac{v_P - v_N}{V_T}\right) \tag{5.4}$$

where we have used $v_{EB1} - v_{EB2} = v_{E1} - v_{B1} - (v_{E2} - v_{B2}) = v_{B2} - v_{B1} = v_P - v_N$.

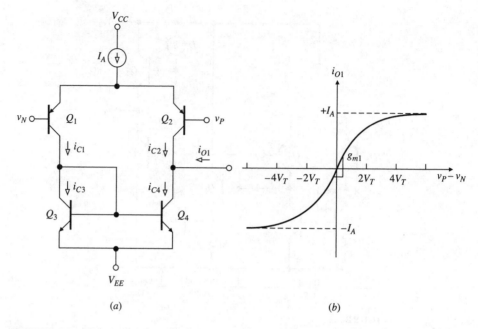

FIGURE 5.2
Input stage and its transfer characteristic.

In response to i_{C1}, Q_3 develops a certain base-emitter voltage drop v_{BE3}. Since $v_{BE4} = v_{BE3}$, Q_4 is forced to draw the same current as Q_3, or $i_{C4} = i_{C3}$; hence the designation current mirror. But, $i_{C3} = i_{C1}$, so the first-stage output current is, by KCL, $i_{O1} = i_{C4} - i_{C2} = i_{C1} - i_{C2}$. Solving Eqs. (5.2) and (5.4) for i_{C1} and i_{C2}, and then taking their difference, we get

$$i_{O1} = I_A \tanh \frac{v_P - v_N}{2V_T} \tag{5.5}$$

This function is plotted in Fig. 5.2b.

We observe that under the balanced condition $v_P = v_N$, I_A splits equally between Q_1 and Q_2, thus yielding $i_{O1} = 0$. However, any imbalance between v_P and v_N will divert more of I_A through Q_1 and less through Q_2, or vice versa, thus yielding $i_{O1} \neq 0$. For sufficiently small imbalances, also referred to as *small-signal conditions,* the transfer characteristic is approximately linear and is expressed by Eq. (5.1). The slope, or transconductance, is found as $g_{m1} = di_{O1}/d(v_P - v_N)|_{v_P = v_N}$. The result is

$$g_{m1} = \frac{I_A}{2V_T} \tag{5.6}$$

Overdriving the input stage will eventually force all of I_A through Q_1 and none through Q_2, or vice versa, thus causing i_{O1} to saturate at $\pm I_A$. The overdrive conditions are referred to as *large-signal conditions.* From the figure we see that the onset of saturation occurs for $v_P - v_N \cong \pm 4V_T \cong \pm 100$ mV. As we know, an op amp with negative feedback normally forces v_N to closely track v_P, indicating small-signal operation.

The Second Stage

This stage is made up of the Darlington pair Q_5 and Q_6, and the frequency-compensation capacitance C_c. The Darlington pair is designed to provide additional gain as well as a wider signal swing. The capacitance is designed to stabilize the op amp against unwanted oscillations in negative-feedback applications, a subject to be addressed in Chapter 8. Since C_c is fabricated on-chip, the op amp is said to be *internally compensated*. By contrast, uncompensated op amps require that the compensation network be supplied externally by the user. The 741 op amp is internally compensated. A popular uncompensated contemporary is the 301 op amp.

The Output Stage

This stage, based on the emitter followers Q_7 and Q_8, is designed to provide low output impedance. Though its voltage gain is only approximately unity, its current gain is fairly high, indicating that this stage acts as a power booster for the second-stage output.

Transistors Q_7 and Q_8 are referred to as a *push-pull pair* because in the presence of a grounded output load, Q_7 will source (or push) current to the load during positive output voltage swings, whereas Q_8 will sink (or pull) current from the load during negative swings. The function of the diodes D_1 and D_2 is to develop a pair of *pn*-junction voltage drops suitable for biasing Q_7 and Q_8 in the forward-active region and thus minimize crossover distortion at the output.

The Input Stage of the 741 Op Amp

Figure 5.3 shows a more detailed diagram of the 741 input stage.[2] To cope with the notoriously low current-gain β_F of lateral *pnp* BJTs, the input drive is provided via the *npn* BJTs Q_1 and Q_2, whose much higher β_Fs ensure a higher input impedance r_d and lower input currents I_P and I_N. These BJTs operate as voltage followers, and the *pnp* BJTs Q_3 and Q_4 form a common-base differential pair. The addition of the voltage followers halves the transconductance g_{m1}, which is now

$$g_{m1} = \frac{I_A}{4V_T} \tag{5.7}$$

Moreover, the large-signal transfer characteristic becomes

$$i_{O1} = I_A \tanh \frac{v_P - v_N}{4V_T} \tag{5.8}$$

As we proceed we shall use the following working values for the 741 op amp: $I_A = 19.6\ \mu A$ and $V_T = 25.9$ mV, so $g_{m1} = 189\ \mu A/V$.

SPICE Models

There are various levels at which an op amp can be simulated. In IC design, op amps are simulated at the transistor level,[3] also called the *micromodel* level. Such a simulation requires a detailed knowledge of both the circuit schematic and the fabrication

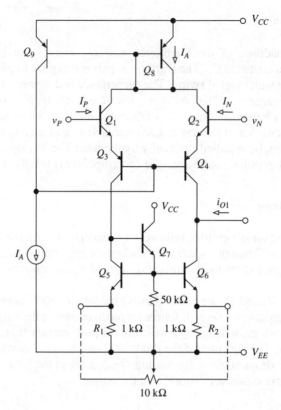

FIGURE 5.3
Detailed diagram of the input stage of the 741 op amp.

process parameters. However, this proprietary information is not easily accessible to the user. Even so, the level of detail may require excessive computation time or may even cause convergence problems, especially in more complex circuit systems.

To cope with these difficulties, simulations by the user are usually carried out at the *macromodel* level.[4] A macromodel uses a much reduced set of circuit elements to closely match the measured behavior of the finished device while saving considerable simulation time. Like any model, a macromodel comes with limitations, and the user need be aware of the parameters the particular macromodel fails to simulate. Macromodels are available from a number of manufacturers (Analog Devices, Burr-Brown, Comlinear, Linear Technology, Maxim, National Semiconductor, Texas Instruments), and can usually be downloaded via the World Wide Web.

The library file EVAL.LIB that comes with the student version of PSpice includes a 741 op amp macromodel based on the so-called Boyle model[5] of Fig. 5.4. This macromodel has been coded as a subcircuit named μA741. The user need not be concerned with the actual subcircuit code, though if desired it can be printed out. The user activates the macromodel via the following commands:

```
.lib eval.lib
XOA vP vN VCC VEE vO uA741
```

The first command instructs PSpice to look up the subcircuit in the EVAL.LIB file, and appears only once. The second command activates the μA741 subcircuit.

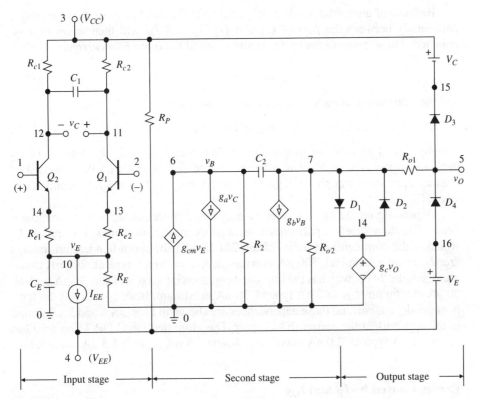

FIGURE 5.4
741 op amp macromodel.

At times we may wish to focus on just one particular op amp feature and thus develop an even simpler model on our own. A typical example is offered by the frequency response, to be studied in Chapter 6. Regardless of the model used, a circuit must eventually be breadboarded and tried out in the lab, where its behavior is evaluated in the presence of parasitics and other factors related to actual circuit construction, which computer simulation, unless properly instructed, fails to account for.

5.2
INPUT BIAS AND OFFSET CURRENTS

Practical op amps do draw small currents at their input pins. These currents cause errors that may be of concern, depending on the application. The 741 input stage of Fig. 5.3 reveals that I_P and I_N are the base currents needed to bias Q_1 and Q_2 in the forward-active region. Q_1 and Q_2 draw these currents automatically from the external circuitry. In fact, for the op amp to function, each input terminal must be provided with a series dc path through which current can flow (we have seen an example in connection with the GIC of Chapter 4). In the case of purely capacitive termination the input current will charge or discharge the capacitor, making a periodic reinitialization necessary. Barring exceptions to be addressed in the next section, I_P and I_N flow *into* the op amp if its input transistors are *npn* BJTs or *p*-channel JFETs, and *out* of the op amp for *pnp* BJTs or *n*-channel JFETs.

Because of unavoidable mismatches between the two halves of the input stage, particularly between the β_Fs of Q_1 and Q_2, I_P and I_N will themselves be mismatched. The average of the two currents is called the *input bias current,*

$$I_B = \frac{I_P + I_N}{2} \tag{5.9}$$

and their difference is called the *input offset current,*

$$I_{OS} = I_P - I_N \tag{5.10}$$

Usually I_{OS} is an order of magnitude smaller than I_B. While the polarity of I_B depends on the type of input transistors, that of I_{OS} depends on the direction of mismatch, so some samples of a given op amp family will have $I_{OS} > 0$, and others $I_{OS} < 0$.

Depending on the op amp type, I_B may range from nanoamperes to femtoamperes. The data sheets report typical as well as maximum values. For the 741C, which is the commercial version of the 741 family, the room-temperature ratings are: $I_B = 80$ nA typical, 500 nA maximum; $I_{OS} = 20$ nA typical, 200 nA maximum. For the 741E, which is the improved commercial version, $I_B = 30$ nA typical, 80 nA maximum; $I_{OS} = 3$ nA typical, 30 nA maximum. Both I_B and I_{OS} are temperature dependent, and these dependences are shown in Figs. 5A.8 and 5A.9, found in the appendix at the end of this chapter. The aforementioned OP-77 op amp has $I_B = 1.2$ nA typical, 2.0 nA maximum; $I_{OS} = 0.3$ nA typical, 1.5 nA maximum.

Errors Caused by I_B and I_{OS}

A straightforward way of assessing the effect of the input currents is to find the output with all input signals set to zero. We shall illustrate for two representative cases, namely, the cases of resistive and capacitive feedback shown in Fig. 5.5. Once we understand these cases, we can readily generalize to other circuits. Our analysis assumes that the op amp, aside from the presence of I_P and I_N, is ideal.

There are many circuits that, once their active inputs are set to zero, reduce to an equivalent circuit of the type of Fig. 5.5a, including the inverting and noninverting

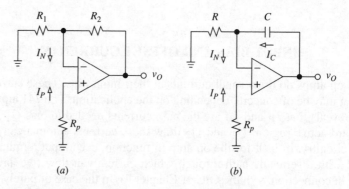

(a)   (b)

FIGURE 5.5
Estimating the output error due to the input bias currents for the case of resistive and capacitive feedback.

amplifiers, the summing and difference amplifiers, I-V converters, and others. By Ohm's law, the voltage at the noninverting input is $V_P = -R_p I_P$. Using the superposition principle, we have $v_O = (1 + R_2/R_1)V_P + R_2 I_N = R_2 I_N - (1 + R_2/R_1)R_p I_P$, or $v_O = E_O$, where

$$E_O = \left(1 + \frac{R_2}{R_1}\right)[(R_1 \parallel R_2)I_N - R_p I_P] \qquad (5.11)$$

This insightful form elicits a number of observations. First, in spite of the absence of any input signal, the circuit yields some output E_O. We regard this unwanted output as an error or, more properly, as *output dc noise*. Second, the circuit produces E_O by taking an input error, or *input dc noise,* and amplifying it by $(1 + R_2/R_1)$, which is aptly called the *dc noise gain*. Third, this input error consists of two terms, the voltage drop $-R_p I_P$ due to I_P flowing through R_p, and the voltage drop $(R_1 \parallel R_2)I_N$ due to I_N flowing through the combination $R_1 \parallel R_2$. Fourth, the two terms tend to compensate for each other since they have opposite polarities.

Depending on the application, the error E_O may be unacceptable and one must devise suitable means to reduce it to a tolerable level. Putting Eq. (5.11) in the form

$$E_O = \left(1 + \frac{R_2}{R_1}\right)\{[(R_1 \parallel R_2) - R_p]I_B - [(R_1 \parallel R_2) + R_p]I_{OS}/2\}$$

reveals that if we install a dummy resistance R_p, as shown, and we impose

$$R_p = R_1 \parallel R_2 \qquad (5.12)$$

then the term involving I_B will be eliminated, leaving

$$E_O = \left(1 + \frac{R_2}{R_1}\right)(-R_1 \parallel R_2)I_{OS} \qquad (5.13)$$

The error is now proportional to I_{OS}, which is typically an order of magnitude smaller than either I_P or I_N.

E_O can be reduced further by scaling down all resistances. For instance, reducing all resistance by a factor of 10 will leave gain unaffected, but will cause a tenfold reduction in the input error $-(R_1 \parallel R_2)I_{OS}$. Reducing resistances, however, increases power dissipation, so a compromise will have to be reached. If E_O is still unacceptable, selecting an op amp type with a lower I_{OS} rating is the next logical step. Other techniques for reducing E_O will be discussed in Section 5.6.

EXAMPLE 5.1. In the circuit of Fig. 5.5a let $R_1 = 22$ kΩ and $R_2 = 2.2$ MΩ, and let the op amp ratings be $I_B = 80$ nA and $I_{OS} = 20$ nA. (a) Calculate E_O for the case $R_p = 0$. (b) Repeat, but with $R_p = R_1 \parallel R_2$ in place. (c) Repeat part (b), but with all resistances simultaneously reduced by a factor of 10. (d) Repeat part (c), but with the op amp replaced by one with $I_{OS} = 3$ nA. Comment.

Solution.

(a) The dc noise gain is $1 + R_2/R_1 = 101$V/V; moreover, $(R_1 \parallel R_2) \cong 22$ kΩ. With $R_p = 0$, we have $E_O = 101 \times (R_1 \parallel R_2)I_N \cong 101 \times (R_1 \parallel R_2)I_B \cong 101 \times 22 \times 10^3 \times 80 \times 10^{-9} \cong 175$ mV.

(b) With $R_p = R_1 \parallel R_2 \cong 22 \text{ k}\Omega$ in place, $E_O \cong 101 \times 22 \times 10^3 \times (\pm 20 \times 10^{-9}) = \pm 44$ mV, where we write "$\pm$" to reflect the fact that I_{OS} may be of either polarity.

(c) With $R_1 = 2.2 \text{ k}\Omega$, $R_2 = 220 \text{ k}\Omega$, and $R_p = 2.2 \text{ k}\Omega$, we get $E_O = 101 \times 2.2 \times 10^3 \times (\pm 20 \times 10^{-9}) \cong \pm 4.4$ mV.

(d) $E_O = 101 \times 2.2 \times 10^3 \times (\pm 3 \times 10^{-9}) \cong \pm 0.7$ mV. Summarizing, with R_p in place, E_O is reduced by 4; scaling the resistances reduces E_O by an additional factor of 10; finally, using a better op amp reduces it by yet another factor of 7.

Turning next to the circuit of Fig. 5.5b, we note that we still have $V_N = V_P = -R_p I_P$. Summing currents at the inverting-input node yields $V_N/R + I_N - I_C = 0$. Eliminating V_N, we get

$$I_C = \frac{1}{R}(RI_N - R_p I_P) = \frac{1}{R}[(R - R_p)I_B - (R + R_p)I_{OS}/2] \qquad (5.14)$$

Applying the capacitance law $v = (1/C) \int i \, dt$, we readily get

$$v_O(t) = E_O(t) + v_O(0) \qquad (5.15)$$

$$E_O(t) = \frac{1}{RC} \int_0^t [(R - R_p)I_B - (R + R_p)I_{OS}/2] \, d\xi \qquad (5.16)$$

where $v_O(0)$ is the initial value of v_O. In the absence of any input signal, we expect the circuit to yield a constant output, or $v_O(t) = v_O(0)$. In practice, besides $v_O(0)$, it yields the *output error* $E_O(t)$, which is the result of integrating the *input error* $[(R - R_p)I_B - (R + R_p)I_{OS}/2]$ over time. Since I_B and I_{OS} are relatively constant, we can write $E_O(t) = [(R - R_p)I_B - (R + R_p)I_{OS}/2]t/RC$. The error is thus a voltage ramp, whose tendency is to drive the op amp into saturation.

It is apparent that installing a dummy resistance R_p such that

$$R_p = R \qquad (5.17)$$

will reduce the error to

$$E_O(t) = \frac{1}{RC} \int_0^t -RI_{OS} \, d\xi \qquad (5.18)$$

This error can be reduced further by component scaling, or by using an op amp with a lower I_{OS} rating.

EXAMPLE 5.2. In the circuit of Fig. 5.5b let $R = 100 \text{ k}\Omega$, $C = 1$ nF, and $v_O(0) = 0$ V. Assuming an op amp with $I_B = 80$ nA, $I_{OS} = 20$ nA, and $\pm V_{\text{sat}} = \pm 13$ V, find how long it takes for the op amp to enter saturation if (a) $R_p = 0$, and (b) $R_p = R$.

Solution.

(a) The input error is $RI_N \cong RI_B = 10^5 \times 80 \times 10^{-9} = 8$ mV. So, $v_O(t) = (RI_N/RC)t = 80t$, which represents a positive voltage ramp. Imposing $13 = 80t$ yields $t = 13/80 = 0.1625$ s.

(b) The input error is now $-RI_{OS} = \pm 2$ mV, indicating that the op amp may saturate at either rail. The time it takes to saturate is now extended in proportion to $0.1625 \times 80/20 = 0.65$ s.

Summarizing, to minimize the errors due to I_B and I_{OS}, adhere to the following rules whenever possible: (a) modify the circuit so that the resistances seen by I_P and

I_N with all sources suppressed are equal, that is, impose $R_p = R_1 \parallel R_2$ in Fig. 5.5a and $R_p = R$ in Fig. 5.5b; (b) keep resistances as low as the application will allow; (c) use op amps with adequately low I_{OS} ratings.

5.3
LOW-INPUT-BIAS-CURRENT OP AMPS

Op amp designers strive to keep I_B and I_{OS} as small as other design constraints allow. Following are the most common techniques.

Superbeta-Input Op Amps

One way of achieving low I_Bs is by using input BJTs with extremely high current gains. Known as *superbeta transistors,* these BJTs achieve β_Fs in excess of 10^3 A/A by utilizing very thin base regions to minimize the recombination component[2] of the base current. This technique was pioneered with the LM308 op amp (National Semiconductor), whose input stage is shown in Fig. 5.6a. The heart of the circuit is the superbeta differential pair Q_1 and Q_2. These BJTs are connected in cascode with the standard-beta BJTs Q_3 and Q_4 to form a composite structure with high current gain as well as high breakdown voltage. Q_5 and Q_6 provide a bootstrapping function to bias Q_1 and Q_2 at zero base-collector voltage regardless of the input

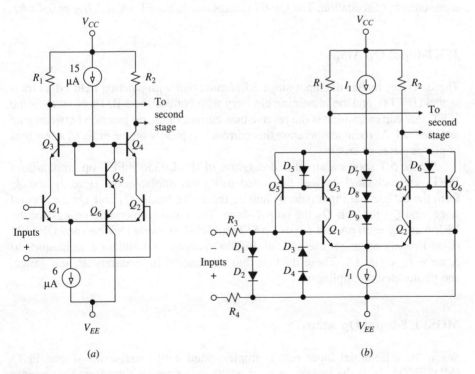

FIGURE 5.6
(a) Superbeta input stage, and (b) input-bias-current cancellation.

common-mode voltage. This avoids the low-breakdown limitations of the superbeta BJTs and also reduces collector-base leakage. Superbeta op amps have typically $I_B \cong 1$ nA or less.

Input-Bias-Current Cancellation

Another popular technique for achieving low $I_B s$ is current cancellation.[2] Special circuitry anticipates the base currents needed to bias the input transistors, then itself supplies these currents internally, making the op amp appear to an outsider as if it were capable of operating without any input bias current.

Figure 5.6b shows the cancellation scheme utilized by the OP-07 op amp (Analog Devices). Once again, the heart of the circuit is the differential pair Q_1 and Q_2. The base currents of Q_1 and Q_2 are duplicated at the bases of common-base transistors Q_3 and Q_4, where they are sensed by current mirrors Q_5-D_5 and Q_6-D_6. The mirrors reflect these currents and then reinject them into the bases of Q_1 and Q_2, thus providing input-bias-current cancellation.

In practice, because of device mismatches, cancellation is not perfect, so the input pins will still draw residual currents. However, since these currents are now the result of a mismatch, they are typically an order of magnitude less than the actual base currents. We observe that I_P and I_N may flow either into or out of the op amp, depending on the direction of the mismatch. Moreover, I_{OS} is of the same order of magnitude as I_B, so there is no use installing a dummy resistance R_p in op amps with input-current cancellation. The OP-07 ratings are $I_B = \pm 1$ nA and $I_{OS} = 0.4$ nA.

JFET-Input Op Amps

These devices realize the input-stage differential pair with junction field-effect transistors (JFETs), and the remaining circuitry with conventional BJTs. Now I_B is the JFET gate current, which is the reverse bias current of the *pn* junction between gate and channel. At room temperature this current is typically on the order of a few tens of picoamperes or less.

Figure 5.7 shows a simplified diagram of the LF356 biFET op amp, whose JFETs are *p*-channel devices fabricated using ion implantation. Here J_1 and J_2 form the differential input pair, J_3 and J_4 the active loads, Q_1 and Q_2 the second stage, and Q_3 through Q_5 the output stage. The room-temperature ratings for the LF356 are $I_B = 30$ pA and $I_{OS} = 3$ pA. The AD549 (Analog Devices) and OPA129 (Burr-Brown) op amps use special JFET structures and isolation techniques to achieve $I_B < 100$ fA. These devices find application in electrometer, ion gauge, and photodetector amplifiers.

MOSFET-Input Op Amps

When the differential input pair is implemented with metal-oxide-silicon FETs (MOSFETs), I_B is the leakage current of the gate-channel capacitor. This current is typically in the range of a few picoamperes. In BiMOS op amps the input pair is in MOS technology and the rest of the circuitry in bipolar. However, op amps are

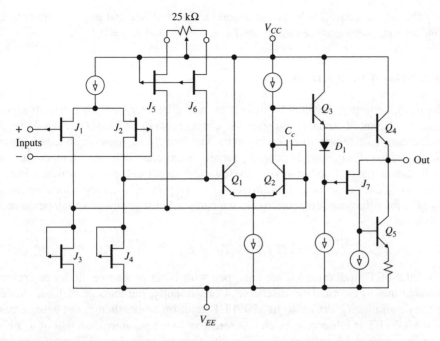

FIGURE 5.7
Circuit diagram of the LF356 biFET op amp. (Courtesy of National Semiconductor.)

also available entirely in MOSFET technology, either as stand-alone devices, or as part of complex systems such as switched-capacitor filters. The stand-alone types are usually implemented in complementary MOS (CMOS) technology.

Figure 5.8 shows a simplified diagram of the TLC279 CMOS op amp, which uses p-channel transistors M_1 and M_2 as the differential input pair, n-channel transistors

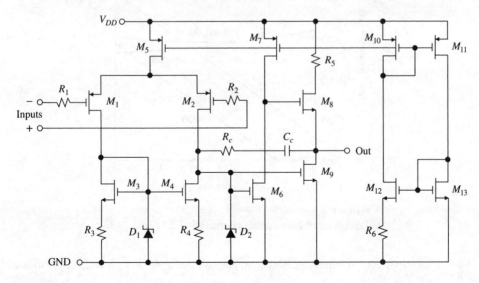

FIGURE 5.8
Circuit diagram of the TLC279 CMOS op amp. (Reprinted by permission of Texas Instruments.)

M_2 and M_4 as active loads, M_6 as second stage, and M_8 and M_9 as output stage. Typical room-temperature ratings are $I_B = 0.7$ pA and $I_{OS} = 0.1$ pA.

Input-Bias-Current Drift

Figure 5.9 compares typical input-bias-current characteristics for different input-stage arrangements and technologies. We observe that in BJT-input devices I_B tends to decrease with temperature, owing to the fact that β_F increases with temperature. However, for JFET-input devices, I_B increases exponentially with temperature. A well-known rule of thumb states that the reverse-bias current of a *pn* junction, whether it is that of a diode or of a JFET, *doubles for every 10 °C increase*. Once we know I_B at some reference temperature T_0, we can predict it at any other temperature T using

$$I_B(T) \cong I_B(T_0) \times 2^{(T-T_0)/10} \qquad (5.19)$$

MOSFET-input op amps are equipped with input protective diodes to prevent damage due to electrostatic discharge. Consequently, the leakage of these diodes causes a similar I_B drift also in MOSFET-input op amps, though the gate current of a MOSFET is inherently much less sensitive to temperature than that of a JFET. The low-current advantages of FET-input op amps over their BJT-input counterparts tend to disappear at higher temperatures. Knowledge of the intended operating temperature range is an important factor when selecting the optimal device.

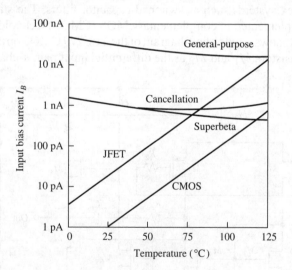

FIGURE 5.9
Typical input-bias-current characteristics.

EXAMPLE 5.3. A certain FET-input op amp is rated at $I_B = 1$ pA at 25 °C. Estimate I_B at 100 °C.

Solution. $I_B(100 \text{ °C}) \cong 10^{-12} \times 2^{(100-25)/10} = 0.18$ nA.

Input Guarding

When applying op amps with ultralow input bias current, special attention must be paid to wiring and circuit construction in order to fully realize the capabilities of these devices. Data sheets usually provide helpful guidelines in this respect. Of special concern are leakage currents across the printed-circuit board. They can easily exceed I_B itself and thus defeat what has been so painstakingly achieved in terms of circuit design.

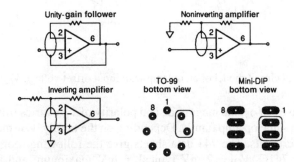

FIGURE 5.10
Guard-ring layout and connections.

The effect of leakage can be reduced significantly by using guard rings around the input pins. As shown in Fig. 5.10, a guard consists of a conductive pattern held at the same potential as v_P and v_N. This pattern will absorb any leakages from other points on the board and thus prevent them from reaching the input pins. Guard rings also act as shields against noise pickup. For best results, board surfaces should be kept clean and moisture-free. If sockets are required, best results are obtained by using Teflon sockets or standoffs.

5.4
INPUT OFFSET VOLTAGE

Shorting together the inputs of an op amp should yield $v_O = a(v_P - v_N) = a \times 0 = 0$ V. However, because of inherent mismatches between the input-stage halves processing v_P and v_N, a practical op amp will generally yield $v_O \neq 0$. To force v_O to zero, a suitable correction voltage must be applied between the input pins. This is tantamount to saying that the open-loop VTC does not go through the origin, but is shifted either to the left or to the right, depending on the direction of the mismatch. This shift is called the *input offset voltage V_{OS}*. As shown in Fig. 5.11, we can model a practical op amp with an ideal or offsetless op amp having a tiny source V_{OS} in series with one of its inputs. The VTC is now

$$v_O = a[v_P + V_{OS} - v_N] \qquad (5.20)$$

To drive the output to zero, we need $v_P + V_{OS} - v_N = 0$, or

$$v_N = v_P + V_{OS} \qquad (5.21)$$

Note that because of V_{OS}, we now have $v_N \neq v_P$.

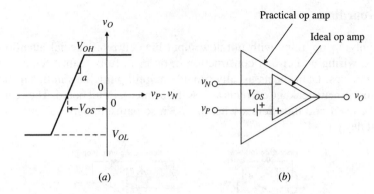

(a) (b)

FIGURE 5.11
VTC and circuit model of an op amp with input offset voltage V_{OS}.

As in the case of I_{OS}, the magnitude and polarity of V_{OS} varies from one sample to another of the same op amp family. Depending on the family, V_{OS} may range from millivolts to microvolts. The 741 data sheets give the following room-temperature ratings: for the 741C, $V_{OS} = 2$ mV typical, 6 mV maximum; and for the 741E, $V_{OS} = 0.8$ mV typical, 3 mV maximum. The OP-77 ultralow offset voltage op amp has $V_{OS} = 10$ μV typical, 50 μV maximum.

Errors Caused by V_{OS}

As in Section 5.2, we shall examine the effect of V_{OS} for the resistive-feedback and capacitive-feedback cases of Fig. 5.12. Note that we are omitting the dummy resistance R_P since the present analysis deliberately ignores I_B and I_{OS} to focus on V_{OS} alone. In Section 5.6 we shall address the general case in which I_B, I_{OS}, and V_{OS} are present simultaneously.

In Fig. 5.12a, the offset-free op amp acts as a noninverting amplifier with respect to V_{OS}, so $v_O = E_O$, where

$$E_O = \left(1 + \frac{R_2}{R_1}\right) V_{OS} \tag{5.22}$$

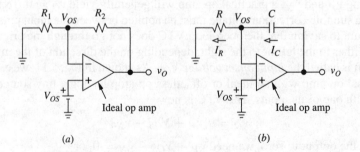

(a) (b)

FIGURE 5.12
Estimating the output error due to V_{OS} for the case of resistive and capacitive feedback.

is the output error, and $(1 + R_2/R_1)$ is again the dc noise gain. Clearly, the larger the noise gain, the larger the error. For instance, with $R_1 = R_2$, a 741C op amp yields $E_O = (1+1) \times (\pm 2 \text{ mV}) = \pm 4 \text{ mV}$ typical, $(1+1) \times (\pm 6 \text{ mV}) = \pm 12 \text{ mV}$ maximum. However, with $R_2 = 10^3 R_1$, it yields $E_O = (1 + 10^3) \times (\pm 2 \text{ mV}) \cong \pm 2 \text{ V}$ typical, $\pm 6 \text{ V}$ maximum—quite an error! Conversely, we can use the present circuit to measure V_{OS}. For instance, let $R_1 = 10 \, \Omega$ and $R_2 = 10 \text{ k}\Omega$, so that the dc noise gain is 1001 V/V and the combination $R_1 \parallel R_2$ is sufficiently small to make the effect of I_N negligible. Suppose we measure the output and find $E_O = -0.5 \text{ V}$. Then, $V_{OS} \cong E_O/1001 \cong -0.5 \text{ mV}$, a negative offset in this specific example.

In the circuit of Fig. 5.12b we note that since the offset-free op amp keeps $V_N = V_{OS}$, we have $I_C = I_R = V_{OS}/R$. Using again the capacitance law, we get $v_O(t) = E_O(t) + v_O(0)$, where the output error is now

$$E_O(t) = \frac{1}{RC} \int_0^t V_{OS} \, d\xi \qquad (5.23)$$

or $E_O(t) = (V_{OS}/RC)t$. This voltage ramp, resulting from the integration of V_{OS} over time, tends, as we know, to drive the op amp into saturation.

Thermal Drift

Like most other parameters, V_{OS} is temperature-dependent, a feature expressed in terms of the *temperature coefficient*

$$\text{TC}(V_{OS}) = \frac{\partial V_{OS}}{\partial T} \qquad (5.24)$$

where T is absolute temperature, in kelvins, and $\text{TC}(V_{OS})$ is in microvolts per degree Celsius. For low-cost, general-purpose op amps such as the 741, $\text{TC}(V_{OS})$ is typically on the order of 5 μV/°C. Thermal drift stems from inherent mismatches as well as thermal gradients across the two halves of the input stage. Op amps specifically designed for low-input offset also tend to exhibit lower thermal drifts, thanks to superior matching and thermal tracking at the input stage. The OP-77 has $\text{TC}(V_{OS}) = 0.1 \, \mu$V/°C typical, 0.3 μV/°C maximum.

Using the average value of the temperature coefficient, one can estimate V_{OS} at a temperature other than 25 °C as

$$V_{OS}(T) \cong V_{OS}(25 \text{ °C}) + \text{TC}(V_{OS})_{\text{avg}} \times (T - 25 \text{ °C}) \qquad (5.25)$$

For instance, an op amp with $V_{OS}(25 \text{ °C}) = 1 \text{ mV}$ and $\text{TC}(V_{OS})_{\text{avg}} = 5 \, \mu$V/°C would have $V_{OS}(70 \text{ °C}) = 1 \text{ mV} + (5 \, \mu\text{V}) \times (70 - 25) = 1.225 \text{ mV}$.

Common-Mode Rejection Ratio (CMRR)

In the absence of input offset, an op amp should respond only to the voltage difference between its inputs, or $v_O = a(v_P - v_N)$. A practical op amp is somewhat sensitive also to the common-mode input voltage $v_{CM} = (v_P + v_N)/2$. Its transfer characteristic is thus $v_O = a(v_P - v_N) + a_{cm}v_{CM}$, where a is the differential-mode gain, and a_{cm} is the common-mode gain. Rewriting as $v_O = a[v_P - v_N + (a_{cm}/a)v_{CM}]$, and recalling that the ratio a/a_{cm} is the common-mode rejection ratio CMRR, we have

$$v_O = a\left(v_P + \frac{v_{CM}}{CMRR} - v_N\right)$$

Comparison with Eq. (5.20) indicates that the sensitivity to v_{CM} can be modeled with an input-offset-voltage term of value $v_{CM}/CMRR$. The common-mode sensitivity stems from the fact that a change in v_{CM} will alter the operating points of the input-stage transistors and cause a change at the output. It is comforting to know that such a complex phenomenon can be reflected to the input in the form of a mere offset error! We thus redefine the CMRR as

$$\frac{1}{CMRR} = \frac{\partial V_{OS}}{\partial v_{CM}} \tag{5.26}$$

and interpret it as the change in V_{OS} brought about by a 1-V change in v_{CM}. We express 1/CMRR in microvolts per volt. Because of stray capacitances, the CMRR deteriorates with frequency. Typically, it is high from dc to a few tens or a few hundreds of hertz, after which it rolls off with frequency at the rate of -20 dB/dec.

Data sheets usually give CMRR in decibels. As we know, the conversion to microvolts per volt is readily accomplished via

$$\frac{1}{CMRR} = 10^{-CMRR_{dB}/20} \tag{5.27}$$

where $CMRR_{dB}$ represents the decibel value of CMRR. From Fig. 5A.4, the dc ratings for the 741 op amp are $CMRR_{dB} = 90$ dB typical, 70 dB minimum, indicating that V_{OS} changes with v_{CM} at the rate of $1/CMRR = 10^{-90/20} = 31.6 \mu$V/V typical, and $10^{-70/20} = 316 \mu$V/V maximum. The OP-77 op amp has $1/CMRR = 0.1 \mu$V/V typical, 1 μV/V maximum. Figure 5A.6 shows that the CMRR of the 741 starts to roll off just above 100 Hz.

Since op amps keep v_N fairly close to v_P, we can write $v_{CM} \cong v_P$. The CMRR is of no concern in inverting applications, where $v_P = 0$. However, it may pose problems when v_P is allowed to swing, as in an instrumentation amplifier.

EXAMPLE 5.4. The difference amplifier of Fig. 2.13 uses a 741 op amp and a perfectly matched resistance set with $R_1 = 10$ kΩ and $R_2 = 100$ kΩ. Suppose the inputs are tied together and driven with a common signal v_I. Estimate the typical change in v_O if (a) v_I is slowly changed from 0 to 10 V, and (b) v_I is a 10-kHz, 10-V peak-to-peak sine wave.

Solution.

(a) At dc we have $1/CMRR = 10^{-90/20} = 31.6 \mu$V/V, typical. The common-mode change at the op amp input pins is $\Delta v_P = [R_2/(R_1 + R_2)]\Delta v_I = [100/(10 + 100)]10 = 9.09$ V. Thus, $\Delta V_{OS} = (1/CMRR)\Delta v_P = 31.6 \times 9.09 = 287 \mu$V. The dc noise gain is $1 + R_2/R_1 = 11$ V/V. Hence, $\Delta v_O = 11 \times 287 = 3.16$ mV.

(b) From the CMRR curve of Fig. 5A.6 we find $CMRR_{dB}(10$ kHz$) \cong 57$ dB. So, $1/CMRR = 10^{-57/20} = 1.41$ mV/V, $\Delta V_{OS} = 1.41 \times 9.09 = 12.8$ mV (peak-to-peak), and $\Delta v_O = 11 \times 12.8 = 0.141$ V (peak-to-peak). The output error at 10 kHz is much worse than at dc.

Power-Supply Rejection Ratio (PSRR)

If we change one of the op amp supply voltages V_S by a given amount ΔV_S, the operating points of the internal transistors will be altered, generally causing a small change in v_O. By analogy with the CMRR, we model this phenomenon with a change in the input offset voltage, which we express in terms of the *power-supply rejection ratio* (PSRR) as $(1/PSRR) \times \Delta V_S$. The parameter

$$\frac{1}{PSRR} = \frac{\partial V_{OS}}{\partial V_S} \qquad (5.28)$$

represents the change in V_{OS} brought about by a 1-V change in V_S, and is expressed in microvolts per volt. Like the CMRR, the PSRR deteriorates with frequency.

Some data sheets give separate PSRR ratings, one for changes in V_{CC} and the other for changes in V_{EE}. Others specify the PSRR for V_{CC} and V_{EE} changing symmetrically. The $PSRR_{dB}$ ratings of most op amps fall in the range of 80 dB to 120 dB. The devices of superior matching usually offer the highest PSRRs. From Fig. 5A.4, the 1/PSRR ratings for the 741C, which are given for symmetric supply changes, are 30 μV/V typical, 150 μV/V maximum. This means that changing, for instance, the supply voltages from ±15 V to ±12 V yields $\Delta V_{OS} = (1/PSRR)\Delta V_S = (30 \ \mu V)(15 - 12) = \pm90 \ \mu V$ typical, $\pm450 \ \mu V$ maximum. The OP-77 op amp has $1/PSRR = 0.7 \ \mu V/V$ typical, 3 $\mu V/V$ maximum.

When the op amp is powered from well-regulated and properly bypassed supplies, the effect of the PSRR is usually negligible. Otherwise, any variation on the supply busses will induce a corresponding variation in V_{OS}, which in turn is amplified by the noise gain. A classical example is offered by audio preamplifiers, where the residual 60 Hz (or 120 Hz) ripple on the supply rails may cause intolerable hum at the output. Another case in point is offered by switchmode power supplies, whose high-frequency ripple is usually inadequately rejected by op amps, indicating that these supplies are unsuited to high-precision analog circuitry.

EXAMPLE 5.5. A 741 op amp is connected as in Fig. 5.12a with $R_1 = 100 \ \Omega$ and $R_2 = 100 \ k\Omega$. Predict the typical as well as the maximum ripple at the output for a power-supply ripple of 0.1 V (peak-to-peak) at 120 Hz.

Solution. The 741 data sheets do not show the PSRR rolloff with frequency, so let us use the ratings given at dc, keeping in mind that the results will be optimistic. The induced ripple at the input is $\Delta V_{OS} = (30 \ \mu V)0.1 = 3 \ \mu V$ typical, 15 μV maximum (peak-to-peak). The noise gain is $1 + R_2/R_1 \cong 1000$ V/V, so the output ripple is $\Delta v_O = 3$ mV typical, 15 mV maximum (peak-to-peak).

Change of V_{OS} with the Output Swing

In a practical op amp the open-loop gain a is finite, so the difference $v_P - v_N$ changes also with the output swing Δv_O by the amount $\Delta v_O/a$. This effect can conveniently be regarded as an effective offset voltage change $\Delta V_{OS} = \Delta v_O/a$. Even an op amp with $V_{OS} = 0$ for $v_O = 0$ will exhibit some input offset for $v_O \neq 0$. For instance, to sustain $v_O = 10$ V with $a = 10^5$ V/V, such an op amp requires

$V_{OS} = 10/10^5 = 100 \, \mu V$. This must be taken into account if we wish to continue using the model of Fig. 5.11b.

We summarize this section by writing a general expression for V_{OS} in terms of the various operating changes affecting it,

$$V_{OS} = V_{OS0} + TC(V_{OS})\Delta T + \frac{\Delta v_P}{CMRR} + \frac{\Delta V_S}{PSRR} + \frac{\Delta v_O}{a} \qquad (5.29)$$

where V_{OS0}, the *initial input offset voltage,* is the value of V_{OS} at some reference operating point, such as ambient temperature, nominal supply voltages, and v_p and v_O halfway between the supply voltages. This parameter itself drifts with time. As an example, the OP-77 has a long-term stability of 0.2 μV/month. In error-budget analysis, the various offset changes are combined *additively* when we wish to estimate the *worst-case* change, and in *root-sum-square (rss)* fashion when we are interested in the *most probable* change.

> **EXAMPLE 5.6.** An op amp has the following ratings: $a = 10^5$ V/V typical, 10^4 V/V minimum, $TC(V_{OS})_{avg} = 3 \, \mu V/°C$, and $CMRR_{dB} = PSRR_{dB} = 100$ dB typical, 80 dB minimum. Estimate the worst-case as well as the most probable change in V_{OS} over the following range of operating conditions: $0 \, °C \le T \le 70 \, °C$, $V_S = \pm 15$ V $\pm 5\%$, -1 V $\le v_P \le +1$ V, and -5 V $\le v_O \le +5$ V.
>
> **Solution.** The thermal change from room temperature is $\Delta V_{OS1} = (3 \, \mu V/°C)(70 - 25)°C = 135 \, \mu V$. With $1/CMRR = 1/PSRR = 10^{-100/20} = 10 \, \mu V/V$ typical, 100 μV/ V maximum, the changes with v_P and V_S are $\Delta V_{OS2} = (\pm 1 \text{ V})/CMRR = \pm 10 \, \mu V$ typical, $\pm 100 \, \mu V$ maximum; $\Delta V_{OS3} = 2 \times (\pm 0.75 \text{ V})/PSRR = \pm 15 \, \mu V$ typical, $\pm 150 \, \mu V$ maximum. Finally, the change with v_O is $\Delta V_{OS4} = (\pm 5 \text{ V})/a = \pm 50 \, \mu V$ typical, $\pm 500 \, \mu V$ maximum. The worst-case change in V_{OS} is $\pm(135 + 100 + 150 + 500) = \pm 885 \, \mu V$. The most probable change is $\pm(135^2 + 10^2 + 15^2 + 50^2)^{1/2} = \pm 145 \, \mu V$.

5.5
LOW-INPUT-OFFSET-VOLTAGE OP AMPS

The initial input offset voltage V_{OS0} is due primarily to device mismatches and bias imbalances in the input stage.

Bipolar Op Amps

Let us return to the simplified input stage of Fig. 5.2a. Taking mismatches between Q_1 and Q_2 into account, we rewrite Eq. (5.4) as $i_{C1}/i_{C2} = (I_{s1}/I_{s2}) \exp[(v_P - v_N)/V_T]$, or $v_P - v_N = V_T \ln[(i_{C1}/i_{C2})(I_{s2}/I_{s1})]$. Similarly, $i_{C3}/i_{C4} = I_{s3}/I_{s4}$. In order to drive i_{O1} to zero, we need, by definition, $v_N = v_P + V_{OS}$. But, $v_N = v_P + V_T \ln[(I_{s4}/I_{s3})(I_{s1}/I_{s2})]$, where we have used $i_{C3} = i_{C1}$ and $i_{C4} = i_{C2}$ to let $i_{C1}/i_{C2} = i_{C3}/i_{C4} = I_{s3}/I_{s4}$. Thus,

$$V_{OS} = V_T \ln \frac{I_{s1}}{I_{s2}} \frac{I_{s4}}{I_{s3}} \qquad (5.30)$$

With $V_T \cong 26$ mV and I_s mismatches on the order of 5%, V_{OS} is typically in the range of 1 mV to 2 mV at room temperature. Moreover, given that $V_T = kT/q$, where k is Boltzmann's constant, q the electron charge, and T absolute temperature, we readily find

$$\text{TC}(V_{OS}) = \frac{V_{OS}}{T} \tag{5.31}$$

Thus, at room temperature ($T \cong 300$ K), a bipolar input stage exhibits a TC(V_{OS}) of about 3.3 μV/°C for every millivolt of offset voltage.

Further insight can be gained by examining the expression for the BJT saturation current,[2]

$$I_s = \frac{qD_B}{N_B} \times n_i^2(T) \times \frac{A_E}{W_B} \tag{5.32}$$

where D_B and N_B are the minority-carrier diffusion constant and the doping concentration in the base region; $n_i(T)$ is the intrinsic carrier concentration, a strong function of temperature; and A_E and W_B are the emitter-junction area and the base width.

The first class of mismatches stems from fabrication process variations, such as mask resolution, which affects A_E, and diffusion process nonuniformities, which affect N_B and W_B. In the design of low-offset op amps, these mismatches are reduced by increasing input-stage device geometries and sizes[2] to make the above parameters less sensitive to edge resolution and diffusion irregularities. In the case of MOSFET-input op amps, large transistor sizes also improve noise performance, an issue we shall address in Chapter 7.

The second class of mismatches stems from thermal gradients and process-related gradients across the chip. Thermal gradients, in particular, tend to affect $n_i(T)$ significantly. The input-stage sensitivity to gradients is reduced by a symmetrical device placement technique known as *common-centroid layout*.[2] As exemplified in Fig. 5.13 for a differential input pair, each transistor is made up of two identical halves connected in parallel, but laid out diagonally opposite to each other. The resulting quad structure provides a multifold symmetry that tends to cancel out the effects of gradient-induced mismatches.

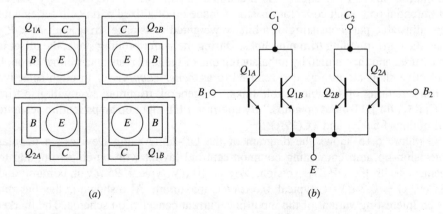

(a)   (b)

FIGURE 5.13
Common-centroid topology: (a) layout and (b) interconnections.

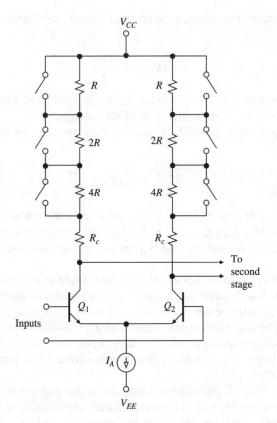

FIGURE 5.14
On-chip V_{OS} trimming using shortable links.

Another method of reducing the initial offset is *on-chip trimming,* which is carried out by means of a laser trim or by selectively shorting or opening suitable trimming links in the circuit. As illustrated in Fig. 5.14 for a resistively loaded differential pair, each collector resistor is made up of a fixed part R_c in series with an adjustable part consisting of a binary-weighted resistance string with $R \ll R_c$, and the corresponding trimming links. During the wafer probing stage, the offset is measured and then nulled by unbalancing one of the load resistances either through selective short-circuiting, also referred to as *Zener zapping,* or through selective open-circuiting of suitable fusible links.[2] In general, trimming V_{OS} will also trim $TC(V_{OS})$ for BJT-input op amps.[6] By contrast, FET-input op amps require separate trimmings for V_{OS} and $TC(V_{OS})$.

Figure 5.15 shows the diagram of the OP-27 (Analog Devices), a popular precision op amp combining common-centroid layout with on-chip trimming to achieve, with the OP-27E version, $V_{OS} = 10\ \mu V$ typical, 25 μV maximum; and $TC(V_{OS}) = 0.2\ \mu V/°C$ typical, 0.6 $\mu V/°C$ maximum. Also shown in the diagram is an interesting variant of the input-bias-current cancellation scheme. The market offers a number of other bipolar products with comparable characteristics.

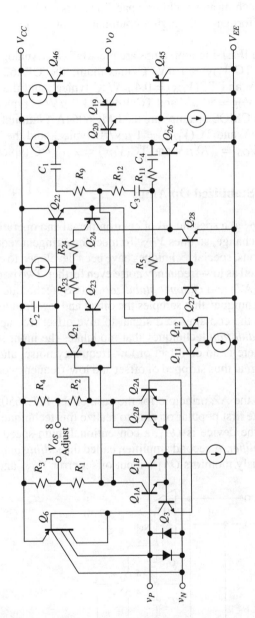

FIGURE 5.15

Simplified circuit diagram of the OP-27 op amp: R_1 and R_2 are adjusted at wafer test for minimum offset voltage. (Courtesy of Analog Devices.)

FET-Input Op Amps

Though in the past FET-input op amps were considered inferior to their BJT-input counterparts in terms of matching and tracking capabilities, it is nevertheless possible to achieve respectable performance through a combination of design, layout, and on-chip trimming.

Examples of precision JFET-input op amps are the AD547L (Analog Devices), with $V_{OS} = 250 \, \mu V$ and $TC(V_{OS}) = 1 \, \mu V/^\circ C$ maximum; the OPA627B (Burr-Brown), with $V_{OS} = 40 \, \mu V$ and $TC(V_{OS}) = 0.4 \, \mu V/^\circ C$ typical; and the LT1055A (Linear Technology) with $V_{OS} = 50 \, \mu V$ and $TC(V_{OS}) = 1.2 \, \mu V/^\circ C$ typical.

Examples of precision CMOS op amps are the LMC6064A (National Semiconductor) with $V_{OS} = 100 \, \mu V$ and $TC(V_{OS}) = 1 \, \mu V/^\circ C$ typical, and the TLC279C (Texas Instruments) with $V_{OS} = 370 \, \mu V$ and $TC(V_{OS}) = 2 \, \mu V/^\circ C$ typical.

Autozero and Chopper-Stabilized Op Amps

On-chip trimming nulls V_{OS} at a specific set of environmental and operating conditions. As these conditions change, so does V_{OS}. To meet the stringent requirements of high-precision applications, special techniques have been developed to effectively reduce the input offset as well as low-frequency noise even further. Two popular such methods are the *autozero* (AZ) and *chopper stabilization* (CS) techniques. The AZ technique is a *sampling* technique[7] that samples the offset and low-frequency noise and then subtracts it from the contaminated signal to give offset-free appearance. The CS technique is a *modulation* technique[7] that modulates the input signal to a higher frequency where there is no dc offset or low-frequency noise, and then demodulates the amplified signal thus stripped of offset and low-frequency errors back to the baseband.

Figure 5.16 illustrates the AZ principle for the case of the ICL7650S op amp (Harris Semiconductor), the first popular op amp to realize this technique in monolithic form. The heart of the device is OA_1, a conventional, high-speed amplifier referred to as the *main amplifier*. A second amplifier, called the *nulling amplifier* and denoted as OA_2, continuously monitors OA_1's input offset error V_{OS1} and drives it

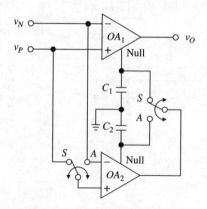

FIGURE 5.16
Chopper-stabilized op amp (CSOA).

to zero by applying a suitable correcting voltage at OA_1's null pin. This mode of operation is called the *sampling mode.*

Note, however, that OA_2 too has an input offset V_{OS2}, so it must correct its own error before attempting to improve OA_2's error. This is achieved by momentarily disconnecting OA_2 from the main amplifier, shorting its inputs together, and coupling its output to its own null pin. This mode, referred to as the *autozero mode,* is activated by flipping the MOS switches from the S (sampling) position to the A (autozero) position. During the autozero mode, the correction voltage for OA_1 is momentarily held by C_1, which therefore acts as an analog memory for this voltage. Similarly, C_2 holds the correction voltage for OA_2 during the sampling mode.

Alternation between the two modes takes place at a typical rate of a few hundred cycles per second, and is controlled by an on-chip oscillator, making the AZ operation transparent to the user. The error-holding capacitors (0.1 μF for the aforementioned ICL7650S) are supplied off-chip by the user. The room-temperature rating for the ICL7650S is $V_{OS} = \pm 0.7$ μV.

Like AZ op amps, CS op amps also utilize a pair of capacitors to realize the modulation/demodulation function. In some devices these capacitors are encapsulated in the IC package itself to save space. Examples of this type of CS op amp are the LTC1050 (Linear Technology) with $V_{OS} = 0.5$ μV and $TC(V_{OS}) = 0.01$ μV/°C typical, and the MAX420 (Maxim) with $V_{OS} = 1$ μV and $TC(V_{OS}) = 0.02$ μV/°C.

The impressive dc specifications of AZ and CS op amps do not come for free, however. Since the nulling circuit is a sampled-data system, clock-feedthrough noise and frequency aliasing problems arise, which need be taken into consideration when selecting the device best suited to the application.

AZ and CS op amps can be used either alone or as part of composite amplifiers to improve existing input specifications.[8,9] To fully realize these specifications, considerable attention must be paid to circuit board layout and construction.[8,9] Of particular concern are input leakage currents and thermocouple effects arising at the junction of dissimilar metals. They can grossly degrade the input specifications of the device and completely defeat what has been so painstakingly achieved in terms of circuit design. Consult the data sheets for valuable hints in this regard.

5.6
INPUT OFFSET-ERROR COMPENSATION

We are now ready to investigate the effect of I_{OS} and V_{OS} acting simultaneously. We begin with the familiar amplifiers of Fig. 5.17 (ignore the 10-kΩ potentiometers for the time being).

Using Eqs. (5.13) and (5.22), along with the superposition principle, it is readily seen that both circuits yield

$$v_O = A_s v_I + E_O \tag{5.33a}$$

$$E_O = \left(1 + \frac{R_2}{R_1}\right)[V_{OS} - (R_1 \parallel R_2)I_{OS}] = \frac{1}{\beta}E_I \tag{5.33b}$$

where $A_s = -R_2/R_1$ for the inverting amplifier, and $A_s = 1 + R_2/R_1$ for the noninverting one. We call A_s the *signal gain* to distinguish it from the *dc noise gain,*

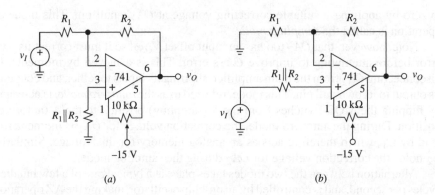

FIGURE 5.17
Inverting and noninverting amplifiers with internal offset-error nulling.

which is $1/\beta = 1 + R_2/R_1$ for *both* circuits. Moreover, $E_I = V_{OS} - (R_1 \parallel R_2)I_{OS}$ is the *total offset error referred to the input,* and E_O the *total offset error referred to the output.* The negative sign does not necessarily imply a tendency by the two terms to compensate for each other, since V_{OS} and I_{OS} may be of either polarity. A prudent designer will take a conservative viewpoint and combine them additively.

The presence of the output error E_O may or may not be a drawback, depending on the application. In audio applications, where dc voltages are usually blocked out through capacitive coupling, offset voltages are seldom of major concern. Not so in low-level signal detection, such as thermocouple or strain-gauge amplification, or in wide dynamic-range applications, such as logarithmic compression and high-resolution data conversion. Here v_I may be of comparable magnitude to E_I, so its information content may easily be obliterated. The problem then arises of reducing E_I below a tolerable level.

Turning next to the integrator of Fig. 5.18, we use Eqs. (5.18) and (5.23) and the superposition principle to write

$$v_O(t) = -\frac{1}{RC} \int_0^t [v_I(\xi) + E_I]\, d\xi + v_O(0) \tag{5.34a}$$

$$E_I = RI_{OS} - V_{OS} \tag{5.34b}$$

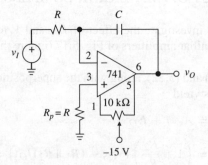

FIGURE 5.18
Integrator with internal offset-error nulling.

Now the effect of V_{OS} and I_{OS} is to offset v_I by the error E_I. Even with $v_I = 0$, the output will ramp up or down until saturation is reached.

The input-referred error E_I in Eqs. (5.33b) and (5.34b) can be nulled by means of a suitable trimmer, as we are about to see. However, as we know, trimmers increase production costs and drift with temperature and time. A wise designer will try minimizing E_I by a combination of circuit tricks, such as resistance scaling and op amp selection. Only as a last resort should one turn to trimmers. Offset nulling techniques are classified as *internal* and *external*.

Internal Offset Nulling

Internal nulling is based on the deliberate unbalancing of the input stage to make up for inherent mismatches and drive the error to zero. This imbalance is introduced by means of an external trimmer, as recommended in the data sheets. Figure 5.3 shows the trimmer connection for the internal nulling for the 741 op amp. The input stage consists of two nominally identical halves: the Q_1-Q_3-Q_5-R_1 half to process v_P and the Q_2-Q_4-Q_6-R_2 half to process v_N. Varying the wiper away from its center position will place more resistance in parallel with one side and less with the other, thus unbalancing the circuit. To calibrate the amplifiers of Fig. 5.17, we set $v_I = 0$ and we adjust the wiper for $v_O = 0$. To calibrate the integrator of Fig. 5.18, we set $v_I = 0$ and we adjust the wiper for v_O as steady as possible in the vicinity of 0 V.

From the 741C data sheets of Fig. 5A.3, we note that the *offset-voltage adjustment range* is typically ± 15 mV, indicating that for this compensating scheme to succeed we must have $|E_I| < 15$ mV. Since the 741C has $V_{OS} = 6$ mV maximum, this leaves 9 mV for the offset term due to I_{OS}. If this term exceeds 9 mV, we must either scale down the external resistances or resort to external nulling, to be discussed below.

> **EXAMPLE 5.7.** A 741C op amp is to be used in the circuit of Fig. 5.17a to yield $A_s = -10$ V/V. Specify suitable resistances that will maximize the input resistance R_i of the circuit.
>
> **Solution.** Since $R_i = R_1$, we need to maximize R_1. Imposing $R_2 = 10R_1$, and $V_{OS(max)} + (R_1 \| R_2)I_{OS(max)} \le 15$ mV, we get $R_1 \| R_2 \le (15 \text{ mV} - 6 \text{ mV})/(200 \text{ nA}) = 45$ kΩ, or $1/R_1 + 1/10R_1 \ge 1/(45 \text{ kΩ})$. Solving yields $R_1 \le 49.5$ kΩ. Use the standard values $R_1 = 47$ kΩ, $R_2 = 470$ kΩ, and $R_p = 43$ kΩ.

Internal offset nulling can be applied to any of the circuits studied so far. In general, the nulling scheme varies from one op amp family to another. For instance, Fig. 5.7 indicates that internal nulling of the LF356 op amp is accomplished with a 25-kΩ potentiometer with the wiper at V_{CC}. To find the recommended nulling scheme for a given device, consult the data sheets. We observe that dual- and quad-op-amp packages usually do not have provisions for internal nulling due to lack of available pins.

External Offset Nulling

External nulling is based on the injection of an adjustable voltage or current into the circuit to compensate for its offset error. This scheme does not introduce any additional imbalances in the input stage, so there is no degradation in drift, CMRR, or PSRR.

The most convenient point of injection of the correcting signal depends on the particular circuit. For inverting-type configurations like the amplifier and integrator of Fig. 5.19, we simply lift R_p off ground and return it to an adjustable voltage V_X. By the superposition principle, we now have an apparent input error of $E_I + V_X$, and we can always adjust V_X to neutralize E_I. V_X is obtained from a dual reference source, such as the supply voltages if they are adequately regulated and filtered. In the circuits shown, we impose $R_B \gg R_C$ to avoid excessive loading at the wiper, and $R_A \ll R_p$ to avoid perturbing the existing resistance levels. The calibration procedure is similar to that for internal nulling.

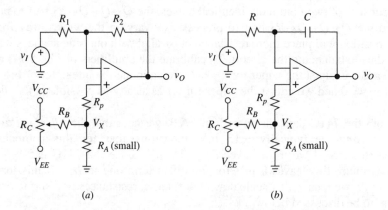

(a)   (b)

FIGURE 5.19
External offset-error nulling for the inverting amplifier and integrator.

EXAMPLE 5.8. A 741C op amp is to be used in the circuit of Fig. 5.19a to yield $A_s = -5$ V/V and $R_i = 30$ kΩ. Specify suitable resistances.

Solution. $R_1 = 30$ kΩ, $R_2 = 5R_1 = 150$ kΩ, and $R_p = R_1 \parallel R_2 = 25$ kΩ. Use the standard value $R_p = 24$ kΩ, and impose $R_A = 1$ kΩ to make up for the difference. We have $E_{I(\text{max})} = V_{OS(\text{max})} + (R_1 \parallel R_2)I_{OS(\text{max})} = 6$ mV $+ (25$ k$\Omega) \times (200$ nA$) = 11$ mV. To be on the safe side, impose -15 mV $\leq V_X \leq 15$ mV. Thus, with the wiper all the way up, we want $R_A/(R_A + R_B) = (15$ mV$)/(15$ V$)$, or $R_B \cong 10^3 R_A = 1$ MΩ. Finally, choose $R_C = 100$ kΩ.

In principle, the foregoing scheme can be applied to any circuit that comes with a dc return to ground. In the circuit of Fig. 5.20, R_1 has been lifted off ground and returned to the adjustable voltage V_X. To avoid upsetting the signal gain we must impose $R_{\text{eq}} \ll R_1$, where R_{eq} is the equivalent resistance of the nulling network as seen by R_1 (for $R_A \ll R_B$ we have $R_{\text{eq}} \cong R_A$.) Alternatively, we must decrease R_1 to the value $R_1 - R_{\text{eq}}$.

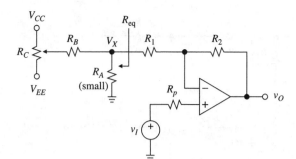

FIGURE 5.20
External offset-error nulling for the noninverting amplifier.

EXAMPLE 5.9. Assuming a 741C op amp in Fig. 5.20, specify suitable resistances for
(a) $A_s = 5$ V/V, and (b) $A_s = 100$ V/V.

Solution.

(a) We want $A_s = 1 + R_2/R_1 = 5$, or $R_2 = 4R_1$. Pick $R_1 = 25.5$ kΩ, 1%, and $R_2 = 102$ kΩ, 1%. Then $R_p \cong 20$ kΩ. Moreover, $E_{O(\text{max})} = (1/\beta)E_{I(\text{max})} = 5[6$ mV $+ (20$ kΩ$) \times (200$ nA$)] = 50$ mV. To balance this out we need $V_X = E_{O(\text{max})}/(-R_2/R_1) = 50/(-4) = -12.5$ mV. Pick a range of ± 15 mV to make sure. To avoid upsetting A_s, choose $R_A \ll R_1$, say, $R_A = 100$ Ω. Then, imposing $R_A/(R_A + R_B) = (15$ mV$)/(15$ V$)$ yields $R_B \cong 10^3 R_A = 100$ kΩ. Finally, let $R_C = 100$ kΩ.

(b) Now $1 + R_2/R_1 = 100$, or $R_2 = 99R_1$. Let $R_2 = 100$ kΩ, so $R_1 = 1010$ Ω. If we were to use $R_A = 100$ Ω as before, R_A would no longer be negligible compared to R_1. So let $R_1 = 909$ Ω, 1%, and $R_A = 1010 - 909 = 101$ Ω (use 102 Ω, 1%), so that the series $(R_1 + R_A)$ still ensures $A_s = 100$ V/V. Moreover, let $R_p \cong 1$ kΩ. Then, $E_{O(\text{max})} = 100[6$ mV $+ (1$ kΩ$) \times (200$ nA$)] = 620$ mV, and $V_X = E_{O(\text{max})}/(-R_2/R_1) = 620/(-10^5/909) = -5.6$ mV. Pick a range of ± 7.5 mV to make sure. Imposing $R_A/(R_A + R_B) = (7.5$ mV$)/(15$ V$)$ gives $R_B \cong 2000R_A \cong 200$ kΩ. Finally, let $R_C = 100$ kΩ.

In multiple-op-amp circuits it is worth seeking ways of nulling the cumulative offset error with just one adjustment. A classic example is offered by the triple-op-amp IA, where other critical parameters may also need adjustment, such as gain and CMRR.

In the circuit of Fig. 5.21, the voltage V_X is buffered by the low-output-impedance follower OA_4 to avoid upsetting bridge balance. The overall CMRR is the combined result of resistance mismatches and finite CMRRs of the individual op amps. At dc, where C_1 acts as an open circuit and R_9 has thus no effect, we adjust R_{10} to optimize the dc CMRR. At some high frequency, where C_1 provides a conductive path from R_9's wiper to ground, we adjust R_9 to deliberately unbalance the second stage and thus optimize the ac CMRR. The circuit is calibrated as follows:

1. With v_1 and v_2 grounded, adjust R_C for $v_O = 0$.
2. Adjust R_8 for the desired gain of 1000 V/V.
3. With the inputs tied together to a common source v_I, adjust R_{10} for the minimum change in v_O as v_I is switched from -10 V dc to $+10$ V dc.
4. With v_I a 10-kHz, 20-V peak-to-peak sine wave, adjust R_9 for the minimum ac component at the output.

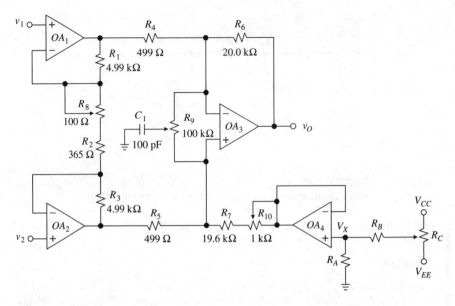

FIGURE 5.21
Instrumentation amplifier with $A = 1$ V/mV. (OA_1, OA_2, and OA_3: OP-37C; OA_4: OP-27; fixed resistances are 0.1%.)

EXAMPLE 5.10. Specify R_A, R_B, and R_C in Fig. 5.21, given the following maximum ratings for the OP-37C low-noise precision high-speed op amp at $T = 25\,°C$: $I_B = 75$ nA, $I_{OS} = \pm 80$ nA, and $V_{OS} = 100\,\mu$V. Assume ± 15-V supplies.

Solution. $E_{I1} = E_{I2} = V_{OS} + [R_1 \,\|\, (R_2 + R_8/2)]I_B = 10^{-4} + (5000 \,\|\, 208)75 \times 10^{-9} \cong 115\,\mu$V; $E_{I3} = 10^{-4} + (500 \,\|\, 20{,}000)80 \times 10^{-9} \cong 139\,\mu$V; $E_O = A(E_{I1} + E_{I2}) + (1/\beta_3)E_{I3} = 10^3 \times 2 \times 115 + (1 + 20/0.5)139 \cong 230$ mV $+ 5.7$ mV $= 236$ mV. According to Eq. (2.40) we need -236 mV $\leq V_X \leq +236$ mV. Use 300 mV to make sure. Then, $R_A = 2\,k\Omega$, $R_B = 100\,k\Omega$, $R_C = 100\,k\Omega$.

Whether internal or external, nulling compensates only for the initial offset error V_{OS0}. As the operating conditions change, the error will reemerge, and if it rises above an intolerable level, it must be nulled periodically. The use of AZ or CS op amps may then be a preferable alternative.

5.7
MAXIMUM RATINGS

Like all electronic devices, op amps require that the user respect certain electrical and environmental limits. Exceeding these limits will generally result in malfunction or even damage. The range of operating temperatures over which op amp ratings are given are the *commercial range* (0 °C to + 70 °C), the *industrial range* (−25 °C to +85 °C), and the *military range* (−55 °C to +125 °C).

Absolute Maximum Ratings

These are the ratings that, if exceeded, are likely to cause permanent damage. The most important ones are the *maximum supply voltages,* the *maximum differential-mode* and *common-mode input voltages,* and the *maximum internal power dissipation* P_{max}.

Figure 5A.1 indicates that for the 741C the maximum voltage ratings are, respectively, ± 18 V, ± 30 V, and ± 15 V. (The large differential-mode rating of the 741 is made possible by the lateral *pnp* BJTs Q_3 and Q_4.) Exceeding these limits may trigger internal reverse-breakdown phenomena and other forms of electrical stress, whose consequences are usually detrimental, such as irreversible degradation of gain, input bias and offset currents, and noise, or permanent damage to the input stage. It is the user's responsibility to ensure that the device operates below its maximum ratings under all possible circuit and signal conditions.

Potentially deleterious conditions may arise during power turn-on and turn-off. Since different parts of a system may go on or off at different times, especially if large capacitors are present, the voltages at the input pins may momentarily exceed those at the supply pins. To prevent damage, the inputs must be equipped with suitable diode clamps to limit the input voltages, and series resistances to limit current during clamping.[9] For example, the op amp of Fig. 5.15 comes with input clamps already on-chip.

Exceeding P_{max} will raise the chip temperature to intolerable levels and cause internal component damage. The value of P_{max} depends on the package type as well as the ambient temperature. The popular mini DIP package has $P_{\text{max}} = 310$ mW up to 70 °C of ambient temperature, and derates linearly by 5.6 mW/°C beyond 70 °C.

> **EXAMPLE 5.11.** What is the maximum current that a mini DIP 741C op amp is allowed to source at 0 V if $T \leq 70$ °C? If $T = 100$ °C?
>
> **Solution.** From Fig. 5A.3 we find the supply current to be $I_Q = 2.8$ mA maximum. Recall from Section 1.8 that an op amp sourcing current dissipates $P = (V_{CC} - V_{EE})I_Q + (V_{CC} - V_O)I_O = 30 \times 2.8 + (15 - V_O)I_O$. Imposing $P \leq 310$ mV gives $I_O(V_O = 0) \leq (310 - 84)/15 \cong 15$ mA for $T \leq 70$ °C. For $T = 100$ °C we have $P_{\text{max}} = 310 - (100 - 70)5.6 = 142$ mW, so now $I_O(V_O = 0) = (142 - 84)/15 \cong 3.9$ mA.

Input Voltage Range

This is the range of input voltages over which the op amp will still operate properly. From Fig. 5A.3 we find that for the 741C this range is typically ± 13 V. Operating the device outside this range, but still below its maximum input voltage rating (between ± 13 V and ± 15 V for the 741C), does not necessarily cause damage; it only results in malfunction, such as causing output saturation or output polarity reversal.

Even though the data sheets provide all the information the user needs to know about the input voltage range, it is instructive to investigate its origin. For bipolar devices such as the 741 op amp, this is the range of input voltages for which each BJT still operates in the forward-active (FA) region, all the way to the edge of saturation (EOS). This type of operation is defined as $v_{BE} = V_{BE(\text{on})} \cong 0.7$ V and $v_{CE} \geq V_{CE(\text{EOS})} \cong 0.1$ V for *npn* BJTs, $v_{EB} = V_{EB(\text{on})} \cong 0.7$ V and $v_{EC} \geq V_{EC(\text{EOS})} \cong 0.1$ V for *pnp* BJTs.

With reference to the 741 diagram of Fig. 5A.2, we observe that to keep Q_2 and Q_8 in the FA region, we need $v_N \leq V_{CC} - V_{EB8(on)} - V_{CB2(EOS)} \cong V_{CC} - 0.7 - (-0.6) = V_{CC} - 0.1$ V; to keep Q_2, Q_4, Q_{16}, and Q_{17} in the FA region, we need $v_N \geq V_{EE} + V_{BE17(on)} + V_{BE16(on)} + V_{EC4(EOS)} + V_{BE2(on)} \cong V_{EE} + 0.7 + 0.7 + 0.1 + 0.7 = V_{EE} + 2.2$ V. Since v_N tracks v_P, the permissible input range is from $V_{EE} + 2.2$ V to $V_{CC} - 0.1$ V. This range depends on V_{CC} and V_{EE}; the higher the supply voltages, the wider this range. Figure 5A.6 shows the 741 input range as a function of the supply voltages.

Op amps specifically designed for an input range extending all the way down to V_{EE} are called *single-supply* op amps because they can be powered between $V_{CC} = V_S$ and $V_{EE} = 0$ V and still provide a virtual ground at the inverting input. These devices find application in battery-operated equipment and single-supply digital systems. A popular example is the LM324 (National Semiconductor), whose input range for single-supply operation extends from $(V_S - 1.5$ V) all the way down to 0 V.

Output Voltage Swing

As we know, this is the range $V_{OL} \leq v_O \leq V_{OH}$, and is usually specified for a 2-kΩ output load. It is again instructive to estimate this range directly from the circuit diagram of Fig. 5A.2. Thus, $V_{OH} = V_{CC} - V_{EC13(sat)} - V_{BE14(on)} - V_{R_6} \cong V_{CC} - 0.1 - 0.7 - 0 = V_{CC} - 0.8$ V. Likewise, $V_{OL} = V_{EE} + V_{CE17(min)} + V_{EB22(on)} + V_{EB20(on)} + V_{R_7} \cong V_{EE} + 0.7 + 0.7 + 0.7 + 0 = V_{EE} + 2.1$ V. For ± 15-V supplies this gives $V_{OH} \cong 14.2$ V and $V_{OL} \cong -12.9$ V, in reasonable agreement with the data sheets. As with the input range, the higher the supply voltages, the wider the output swing. This is illustrated in Fig. 5A.6.

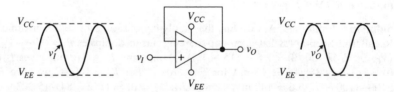

FIGURE 5.22
Waveforms for a voltage follower with rail-to-rail input and output capabilities.

Op amps specifically designed for an output range extending all the way up to V_{CC} and down to V_{EE} are called *rail-to-rail op amps*. As we know, CMOS op amps belong to this class of devices, though rail-to-rail op amps are available also in bipolar technology. The LMC6464 CMOS op amp (National Semiconductor) offers rail-to-rail capabilities both at the input and at the output. Figure 5.22 shows the input and output waveforms of a voltage follower implemented with an op amp possessing such capabilities.

Overload Protection

To prevent excessive power dissipation in case of output overload, op amps are equipped with protective circuitry designed to limit the output current below a safety level called the *output short-circuit current* I_{sc}. The 741C has typically $I_{sc} \cong 25$ mA.

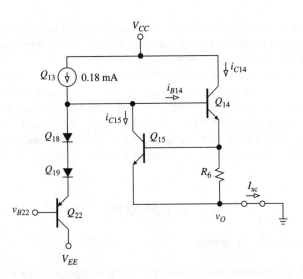

FIGURE 5.23
Partial illustration of overload protection circuitry for the
741 op amp.

In the 741 diagram of Fig. 5A.2, overload protection is provided by the watchdog BJTs Q_{15} and Q_{21} and the current-sensing resistors R_6 and R_7. Under normal conditions these BJTs are off. However, should an output overload condition arise, such as an accidental short circuit, the resistance sensing the overload current will develop enough voltage to turn on the corresponding watchdog BJT; this, in turn, will limit the current through the corresponding output-stage BJT.

To illustrate with an example, suppose the op amp is designed to output a positive voltage, but an inadvertent output short forces v_O to 0 V, as depicted in Fig. 5.23. In response to this short, the second stage of the op amp will drive v_{B22} as positive as it can in a futile attempt to raise v_O. Consequently, Q_{22} will go off and let the entire bias current of 0.18 mA flow toward the base of Q_{14}. Were it not for the presence of Q_{15}, Q_{14} would amplify this current by β_{14} while sustaining $V_{CE} = V_{CC}$; the resulting power dissipation would most likely destroy it. However, with Q_{15} in place, only the current $i_{B14(\text{max})} = i_{C14(\text{max})}/\beta_{14} \cong [V_{BE15(\text{on})}/R_6]/\beta_{14}$ is allowed to reach the base of Q_{14}, the remainder being diverted to the output short via Q_{15}; hence, Q_{14} is protected.

With reference to Fig. 5A.2, we observe that just like Q_{15} protects Q_{14} when the op amp is *sourcing* current, Q_{21} protects Q_{20} during current *sinking*. However, since the base of Q_{20} is a low-impedance node because it is driven by emitter-follower Q_{22}, the action of Q_{21} is applied further upstream, via Q_{23}.

EXAMPLE 5.12. Find all currents in the circuit of Fig. 5.23 if $R_6 = 27\ \Omega$, $\beta_{14} = \beta_{15} = 250$, and $V_{BE15(\text{on})} = 0.7$ V.

Solution. Q_{14} is limited to $I_{C14} = \alpha_{14}I_{E14} = \alpha_{14}[I_{R_6} + I_{B15}] \cong I_{R_6} = V_{BE15(\text{on})}/R_6 = 0.7/27 \cong 26$ mA. The current reaching the base of Q_{14} is $I_{B14} = I_{C14}/\beta_{14} = 26/250 \cong 0.104$ mA; the remainder, $I_{C15} = 0.18 - 0.104 \cong 76\ \mu$A, is diverted to the short. Hence, $I_{sc} \cong I_{C14} + I_{C15} \cong 26$ mA.

It is important to realize that during overload the actual output voltage is not what it should be: the protection circuitry prevents the op amp from properly influencing v_N, so during overload we generally have $v_N \neq v_P$.

Op amp types are available with much higher output current capabilities than the 741. Aptly referred to as *power op amps*, they are similar to their low-power counterparts except for the presence of heftier output stages and proper power packaging to handle the increased dissipation of heat. These op amps usually require heatsink mounting. Examples of power op amps are the PA04 (Apex Microtechnology) and the OPA501 (Burr-Brown), with peak output-current capabilities of 20 A and 10 A, respectively.

PROBLEMS

5.1 Simplified op amp circuit diagram

5.1 If the 741 op amp were redesigned with I_A twice as large, which of the op amp parameters discussed in Section 5.1 would be affected, and how?

5.2 Input bias and offset currents

5.2 The circuit of Fig. 5.5a is to be used as an inverting amplifier with a gain of 10 V/V and is to employ the μA741C op amp. Specify suitable component values to ensure a maximum output error of 10 mV with minimum power dissipation in the resistors.

5.3 (a) Investigate the effect of I_B on the performance of the inverting amplifier of Fig. P1.54 if $I_B = 10$ nA and all resistances are 100 kΩ. (b) What dummy resistance R_p must be installed in series with the noninverting input to minimize E_O?

5.4 Investigate the effect of I_B and I_{OS} on the performance of the circuit of Fig. P1.17 if $I_B = 100$ nA and $I_{OS} = 10$ nA.

5.5 The circuit of Fig. P5.5 exploits the matching properties of dual op amps to minimize the overall input current I_I. (a) Find the condition between R_2 and R_1 that yields $I_I = 0$ when the op amps are perfectly matched. (b) What if there is a 10% mismatch between the I_Bs of the op amps?

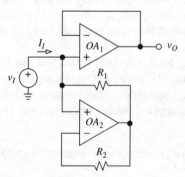

FIGURE P5.5

5.6 (a) Investigate the effect of I_{OS} on the performance of the Deboo integrator. (b) Assuming $C = 1$ nF and 100-kΩ resistances throughout, find $v_O(t)$ if $I_{OS} = \pm 1$ nA and $v_O(0) = 1$ V.

5.7 Investigate the effect of using an op amp with $I_B = 1$ nA and $I_{OS} = 0.1$ nA in the high-sensitivity I-V converter of Example 2.2. What dummy resistance R_p would you install in series with the noninverting input?

5.8 If $R_4/R_3 = R_2/R_1$, the circuit of Fig. P2.14 is a true V-I converter with $i_O = (R_2/R_1R_5) \times (v_2 - v_1)$ and $R_o = \infty$. What if the op amps have input bias currents I_{B1} and I_{B2}, and input offset currents I_{OS1} and I_{OS2}? Is i_O affected? Is R_o affected? How would you modify the circuit to optimize its dc performance?

5.9 Investigate the effect of I_B and I_{OS} in the current amplifier of Fig. 2.11. How would you modify the circuit to minimize its dc error?

5.10 Assuming the multiple-feedback low-pass filter of Fig. 3.32 is in dc steady state (i.e., all transients have died out), investigate the effect of $I_B = 50$ nA if all resistances are 100 kΩ. What dummy resistor would you use to optimize the dc performance of the circuit? *Hint:* Assume a zero input

5.3 Low-input-bias-current op amps

5.11 The bottom plate of a low-leakage 10-nF charged capacitor is at ground, and the top plate at 10 V. Next, a voltage follower is connected to the top plate, and the follower output is monitored with a voltmeter to observe how the input bias current discharges the capacitor. (*a*) If it is found that the output decreases at the rate of 1 mV/s, what do you conclude about the technology of the input stage? (*b*) Estimate the temperature rise needed for a discharge rate of 0.1 V/s.

5.4 Input offset voltage

5.12 A FET-input op amp is connected as in Fig. 5.12*a* with $R_1 = 100$ Ω and $R_2 = 33$ kΩ, and gives $v_O = -0.5$ V. The same op amp is then moved to the circuit of Fig. 5.12*b* with $R = 100$ kΩ and $C = 1$ nF. Assuming $v_O(0) = 0$ and symmetric saturation voltages of ± 14 V, find the time it takes for the output to saturate.

5.13 If $R_4/R_3 = R_2/R_1$, the circuit of Fig. P2.15 is a true V-I converter with $i_O = R_2v_I/R_1R_5$ and $R_o = \infty$. What if the op amps have input offset voltages V_{OS1} and V_{OS2}, but are otherwise ideal? Is i_O affected? Is R_o affected?

5.14 In the circuit of Fig. 5.12*a* let $R_1 = 10$ Ω and $R_2 = 100$ kΩ, and let the op amp have an offset drift of 5 μV/°C. (*a*) If the op amp has been trimmed for $v_O(25\,°C) = 0$, estimate $v_O(0\,°C)$ and $v_O(70\,°C)$. What do you expect their relative polarities to be? (*b*) If the same op amp is moved to the circuit of Fig. 5.12*b* with $R = 100$ kΩ and $C = 1$ nF, find $v_O(t)$ both at 0 °C and at 70 °C.

5.15 Investigate the effect of using an op amp with $CMRR_{dB} = 100$ dB on the output resistance of a Howland current pump made up of four perfectly matched 10-kΩ resistances. Except for CMRR, the op amp is ideal.

5.16 Investigate the effect of using an op amp with $V_{OS0} = 100$ μV and $CMRR_{dB} = 100$ dB in a Deboo integrator which uses four perfectly matched 100-kΩ resistances and a 1-nF capacitance. Except for V_{OS0} and CMRR, the op amp is ideal.

5.17 Assuming perfectly matched resistances in the difference amplifier of Fig. 2.13*a*, show that if we define the CMRR of the op amp as $1/CMRR_{OA} = \partial V_{OS}/\partial v_{CM(OA)}$ and that of the difference amplifier as $1/\mathrm{CMRR_{DA}} = A_{cm}/A_{dm}$, where $A_{cm} = \partial v_O/\partial v_{CM(DA)}$ and $A_{dm} = R_2/R_1$, then we have $\mathrm{CMRR_{DA}} = \mathrm{CMRR_{OA}}$.

5.18 The difference amplifier of Problem 5.17 uses a 741 op amp with $R_1 = 1\ \mathrm{k\Omega}$ and $R_2 = 100\ \mathrm{k\Omega}$. Find the worst-case CMRR of the circuit for the case of (a) perfectly matched resistances, and (b) 1% resistances. Comment.

5.19 In the difference amplifier of Problem 5.18 the inputs are tied together and are driven by $v_{CM} = 1 \sin 2\pi ft$ V. Using the CMRR plot of Fig. 5A.6, predict the output at $f = 1$ Hz, 1 kHz, and 10 kHz.

5.20 (a) Assuming perfectly matched op amps and resistances in the dual-op-amp IA of Fig. 2.23, show that if we define the CMRR of each op amp as $1/\mathrm{CMRR_{OA}} = \partial V_{OS}/\partial v_{CM(OA)}$ and that of the IA as $1/\mathrm{CMRR_{IA}} = A_{cm}/A_{dm}$, where $A_{cm} = \partial v_O/\partial v_{CM(DA)}$ and $A_{dm} = 1 + R_2/R_1$, then we have $\mathrm{CMRR_{IA(min)}} = 0.5 \times \mathrm{CMRR_{OA(min)}}$. (b) If an IA with a gain of 100 V/V is implemented with perfectly matched resistances and a dual OP-227A op amp, ($\mathrm{CMRR_{dB}} = 126$ dB typical, 114 dB minimum), find the worst-case output change for a 10-V common-mode input change. What is the corresponding A_{cm}?

5.21 Assuming perfectly matched op amps and resistances in the triple-op-amp IA of Fig. 2.20, derive a relationship between $\mathrm{CMRR_{IA(min)}}$ and $\mathrm{CMRR_{OA(min)}}$, where $1/\mathrm{CMRR_{OA}} = \partial V_{OS}/\partial v_{CM(OA)}$, and $1/\mathrm{CMRR_{IA}} = A_{cm}/A_{dm}$.

5.22 In the inverting integrator of Fig. 1.19 let $R = 100\ \mathrm{k\Omega}$, $C = 10$ nF, and $v_I = 0$, and let the capacitor be initially charged such that $v_O(t = 0) = 10$ V. Except for a finite open-loop gain of 10^5 V/V, the op amp is ideal. Find $v_O(t > 0)$.

5.23 An op amp with $a_{min} = 10^4$ V/V, $V_{OS0(max)} = 2$ mV, and $\mathrm{CMRR_{dB(min)}} = \mathrm{PSRR_{dB(min)}} = 74$ dB is configured as a voltage follower. (a) Estimate the worst-case departure of v_O from the ideal for $v_I = 0$ V. (b) Repeat with $v_I = 10$ V. (c) Repeat if the supply voltages are lowered from ± 15 V to ± 12 V.

5.5 Low-input-offset-voltage op amps

5.24 With reference to the op amp of Fig. 5.1, investigate the effect of (a) a 10% mismatch between the emitter areas of Q_1 and Q_2, and (b) a 1 °C temperature gradient across Q_1 and Q_2.

5.6 Input offset-error compensation

5.25 Repeat Example 5.8, but for the integrator of Fig. 5.19b for the case $R = 100\ \mathrm{k\Omega}$.

5.26 In the noninverting amplifier of Fig. 1.14a let $R_1 = 10\ \Omega$, $R_2 = 10\ \mathrm{k\Omega}$, and $v_I = 0$. The output v_O is monitored with a voltmeter and is found to be 0.480 V. If adding a 1-MΩ resistor in series with the noninverting input pin gives $v_O = 0.780$ V, but adding it in series with the inverting input pin gives $v_O = 0.230$ V, find I_B, I_{OS}, and V_{OS}. What is the direction of I_B?

5.27 Figure P5.27 shows a widely used test fixture to characterize the op amp referred to as *device under test* (DUT). The purpose of OA_2, which is assumed ideal, is to keep DUT's output near 0 V, or in the middle of the linear region. Find V_{OS0}, I_P, I_N, I_B, I_{OS}, and the gain a for the DUT, given the following measurements: (a) $v_2 = -0.75$ V with SW_1 and SW_2 closed and $v_1 = 0$ V; (b) $v_2 = +0.30$ V with SW_1 closed, SW_2

open, and $v_1 = 0$ V; (c) $v_2 = -1.70$ V with SW_1 open, SW_2 closed, and $v_1 = 0$ V; (d) $v_2 = -0.25$ V with SW_1 and SW_2 closed, and $v_1 = -10$ V.

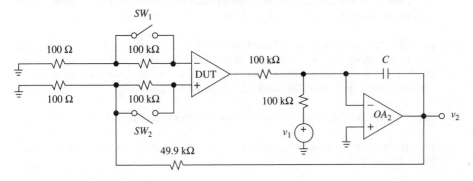

FIGURE P5.27

5.28 (a) In the circuit of Fig. P1.15 obtain an expression for the output error E_O as a function of I_P, I_N, and V_{OS}. (b) Repeat, but for the circuit of Fig. P1.16. *Hint:* In each case set the independent source to zero.

5.29 Repeat Problem 5.28, but for the circuits of Figs. P1.18 and P1.19.

5.30 In the circuit of Fig. P1.60 obtain an expression for the output error E_O as a function of I_P, I_N, and V_{OS}. *Hint:* Assume a zero input.

5.31 (a) Find output error E_O for the *I-V* converter of Fig. 2.1. (b) Repeat if the noninverting input pin is returned to ground via a dummy resistance $R_p = R$. (c) Devise a scheme for the external nulling of E_O if $R = 1$ MΩ, $I_{OS} = 1$ nA maximum, and $V_{OS} = 1$ mV maximum.

5.32 What input-stage technology would you choose for the op amp of the high-sensitivity *I-V* converter of Example 2.2? How would you modify the circuit for a minimum output error E_O? How would you make provision for the external nulling of E_O?

5.33 Using the OP-227A dual-precision op amp ($V_{OS(max)} = 80$ μV, $I_{B(max)} = \pm 40$ nA, $I_{OS(max)} = 35$ nA, and CMRR$_{dB(min)} = 114$ dB), design a dual-op-amp IA with a gain of 100 V/V. Assuming perfectly matched resistances, what is the maximum output error for $v_1 = v_2 = 0$? For $v_1 = v_2 = 10$ V?

5.34 If $R_2 + R_3 = R_1$, the circuit of Fig. P2.16 is a true *V-I* converter with $i_O = v_I/R_3$ and $R_o = \infty$. What if the op amps have nonzero input bias and offset currents and offset voltages? Is i_O affected? Is R_o affected? How would you make provisions for minimizing the total error? For externally nulling it?

5.35 (a) Investigate the effect of the offset voltages V_{OS1} and V_{OS2} on the performance of the dual-op-amp transducer amplifier of Fig. 2.40 for the case $\delta = 0$. (b) Devise a scheme to externally null the output offset error, and illustrate how it works.

5.36 Repeat Problem 5.35, but for the transducer amplifier of Fig. P2.54.

5.37 An *I-V* converter with a sensitivity of 1 V/μA is to be designed using an op amp with $V_{OS(max)} = 1$ mV and $I_{OS(max)} = 2$ nA. Two alternatives are being evaluated, namely, the circuit of Fig. 2.1 with $R = 1$ MΩ, and the circuit of Fig. 2.2 with $R = 100$ kΩ, $R_1 = 2.26$ kΩ, and $R_2 = 20$ kΩ; both circuits use an appropriate dummy resistance R_p to minimize the error due to I_B. Which circuit is preferable from the viewpoint of minimizing the untrimmed output error? What is the main reason?

5.38 Assuming the multiple-feedback band-pass filter of Example 3.15 is in dc steady state (i.e., all transients have died out), investigate the effect of $I_B = 50$ nA, $I_{OS} = 5$ nA, and $V_{OS} = 1$ mV upon the circuit's performance. How would you modify the circuit to minimize the output error? To null it? *Hint:* Assume a zero input.

5.39 Repeat Problem 5.38, but for the low-pass *KRC* filter of Example 3.8.

5.40 Repeat Problem 5.38, but for the band-bass and band-reject *KRC* filters of Examples 3.13 and 3.14.

5.41 The biquad filter of Example 3.19 is implemented with FET-input op amps having maximum input offset voltages of 5 mV. Investigate the effect on circuit performance and devise a method to trim the output dc error for the low-pass output.

5.7 Maximum ratings

5.42 Let the inverting amplifier in the single-supply system of Fig. 1.40 be a rail-to-rail op amp with a gain of -2 V/V. (*a*) Sketch and label v_I, v_D, and v_O if v_I is a 1-kHz sine wave with 1.5-V peak amplitude. (*b*) Find the expression for the input sine wave that will result in a rail-to-rail output.

5.43 A 741 op amp is connected as a voltage follower and programmed to give $v_O = 10$ V. Using the simplified circuit of Fig. 5.23 with $R_6 = 27$ Ω, β_{FS} of 250, and base-emitter junction drops of 0.7 V, find v_{B22}, i_{C14}, i_{C15}, $P_{Q_{14}}$, and v_O if the output load is (*a*) $R_L = 2$ kΩ, and (*b*) $R_L = 200$ Ω.

REFERENCES

1. J. E. Solomon, "The Monolithic Operational Amplifier: A Tutorial Study," *IEEE J. Solid-State Circuits,* Vol. SC-9, Dec. 1974, pp. 314–332.
2. P. R. Gray and R. G. Meyer, *Analysis and Design of Analog Integrated Circuits,* 3d ed., John Wiley & Sons, New York, 1993.
3. G. W. Roberts and A. Sedra, *SPICE,* 2d ed., Oxford University Press, New York, 1997.
4. J. Buxton, "Analog Circuit Simulation," *Amplifier Applications Guide,* Analog Devices, Norwood, MA, 1992.
5. G. R. Boyle, B. M. Cohn, D. O. Pederson, and J. E. Solomon, "Macromodeling of Integrated Circuit Operational Amplifiers," *IEEE J. Solid-State Circuits,* Vol. SC-9, Dec. 1974, pp. 353–363.
6. J. Dostál, *Operational Amplifiers,* 2d ed., Butterworth-Heinemann, Stoneham, MA, 1993.
7. C. C. Enz and G. C. Temes, "Circuit Techniques for Reducing the Effects of Op-Amp Imperfections: Autozeroing, Correlated Double Sampling, and Chopper Stabilization," *IEEE Proceedings,* Vol. 84, No. 11, Nov. 1996, pp. 1584–1614.
8. J. Williams, "Chopper-Stabilized Monolithic Op Amp Suits Diverse Uses," *EDN,* February 21, 1985, pp. 305–312, and "Chopper Amplifier Improves Operation of Diverse Circuits," *EDN,* March 7, 1985, pp. 189–207.
9. J. Bryant, J. Buxton, A. Garcia, and J. Wong, "Precision Sensor Signal Conditioning and Transmission," *System Applications Guide,* Analog Devices, Norwood, MA, 1993.

APPENDIX 5A
DATA SHEETS OF THE μA741 OP AMP*

249

APPENDIX 5A
Data Sheets of the
μA741 Op Amp

A Schlumberger Company

μA741
Operational Amplifier

Linear Division Operational Amplifiers

Description

The μA741 is a high performance monolithic operational amplifier constructed using the Fairchild Planar Epitaxial process. It is intended for a wide range of analog applications. High common mode voltage range and absence of latch up tendencies make the μA741 ideal for use as a voltage follower. The high gain and wide range of operating voltage provide superior performance in integrator, summing amplifier, and general feedback applications.

- **No Frequency Compensation Required**
- **Short Circuit Protection**
- **Offset Voltage Null Capability**
- **Large Common Mode And Differential Voltage Ranges**
- **Low Power Consumption**
- **No Latch Up**

Absolute Maximum Ratings

Storage Temperature Range	
Metal Can and Ceramic DIP	−65°C to +175°C
Molded DIP and SO-8	−65°C to +150°C
Operating Temperature Range	
Extended (μA741AM, μA741M)	−55°C to +125°C
Commercial (μA741EC, μA741C)	0°C to +70°C
Lead Temperature	
Metal Can and Ceramic DIP	
(soldering, 60 s)	300°C
Molded DIP and SO-8	
(soldering, 10 s)	265°C
Internal Power Dissipation[1, 2]	
8L-Metal Can	1.00 W
8L-Molded DIP	0.93 W
8L-Ceramic DIP	1.30 W
SO-8	0.81 W
Supply Voltage	
μA741A, μA741, μA741E	± 22 V
μA741C	± 18 V
Differential Input Voltage	± 30 V
Input Voltage[3]	± 15 V
Output Short Circuit Duration[4]	Indefinite

Notes
1. $T_{J \, Max}$ = 150°C for the Molded DIP and SO-8, and 175°C for the Metal Can and Ceramic DIP.
2. Ratings apply to ambient temperature at 25°C. Above this temperature, derate the 8L-Metal Can at 6.7 mW/°C, the 8L-Molded DIP at 7.5 mW/°C, the 8L-Ceramic DIP at 8.7 mW/°C, and the SO-8 at 6.5 mW/°C.
3. For supply voltages less than ± 15 V, the absolute maximum input voltage is equal to the supply voltage.
4. Short circuit may be to ground or either supply. Rating applies to 125°C case temperature or 75°C ambient temperature.

Connection Diagram
8-Lead Metal Package
(Top View)

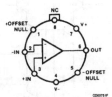

Lead 4 connected to case.

Order Information

Device Code	Package Code	Package Description
μA741HM	5W	Metal
μA741HC	5W	Metal
μA741AHM	5W	Metal
μA741EHC	5W	Metal

Connection Diagram
8-Lead DIP and SO-8 Package
(Top View)

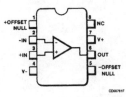

Order Information

Device Code	Package Code	Package Description
μA741RM	6T	Ceramic DIP
μA741RC	6T	Ceramic DIP
μA741SC	KC	Molded Surface Mount
μA741TC	9T	Molded DIP
μA741ARM	6T	Ceramic DIP
μA741ERC	6T	Ceramic DIP
μA741ETC	9T	Molded DIP

FIGURE 5A.1

μA741

Equivalent Circuit

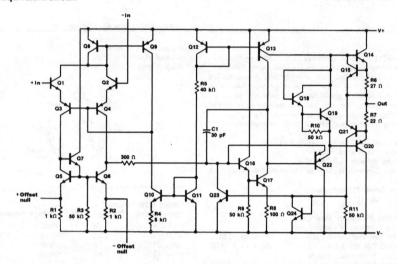

FIGURE 5A.2

μA741

μA741 and μA741C
Electrical Characteristics $T_A = 25°C$, $V_{CC} = \pm 15$ V, unless otherwise specified.

Symbol	Characteristic		Condition	μA741			μA741C			Unit
				Min	Typ	Max	Min	Typ	Max	
V_{IO}	Input Offset Voltage		$R_S \leqslant 10$ kΩ		1.0	5.0		2.0	6.0	mV
$V_{IO\ adj}$	Input Offset Voltage Adjustment Range				± 15			± 15		mV
I_{IO}	Input Offset Current				20	200		20	200	nA
I_{IB}	Input Bias Current				80	500		80	500	nA
Z_I	Input Impedance			0.3	2.0		0.3	2.0		MΩ
I_{CC}	Supply Current				1.7	2.8		1.7	2.8	mA
P_c	Power Consumption				50	85		50	85	mW
CMR	Common Mode Rejection			70			70	90		dB
V_{IR}	Input Voltage Range			± 12	± 13		± 12	± 13		V
PSRR	Power Supply Rejection Ratio				30	150				μV/V
			$V_{CC} = \pm 5.0$ V to ± 18 V					30	150	
I_{OS}	Output Short Circuit Current				25			25		mA
A_{VS}	Large Signal Voltage Gain		$R_L \geqslant 2.0$ kΩ, $V_O = \pm 10$ V	50	200		20	200		V/mV
V_{OP}	Output Voltage Swing		$R_L = 10$ kΩ	± 12			± 12	± 14		V
			$R_L = 2.0$ kΩ	± 10			± 10	± 13		
TR	Transient Response	Rise time	$V_I = 20$ mV, $R_L = 2.0$ kΩ, $C_L = 100$ pF, $A_V = 1.0$		0.3			0.3		μs
		Overshoot			5.0			5.0		%
BW	Bandwidth				1.0			1.0		MHz
SR	Slew Rate		$R_L \geqslant 2.0$ kΩ, $A_V = 1.0$		0.5			0.5		V/μs

FIGURE 5A.3

μA741

μA741 and μA741C (Cont.)
Electrical Characteristics Over the range of −55°C ⩽ T$_A$ ⩽ +125°C for μA741, 0°C ⩽ T$_A$ ⩽ +70°C for μA741C, unless otherwise specified.

Symbol	Characteristic	Condition	μA741			μA741C			Unit
			Min	Typ	Max	Min	Typ	Max	
V$_{IO}$	Input Offset Voltage							7.5	mV
		R$_S$ ⩽ 10 kΩ		1.0	6.0				
V$_{IO\,adj}$	Input Offset Voltage Adjustment Range			±15			±15		mV
I$_{IO}$	Input Offset Current							300	nA
		T$_A$ = +125°C		7.0	200				
		T$_A$ = −55°C		85	500				
I$_{IB}$	Input Bias Current							800	nA
		T$_A$ = +125°C		0.03	0.5				μA
		T$_A$ = −55°C		0.3	1.5				
I$_{CC}$	Supply Current	T$_A$ = +125°C		1.5	2.5				mA
		T$_A$ = −55°C		2.0	3.3				
P$_c$	Power Consumption	T$_A$ = +125°C		45	75				mW
		T$_A$ = −55°C		60	100				
CMR	Common Mode Rejection	R$_S$ ⩽ 10 kΩ	70	90					dB
V$_{IR}$	Input Voltage Range		±12	±13					V
PSRR	Power Supply Rejection Ratio			30	150				μV/V
A$_{VS}$	Large Signal Voltage Gain	R$_L$ ⩾ 2.0 kΩ, V$_O$ = ±10 V	25			15			V/mV
V$_{OP}$	Output Voltage Swing	R$_L$ = 10 kΩ	±12	±14					V
		R$_L$ = 2.0 kΩ	±10	±13		±10	±13		

FIGURE 5A.4

253

APPENDIX 5A
Data Sheets of the
μA741 Op Amp

μA741

μA741A and μA741E
Electrical Characteristics $T_A = 25°C$, $V_{CC} = \pm 15$ V, unless otherwise specified.

Symbol	Characteristic		Condition	Min	Typ	Max	Unit
V_{IO}	Input Offset Voltage		$R_S \leqslant 50$ Ω		0.8	3.0	mV
I_{IO}	Input Offset Current				3.0	30	nA
I_{IB}	Input Bias Current				30	80	nA
Z_I	Input Impedance		$V_{CC} = \pm 20$ V	1.0	6.0		MΩ
P_c	Power Consumption		$V_{CC} = \pm 20$ V		80	150	mW
PSRR	Power Supply Rejection Ratio		$V_{CC} = +10$ V, -20 V to $V_{CC} = +20$ V, -10 V, $R_S = 50$ Ω		15	50	μV/V
I_{OS}	Output Short Circuit Current			10	25	40	mA
A_{VS}	Large Signal Voltage Gain		$V_{CC} = \pm 20$ V, $R_L \geqslant 2.0$ kΩ, $V_O = \pm 15$ V	50	200		V/mV
TR	Transient Response	Rise time	$A_V = 1.0$, $V_{CC} = \pm 20$ V, $V_I = 50$ mV, $R_L = 2.0$ kΩ, $C_L = 100$ pF		0.25	0.8	μs
		Overshoot			6.0	20	%
BW	Bandwidth			0.437	1.5		MHz
SR	Slew Rate		$V_I = \pm 10$ V, $A_V = 1.0$		0.3	0.7	V/μs

The following specifications apply over the range of $-55°C \leqslant T_A \leqslant +125°C$ for the μA741A, and $0°C \leqslant T_A \leqslant +70°C$ for the μA741E.

Symbol	Characteristic	Condition			Min	Typ	Max	Unit
V_{IO}	Input Offset Voltage						4.0	mV
$\Delta V_{IO}/\Delta T$	Input Offset Voltage Temperature Sensitivity						15	μV/°C
$V_{IO \, adj}$	Input Offset Voltage Adjustment Range	$V_{CC} = \pm 20$ V			10			mV
I_{IO}	Input Offset Current						70	nA
$\Delta I_{IO}/\Delta T$	Input Offset Current Temperature Sensitivity						0.5	nA/°C
I_{IB}	Input Bias Current						210	nA
Z_I	Input Impedance				0.5			MΩ
P_c	Power Consumption	$V_{CC} = \pm 20$ V	μA741A	$-55°C$			165	mW
				$+125°C$			135	
			μA741E				150	
CMR	Common Mode Rejection	$V_{CC} = \pm 20$ V, $V_I = \pm 15$ V, $R_S = 50$ Ω			80	95		dB
I_{OS}	Output Short Circuit Current				10		40	mA
A_{VS}	Large Signal Voltage Gain	$V_{CC} = \pm 20$ V, $R_L \geqslant 2.0$ kΩ, $V_O = \pm 15$ V			32			V/mV
		$V_{CC} = \pm 5.0$ V, $R_L \geqslant 2.0$ kΩ, $V_O = \pm 2.0$ V			10			
V_{OP}	Output Voltage Swing	$V_{CC} = \pm 20$ V	$R_L = 10$ kΩ		± 16			V
			$R_L = 2.0$ kΩ		± 15			

FIGURE 5A.5

µA741

Typical Performance Curves

**Voltage Gain vs
Supply Voltage for µA741/A**

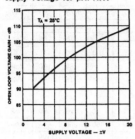

**Output Voltage Swing vs
Supply Voltage for µA741/A**

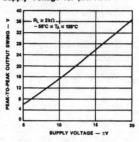

**Input Common Mode Voltage
vs Supply Voltage for µA741/A**

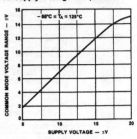

**Voltage Gain vs
Supply Voltage for µA741C/E**

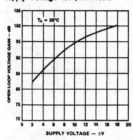

**Output Voltage Swing vs
Supply Voltage for µA741C/E**

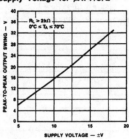

**Input Common Mode Voltage
Range vs Supply Voltage
for µA741C/E**

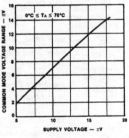

Transient Response for µA741C/E

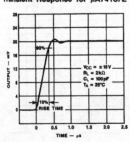

**Transient Response Test Circuit
for µA741C/E**

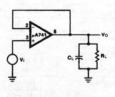

Lead numbers are shown
for metal package only

**Common Mode Rejection Ratio
vs Frequency for µA741C/E**

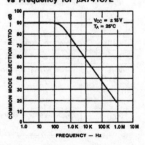

FIGURE 5A.6

μA741

Typical Performance Curves (Cont.)

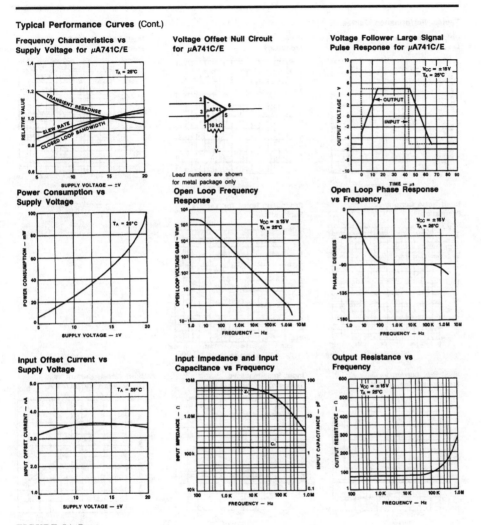

Frequency Characteristics vs
Supply Voltage for μA741C/E

Voltage Offset Null Circuit
for μA741C/E

Lead numbers are shown
for metal package only

Voltage Follower Large Signal
Pulse Response for μA741C/E

Power Consumption vs
Supply Voltage

Open Loop Frequency
Response

Open Loop Phase Response
vs Frequency

Input Offset Current vs
Supply Voltage

Input Impedance and Input
Capacitance vs Frequency

Output Resistance vs
Frequency

FIGURE 5A.7

μA741

Typical Performance Curves (Cont.)

**Output Voltage Swing vs
Load Resistance**

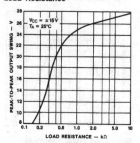

**Output Voltage Swing vs
Frequency**

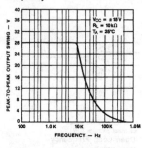

**Input Noise Voltage vs
Frequency**

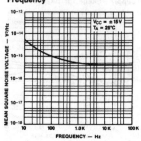

**Input Noise Current vs
Frequency**

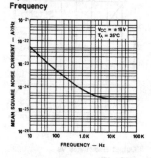

**Broadband Noise for Various
Bandwidths**

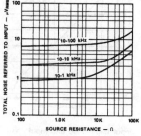

**Input Bias Current vs
Temperature for μA741/A**

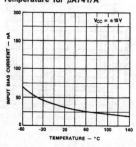

**Input Impedance vs
Temperature for μA741/A**

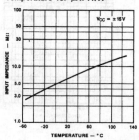

**Short Circuit Current vs
Temperature for μA741/A**

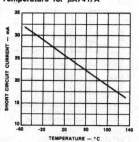

FIGURE 5A.8

μA741

Typical Performance Curves (Cont.)

Input Offset Current vs Temperature for μA741/A

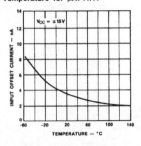

Power Consumption vs Temperature for μA741/A

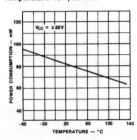

Frequency Characteristics vs Temperature for μA741/A

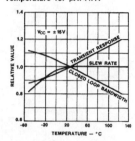

Input Bias Current vs Temperature for μA741C/E

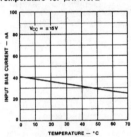

Input Impedance vs Temperature for μA741C/E

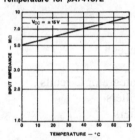

Input Offset Current vs Temperature for μA741C/E

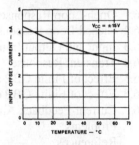

Power Consumption vs Temperature for μA741C/E

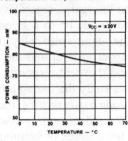

Short Circuit Current vs Temperature for μA741C/E

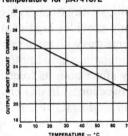

Frequency Characteristics vs Temperature for μA741C/E

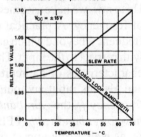

FIGURE 5A.9

6

DYNAMIC OP AMP LIMITATIONS

6.1 Open-Loop Response
6.2 Closed-Loop Response
6.3 Input and Output Impedances
6.4 Transient Response
6.5 Effect of Finite GBP on Integrator Circuits
6.6 Effect of Finite GBP on Filters
6.7 Current-Feedback Amplifiers
     Problems
     References

Up to now we have assumed op amps with extremely high open-loop gains, regardless of frequency. A practical op amp provides high gain only from dc up to a given frequency, beyond which gain decreases with frequency and the output is also delayed with respect to the input. These limitations have a profound impact on the closed-loop characteristics of a circuit: they affect both its frequency and transient responses, and also its input and output impedances. In this chapter we study the *unity-gain frequency* f_t, the *gain-bandwidth product* (GBP), the *closed-loop bandwidth* f_B, the *full-power bandwidth* (FPB), the *rise time* t_R, the *slew rate* (SR), and the *settling time* t_S, as well as the impact on the responses and the terminal impedances of familiar circuits such as the four amplifier types, and filters. We also take the opportunity to discuss *current-feedback amplifiers* (CFAs), a class of op amps designed specifically for high-speed applications.

Since data sheets show frequency responses in terms of the *cyclical frequency f*, we shall work with this frequency rather than with the *angular frequency* ω. One can readily convert from one frequency to the other via $\omega \leftrightarrow 2\pi f$. Moreover, a frequency response $H(jf)$ is readily converted to the *s*-domain by letting $jf \rightarrow s/2\pi$.

The open-loop response $a(jf)$ of an op amp can be quite complex and will be investigated in general terms in Chapter 8. In the present chapter we limit ourselves to the particular but most common case of *internally compensated op amps,* that is, op amps incorporating on-chip components to stabilize their behavior against unwanted oscillations. Most op amps are compensated so that $a(jf)$ is dominated by a single low-frequency pole.

6.1
OPEN-LOOP RESPONSE

The most common open-loop response is the *dominant-pole response,* so called because its frequency profile is primarily controlled by a single pole. To understand its origin, refer to Fig. 6.1, which provides a block diagram of the three-stage op amp circuit of Fig. 5.1. Here g_{m1} is the transconductance gain of the first stage, and $-a_2$ is the voltage gain of the second stage, which is an inverting stage. Moreover, R_{eq} and C_{eq} represent the net equivalent resistance and capacitance between the node common to the first and second stage, and ground.

At low frequencies, where C_c acts as an open circuit, we have $v_O = 1 \times (-a_2) \times (-R_{eq}i_{O1}) = g_{m1}R_{eq}a_2(v_P - v_N)$. The low-frequency gain, called the *dc gain* and denoted as a_0, is thus

$$a_0 = g_{m1}R_{eq}a_2 \tag{6.1}$$

As we know, this is a fairly large number. For the 741 op amp we shall assume the following working values: $g_{m1} = 189\ \mu\text{A/V}$, $R_{eq} = 1.95\ \text{M}\Omega$, and $a_2 = 544\ \text{V/V}$. Substituting into Eq. (6.1) yields the familiar typical value $a_0 = 200\ \text{V/mV}$, or 106 dB.

Increasing the operating frequency will bring the impedance of C_{eq} into play, causing gain to roll off with frequency because of the low-pass filter action provided by R_{eq} and C_{eq}. Gain starts to roll off at the frequency f_b that makes $|Z_{C_{eq}}| = R_{eq}$, or $1/2\pi f_b C_{eq} = R_{eq}$. This frequency, called the *dominant-pole frequency,* is thus

$$f_b = \frac{1}{2\pi R_{eq}C_{eq}} \tag{6.2}$$

From the data sheets we find that the 741 op amp has typically $f_b = 5$ Hz, indicating a dominant pole at $s = -2\pi f_b = -10\pi$ Np/s. Such a low-frequency pole requires that for a given R_{eq}, C_{eq} be suitably large. For the 741 op amp $C_{eq} = 1/2\pi f_b R_{eq} = 1/(2\pi 5 \times 1.95 \times 10^6) = 16.3$ nF. The on-chip fabrication of such a large capacitance would be prohibitive in terms of the chip area needed. This drawback is ingeniously

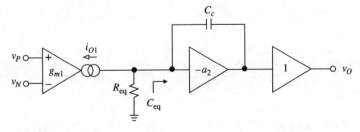

FIGURE 6.1
Simplified op-amp block diagram.

avoided by starting out with an acceptable value for C_c, and then exploiting the multiplicative property of the Miller effect to increase its effective value to $C_{eq} = (1 + a_2)C_c$. The 741 uses $C_c = 30$ pF to achieve $C_{eq} = (1 + 544)30 = 16.3$ nF.

Expression for the Open-Loop Gain

In addition to the dominant pole created by R_{eq} and C_{eq}, the open-loop response will generally include higher-order zeros and poles due to the transistors making up the different stages. The dominant-pole frequency is chosen deliberately low (5 Hz for the 741) to ensure that gain has dropped well below unity at the higher-order root frequencies, and their effect can thus be ignored. With this in mind, the open-loop response of an internally compensated op amp can be approximated as

$$a(jf) = \frac{a_0}{1 + jf/f_b} \tag{6.3}$$

where j is the imaginary unit ($j^2 = -1$), a_0 is the *open-loop dc gain*, and f_b is the *open-loop −3-dB frequency*, also called the *open-loop bandwidth*. Magnitude and phase are calculated as

$$|a(jf)| = \frac{a_0}{\sqrt{1 + (f/f_b)^2}} \qquad \sphericalangle a(jf) = -\tan^{-1}(f/f_b) \tag{6.4}$$

and are plotted in Fig. 6.2. We observe that gain is high and approximately constant only from dc up to f_b. Past f_b it rolls off at the approximately constant rate of

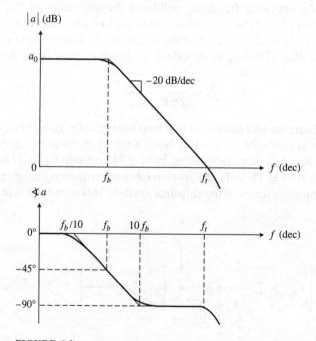

FIGURE 6.2
Typical open-loop response of an internally compensated op amp. (The 741 op amp has typically $a_0 = 200$ V/mV, $f_b = 5$ Hz, and $f_t = 1$ MHz.)

−20 dB/dec, until it drops to 0 dB (or 1 V/V) for $f = f_t$. This frequency is called the *unity-gain frequency,* or also the *transition frequency,* because it marks the transition from amplification (positive decibels) to attenuation (negative decibels). Imposing $1 = a_0/\sqrt{1 + (f_t/f_b)^2}$ in Eq. (6.4) and using the fact that $f_t \gg f_b$, we get

$$f_t = a_0 f_b \qquad (6.5)$$

The 741 op amp has typically $f_t = 200{,}000 \times 5 = 1$ MHz. We wish to emphasize the following special cases:

$$a(jf)\big|_{f \ll f_b} \to a_0\ \underline{/0°} \qquad (6.6a)$$

$$a(jf)\big|_{f = f_b} = \frac{a_0}{\sqrt{2}}\ \underline{/{-45°}} \qquad (6.6b)$$

$$a(jf)\big|_{f \gg f_b} \to \frac{f_t}{f}\ \underline{/{-90°}} \qquad (6.6c)$$

We observe that over the frequency region $f \gg f_b$ the op amp behaves as an integrator, and that its *gain-bandwidth product,* defined as GBP $= |a(jf)| \times f$, is constant

$$\text{GBP} = f_t \qquad (6.7)$$

For this reason, op amps with dominant-pole compensation are also referred to as *constant-GBP op amps:* increasing (or decreasing) f by a given amount in the region of integrator behavior will decrease (or increase) $|a|$ by the same amount. This can be exploited to estimate gain at any frequency above f_b. Thus, at $f = 100$ Hz the 741 has $|a| = f_t/f = 10^6/10^2 = 10{,}000$ V/V; at $f = 1$ kHz it has $|a| = 1000$ V/V; at $f = 10$ kHz it has $|a| = 100$ V/V; at $f = 100$ kHz it has $|a| = 10$ V/V, and so forth. Browsing through linear databooks will reveal quite a few op amp families with a gain response of the type of Fig. 6.2. Most general-purpose types tend to have GBPs between 500 kHz and 20 MHz, with 1 MHz being one of the most frequent values. However, for wideband applications, op amp types are available with much higher GBPs. Current-feedback amplifiers, to be discussed in Section 6.7, are an example.

Though a_0 and f_b may be useful for mathematical manipulations, in practice they are very ill-defined parameters because so are R_{eq} and a_2, due to manufacturing process variations. We shall instead focus on the unity-gain frequency f_t, which turns out to be a more predictable parameter. To justify this claim, we note that at high frequencies the circuit of Fig. 6.1 yields $V_o \cong 1 \times Z_{C_c} I_{o1} = (1/j2\pi f C_c)g_{m1} \times (V_p - V_n)$, or $a = g_{m1}/j2\pi f C_c$. Comparing with Eq. (6.6c) gives

$$f_t = \frac{g_{m1}}{2\pi C_c} \qquad (6.8)$$

By Eq. (5.7), $g_{m1} = I_A/4V_T$. Substituting into Eq. (6.8) gives, for the 741 op amp,

$$f_t = \frac{I_A}{8\pi V_T C_c} \qquad (6.9)$$

It is possible to design for reasonably stable and predictable values of I_A and C_c, thus resulting in a dependable value for f_t. For the 741, $f_t = (19.6 \times 10^{-6})/(8\pi \times 0.026 \times 30 \times 10^{-12}) = 1$ MHz.

Graphical Visualization of the Loop Gain T

We are well aware of the central role played by the loop gain $T = a\beta$ in negative feedback. Since both a and β are generally frequency-dependent, so is T, and we seek a quick means for visualizing this dependence. Letting $T = a\beta = a/(1/\beta)$ allows us to write $|T|_{dB} = 20\log_{10}|T| = 20\log_{10}|a| - 20\log_{10}(1/\beta)$, or

$$|T|_{dB} = |a|_{dB} - |1/\beta|_{dB} \qquad (6.10a)$$

$$\angle T = \angle a - \angle(1/\beta) \qquad (6.10b)$$

indicating that the Bode plots of T can be found graphically as the *difference* between the individual plots of a and $1/\beta$.

Figure 6.3 depicts the magnitude plot. To construct it, we first obtain the open-loop curve from the data sheets. Next, we find β using the techniques of Section 1.7, take its reciprocal $1/\beta$, and then plot $|1/\beta|$. Since usually $|\beta| \leq 1$ V/V, or $|\beta| \leq 0$ dB, it follows that $|1/\beta| \geq 1$ V/V, or $|1/\beta| \geq 0$ dB; that is, the $|1/\beta|$ curve extends above the 0-dB axis. This curve will generally have some breakpoints, though in many cases it is flat. As shown, its low-frequency and high-frequency asymptotes are denoted as $|1/\beta_0|$ and $|1/\beta_\infty|$. Finally, we visualize $|T|$ as the *difference* between the $|a|$ and $|1/\beta|$ curves. The $|T|$ curve is shown explicitly at the bottom, but you should learn to visualize it directly from the diagram at the top.

The frequency f_x at which the two curves meet is called the *crossover frequency*. Clearly, $|T(jf_x)|_{dB} = 0$ dB, or $|T(jf_x)| = 1$. In the example shown, for $f \ll f_x$ we have $|T| \gg 1$, indicating a closed-loop behavior nearly ideal there. However, for $f > f_x$ we have $|T|_{dB} < 0$ dB, or $|T| < 1$, indicating a significant departure from the ideal. Thus, the useful frequency range for the op amp circuit is to the left of f_x. In Chapter 8 we shall find that $\angle T(jf_x)$, the phase angle of T at f_x, determines whether a circuit is stable as opposed to oscillatory.

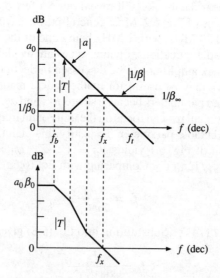

FIGURE 6.3
In a Bode plot, the loop gain $|T|$ is the difference between the $|a|$ and $|1/\beta|$ curves.

Dominant-Pole PSpice Model

Though an op amp can be simulated at either the transistor or the macromodel level, at times we wish to use an even simpler model to focus on just one feature, such as the effect of the dominant pole. The circuit of Fig. 6.4 uses R_{eq} and C_{eq} to create a pole frequency at $f_b = 1/2\pi R_{eq}C_{eq}$. The following subcircuit file reflects typical 741 parameters:

```
*Simple one-pole op amp: a0 = 200V/mV, fb = 5Hz:
.subckt OA1 vP vN vO
rd vP vN 2Meg          ;input resistance
ea0 1 0 vP vN 200k     ;dc gain
Req 1 2 1.95Meg        ;with Ceq, it sets fb as
Ceq 2 0 16.32nF        ;fb=1/2*pi*Req*Ceq
ebuf 3 0 2 0 1         ;output buffer
ro 3 vO 75             ;output resistance
.ends OA1
```

As we proceed, we shall make frequent use of this subcircuit.

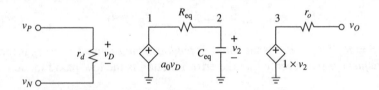

FIGURE 6.4
Simple PSpice model for a one-pole op amp.

6.2
CLOSED-LOOP RESPONSE

The fact that the loop gain T is frequency-dependent will make the closed-loop response A depend on frequency even when A_{ideal} is designed to be frequency-independent, as in the case of purely resistive feedback. To stress this fact, we write

$$A(jf) = A_{ideal} \times \frac{1}{1 + 1/T(jf)} \tag{6.11}$$

The deviation of the error function $1/(1+1/T)$ from $1\ \underline{/0°}$ is now specified in terms of two parameters, namely, the *magnitude error*

$$\epsilon_m = \left| \frac{1}{1 + 1/T(jf)} \right| - 1 \tag{6.12a}$$

and the *phase error*

$$\epsilon_\phi = -\sphericalangle[1 + 1/T(jf)] \tag{6.12b}$$

The Noninverting Amplifier

The closed-loop response of the noninverting amplifier of Fig. 6.5a was given in Eq. (1.12). Substituting $a = a_0/(1 + jf/f_b)$,

$$A(jf) = \left(1 + \frac{R_2}{R_1}\right) \frac{1}{1 + (1 + R_2/R_1)(1 + jf/f_b)/a_0}$$

Using straightforward algebra, we can readily put this in the form $A(jf) = A_0/(1 + jf/f_B)$, where

$$A_0 = \left(1 + \frac{R_2}{R_1}\right) \frac{1}{1 + (1 + R_2/R_1)/a_0} \qquad f_B = f_b\left(1 + a_0\frac{R_1}{R_1 + R_2}\right)$$

Note again the use of lowercase letters for open-loop parameters, and uppercase letters for closed-loop parameters. Exploiting the fact that $a_0 \gg 1 + R_2/R_1$, we can write

$$A(jf) = A_0 \times \frac{1}{1 + jf/f_B} \tag{6.13a}$$

$$A_0 \cong 1 + \frac{R_2}{R_1} \qquad f_B \cong \beta f_t \tag{6.13b}$$

where $\beta = R_1/(R_1 + R_2)$ is the familiar *feedback factor* and $f_t = a_0 f_b$ is the *unity-gain frequency*. We observe that the error function is the low-pass function $1/(1 + jf/f_B)$.

As depicted in Fig. 6.5b, the closed-loop gain $A(jf)$ has A_0 as the dc gain and f_B as the -3-dB frequency, also called the *closed-loop bandwidth*. We note the following:

1. At low frequencies, where $|T| \gg 1$, Eq. (6.11) predicts $A \rightarrow A_{\text{ideal}} = 1 + R_2/R_1$.
2. At the crossover frequency, where $|T| = 1$, we have $\sphericalangle T = \sphericalangle a - \sphericalangle(1/\beta) \cong -90° - 0° = -90°$, or $T = 1\underline{/-90°} = -j1$. By Eq. (6.11), $A = A_{\text{ideal}}/(1 + j1) = (A_{\text{ideal}}/\sqrt{2})\underline{/-45°}$. Clearly, this is the -3-dB frequency f_B, and we can find it graphically by plotting $|1/\beta|$ on the data-sheet graph of $|a|$, and then locating the frequency at which the two curves intersect.

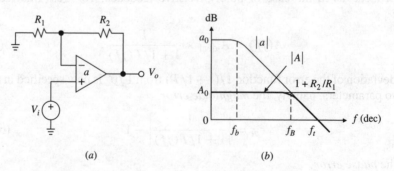

(a)   (b)

FIGURE 6.5
The noninverting amplifier and its frequency response $|A|$.

3. At high frequencies, where $|T| \ll 1$, Eq. (6.11) predicts $A \to A_{\text{ideal}} \times T = (1 + R_2/R_1) \times a \times R_1/(R_1 + R_2) = a$, indicating that $|A|$ will roll off with $|a|$ there.

It is apparent that negative feedback reduces gain from a_0 to A_0 ($A_0 \ll a_0$), but widens bandwidth from f_b to f_B ($f_B \gg f_b$). This curative property, referred to as *broadbanding,* constitutes another important advantage of negative feedback. It also benefits phase since $\sphericalangle a = -45°$ at f_b, but $\sphericalangle A = -45°$ at f_B, $f_B \gg f_b$.

EXAMPLE 6.1. A 741 op amp is configured as a noninverting amplifier with $R_1 = 2\,k\Omega$ and $R_2 = 18\,k\Omega$. Find (a) the 1% magnitude error and (b) the 5° phase error bandwidths, defined, respectively, as the frequency ranges over which $|\epsilon_m| \leq 0.01$ and $|\epsilon_\phi| \leq 5°$.

Solution.

(a) We have $\beta = 0.1$ V/V, so $f_B = \beta f_t = 100$ kHz. By Eq. (6.12a), $\epsilon_m = 1/\sqrt{1 + (f/f_B)^2} - 1$. Imposing $|\epsilon_m| \leq 0.01$ yields $1/\sqrt{1 + (f/10^5)^2} \geq 0.99$, or $f \leq 14.2$ kHz.

(b) By Eq. (6.12b), $\epsilon_\phi = -\tan^{-1}(f/f_B)$. Imposing $|\epsilon_\phi| \leq 5°$ gives $\tan^{-1}(f/10^5) \leq 5°$, or $f \leq 8.75$ kHz.

Gain-Bandwidth Tradeoff

By Eq. (6.13), the gain-bandwidth product of the noninverting amplifier is

$$\text{GBP} = A_0 f_B = f_t \tag{6.14}$$

indicating a *gain-bandwidth tradeoff.* For instance, a 741 op amp configured for $A_0 = 1000$ V/V will have $f_B = f_t/A_0 = 10^6/10^3 = 1$ kHz. Reducing A_0 by a decade, to 100 V/V, will increase f_B also by a decade, to 10 kHz. The amplifier with the lowest gain has also the widest bandwidth: this is the voltage follower, for which $A_0 = 1$ V/V and $f_B = f_t = 1$ MHz. It is apparent that f_t represents a figure of merit for op amps. The gain-bandwidth tradeoff can be exploited to meet specific bandwidth requirements, as illustrated in the following example.

EXAMPLE 6.2. (a) Using 741 op amps, design an audio amplifier with a gain of 60 dB. (b) Sketch its magnitude plot. (c) Find its actual bandwidth.

Solution.

(a) Since $10^{60/20} = 10^3$, the design calls for an amplifier with $A_0 = 10^3$ V/V and $f_B \geq 20$ kHz. A single 741 op amp will not do, because it would have $f_B = 10^6/10^3 = 1$ kHz. Let us try cascading two noninverting stages with lesser individual gains but wider bandwidths, as depicted in Fig. 6.6a. Denoting the individual gains as A_1 and A_2, the overall gain is then $A = A_1 \times A_2$. One can easily prove that the widest bandwidth for A is achieved when A_1 and A_2 are made equal, or $A_{10} = A_{20} = \sqrt{1000} = 31.62$ V/V, or 30 dB. Then, $f_{B1} = f_{B2} = 10^6/31.62 = 31.62$ kHz.

(b) To construct the magnitude plot we note that since $A = A_1^2$, we have $|A|_{dB} = 2|A_1|_{dB}$, indicating that the magnitude plot of A is obtained by multiplying that of A_1 by 2, point by point. The plot of $|A_1|$ is in turn obtained via the graphical technique of Fig. 6.5b. The final result is shown in Fig. 6.6b.

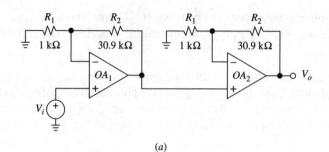

(a)

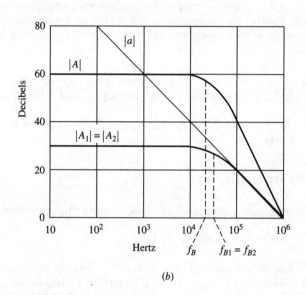

(b)

FIGURE 6.6
Cascading two amplifiers, and the resulting frequency
response $|A|$.

(*c*) Note that at 31.62 kHz both $|A_1|$ and $|A_2|$ are 3 dB below their dc values, making $|A|$
in turn 6 dB below its dc value. The -3-dB frequency f_B is such that $|A(jf_B)| = 10^3/\sqrt{2}$. But, $|A(jf)| = |A_1(jf)|^2 = 31.62^2/[1 + (f/f_B)^2]$. We thus impose

$$\frac{10^3}{\sqrt{2}} = \frac{31.62^2}{1 + [f_B/(31.62 \times 10^3)]^2}$$

to obtain $f_B = 31.62\sqrt{\sqrt{2} - 1} = 20.35$ kHz, which indeed meets the audio
bandwidth requirement.

The Inverting Amplifier

Applying similar reasoning to the inverting amplifier of Fig. 6.7*a*, whose gain was
obtained in Eq. (1.19), we get

$$A(jf) = A_0 \times \frac{1}{1 + jf/f_B} \tag{6.15a}$$

$$A_0 \cong -\frac{R_2}{R_1} \qquad f_B \cong \beta f_t \tag{6.15b}$$

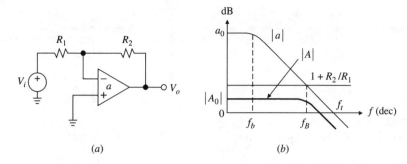

(a)   (b)

FIGURE 6.7
The inverting amplifier and its frequency response $|A|$.

where the feedback factor is again $\beta = R_1/(R_1 + R_2)$. As before, these expressions hold as long as $a_0 \gg 1 + R_2/R_1$. We note that the bandwidth f_B is the same as for the noninverting amplifier; this is not surprising, since f_B depends on T, which in turn depends on the op amp and its feedback network, regardless of the point of injection of the external signal. As depicted in Fig. 6.7b, we can still find f_B graphically as the crossover frequency. However, since we now have $|A_0| = R_2/R_1 < (1 + R_2/R_1)$, the plot of $|A|$ will be shifted downward.

The gain-bandwidth product of the inverting amplifier is GBP $= |A_0| \times f_B = (R_2/R_1) \times f_t R_1/(R_1 + R_2)$, or

$$\text{GBP} = \frac{R_2}{R_1 + R_2} f_t = (1 - \beta) f_t \qquad (6.16)$$

This is less than the GBP of the noninverting counterpart, which coincides with f_t. The difference is more noticeable at low dc gains. For instance, a unity-gain noninverting amplifier ($R_1 = \infty$, $R_2 = 0$) has GBP $= f_t$, whereas a unity-gain inverting amplifier ($R_1 = R_2$) has GBP $= 0.5 f_t$. From the viewpoint of maximizing bandwidth, the former is obviously preferable.

If we observe and plot $|A|$ experimentally, we are likely to find a high-frequency departure from the curve predicted by Eq. (6.15). The high-frequency asymptotic value of A is the feedthrough gain of Eq. (1.65), which we rephrase as

$$\lim_{a \to 0} A = \frac{z_o}{R_1 + (R_2 + z_o)(1 + R_1/z_d)} \qquad (6.17)$$

to account for the fact that at high frequencies the input and output impedances z_d and z_o are generally no longer resistive. Depending on the op amp and its feedback network, as well as the application at hand, this asymptotic departure may be of concern.

EXAMPLE 6.3. Use PSpice to investigate the high-frequency behavior of a 741 inverting amplifier with $R_1 = R_2 = 1\ k\Omega$.

Solution. As shown in Fig. 6.8, we use the test source V_i to excite both the circuit, shown at the right, and a replication of its feedback network, shown at the left and identified by subscripts f. Using the subcircuit OA1 discussed at the end of Section 6.1,

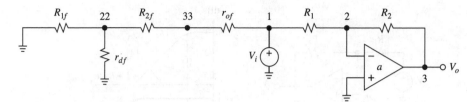

FIGURE 6.8
PSpice circuit of Example 6.3.

we write the following circuit file:

```
Plotting a, 1/beta, and A for the inverting amp:
Vi 1 0 ac 1V
*Circuit to plot A:
R1 1 2 1k
R2 2 3 1k
X1 0 2 3 OA1
*Circuit to plot 1/beta:
rof 1 33 75
R2f 33 22 1k
R1f 22 0 1k
rdf 22 0 2Meg
*Circuit to plot a:
X2 1 0 6 OA1
RL 6 0 2k ;avoids floating nodes
.ac dec 10 1Hz 100MegHz
.probe ;a=V(6)/V(1), 1/beta=V(1)/V(22), A=V(3)/V(1)
.end
```

The plot of Fig. 6.9 confirms that $A|_{f\to\infty} \cong 75/(10^3 + 10^3 + 75) = 36.14$ mV/V $= -28.8$ dB. This value is affected by our choice of R_1 and R_2. For instance, increasing them to $R_1 = R_2 = 10$ kΩ will push the high-frequency asymptote further down, to -48.6 dB. The feedforward gain is of less concern in the noninverting configuration because the signal has to propagate through z_d, which is usually very large and thus causes a great amount of attenuation.

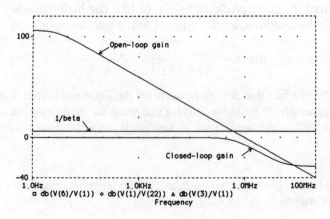

FIGURE 6.9
Frequency plots for the circuit of Fig. 6.8.

6.3
INPUT AND OUTPUT IMPEDANCES

Figure 5A.7 indicates that at high frequencies the *differential input impedance* z_d and the *output impedance* z_o of the 741 op amp become, respectively, capacitive and inductive. This behavior is typical of most op amps, and is due primarily to the stray capacitances of the input transistors and to the frequency limitations of the output transistors. Moreover, if the inputs of a practical op amp are tied together and the impedance to ground is measured, the result is the *common-mode input impedance* z_c. In the op amp model of Fig. 6.10, z_c has been split equally between the two inputs in order to yield $(2z_c) \parallel (2z_c) = z_c$ when they are tied together.

Data sheets usually specify only the resistive portion of these impedances, namely, r_d, r_c, and r_o. For BJT-input op amps, r_d and r_c are typically in the megaohm and gigaohm range, respectively. Since $r_c \gg r_d$, the specification of r_c is often omitted, and only r_d is given. For FET-input devices, r_d and r_c are of the same order of magnitude and in the range of 100 GΩ or higher.

A few manufacturers specify the reactive portions of z_d and z_c, namely, the *differential input capacitance* C_d, and the *common-mode input capacitance* C_c. For example, the AD705 op amp (Analog Devices) has typically $z_d = r_d \parallel C_d = (40 \text{ M}\Omega) \parallel (2 \text{ pF})$ and $z_c = r_c \parallel C_c = (300 \text{ G}\Omega) \parallel (2 \text{ pF})$. In general, it is safe to assume values on the order of few picofarads for both C_d and C_c. Though irrelevant at low frequencies, these capacitances may cause significant degradation at high frequencies. For instance, at dc the AD705 op amp has $z_c = r_c = 300 \text{ G}\Omega$; however, at 1 kHz, where $Z_{C_c} = 1/(j2\pi \times 10^3 \times 2 \times 10^{-12}) \cong -j80 \text{ M}\Omega$, it has $z_c = (300 \text{ G}\Omega) \parallel (-j80 \text{ M}\Omega) \cong -j80 \text{ M}\Omega$, a drastically reduced magnitude.

Let us now investigate how open-loop gain rolloff affects impedances. As we know, a closed-loop impedance Z can often be expressed in terms of its open-loop counterpart z as

$$Z \cong z(1 + T)^{\pm 1} \tag{6.18}$$

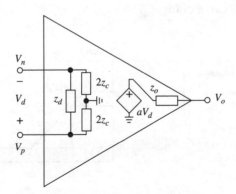

FIGURE 6.10
Modeling the input and output impedances
of a practical op amp.

where we use $+1$ for *series* topologies and -1 for *shunt* topologies. Given that $|T|$ decreases with frequency, we expect the impedance of a *series* topology to decrease with frequency and be thus *capacitive*, and that of a *shunt* topology to increase with frequency and be thus *inductive*. To simplify our calculations, we assume the open-loop impedances to be purely resistive.

Series Impedances

According to Eq. (1.59), negative feedback raises z_d to $Z_d \cong z_d(1 + a\beta)$, where $\beta \cong R_1/(R_1 + R_2)$. Substituting $a = a_0/(1 + jf/f_b)$ and letting $z_d \cong r_d$, we get

$$Z_d \cong R_d \frac{1 + jf/f_B}{1 + jf/f_b} \tag{6.19}$$

where $R_d = r_d(1 + a_0\beta)$ is the closed-loop dc differential input resistance, and $f_B = \beta f_t$. The logarithmic plot of $|Z_d|$ is shown as a broken curve in Fig. 6.11b. Physically, we justify it as follows.

At low frequencies, where $T \gg 1$, negative feedback raises the impedance significantly, so $Z_d \to r_d(1 + a_0\beta)$. At high frequencies, where $T \to 0$, feedback no longer has any effect, so $Z_d \to r_d + [R_1 \parallel (R_2 + r_o)] \cong r_d$. The plot has two breakpoints: one at f_b, where $|T|$ starts to roll off with frequency, and the other at f_B, where $|T|$ drops below unity. Moreover, f_B can be found via the graphical method of Fig. 6.5b. Clearly, $Z_d(s)$ has a pole at $s = -2\pi f_b$ and a zero at $s = -2\pi f_B$.

The overall impedance Z_i of the input series topology of Fig. 6.11a is

$$Z_i = (2z_c) \parallel Z_d \tag{6.20}$$

where $2z_c$ is the common-mode impedance component of the noninverting input. This component is impervious to the curative properties of negative feedback because it is outside the loop. Its effect is to reduce the low-frequency asymptotic value, as indicated by the solid curve of Fig. 6.11b.

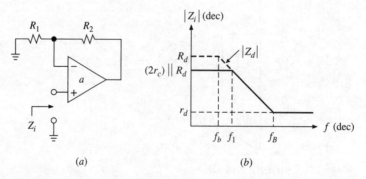

(a) $\qquad\qquad\qquad\qquad\qquad\qquad$ (b)

FIGURE 6.11
Z_i for the input-series topology.

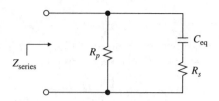

FIGURE 6.12
Equivalent circuit of a series-topology
impedance.

Since Z_i is capacitive, we can model it as in Fig. 6.12. To find its elements, we match their values to the asymptotic values of the plot as follows: At low frequencies, where C_{eq} acts as an open circuit, we impose $R_p = (2z_c) \parallel R_d \cong (2r_c) \parallel [r_d(1 + a_0\beta)]$. At high frequencies, where C_{eq} acts as a short, impose $R_p \parallel R_s = r_d$; but, $R_p \gg r_d$, so $R_s \cong r_d$. Finally, C_{eq} can be computed at f_B by imposing $|Z_{C_{eq}}(jf_B)| = r_d$, or $C_{eq} = 1/2\pi f_B r_d$. For a more realistic picture, one ought to take into account also C_d and C_c, especially at high frequencies. Clearly, Eq. (6.20) provides only a starting point, which can be refined via proper computer simulation.

EXAMPLE 6.4. A certain op amp has $r_d = 1$ MΩ, $r_c = 1$ GΩ, $a_0 = 10^5$ V/V, $r_o = 100$ Ω, and $f_t = 1$ MHz. If it is used in the circuit of Fig. 6.11 with $R_1 = 2$ kΩ and $R_2 = 18$ kΩ, find the element values in the equivalent circuit of Z_i, as well as the breakpoint frequencies of its magnitude plot.

Solution. We have $\beta = 1/10$, $f_B = \beta f_t = 100$ kHz, $R_s = r_d = 1$ MΩ, $R_d = 10^6(1 + 10^5/10) \cong 10$ GΩ, $R_p = (2r_c) \parallel R_d = 2 \parallel 10 = 1.67$ GΩ, and $C_{eq} = 1/(2\pi 10^5 \times 10^6) = 1.59$ pF. Exploiting the constancy of the magnitude-frequency product, we write $R_p f_1 = R_s f_B$, so $f_1 = (R_s/R_p)f_B = 60$ Hz.

The impedance of the output-series topology of Fig. 6.13a is, by Eq. (2.7), $Z_o \cong R(1 + a)$. Substituting the expression for $a(jf)$ gives

$$Z_o \cong R_o \frac{1 + jf/f_t}{1 + jf/f_b} \qquad (6.21)$$

where $R_o \cong R(1 + a_0)$. Its logarithmic plot, shown in Fig. 6.13b, can easily be justified via physical insight. Moreover, Z_o can be modeled as in Fig. 6.12 with $R_p \cong R(1 + a_0)$, $R_s \cong R$, and $C_{eq} = 1/2\pi f_t R$.

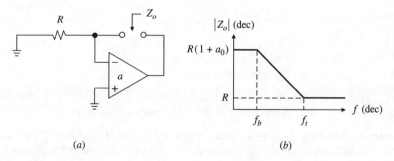

(a)   (b)

FIGURE 6.13
Z_o for the output-series topology.

Shunt Impedances

The impedance of the output-shunt topology of Fig. 6.14a is, by Eq. (1.61), $Z_o \cong r_o/(1 + a\beta)$, where $\beta \cong R_1/(R_1 + R_2)$. Proceeding in the usual manner, we get

$$Z_o \cong R_o \frac{1 + jf/f_b}{1 + jf/f_B} \qquad (6.22)$$

where $R_o = r_o/(1 + a_0\beta)$. Clearly, $Z_o(s)$ has a zero at $s = -2\pi f_b$, and a pole at $s = -2\pi f_B$. Referring to its plot of Fig. 6.14b we observe that the benefits of negative feedback are realized only at low frequencies, where T is fairly large.

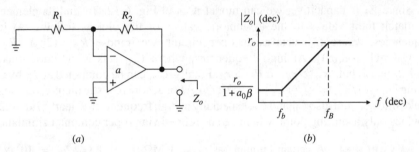

(a)   (b)

FIGURE 6.14
Z_o for the output-shunt topology.

This inductive impedance can be modeled as in Fig. 6.15. At high frequencies, where L_{eq} acts as an open, we impose $R_p = r_o$. At low frequencies, where L_{eq} acts as a short, we impose $R_p \parallel R_s \cong R_s = r_o/(1 + a_0\beta)$. Finally, at f_B we impose $|Z_{L_{eq}}(jf_B)| = r_o$, or $L_{eq} = r_o/2\pi f_B$. Note that Eq. (6.22) ignores the inductive behavior of z_o. One can obtain a more realistic picture via computer simulation, provided the macromodel being used duly reflects the actual behavior of z_o with frequency.

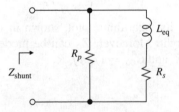

FIGURE 6.15
Equivalent circuit of a shunt-topology impedance.

EXAMPLE 6.5. Repeat Example 6.4, but for the output impedance Z_o.

Solution. The frequency breakpoints are $f_b = f_t/a_0 = 10$ Hz and $f_B = \beta f_t = 100$ kHz. The element values are $R_p = 100\ \Omega$, $R_s = 100/(1 + 10^4) \cong 10$ mΩ, and $L_{eq} = 100/2\pi 10^5 = 159\ \mu$H.

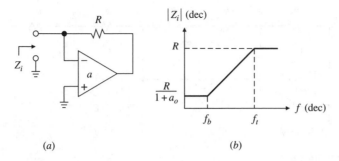

(a)   (b)

FIGURE 6.16
Z_i for the input-shunt topology.

The impedance of the input-shunt topology of Fig. 6.16a is, by Eq. (1.67), $Z_i \cong R/(1 + a)$. This now becomes

$$Z_i \cong R_i \frac{1 + jf/f_b}{1 + jf/f_t} \tag{6.23}$$

where $R_i \cong R/(1 + a_0)$. Its plot is shown in Fig. 6.16b. The inductive equivalent of Fig. 6.15 has now $R_p \cong R$, $R_s \cong R/(1 + a_0)$, and $L_{eq} = R/2\pi f_t$. It is apparent that the virtual-ground concept holds reasonably well only as long as $|T| \gg 1$. As frequency is increased, its impedance deteriorates, thus leading to an increase of the inverting-input voltage V_n with frequency.

Given the tendency of shunt topologies to be inductive, terminating them on capacitive loads may cause instability. The capacitance of the load tends to form a resonant circuit with the equivalent inductance presented by the shunt topology, and this may cause undesirable peaking and ringing, unless the termination is properly damped. Examples of capacitive termination are the stray capacitance of the inverting-input pin and the load capacitance when the op amp drives a long cable. These issues will be studied in Chapter 8.

EXAMPLE 6.6. (a) Estimate $A(jf)$, $Z_i(jf)$, and $Z_o(jf)$ for the high-sensitivity I-V converter of Fig. 2.2 if it is implemented with $R = 100\,\text{k}\Omega$, $R_1 = 2\,\text{k}\Omega$, $R_2 = 18\,\text{k}\Omega$, and a 741 op amp. (b) Compare with PSpice using the μA741 subcircuit of the EVAL.LIB file.

Solution. Since $r_d \gg R$ and $r_o \ll R_2$, we can write $\beta \cong R_1/(R_1 + R_2) = 0.1$ V/V. Then, $A_0 \cong -(1 + R_2/R_1)R = -1$ V/μA, $a_0\beta = 20 \times 10^3$, $f_B = \beta f_t = 100$ kHz, $R_i \cong [R + (R_1 \parallel R_2)]/(1 + a_0\beta) \cong 5\,\Omega$, and $R_o \cong r_o/(1 + a_0\beta) = 3.75$ mΩ. Based on the above approximations, we estimate

$$A(jf) \cong \frac{-10^6 \text{ V/A}}{1 + jf/10^5}$$

$$Z_i(jf) \cong 5\frac{1 + jf/5}{1 + jf/10^5}\,\Omega \qquad Z_o(jf) \cong 3.75\frac{1 + jf/5}{1 + jf/10^5}\,\text{m}\Omega$$

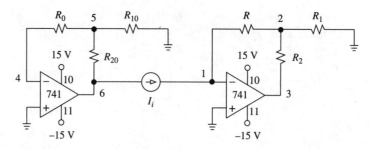

FIGURE 6.17
PSpice circuit of Example 6.6.

With reference to Fig. 6.17, we write the following circuit file using the same test source at both the input and the output:

```
High-sensitivity I-V converter:
*Using library subckt uA741 to find A, Zi, and Zo:
.lib eval.lib
VCC 10 0 dc 15V
VEE 11 0 dc -15V
Ii 6 1 ac 100nA
*Circuit to find A and Zi:
R 1 2 100k
R1 2 0 2k
R2 2 3 18k
XOA1 0 1 10 11 3 uA741
*Circuit to find Zo:
R0 4 5 100k
R10 5 0 2k
R20 5 6 18k
XOA2 0 4 10 11 6 uA741
.ac dec 10 1Hz 10MegHz
.probe ;A=Vm(3)/Ii, Zi=Vm(1)/Ii, Zo=Vm(6)/Ii
.end
```

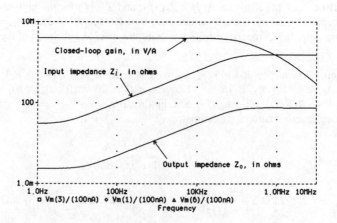

FIGURE 6.18
Frequency plots for the *I-V* converter of Fig. 6.17.

The results of the simulation, shown in Fig. 6.18, are in reasonable agreement with our predictions. The minor discrepancies are due to differences between the simplified model used in our calculations and the Boyle model used by PSpice.

6.4
TRANSIENT RESPONSE

So far we have investigated the effect of the open-loop dominant pole in the frequency domain. We now turn to the time domain by examining the transient response, that is, the response to an input step as a function of time. This response, like its frequency-domain counterpart, varies with the amount of feedback applied. In the data sheets it is usually specified for unity feedback, that is, for the voltage follower configuration; however, the results can readily be generalized to other feedback factors.

The Rise Time t_R

As we know, the small-signal bandwidth of the voltage follower is f_t, so its frequency response can be written as

$$A(jf) = \frac{1}{1 + jf/f_t} \tag{6.24}$$

indicating a pole at $s = -2\pi f_t$. Subjecting the voltage follower of Fig. 6.19a to an input voltage step of sufficiently small amplitude V_m will result in the well-known exponential response

$$v_O(t) = V_m(1 - e^{-t/\tau}) \tag{6.25a}$$

$$\tau = \frac{1}{2\pi f_t} \tag{6.25b}$$

The time t_R it takes for v_O to swing from 10% to 90% of V_m is called the *rise time*, and it provides an indication of how rapid the exponential swing is. We easily find

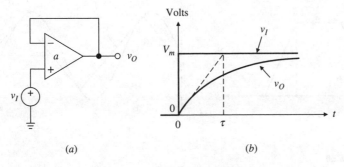

(a)   (b)

FIGURE 6.19
Voltage follower and its small-signal step response.

$t_R = \tau(\ln 0.9 - \ln 0.1)$, or

$$t_R = \frac{0.35}{f_t} \tag{6.26}$$

This provides a link between the frequency-domain parameter f_t and the time-domain parameter t_R; clearly, the higher f_t, the lower t_R.

The 741 op amp has $\tau = 1/(2\pi \times 10^6) \cong 159$ ns and $t_R \cong 350$ ns. A closer look at its small-signal-step response of Fig. 5A.6 indicates a small amount of ringing. This is due to higher-order complex pole pairs, which we have neglected in our dominant-pole approximation.

Slew-Rate Limiting

The rate at which v_O changes with time is highest at the beginning of the exponential transition. Using Eq. (6.25a), we find $dv_O/dt|_{t=0} = V_m/\tau$, which is also illustrated in Fig. 6.19b. If we increase V_m, the rate at which the output slews will have to increase accordingly in order to complete the 10%-to-90% transition within the time t_R. In practice it is observed that above a certain step amplitude the output slope saturates at a constant value called the *slew rate* (SR). The output waveform, rather than an exponential curve, is now a ramp. Figure 6.20a shows the slew-rate limited response to a pulse. As we shall see in greater detail shortly, slew-rate limiting is a nonlinear effect that stems from the limited ability by the internal circuitry to charge or discharge the frequency-compensation capacitance C_c.

The SR is expressed in volts per microsecond. The data sheets give SR = 0.5 V/μs for the 741C op amp version and SR = 0.7 V/μs for the 741E version. This means that to complete a 10-V output swing, a 741C voltage follower takes approximately $(10 \text{ V})/(0.5 \text{ V}/\mu\text{s}) = 20 \ \mu\text{s}$.

When an op amp is operated in the inverting mode, the slew rate during a positive-going swing is usually the same as that during a negative-going swing. However, when operation is in the noninverting mode, the common-mode input swing brings

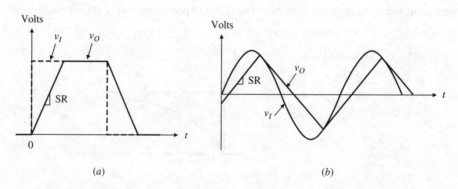

(a)   (b)

FIGURE 6.20
Effect of slew-rate limiting for (a) a pulse input, and (b) a sine wave input.

additional parasitic capacitances into play, which result in asymmetric SR values as well as other second-order effects such as discontinuities at the onset of the step.[1] This is shown in Fig. 5A.7 for the 741 op amp. Unless stated to the contrary, we shall assume symmetric SR values for simplicity.

We stress that SR is a nonlinear large-signal parameter, while t_R is a linear small-signal parameter. The critical output-step magnitude corresponding to the onset of slew-rate limiting is such that $V_{om(\text{crit})}/\tau = \text{SR}$. Using Eq. (6.25b), this gives

$$V_{om(\text{crit})} = \frac{\text{SR}}{2\pi f_t} \qquad (6.27)$$

For the 741C, $V_{om(\text{crit})} = 0.5 \times 10^6/(2\pi \times 10^6) = 80$ mV. This means that as long as the input step is less than 80 mV, a 741C voltage follower responds with an exponential transition governed by $\tau = 159$ ns. However, for a greater input step, the output slews at a constant rate of 0.5 V/μs until it comes within 80 mV of the final value, after which it performs the remainder of the transition in exponential fashion. The above results can be generalized to circuits with $\beta < 1$ by replacing f_t with βf_t.

EXAMPLE 6.7. With an input-stage bias current I_A of 19.6 μA and a compensation capacitance C_c of 30 pF, a 741-type op amp gives SR = 0.633 V/μs. (a) If such an op amp is configured as in Fig. 6.21, find its response $v_O(t)$ to an input step of -0.5 V. (b) Verify with PSpice.

Solution.

(a) By inspection, $A_0 = -4$ V/V and $\beta = 0.2$ V/V. So, $\tau = 1/2\pi\beta f_t = 1/(2\pi \times 0.2 \times 10^6) = 796$ ns and $V_{om(\text{crit})} = \text{SR} \times \tau = 0.504$ V. Once the transient has died out, we have $v_O(\infty) = A_0 V_{im} = -4(-0.5) = 2$ V. Since this is greater than 0.504 V, $v_O(t)$ will be a slew-rate limited ramp until it reaches $2 - 0.504 = 1.496$ V and be an exponential transient thereafter.

Let $v_I(t) = -0.5u(t)$ V, where $u(t)$ is the unit step function. As long as $v_O < 1.496$ V we have $v_O(t) = \text{SR} \times t = 0.633 \times 10^6 t$ V. The instant at which v_O reaches 1.496 V is $t_1 = 1.496/(0.633 \times 10^6) = 2.36$ μs. For $t > t_1$ we can write[2] $v_O(t) = v_O(\infty) + [v_O(t_1) - v(\infty)]\exp[-(t - t_1)/\tau] = 2 - 9.81 \times \exp[-t/(796 \text{ ns})]$ V.

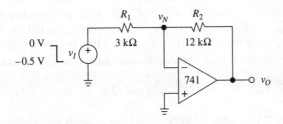

FIGURE 6.21
Circuit of Example 6.7.

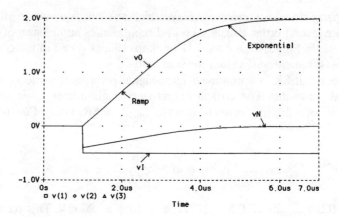

FIGURE 6.22
Step response of the circuit of Fig. 6.21.

(*b*) The input circuit file is the following.

```
Large-signal step response:
vI 1 0 pulse (0 -0.5 1us 10ns 10ns 6us 12us)
R1 1 2 3k
R2 2 3 12k
*Input stage: iO1 = IA*tanh [vD/(4*VT)]:
gm1 4 0 value = {19.6E-6*((exp(19.3*v(0,2))-1)/(exp(19.3*v(0,2))+1))}
Req 4 0 1.95Meg
*Second stage: a2 = -544 V/V
ea2 3 0 4 0 -544
Cc 4 3 30pF
.tran 10ns 6us
.probe ;vI=v(1), vO=v(3), vN=v(2)
.end
```

The results of the simulation are shown in Fig. 6.22, which will be discussed further in the subsection on slew-rate limiting's causes and cures.

Full-Power Bandwidth

The effect of slew-rate limiting is to distort the output signal whenever an attempt is made to exceed the SR capabilities of the op amp. This is illustrated in Fig. 6.20*b* for a sinusoidal signal. In the absence of slew-rate limiting, the output would be $v_O = V_{om} \sin 2\pi ft$. Its rate of change is $dv_O/dt = 2\pi f V_{om} \cos 2\pi ft$, whose maximum is $2\pi f V_{om}$. To prevent distortion we must require $(dv_O/dt)_{\max} \le SR$, or

$$f V_{om} \le SR/2\pi \tag{6.28}$$

indicating a tradeoff between frequency and amplitude. If we want to operate at high frequencies, then we must keep V_{om} suitably small to avoid slew-rate distortion. In particular, if we want to exploit the full small-signal bandwidth f_t of a 741C voltage follower, then we must keep $V_{om} \le SR/2\pi f_t \cong 80$ mV. Conversely, if we want to ensure an undistorted output with $V_{om} > V_{om(\text{crit})}$, then we must keep $f \leq \text{SR}/2\pi V_{om}$. For instance, for an undistorted ac output with $V_{om} = 1$ V, a 741C follower must be operated below $0.5 \times 10^6/2\pi 1 = 80$ kHz, which is way below $f_t = 1$ MHz.

The *full-power bandwidth* (FPB) is the maximum frequency at which the op amp will yield an undistorted ac output with the largest possible amplitude. This amplitude depends on the particular op amp as well as its power supplies. Assuming symmetric output saturation values of $\pm V_{\text{sat}}$, we can write

$$\text{FPB} = \frac{\text{SR}}{2\pi V_{\text{sat}}} \tag{6.29}$$

Thus, a 741C with $V_{\text{sat}} = 13$ V has FPB $= 0.5 \times 10^6/2\pi 13 = 6.1$ kHz. Exceeding this frequency will yield a distorted as well as reduced output. When applying an amplifier we must make sure that neither its slew-rate limit SR nor its -3-dB frequency f_B is exceeded.

> **EXAMPLE 6.8.** A 741C op amp with ± 15-V supplies is configured as a noninverting amplifier with a gain of 10 V/V. (a) If the ac input amplitude is $V_{im} = 0.5$ V, what is the maximum frequency before the output distorts? (b) If $f = 10$ kHz, what is the maximum value of V_{im} before the output distorts? (c) If $V_{im} = 40$ mV, what is the useful frequency range of operation? (d) If $f = 2$ kHz, what is the useful input amplitude range?
>
> **Solution.**
>
> (a) $V_{om} = A V_{im} = 10 \times 0.5 = 5$ V; $f_{\text{max}} = \text{SR}/2\pi V_{om} = 0.5 \times 10^6/2\pi 5 \cong 16$ kHz.
> (b) $V_{om(\text{max})} = \text{SR}/2\pi f = 0.5 \times 10^6/2\pi 10^4 = 7.96$ V; $V_{im(\text{max})} = V_{om(\text{max})}/A = 7.96/10 = 0.796$ V.
> (c) To avoid slew-rate limiting, keep $f \leq 0.5 \times 10^6/(2\pi \times 10 \times 40 \times 10^{-3}) \cong 200$ kHz. Note, however, that $f_B = f_t/A_0 = 10^6/10 = 100$ kHz. The useful range is thus $f \leq 100$ kHz, and is dictated by small-signal considerations, rather than slew-rate limiting.
> (d) $V_{om(\text{max})} = 0.5 \times 10^6/(2\pi \times 2 \times 10^3) = 39.8$ V. Since this is greater than V_{sat}, or 13 V, the limiting factor is in this case output saturation. Thus, the useful input amplitude range is $V_{im} \leq V_{\text{sat}}/A = 13/10 = 1.3$ V.

The Settling Time t_S

The rise time t_R and slew rate SR give an indication of how rapidly the output changes, respectively, under small-signal and large-signal conditions. The parameter of greatest concern in many applications is the *settling time* t_S, defined as the time it takes for the response to a large input step to settle and remain within a specified error-band, usually symmetric about its final value. Settling times are typically specified to accuracies of 0.1% and 0.01% of a 10-V input step. As an example, the AD843 op amp (Analog Devices) has typically $t_S = 135$ ns to 0.01% of a 10-V step.

As shown in Fig. 6.23a, t_S is comprised of an initial propagation delay due to higher-order poles, followed by an SR-limited transition to the vicinity of the final value, followed by a period to recover from the overload conditon associated with the SR, and finally settle toward the final equilibrium value. The settling time depends on both linear and nonlinear factors, and is generally a complex phenomenon.[3,4] A

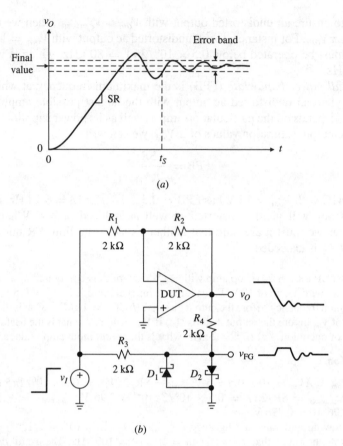

(a)

(b)

FIGURE 6.23
Settling time t_S, and circuit to measure t_S. (D_1 and D_2 are HP2835
Schottky diodes.)

fast t_R or a high SR does not necessarily guarantee a fast t_S. For instance, an op amp
may settle quickly within 0.1%, but may take considerably longer to settle within
0.01% due to excessively long ringing.

Figure 6.23b shows a popular test circuit[5] for the measurement of t_S. The device under test (DUT) is configured as a unity-gain inverting amplifier, while the
equal-valued resistors R_3 and R_4 synthesize what is commonly referred to as a *false
ground*. Since $v_{FG} = \frac{1}{2}(v_I + v_O)$, with $v_O = -v_I$ we expect $v_{FG} = 0$ V. In practice,
because of the transient due to the op amp, v_{FG} will momentarily deviate from zero
and we can observe this deviation to measure t_S. For an error band of ±0.01% of a
10-V step, v_{FG} will have to settle within ±0.5 mV of its final value. The purpose of
the Schottky diodes is to prevent overloading the oscilloscope's input amplifier. To
avoid loading by the probe's stray capacitance, v_{FG} can be buffered by means of a
JFET source-follower. Consult the data sheets for the recommended test circuit to
measure t_S.

In order to fully realize the settling-time capabilities of the op amp, one must
pay proper attention to component selection, layout, and grounding; otherwise,

the painstaking process of amplifier design can easily be defeated.[5] This includes keeping component leads extremely short, using metal-film resistors, orienting components so as to minimize stray capacitances and connection inductances, properly bypassing the power supplies, and providing separate ground returns for the input, the load, and the feedback network. Fast settling times are particularly desirable in high-speed, high-accuracy D-A converters, sample-and-hold amplifiers, and multiplexed amplifiers.

Slew-Rate Limiting: Causes and Cures

It is instructive to investigate the causes of slew-rate limiting since even a qualitative understanding can better help the user in the op amp selection process. Referring to the block diagram[1] of Fig. 6.24, we observe that as long as the input step amplitude V_m is sufficiently small, the input stage will respond in proportion and yield $i_{O1} = g_{m1}V_m$. By the capacitance law, $dv_O/dt = i_{O1}/C_c = g_{m1}V_m/C_c$, thus confirming that the output rate of change is also propotional to V_m. However, if we overdrive the input stage, i_{O1} will saturate at $\pm I_A$, as depicted in Fig. 5.2b. The capacitor C_c will become current-starved, and $(dv_O/dt)_{max} = I_A/C_c$. This is precisely the slew rate,

$$SR = \frac{I_A}{C_c} \tag{6.30}$$

Using the 741 op amp working values of Section 5.1, namely, $I_A = 19.6\ \mu A$ and $C_c = 30$ pF, we estimate SR $= 0.653$ V/μs, in reasonable agreement with the data sheets.

It is important to realize that during slew-rate limiting v_N may depart from v_P significantly because of the drastic drop in the open-loop gain brought about by input-stage saturation. During limiting the circuit is insensitive to any high-frequency

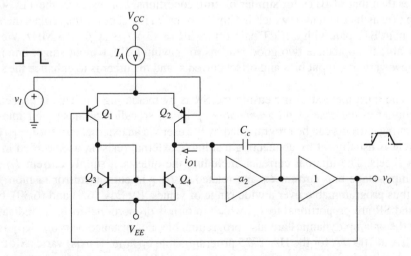

FIGURE 6.24
Op amp model to investigate slew-rate limiting.

components at the input. In particular, the virtual-ground condition of the inverting configuration does not hold during limiting. This is confirmed by the shape of v_N in Fig. 6.22.

We can gain additional insight by relating large-signal and small-signal behavior.[1,6] In Eq. (6.8) it was found that $f_t = g_{m1}/2\pi C_c$. Solving for C_c and substituting into Eq. (6.30) gives

$$SR = \frac{2\pi I_A f_t}{g_{m1}} \tag{6.31}$$

This expression points to three different ways of increasing the SR, namely, (a) by increasing f_t, (b) by reducing g_{m1}, or (c) by increasing I_A.

In general, an op amp with a high f_t tends to exhibit also a high SR. By Eq. (6.8), f_t can be increased by reducing C_c. This is especially useful in the case of uncompensated op amps, for then the user can specify a compensation network that will also maximize the SR. A popular example is offered by the 301 and 748 op amps, which, when used in high-gain configurations, can be compensated with a smaller C_c value to achieve a higher f_t as well as a higher SR. Even in low-gain applications, other frequency-compensation schemes than the dominant pole are possible, which may improve the SR significantly. Popular examples are the so-called *input-lag* and *feedforward* compensation methods, to be addressed in Chapter 8. For instance, with dominant-pole compensation, the 301 op amp offers dynamic characteristics similar to those of the 741; however, with feedforward compensation, it achieves $f_t = 10$ MHz and SR = 10 V/μs.

The second method of increasing the SR is by reducing the input-stage transconductance g_{m1}. For BJT input stages, g_{m1} can be reduced via *emitter degeneration,* which is obtained by including suitable resistances in series with the emitters in the differential input pair to deliberately reduce, or degenerate, transconductance. The LM318 op amp (National Semiconductor) utilizes this technique to achieve SR = 70 V/μs with $f_t = 15$ MHz. Alternatively, g_{m1} can be reduced by implementing the differential input pair with FETs, whose transconductance is notoriously lower than that of BJTs for similar biasing conditions. For instance, the TL080 op amp (Texas Instruments), which is similar to the 741 except for the replacement of the input BJT pair with a JFET pair, offers SR = 13 V/μs at $f_t = 3$ MHz. We are now able to appreciate two good reasons for having a JFET input stage: one is to achieve very low input bias and offset currents, and the other is to enhance the slew rate.

The third method of increasing the SR is by increasing I_A. This is especially important in the case of *programmable op amps,* so called because their internal operating currents can be programmed by the user via an external current I_{SET}. This current is usually set by connecting a suitable external resistor, as specified in the data sheets. The internal currents, including the quiescent supply current I_Q and the input-stage bias current I_A, are related to I_{SET} in current-mirror fashion, and are thus programmable over a wide range of values. By Eqs. (6.9) and (6.30), both f_t and SR are proportional to I_A, which in turn is proportional to I_{SET}, indicating that the op amp dynamics are also programmable. For instance, varying I_{SET} from 0.1 μA to 100 μA for the HA-2725 programmable op amp (Harris) varies SR from 0.06 V/μs to 6 V/μs and f_t from 5 kHz to 10 MHz, providing the user with the ability to tailor the dynamics to a wide variety of situations.

6.5
EFFECT OF FINITE GBP ON INTEGRATOR CIRCUITS

283

SECTION 6.5
Effect of Finite
GBP on Integrator
Circuits

As we know, the integrator of Fig. 6.25a yields

$$H_{ideal}(jf) = -\frac{1}{jf/f_0} \tag{6.32}$$

where $f_0 = 1/2\pi RC$ is the unity-gain frequency. To investigate the effect of the open-loop gain rolloff, we calculate the feedback factor $\beta = R/(R + 1/j2\pi fC)$. Expanding, we get

$$\frac{1}{\beta} = \frac{1 + jf/f_0}{jf/f_0}$$

As shown in Fig. 6.25b, $|1/\beta|$ has the low-frequency and high-frequency asymptotes $|1/\beta_0| = 1/(f/f_0)$ and $|1/\beta_\infty| = 1$ V/V = 0 dB, and it intercepts the $|a|$ curve at $f = f_0/a_0$ and at $f = f_t$.

The frequency region of nearly ideal behavior is $f_0/a_0 \ll f \ll f_t$, where $|T| \gg 1$. Below f_0/a_0, C acts as an open circuit compared to R, so the circuit amplifies with the full open-loop gain there, giving $H = -a_0$. Above f_t, $|T|$ drops below unity to give $H \cong H_{ideal} \times T$ there, indicating a frequency rolloff of -40 dB/dec. It is apparent that for a single-pole op amp, $H(s)$ has $-a_0$ as dc gain and two real poles at $s = -2\pi f_0/a_0$ and $s = -2\pi f_t$, so we write

$$H(jf) \cong \frac{-a_0}{[1 + jf/(f_0/a_0)][1 + jf/f_t]} \tag{6.33}$$

Compared to Eq. (6.32), the actual response is of the second order because of the presence of two reactive elements, namely, the external capacitance C and the internal compensation capacitance C_c.

According to Eq. (6.32), the integrator should provide a phase shift of 90°. In practice, because of the two breakpoints, the shift will depart from 90° at both the

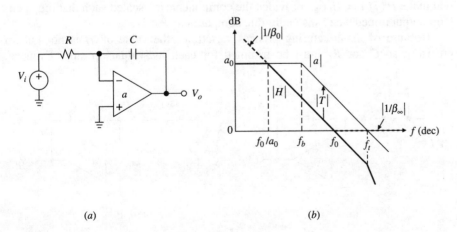

(a)　　　　　　　　　　　　　　(b)

FIGURE 6.25
The inverting integrator and its transfer function $|H|$.

low and high ends of the frequency spectrum. We shall see soon that the latter is a source of concern in integrator-based filters such as dual-integrator loops. At high frequencies Eq. (6.33) simplifies to

$$H(jf) = -\frac{1}{jf/f_0} \times \frac{1}{1 + jf/f_t} \tag{6.34}$$

indicating that the error function is the usual low-pass function $1/(1 + jf/f_t)$. As we know from Eq. (6.12b), the corresponding phase error is $\epsilon_\phi = -\tan^{-1}(f/f_t)$. We are particularly interested in ϵ_ϕ in the vicinity of f_0. Since a well-designed integrator has $f_0 \ll f_t$, we can approximate, for $f \ll f_t$,

$$\epsilon_\phi \cong -f/f_t \tag{6.35}$$

We can reduce ϵ_ϕ by introducing a suitable amount of phase lead to counteract the phase lag due to the pole frequency f_t. This process is called *phase-error compensation.*

Passive Compensation of Integrators

The integrator of Fig. 6.26a is compensated by means of an input parallel capacitance C_c. If we specify its value so that $|Z_{C_c}(jf_t)| = R$, or $1/2\pi f_t C_c = R$, then the phase lead due to the high-pass action by C_c will compensate for the phase lag due to the low-pass term $1/(1 + jf/f_t)$, thus expanding the frequency range of negligible phase error. This technique, also referred to as *zero-pole cancellation,* requires that

$$C_c = 1/2\pi R f_t \tag{6.36}$$

The scheme of Fig. 6.26b achieves a similar result, but using a feedback series resistance R_c and decreasing the input resistance from R to $R - R_c$. This method offers better trimming capabilities than capacitive compensation. It can be shown (see Problem 6.45) that letting

$$R_c = 1/2\pi C f_t \tag{6.37}$$

will make $H(jf) = H_{\text{ideal}}$, provided the components are scaled such that the open-loop output impedance z_o is negligible compared to R_c.

Because of manufacturing process variations, the value of f_t is not known precisely, so C_c or R_c must be trimmed for each individual op amp. Even so,

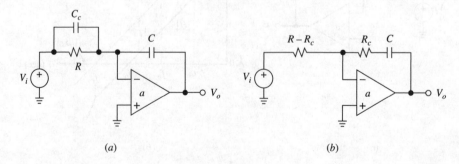

(a)   (b)

FIGURE 6.26
Passive compensation of the integrator: (a) capacitive, and (b) resistive.

compensation is difficult to maintain because f_t is sensitive to temperature and power-supply variations.

Active Compensation of Integrators

The drawbacks of passive compensation are ingeniously avoided with active compensation,[7] so called because it exploits the matching and tracking properties of dual op amps to compensate for the frequency limitations of one device using the very same limitations of the other. Although this technique is general and will be readdressed in Section 8.6, we are presently focusing on the compensation of integrators.

Applying the superposition principle to the circuit of Fig. 6.27a, we can write

$$V_o = -a_1 \left(\frac{1}{1 + jf/f_0} V_i + \frac{jf/f_0}{1 + jf/f_0} A_2 V_o \right) \qquad A_2 = \frac{1}{1 + jf/f_{t2}}$$

where $f_0 = 1/2\pi RC$. To find $H = V_o/V_i$, we eliminate A_2, substitute $a_1 \cong f_{t1}/jf$, and let $f_{t2} = f_{t1} = f_t$ to reflect matching. This gives $H(jf) = H_{ideal} \times 1/(1+1/T)$, where the error function is now

$$\frac{1}{1 + 1/T} = \frac{1 + jf/f_t}{1 + jf/f_t - (f/f_t)^2} = \frac{1 - j(f/f_t)^3}{1 - (f/f_t)^2 - (f/f_t)^4} \qquad (6.38)$$

The last step reveals an interesting property: the rationalization process leads to the mutual cancellation of the first- and second-order terms in f/f_t in the numerator, leaving only the third-order term. We thus approximate, for $f \ll f_t$,

$$\epsilon_\phi \cong -(f/f_t)^3 \qquad (6.39)$$

indicating a much smaller error than in Eq. (6.35).

In Fig. 6.27b, OA_1 contains the inverting op amp OA_2 in its feedback path, so its input polarities have been interchanged to keep feedback negative. One can prove (see Problem 6.46) that

$$\frac{1}{1 + 1/T} = \frac{1 + jf/0.5f_t}{1 - jf/f_t - (f/0.5f_t)^2} \cong \frac{1 + jf/f_t}{1 - 3(f/f_t)^2}$$

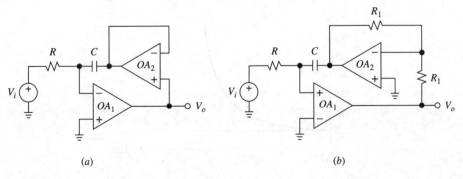

(a)   (b)

FIGURE 6.27
Active compensation of the integrator: (a) $\epsilon_\phi = -(f/f_t)^3$, and (b) $\epsilon_\phi = +f/f_t$.

where we have ignored higher-order terms in f/f_t. We now have

$$\epsilon_\phi \cong +f/f_t \qquad (6.40)$$

Though not as small as in Eq. (6.39), this phase error has the advantage of being positive, a feature we shall exploit shortly.

Q-Enhancement Compensation

It has been found that the effect of nonideal op amps on dual-integrator-loop filters such as the state-variable and biquad varieties is to raise the actual value of Q above the design value predicted under ideal op amp assumptions. This effect, aptly referred to as *Q enhancement*, has been analyzed[8] for the case of the biquad configuration in terms of the phase errors introduced by the two integrators and the third amplifier. The result is

$$Q_{\text{actual}} \cong \frac{Q}{1 - 4Qf_0/f_t} \qquad (6.41)$$

where f_0 is the integrator unity-gain frequency, f_t is the op amp transition frequency, and Q is the quality factor in the ideal op amp limit $f_t \to \infty$. As pictured in Fig. 6.28 for a design value of $Q = 25$ and op amps with $f_t = 1$ MHz, Q_{actual} increases with f_0 until it becomes infinite for $f_0 \cong f_t/4Q = 10^6/100 = 10$ kHz. At this point the circuit becomes oscillatory.

Besides Q enhancement, the finite GBP of the op amps causes also a shift in the characteristic frequency f_0 of the filter,[9]

$$\frac{\Delta f_0}{f_0} \cong -(f_0/f_t) \qquad (6.42a)$$

For small Q deviations, Eq. (6.41) gives

$$\frac{\Delta Q}{Q} \cong 4Qf_0/f_t \qquad (6.42b)$$

Together, these equations indicate the GBP that is needed to contain $\Delta f_0/f_0$ and $\Delta Q/Q$ within specified limits.

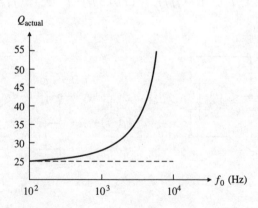

FIGURE 6.28
Q enhancement.

EXAMPLE 6.9. Specify suitable components in the biquad filter of Fig. 3.36 to achieve $f_0 = 10$ kHz, $Q = 25$, and $H_{0BP} = 0$ dB, under the constraint that the deviations of f_0 and Q from their design values because of finite GBPs be within 1%.

Solution. Use $R_1 = R_2 = R_5 = R_6 = 10$ kΩ, $R_3 = R_4 = 250$ kΩ, $C_1 = C_2 = 5/\pi$ nF. To meet the f_0 and Q specifications, we need, respectively, $f_t \geq f_0/(\Delta f_0/f_0) = 10^4/0.01 = 1$ MHz, and $f_t \geq 4 \times 25 \times 10^4/0.01 = 100$ MHz. The Q specification is the most demanding, so we need GBP ≥ 100 MHz.

The onerous GBP requirements imposed by the Q specification can be relaxed dramatically if we use phase-error compensation to eliminate the Q-enhancement effect. Figure 6.29a shows a passively compensated realization of the filter of Example 6.9, but using 1-MHz op amps. To compensate for the phase errors of the integrators as well as the inverting amplifier, whose pole frequency f_B is half

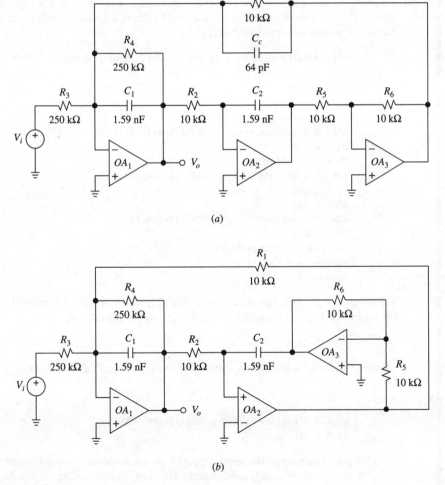

(a)

(b)

FIGURE 6.29
Biquad filter with (a) passive and (b) active compensation.

the pole frequency f_t of each integrator, we use a single capacitance, but four times as large as that predicted by Eq. (6.36), or $C_c = 2/(\pi R f_t) \cong 64$ pF.

Figure 6.29b eliminates the Q-enhancement effect using the active compensation scheme of Fig. 6.27b. In this case the phase error of the inverting amplifier is used to change the error of the rightmost integrator from negative to positive, and thus cancel out the negative error of the leftmost integrator. It is intriguing that just a few rewirings can accomplish so much!

Whether actively or passively compensated, the filter still exhibits the frequency shift of Eq. (6.42a). We eliminate it by altering the design values so as to make actual values coincide with desired values, a technique referred to as *predistortion*.

EXAMPLE 6.10. (*a*) Verify the circuits of Fig. 6.29 with PSpice. (*b*) Predistort the component values so that $f_0 = 10$ kHz.

Solution.

(*a*) Let nodes be numbered sequentially from left to right, and let $a_0 = 1$ V/μV and $f_b = 1$ Hz. The following circuit file uses the LAPLACE facility of PSpice as an alternative method for simulating $a(jf)$.

```
Biquad filter with f0 = 10 kHz, Q = 25, H0BP = 0 dB:
Vi 1 0 ac 1V
R3 1 2 250k
R4 2 3 250k
C1 2 3 1.5915nF
eOA1 3 0 Laplace {V(0,2)}={1E6/(1+s/6.283)}
R2 3 4 10k
C2 4 5 1.5915nF
eOA2 5 0 Laplace {V(0,4)}={1E6/(1+s/6.283)}
R5 5 6 10k
R6 6 7 10k
eOA3 7 0 Laplace {V(0,6)}={1E6/(1+s/6.283)}
R1 7 2 10k
.ac lin 100 9.1kHz 11kHz
.probe ;H = V(3)/V(1)
.end
```

The results of the simulation, shown in Fig. 6.30 (top) reveal an intolerable Q enhancement. To provide passive compensation, we simply add the statement

```
Cc 2 7 64pF
```

whereas to provide active compensation, we change the node connections for C_2, OA_2, and R_1, as

```
C2 4 7 1.5915nF
eOA2 5 0 Laplace {V(4,0)}={1E6/(1+s/6.283)}
R1 5 2 10k
```

everything else remaining the same. The plot shows that compensation, whether active or passive, eliminates Q enhancement. However, by Eq. (6.42a), we still have a frequency downshift compared to the ideal response. The ideal response has been obtained by changing the value of each Laplace source from 1E6 to 1E9.

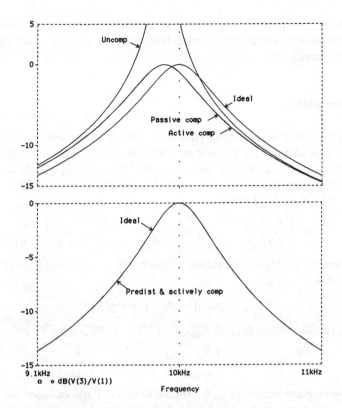

FIGURE 6.30
Frequency plots for the circuit of Example 6.10.

(b) To obtain $f_0 = 10$ kHz, we reduce either all capacitances or all resistances by the amount $f_0/f_t = 10^4/10^6 = 0.01$, or 1%. For instance, changing both capacitances from 1.5915 nF to 1.5756 nF in the circuit of Fig. 6.29b gives a response hardly distinguishable from the ideal one, as shown in Fig. 6.30 (bottom).

Before concluding, we wish to point out that the results derived above for the active compensation schemes are based on a single-pole open-loop response. Practical op amps exhibit additional higher-order roots whose effect is to increase the amount of phase lag at f_t sometimes well above that associated with a single pole. Additional lag is introduced also by inverting-input parasitics, a subject that will be addressed in Section 8.2. Consequently, a practical filter circuit may exhibit localized oscillations, thus requiring additional compensation measures in order to function properly.

6.6
EFFECT OF FINITE GBP ON FILTERS

To assess the effect of finite GBPs on filter performance we must take into account the open-loop gain $a(jf)$ when deriving transfer functions. It is fair to say that because of the additional reactive element provided by the internal compensation capacitance C_c, the order of the transfer function will generally be increased by the number of op amps present. The amount of algebra involved is still manageable in the case of first-order filters, but becomes prohibitive as the order and complexity of the filter are increased.

First-Order Filters

First-order filters can be solved analytically, as already exemplified by the integrators of the previous section. We present an additional example in the high-pass filter of Fig. 6.31a. Ideally, this filter has the cutoff frequency $f_0 = 1/2\pi R_1 C$ and the high-frequency gain $H_0 = -R_2/R_1$. To investigate the effect of GBP, we first find $1/\beta = 1 + R_2/(R_1 + 1/j2\pi f C)$, or

$$\frac{1}{\beta} = \frac{1 + jf/f_z}{1 + jf/f_p} \qquad f_p = \frac{1}{2\pi R_1 C} \qquad f_z = \frac{f_p}{1 + R_2/R_1}$$

This function has the high-frequency asymptote $1/\beta_\infty = 1 + R_2/R_1$, so it intercepts the $|a|$ curve at

$$f_x = \beta_\infty f_t = f_t/(1 + R_2/R_1) \tag{6.43}$$

The actual transfer function $H = H_{ideal} \times 1/(1 + 1/T)$ is thus

$$H(jf) = -\left(\frac{R_2}{R_1}\right)\frac{jf/f_0}{1 + jf/f_0} \times \frac{1}{1 + jf/f_x} \tag{6.44}$$

As also shown graphically in Fig. 6.31b, the finite GBP has changed the filter from a first-order high-pass to a second-order wideband band-pass. As usual, the region of nearly ideal high-pass behavior is $f \ll f_x$, where $|T| \gg 1$.

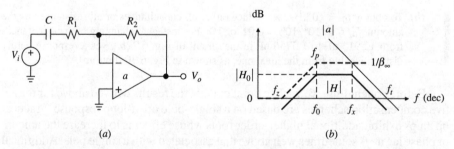

$$(a) \qquad\qquad\qquad\qquad (b)$$

FIGURE 6.31
High-pass filter with gain.

EXAMPLE 6.11. In the circuit of Fig. 6.31a let $C = 5/\pi$ nF, $R_1 = 10$ kΩ, $R_2 = 30$ kΩ, and GBP $= 1$ MHz. Find the frequency range over which the departure of $|H|$ from $|H_{ideal}|$ is less than 1%. How does the finite GBP affect the cutoff frequency?

Solution. We have $f_x = 10^6/(1 + 30/10) = 250$ kHz. Imposing $1/\sqrt{1 + (f/f_x)^2} \geq 0.99$ yields $f \leq 36.6$ kHz. Ideally, $f_{-3\,dB} = f_0 = 10$ kHz. To find the actual value, impose $[1 + (f_{-3\,dB}/f_0)^2][1 + (f_{-3\,dB}/f_x)^2] = 2(f_{-3\,dB}/f_0)^2$. This gives $f_{-3\,dB} = 10.016$ kHz, so $\Delta f_{-3\,dB}/f_{-3\,dB} = 0.16\%$.

Additional first-order filter examples are covered in the end-of-chapter problems.

Second-Order Filters

The analysis of second-order filters is more contrived than that of first-order circuits. In a single-op-amp configuration the actual transfer function $H(s)$ will have three poles; in a three-op-amp structure such as the SV and biquad filters, $H(s)$ will have five poles. In general, the effect of finite GBPs is to create new poles as well as rearrange the existing ones, thus altering the frequency response. In some cases the poles may spill into the right half of the s plane and lead to instability; the biquad filter of Section 6.5 is an example.

To gain a qualitative feel, we investigate the multiple-feedback filter of Fig. 6.32 for the equal-C case. To find $H(s)$, we first obtain an expression for V_n in terms of V_i and V_o by applying KCL at nodes 2 and 4. Then, we let $V_o = -a(s)V_n$ and solve for the ratio V_o/V_i. The result is

$$H(s) = H_{0BP} \frac{(s/\omega_0)/Q}{\dfrac{s^2}{\omega_0^2} + \dfrac{1}{Q}\dfrac{s}{\omega_0} + 1 + \dfrac{1}{a}\left(\dfrac{s^2}{\omega_0^2} + \dfrac{2Q^2+1}{Q}\dfrac{s}{\omega_0} + 1\right)} \qquad (6.45)$$

where H_{0BP}, ω_0, and Q are as in Eq. (3.71). It is apparent that once we substitute $a \cong \omega_t/s = 2\pi f_t/s$, we end up with a third-order function, whereas for $a = \infty$ the order is only two.

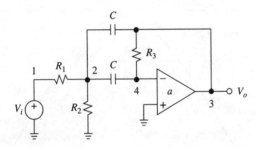

FIGURE 6.32
PSpice circuit of a multiple-feedback band-pass filter.

We are primarily interested in the deviations of the resonance frequency and the -3-dB bandwidth from their design values. It has been shown[9] that as long as $Qf_0 \ll f_t$, we have

$$\frac{\Delta f_0}{f_0} \cong -\frac{\Delta Q}{Q} \cong -Qf_0/f_t \qquad (6.46)$$

Evidently the product $Q \times f_0$ provides an indication of how demanding the filter specifications are in terms of the GBP.

> **EXAMPLE 6.12.** Using 10-nF capacitances, specify suitable components in the circuit of Fig. 6.32 for $H_{0BP} = 0$ dB, $f_0 = 10$ kHz, $Q = 10$, and a BW deviation from its design value due to finite GBP of 1% or less.
>
> **Solution.** Using Eqs. (3.72) and (3.73), we get $R_1 = 15.92$ kΩ, $R_2 = 79.98$ Ω, and $R_3 = 31.83$ kΩ. Since BW $= f_0/Q$, Eq. (6.46) gives ΔBW/BW $\cong -2Qf_0/f_t$. Consequently, GBP $\geq 2 \times 10 \times 10^4/0.01 = 20$ MHz.

An alternative to using high-GBP op amps is to predistort the filter parameters so as to make the actual values coincide with those given in the specifications. In this respect, PSpice simulation provides an invaluable tool in determining the amount of predistortion required for a given value of f_t.

EXAMPLE 6.13. Design a filter meeting the specifications of Example 6.12 with a 1-MHz op amp.

Solution. With $f_t = 1$ MHz we get $Qf_0/f_t = 0.1$, so by Eq. (6.46) we expect a decrease in f_0 and an increase in Q on the order of 10%. For more accurate estimates we use PSpice with the following input file.

```
MF band-pass filter with f0 = 10 kHz and Q = 10:
Vi 1 0 ac 1V
R1 1 2 15.92k
R2 2 0 79.98
R3 4 3 31.83k
C1 2 3 10nF
C2 2 4 10nF
*Op amp with GBP = 1 MHz:
eOA 3 0 Laplace {V(0,4)}={1E6/(1+s/6.283)}
.ac lin 1000 5kHz 20kHz
.probe ;H = V(3)/V(1)
.end
```

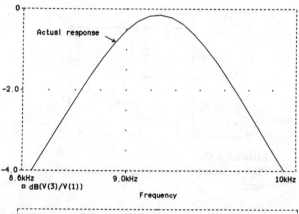

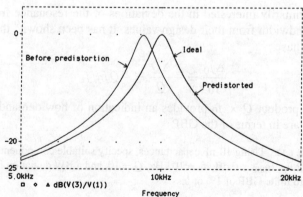

FIGURE 6.33
Frequency plots for the band-pass filter of Example 6.13.

With reference to Fig. 6.33 (top), we use the cursor facility of the Probe postprocessor to measure $f_0 = 9.12$ kHz, $H_{0BP} = 0.983$ V/V, $f_L = 8.71$ kHz, and $f_H = 9.56$ kHz. So, $Q = f_0/(f_H - f_L) = 10.8$.

To achieve the desired parameter values, we redesign the circuit for $f_0 = 10/9.12 = 10.9$ kHz, $Q = 10/10.8 = 9.29$, and $H_{0BP} = 1/0.983 = 1.02$ V/V. Using again Eqs. (3.72) and (3.73), we find that the following changes need to be made.

```
R1 1 2 13.3k
R2 2 0 78.6
R3 4 3 27.0k
```

The responses before and after predistortion are compared in Fig. 6.33 (bottom). Also shown for reference is the ideal response, obtained by changing the value of the eOA source from 1E6 to 1E9 in the original file.

The interested reader is referred to the literature[9] for detailed studies of the effect of finite GBP on filters. Within the scope of this book, we limit ourself to finding the actual response via computer simulation, using the more realistic SPICE macromodels provided by the manufacturers, and then applying predistortion in the manner of Examples 6.10 and 6.13. As rule of thumb, one should select an op amp with a GBP at least an order of magnitude higher than the filter product Qf_0 in order to reduce the effect of GBP variations due to environmental and manufacturing process variations.

6.7
CURRENT-FEEDBACK AMPLIFIERS[10]

The op amps considered so far are also referred to as *voltage-feedback amplifiers* (VFAs) because they respond to voltages. As we know, their dynamics are limited by the gain-bandwidth product and the slew rate. By contrast, *current-feedback amplifiers* (CFAs) exploit a circuit topology that emphasizes current-mode operation, which is inherently much faster than voltage-mode operation because it is less prone to the effect of stray node-capacitances. Fabricated using high-speed complementary bipolar processes, CFAs can be orders of magnitude faster than VFAs.

As shown in the simplified diagram of Fig. 6.34, a CFA consists of three stages: (a) a *unity-gain input buffer,* (b) a pair of *current mirrors,* and (c) an *output buffer.* The input buffer is based on the push-pull pair Q_1 and Q_2, whose purpose is to provide very low impedance at its output node v_N, which also acts as the inverting input of the CFA. In the presence of an external network, the push-pull pair can easily source or sink a substantial current i_N, though we shall see that in steady state i_N approaches zero. Q_1 and Q_2 are driven by the emitter followers Q_3 and Q_4, whose purpose is to raise the impedance and lower the bias current at the noninverting input v_P. The followers also provide suitable *pn*-junction voltage drops to bias Q_1 and Q_2 in the forward-active region and thus reduce crossover distortion. By design, the input buffer forces v_N to track v_P. This is similar to ordinary VFAs, except that the latter force v_N to track v_P via negative feedback.

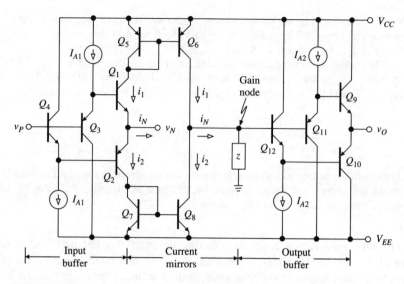

FIGURE 6.34
Simplified circuit diagram of a current-feedback amplifier.

Any current drawn at node v_N by the external network causes an imbalance between the currents of the push-pull pair,

$$i_1 - i_2 = i_N \qquad (6.47)$$

The current mirrors Q_5-Q_6 and Q_7-Q_8 replicate i_1 and i_2 and sum them at a common node called the *gain node*. The voltage of this node is buffered to the outside by another unity-gain buffer made up of Q_9 through Q_{12}. Ignoring the input bias current of this buffer, we can write, by Ohm's law,

$$V_o = z(jf)I_n \qquad (6.48)$$

where $z(jf)$, the net equivalent impedance of the gain node toward ground, is called the *open-loop transimpedance gain*. This transfer characteristic is similar to that of a VFA, except that the error signal i_N is a current rather than a voltage, and the gain $z(jf)$ is in volts per ampere rather than volts per volt. For this reason CFAs are also called *transimpedance amplifiers*.

The relevant CFA features are summarized in the block diagram of Fig. 6.35, where z has been decomposed into the *transresistance* component R_{eq} and *trans-capacitance* component C_{eq}. Letting $z(jf) = R_{eq} \parallel (1/j2\pi f C_{eq})$ and expanding, we get

$$z(jf) = \frac{z_0}{1 + jf/f_b} \qquad (6.49)$$

$$f_b = \frac{1}{2\pi R_{eq}C_{eq}} \qquad (6.50)$$

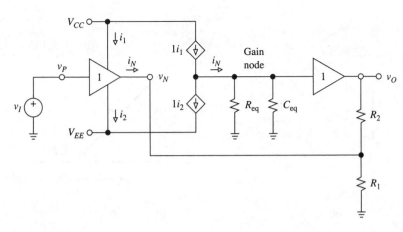

FIGURE 6.35
Block diagram of a CFA configured as a noninverting amplifier.

where $z_0 = R_{eq}$ is the dc value of $z(jf)$. The gain $z(jf)$ is approximately constant from dc to f_b; thereafter it rolls off with frequency at the rate of -10 dec/dec. Typically, R_{eq} is on the order of $10^6\ \Omega$ (which makes z_0 on the order of 1 V/μA), C_{eq} on the order of 10^{-12} F, and f_b on the order of 10^5 Hz.

EXAMPLE 6.14. The CLC401 CFA (Comlinear) has $z_0 \cong 0.71$ V/μA and $f_b \cong$ 350 kHz. (a) Find C_{eq}. (b) Find i_N for $v_O = 5$ V (dc).

Solution.

(a) $R_{eq} \cong 710\ k\Omega$, so $C_{eq} = 1/(2\pi R_{eq} f_b) \cong 0.64$ pF.
(b) $i_N = v_O/R_{eq} \cong 7.04\ \mu A$.

Closed-Loop Gain

Figure 6.36a shows a simplified CFA model, along with a negative-feedback network. Whenever an external signal V_i tries to unbalance the CFA inputs, the input buffer begins sourcing (or sinking) an imbalance current I_n. By Eq. (6.48), this current causes V_o to swing in the positive (or negative) direction until the original imbalance is neutralized via the negative-feedback loop, thus confirming the role of I_n as error signal.

Applying the superposition principle, we can write

$$I_n = \frac{V_i}{R_1 \parallel R_2} - \frac{V_o}{R_2} \qquad (6.51)$$

Clearly, the feedback signal V_o/R_2 is a current, and the feedback factor $\beta = 1/R_2$ is now in amperes per volt. Substituting into Eq. (6.48) and collecting gives the

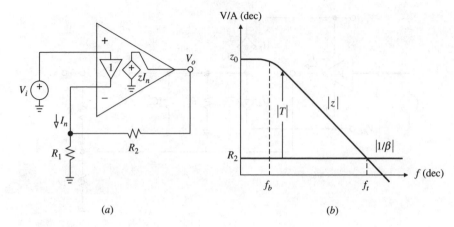

FIGURE 6.36
Noninverting CFA amplifier, and graphical method to visualize the loop gain $|T|$.

closed-loop gain

$$A(jf) = \frac{V_o}{V_i} = \left(1 + \frac{R_2}{R_1}\right)\frac{1}{1 + 1/T(jf)} \tag{6.52}$$

$$T(jf) = \frac{z(jf)}{R_2} \tag{6.53}$$

where $T(jf)$ is called the *loop gain*. This name stems from the fact that a current flowing around the loop is first multiplied by $z(jf)$ to be converted to a voltage, and then divided by R_2 to be converted back to a current, thus experiencing an overall gain of $T(jf) = z(jf)/R_2$. In the decade plot of $|z|$ and $|1/\beta|$ of Fig. 6.36b we can visualize the decade value of $|T|$ as the decade difference between the two curves. For instance, if at a given frequency $|z| = 10^5$ V/A and $|1/\beta| = 10^3$ V/A, then $|T| = 10^{5-3} = 10^2$.

In their effort to maximize $T(jf)$ and thus reduce the gain error, manufacturers strive to maximize $z(jf)$ relative to R_2. Consequently, the inverting-input current $I_n = V_o/z$ will be very small, even though this is the low-impedance output node of a buffer. In the limit $z \to \infty$ we obtain $I_n \to 0$, indicating that a *CFA will ideally provide whatever output is needed to drive I_n to zero*. Thus, the *input voltage constraint*

$$V_n \to V_p \tag{6.54a}$$

and the *input current constraints*

$$I_p \to 0 \qquad I_n \to 0 \tag{6.54b}$$

hold also for CFAs, though for different reasons than for VFAs. Equation (6.54a) holds by design in a CFA, and by negative-feedback action in a VFA; Eq. (6.54b) holds by negative-feedback action in a CFA, and by design in a VFA. We can apply these constraints to the analysis of CFA circuits, very much like in the analysis of conventional VFAs.[11]

CFA Dynamics

To investigate the dynamics of the CFA of Fig. 6.35, we substitute Eq. (6.49) into Eq. (6.53), and then into Eq. (6.52). This gives, for $z_0/R_2 \gg 1$,

$$A(jf) = A_0 \times \frac{1}{1 + jf/f_t} \tag{6.55}$$

$$A_0 = 1 + \frac{R_2}{R_1} \qquad f_t = \frac{1}{2\pi R_2 C_{eq}} \tag{6.56}$$

where A_0 and f_t are, respectively, the *closed-loop dc gain* and *bandwidth*. With R_2 in the kiloohm range and C_{eq} in the picofarad range, f_t is typically in the range of 10^8 Hz. We observe that for a given CFA, the closed-loop bandwidth depends on only R_2. We can thus use R_2 to set f_t, and then adjust R_1 to set A_0. The ability to control gain independently of bandwidth constitutes the first major advantage of CFAs over conventional op amps. Bandwidth constancy is illustrated in Fig. 6.37a.

Next, we investigate the transient response. Applying a step $v_I = V_{im}u(t)$ to the circuit of Fig. 6.36a will, by Eq. (6.51), result in the current $i_N = V_{im}/(R_1 \parallel R_2) - v_O/R_2$. With reference to Fig. 6.35, we can also write $i_N = v_O/R_{eq} + C_{eq}dv_O/dt$. Eliminating i_N, we get, for $R_2 \ll R_{eq}$,

$$R_2 C_{eq} \frac{dv_O}{dt} + v_O = A_0 V_{im}$$

whose solution is $v_O = A_0 V_{im}[1 - \exp(t/\tau)]u(t)$,

$$\tau = R_2 C_{eq} \tag{6.57}$$

The response is an exponential transient regardless of the input step magnitude, and the time constant governing it is set by R_2 regardless of A_0. For instance, a CLC401 op amp with $R_2 = 1.5$ kΩ has $\tau = 1.5 \times 10^3 \times 0.64 \times 10^{-12} \cong 1$ ns. The rise time is $t_R = 2.2\tau \cong 2.2$ ns, and the settling time within 0.1% of the final value is $t_S \cong 7\tau \cong 7$ ns, in reasonable agreement with the data-sheet values $t_R = 2.5$ ns and $t_S = 10$ ns.

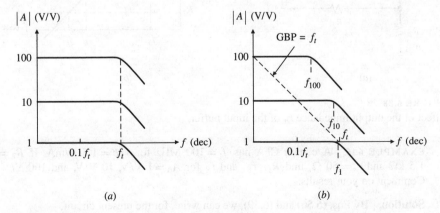

(a) (b)

FIGURE 6.37
Closed-loop bandwidth as a function of gain for (a) an ideal CFA and (b) a practical CFA.

Since R_2 controls the closed-loop dynamics, data sheets usually recommend an optimum value, typically in the range of 10^3 Ω. For voltage follower operation R_1 is removed, but R_2 must be left in place to set the dynamics of the device.

Second-Order Effects

According to the above analysis, once R_2 has been set, the dynamics appear to be unaffected by the closed-loop gain setting. However, the bandwidth and rise time of a practical CFA do vary with A_0 somewhat, though not as drastically as in VFAs. The main reason is the nonzero output resistance r_n of the input buffer, whose effect is to reduce the loop gain somewhat, degrading the closed-loop dynamics in proportion. Using the more realistic CFA model of Fig. 6.38a we get, by the superposition principle, $I_n = V_i/[r_n + (R_1 \parallel R_2)] - \beta V_o$, where the feedback factor β is found using the current-divider formula and Ohm's law,

$$\beta = \frac{R_1}{R_1 + r_n} \times \frac{1}{R_2 + (r_n \parallel R_1)} = \frac{1}{R_2 + r_n(1 + R_2/R_1)} \tag{6.58}$$

Clearly, the effect of r_n is to shift the $|1/\beta|$ curve upward, from R_2 to $R_2 + r_n(1 + R_2/R_1)$. As pictured in Fig. 6.38b, this causes a decrease in the crossover frequency, which we shall now denote as f_B. This frequency is obtained by letting $f_t \rightarrow f_B$ and $R_2 \rightarrow R_2 + r_n(1 + R_2/R_1)$ in Eq. (6.56). The result can be put in the form

$$f_B = \frac{f_t}{1 + r_n/(R_1 \parallel R_2)} \tag{6.59}$$

where now f_t is the extrapolated value of f_B in the limit $r_n \rightarrow 0$.

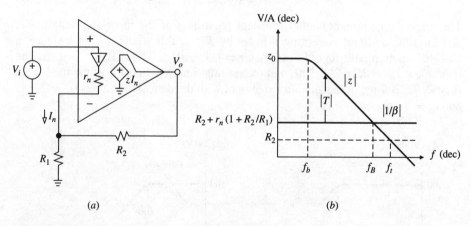

(a)   (b)

FIGURE 6.38
Effect of the output impedance r_n of the input buffer.

EXAMPLE 6.15. A certain CFA has $f_t = 100$ MHz for $1/\beta = 1.5$ V/mA. If $R_2 = 1.5$ kΩ and $r_n = 50$ Ω, find R_1, f_B, and t_R for $A_0 = 1$ V/V, 10 V/V, and 100 V/V. Comment on your results.

Solution. By Eqs. (6.56) and (6.59), we can write, for the present circuit,

$$R_1 = R_2/(A_0 - 1)$$

$$f_B = 10^8/(1 + A_0/30)$$

Moreover, $t_R \cong 2.2/2\pi f_B$. For $A_0 = 1$, 10, and 100 V/V we get, respectively, $R_1 = \infty$, 166.7 Ω, and 15.15 Ω; $f_B = 96.8$ MHz, 75.0 MHz, and 23.1 MHz; $t_R = 2.2/(2\pi \times 96.8 \times 10^6) = 3.6$ ns, 4.7 ns, and 15.2 ns. The bandwidth reductions, depicted in Fig. 6.37b, still compare favorably with those of a VFA, whose bandwidth would be reduced, respectively, by 1, 10, and 100.

The values of R_1 and R_2 can be predistorted to compensate for bandwidth reduction. We first find R_2 for a given f_B at a given A_0, then we find R_1 for the given A_0.

EXAMPLE 6.16. (a) Redesign the amplifier of Example 6.15 so that with $A_0 = 10$ V/V it has $f_B = 100$ MHz rather than 75 MHz. (b) Assuming $z_0 = 0.75$ V/μA, find the dc gain error.

Solution.

(a) For $f_B = 100$ MHz we need $R_2 + r_n(1 + R_2/R_1) = 1.5$ V/mA, or $R_2 = 1500 - 50 \times 10 = 1$ kΩ. Then, $R_1 = R_2/(A_0 - 1) = 10^3/(10 - 1) = 111$ Ω.

(b) $T_0 = \beta z_0 = (1/1500)0.75 \times 10^6 = 500$. The dc gain error is $\epsilon \cong -100/T_0 = -0.2\%$.

Applying CFAs

Though we have focused on the noninverting amplifier, we can configure a CFA for other familiar topologies.[11,12] For instance, if we lift R_1 off ground in Fig. 6.36a, and apply V_i via R_1 with the noninverting input at ground, we obtain the familiar inverting amplifier. Its dc gain is $A_0 = -R_2/R_1$, and its bandwidth is given by Eq. (6.59). Likewise, we can configure CFAs as summing or difference amplifiers, I-V converters, and so forth. Except for its much faster dynamics, a CFA works much like a VFA, but with one notorious exception that will be explained in Chapter 8: it must never include a direct capacitance between its output and inverting-input pins, since this tends to make the circuit oscillatory. In fact, stable amplifier operation requires that $1/\beta \geq (1/\beta)_{min}$, where $(1/\beta)_{min}$ is also given in the data sheets.

Compared to VFAs, CFAs generally suffer from poorer input-offset-voltage and input-bias-current characteristics. Moreover, they afford lower dc loop gains, usually on the order of 10^3 or less. Finally, having much wider bandwidths, they tend to be noisier. CFAs are suited to moderately accurate but very high-speed applications.

PSpice Models

CFA manufacturers provide macromodels to facilitate the application of their products. Alternatively, the user can create simplified models for a quick test of such characteristics as noise and stability. Figure 6.39 shows one such model.

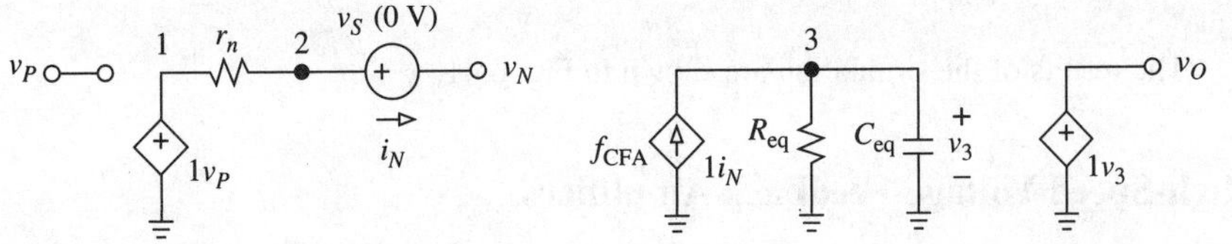

FIGURE 6.39
Simple PSpice model for a one-pole CFA.

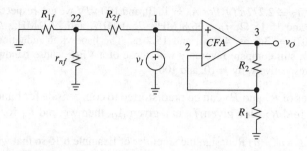

FIGURE 6.40
PSpice circuit of Example 6.16.

EXAMPLE 6.17. Use PSpice to verify the case $A_0 = 10$ V/V in Example 6.16.

Solution. With reference to Fig. 6.40, we write the following file.

```
Noninverting CFA Amp with A0 = 10 V/V and fB = 100 MHz:
*One-pole CFA: z0 = 0.75 V/µA, fb = 200 kHz
.subckt CFA vP vN vO
ein 1 0 vP 0 1        ;input buffer
rn 1 2 50             ;buffer's output resistance
vS 2 vN dc 0          ;0-V source to sense iN
fCFA 0 3 vS 1         ;CCCS
Req 3 0 750k          ;dc gain
Ceq 3 0 1.061pF       ;fB=200kHz
eout vO 0 3 0 1       ;output buffer
.ends CFA
*Circuit to plot A and vo(t)
vI 1 0 ac 1V pulse(0 1V 0 0.1ns 0.1ns 10ns 20ns)
R1 0 2 111.1
R2 2 3 1k
XCFA1 1 2 3 CFA
*Circuit to plot 1/beta
R2f 1 22 1k
R1f 22 0 111.1
rnf 0 22 50
*Circuit to plot z:
Vs 5 0 dc 0V
XCFA2 1 5 6 CFA
RL 6 0 2k             ;avoids floating nodes
.ac dec 10 100kHz 1GHz
.tran 0.5ns 10ns
.probe ;z=V(6)/I(Vs), 1/beta=V(1)/I(rnf)
       ;A=V(3)/V(1), vo(t)=v(3)
.end
```

The results of the simulation are shown in Fig. 6.41.

High-Speed Voltage-Feedback Amplifiers

The availability of high-speed complementary bipolar processes and the emergence
of applications requiring increased speeds have led to the development of faster

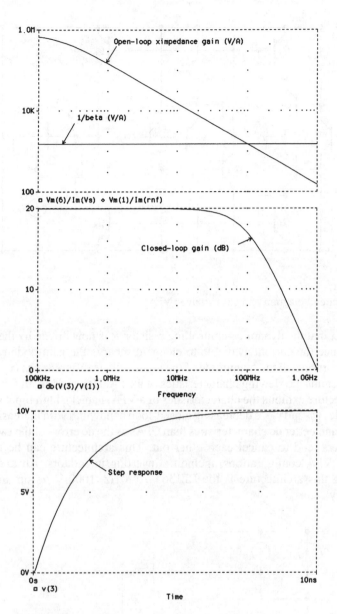

FIGURE 6.41
Frequency plots and step response for the CFA circuit of Example 6.17.

voltage-feedback amplifiers (VFAs),[13] alongside the current-feedback amplifiers (CFAs) just discussed. Though the borderline between standard and high-speed VFAs keeps changing, at the time of writing we can take a high-speed VFA as one having[14] GBP > 50 MHz and SR > 100 V/μs. Two of the most popular high-speed VFA architectures in current use are illustrated in Figs. 6.42 and 6.43.

The VFA of Fig. 6.42 is similar to the CFA of Fig. 6.34, except for the addition of a unity-gain buffer (Q_{13} through Q_{16}) to raise the input impedance at node v_N,

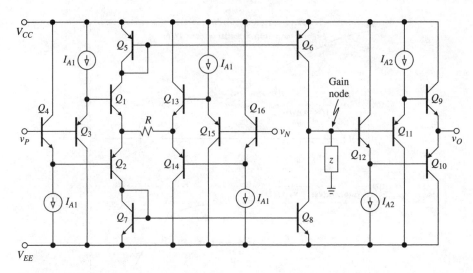

FIGURE 6.42
Simplified circuit diagram of a CFA-derived VFA.

and the fact that the dynamics-controlling resistor R is now driven by the two input buffers. Since the current available to charge/discharge the gain-node capacitance C_{eq} is proportional to the magnitude of the input voltage difference as $(v_P - v_N)/R$, this VFA retains the slewing characteristics of a CFA. However, in all other respects this architecture exhibits the characteristics of a VFA, namely, high input impedance at both nodes v_P and v_N, a decreasing closed-loop bandwidth with increasing closed-loop gain, and better dc characteristics than CFAs as the dc errors of the two matched input buffers tend to cancel each other out. This architecture can be used in all traditional VFA configurations, including inverting integrators. An example of a VFA using this architecture is the LT1363 70-MHz, 1000-V/μs op amp (Linear Technology).

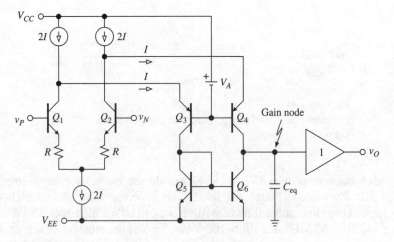

FIGURE 6.43
Simplified circuit diagram of a folded cascode bipolar VFA.

The trend toward high speed as well as low-power-supply voltages has inspired the folded cascode architecture, which finds wide use both in complementary bipolar processes and CMOS processes. In the bipolar illustration[15] of Fig. 6.43, any imbalance between v_P and v_N will cause an imbalance in the collector currents of the common-emitter *npn* pair Q_1 and Q_2, and this current imbalance is in turn fed to the emitters of the common-base *pnp* pair Q_3 and Q_4 (hence the term *folded cascode*). The latter pair is actively loaded by the current mirror Q_5 and Q_6 to provide high voltage gain at the gain node, whence the signal is buffered to the outside via a suitable unity-gain stage. Product examples utilizing this architecture are the EL2044C low-power/low-voltage 120-MHz unity-gain stable op amp (Elantec), and the THS4401 high-speed VFA (Texas Instruments) offering a unity-gain bandwidth of 300 MHz, SR $= 400$ V/μs, and $t_S = 30$ ns to 0.1%.

PROBLEMS

6.1 Open-loop response

6.1 (*a*) Because of manufacturing process variations, the second-stage gain of a certain 741 op amp version is $-a_2 = -544$ V/V $\pm 20\%$. How does this affect a_0, f_b, and f_t? (*b*) Repeat, but for $C_c = 30$ pF $\pm 10\%$.

6.2 The open-loop response of a constant-GBP op amp is measured in the lab. If $\angle a(j80\text{ Hz}) = -58°$ and $|a(j1\text{ Hz})| = 100$ V/mV, find a_0, f_b, and f_t.

6.3 Given that a constant-GBP op amp has $|a(j100\text{ Hz})| = 1$ V/mV and $|a(j1\text{ MHz})| = 10$ V/V, find (*a*) the frequency at which $\angle a = -60°$, and (*b*) the frequency at which $|a| = 2$ V/V. *Hint:* Start out with the linearized magnitude plot.

6.2 Closed-loop response

6.4 Show that the circuit of Example 6.2 yields $A(jf) = H_{0\text{LP}} \times H_{\text{LP}}$. What are the values of $H_{0\text{LP}}$, f_0, and Q?

6.5 (*a*) Show that cascading n identical noninverting amplifiers with individual dc gains A_0 yields a composite amplifier with the overall bandwidth $f_B = (f_t/A_0)\sqrt{2^{1/n} - 1}$. (*b*) Develop a similar expression for the case of n inverting amplifiers with individual dc gains $-A_0$.

6.6 (*a*) Repeat Example 6.2, but for a cascade of three 741 noninverting amplifiers with individual dc gains of 10 V/V. (*b*) Compare the -3-dB bandwidths of the one-op-amp, two-op-amp, and three-op-amp designs, and comment.

6.7 (*a*) Consider the cascade connection of a noninverting amplifier with $A_0 = 2$ V/V, and an inverting amplifier with $A_0 = -2$ V/V. If both amplifiers use op amps with GBP $= 5$ MHz, find the -3-dB frequency of the composite amplifier. (*b*) Find the 1% magnitude error and the 5° phase-error bandwidths.

6.8 (*a*) Find the closed-loop GBP of the inverting amplifier of Fig. P1.54 if $R_1 = R_2 = \cdots = R_6 = R$, $r_d \gg R$, $r_o \ll R$, and $f_t = 4$ MHz. (*b*) Repeat if the source

v_I is applied at the noninverting input and the left terminal of R_1 is connected to ground. (c) Repeat part (b), but with the left terminal of R_1 left floating. Comment.

6.9 (a) Using a 741 op amp, design a two-input summing amplifier such that $v_O = -10(v_1 + v_2)$; hence, find its -3-dB frequency. (b) Repeat, but for five inputs, or $v_O = -10(v_1 + \cdots + v_5)$. Compare with the amplifier of part (a) and comment.

6.10 Assuming 741 op amps, find the -3-dB frequency of the circuits of (a) Fig. P1.17, (b) Fig. P1.19, (c) Fig. P1.21, and (d) Fig. P1.65.

6.11 Find the -3-dB frequency of the triple-op-amp IA of Fig. 2.21, given that all op amps have GBP $= 8$ MHz. Calculate with the wiper all the way down and all the way up.

6.12 In the dual-op-amp IA of Fig. 2.23 let $R_3 = R_1 = 1$ kΩ, $R_4 = R_2 = 9$ kΩ, and $f_{t1} = f_{t2} = 1$ MHz. Find the -3-dB frequency with which the IA processes V_2, and that with which it processes V_1.

6.13 Sketch and label the frequency plot of the $CMRR_{dB}$ of the IA of Problem 6.12. Except for the finite f_t, the op amps are ideal and the resistance ratios are perfectly matched.

6.14 A triple-op-amp instrumentation amplifier with $A = 10$ V/V is to be designed using three constant-GBP, JFET-input op amps of the same family. Letting $A = A_{\mathrm{I}} \times A_{\mathrm{II}}$, how would you choose A_{I} and A_{II} in order to minimize the worst-case output dc error E_0? Maximize the overall -3-dB bandwidth?

6.15 Three signals v_1, v_2, and v_3 are to be summed using the topology of Fig. P1.31, and two alternatives are being considered: $v_O = v_1 + v_2 + v_3$ and $v_O = -(v_1 + v_2 + v_3)$. Which option is most desirable from the viewpoint of minimizing the untrimmed dc output error E_O? Maximizing the -3-dB frequency?

6.16 A unity-gain buffer is needed and the following options are being considered, each offering advantages and disadvantages in the event that the circuit must subsequently need to be altered: (a) a voltage follower, (b) a noninverting amplifier with $A_0 = 2$ V/V followed by a 2:1 voltage divider, and (c) a cascade of two unity-gain inverting amplifiers. Assuming constant-GBP op amps, compare the advantages and disadvantages of the three alternatives.

6.17 Assuming the op amp of Fig. P1.60 has a constant GBP of 3 MHz, find the closed-loop parameters A_0 and f_B. Except for the GBP, the op amp is ideal.

6.18 Find the closed-loop GBP of the inverting amplifier of Fig. 1.32a, given that $R_1 = 10$ kΩ, $R_2 = 20$ kΩ, $R_3 = 120$ kΩ, $R_4 = 30$ kΩ, $R_L = \infty$, and $f_t = 27$ MHz. Except for its finite f_t, the op amp can be considered ideal.

6.19 Find the closed-loop gain and bandwidth of the high-sensitivity I-V converter of Fig. 2.2 if $R = 200$ kΩ, $R_1 = R_2 = 100$ kΩ, and the input source has a 200-kΩ parallel resistance toward ground. The op amp is ideal, except for a constant GBP such that at 1.8 kHz the open-loop gain is 80 dB.

6.20 The circuit of Fig. P1.21 is implemented with three 10-kΩ resistances and an op amp with $a_0 = 50$ V/mV, $I_B = 50$ nA, $I_{OS} = 10$ nA, $V_{OS} = 0.75$ mV, $CMRR_{dB} = 100$ dB, and $f_t = 1$ MHz. Assuming $v_I = 5$ V, find the maximum dc output error as well as the small-signal bandwidth with both the switch open and the switch closed.

6.3 Input and output impedances

6.21 If the floating-load *V-I* converter of Fig. 2.4*a* is implemented with an op amp having $a_0 = 10^5$ V/V, $f_b = 10$ Hz, $r_c \gg r_d \gg R$, $r_o \ll R$, and $R = 10$ kΩ, sketch and label the magnitude Bode plot of the impedance $Z_o(jf)$ seen by the load; hence, find the element values of its equivalent circuit.

6.22 Find the impedance $Z_o(jf)$ seen by the load in the *V-I* converter of Fig. P2.5 if the op amp has $a_0 = 10^5$ V/V, $f_t = 1$ MHz, $r_d = \infty$, $r_o = 0$, $R_1 = R_2 = 18$ kΩ, and $R_3 = 2$ kΩ.

6.23 If the Howland current pump of Fig. 2.6*a* is implemented with four 10-kΩ resistances and an op amp having $a_0 = 10^5$ V/V, $f_t = 1$ MHz, $r_d = \infty$, and $r_o = 0$, sketch and label the magnitude plot of the impedance Z_o seen by the load. Justify using physical insight.

6.24 The negative-resistance converter of Fig. 1.20*b* is implemented with three 10-kΩ resistances and an op amp with GBP = 1 MHz. Find its input impedance Z_{eq}. How does it change as f is swept from 0 to ∞?

6.25 The grounded-load current amplifier of Fig. 2.12 is implemented with $R_1 = R_2 = 10$ kΩ and an op amp having $f_t = 10$ MHz, $r_d = \infty$, and $r_o = 0$. If the amplifier is driven by a source with a parallel resistance of 30 kΩ and drives a load of 2 kΩ, sketch and label the magnitude plots of the gain, the impedance seen by the source, and the impedance seen by the load.

6.26 A constant-GBP JFET-input op amp with $a_0 = 10^5$ V/V, $f_t = 4$ MHz, and $r_o = 100$ Ω is configured as an inverting amplifier with $R_1 = 10$ kΩ and $R_2 = 20$ kΩ. What is the frequency at which resonance with a 0.1-μF load capacitance will occur? What is the value of Q?

6.27 In the circuit of Fig. 1.32*a* let $R_1 = R_2 = R_3 = 30$ kΩ, $R_4 = R_L = \infty$, and let the op amp have $a_0 = 300$ V/mV and $f_b = 10$ Hz. Assuming $r_d = \infty$ and $r_o = 0$, sketch and label the magnitude plot of the impedance $Z(jf)$ between node v_1 and ground; use log-log scales.

6.28 In the circuit of Fig. 1.13*b* let both the 10-kΩ and 30-kΩ resistances be changed to 1 kΩ, and let the 20-kΩ resistance be changed to 18 kΩ. Assuming $r_d = \infty$, $r_o = 0$, and $f_t = 1$ MHz, sketch and label the magnitude plot of the impedance $Z(jf)$ seen by the input source; use log-log scales.

6.29 Let the inverting integrator of Fig. 6.25*a* be implemented with a 741 op amp, and with $R = 158$ kΩ and $C = 1$ nF. Sketch and label the magnitude plot of its output impedance $Z_o(jf)$; use log-log scales. *Hint:* First plot T.

6.4 Transient response

6.30 Investigate the response of the high-sensitivity *I-V* converter of Example 2.2 to an input step of 10 nA. Except for $f_t = 1$ MHz and SR $= 5$ V/μs, the op amp is ideal.

6.31 Investigate the response of a Howland current pump to an input step of 1 V. The circuit is implemented with four 10-kΩ resistances and a 741C op amp, and it drives a 2-kΩ load.

6.32 (a) Using a 741C op amp powered from ±15-V regulated supplies, design a circuit that gives $v_O = -(v_I + 5 \text{ V})$ with the maximum small-signal bandwidth possible. (b) What is this bandwidth? What is the FPB?

6.33 An inverting amplifier with $A_0 = -2$ V/V is driven with a square wave of peak values $\pm V_{im}$ and frequency f. With $V_{im} = 2.5$ V, it is observed that the output turns from trapezoidal to triangular when f is raised to 250 kHz; with $f = 100$ kHz, it is found that slew-rate limiting ceases when V_{im} is lowered to 0.4 V. If the input is changed to a 3.5-V (rms) ac signal, what is the useful bandwidth of the circuit? Is it small-signal or large-signal limited?

6.34 Find the response of the cascaded amplifier of Example 6.2 to a 1-mV input step.

6.35 A cascaded amplifier consists of an op amp OA_1, operating as a noninverting amplifier with $A_0 = +20$ V/V, followed by an op amp OA_2, operating as an inverting amplifier with $A_0 = -10$ V/V. Sketch the circuit; then find the minimum values of f_{t1}, SR_1, f_{t2}, and SR_2 needed to ensure an overall bandwidth of 100 kHz with a full-power output signal of 5 V (rms).

6.36 In the dual-op-amp IA of Fig. 2.23 let $R_3 = R_1 = 1$ kΩ, $R_4 = R_2 = 9$ kΩ, and $f_{t1} = f_{t2} = 1$ MHz. Find the small-signal step response if (a) $v_1 = 0$ and the step is applied at v_2, (b) $v_2 = 0$ and the step is applied at v_1, and (c) the step is applied at v_1 and v_2 tied together.

6.37 Using the LF353 dual JFET-input op amp, whose ratings are $V_{OS(\text{max})} = 10$ mV, GBP = 4 MHz, and SR = 13 V/μs, design a cascaded amplifier having an overall gain of 100 V/V as well as provision for overall offset-error nulling. (b) Find the small-signal bandwidth as well as the FPB. (c) If the circuit is to operate with a 50-mV (rms) ac input, what is its useful frequency range of operation? Is it small-signal or large-signal limited?

6.38 A TL071 JFET-input op amp is configured as an inverting amplifier with $A_0 = -10$ V/V and is driven by a 1-V (peak-to-peak) ac signal. Assuming $a_0 = 200$ V/mV, $f_t = 3$ MHz, and SR = 13 V/μs, estimate the peak-to-peak amplitude of the inverting input voltage v_N for $f = 1$ Hz, 10 Hz, . . . , 10 MHz. Comment.

6.39 In the high-sensitivity I-V converter of Fig. 2.2 let $R = 100$ kΩ, $R_1 = 10$ kΩ, $R_2 = 30$ kΩ, and let the op amp have $f_t = 4$ MHz and SR = 15 V/μs. Except for these limitations, the op amp can be considered ideal. If $i_I = 20 \sin(2\pi f t)$ μA, what is the useful bandwidth of the circuit? Is it small-signal or large-signal limited?

6.40 Equation (6.27) indicates that if we want to avoid slew-rate limiting in a voltage follower implemented with an op amp having SR = 0.5 V/μs and $f_t = 1$ MHz, we must limit the input step magnitude below about 80 mV. What is the maximum allowed input step if the same op amp is configured as: (a) An inverting amplifier with a gain of -1 V/V? (b) A noninverting amplifier with a gain of $+2$ V/V? (c) An inverting amplifier with a gain of -2 V/V?

6.41 Assuming equal resistors in the circuit of Fig. P1.54, find the minimum values of SR and f_t required for a useful bandwidth of 1 MHz for a sinusoidal input with a peak amplitude of 1 V.

6.42 The wideband band-pass filter of Example 3.5 is to be implemented with a constant GBP op amp. Find the minimum f_t and SR for an undistorted full-power output with a magnitude error of less than 1% over the entire audio range (that is, 20 Hz to 20 kHz).

6.5 Effect of finite GBP on integrator circuits

6.43 (a) Using a 741 op amp with four equal resistances and a 10-nF capacitance, design a Deboo integrator with $f_0 = 1$ kHz. (b) Sketch the linearized Bode plots of $|a|$, $|1/\beta|$, and $|H|$. (c) Find an expression for $H(jf)$.

6.44 (a) Assuming $r_d = \infty$, $r_o = 0$, and $a(jf) \cong f_t/jf$, find $H(jf)$ for the compensated integrator of Fig. 6.26a. (b) Show that letting $C_c = C/(f_t/f_0 - 1)$ makes $H \cong H_{\text{ideal}}$. (c) Specify suitable components for $f_0 = 10$ kHz, and verify with PSpice for $f_t = 1$ MHz.

6.45 (a) Assuming $r_d = \infty$, $r_o = 0$, and $a(jf) \cong f_t/jf$, find $H(jf)$ for the compensated integrator of Fig. 6.26b. (b) Show that letting $R_c = 1/2\pi C f_t$ makes $H = H_{\text{ideal}}$. (c) Specify suitable components for $f_0 = 10$ kHz if $r_o = 100\ \Omega$, and verify with PSpice for $f_t = 1$ MHz.

6.46 (a) Find $H(jf)$ for the circuit of Fig. 6.27b, rationalize it, and discard higher-order terms to show that $\epsilon_\phi = +f/f_t$ for $f \ll f_t$. (b) Verify with PSpice for the case $f_0 = 10$ kHz and $f_t = 1$ MHz.

6.47 (a) Find an expression for the phase error of the Deboo integrator of Problem 6.43. (b) Find a suitable resistance R_c that, when placed in series with the capacitance, will provide phase-error compensation.

6.48 The active compensation scheme of Fig. P6.48 (see *Electronics and Wireless World,* May 1987) is a generalization of that of Fig. 6.27a, in that it allows for phase-error control. Verify that the error function of this circuit is $(1 + jf/\beta_2 f_{t2})/(1 + jf/f_{t1} - f^2/\beta_2 f_{t1} f_{t2})$, $\beta_2 = R_1/(R_1 + R_2)$. What happens if the op amps are matched and $R_1 = R_2$? Would you have any use for this circuit?

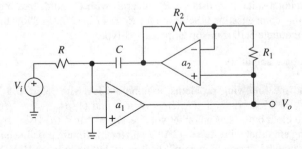

FIGURE P6.48

6.49 The active compensation method of Problem 6.48 can also be applied to the Deboo integrator, as shown in Fig. P6.49 (see *Proceedings of the IEEE*, Feb. 1979, pp. 324–325). Show that for matched op amps and $f \ll f_t$ we have $\epsilon_\phi \cong -(f/0.5f_t)^3$.

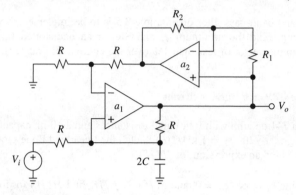

FIGURE P6.49

6.6 Effect of finite GBP on filters

6.50 The inverting amplifier of Fig. 1.10a is implemented with $R_1 = 10\,\text{k}\Omega$, $R_2 = 100\,\text{k}\Omega$, and a 741 op amp. Sketch and label the magnitude Bode plot of its closed-loop gain if the circuit contains also a 100-pF capacitance in parallel with R_2.

6.51 Investigate the effect of finite GBP on the phase-shifter circuit of Fig. 3.12a.

6.52 Investigate the effect of using op amps with GBP = 1 MHz in the inductance simulator of Example 4.8.

6.53 Obtain an expression of the type of Eq. (6.45) for the low-pass *KRC* filter of Fig. 3.23.

6.54 Use the μA741 macromodel of PSpice to assess the departure from ideality of the band-pass response of the state-variable filter of Example 3.18. If needed, compensate and predistort to improve accuracy.

6.55 Investigate the effect of using an op amp with GBP = 1 MHz in the notch filter of Example 3.14.

6.56 The effect of finite GBP on the unity-gain *KRC* filter of Fig. 3.25 can be compensated for by placing a suitable resistance R_c in series with C and decreasing R to $R - R_c$. (a) Show that compensation is achieved for $R_c = 1/2\pi C f_t$. (b) Show the compensated circuit of Example 3.10 if the op amp is a 741 type.

6.7 Current-feedback amplifiers

6.57 In this and the following problems, assume a CFA with $z_0 = 0.5\ \text{V}/\mu\text{A}$, $C_{eq} = 1.59\ \text{pF}$, $r_n = 25\ \Omega$, $I_P = 1\ \mu\text{A}$, $I_N = 2\ \mu\text{A}$, and $(1/\beta)_{\min} = 1\ \text{V/mA}$. Moreover, assume the input buffer has an offset voltage $V_{OS} = 1\ \text{mV}$. (a) Using this CFA, design an inverting amplifier with $A_0 = -2\ \text{V/V}$ and the maximum possible bandwidth. What is this bandwidth? The dc loop gain? (b) Repeat, but for $A_0 = -10\ \text{V/V}$ and the same bandwidth as in part (a). (c) Repeat (a), but for a difference amplifier with a dc gain of 1 V/V.

6.58 (*a*) Using the CFA of Problem 6.57, design a voltage follower with the widest possible bandwidth. (*b*) Repeat, but for a unity-gain inverting amplifier. How do the closed-loop GBPs compare? (*c*) Modify both circuits so that the closed-loop bandwidth is reduced in half. (*d*) How do the maximum dc output errors compare in the various circuits?

6.59 (*a*) Using the CFA of Problem 6.57, provide two designs for an *I-V* converter with a dc sensitivity of -10 V/mA. (*b*) How do the closed-loop bandwidths compare? How do the maximum output errors compare?

6.60 The data sheets recommend the circuit of Fig. P6.60 to adjust the closed-loop dynamics. Assuming the CFA data of Problem 6.57, estimate the closed-loop bandwidth and rise time as the wiper is varied from end to end.

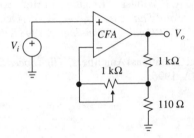

FIGURE P6.60

6.61 Using the CFA of Problem 6.57, design a second-order 10-MHz low-pass filter with $Q = 5$.

6.62 (*a*) Sketch a block diagram of the type of Fig. 6.35, but for the CFA-derived VFA of Fig. 6.42. Hence, denoting the output resistance of each input buffer as r_o, obtain expressions for the open-loop gain $a(jf)$ and the slew-rate SR. (*b*) Assuming $z(jf)$ can be modeled with a 1-MΩ resistance in parallel with a 2-pF capacitance, and $R = 500\ \Omega$ and $r_o = 25\ \Omega$, find a_0, f_b, f_t, β, T_0, A_0, and f_B, if $R_1 = R_2 = 1\ \text{k}\Omega$. (*c*) What is the SR for the case of a 1-V input step?

REFERENCES

1. J. E. Solomon, "The Monolithic Operational Amplifier: A Tutorial Study," *IEEE J. Solid-State Circuits,* vol. SC-9, Dec. 1974, pp. 314–332.
2. S. Franco, *Electric Circuits Fundamentals,* Oxford University Press, New York, 1995.
3. R. I. Demrow, "Settling Time of Operational Amplifiers," Application Note AN-359, *Applications Reference Manual,* Analog Devices, Norwood, MA, 1993.
4. C. T. Chuang, "Analysis of the Settling Behavior of an Operational Amplifier," *IEEE J. Solid-State Circuits,* vol. SC-17, Feb. 1982, pp. 74–80.
5. J. Williams, "Settling Time Measurements Demand Precise Test Circuitry," *EDN,* Nov. 15, 1984, p. 307.
6. P. R. Gray and R. G. Meyer, *Analysis and Design of Analog Integrated Circuits,* 3d ed., John Wiley & Sons, New York, 1993.
7. P. O. Brackett and A. S. Sedra, "Active Compensation for High-Frequency Effects in Op Amp Circuits with Applications to Active RC Filters," *IEEE Trans. Circuits Syst.,* vol. CAS-23, Feb. 1976, pp. 68–72.

8. L. C. Thomas, "The Biquad: Part I—Some Practical Design Considerations," and "Part II—A Multipurpose Active Filtering System," *IEEE Trans. Circuit Theory,* vol. CT-18, May 1971, pp. 350–361.

9. A. Budak, *Passive and Active Network Analysis and Synthesis,* Waveland Press, Prospect Heights, IL, 1991.

10. Based on the author's article "Current-Feedback Amplifiers Benefit High-Speed Designs," *EDN,* January 5, 1989, pp. 161–172. © Cahners Publishing Company, a Div. of Reed Elsevier Inc., 1997.

11. R. Mancini, "Converting from Voltage-Feedback to Current-Feedback Amplifiers," *Electronic Design Special Analog Issue,* June 26, 1995, pp. 37–46.

12. S. Evans, "Current-Feedback Op Amp Applications Circuit Guide," Application Note OA-07, *Comlinear Corporation Databook,* Fort Collins, CO, 1993-94.

13. D. Smith, M. Koen, and A. F. Witulski, "Evolution of High-Speed Operational Amplifier Architectures," *IEEE J. Solid-State Circuits,* vol. SC-29, Oct. 1994, pp. 1166–1179.

14. Texas Instruments Staff, *DSP/Analog Technologies,* 1998 Seminar Series, Texas Instruments, Dallas, TX, 1998.

15. W. Kester, "High Speed Operational Amplifiers," *High Speed Design Techniques,* Analog Devices, Norwood, MA, 1996.

7

NOISE

7.1 Noise Properties
7.2 Noise Dynamics
7.3 Sources of Noise
7.4 Op Amp Noise
7.5 Noise in Photodiode Amplifiers
7.6 Low-Noise Op Amps
     Problems
     References

Any unwanted disturbance that obscures or interferes with a signal of interest is generally referred to as *noise*.[1,2] The offset error due to the input bias current and input offset voltage is a familiar example of noise, dc noise in this case. However, there are many other forms of noise, particularly ac noise, which can significantly degrade the performance of a circuit unless proper noise reduction measures are taken. Depending on its origin, ac noise is classified as *external,* or *interference, noise,* and *internal,* or *inherent, noise.*

Interference Noise

This type of noise is caused by unwanted interaction between the circuit and the outside, or even between different parts of the circuit itself. This interaction can be electric, magnetic, electromagnetic, or even electromechanical, such as microphonic and piezoelectric noise. Electric and magnetic interaction takes place through the parasitic capacitances and mutual inductances between adjacent circuits or adjacent parts of the same circuit. Electromagnetic interference stems from the fact that each wire and trace constitutes a potential antenna. External noise can inadvertently be injected into a circuit also via the ground and power-supply busses.

Interference noise can be periodic, intermittent, or completely random. Usually it is reduced or forestalled by minimizing electrostatic and electromagnetic pickup from line frequency and its harmonics, radio stations, mechanical switch arching, reactive component voltage spikes, etc. These precautions may include filtering, decoupling, guarding, electrostatic and electromagnetic shielding, physical reorientation of components and leads, use of snubber networks, ground-loop elimination, and use of low-noise power supplies. Though often misconceived as "black magic," interference noise can be explained and dealt with in a rational manner.[3,4]

Inherent Noise

Even if we manage to remove all interference noise, a circuit will still exhibit inherent noise. This form of noise is random in nature and is due to random phenomena, such as the thermal agitation of electrons in resistors and the random generation and recombination of electron-hole pairs in semiconductors. Because of thermal agitation, each vibrating electron inside a resistor constitutes a minuscule current. These currents add up algebraically to originate a net current and, hence, a net voltage that, though zero on average, is constantly fluctuating because of the random distribution of the instantaneous magnitudes and directions of the individual currents. These fluctuations occur even if the resistor is sitting in a drawer. Thus, it is quite appropriate to assume that each node voltage and each branch current in a circuit are constantly fluctuating around their desired values.

Signal-to-Noise Ratio

The presence of noise degrades the quality of a signal and poses the ultimate limit on the size of signals that can be successfully detected, measured, and interpreted. The quality of a signal in the presence of noise is specified by means of the *signal-to-noise ratio* (SNR)

$$ \text{SNR} = 10 \log_{10} \frac{X_s^2}{X_n^2} \qquad (7.1) $$

where X_s is the rms value of the signal, and X_n is that of its noise component. The poorer the SNR, the more difficult it is to rescue the useful signal from noise. Even though a signal buried in noise can be rescued by suitable signal processing, such as signal averaging, it always pays to keep the SNR as high as other design constraints allow.

The degree to which circuit designers should be concerned about noise ultimately depends on the performance requirements of the application. With the tremendous improvements in op amp input offset-error characteristics, as well A-D and D-A converter resolution, noise is an increasingly important factor in the error budget analysis of high-performance systems. Taking a 12-bit system as an example, we note that with a 10-V full scale, $\frac{1}{2}$ LSB corresponds to $10/2^{13} = 1.22$ mV, which by itself may pose problems in converter design. In the real world, the signal may be produced by a transducer and require considerable amplification to achieve a

10-V full scale. Taking 10 mV as a typical full-scale transducer output, $\frac{1}{2}$ LSB now corresponds to 1.22 μV. If the amplifier generates only 1 μV of input-referred noise, the LSB resolution would be invalidated!

To take full advantage of sophisticated devices and systems, the designer must be able to understand noise mechanisms; perform noise calculations, simulations, and measurements; and minimize noise as required. These are the topics to be addressed in this chapter.

7.1
NOISE PROPERTIES

Since noise is a random process, the instantaneous value of a noise variable is unpredictable. However, we can deal with noise on a statistical basis. This requires introducing special terminology as well as special calculation and measurement.

Rms Value and Crest Factor

Using subscript n to denote noise quantities, we define the *root-mean-square (rms) value* X_n of a noise voltage or current $x_n(t)$ as

$$X_n = \left(\frac{1}{T} \int_0^T x_n^2(t)\, dt \right)^{1/2} \tag{7.2}$$

where T is a suitable averaging time interval. The square of the rms value, or X_n^2, is called the *mean square value*. Physically, X_n^2 represents the average power dissipated by $x_n(t)$ in a 1-Ω resistor.

In voltage-comparator applications, such as A-D converters and precision multivibrators, accuracy and resolution are affected by the instantaneous rather than the rms value of noise. In these situations, expected peak values of noise are of more concern. Most noise has a Gaussian, or normal, distribution as shown in Fig. 7.1, so it is possible to predict instantaneous values in terms of probabilities. The *crest factor* (CF) is defined as the *ratio of the peak value to the rms value* of noise. Though all

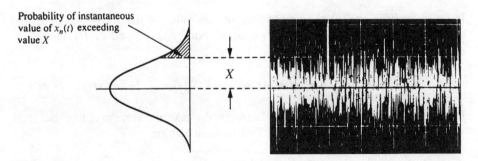

Probability of instantaneous value of $x_n(t)$ exceeding value X

X

FIGURE 7.1
Voltage noise (right), and Gaussian distribution of amplitude.

CF values are possible in principle, the likelihood of $x_n(t)$ exceeding a given value X decreases very rapidly with X, as indicated by the residual area under the distribution curve. Suitable calculations[5] reveal that for Gaussian noise the probability of CF exceeding 1 is 32%, that of exceeding 2 is 4.6%, that of exceeding 3 is 0.27%, that of exceeding 3.3 is 0.1%, and that of exceeding 4 is 0.0063%. It is common practice to take the *peak-to-peak value* of Gaussian noise to be 6.6 times the rms value, since the instantaneous value is within this range 99.9% of the time, which is close to 100%.

Noise Observation and Measurement

Voltage noise can readily be observed with an oscilloscope of adequate sensitivity. An advantage of this instrument is that it allows us to actually see the signal and thus make sure it is internal noise and not externally induced noise, such as 60-Hz pickup. One way of estimating the rms value is by observing the maximum peak-to-peak fluctuation, and then dividing by 6.6. A less subjective alternative[6] is to observe noise with two equally calibrated channels, and adjust the offset of one channel until the two noisy traces just merge; if we then remove both noise sources and measure the difference between the two clean traces, the result is approximately *twice the rms value*.

Noise can be measured with a multimeter. Ac meters fall into two categories: *true rms meters* and *averaging-type meters*. The former yield the correct rms value regardless of the waveform, provided that the CF specifications of the instrument are not exceeded. The latter are calibrated to give the rms value of a sine wave. They first rectify the signal and compute its average, which for ac signals is $2/\pi$ times the peak value; then they synthesize the rms value, which for ac signals is $1/\sqrt{2}$ times the peak value, by amplifying the average value by $(1/\sqrt{2})/(2/\pi) = 1.11$. For Gaussian noise the rms value is $\sqrt{\pi/2} = 1.25$ times the average value,[2] so the noise reading provided by an averaging-type meter must be multiplied by $1.25/1.11 = 1.13$, or, equivalently, it must be increased by $20 \log_{10} 1.13 \cong 1$ dB to obtain the correct value.

Noise Summation

In noise analysis one often needs to find the rms value of noise voltages in series or noise currents in parallel. Given two noise sources $x_{n1}(t)$ and $x_{n2}(t)$, the mean square value of their sum is

$$X_n^2 = \frac{1}{T} \int_0^T [x_{n1}(t) + x_{n2}(t)]^2 \, dt = X_{n1}^2 + X_{n2}^2 + \frac{2}{T} \int_0^T x_{n1}(t)x_{n2}(t) \, dt$$

If the two signals are uncorrelated, as is usually the case, the average of their product vanishes, so the rms values add up in Pythagorean fashion,

$$X_n = \sqrt{X_{n1}^2 + X_{n2}^2} \tag{7.3}$$

This indicates that if the sources are of uneven strengths, minimization efforts should be directed primarily at the strongest one. For instance, two noise sources with rms values of 10 μV and 5 μV combine to give an overall rms value of $\sqrt{10^2 + 5^2} = 11.2$ μV, which is only 12% higher than that of the dominant source. It is readily seen that reducing the dominant source by 13.4% has the same effect as eliminating the weaker source altogether!

As mentioned, the dc error referred to the input is also a form of noise, so when performing budget-error analysis we must add dc noise and rms ac noise *quadratically*.

Noise Spectra

Since X_n^2 represents the average power dissipated by $x_n(t)$ in a 1-Ω resistor, the physical meaning of mean square value is the same as for ordinary ac signals. However, unlike an ac signal, whose power is concentrated at just one frequency, noise power is usually spread all over the frequency spectrum because of the random nature of noise. Thus, when referring to rms noise, we must always specify the *frequency band* over which we are making our observations, measurements, or calculations.

In general, noise power depends on both the width of the frequency band and the band's location within the frequency spectrum. The rate of change of noise power with frequency is called the *noise power density*, and is denoted as $e_n^2(f)$ in the case of voltage noise, and $i_n^2(f)$ in the case of current noise. We have

$$e_n^2(f) = \frac{dE_n^2}{df} \qquad i_n^2(f) = \frac{dI_n^2}{df} \tag{7.4}$$

where E_n^2 and I_n^2 are the mean square values of voltage noise and current noise. Note that the units of $e_n^2(f)$ and $i_n^2(f)$ are volts squared per hertz (V^2/Hz) and amperes squared per hertz (A^2/Hz). Physically, noise power density represents the average noise power over a 1-Hz bandwidth as a function of frequency. When plotted versus frequency, it provides a visual indication of how power is distributed over the frequency spectrum. In integrated circuits, the two most common forms of power density distribution are white noise and $1/f$ noise.

The quantities $e_n(f)$ and $i_n(f)$ are called the *spectral noise densities*, and are expressed in volts per square root of hertz (V/$\sqrt{\text{Hz}}$) and amperes per square root of hertz (A/$\sqrt{\text{Hz}}$). Some manufacturers specify noise in terms of noise power densities, others in terms of spectral noise densities. Conversion between the two is accomplished by squaring or by extracting the square root.

Multiplying both sides in Eq. (7.4) by df and integrating from f_L to f_H, the lower and upper limits of the frequency band of interest, allows us to find the rms values in terms of the power densities,

$$E_n = \left(\int_{f_L}^{f_H} e_n^2(f)\,df \right)^{1/2} \qquad I_n = \left(\int_{f_L}^{f_H} i_n^2(f)\,df \right)^{1/2} \tag{7.5}$$

Once again it is stressed that the concept of rms cannot be separated from that of frequency band: in order to find the rms value we need to know the lower and upper limits of the band as well as the density within the band.

White Noise and 1/f Noise

White noise is characterized by a uniform spectral density, or $e_n = e_{nw}$ and $i_n = i_{nw}$, where e_{nw} and i_{nw} are suitable constants. It is so called by analogy with white light, which consists of all visible frequencies in equal amounts. When played through a loudspeaker it produces a waterfall sound. Applying Eq. (7.5) we get

$$E_n = e_{nw} \sqrt{f_H - f_L} \qquad I_n = i_{nw} \sqrt{f_H - f_L} \qquad (7.6)$$

indicating that the rms value of white noise increases with the square root of the frequency band. For $f_H \geq 10 f_L$ we can approximate as $E_n \cong e_{nw} \sqrt{f_H}$ and $I_n \cong i_{nw} \sqrt{f_H}$ at the risk of an error of about 5% or less.

Squaring both sides in Eq. (7.6) yields $E_n^2 = e_{nw}^2 (f_H - f_L)$ and $I_n^2 = i_{nw}^2 (f_H - f_L)$, indicating that white-noise power is *proportional to the bandwidth,* regardless of the band's location within the frequency spectrum. Thus, the noise power within the 10-Hz band between 20 Hz and 30 Hz is the same as that within the band between 990 Hz and 1 kHz.

The other common form of noise is $1/f$ noise, so called because its power density varies with frequency as $e_n^2(f) = K_v^2/f$ and $i_n^2(f) = K_i^2/f$, where K_v and K_i are suitable constants. The spectral densities are $e_n = K_v/\sqrt{f}$ and $i_n = K_i/\sqrt{f}$, indicating that when plotted versus frequency on logarithmic scales, power densities have a slope of -1 dec/dec, and spectral densities a slope of -0.5 dec/dec. Substituting into Eq. (7.5) and integrating yields

$$E_n = K_v \sqrt{\ln(f_H/f_L)} \qquad I_n = K_i \sqrt{\ln(f_H/f_L)} \qquad (7.7)$$

Squaring both sides in Eq. (7.7) yields $E_n^2 = K_v^2 \ln(f_H/f_L)$ and $I_n^2 = K_i^2 \ln(f_H/f_L)$, indicating that $1/f$-noise power is *proportional to the log ratio of the frequency band extremes,* regardless of the band's location within the frequency spectrum. Consequently, $1/f$ noise is said to have the same power content in each frequency decade (or octave). Once the noise rms of a particular decade (or octave) is known, the noise rms over m decades (or octaves) is obtained by multiplying the former by $\sqrt{m}$. For example, if the rms value within the decade $1 \text{ Hz} \leq f \leq 10 \text{ Hz}$ is $1 \mu\text{V}$, then the noise rms in the 9-decade span below 1 Hz, that is, down to about 1 cycle per 32 years, is $\sqrt{9} \times 1 \mu\text{V} = 3 \mu\text{V}$.

Integrated-Circuit Noise

Integrated-circuit noise is a mixture of white and $1/f$ noise, as shown in Fig. 7.2. At high frequencies, noise is predominantly white, while at low frequencies $1/f$ noise dominates. The borderline frequency, or *corner frequency,* is found graphically as the intercept of the $1/f$ asymptote and the white-noise floor. Power densities are expressed analytically as

$$e_n^2 = e_{nw}^2 \left(\frac{f_{ce}}{f} + 1 \right) \qquad i_n^2 = i_{nw}^2 \left(\frac{f_{ci}}{f} + 1 \right) \qquad (7.8)$$

where e_{nw} and i_{nw} are the *white-noise floors,* and f_{ce} and f_{ci} the *corner frequencies.* The μA741 data sheets of Fig. 5A.8 indicate $e_{nw} \cong 20 \text{ nV}/\sqrt{\text{Hz}}$, $f_{ce} \cong 200 \text{ Hz}$, $i_{nw} \cong 0.5 \text{ pA}/\sqrt{\text{Hz}}$, and $f_{ci} \cong 2 \text{ kHz}$. Inserting Eq. (7.8) into Eq. (7.5) and

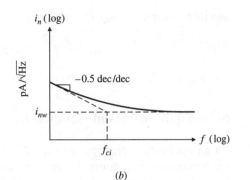

FIGURE 7.2
Typical IC noise densities.

integrating, we get

$$E_n = e_{nw}\sqrt{f_{ce}\ln(f_H/f_L) + f_H - f_L} \tag{7.9a}$$

$$I_n = i_{nw}\sqrt{f_{ci}\ln(f_H/f_L) + f_H - f_L} \tag{7.9b}$$

EXAMPLE 7.1. Estimate the rms input voltage noise of the 741 op amp over the following frequency bands: (a) 0.1 Hz to 100 Hz (instrumentation range), (b) 20 Hz to 20 kHz (audio range), and (c) 0.1 Hz to 1 MHz (wideband range).

Solution.

(a) Equation (7.9a) gives $E_n = 20 \times 10^{-9}\sqrt{200\ln(10^2/0.1) + 10^2 - 0.1} = 20 \times 10^{-9}\sqrt{1382 + 98.9} = 0.770\ \mu V$

(b) $E_n = 20 \times 10^{-9}\sqrt{1382 + 19{,}980} = 2.92\ \mu V$

(c) $E_n = 20 \times 10^{-9}\sqrt{3224 + 10^6} = 20.0\ \mu V$

We observe that $1/f$ noise dominates at low frequencies, white noise dominates at high frequencies—and the wider the frequency band, the higher the noise. Consequently, to minimize noise one should *limit the bandwidth to the strict minimum required.*

7.2
NOISE DYNAMICS

A common task in noise analysis is finding the total rms noise at the output of a circuit, given the noise density at its input as well as its frequency response. A typical example is offered by the voltage amplifier. The noise density at the output is $e_{no}(f) = |A_n(jf)|e_{ni}(f)$, where $e_{ni}(f)$ is the noise density at the input and $A_n(jf)$ is the *noise gain*. The *total output rms noise* is then $E_{no}^2 = \int_0^\infty e_{no}^2(f)\,df$, or

$$E_{no} = \left(\int_0^\infty |A_n(jf)|^2 e_{ni}^2(f)\,df\right)^{1/2} \tag{7.10}$$

Similar considerations hold for current amplifiers. Another common example is offered by the transimpedance amplifier. Denoting its input noise density as $i_n(f)$

and its noise gain as $Z_n(jf)$, we have

$$E_{no} = \left(\int_0^\infty |Z_n(jf)|^2 i_{ni}^2(f)\, df \right)^{1/2} \tag{7.11}$$

Similar considerations apply to transadmittance and current amplifiers.

Noise Equivalent Bandwidth (NEB)

As an application example of Eq. (7.10), consider the case of white noise with spectral density e_{nw} going through a simple RC filter as in Fig. 7.3a. Since $|A_n|^2 = 1/[1 + (f/f_0)^2]$, where f_0 is the -3-dB frequency, Eq. (7.10) gives

$$E_{no} = e_{nw} \left(\int_0^\infty \frac{df}{1 + (f/f_0)^2} \right)^{1/2} = e_{nw}\sqrt{\pi f_0/2} = e_{nw}\sqrt{1.57 f_0} \tag{7.12}$$

Comparing with Eq. (7.6), we observe that white noise is passed as if the filter were a brick-wall type, but with a cutoff frequency 1.57 times as large. As depicted in Fig. 7.3b, the fraction 0.57 accounts for the transmitted noise above f_0 as a consequence of the gradual rolloff, or *skirt*. This property holds for all first-order low-pass functions, not just for RC networks. As we know, the closed-loop response of many amplifiers is a first-order function with $f_B = \beta f_t$ as the -3-dB frequency. These amplifiers pass white noise with a cutoff frequency of $1.57 f_B$.

The quantity $1.57 f_0$ is called the *noise equivalent bandwidth* (NEB) of the given circuit. More generally, the NEB of a circuit with noise gain $A_n(jf)$ is defined as[2]

$$\text{NEB} = \frac{1}{A_{n(\text{max})}^2} \int_0^\infty |A_n(jf)|^2\, df \tag{7.13}$$

where $A_{n(\text{max})}$ is the peak magnitude of the noise gain. The NEB represents the *frequency span of a brick-wall power gain response having the same area as the power gain response of the original circuit.*

The NEB can be computed analytically for higher-order responses. For instance, for an nth-order maximally flat low-pass response we have

$$\text{NEB}_{\text{MF}} = \int_0^\infty \frac{df}{1 + (f/f_0)^{2n}} \tag{7.14a}$$

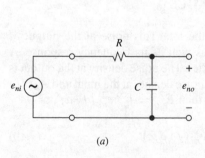

(a)

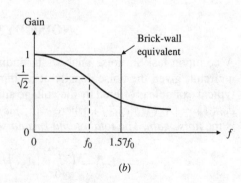

(b)

FIGURE 7.3
Noise equivalent bandwidth (NEB).

The results are[2] $\text{NEB}_{\text{MF}} = 1.57 f_0$ for $n = 1$, $1.11 f_0$ for $n = 2$, $1.05 f_0$ for $n = 3$, and $1.025 f_0$ for $n = 4$, indicating that NEB_{MF} rapidly approaches f_0 as n is increased.

Likewise, one can prove (see Problem 7.3) that the noise equivalent bandwidths of the standard second-order low-pass and band-pass functions H_{LP} and H_{BP} defined in Section 3.4 are, respectively,

$$\text{NEB}_{\text{LP}} = Q^2 \text{NEB}_{\text{BP}} = Q\pi f_0/2 \qquad (7.14b)$$

When the NEB cannot be calculated analytically, it can be estimated by piecewise graphical integration, or it can be found by computer via numerical integration. Numerical integration can be carried out with PSpice using the "s" function available with the Probe postprocessor.

EXAMPLE 7.2. Use PSpice to find the NEB of the circuit of Fig. 7.4, given that the op amp has GBP = 1 MHz.

Solution. The input circuit file is as follows.

```
Finding the NEB:
Vi 1 0 ac 1V
Rx 1 0 1k              ;avoids floating nodes
C1 0 4 1uF
R1 4 2 1k
R2 2 3 100k
C2 2 3 10nF
ea0 5 0 1 2 100k      ;a0 = 100 V/mV
Req 5 6 1Meg
Ceq 6 0 15.72nF       ;fb = 10 Hz
ebuf 3 0 6 0 1
.ac dec 10 1Hz 100MegHz
.probe ;A=V(3)/V(1), NEB=s(Vm(3)*Vm(3))/2601
.end
```

The plot of Fig. 7.5 (top) indicates that $A_{n(\text{max})} \cong 51$ V/V, so we direct the Probe postprocessor to display $s(Vm(3)*Vm(3))/2601$. The resulting curve, shown in Fig. 7.5 (bottom), tends asymptotically to the value $\text{NEB} \cong 1.1$ kHz.

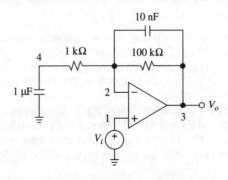

FIGURE 7.4
PSpice circuit of Example 7.2.

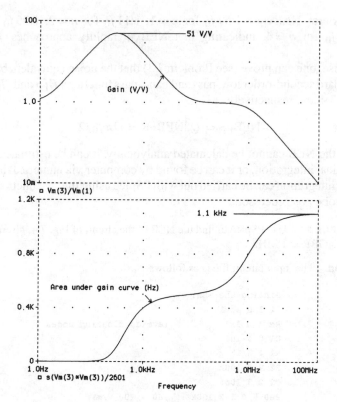

FIGURE 7.5
Finding the NEB of the circuit of Fig. 7.4.

Piecewise Graphical Integration

Noise densities and noise gains are often available only in graphical form. When this is the case, E_{no} is estimated by graphical integration, as illustrated in the following example.

EXAMPLE 7.3. Estimate the total rms output noise above 1 Hz for noise with the spectral density of Fig. 7.6 (top) going through an amplifier with the noise-gain characteristic of Fig. 7.6 (center).

Solution. To find the output density e_{no} we multiply out the two curves point by point and obtain the curve of Fig. 7.6 (bottom). Clearly, the use of linearized Bode plots simplifies graphical multiplications considerably. Next, we integrate e_{no}^2 from $f_L = 1$ Hz to $f_H = \infty$. To facilitate our task, we break down the integration interval into three parts, as follows.

For 1 Hz $\leq f \leq 1$ kHz we can apply Eq. (7.9a) with $e_{nw} = 20$ nV/$\sqrt{\text{Hz}}$, $f_{ce} = 100$ Hz, $f_L = 1$ Hz, and $f_H = 1$ kHz. The result is $E_{no1} = 0.822$ μV.

For 1 kHz $\leq f \leq 10$ kHz the density e_{no} increases with f at the rate of $+1$ dec/dec, so we can write $e_{no}(f) = (20$ nV/$\sqrt{\text{Hz}}) \times (f/10^3) = 2 \times 10^{-11} f$ V/$\sqrt{\text{Hz}}$. Then,

$$E_{no2} = 2 \times 10^{-11} \left(\int_{10^3}^{10^4} f^2 \, df \right)^{1/2} = 2 \times 10^{-11} \left(\frac{1}{3} f^3 \Big|_{10^3}^{10^4} \right)^{1/2} = 11.5 \, \mu\text{V}$$

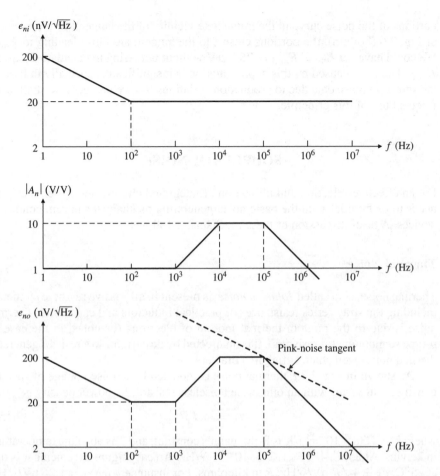

FIGURE 7.6
Noise spectra of Example 7.3.

For 10 kHz $\leq f \leq \infty$ we have white noise with $e_{nw} = 200$ nV/$\sqrt{Hz}$ going through a low-pass filter with $f_0 = 100$ kHz. By Eq. (7.12), $E_{no3} = 200 \times 10^{-9}(1.57 \times 10^5 - 10^4)^{1/2} = 76.7$ μV.

Finally, we add up all components in rms fashion to obtain $E_{no} = \sqrt{E_{no1}^2 + E_{no2}^2 + E_{no3}^2} = \sqrt{0.822 + 11.5^2 + 76.7^2} = 77.5$ μV.

The Pink-Noise Tangent Principle

Looking at the result of the foregoing example, we note that the largest contribution comes from E_{no3}, which represents noise above 10 kHz. We wonder if there is a quick method of predicting this, without having to go through all calculations. Such a method exists; it is offered by the *pink-noise tangent principle.*[5]

The pink-noise curve is the locus of points contributing equal-per-decade (or equal-per-octave) noise power. Its noise density slope is −0.5 dec/dec. The pink-noise principle states that if we lower the pink-noise curve until it becomes tangent to the noise curve $e_{no}(f)$, then the main contribution to E_{no} will come from the portions of the noise curve in the immediate vicinity of the tangent. In the example of Fig. 7.6 (bottom), the portions closest to the tangent are those leading to E_{no3}. We could have set $E_{no} \cong E_{no3} = 76.7$ μV without bothering to calculate E_{no1} and E_{no2}. The error caused by this approximation is insignificant, especially in light of the spread in noise data due to production variations. As we proceed, we shall make frequent use of this principle.

7.3
SOURCES OF NOISE

For an effective selection and utilization of integrated circuits, the system designer needs to be familiar with the basic noise-generating mechanisms in semiconductor devices. A brief discussion of these mechanisms follows.

Thermal Noise

Thermal noise, also called *Johnson noise,* is present in all passive resistive elements, including the stray series resistances of practical inductors and capacitors. Thermal noise is due to the random thermal motion of electrons (or holes, in the case of p-type semiconductor resistors). It is unaffected by dc current, so a resistor generates thermal noise even when sitting in a drawer.

As shown in Fig. 7.7a, thermal noise is modeled by a noise voltage of spectral density e_R in *series* with an otherwise noiseless resistor. Its power density is

$$e_R^2 = 4kTR \tag{7.15a}$$

where $k = 1.38 \times 10^{-23}$ J/K is Boltzmann's constant, and T is absolute temperature, in kelvins. At 25°C, $4kT = 1.65 \times 10^{-20}$ W/Hz. An easy figure to remember is that at 25°C, $e_R \cong 4\sqrt{R}$ nV/$\sqrt{\text{Hz}}$, R in kiloohms. For instance, $e_{100\ \Omega} = 1.26$ nV/$\sqrt{\text{Hz}}$, and $e_{10\ k\Omega} = 12.6$ nV/$\sqrt{\text{Hz}}$.

Converting from Thévenin to Norton, we can model thermal noise also with a noise current i_R in *parallel* with an otherwise noiseless resistor, as shown in Fig. 7.7b. We have $i_R^2 = e_R^2/R^2$, or

$$i_R^2 = 4kT/R \tag{7.15b}$$

The preceding equations indicate that thermal noise is of the white type. Purely reactive elements are free from thermal noise.

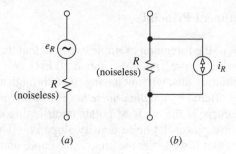

FIGURE 7.7
Thermal noise models.

EXAMPLE 7.4. Consider a 10-kΩ resistor at room temperature. Find (a) its voltage and (b) current spectral densities, and (c) its rms noise voltage over the audio range.

Solution.

(a) $e_R = \sqrt{4kTR} = \sqrt{1.65 \times 10^{-20} \times 10^4} = 12.8 \text{ nV}/\sqrt{\text{Hz}}$
(b) $i_R = e_R/R = 1.28 \text{ pA}/\sqrt{\text{Hz}}$
(c) $E_R = e_R\sqrt{f_H - f_L} = 12.8 \times 10^{-9} \times \sqrt{20 \times 10^3 - 20} = 1.81 \text{ μV}$

Shot Noise

This type of noise arises whenever charges cross a potential barrier, such as in diodes or transistors. Barrier crossing is a purely random event and the dc current we observe microscopically is actually the sum of many random elementary current pulses. Shot noise has a uniform power density,

$$i_n^2 = 2qI \qquad (7.16)$$

where $q = 1.602 \times 10^{-19}$ C is the electron charge, and I is the dc current through the barrier. Shot noise is present in BJT base currents as well as in current-output D-A converters.

EXAMPLE 7.5. Find the signal-to-noise ratio for diode current over a 1-MHz bandwidth if (a) $I_D = 1 \text{ μA}$ and (b) $I_D = 1 \text{ nA}$.

Solution.

(a) $I_n = \sqrt{2qI_Df_H} = \sqrt{2 \times 1.62 \times 10^{-19} \times 10^{-6} \times 10^6} = 0.57 \text{ nA (rms)}$. Thus, SNR $= 20 \log_{10}[(1 \text{ μA})/(0.57 \text{ nA})] = 64.9 \text{ dB}$.
(b) By similar procedure, SNR $= 34.9 \text{ dB}$. We observe that the SNR deteriorates as the operating current is lowered.

Flicker Noise

Flicker noise, also called $1/f$ *noise,* or *contact noise,* is present in all active as well as in some passive devices and has various origins, depending on device type. In active devices it is due to traps, which, when current flows, capture and release charge carriers randomly, thus causing random fluctuations in the current itself. In BJTs these traps are associated with contamination and crystal defects at the base-emitter junction. In MOSFETs they are associated with extra electron energy states at the boundary between silicon and silicon dioxide. Among active devices, MOSFETs suffer the most, and this can be a source of concern in low-noise MOS applications.

Flicker noise is always associated with a dc current, and its power density is of the type

$$i_n^2 = K\frac{I^a}{f} \qquad (7.17)$$

where K is a device constant, I is the dc current, and a is another device constant in the range $\frac{1}{2}$ to 2.

Flicker noise is also found in some passive devices, such as carbon composition resistors, in which case it is called *excess noise* because it appears in addition to the thermal noise already there. However, while thermal noise is also present without a dc current, flicker noise requires a dc current in order to exist. Resistors of the wire-wound type are the quietest in terms of $1/f$ noise, while the carbon composition types can be noisier by as much as an order of magnitude, depending on operating conditions. Carbon-film and metal-film types fall in between. However, if the application requires that a given resistor carry a fairly small current, thermal noise will predominate and it will make little difference which resistor type one uses.

Avalanche Noise

This form of noise is found in *pn* junctions operated in the reverse breakdown mode. Avalanche breakdown occurs when electrons, under the influence of the strong electric field inside the space-charge layer, acquire enough kinetic energy to create additional electron-hole pairs by collision against the atoms of the crystal lattice. These additional pairs can, in turn, create other pairs in avalanche fashion. The resulting current consists of randomly distributed noise spikes flowing through the reverse biased junction. Like shot noise, avalanche noise requires current flow. However, avalanche noise is usually much more intense than shot noise, making Zener diodes notoriously noisy. This is one of the reasons why voltage references of the bandgap type are preferable to Zener-diode references.

Noise in BJTs

With the exception of avalanche noise, transistors generally exhibit all forms of noise just discussed. A feel for transistor noise mechanisms will help the user better understand the noise characteristics of op amps. As shown in Fig. 7.8, transistor noise is characterized by a pair of equivalent input noise sources with spectral densities e_n and i_n.

The noise power densities for BJTs are[7]

$$e_n^2 = 4kT \left(r_b + \frac{1}{2g_m} \right) \tag{7.18a}$$

$$i_n^2 = 2q \left(I_B + K_1 \frac{I_B^a}{f} + \frac{I_C}{|\beta(jf)|^2} \right) \tag{7.18b}$$

where r_b is the intrinsic base resistance, I_B and I_C are the dc base and collector

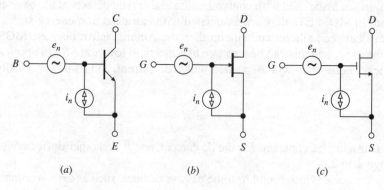

(a) $\qquad\qquad\qquad$ (b) $\qquad\qquad\qquad$ (c)

FIGURE 7.8
Transistor noise models.

currents, $g_m = qI_C/kT$ is the transconductance, K_1 and a are appropriate device constants, and $\beta(jf)$ is the forward current gain, which decreases at high frequencies.

In the expression for e_n^2, the first term represents thermal noise from r_b and the second term represents the effect of collector-current shot noise referred to the input. In the expression for i_n^2, the first two terms represent base-current shot and flicker noise, and the last term represents collector-current shot noise reflected to the input.

To achieve a high β, the base region of a BJT is doped lightly and fabricated very thin. This, however, increases the intrinsic base resistance r_b. Moreover, the transconductance g_m and the base current I_B are directly proportional to I_C. Thus, what works to minimize voltage noise (low r_b and high I_C) is the opposite of what is good for low current noise (high β and low I_C). This represents a fundamental tradeoff in bipolar-op-amp design.

Noise in JFETs

The noise power densities for JFETs are[7]

$$e_n^2 = 4kT \left(\frac{2}{3g_m} + K_2 \frac{I_D^a/g_m^2}{f} \right) \tag{7.19a}$$

$$i_n^2 = 2qI_G + \left(\frac{2\pi f C_{gs}}{g_m} \right)^2 \left(4kT \frac{2}{3} g_m + K_3 \frac{I_D^a}{f} \right) \tag{7.19b}$$

where g_m is the transconductance; I_D is the dc drain current; I_G is the gate leakage current; K_2, K_3, and a are appropriate device constants; and C_{gs} is the gate-to-source capacitance.

In the expression for e_n^2, the first term represents thermal noise in the channel, and the second represents drain-current flicker noise. At room temperature and at moderate frequencies, all terms in the expression for i_n^2 are negligible, making JFETs virtually free of input current noise. Recall, however, that gate leakage increases very rapidly with temperature, so i_n^2 may no longer be neglected at higher temperatures.

Compared to BJTs, FETs have notoriously low g_m values, indicating that FET-input op amps tend to exhibit higher voltage noise than BJT-input types for similar operating conditions. Moreover, e_n^2 in the JFET contains flicker noise. These disadvantages are offset by better current noise performance, at least near room temperature.

Noise in MOSFETs

The noise power densities for MOSFETs are[7]

$$e_n^2 = 4kT \frac{2}{3g_m} + K_4 \frac{1}{WLf} \tag{7.20a}$$

$$i_n^2 = 2qI_G \tag{7.20b}$$

where g_m is the transconductance, K_4 is a device constant, and W and L are the channel width and length. As in the JFET case, i_n^2 is negligible at room temperature, but increases with temperature.

In the expression for e_n^2, the first term represents thermal noise from the channel resistance and the second represents flicker noise. It is the latter that is of most concern in MOSFET-input op amps. Flicker noise is inversely proportional to the transistor area $W \times L$, so this type of noise is reduced by using input-stage transistors with large geometries. As discussed in Chapter 5, when large geometries are combined with common-centroid layout techniques, the input offset voltage and offset drift characteristics are also improved significantly.

Noise Modeling in PSpice

When performing noise analysis SPICE calculates the thermal-noise density for each resistor in the circuit, as well as the shot-noise and flicker-noise densities for each diode and transistor. When using op amp macromodels, the need arises for noise sources with spectral densities of the type of Fig. 7.2. We shall synthesize these sources[2] by exploiting the fact that SPICE calculates the noise current of a diode according to

$$i_d^2 = \text{KF}\frac{I_D^{\text{AF}}}{f} + 2q I_D = 2q I_D \left(\frac{\text{KF} \times I_D^{\text{AF}-1}/2q}{f} + 1 \right)$$

where I_D is the diode bias current, q the electron charge, and KF and AF are parameters that can be specified by the user. This is a power density with white-noise floor $i_w^2 = 2q I_D$ and corner frequency $f_c = \text{KF} \times I_D^{\text{AF}-1}/2q$. If we let $\text{AF} = 1$ for mathematical convenience, then the required I_D and KF for given i_w^2 and f_c are

$$I_D = i_w^2/2q \qquad \text{KF} = 2q f_c \tag{7.21}$$

Once we have a source of current noise, we can readily convert it to a source of voltage noise via a CCVS.

> **EXAMPLE 7.6.** Verify Example 7.1 using PSpice.
>
> **Solution.** We need to create a source e_n with $e_{nw} = 20 \text{ nV}/\sqrt{\text{Hz}}$ and $f_{ce} = 200$ Hz. First we create a noise current source with $i_w = 1 \text{ pA}/\sqrt{\text{Hz}}$ and $f_c = 200$ Hz, then we use an H-type source of value 20 nV/pA to convert to e_n. As shown in Fig. 7.9, we bias the diode with $I_D = (1 \times 10^{-12})^2/(2 \times 1.602 \times 10^{-19}) = 3.12 \ \mu\text{A}$, and we impose $\text{KF} = 2 \times 1.602 \times 10^{-19} \times 200 = 6.41 \times 10^{-17}$ A. The 1-GF capacitor couples the ac

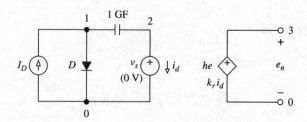

FIGURE 7.9
Using diode noise current to create a source of voltage noise with spectral density e_n.

noise current generated by the diode to the current-sensing source v_s, which then controls the CCVS to produce e_n. The input circuit file follows.

```
Calculating rms noise:
ID 0 1 dc 3.12uA
D 1 0 Dnoise
.model Dnoise D (KF=6.41E-17,AF=1)
C 1 2 1GF
vs 2 0 dc 0V
he 3 0 vs 20k
Rx 3 0 1            ;avoids floating nodes
.ac dec 10 0.1Hz 1MegHz
.noise v(3) vs 10
.probe ;en(f)=v(onoise), En=sqrt(s(v(onoise)*v(onoise)))
.end
```

Figure 7.10 shows the plots of the spectral density $e_n =$ v(onoise) and the rms value $E_n =$ sqrt(s(v(onoise)*v(onoise))), where "sqrt" and "s" stand for the square root and integral functions available with the Probe postprocessor. Using the cursor facility to measure specific values, we find that for 0.1 Hz $\leq f \leq$ 100 Hz, $E_n \cong 0.77\ \mu$V; for 20 Hz $\leq f \leq$ 20 kHz, $E_n \cong 3\ \mu$V; and for 0.1 Hz $\leq f \leq$ 1 MHz, $E_n = 20\ \mu$V. This corroborates the results of the hand calculations of Example 7.1.

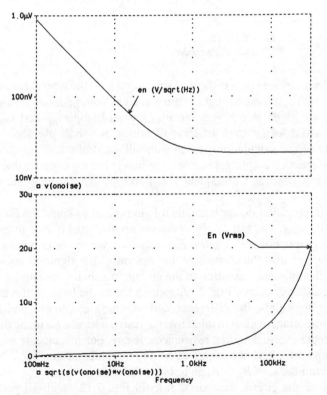

FIGURE 7.10
Using PSpice to generate IC noise and calculate rms noise.

7.4
OP AMP NOISE

Op amp noise is characterized by three equivalent noise sources: a voltage source with spectral density e_n, and two current sources with densities i_{np} and i_{nn}. As shown in Fig. 7.11, a practical op amp can be regarded as a noiseless op amp equipped with these sources at the input. This model is similar to that used to account for the input offset voltage V_{OS} and the input bias currents I_P and I_N. This is not surprising since these parameters are themselves special forms of noise, namely, dc noise. Note, however, that the magnitudes and directions of $e_n(t)$, $i_{np}(t)$, and $i_{nn}(t)$ are constantly changing due to the random nature of noise and that noise terms must be added up in rms rather than algebraic fashion.

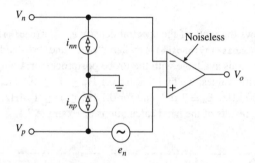

FIGURE 7.11
Op amp noise model.

Noise densities are given in the data sheets and have the typical forms of Fig. 7.2. For devices with symmetric input circuitry, such as voltage-mode op amps (VFAs), i_{np} and i_{nn} are given as a single density i_n, even though i_{np} and i_{nn} are uncorrelated. To avoid losing track of their identities, we shall use separate symbols until the end of our calculations, when we shall substitute i_n for both i_{np} and i_{nn}. For current-feedback amplifiers (CFAs), the inputs are asymmetric due to the presence of the input buffer. Consequently, i_{np} and i_{nn} are different and are graphed separately.

Just as in precision dc applications it is important to know the dc output error E_O caused by V_{OS}, I_P, and I_N, in low-noise applications it is of interest to know the total rms output noise E_{no}. Once E_{no} is known, we can refer it back to the input and compare it against the useful signal to determine the signal-to-noise ratio SNR and, hence, the ultimate resolution of the circuit. We shall illustrate for the familiar resistive-feedback circuit of Fig. 7.12a, which forms the basis of the inverting and noninverting amplifiers, the difference and summing amplifiers, and a variety of others. It is important to keep in mind that the resistances shown in the diagram must include also the external source resistances, if any. For instance, if we lift node A off ground and drive it with a source v_S having internal resistance R_s, then we must replace R_1 with the sum $R_s + R_1$ in our calculations.

To analyze the circuit we redraw it as in Fig. 7.12b with all pertinent noise sources in place, including the thermal noise sources of the resistors. As we know,

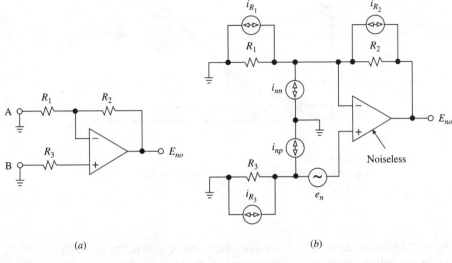

FIGURE 7.12
Resistive-feedback op amp circuit and its noise model.

resistor noise can be modeled with either a series voltage source or a parallel current source. The reason for choosing the latter will become apparent shortly.

Overall Input Spectral Density

The first task is to find the overall spectral density e_{ni} referred to the input of the op amp. We can apply the superposition principle as when we calculate the overall input error E_I due to V_{OS}, I_P, and I_N, except that now the individual terms must be added up in rms fashion. Thus, the noise voltage e_n contributes the term e_n^2. The noise currents i_{np} and i_{R_3} are flowing through R_3, so their combined contribution is, by Eq. (7.15), $(R_3 i_{np})^2 + (R_3 i_{R_3})^2 = R_3^2 i_{np}^2 + 4kT R_3$. The noise currents i_{nn}, i_{R_1}, and i_{R_2} are flowing through the parallel combination $R_1 \| R_2$, so their contribution is $(R_1 \| R_2)^2 (i_{nn}^2 + i_{R_1}^2 + i_{R_2}^2) = (R_1 \| R_2)^2 i_{nn}^2 + 4kT(R_1 \| R_2)$. Combining all terms gives the overall input spectral density

$$e_{ni}^2 = e_n^2 + R_3^2 i_{np}^2 + (R_1 \| R_2)^2 i_{nn}^2 + 4kT[R_3 + (R_1 \| R_2)] \qquad (7.22)$$

For op amps with symmetric inputs and uncorrelated noise currents we have $i_{np} = i_{nn} = i_n$, where i_n is the noise current density given in the data sheets.

To gain better insight into the relative weights of the various terms, consider the special but familiar case in which $R_3 = R_1 // R_2$. Under this constraint, Eq. (7.22) simplifies as

$$e_{ni}^2 = e_n^2 + 2R^2 i_n^2 + 8kTR \qquad (7.23a)$$

$$R = R_1 // R_2 = R_3 \qquad (7.23b)$$

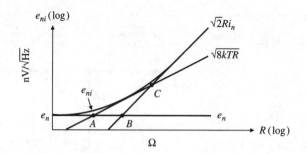

FIGURE 7.13
Op-amp input spectral noise e_{ni} as a function of R of Eq. (7.23b).

Figure 7.13 shows e_{ni} as well as its three individual components as a function of R. While the voltage term e_n is independent of R, the current term $\sqrt{2}Ri_n$ increases with R at the rate of 1 dec/dec, and the thermal term $\sqrt{8kTR}$ increases at the rate of 0.5 dec/dec.

We observe that for R sufficiently small, voltage noise dominates. In the limit $R \to 0$ we get $e_{ni} \to e_n$, so e_n is aptly called the *short-circuit* noise: this is the noise produced by the internal components of the op amp, regardless of the external circuitry. For R sufficiently large, current noise dominates. In the limit $R \to \infty$ we get $e_{ni} \to \sqrt{2}Ri_n$, so i_n is aptly called the *open-circuit* noise. This form of noise stems from input-bias-current flow through the external resistors. For intermediate values of R_s, thermal noise may also come into play, depending on the relative magnitudes of the other two terms. In the example pictured, point A is where thermal noise overtakes voltage noise, point B where current noise overtakes voltage noise, and point C where current noise overtakes thermal noise. The relative positions of A, B, and C vary from one op amp to another, and can be used to compare different devices.

We note that while it is desirable to install a dummy resistance $R_3 = R_1 \parallel R_2$ in order to provide bias-current compensation, in terms of noise it is preferable to have $R_3 = 0$ since this resistor only contributes additional noise. When the presence of R_3 is mandatory, the corresponding thermal noise can be filtered out by connecting a suitably large capacitance in parallel with R_3. This will also suppress any external noise that might be accidentally injected into the noninverting input pin.

Rms Output Noise

Like offsets and drift, e_{ni} is amplified by the noise gain of the circuit. This gain is not necessarily the same as the signal gain, so we shall denote *signal gain* as $A_s(jf)$ and *noise gain* as $A_n(jf)$ to avoid confusion. Recall that the dc value of $A_n(jf)$ is $A_{n0} = 1/\beta = 1 + R_2/R_1$. Moreover, for a constant-GBP op amp, the closed-loop bandwidth of $A_n(jf)$ is $f_B = \beta f_t = f_t/(1 + R_2/R_1)$, where f_t is the unity-gain frequency of the op amp. The output spectral density can thus be expressed as

$$e_{no} = \frac{1 + R_2/R_1}{\sqrt{1 + (f/f_B)^2}} e_{ni} \qquad (7.24)$$

Noise is observed or measured over a finite time interval T_{obs}. The *total rms output noise* is found by integrating e_{no}^2 from $f_L = 1/T_{obs}$ to $f_H = \infty$. Using Eqs. (7.9), (7.12), and (7.22) we get

$$
\begin{aligned}
E_{no} = \Bigg[&1 + \frac{R_2}{R_1} \Bigg] \times \Bigg[e_{nw}^2 \left(f_{ce} \ln \frac{f_B}{f_L} + 1.57 f_B - f_L \right) \\
&+ R_3^2 i_{npw}^2 \left(f_{cip} \ln \frac{f_B}{f_L} + 1.57 f_B - f_L \right) \\
&+ (R_1 \parallel R_2)^2 i_{nnw}^2 \left(f_{cin} \ln \frac{f_B}{f_L} + 1.57 f_B - f_L \right) \\
&+ 4kT(R_3 + R_1 \parallel R_2)(1.57 f_B - f_L) \Bigg]^{1/2} \qquad (7.25)
\end{aligned}
$$

This expression indicates the considerations in low-noise design: (*a*) select op amps with low-noise floors e_{nw} and i_{nw} as well as low corner frequencies f_{ce} and f_{ci}; (*b*) keep the external resistances sufficiently small to make current noise and thermal noise negligible compared to voltage noise (if possible, make $R_3 = 0$); (*c*) limit the noise-gain bandwidth to the strict minimum required.

The popular OP-27 op amp has been specifically designed for low-noise applications. Its characteristics are $f_t = 8$ MHz, $e_{nw} = 3$ nV/$\sqrt{\text{Hz}}$, $f_{ce} = 2.7$ Hz, $i_{nw} = 0.4$ pA/$\sqrt{\text{Hz}}$, and $f_{ci} = 140$ Hz.

EXAMPLE 7.7. A 741 op amp is configured as an inverting amplifier with $R_1 = 100\,\text{k}\Omega$, $R_2 = 200$ kΩ, and $R_3 = 68$ kΩ. (*a*) Assuming $e_{nw} = 20$ nV/$\sqrt{\text{Hz}}$, $f_{ce} = 200$ Hz, $i_{nw} = 0.5$ pA/$\sqrt{\text{Hz}}$, and $f_{ci} = 2$ kHz, find the total output noise above 0.1 Hz, both rms and peak-to-peak. (*b*) Verify with PSpice.

Solution.

(*a*) We have $R_1 \parallel R_2 = 100 \parallel 200 \cong 67\,\text{k}\Omega$, $A_{n0} = 1 + R_2/R_1 = 3$ V/V, and $f_B = 10^6/3 = 333$ kHz. The noise voltage component is $E_{noe} = 3 \times 20 \times 10^{-9}[200 \ln(333 \times 10^3/0.1) + 1.57 \times 333 \times 10^3 - 0.1]^{1/2} = 43.5\ \mu$V. The current noise component is $E_{noi} = 3[(68 \times 10^3)^2 + (67 \times 10^3)^2]^{1/2} \times 0.5 \times 10^{-12} \times [2 \times 10^3 \ln(333 \times 10^4) + 523 \times 10^3]^{1/2} = 106.5\ \mu$V. The thermal noise component is $E_{noR} = 3[1.65 \times 10^{-20}(68 + 67) \times 10^3 \times 523 \times 10^3]^{1/2} = 102.4\ \mu$V. Finally,

$$E_{no} = \sqrt{E_{noe}^2 + E_{noi}^2 + E_{noR}^2} = \sqrt{43.5^2 + 106.5^2 + 102.4^2} = 154\ \mu\text{V (rms)}$$

or $6.6 \times 154 = 1.02$ mV (peak-to-peak).

(*b*) As shown in Fig. 7.14, we model e_n with an *H*-type source, and i_{np} and i_{nn} with *F*-type sources. The corresponding diode noise generators, omitted for simplicity, are as in Fig. 7.9. To ensure statistical independency, we must use three different such generators. Moreover, to model a noiseless op amp, we use the LAPLACE facility of PSpice. The input circuit file follows.

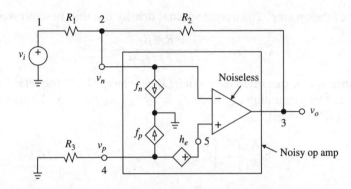

FIGURE 7.14
PSpice circuit to find the rms output noise E_{no}.

```
Finding the total rms output noise Eno:
*Input noise sources:
IDe 0 11 dc 3.12uA
De 11 0 De
.model De D (KF=6.41E-17,AF=1)
Ce 11 12 1GF
vse 12 0 dc 0V
he 5 4 vse 20k      ;enw = 20 nV/sqrt(Hz), fce = 200 Hz
ix 5 0 dc 0         ;avoids floating nodes
IDp 0 21 dc 3.12uA
Dp 21 0 Dp
.model Dp D (KF=6.41E-16,AF=1)
Cp 21 22 1GF
vsp 22 0 dc 0V
fp 4 0 vsp 0.5      ;inpw = 0.5 pA/sqrt(Hz), fcip = 2 kHz
IDn 0 31 dc 3.12uA
Dn 31 0 Dn
.model Dn D (KF=6.41E-16,AF=1)
Cn 31 32 1GF
vsn 32 0 dc 0V
fn 2 0 vsn 0.5      ;innw = 0.5 pA/sqrt(Hz), fcin = 2 kHz
*Noiseless op amp with a0 = 200 V/mV and fb = 5 Hz:
eOA 3 0 Laplace {V(5,2)}={2E5/(1+s/31.42)}
*Main circuit:
vi 1 0 ac 1V
R1 1 2 100k
R2 2 3 200k
R3 4 0 68k
.ac dec 10 0.1Hz 100MegHz
.noise v(3) vi 10
.probe ;eno = v(onoise), En = sqrt(s(v(onoise)*v(onoise)))
.end
```

The results of Fig. 7.15 agree with our calculations. We also note that we could have used the pink-noise tangent principle to estimate $E_{no} \cong (0.21 \ \mu\text{V}) \times \sqrt{1.57 \times 333 \text{ kHz}} = 152 \ \mu\text{V (rms)}$.

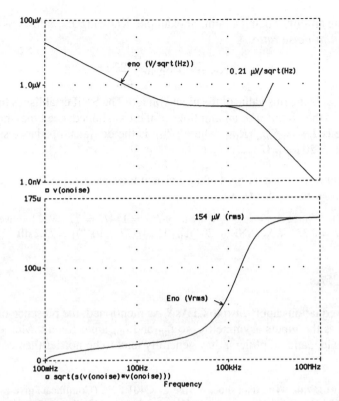

FIGURE 7.15
Finding the total rms output noise.

In terms of noise, the circuit of Example 7.7 is poorly designed because E_{noi} and E_{noR} far exceed E_{noe}. This can be improved by scaling down all resistances. A good rule of thumb is to impose $E_{noi}^2 + E_{noR}^2 \leq E_{noe}^2/3^2$, since this raises E_{no} only by about 5%, or less, above E_{noe}.

EXAMPLE 7.8. Scale the resistances of the circuit of Example 7.7 so that $E_{no} = 50\,\mu\text{V}$.

Solution. We want $E_{noi}^2 + E_{noR}^2 = E_{no}^2 - E_{noe}^2 = 50^2 - 43.5^2 = (24.6\,\mu\text{V})^2$. Letting $R = R_3 + R_1 \parallel R_2$, we have $E_{noi}^2 = 3^2 \times R^2(0.5 \times 10^{-12})^2 \times [2 \times 10^3 \ln(333 \times 10^4) + 523 \times 10^3] = 1.24 \times 10^{-18} R^2$, and $E_{noR}^2 = 3^2 \times 1.65 \times 10^{-20} \times R \times 523 \times 10^3 = 7.77 \times 10^{-14} R$. We want $1.24 \times 10^{-18} R^2 + 7.77 \times 10^{-14} R = (24.6\,\mu\text{V})^2$, which gives $R = 7\,\text{k}\Omega$. Thus, $R_3 = R/2 = 3.5\,\text{k}\Omega$, and $1/R_1 + 1/R_2 = 1/(3.5\,\text{k}\Omega)$. Since $R_2 = 2R_1$, this yields $R_1 = 5.25\,\text{k}\Omega$ and $R_2 = 10.5\,\text{k}\Omega$.

Signal-to-Noise Ratio

Dividing E_{no} by the dc signal gain $|A_{s0}|$ yields the *total rms input noise*,

$$E_{ni} = \frac{E_{no}}{|A_{s0}|} \tag{7.26}$$

We again stress that the signal gain A_s may be different from the noise gain A_n, the inverting amplifier being a familiar example. Knowing E_{ni} allows us to find the *input signal-to-noise ratio,*

$$\text{SNR} = 20 \log_{10} \frac{V_{i\,(\text{rms})}}{E_{ni}} \tag{7.27}$$

where $V_{i\,(\text{rms})}$ is the rms value of the input voltage. The SNR establishes the ultimate resolution of the circuit. For an amplifier of the transimpedance type, the total rms input noise is $I_{ni} = E_{no}/|R_{s0}|$, where $|R_{s0}|$ is the dc transimpedance signal gain. Then, $\text{SNR} = 20 \log_{10}(I_{i\,(\text{rms})}/I_{ni})$.

> **EXAMPLE 7.9.** Find the SNR of the circuit of Example 7.7 if the input is an ac signal with a peak amplitude of 0.5 V.
>
> **Solution.** Since $A_{s0} = -2$ V/V, we have $E_{ni} = 154/2 = 77$ μV. Moreover, $V_{i\,(\text{rms})} = 0.5/\sqrt{2} = 0.354$ V. So, $\text{SNR} = 20 \log_{10}[0.354/(77 \times 10^{-6})] = 73.2$ dB.

Noise in CFAs

The above equations apply also to CFAs.[8] As mentioned, the presence of the input buffer makes the inputs asymmetric, so i_{np} and i_{nn} are different. Moreover, since CFAs are wideband amplifiers, they generally tend to be noisier than conventional op amps.[9]

> **EXAMPLE 7.10.** The data sheets of the CLC401 CFA (Comlinear) give $z_0 \cong 710$ kΩ, $f_b \cong 350$ kHz, $r_n \cong 50$ Ω, $e_{nw} \cong 2.4$ nV/$\sqrt{\text{Hz}}$, $f_{ce} \cong 50$ kHz, $i_{npw} \cong 3.8$ pA/$\sqrt{\text{Hz}}$, $f_{cip} \cong 100$ kHz, $i_{nnw} \cong 20$ pA/$\sqrt{\text{Hz}}$, and $f_{cin} \cong 100$ kHz. Find the total rms output noise above 0.1 Hz if the CFA is configured as a noninverting amplifier with $R_1 = 166.7$ Ω and $R_2 = 1.5$ kΩ, and is driven by a source with an internal resistance of 100 Ω.
>
> **Solution.** Since $f_t = z_0 f_b / R_2 = 166$ MHz, we have $f_B = f_t/[1 + r_n/(R_1 \| R_2)] = 124$ MHz. Applying Eq. (7.25) gives $E_{no} = 10[(33.5\,\mu\text{V})^2 + (3.6\,\mu\text{V})^2 + (35.6\,\mu\text{V})^2 + (28.4\,\mu\text{V})^2]^{1/2} \cong 566$ μV (rms), or $6.6 \times 566 \cong 3.7$ mV (peak-to-peak).

Noise Filtering

Since broadband noise increases with the square root of the noise-gain bandwidth, noise can be reduced through narrowbanding. The most common technique is to pass the signal through a simple *R-C* network with *R* small enough to avoid adding appreciably to the existing noise. This filter is susceptible to output loading, so we may want to buffer it with a voltage follower. However, this would add the noise of the follower, whose equivalent bandwidth NEB $= (\pi/2) f_t$ is quite wide.

The topology[10] of Fig. 7.16 places the op amp upstream of the *R-C* network so that the noise of the op amp itself is filtered. Moreover, *R* is placed within the feedback loop to reduce its effective value by $1 + T$ and thus reduce output loading significantly. Even though *T* decreases with frequency, the presence of *C* helps maintain a low output impedance well into the upper frequency range. The purpose of *mR* and *nC* is to provide frequency compensation, an issue addressed in Section 8.2. Suffice it to say here that the circuit exhibits a good tolerance to capacitive loads.

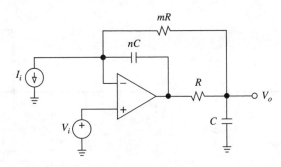

FIGURE 7.16
Low-pass noise filter. Input may be either a current
or a voltage.

The circuit lends itself to filtering both voltages and currents. It can be shown
(see Problem 7.26) that

$$V_o = H_{\text{LP}} m R I_i + (H_{\text{LP}} + H_{\text{BP}}) V_i \tag{7.28}$$

$$f_0 = \frac{1}{2\pi \sqrt{mn} RC} \qquad Q = \frac{\sqrt{m/n}}{m+1} \tag{7.29}$$

where H_{LP} and H_{BP} are the standard second-order low-pass and band-pass functions defined in Section 3.3. This filter finds application in voltage-reference and
photodiode-amplifier noise reduction.

7.5
NOISE IN PHOTODIODE AMPLIFIERS

An area in which noise is of concern is low-level signal detection, such as instrumentation applications and high-sensitivity I-V conversion. In particular, photodiode
amplifiers have been at the center of considerable attention,[11] so we examine this
class of amplifiers in some detail.

The photodiode of Fig. 7.17a responds to incident light with a current i_S that
the op amp subsequently converts to a voltage v_O. For a realistic analysis we use
the model of Fig. 7.17b, where R_1 and C_1 represent the combined resistance and
capacitance toward ground of the diode and the inverting-input pin of the op amp,
and C_2 represents the stray capacitance of R_2. With careful printed-circuit board
layout, C_2 can be kept in the range of 1 pF or less. Usually $C_1 \gg C_2$ and $R_1 \gg R_2$.

We are interested in the signal gain $A_s = V_o/I_s$ as well as the noise gain
$A_n = e_{no}/e_{ni}$. To this end, we need to find the feedback factor $\beta = Z_1/(Z_1 + Z_2)$,
$Z_1 = R_1 \parallel (1/j2\pi f C_1)$, $Z_2 = R_2 \parallel (1/j2\pi f C_2)$. Expanding gives

$$\frac{1}{\beta} = \left(1 + \frac{R_2}{R_1}\right) \frac{1 + jf/f_z}{1 + jf/f_p} \tag{7.30a}$$

$$f_z = \frac{1}{2\pi(R_1 \parallel R_2)(C_1 + C_2)} \qquad f_p = \frac{1}{2\pi R_2 C_2} \tag{7.30b}$$

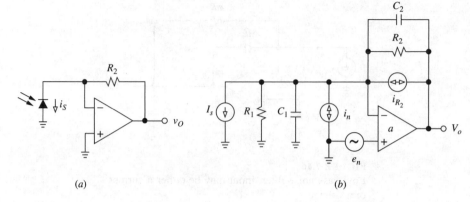

FIGURE 7.17
Photodiode amplifier and its noise model.

The $1/\beta$ function has the low-frequency asymptote $1/\beta_0 = 1 + R_2/R_1$, the high-frequency asymptote $1/\beta_\infty = 1 + C_1/C_2$, and two breakpoints at f_z and f_p. As shown in Fig. 7.18a, the crossover frequency is $f_x = \beta_\infty f_t$, so the noise gain is $A_n = (1/\beta)/(1 + jf/f_x)$, or

$$A_n = \left(1 + \frac{R_2}{R_1}\right) \frac{1 + jf/f_z}{(1 + jf/f_p)(1 + jf/f_x)} \tag{7.31}$$

We also observe that for $a \to \infty$ we have $A_{s(\text{ideal})} = R_2/(1 + jf/f_p)$, so the signal gain is

$$A_s = \frac{R_2}{(1 + jf/f_p)(1 + jf/f_x)} \tag{7.32}$$

and is shown in Fig. 7.18b. With $C_1 \gg C_2$, the noise-gain curve exhibits significant peaking, a notorious feature of photodiode amplifiers. This can be reduced by

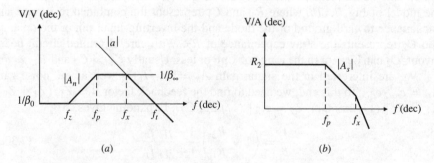

FIGURE 7.18
Noise gain A_n and signal gain A_s for the photodiode amplifier.

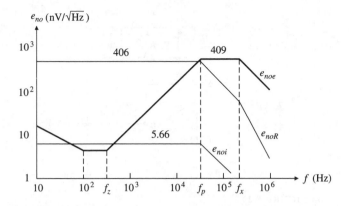

FIGURE 7.19
Output spectral densities of the photodiode amplifier of Example 7.11.

adding a capacitance in parallel with R_2; however, this also reduces the signal-gain bandwidth f_p.

EXAMPLE 7.11. In the circuit[11] of Fig. 7.17 let the op amp be the OPA627 JFET-input op amp (Burr-Brown), for which $f_t = 16$ MHz, $e_{nw} = 4.5$ nV/$\sqrt{\text{Hz}}$, $f_{ce} = 100$ Hz, and $I_B = 1$ pA. Estimate the total output noise E_{no} above 0.01 Hz if $R_1 = 100$ GΩ, $C_1 = 45$ pF, $R_2 = 10$ MΩ, and $C_2 = 0.5$ pF.

Solution. With the above data we have $1/\beta_0 \cong 1$ V/V, $1/\beta_\infty = 91$ V/V, $f_z = 350$ Hz, $f_p = 31.8$ kHz, and $f_x = 176$ kHz. Moreover, by Eqs. (7.15b) and (7.16), $i_{R_2} = 40.6$ fA/$\sqrt{\text{Hz}}$ and $i_n = 0.566$ fA/$\sqrt{\text{Hz}}$. We observe that the noise gain for e_n is A_n, whereas the noise gains for i_n and i_{R_2} coincide with the signal gain A_s. The output densities, obtained as $e_{noe} = |A_n|e_n$, $e_{noi} = |A_s|i_n$, and $e_{noR} = |A_s|i_{R_2}$, are plotted in Fig. 7.19.

The pink-noise tangent principle reveals that the dominant components are the voltage noise e_{noe} in the vicinity of f_x, and the thermal noise e_{noR} in the vicinity of f_p. Current noise is negligible because we are using a JFET-input op amp. Thus, $E_{noe} \cong (1/\beta_\infty)e_n\sqrt{(\pi/2)f_x - f_p} = 91 \times 4.5 \times 10^{-9}\sqrt{(1.57 \times 176 - 31.8)10^3} = 202$ μV (rms), and $E_{noR} \cong R_2 i_{R_2} \times \sqrt{(\pi/2)f_p} \cong 91$ μV. Finally, $E_{no} \cong \sqrt{202^2 + 91^2} = 222$ μV (rms). A PSpice simulation (see Problem 7.30) gives $E_{no} = 230$ μV (rms), indicating that our hand-calculation approximations are quite reasonable.

Noise Filtering

The modified photodiode amplifier of Fig. 7.20 incorporates the current-filtering option of Fig. 7.16 to reduce noise. In choosing the filter cutoff frequency f_0 we must be careful that the signal-gain bandwidth is not reduced unnecessarily. Moreover, the optimum value of Q is the result of a compromise between noise and response characteristics such as peaking and ringing. A reasonable approach is to start with $C_c = C_2$ and $R_3 C_3 = R_2 C_c$, so that $m = 1/n$ and $Q \cong 1$ for $m \gg 1$. Then we fine-tune C_c and R_3 for a best compromise between noise and response characteristics.

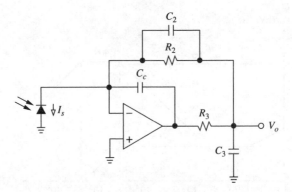

FIGURE 7.20
Photodiode amplifier with noise filtering.

EXAMPLE 7.12. Assuming the parameters of Example 7.11, find suitable values for C_c, R_3, and C_3 in the circuit of Fig. 7.20.

Solution. Let $C_c = C_2 = 0.5$ pF. Pick $C_3 = 10$ nF as a convenient value. Then, $R_3 = R_2 C_c / C_3 = 500\ \Omega$.

PSpice simulations for different values of R_3 give a good compromise for $R_3 = 1$ kΩ, which results in a signal-gain bandwidth of about 24 kHz and $E_{no} \cong 80\ \mu$V (rms). Thus, filtering has reduced noise to about one third of the original value of 230 μV (rms). When the circuit is tried out in the lab, empirical tuning is necessary because of parasitics not accounted for by our PSpice model.

T-Feedback Photodiode Amplifiers

As we know, the use of a *T*-network makes it possible to achieve extremely high sensitivities using moderately high resistances. To assess its impact on dc as well as noise, we use the model of Fig. 7.21. The *T*-network is usually implemented with

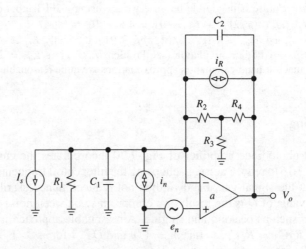

FIGURE 7.21
T-network photodiode amplifier.

$R_3 \parallel R_4 \ll R_2$, so R_2 is raised to the equivalent value $R_{eq} \cong (1 + R_4/R_3)R_2$, and $i_R^2 \cong i_{R_2}^2 = 4kT/R_2$. One can show (see Problem 7.33) that the noise and signal gains are now

$$A_n \cong \left(1 + \frac{R_2}{R_1}\right)\left(1 + \frac{R_4}{R_3}\right)\frac{1 + jf/f_z}{(1 + jf/f_p)(1 + jf/f_x)} \qquad (7.33a)$$

$$A_s \cong \frac{(1 + R_4/R_3)R_2}{(1 + jf/f_p)(1 + jf/f_x)} \qquad (7.33b)$$

$$f_z = \frac{1}{2\pi(R_1 \parallel R_2)(C_1 + C_2)} \qquad f_p = \frac{1}{2\pi(1 + R_4/R_3)R_2 C_2} \qquad (7.34)$$

indicating that the dc values of both gains are raised by a factor of $1 + R_4/R_3$. In particular, we observe that $E_{noR} \cong (1 + R_4/R_3) \times R_2 i_R \sqrt{\pi f_p/2} = [(1 + R_4/R_3)kT/C_2]^{1/2}$, indicating that thermal noise increases with the square root of the factor $1 + R_4/R_3$. Consequently, we must suitably limit this factor in order to avoid raising noise unnecessarily. As it turns out, the T-network option is worthwhile[11] when high-sensitivity amplifiers are used in connection with large-area photodiodes. The large capacitances of these devices cause enough noise-gain peaking to allow for thermal noise increase without jeopardizing the overall noise performance.

EXAMPLE 7.13. In the circuit[11] of Fig. 7.21 let the op amp be the OPA627 of Example 7.11, and let the diode be a large-area photodiode such that $C_1 = 2$ nF, everything else remaining the same. (a) Specify a T-network for a dc sensitivity of 1 V/nA. (b) Find the total rms output noise and the signal bandwidth.

Solution.

(a) We now have $1/\beta_0 \cong 1 + R_4/R_3$, $1/\beta_\infty = 1 + C_1/C_2 = 4000$ V/V and $f_x = \beta_\infty f_t = 4$ kHz. To avoid increasing voltage noise unnecessarily, impose $1/\beta_0 < 1/\beta_\infty$, or $1 + R_4/R_3 < 4000$. Then, $E_{noe} \cong (1/\beta_\infty)e_n\sqrt{\pi f_x/2} = 1.43$ mV. To avoid increasing thermal noise unnecessarily, impose $E_{noR} \leq E_{noe}/3$, or $[(1 + R_4/R_3)kT/C_2]^{1/2} \leq E_{noe}/3$. This yields $1 + R_4/R_3 \leq 27$ (< 4000). Then $R_2 = 10^9/27 = 37$ MΩ. Pick $R_2 = 36.5$ MΩ, $R_3 = 1.00$ kΩ, $R_4 = 26.7$ kΩ.

(b) The signal bandwidth is $f_B = f_p = 1/(2\pi \times 10^9 \times 0.5 \times 10^{-12}) = 318$ Hz. Moreover, $E_{noR} \cong 0.5$ mV, $E_{noi} = 10^9 \times 0.566 \times 10^{-15}\sqrt{1.57 \times 318} = 12.6$ μV, and $E_{no} \cong \sqrt{1.43^2 + 0.5^2} = 1.51$ mV (rms).

7.6
LOW-NOISE OP AMPS

As discussed in Section 7.4, the figures of merit in op amp noise performance are the white-noise floors e_{nw} and i_{nw}, and the corner frequencies f_{ce} and f_{ci}. The lower their values, the quieter the op amp. In wideband applications, usually only the white-noise floors are of concern; however, in instrumentation applications also the corner frequencies may be crucial.

Figure 7.22a and b shows the noise characteristics of the industry-standard OP-27 low-noise precision op amp (Analog Devices), whose typical ratings are $e_{nw} = 3$ nV/$\sqrt{\text{Hz}}$ (the same spectral density as a 545-Ω resistor), $f_{ce} = 2.7$ Hz,

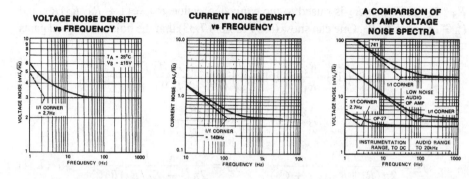

FIGURE 7.22

(*a*) Noise-voltage and (*b*) noise-current characteristics of the OP-27/37 op amp. (*c*) Noise-voltage comparison of three popular op amps. (Courtesy of Analog Devices.)

$e_{nw} = 0.4$ pA/$\sqrt{\text{Hz}}$, and $f_{ci} = 140$ Hz. Another low-noise op amp is the LT1028 (Linear Technology), with $e_{nw} = 0.9$ nV/$\sqrt{\text{Hz}}$.

Figure 7.22c compares the noise-voltage characteristic of the OP-27 low-noise op amp, the NE5533/5534 low-noise audio op amp (Signetics), and the μA741 general-purpose op amp.

Except for programmable op amps, the user has no control over the noise characteristics; however, a basic understanding of how these characteristics originate will help in the device selection process. As with the input offset voltage and bias current, both voltage and current noise depend very heavily on the technology and operating conditions of the differential transistor pair of the input stage. Voltage noise is also affected by the load of the input pair and by the second stage. The noise produced by the subsequent stages is usually insignificant since, when referred to the input, it is divided by the gains of all preceding stages.

Differential Input-Pair Noise

The noise contributed by the differential input pair can be minimized by proper choice of transistor type, geometry, and operating current. Consider BJT-input op amps first. Recall from Eq. (7.18*a*) that BJT voltage noise depends on the base-spreading resistance r_b and transconductance g_m. In the OP-27 the differential-pair BJTs are realized in the *striped geometry* (long and narrow emitters surrounded by base contacts on both sides) to minimize r_b, and are biased at substantially higher than normal collector currents (120 μA per side) to increase g_m.[12] The increase in operating current, however, has an adverse effect on the input bias current I_B and the input noise current i_n. In the OP-27, shown in Fig. 5.15, I_B is reduced by the current-cancellation technique. Noise densities, however, do not cancel but add up in rms fashion, so in current-cancellation schemes i_{nw} is higher than the shot-noise value predicted by Eq. (7.18*b*).

When the application requires large external resistances, FET-input op amps offer a better alternative since their noise current levels are orders of magnitude lower than those of BJT-input devices, at least near room temperature. FETs, on the other hand, tend to exhibit higher voltage noise, mainly because they have lower

$g_m s$ than BJTs. As an example of a JFET-input op amp, the OPA627 (Burr-Brown) has $e_{nw} = 4.5$ nV/$\sqrt{\text{Hz}}$ and $i_n = 1.6$ fA/$\sqrt{\text{Hz}}$ at 100 Hz.

In the case of MOSFETs, $1/f$ noise is also a critical factor. By Eq. (7.20a), the $1/f$ component can be reduced by using large-area devices. Moreover, the empirical observation that p-channel devices tend to display less $1/f$ noise than n-channel types indicates that, in general, the best noise performance in CMOS op amps is achieved by using p-channel input transistors with n-channel active loads.[7] As an example of a MOSFET-input op amp, the TLC279 (Texas Instruments) has $e_{nw} = 25$ nV/$\sqrt{\text{Hz}}$.

Input-Pair Load Noise

Another critical source of noise is the load of the differential input pair. In general-purpose op amps such as the 741, this load is implemented with a current-mirror active load to maximize gain. Active loads, however, are notoriously noisy since they amplify their own noise current. Once divided by the first-stage transconductance and converted to an equivalent input noise voltage, this component can degrade the noise characteristics significantly. In fact, in the 741, noise from the active load exceeds noise from the differential input pair itself.[7]

The OP-27 avoids this problem by using a resistively loaded input stage,[12] as shown in Fig. 5.15. In CMOS op amps, the noise contribution from the active load, when reflected back to the input, is multiplied by the ratio of the g_m of the load to the g_m of the differential pair.[7] Thus, using a load with low g_m reduces this component significantly.

Second-Stage Noise

The last potentially critical contributor to e_n is the second stage, particularly when this is implemented with pnp transistors to provide level shifting as well as additional gain (see Q_{23} and Q_{24} in Fig. 5.15). Being surface devices, pnp transistors suffer from large $1/f$ noise and poor β. Once this noise is reflected back to the input, it can increase f_{ce} significantly. The OP-27 avoids this drawback by using emitter followers Q_{21} and Q_{22} (see again Fig. 5.15) to *isolate* the first stage from the pnp pair.[12]

Ultralow-Noise Op Amps

High-precision instrumentation often requires ultrahigh open-loop gains to achieve the desired degree of linearity, together with ultralow noise to ensure an adequate SNR. In these situations, considerations of cost and availability may justify the development of specialized circuits to meet the requirements.

Figure 7.23 shows an example of specialized op amp design whose dc specifications are compatible with high-precision transducer requirements and ac specifications are suitable for professional audio work.[13] The circuit uses the low-noise OP-27 op amp with a differential front end to simultaneously increase the open-loop gain and reduce voltage noise. The front end consists of three parallel-connected MAT-02 low-noise dual BJTs operating at moderately high collector currents

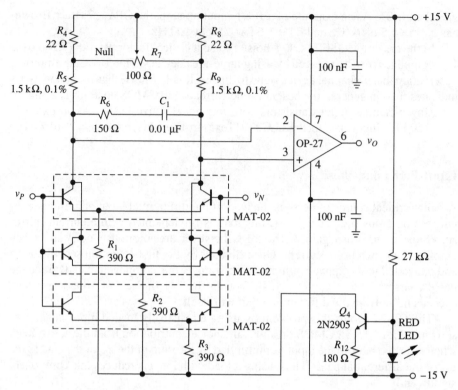

FIGURE 7.23
Ultralow-noise op amp. (Courtesy of Analog Devices.)

(1 mA per transistor). The parallel arrangement reduces the base spreading resistance of the composite device by $\sqrt{3}$, while the high collector current increases g_m. This yields an equivalent input noise voltage with $e_{nw} = 0.5$ nV/$\sqrt{\text{Hz}}$ and $f_{ce} = 1.5$ Hz.

Transistor Q_4, in conjunction with R_{12} and the LED, forms a temperature-stable 6-mA current sink that R_1 through R_3 then split evenly among the three differential pairs. R_6 and C_1 provide frequency compensation for closed-loop gains greater than 10, and R_7 nulls the input offset voltage.

The additional gain provided by the front end increases the overall dc gain to $a_0 = 3 \times 10^7$ V/V. Other measured parameters are $i_{nw} = 1.5$ pA/$\sqrt{\text{Hz}}$, TC(V_{OS}) = $0.1 \ \mu$V/°C (max), GBP = 150 MHz with $A_0 = 10^3$ V/V, and CMRR$_{\text{dB}} = 130$ dB. Similar front-end designs can be used to improve the noise characteristics of other critical circuits, such as instrumentation amplifiers and audio preamps.

PROBLEMS

7.1 Noise properties

7.1 Two IC noise spot measurements, performed respectively at $f_1 = 10$ Hz and $f_2 \gg f_{ce}$, yield $e_n(f_1) = 20$ nV/$\sqrt{\text{Hz}}$ and $e_n(f_2) = 6$ nV/$\sqrt{\text{Hz}}$. Find the rms noise from 1 mHz to 1 MHz.

7.2 Find the NEB of a composite amplifier consisting of two identical stages cascaded as in Fig. 6.6, each stage having a gain of the type $A(jf) = A_0/(1 + jf/f_B)$.

7.3 Show that the standard second-order low-pass and band-pass functions H_{LP} and H_{BP} defined in Section 3.4 have $NEB_{LP} = Q^2 NEB_{BP} = Q\pi f_0/2$. Can you justify the similarity intuitively?

7.4 (*a*) Find the NEB of a filter consisting of an R-C network, followed by a buffer, followed by another R-C network. (*b*) Repeat, but for a filter consisting of a C-R network, followed by a buffer, followed by an R-C network. (*c*) Repeat, but for a filter consisting of an R-C network, followed by a buffer, followed by a C-R network. (*d*) Rank the three filters in terms of noise minimization.

7.5 Confirm the results of Example 7.2 using piecewise graphical integration.

7.6 Estimate the NEB of the RIAA response of Fig. 3.13. Confirm with PSpice.

7.7 Find the NEB if $A_n(s)$ has two zeros at $s = -20\pi$ rad/s and $s = -2\pi 10^3$ rad/s, and four poles at $s = -200\pi$ rad/s, $s = -400\pi$ rad/s, $s = -2\pi 10^4$ rad/s, and $s = -2\pi 10^4$ rad/s.

7.8 Find the total output noise when a noise source with $f_{ce} = 100$ Hz and $e_{nw} = 10$ nV/$\sqrt{\text{Hz}}$ is played through a noiseless wideband band-pass filter with a mid-frequency gain of 40 dB, $f_L = 10$ Hz and $f_H = 1$ kHz. Confirm using the pink-noise tangent principle.

7.9 The spectral noise e_{no} of a certain amplifier below 100 Hz consists of $1/f$ noise with $f_{ce} = 1$ Hz and $e_{nw} = 10$ nV/$\sqrt{\text{Hz}}$; from 100 Hz to 1 kHz it rolls off at the rate of -1 dec/dec; from 1 kHz to 10 kHz it is again constant at 1 nV/$\sqrt{\text{Hz}}$; and past 10 kHz it rolls off at the rate of -1 dec/dec. Sketch and label e_{no}, estimate the total rms noise above 0.01 Hz, and confirm using the pink-noise tangent principle.

7.10 The LT1009 2.5-V reference diode (Linear Technology), when suitably biased, acts as a 2.5-V source with superimposed noise of the type $e_n^2 \cong (118 \text{ nV}/\sqrt{\text{Hz}})^2(30/f + 1)$. If the diode voltage is sent through an R-C filter with $R = 10$ kΩ and $C = 1$ μF, estimate the peak-to-peak noise that one would observe at the output over a 1-minute interval.

7.3 Sources of noise

7.11 Find a resistance that will produce the same amount of room-temperature noise as a diode operating with (*a*) a forward-bias current of 50 μA, and (*b*) a reverse-bias current of 1 pA.

7.12 (*a*) Show that the total rms noise voltage across the parallel combination of a resistance R and capacitance C is $E_n = \sqrt{kT/C}$, regardless of R. (*b*) Find an expression for the total rms value of the noise current flowing through a resistance R in series with an inductance L.

7.13 (*a*) Find a resistance that produces the same e_{nw} as a 741 op amp at room temperature. (*b*) Find a reverse-biased diode current that produces the same i_{nw} as a 741 op amp. How does this current compare with the input bias current of the 741?

7.4 Op amp noise

7.14 In the difference amplifier of Fig. 1.17 let $R_1 = R_3 = 10\,\text{k}\Omega$ and $R_2 = R_4 = 100\,\text{k}\Omega$. Find the total output noise E_{no} above 0.1 Hz if the op amp is (a) the 741 type, and (b) the OP-27 type. Compare also the individual components E_{noe}, E_{noi}, and E_{noR}, and comment. For the 741 assume $f_t = 1\,\text{MHz}$, $e_{nw} = 20\,\text{nV}/\sqrt{\text{Hz}}$, $f_{ce} = 200\,\text{Hz}$, $i_{nw} = 0.5\,\text{pA}/\sqrt{\text{Hz}}$, and $f_{ci} = 2\,\text{kHz}$; for the OP-27 assume $f_t = 8\,\text{MHz}$, $e_{nw} = 3\,\text{nV}/\sqrt{\text{Hz}}$, $f_{ce} = 2.7\,\text{Hz}$, $i_{nw} = 0.4\,\text{pA}/\sqrt{\text{Hz}}$, and $f_{ci} = 140\,\text{Hz}$.

7.15 Using a 741 op amp, design a circuit that accepts three inputs v_1, v_2, and v_3, and yields $v_O = 2(v_1 - v_2 - v_3)$; hence, estimate its total output noise above 1 Hz.

7.16 In the bridge amplifier of Fig. P1.74 let $R = 100\,\text{k}\Omega$ and $A = 2\,\text{V}/\text{V}$, and let the op amps be 741 types. Estimate the total output noise above 1 Hz.

7.17 (a) Find the total rms output noise above 0.1 Hz for the *I-V* converter of Fig. 2.2 if $R = 10\,\text{k}\Omega$, $R_1 = 2\,\text{k}\Omega$, $R_2 = 18\,\text{k}\Omega$, and the op amp is the OP-27, whose characteristics are given in Problem 7.14. (b) Find the SNR if i_I is a triangular wave with peak values of $\pm 10\,\mu\text{A}$.

7.18 (a) Find the total output noise above 0.1 Hz for the inverting amplifier of Fig. P1.54 if all resistances are 10 kΩ and the op amp is the 741 type. (b) Find the SNR if $v_I = 0.5\cos 100t + 0.25\cos 300t$ V.

7.19 A JFET-input op amp with $e_{nw} = 18\,\text{nV}/\sqrt{\text{Hz}}$, $f_{ce} = 200\,\text{Hz}$, and $f_t = 3\,\text{MHz}$ is configured as an inverting integrator with $R = 159\,\text{k}\Omega$ and $C = 1\,\text{nF}$. Estimate the total output noise above 1 Hz.

7.20 It is required to design an amplifier with $A_0 = 60\,\text{dB}$ using op amps with GBP = 1 MHz. Two alternatives are being evaluated, namely, a single-op-amp realization and a two-op-amp cascade realization of the type of Example 6.2. Assuming the resistances are sufficiently low to render current and resistor noise negligible, which of the two configurations is noisier and by how much?

7.21 Using the OP-227 dual op amp, design a dual-op-amp instrumentation amplifier with a gain of 10^3 V/V, and find its total output noise above 0.1 Hz. Try keeping noise as low as practical. The OP-227 consists of two OP-27 op amps in the same package, so use the data of Problem 7.14.

7.22 With reference to the triple-op-amp instrumentation amplifier of Fig. 2.20, consider the first stage, whose outputs are v_{O1} and v_{O2}. (a) Show that if OA_1 and OA_2 are dual op amps with densities e_n and i_n, the overall input power density of this stage is $e_{ni}^2 = 2e_n^2 + [(R_G \parallel 2R_3)i_n]^2/2 + 4kT(R_G \parallel 2R_3)$. (b) Estimate the total rms noise produced by this stage above 0.1 Hz if $R_G = 100\,\Omega$, $R_3 = 50\,\text{k}\Omega$, and the op amps are from the OP-227 dual-op-amp package, whose characteristics are the same as those of the OP-27 given in Problem 7.14.

7.23 (a) In the triple-op-amp instrumentation amplifier of Fig. 2.21 let the pot be adjusted for a gain of 10^3 V/V. Using the results of Problem 7.22, estimate the total output noise above 0.1 Hz. (b) Find the SNR for a sinusoidal input having a peak amplitude of 10 mV.

7.24 Use PSpice to verify the CFA noise calculations of Example 7.10.

7.25 The circuit of Fig. 7.12a has $R_1 = R_3 = 10$ Ω and $R_2 = 10$ kΩ, and its output is observed through a band-pass filter having NEB $= 100$ Hz. The reading is 0.120 mV (rms), and it can be regarded as being primarily voltage noise since the resistances are so small. Next, a 500-kΩ resistor is inserted in series with each input pin of the op amp to generate substantial current noise. The output reading is now 2.25 mV rms. Find e_n and i_n.

7.26 (a) Derive the transfer function of the noise filter of Fig. 7.16. (b) Modify the circuit so that it works as an inverting voltage amplifier with $H = -10H_{\mathrm{LP}}$.

7.27 Using two 0.1-μF capacitances, specify resistances in the noise filter of Fig. 7.16 for $f_0 = 100$ Hz and $Q = 1/2$. If the op amp is the 741 type, find the total rms noise generated by the filter above 0.01 Hz with V_i and I_i both set to zero.

7.28 Using the voltage-input option of the noise filter of Fig. 7.16, design a circuit to filter the voltage of the LT1009 reference diode of Problem 7.10 for a total output noise above 0.01 Hz of 1 μV (rms) or less. Assume an OP-27 op amp whose characteristics are given in Problem 7.14.

7.29 (a) Find a capacitance C that when connected in parallel with R_2 in the inverting amplifier of Example 7.7 will lower the signal-gain bandwidth to 1 kHz. How does it affect noise? (b) Repeat but also with a 0.1-μF capacitance in parallel with R_3.

7.5 Noise in photodiode amplifiers

7.30 Use PSpice to plot e_{noe}, e_{noi}, e_{noR}, and e_{no} for the circuit of Example 7.11. Hence, use the "s" and "sqrt" Probe functions to find E_{no}.

7.31 Investigate the effect of connecting an additional capacitance $C_f = 2$ pF in parallel with R_2 in the photodiode amplifier of Example 7.11. How does it affect noise? The signal-gain bandwidth?

7.32 Use PSpice to confirm Example 7.12.

7.33 Derive Eqs. (7.33) and (7.34).

7.34 Rework Example 7.11, but with R_2 replaced by a T-network with $R_2 = 1$ MΩ, $R_3 = 2$ kΩ, and $R_4 = 18$ kΩ, everything else staying the same. Comment on your findings.

7.35 Verify Example 7.13 via PSpice.

7.36 Modify the circuit of Example 7.13 to filter noise without significantly reducing the signal bandwidth. What is the total output noise of your circuit?

7.6 Low-noise op amps

7.37 A popular noise reduction technique is to combine N identical voltage sources in the manner of Fig. P7.37. (a) Show that if the noise of the resistors is negligible, the output density e_{no} is related to the individual source densities e_n as $e_{no} = e_n/\sqrt{N}$. (b) Find the maximum value of the resistances in terms of e_n so that the rms noise contributed by the resistances is less than 10% of the rms noise due to the sources.

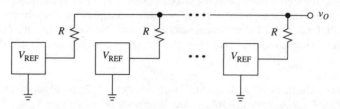

FIGURE P7.37

REFERENCES

1. H. W. Ott, *Noise Reduction Techniques in Electronic Systems,* 2d ed., John Wiley & Sons, New York, 1988.
2. C. D. Motchenbacher and J. A. Connelly, *Low-Noise Electronic System Design,* John Wiley & Sons, New York, 1993.
3. A. P. Brokaw, "An IC Amplifiers User's Guide to Decoupling, Grounding, and Making Things Go Right for a Change," Application Note AN-202, *Applications Reference Manual,* Analog Devices, Norwood, MA, 1993.
4. A. Rich, "Understanding Interference-Type Noise," Application Note AN-346, and "Shielding and Guarding," Application Note AN-347, *Applications Reference Manual,* Analog Devices, Norwood, MA, 1993.
5. A. Ryan and T. Scranton, "Dc Amplifier Noise Revisited," *Analog Dialogue,* Vol. 18, No. 1, Analog Devices, Norwood, MA, 1984.
6. M. E. Gruchalla, "Measure Wide-Band White Noise Using a Standard Oscilloscope," *EDN,* June 5, 1980, pp. 157–160.
7. P. R. Gray and R. G. Meyer, *Analysis and Design of Analog Integrated Circuits,* 3d ed., John Wiley & Sons, New York, 1993.
8. S. Franco, "Current-Feedback Amplifiers," *Analog Circuit Design: Art, Science, and Personalities,* J. Williams ed., Butterworth-Heinemann, Stoneham, MA, 1991.
9. W. Kester, "High Speed Operational Amplifiers," *High-Speed Design Techniques,* Analog Devices, Norwood, MA, 1996.
10. R. M. Stitt, "Circuit Reduces Noise from Multiple Voltage Sources," *Electronic Design,* Nov. 10, 1988, pp. 133–137.
11. J. G. Graeme, *Photodiode Amplifiers–Op Amp Solutions,* McGraw-Hill, New York, 1996.
12. G. Erdi, "Amplifier Techniques for Combining Low Noise, Precision, and High Speed Performance," *IEEE J. Solid-State Circuits,* Vol. SC-16, Dec. 1981, pp. 653–661.
13. A. Jenkins and D. Bowers, "NPN Pairs Yield Ultralow-Noise Op Amp," *EDN,* May 3, 1984, pp. 323–324.

8

STABILITY

8.1 The Stability Problem
8.2 Stability in Constant-GBP Op Amp Circuits
8.3 Internal Frequency Compensation
8.4 External Frequency Compensation
8.5 Stability in CFA Circuits
8.6 Composite Amplifiers
   Problems
   References

Since its conception by Harold S. Black in 1927, negative feedback has become a cornerstone of electronics and control, as well as other areas of applied science, such as biological systems modeling. As seen in the previous chapters, negative feedback results in a number of performance improvements, including gain stabilization against process and environmental variations, reduction of distortion stemming from component nonlinearities, broadbanding, and impedance transformation. These advantages are especially startling if feedback is applied around very high-gain amplifiers such as op amps.

Negative feedback comes at a price, however: the possibility of an oscillatory state. In general, oscillation will result when the system is capable of sustaining a signal around the loop regardless of any applied input. For this to occur, the system must provide enough phase shift around the loop to turn feedback from negative to positive, and enough loop gain to sustain an output oscillation without any applied input.

This chapter provides a systematic investigation of the conditions leading to instability as well as suitable cures, known as *frequency-compensation techniques,* to stabilize a circuit so that the benefits of negative feedback can be fully realized.

8.1
THE STABILITY PROBLEM

The advantages of negative feedback are realized only if the circuit has been stabilized against the possibility of oscillations. For an intuitive discussion,[1] refer again to the feedback system of Fig. 1.21. As we know, whenever the amplifier detects an input error x_d, it tries to reduce it. It takes some time, however, for the amplifier to react and then transmit its response back to the input via the feedback network. The consequence of this combined delay is a tendency on the part of the amplifier to overcorrect the input error, especially if the loop gain is high. If the overcorrection exceeds the original error, a regenerative effect results, whereby the magnitude of x_d diverges, instead of converging, and instability results. Signal amplitudes grow exponentially until inherent circuit nonlinearities limit further growth, forcing the system either to saturate or to oscillate, depending on the order of its system function. By contrast, a circuit that succeeds in making x_d converge is stable.

Gain Margin

Whether a system is stable or unstable is determined by the manner in which its loop gain T varies with frequency. To substantiate, suppose a frequency exists at which the phase angle of T is $-180°$; call this frequency $f_{-180°}$. Then, $T(jf_{-180°})$ is real and negative, indicating that feedback has turned from negative to positive. If $|T(jf_{-180°})| < 1$, then Eq. (1.40), rewritten here as

$$A(jf_{-180°}) = \frac{a(jf_{-180°})}{1 + T(jf_{-180°})}$$

indicates that $A(jf_{-180°})$ is greater than $a(jf_{-180°})$ because the denominator is less than unity. The circuit is nonetheless stable because any signal circulating around the loop will progressively decrease in magnitude and eventually die out; consequently, the poles of $A(s)$ must lie in the left half of the s plane.

If $|T(jf_{-180°})| = 1$, the above equation predicts $A(jf_{-180°}) \to \infty$, indicating that the circuit can now sustain an output signal even with zero input! The circuit is an oscillator, indicating that $A(s)$ must have a conjugate pole pair right on the imaginary axis. Oscillations are initiated by ac noise, which is always present in some form at the amplifier input. An ac noise component x_d right at $f = f_{-180°}$ results in a feedback component $x_f = -x_d$, which is further multiplied by -1 in the summing network to yield x_d itself. Thus, once this ac component has entered the loop, it will be sustained indefinitely.

If $|T(jf_{-180°})| > 1$, mathematical tools other than the foregoing equation are needed to predict circuit behavior. Suffice it to say here that now $A(s)$ may have a conjugate pole pair in the right half of the s plane. Consequently, once started, oscillation will grow in magnitude until some circuit nonlinearity, either inherent, such as a nonlinear VTC, or deliberate, such as an external clamping network, reduces the loop gain to exactly unity. Henceforth, oscillation is of the sustained type.

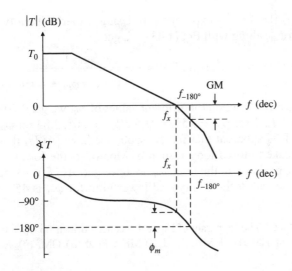

FIGURE 8.1
Visualizing gain margin GM and phase margin ϕ_m.

A quantitative measure of stability is offered by the *gain margin,* defined as

$$GM = 20 \log \frac{1}{|T(jf_{-180°})|} \tag{8.1}$$

The GM represents the number of decibels by which we can increase $|T(jf_{-180°})|$ before it becomes unity and thus leads to instability. For instance, a circuit with $|T(jf_{-180°})| = 1/\sqrt{10}$ has $GM = 20 \times \log_{10} \sqrt{10} = 10$ dB, which is considered a reasonable margin. By contrast, a circuit with $|T(jf_{-180°})| = 1/\sqrt{2}$ has $GM = 3$ dB, not much of a margin: only a modest increase in the gain a because of manufacturing process variations or environmental changes may easily lead to instability! The GM is visualized in Fig. 8.1 (top).

Phase Margin

An alternative and more common way of quantifying stability is via phase. In this case we focus on $\sphericalangle T(jf_x)$, the phase angle of T at the crossover frequency f_x, where $|T| = 1$ by definition, and we define the *phase margin* ϕ_m as the number of degrees by which we can lower $\sphericalangle T(jf_x)$ before it reaches $-180°$ and thus leads to instability. We have $\phi_m = \sphericalangle T(jf_x) - (-180°)$, or

$$\phi_m = 180° + \sphericalangle T(jf_x) \tag{8.2}$$

The phase margin is visualized in Fig. 8.1 (bottom). To investigate its significance, we write $T(jf_x) = 1 \underline{/\phi_m - 180°} = -\exp(j\phi_m)$. The error function is then

$1/[1 + 1/T(jf_x)] = 1/[1 - \exp(-j\phi_m)]$. Using Euler's identity $\exp(-j\phi_m) = \cos\phi_m - j\sin\phi_m$, along with Eq. (1.43), we get

$$|A(jf_x)| = |A_{\text{ideal}}(jf_x)| \times \frac{1}{\sqrt{(1 - \cos\phi_m)^2 + \sin^2\phi_m}}$$

Calculating the error function for different values of ϕ_m we get 0.707 for $\phi_m = 90°$, 1 for $\phi_m = 60°$, 1.31 for $\phi_m = 45°$, 1.93 for $\phi_m = 30°$, 3.83 for $\phi_m = 15°$, and ∞ for $\phi_m = 0°$. It is apparent that for $\phi_m < 60°$ we have $|A(jf_x)| > |A_{\text{ideal}}(jf_x)|$, indicating a peaked closed-loop response. Moreover, the lower ϕ_m is, the more pronounced the peaking. In the limit $\phi_m \to 0$ we get $|A(jf_x)| \to \infty$, or oscillatory behavior. In practical designs, a typical lower limit for ϕ_m is 45°, with 60° being more common.

EXAMPLE 8.1. The loop gain of Fig. 8.1 has been drawn for $T_0 = 10^4$ and three pole frequencies at 100 Hz, 1 MHz, and 10 MHz. Find (a) GM, (b) ϕ_m, and (c) T_0 for $\phi_m = 60°$.

Solution. We have

$$|T(jf)| = \frac{10^4}{\{[1 + (f/10^2)^2][1 + (f/10^6)^2][1 + (f/10^7)^2]\}^{1/2}} \tag{8.3a}$$

$$\sphericalangle T(jf) = -\{\tan^{-1}(f/10^2) + \tan^{-1}(f/10^6) + \tan^{-1}(f/10^7)\} \tag{8.3b}$$

(a) To find GM we need to know $f_{-180°}$. The figure indicates that 1 MHz $\leq f_{-180°} \leq$ 10 MHz. Start out with 5 MHz, as an initial estimate, then use Eq. (8.3b) to find the actual value by trial and error. Letting $f = 5$ MHz in Eq. (8.3b) yields $\sphericalangle T(j5 \times 10^6) = -195.3°$. This is too large, so try $f = 3$ MHz. This gives $\sphericalangle T(j3 \times 10^6) = -178.3°$, which is too small. After a few more trials we find that $\sphericalangle T = -180°$ for $f = 3.16$ MHz. Then, Eq. (8.3a) gives $|T(j3.16 \times 10^6)| = 91.04 \times 10^{-3}$. Finally, Eq. (8.1) gives GM = 20.82 dB.

(b) To find ϕ_m, we need to know f_x. The figure provides the initial estimate $f = 1$ MHz. Substituting into Eq. (8.3a) gives $|T(j10^6)| = 0.7036$, which is too small. So, try $f = 700$ kHz; this yields $|T(j700 \times 10^3)| = 1.167$, which is too large. After a few more iterations we find that $|T| = 1$ for $f = 784$ kHz. Then, Eq. (8.3b) gives $\sphericalangle T(j784 \times 10^3) = -132.6°$, and Eq. (8.2) gives $\phi_m = 47.4°$.

(c) For $\phi_m = 60°$, we want $|T(jf_{-120°})| = 1$. Using Eq. (8.3b) we find, by trial and error, $f_{-120°} = 512$ kHz. The value of the denominator of Eq. (8.3a) at this frequency is 5760. Clearly, for $|T|$ to be unity at this frequency, its dc value T_0 must be lowered from 10^4 to 5760.

Peaking and Ringing

The presence of peaking in the frequency domain is usually accompanied by ringing in the time domain, and vice versa. As illustrated in Fig. 8.2, the two effects are quantified in terms of the *gain peaking GP*, in decibels, and the *overshoot OS*, in percentage. Both effects are absent in first-order systems since it takes a complex pole pair to produce them. For a second-order all-pole system, peaking occurs for $Q > 1/\sqrt{2}$, and ringing for $\zeta < 1$, where the *quality factor Q* and the *damping ratio* ζ are related as $Q = 1/2\zeta$, or $\zeta = 1/2Q$. Second-order systems are well documented

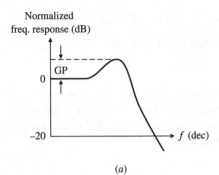

Normalized
freq. response (dB)

Normalized
step response (V)

(a)

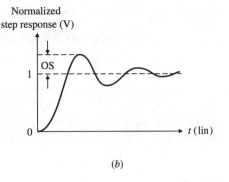

(b)

FIGURE 8.2
Illustrating gain peaking GP and overshoot OS.

in the literature,[2] where it is found that

$$GP = 20 \log_{10} \frac{2Q^2}{\sqrt{4Q^2 - 1}} \qquad \text{for } Q > 1/\sqrt{2} \qquad (8.4)$$

$$OS\,(\%) = 100 \exp \frac{-\pi \zeta}{\sqrt{1 - \zeta^2}} \qquad \text{for } \zeta < 1 \qquad (8.5)$$

$$\phi_m = \cos^{-1}\left(\sqrt{4\zeta^4 + 1} - 2\zeta^2\right) = \cos^{-1}\left(\sqrt{1 + 1/4Q^4} - 1/2Q^2\right) \qquad (8.6)$$

Combining these equations yields the graphs of Fig. 8.3, which relate peaking and ringing to the phase margin. We observe that peaking occurs for $\phi_m \leq \cos^{-1}(\sqrt{2} - 1) = 65.5°$, and ringing for $\phi_m \leq \cos^{-1}(\sqrt{5} - 1) = 76.3°$. It is also worth keeping

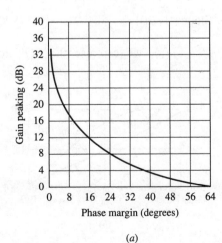

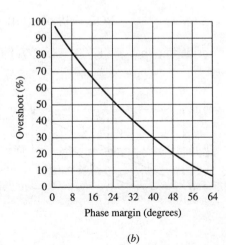

(a)

(b)

FIGURE 8.3
GP and OS as functions of ϕ_m for a second-order all-pole system.

in mind the following frequently encountered values of GP (ϕ_m) and OS (ϕ_m):

$$\text{GP}(60°) \cong 0.3 \text{ dB} \qquad \text{OS}(60°) \cong 8.8\%$$

$$\text{GP}(45°) \cong 2.4 \text{ dB} \qquad \text{OS}(45°) \cong 23\%$$

Depending on the case, a closed-loop response may have a single pole, a pole pair, or a higher number of poles. Mercifully, the response of higher-order circuits is often dominated by a single pole pair, so the graphs of Fig. 8.3 provide a good starting point for a great many circuits of practical interest.

The Rate of Closure (ROC)

We are now ready to develop a quick means for assessing stability from magnitude Bode plots for *minimum-phase systems,* that is, for systems having no roots in the right half of the s plane. To this end, let us first study the plots of Fig. 8.4, which pertain to the single-root function $H(jf) = (1 + jf/f_0)^{\pm 1}$, where -1 holds for a pole frequency, and $+1$ for a zero. Denoting the slope of $|H|$ as Slope($|H|$), we observe that for $f \leq f_0/10$, Slope($|H|$) $\to 0$ dB/dec and $\measuredangle H \to 0°$; for $f \geq 10f_0$, Slope($|H|$) $\to \pm 20$ dB/dec and $\measuredangle H \to \pm 90°$; for $f = f_0$, Slope($|H|$) $\to \pm 10$ dB/dec and $\measuredangle H \to \pm 45°$. We can empirically derive phase (in degrees) from slope (in decibels per decade) as

$$\measuredangle H \cong 4.5 \times \text{Slope}(|H|) \tag{8.7}$$

This correlation holds also if $H(s)$ has more than one root, provided the roots are *real, negative,* and *well separated,* say, at least a decade apart.

Next, suppose both $|a|$ and $|1/\beta|$ have been graphed. Observe the slopes of the two curves at the crossover frequency f_x, and call the magnitude of their difference the *rate of closure,*

$$\text{ROC} = |\text{Slope}(|a|) - \text{Slope}(|1/\beta|)|_{f=f_x} \tag{8.8}$$

Considering that $\measuredangle T(jf_x) = \measuredangle a(jf_x) - \measuredangle \beta^{-1}(jf_x)$, we can use the ROC to

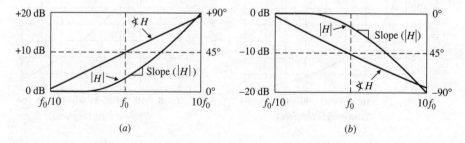

(a) (b)

FIGURE 8.4
Graphical illustration of the relationship $\measuredangle H \cong 4.5 \times \text{Slope}(|H|)$ for (a) a zero and (b) a pole.

estimate ϕ_m via Eq. (8.7). The following cases arise so frequently that it is worth keeping them in mind.

$$\text{ROC} \cong 20 \text{ dB/dec} \Rightarrow \phi_m \cong 90° \qquad (8.9a)$$

$$\text{ROC} \cong 30 \text{ dB/dec} \Rightarrow \phi_m \cong 45° \qquad (8.9b)$$

$$\text{ROC} \cong 40 \text{ dB/dec} \Rightarrow \phi_m \cong 0° \qquad (8.9c)$$

$$\text{ROC} > 40 \text{ dB/dec} \Rightarrow \phi_m < 0° \qquad (8.9d)$$

We shall also make frequent use of the property that for any two frequencies located within a region of constant slope of $\pm n20$ dB/dec, we have

$$|H(jf_1)|/|H(jf_2)| = (f_1/f_2)^{\pm n} \qquad (8.10)$$

For instance, in the region of constant GBP of an op amp, we get the familiar result $|a(jf_1)|/|a(jf_2)| = (f_1/f_2)^{-1} = f_2/f_1$.

Finding T Using PSpice

PSpice is a powerful tool for finding T, especially when complex transistor-level or macromodel-level circuits are involved. A convenient method, developed by S. Rosenstark,[3] requires that we break the loop, inject a test signal in the forward direction, and perform two measurements at the return end, namely, the measurement of the open-circuit voltage V_{ret} and the short-circuit current I_{ret}. Then, we calculate

$$T = \frac{-1}{1/T_{\text{oc}} + 1/T_{\text{sc}}} \qquad (8.11)$$

where $T_{\text{oc}} = V_{\text{ret}}/V_{\text{test}}$ and $T_{\text{sc}} = I_{\text{ret}}/I_{\text{test}}$, V_{test} and I_{test} being the voltage and current at the point of test-signal injection. The advantage of this method is that we can break the loop at any point we wish, without having to worry about the termination issues raised in Section 1.7.

The procedure is illustrated in Fig. 8.5 for a 741 op amp with $\beta = 0.5$. The circuit includes also R_L and C_L to model a typical output load, and C_n to model the stray capacitance of the inverting-input interconnections. Though we have chosen to break the loop at the output of the op amp, we could have broken it at any other point, such as at the inverting-input pin (see Problem 8.8). The only constraint is that as we break the loop we must maintain dc continuity for PSpice to perform the dc bias analysis. In Fig. 8.5a we use the source V_t to inject a test signal, a conveniently large shunt capacitance C_∞ to establish an ac short at the return end, and the source V_r to sense the short-circuit return current. In Fig. 8.5b we use the source G_t to inject a test signal, and a conveniently large series inductance L_∞ to maintain dc continuity

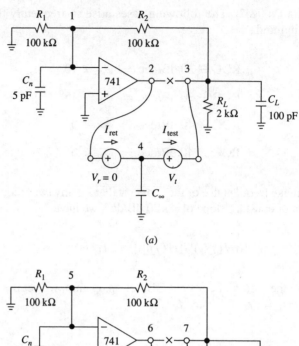

(a)

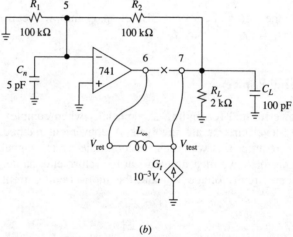

(b)

FIGURE 8.5
PSpice circuits to find T_{sc} and T_{oc}.

while providing an ac open. The PSpice circuit file uses the 741 Byle macromodel as follows.

```
Plotting the Loop Gain T:
.lib eval.lib
VCC 10 0 dc 15V
VEE 11 0 dc -15V
*Circuit to find Tsc:
R1sc 0 1 100k
R2sc 1 3 100k
Cnsc 1 0 5pF
RLsc 3 0 2k
CLsc 3 0 100pF
```

```
XOAsc 0 1 10 11 2 ua741
Vr 2 4 dc 0V
Vt 4 3 ac 1V
C00 4 0 1MegF
*Circuit to find Toc:
R1oc 0 5 100k
R2oc 5 7 100k
Cnoc 5 0 5pF
RLoc 7 0 2k
CLoc 7 0 100pF
XOAoc 0 5 10 11 6 ua741
L00 6 7 1MegH
Gt 0 7 4 3 1m
.ac dec 10 1Hz 10MegHz
.probe ;Tsc = I(Vr)/I(Vt), Toc = V(6)/V(7)
.end
```

The results of the simulation are shown in Fig. 8.6. Using the cursor facility of the Probe postprocessor, we find $f_x \cong 390$ kHz and $\sphericalangle T(jf_x) \cong -134°$, indicating a phase margin $\phi_m \cong 46°$.

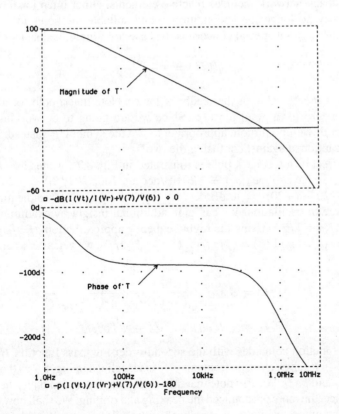

FIGURE 8.6
Bode plots of T for the op amp circuit of Fig. 8.5.

8.2
STABILITY IN CONSTANT-GBP OP AMP CIRCUITS

Op amps with a constant GBP are said to be *unconditionally stable* because with frequency-independent feedback, or $\sphericalangle\beta = 0$, they are stable for any $\beta \leq 1$ V/V. Since we now have $\sphericalangle T = \sphericalangle(a\beta) = \sphericalangle a$, and $\sphericalangle a(jf_x) \cong -90°$, these circuits enjoy $\phi_m = 180° + \sphericalangle a(jf_x) \cong 180° - 90° = 90°$. Look at Figs. 6.5b and 6.7b to appreciate the unconditional stability of the noninverting and inverting amplifiers: in both cases the rate of closure is ROC = 20 dB/dec.

As the transition frequency f_t is approached, constant-GBP op amps exhibit additional phase lag due to higher-order poles. Typically, $\sphericalangle a(jf_t) \cong -120°$, so $60° \leq \phi_m \leq 90°$, depending on the value of β. The circuit with the lowest phase margin is the *voltage follower*, for which $\beta = 1$ V/V and $f_x = f_t$. We observe that an op amp that has been stabilized for voltage-follower operation will be stable also as an *inverting integrator*, since the latter has $\beta(jf_x) = 1$ V/V. Look at Fig. 6.25b to convince yourself.

Feedback Pole

If the feedback network includes reactive elements, either intentional or parasitic, stability may no longer be unconditional, and suitable measures may have to be taken to raise ϕ_m. Of special concern is the case of a single feedback pole, or

$$\beta(jf) = \frac{\beta_0}{1 + jf/f_p} \tag{8.12}$$

where β_0 is the dc value of the feedback factor. Note that a pole (or a zero) of β becomes a zero (or a pole) for $1/\beta$. Since we are going to be working with $1/\beta$ rather than β, we find it more appropriate to use the symbol f_z instead of f_p. (The reader is cautioned against confusing the two!)

The effect of a feedback pole is illustrated in Fig. 8.7 for the case $f_z \ll \beta_0 f_t$. At $f = f_x$ we have Slope $(|a|) \cong -20$ dB/dec and Slope $(|1/\beta|) \cong +20$ dB/dec, so ROC $\cong |-20 - (+20)| = 40$ dB/dec. By Eq. (8.9c), $\phi_m \cong 0°$, indicating a circuit on the verge of oscillation. We can gain additional insight by examining the error function $1/(1 + 1/T)$. Using the high-frequency approximation $a \cong f_t/jf$ and letting $1/T = (1/a) \times (1/\beta) = (jf/f_t) \times (1 + jf/f_z)/\beta_0$, we get, after straightforward algebra,

$$A(jf) = A_{ideal} \times \frac{1}{1 - (f/f_x)^2 + (jf/f_x)/Q} \tag{8.13a}$$

$$f_x = \sqrt{f_z \beta_0 f_t} \qquad Q = \sqrt{\beta_0 f_t/f_z} \tag{8.13b}$$

The error function coincides with the second-order low-pass function H_{LP} defined in Eq. (3.44). Its characteristic frequency f_x is visualized in Fig. 8.7 as the *geometric mean* of f_z and $\beta_0 f_t$. We also note that the lower f_z compared to $\beta_0 f_t$, the higher the Q and, hence, the more pronounced the peaking and ringing. We shall now investigate the most common examples of feedback poles, along with suitable stabilization techniques.

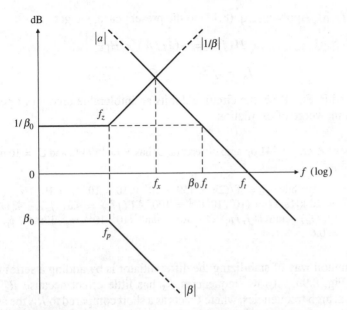

FIGURE 8.7
Illustrating the effect of a pole within the feedback loop of an
internally compensated op amp.

The Differentiator Circuit

As we know, the differentiator of Fig. 8.8a gives, in the limit $a \to \infty$, $H_{ideal} = -(jf/f_0)$, where $f_0 = 1/2\pi RC$ is the *unity-gain frequency*. To find the actual transfer function $H(jf)$, we observe that $\beta = Z_C/(Z_C + R)$, $Z_C = 1/j2\pi fC$, where we have assumed $r_d = \infty$ and $r_o = 0$ for simplicity. Expanding gives $\beta(jf) =$

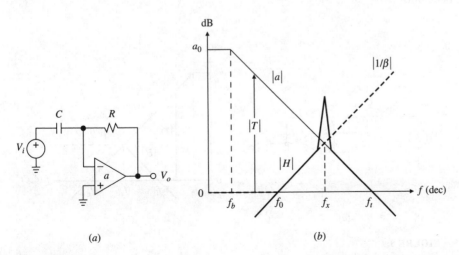

(a)   (b)

FIGURE 8.8
Uncompensated differentiator.

$1/(1 + jf/f_0)$. Applying Eq. (8.13) to the present case, we get

$$H(jf) = -(jf/f_0) \times H_{LP} \qquad (8.14a)$$

$$f_x = \sqrt{f_0 f_t} \qquad Q = \sqrt{f_t/f_0} \qquad (8.14b)$$

As depicted in Fig. 8.8b, the circuit exhibits an intolerable amount of peaking and is thus on the verge of oscillation.

EXAMPLE 8.2. A 741 op amp differentiator has $R = 159\,\mathrm{k\Omega}$ and $C = 10\,\mathrm{nF}$. Find f_x, Q, and ϕ_m.

Solution. We have $f_0 = 1/(2\pi \times 159 \times 10^3 \times 10 \times 10^{-9}) = 10^2\,\mathrm{Hz}$, $f_x = (100 \times 10^6)^{1/2} = 10\,\mathrm{kHz}$, $Q = (10^6/10^2)^{1/2} = 100$, $\sphericalangle T(jf_x) = \sphericalangle a(jf_x) - \sphericalangle[1/\beta(jf_x)] = -\tan^{-1}(f_x/f_a) - \tan^{-1}(f_x/f_0) \cong -90° - \tan^{-1}(10^4/10^2) = -179.4°$, $\phi_m = 180° - 179.4° = 0.6°$.

A common way of stabilizing the differentiator is by adding a series resistance R_s as in Fig. 8.9a. At low frequencies R_s has little effect because $R_s \ll |Z_C|$. However, at high frequencies, where C acts as a short compared to R_s, the asymptotic value becomes $|1/\beta_\infty| = 1 + R/R_s$, indicating the creation of a break frequency past which the $|1/\beta|$ curve flattens out. If we position this breakpoint right on the $|a|$ curve, as in Fig. 8.9b, then we obtain ROC = 30 dB/dec, or $\phi_m = 45°$, by Eq. (8.9b). To find the required R_s, impose $1 + R/R_s = |a(jf_x)| = f_t/f_x = \sqrt{f_t/f_0} \gg 1$. This gives

$$R_s \cong R/\sqrt{f_t/f_0} \qquad (8.15)$$

Thus, for $\phi_m = 45°$ in Example 8.2, use $R_s \cong 159/\sqrt{10^6/100} = 1.59\,\mathrm{k\Omega}$. If a greater phase margin is desired, the second break frequency can be lowered further, but at the price of further reducing the frequency range of near-ideal differentiator behavior.

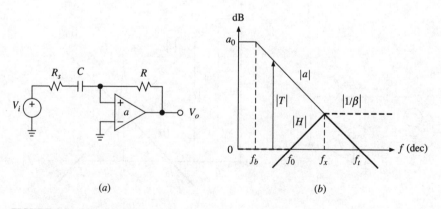

(a) (b)

FIGURE 8.9
Compensated differentiator.

Stray Input Capacitance Compensation

All practical op amps exhibit stray input capacitances. Of special concern is the net capacitance C_n of the inverting input toward ground,

$$C_n = C_d + C_c/2 + C_{ext} \tag{8.16}$$

where C_d is the *differential capacitance* between the input pins; $C_c/2$ is the *common-mode capacitance* of each input to ground, so that when the inputs are tied together the net capacitance is the sum of the two; and C_{ext} is the *external parasitic capacitance* of components, leads, sockets, and printed-circuit traces associated with the inverting input node. Typically, each of the above components is on the order of a few picofarads.

As in the case of the differentiator, C_n creates a feedback pole whose phase lag erodes ϕ_m. A common way of counteracting this lag is by using a feedback capacitance C_f to create feedback phase lead. This is illustrated in Fig. 8.10a for the inverting case. Assuming $r_d = \infty$ and $r_o = 0$, we have $1/\beta = 1 + Z_2/Z_1$, where $Z_1 = R_1 \parallel (1/j2\pi f C_n)$ and $Z_2 = R_2 \parallel (1/j2\pi f C_f)$. Expanding, we get

$$\frac{1}{\beta} = \left(1 + \frac{R_2}{R_1}\right)\frac{1 + jf/f_z}{1 + jf/f_p} \tag{8.17}$$

where $f_z = 1/[2\pi(R_1 \parallel R_2)(C_n + C_f)]$ and $f_p = 1/2\pi R_2 C_f$.

In the absence of C_f we have $1/\beta = (1 + R_2/R_1)\{1 + jf[2\pi(R_1 \parallel R_2)C_n]\}$, indicating that the $|1/\beta|$ curve bends upward. If its break frequency is located well below the crossover frequency, we have ROC $\cong$ 40 dB/dec, or a circuit on the verge of oscillation. This situation corresponds to the curve $\phi_m \cong 0$ in Fig. 8.10b.

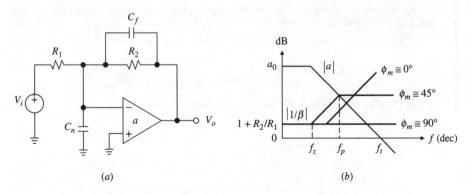

(a)　　　　　　　　　　(b)

FIGURE 8.10
Stray input capacitance compensation.

Inserting C_f creates a second breakpoint at f_p beyond which the $|1/\beta|$ flattens out toward the high-frequency asymptote $1/\beta_\infty = 1 + Z_{C_f}/Z_{C_n} = 1 + C_n/C_f$. By properly positioning this second breakpoint we can increase ϕ_m. For $\phi_m \cong 45°$ we place f_p right on the $|a|$ curve, so $f_p = \beta_\infty f_t$. Rewriting as $1/2\pi R_2 C_f = f_t/(1 +$

C_n/C_f) gives

$$C_f = (1 + \sqrt{1 + 8\pi R_2 C_n f_t})/4\pi R_2 f_t \qquad \text{for } \phi_m \cong 45° \qquad (8.18a)$$

Alternatively, we can compensate for $\phi_m = 90°$. In this case we place f_p right on top of f_z so as to cause a pole-zero cancellation. This makes the $|1/\beta|$ curve flat throughout, or $1/\beta_\infty = |1/\beta_0|$. Rewriting as $1 + C_n/C_f = 1 + R_2/R_1$ yields

$$C_f = (R_1/R_2)C_n \qquad \text{for } \phi_m = 90° \qquad (8.18b)$$

Moreover, the crossover frequency is $\beta_\infty f_t = \beta_0 f_t = f_t/(1 + R_2/R_1)$. This technique, also called *neutral compensation*, is similar to oscilloscope probe compensation.

We observe that the introduction of C_f yields, in the limit $a \to \infty$, $A_{\text{ideal}} = -Z_2/R_1 = (-R_2/R_1)/(1 + jf/f_p)$; that is, A_{ideal} is frequency-dependent with a pole frequency at $f = f_p$. Moreover, the error function $1/(1 + 1/T)$ has a pole frequency at the crossover frequency $\beta_\infty f_t$. Hence, the actual gain $A(jf) = A_{\text{ideal}}/(1 + 1/T)$ has a pole-frequency pair, namely, f_p and $\beta_\infty f_t$.

EXAMPLE 8.3. In Fig. 8.10a let $R_1 = R_2 = 30 \text{ k}\Omega$, and $C_{\text{ext}} = 3 \text{ pF}$. Moreover, let the op amp have GBP = 20 MHz, $C_d = 7 \text{ pF}$, and $C_c/2 = 6 \text{ pF}$. (a) Find ϕ_m with C_f absent. (b) Find C_f for $\phi_m \cong 90°$. (c) Find $A(jf)$ after compensation. (d) Verify with PSpice.

Solution.

(a) We have $1 + R_2/R_1 = 2$, $C_n = 7 + 6 + 3 = 16 \text{ pF}$, $f_z = 1/(2\pi \times 15 \times 10^3 \times 16 \times 10^{-12}) = 663 \text{ kHz}$, and $1/\beta = 2[1 + jf/(663 \text{ kHz})]$. Using Eq. (8.13b) we find $Q = 3.88$, and using Eq. (8.6) we find $\phi_m = 14.7°$—not a very convincing margin.

(b) Use $C_f = (30/30)16 = 16 \text{ pF}$.

(c) We have $f_p = 1/2\pi R_2 C_f = 332 \text{ kHz}$ and $\beta_\infty f_t = (1/2)20 = 10 \text{ MHz}$, so

$$A(jf) = \frac{-1}{[1 + jf/(332 \text{ kHz})][1 + jf/(10 \text{ MHz})]} \text{ V/V}$$

(d) With reference to Fig. 8.11, we write the following file.

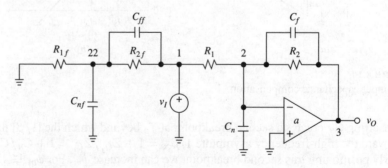

FIGURE 8.11
PSpice circuit of Example 8.3.

```
Stray input capacitance compensation:
vI 1 0 ac 1V pulse (0 1V 0 1ns 1ns 4us 8us)
R1 1 2 30k
Cn 2 0 16pF
R2 2 3 30k
Cf 2 3 16pF
ea0 5 0 0 2 1Meg ;a0 = 1 V/uV
Req 5 6 1Meg
Ceq 6 0 7.958nF ;fb = 20 Hz
eout 3 0 6 0 1 ;output buffer
*Circuit to plot 1/beta:
R2f 1 22 30k
Cff 1 22 16pF
R1f 22 0 30k
Cnf 22 0 16pF
.ac dec 50 100kHz 100MegHz
.tran 10ns 4us 0ns 10ns
.probe ;a=V(3)/V(0,2), 1/beta=V(1)/V(22), A=V(3)/V(1); vO(t)=v(3)
.end
```

The results of the simulation, shown in Fig. 8.12, confirm the stabilizing effect of C_f as well as the closed-loop pole frequencies of 332 kHz and 10 MHz.

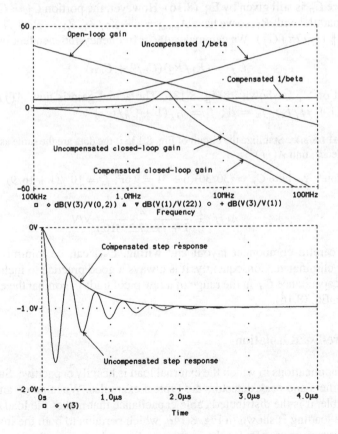

FIGURE 8.12
Frequency and transient responses of the circuit of Fig. 8.11.

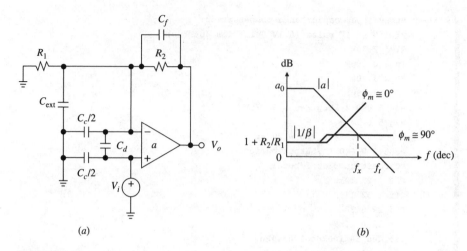

FIGURE 8.13
Stray input capacitance compensation for the noninverting configuration.

We now turn to the noninverting configuration[4] of Fig. 8.13a, where the various stray input capacitances have been shown explicitly. We observe that the overall capacitance C_n is still given by Eq. (8.16). However, the portion $C_1 = C_c/2 + C_{ext}$ is now in parallel with R_1, so we have $A_{ideal} = 1 + Z_2/Z_1$, $Z_1 = R_1 \parallel (1/j2\pi f C_1)$, $Z_2 = R_2 \parallel (1/j2\pi f C_f)$. We can make A_{ideal} frequency-independent by using

$$C_f = (R_1/R_2)(C_c/2 + C_{ext}) \tag{8.19}$$

The effect of C_f is shown in Fig. 8.13b. The actual gain is now $A(jf) \cong (1 + R_2/R_1)/(1 + jf/f_x)$, $f_x = \beta_\infty f_t = f_t/(1 + C_n/C_f)$.

> **EXAMPLE 8.4.** Stabilize the circuit of Fig. 8.13a if the data are the same as in Example 8.3. Hence, find $A(jf)$.
>
> **Solution.** We have $C_f = (30/30)(6 + 3) = 9$ pF, $f_x = 10^7/(1 + 16/9) = 7.2$ MHz, and
>
> $$A(jf) \cong \frac{2}{1 + jf/(7.2 \text{ MHz})} \text{ V/V}$$

With careful component layout and wiring, C_{ext} can be minimized but not altogether eliminated. Consequently, it is always a good practice to include a small feedback capacitance C_f in the range of a few picofarads to combat the effect of C_n as given in Eq. (8.16).

Capacitive-Load Isolation

There are applications in which the external load is heavily capacitive. Sample-and-hold amplifiers and peak detectors are typical examples. When an op amp drives a coaxial cable, it is the distributed cable capacitance that makes the load capacitive. Capacitive loading is shown in Fig. 8.14a, which pertains to both the inverting and the noninverting amplifier: for the former we lift node A off ground and apply the input source there, and for the latter we lift B and use it as the input node.

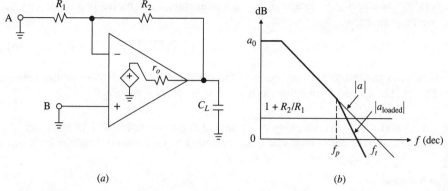

(a)   *(b)*

FIGURE 8.14
Capacitive loading.

The capacitance C_L forms a pole with the open-loop output resistance r_o. Ignoring loading by the feedback network, the loaded gain can be expressed as

$$a_{\text{loaded}} \cong a \frac{1}{1 + jf/f_p}$$

where $f_p = 1/2\pi r_o C_L$. As shown in Fig. 8.14b, the effect of the pole is to increase the ROC and thus invite instability. Looked at from another viewpoint, C_L will tend to resonate with the equivalent inductance L_{eq} of the closed-loop output impedance Z_o investigated in Section 6.3. Hence, intolerable peaking and ringing may ensue.

The popular cure depicted in Fig. 8.15 uses a small series resistance R_s to decouple the amplifier output from C_L, and a small feedback capacitance C_f to provide a high-frequency bypass around C_L as well as to combat the effect of any stray input capacitance C_n. It is possible to specify the compensation network so that the phase lead introduced by C_f exactly neutralizes the phase lag due to C_L. The design equations for neutral compensation are[5]

$$R_s = (R_1/R_2)r_o \qquad C_f = (1 + R_1/R_2)^2 (r_o/R_2)C_L \qquad (8.20a)$$

and the closed-loop bandwidth is $f_B \cong 1/2\pi R_2 C_f$. In the case of voltage-follower

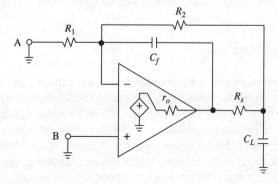

FIGURE 8.15
Stabilizing a capacitively loaded op amp circuit.

operation, where $R_1 = \infty$ and $R_2 = 0$, a convenient alternative is provided by the design equations[6]

$$R_s = 30r_o \qquad C_f = \sqrt{C_L/18\pi r_o \beta f_t} \qquad (8.20b)$$

where f_t is the transition frequency of the op amp and $\beta = 1$ V/V for the voltage follower. The closed-loop bandwidth is now $f_B \cong \sqrt{\beta f_t/18\pi r_o C_L}$.

> **EXAMPLE 8.5.** (a) Assuming the op amp of Fig. 8.14a has GBP = 10 MHz and $r_o = 100\,\Omega$, specify component values for operation as an inverting amplifier with $A_0 = -2$ V/V and $C_L = 5$ nF. (b) Find $A(jf)$.
>
> **Solution.**
>
> (a) For $A_0 = -2$ V/V, use $R_1 = 10$ kΩ and $R_2 = 20$ kΩ, and insert the input source at node A. Then, Eq. (8.20a) yields $R_s = 50\ \Omega$ and $C_f \cong 56$ pF.
> (b) We have $f_{-3\,\text{dB}} = 1/2\pi R_2 C_f \cong 140$ kHz. An additional breakpoint occurs at $f_x = \beta \times$ GBP $= (1/3)10^7 = 3.33$ MHz. Consequently,
>
> $$A(jf) = \frac{-2}{[1 + jf/(140 \times 10^3)][1 + jf/(3.33 \times 10^6)]}\ \text{V/V}$$

We observe that since R_s is inside the feedback loop, its presence does not degrade dc accuracy appreciably. However, R_s should be kept suitably small to avoid excessive output-swing reduction and excessive slew-rate degradation. In a practical op amp the open-loop output impedance tends to behave inductively at high frequencies, so the above equations provide only initial estimates for R_s and C_f. The optimum values must be found empirically once the circuit has been assembled in the lab.[6]

An alternative way of stabilizing a capacitively loaded amplifier is via the input-lag method, to be discussed in Section 8.4. The need to drive capacitive loads arises frequently enough to warrant the design of special op amps with provisions for automatic capacitive-load compensation. The AD817 (Analog Devices) and LT1360 (Linear Technology) op amps are designed to drive unlimited capacitive loads. Special internal circuitry senses the amount of loading and adjusts the open-loop response to maintain an adequate phase margin regardless of the load. The process, completely transparent to the user, is most effective when the load is not fixed or is ill-defined, as in the case of unterminated coaxial cable loads.

Other Sources of Instability

In circuitry incorporating high-gain amplifiers such as op amps and voltage comparators, the specter of instability arises in a number of subtle ways unless proper circuit design and construction rules are followed.[7-10] Two common causes of instability are *poor grounding* and *inadequate power-supply filtering*. Both problems stem from the distributed impedances of the supply and ground busses, which can provide spurious feedback paths around the high-gain device and compromise its stability.

In general, to minimize the ground-bus impedance, it is good practice to use a ground plane, especially in audio and wideband applications. To reduce grounding problems further, it is good practice to provide two separate ground busses: a *signal-ground* bus to provide a return path for critical circuits—such as signal sources, feedback networks, and precision voltage references—and a *power-ground* bus to provide a return path for less critical circuits, such as high-current loads and digital circuits. Every effort is made to keep both dc and ac currents on the signal-ground bus *small* in order to render this bus essentially equipotential. To avoid perturbing this equipotential condition, the two busses are joined only at one point of the circuit.

Spurious feedback paths can also form through the power-supply busses. Because of nonzero bus impedances, any change in supply currents brought about by a load current change will induce a corresponding voltage change across the op amp supply pins. Due to finite PSRR, this change will in turn be felt at the input, thus providing an indirect feedback path. To break this path, each supply voltage must be bypassed with a 0.01-μF to 0.1-μF decoupling capacitor, in the manner already depicted in Fig. 1.36. The best results are obtained with low ESR and ESL ceramic chip capacitors, preferably surface-mounted. For this cure to be effective, the lead lengths must be kept short and the capacitors must be mounted as close as possible to the op amp pins. Likewise, the elements of the feedback network must be mounted close to the inverting-input pin in order to minimize the stray capacitance C_{ext} in Eq. (8.16). Manufacturers often provide evaluation boards to guide the user in the proper construction of the circuit.

8.3
INTERNAL FREQUENCY COMPENSATION

If we were to remove the 30-pF capacitor from the 741 op amp, we would end up with an uncompensated device. Such a device has indeed been marketed as the 748 op amp for those users who prefer custom compensation. Another highly popular uncompensated contemporary is the 301 op amp.

With the low-frequency dominant pole removed, an uncompensated op amp exhibits much higher bandwidth, but also much greater phase shift due to various high-frequency poles and zeros. Such a device is unstable in most applications, so efforts must be made to stabilize it. The overall response of an uncompensated op amp is the result of its individual internal-stage responses, and can be rather complex. For illustration purposes, however, the following three-pole approximation is generally satisfactory,

$$a(jf) = \frac{a_0}{(1 + jf/f_1)(1 + jf/f_2)(1 + jf/f_3)} \tag{8.21}$$

The magnitude plot of Fig. 8.16 (top) shows also important phase values, which have been associated with slope using Eq. (8.7). Note that GBP is constant only for $f_1 < f < f_2$.

Suppose we apply *frequency-independent feedback* around such an op amp. With this type of feedback the $1/\beta$ curve is flat, so we can visualize the $|T|$ curve as the $|a|$ curve, but with the $|1/\beta|$ line as the new 0-dB axis. As long as $1/\beta \geq |a(jf_{-135°})|$, the rate of closure is ROC ≤ 30 dB/dec, indicating a phase margin $\phi_m \geq 45°$. For $|a(jf_{-135°})| \geq 1/\beta \geq |a(jf_{-180°})|$ we have 30 dB/dec $\leq$ ROC ≤ 40 dB/dec, or

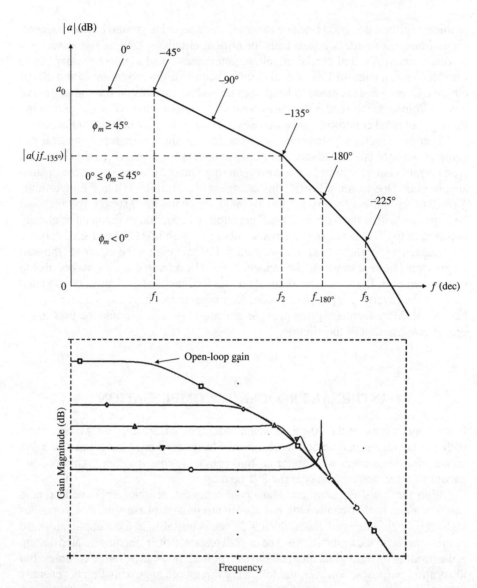

FIGURE 8.16
A three-pole open-loop response, showing a correspondence between phase shift and slope.

$45° \geq \phi_m \geq 0°$, indicating an inadequate degree of stability. Finally, for $1/\beta \leq |a(jf_{-180°})|$ we have ROC $\geq$ 40 dB/dec, or $\phi_m < 0$, indicating oscillatory behavior. Figure 8.16 (bottom) illustrates how peaking increases as we reduce $1/\beta$.

It is apparent that uncompensated op amps provide adequate phase margins only in high-gain applications. For instance, we must have $1/\beta \geq |a(jf_{-135°})|$ for $\phi_m \geq 45°$. To accommodate lower closed-loop gains, frequency compensation is needed. This is achieved by modifying either the open-loop response $a(jf)$ (internal compensation), or the feedback factor $\beta(jf)$ (external compensation), or a combination of both, as in decompensated amplifiers (see Section 8.4).

EXAMPLE 8.6. The μA702, the first monolithic op amp, had[11] $a_0 = 3600$ V/V, $f_1 = 1$ MHz, $f_2 = 4$ MHz, and $f_3 = 40$ MHz. Find (a) $|a(jf_{-135°})|$, and (b) $|a(jf_{-180°})|$.

Solution.

(a) Start out with the estimate $f_{-135°} = 4$ MHz. Then, use the trial-and-error technique of Example 8.1 to find $f_{-135°} = 4.78$ MHz, and $|a(jf_{-135°})| \cong 470$ V/V. An uncompensated 702 circuit is stable with $\phi_m = 45°$ only for $|1/\beta| \geq 470$ V/V.

(b) Similarly, $|a(jf_{-180°})| = |a(j14.3 \text{ MHz})| = 63.7$ V/V, indicating that for $|1/\beta| \leq 63.7$ V/V the circuit oscillates.

Figure 8.17 shows a three-pole op amp model that we shall use as the basis of our discussion as well as PSpice simulations.

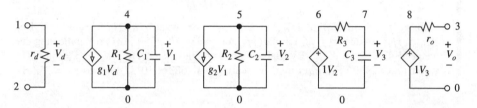

FIGURE 8.17
Three-pole op amp model, consisting of two transconductance stages and a voltage buffer.

Dominant-Pole Compensation

The objective of this method is the deliberate creation of a pole at a sufficiently low frequency f_d to ensure a rolloff rate of -20 dB/dec all the way up to the crossover frequency f_x. Figure 8.18 provides a graphical means for finding f_d. First, we draw the $|1/\beta|$ curve corresponding to the required closed-loop gain. Next, we locate point X corresponding to the desired f_x. For $\phi_m = 45°$, let $f_x = f_1$. From X we draw a line with a slope of -20 dB/dec until it intercepts the dc gain asymptote at point D.

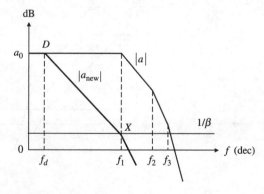

FIGURE 8.18
Dominant-pole compensation.

The abscissa of D is f_d. By the constancy of the GBP we have $a_0 f_d = (1/\beta) f_x$, or

$$f_d = \frac{f_x}{\beta a_0} \tag{8.22}$$

It is apparent that dominant-pole compensation causes a drastic gain reduction above f_d. But, this is the price we are paying for stability!

> **EXAMPLE 8.7.** Find f_d to make the μA702 op amp of Example 8.6 unconditionally stable with $\phi_m = 45°$.
>
> **Solution.** After creating the new pole frequency we have
>
> $$a_{\text{new}}(jf) = \frac{1}{1 + jf/f_d} a(jf)$$
>
> with $a(jf)$ as in Eq. (8.21). For $\phi_m = 45°$ we want $\sphericalangle a_{\text{new}}(jf_x) = -135°$. But, $\sphericalangle a_{\text{new}}(jf_x) = -\tan^{-1}(f_x/f_d) + \sphericalangle a(jf_x)$, or $-135° \cong -90° + \sphericalangle a(jf_x)$, indicating that we need $\sphericalangle a(jf_x) = -45°$. By trial and error we find that $\sphericalangle a = -45°$ at $f = 683$ kHz, where $|a| = 2930$ V/V. Imposing $1 = 2930/\sqrt{1 + (683 \times 10^3/f_d)^2}$ gives $f_d = 233$ Hz.

Shunt-Capacitance Compensation

The above discussion assumes that a fourth pole is added to the open-loop response, and that the existing poles are unaffected by this procedure. For the purpose of maximizing bandwidth, it is more efficient to rearrange the existing poles rather than create a new one. Specifically, if we decrease f_1 until f_x coincides with f_2, as in Fig. 8.19b, then the open-loop bandwidth will be improved by the factor f_2/f_1 compared to Fig. 8.18. A pole frequency is decreased by adding capacitance to the internal node causing it. Referring to Fig. 8.17, we observe that the equivalent resistance and capacitance of node V_1 form a low-pass function with the pole frequency $f_1 = 1/2\pi R_1 C_1$. If we deliberately add an external capacitance C_c as shown for the first-stage model of Fig. 8.19a, then f_1 is changed to $f_{1(\text{new})} = 1/2\pi R_1 (C_1 + C_c)$.

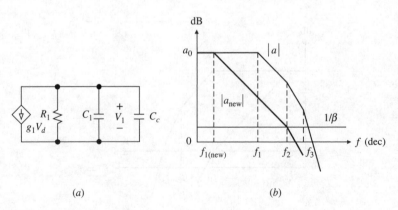

(a)   (b)

FIGURE 8.19
Dominant-pole compensation using a shunt capacitance C_c.

Rewriting Eq. (8.22) as $f_{1(new)} = f_2/\beta a_0$ gives, for $f_{1(new)} \ll f_1$,

$$C_c \cong \frac{\beta a_0}{2\pi R_1 f_2} \qquad (8.23)$$

If the 741 op amp of Fig. 5.1 were not already compensated, a proper place to connect the shunt capacitance would be between the base of Q_5 and the negative-supply rail. Note that adding shunt capacitance to a node usually affects also the other pole frequencies,[11] a feature not explicitly conveyed by the simplified model of Fig. 8.17. Consequently, it may be necessary to calculate or measure the new value of f_2 and perform a few iterations to find the correct value of C_c.

EXAMPLE 8.8. In the op amp model of Fig. 8.17 let $r_d = 1$ MΩ, $g_1 = 2$ mA/V, $R_1 = 100$ kΩ, $g_2 = 10$ mA/V, $R_2 = 50$ kΩ, and $r_0 = 100$ Ω. (a) If the open-loop response has three pole frequencies at $f_1 = 100$ kHz, $f_2 = 1$ MHz, and $f_3 = 10$ MHz, find the dominant pole f_d and shunt capacitance C_c needed for operation as a voltage follower with $\phi_m = 45°$. (b) Repeat, but for operation as a unity-gain inverting amplifier.

Solution.

(a) By inspection, $a_0 = g_1 R_1 g_2 R_2 = 10^5$ V/V, and $C_1 = 1/2\pi R_1 f_1 = 15.9$ pF. For $\beta = 1$ V/V we get $f_{1(new)} = f_2/\beta a_0 = 10$ Hz and $C_c = 159$ nF.

(b) Now $\beta = 0.5$ V/V, so $f_{1(new)} = 20$ Hz and $C_c = 79.6$ nF.

Miller Compensation

Given the low-frequency nature of the dominant pole, the value of the shunt capacitance C_c tends to be too large for monolithic fabrication. As mentioned in Chapters 5 and 6, this drawback is overcome by placing C_c in the feedback path of one of the internal stages to take advantage of the multiplicative action of the Miller effect for capacitance. Luckily, another unexpected benefit accrues from this connection, namely, *pole splitting*.[11,12]

In Fig. 8.20a, C_c has been placed in the feedback path of the second stage, which, for the 741 op amp, is the Darlington stage depicted in Fig. 5.1. In the absence of C_c,

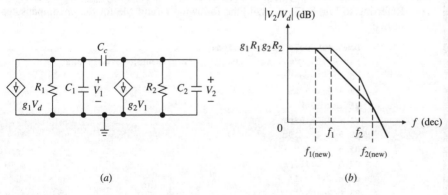

(a)   (b)

FIGURE 8.20
Miller compensation and pole splitting.

the circuit provides the pole frequency $f_1 = 1/2\pi R_1 C_1$ at the input, and the pole frequency $f_2 = 1/2\pi R_2 C_2$ at the output. With C_c present, a detailed ac analysis[12] (see Problem 8.30) yields

$$\frac{V_2}{V_d} \cong g_1 R_1 g_2 R_2 \frac{1 - jf/f_z}{\left(1 + jf/f_{1(\text{new})}\right)\left(1 + jf/f_{2(\text{new})}\right)} \tag{8.24}$$

where $f_z = g_2/2\pi C_c$, and

$$f_{1(\text{new})} \cong \frac{1}{2\pi R_1 g_2 R_2 C_c} \qquad f_{2(\text{new})} \cong \frac{g_2 C_c}{2\pi (C_1 C_2 + C_c C_1 + C_c C_2)} \tag{8.25}$$

Equation (8.24) reveals the presence of a *positive* real zero at $s = 2\pi f_z$, thus providing an example of a circuit that is not a minimum-phase system. This zero stems from direct signal transmission through C_c to the output, and its effect is to reduce ϕ_m. However, in bipolar op amps f_z is usually high enough to warrant approximating $1 - jf/f_z \cong 1$ over the useful frequency range.

Equation (8.25) indicates that increasing C_c lowers $f_{1(\text{new})}$ and raises $f_{2(\text{new})}$, causing the poles to split apart. *Pole splitting,* depicted in Fig. 8.20b, is highly beneficial since the shift in f_2 eases the amount of shift required of f_1, thus allowing for a wider bandwidth. We also note that the dominant-pole frequency is due to the familiar Miller-multiplied capacitance $g_2 R_2 C_c$, which combines with the input node resistance R_1 to form $f_{1(\text{new})}$.

EXAMPLE 8.9. Repeat (*a*) and (*b*) of Example 8.8, but using a feedback capacitance C_c. (*c*) Use PSpice to compare the two compensations for the voltage-follower case.

Solution.

(*a*) $C_1 = 15.9$ pF, $C_2 = 1/2\pi R_2 f_2 = 3.18$ pF. To find $f_{1(\text{new})}$ we need to know $f_{2(\text{new})}$. Assume $C_c \gg C_2$, so we can estimate $f_{2(\text{new})} \cong g_2/[2\pi(C_1 + C_2)] \cong 83$ MHz. Since this is much higher than f_3, which is 10 MHz, we impose $f_x = f_3 = 10$ MHz for $\phi_m \cong 45°$. Then, $f_{1(\text{new})} = f_3/\beta a_0 = 100$ Hz, which gives $C_c = 1/2\pi R_1 g_2 R_2 f_{1(\text{new})} = 31.8$ pF. By Eq. (8.25), $f_{2(\text{new})} \cong 77$ MHz; moreover, $f_z = g_2/2\pi C_c = 50$ MHz, confirming that both f_z and $f_{2(\text{new})}$ are well above f_x. It can be shown (see Problem 8.31) that the actual values of f_x and ϕ_m are 7.9 MHz and 36.7°.

(*b*) Now $\beta = 0.5$ V/V, so $f_{1(\text{new})} = 200$ Hz, $C_c = 15.9$ pF, and $f_{2(\text{new})} \cong 71$ MHz.

(*c*) Referring to Fig. 8.17, we write the following circuit file for the uncompensated device.

```
Dominant-pole compensation:
*a0 = 100 V/mV, f1 = 100 kHz, f2 = 1 MHz, f3 = 10 MHz
rd 1 2 1Meg
g1 4 0 1 2 2m
R1 4 0 100k
C1 4 0 15.92pF
g2 5 0 4 0 10m
R2 5 0 50k
C2 5 0 3.183pF
e3 6 0 5 0 1
R3 6 7 10k
C3 7 0 1.592pF
```

```
e0 8 0 7 0 1
ro 8 3 100
vi 1 0 ac 1 pulse (0 1V 0 10ns 10ns 2us 4us)
Rf 2 3 1k
.ac dec 10 1Hz 100MegHz
.tran 0.1us 2us
.probe ;a = V(3)/V(1,2), vO(t) = v(3)
.end
```

To compensate with a shunt capacitance C_c we add the statement

```
Cc 4 0 159nF
```

whereas to compensate with a feedback capacitance C_c we add

```
Cc 4 5 31.8pF
```

The results of the simulation are shown in Fig. 8.21.

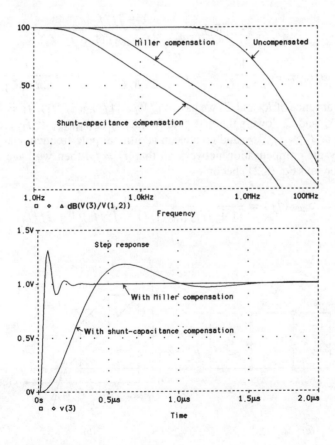

FIGURE 8.21
Frequency and transient responses for Example 8.9.

Unconditionally stable op amps are compensated for $\beta = 1$ V/V. Since this requires the lowest dominant-pole value and, hence, the largest C_c, these op amps are of necessity compensated conservatively. When used with $\beta < 1$ V/V, they tend to be wasteful in terms of bandwidth and slew rate since a smaller value of C_c would suffice. Custom compensation may then prove a better alternative.

Pole-Zero Compensation

An alternative dominant-pole compensation technique is *pole-zero cancellation*. This technique, shown for the first-stage model of Fig. 8.22*a*, uses a capacitance $C_c \gg C_1$ to significantly lower the first-pole frequency f_1, and a resistance $R_c \ll R_1$ to create a zero frequency that is used to cancel the second-pole frequency f_2. The compensated response is then dominated by the lowered first-pole frequency up to f_3, which, for $\phi_m = 45°$, becomes the new crossover frequency. To see how this comes about, note that the transfer function is now $V_1/V_d = -g_1[R_1 \| (1/j2\pi f C_1) \| (R_c + 1/j2\pi f C_c)]$. After expanding (see Problem 8.33), we get, for $C_c \gg C_1$ and $R_c \ll R_1$,

$$\frac{V_1}{V_d} \cong (-g_1 R_1) \frac{1 + jf/f_z}{(1 + jf/f_{1(\text{new})})(1 + jf/f_4)} \tag{8.26}$$

$$f_{1(\text{new})} \cong \frac{1}{2\pi R_1 C_c} \qquad f_z = \frac{1}{2\pi R_c C_c} \qquad f_4 \cong \frac{1}{2\pi R_c C_1} \tag{8.27}$$

In the absence of R_c and C_c we have $V_1/V_d = 1/(1 + jf/f_1)$, $f_1 = 1/2\pi R_1 C_1$. Inserting R_c and C_c lowers the first-pole frequency to $f_{1(\text{new})} \ll f_1$, creates a zero frequency at $f_z \gg f_{1(\text{new})}$, and creates an additional pole frequency at $f_4 \gg f_z$. If we specify the compensation network so that $f_z = f_2$, then we have a zero-pole cancellation and Eq. (8.21) becomes

$$a_{\text{new}}(jf) = \frac{a_0}{(1 + jf/f_{1(\text{new})})(1 + jf/f_3)(1 + jf/f_4)}$$

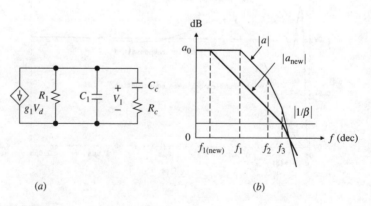

(a) (b)

FIGURE 8.22
Pole-zero compensation.

For $\phi_m = 45°$ we let $f_x = f_3$. Then, rewriting Eq. (8.22) as $f_{1(\text{new})} = f_3/\beta a_0$, and noting that $f_{1(\text{new})} \ll f_1 < f_2$, we get

$$C_c \cong \frac{\beta a_0}{2\pi R_1 f_3}$$ (8.28a)

Moreover, imposing $f_z = f_2$ yields

$$R_c = 1/2\pi C_c f_2$$ (8.28b)

Comparing Eq. (8.28a) with Eq. (8.23) reveals a bandwidth improvement by the factor f_3/f_2 with respect to the shunt-capacitance method.

EXAMPLE 8.10. Use the pole-zero method to compensate the op amp of Example 8.8 for $\phi_m = 45°$ and $\beta = 1$.

Solution. $C_c = 10^5/(2\pi \times 10^5 \times 10^7) = 15.9$ nF, $R_c = 1/(2\pi \times 15.9 \times 10^{-9} \times 10^6) = 10\ \Omega$. Note, incidentally, that $f_4 \cong 1$ GHz, thus justifying our choice of f_3 as the crossover frequency.

Feedforward Compensation

In a multistage amplifier the overall phase shift at f_x is the result of its individual-stage phase contributions. Usually there is one stage that acts as a bandwidth bottleneck by contributing a substantial amount of phase shift. Feedforward compensation creates a high-frequency bypass around this bottleneck stage to suppress its phase contribution in the vicinity of f_x and thus increase the phase margin.

The principle is illustrated in Fig. 8.23a, where the overall gain a of the uncompensated amplifier is expressed as the product of the gain a_1 of the bottleneck stage and the gain a_2 of the remaining stages lumped together. The bypass around

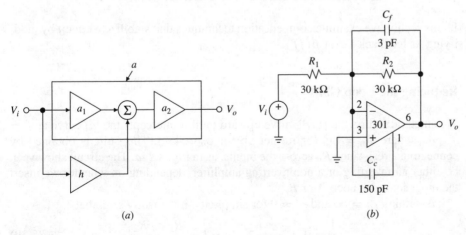

(a)   (b)

FIGURE 8.23
Feedforward compensation and implementation example.

the bottleneck stage is a high-pass function of the type

$$h(jf) = \frac{jf/f_0}{1 + jf/f_0}$$

so that

$$a_{\text{new}}(jf) = [a_1(jf) + h(jf)]a_2(jf)$$

At low frequencies, where $|h| \ll |a_1|$, we have $a_{\text{new}} \cong a_1 a_2 = a$, indicating that the high-gain advantages of the uncompensated response still hold there. However, at high frequencies, where $|a_1| \ll |h|$, we now have $a_{\text{new}} \cong a_2$, indicating a wider bandwidth as well as a lower phase shift because the dynamics are now controlled by a_2 alone.

Problems may arise[1,13] in the frequency region where a_{new} makes its transition from $a_1 a_2$ to a_2. If $\sphericalangle a_{\text{new}}$ approaches $-180°$ before the transition, excessive ringing may develop. Furthermore, a phase shift of $-180°$ at the transition causes signal cancellation at the summing node, thus creating a notch in the compensated response.

Feedforward compensation is implemented with a capacitive bypass around the bottleneck stage. This is shown in Fig. 8.22b for the case of the 301 op amp.[14] In this device the bandwidth bottleneck is the input stage because of the presence of lateral *pnp* transistors, whose frequency characteristics are notoriously poor. Connecting C_c between the inverting input (pin 2) and the input to the second stage (pin 1) bypasses the input stage by creating a high-pass function with $f_0 = 1/2\pi R_{\text{eq}}C_c$, where R_{eq} is the equivalent resistance seen by C_c.

Since only signals at the inverting input are transmitted to the second stage, feedforward compensation provides a much lower bandwidth for signals applied to the noninverting input. Consequently, this compensation is worthwhile only in inverting applications. Note also the presence of the feedback capacitance C_f to combat the effect of stray capacitance at the inverting input.

8.4
EXTERNAL FREQUENCY COMPENSATION

In this section we examine compensation techniques that stabilize a circuit by modifying its feedback factor $\beta(jf)$.

Reducing the Loop Gain

This method[1] shifts the $|1/\beta|$ curve upward until it intercepts the $|a|$ curve at $f = f_{-135°}$, where $\phi_m = 45°$ (or further up for $\phi_m > 45°$). This shift is obtained by connecting a resistance R_c across the inputs, as in Fig. 8.24a. The circuit shown can be either an inverting or a noninverting amplifier, depending on whether we insert the input source at node A or B.

Assuming $r_d = \infty$ and $r_o = 0$ for simplicity, it is readily seen that

$$\frac{1}{\beta} = 1 + \frac{R_2}{(R_1 \parallel R_c)} = 1 + \frac{R_2}{R_1} + \frac{R_2}{R_c} \tag{8.29}$$

By choosing R_c suitably small, we can move the $1/\beta$ curve up until $1/\beta = |a(jf_2)|$,

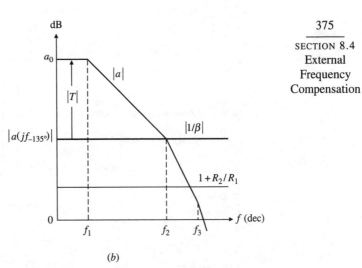

FIGURE 8.24
Frequency compensation via loop-gain reduction.

where $\phi_m = 45°$. This is shown in Fig. 8.24b. Solving for R_c yields

$$R_c = \frac{R_2}{|a(jf_2)| - (1 + R_2/R_1)} \qquad (8.30)$$

If $\phi_m \neq 45°$ is desired, then replace f_2 with $f_{\phi_m - 180°}$, where ϕ_m is the desired phase margin, and $f_{\phi_m - 180°}$ is the frequency at which $\angle a = \phi_m - 180°$.

It should be pointed out that the presence of R_c does not affect A_{ideal} in the relation $A = A_{\text{ideal}}/(1 + 1/T)$; R_c only reduces T, resulting in a larger gain error. Moreover, the much increased dc noise gain may result in an intolerable dc output error E_O. Again, these are the prices we are paying for stability!

> **EXAMPLE 8.11.** An op amp with $a_0 = 10^5$ V/V, $f_1 = 10$ kHz, $f_2 = 3$ MHz, and $f_3 = 30$ MHz is to be used as an inverting amplifier with $R_1 = 10\,\text{k}\Omega$ and $R_2 = 100\,\text{k}\Omega$. Find ($a$) R_c for $\phi_m \cong 45°$, (b) the dc gain error, (c) the dc output error E_O if the total input dc error is $E_I = 1$ mV, and (d) the closed-loop -3-dB frequency.
>
> **Solution.**
>
> (a) We calculate $|a(jf_2)| = 234.5$ V/V. Then, Eq. (8.30) gives $R_c = 447.4\,\Omega$ (use $430\,\Omega$).
> (b) Letting $R_c = 430\,\Omega$ in Eq. (8.29) gives $1/\beta \cong 244$ V/V. Then, $a_0\beta = 10^5/244 \cong 410$, indicating a dc gain error $\epsilon_0 \cong -100/a_0\beta = -0.24\%$.
> (c) $E_O = (1/\beta)E_I = 244 \times 1 = 244$ mV, quite an error!
> (d) $f_{-3\,\text{dB}} = f_2 = 3$ MHz.

Input-Lag Compensation

The high dc noise-gain drawback of the previous method is overcome[1] by placing a capacitance C_c in series with R_c, as in Fig. 8.25a. At high frequencies, where C_c acts as a short compared to R_c, the $|1/\beta|$ curve is unchanged compared to

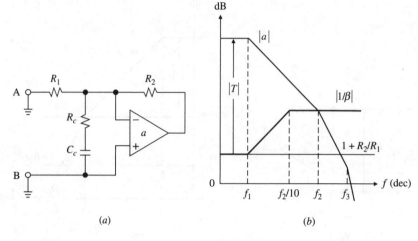

FIGURE 8.25
Input-lag compensation.

Eq. (8.29). However, at low frequencies, where C_c acts as an open, we now have $|1/\beta_0| = 1 + R_2/R_1$. Since this is much lower than the high-frequency value, we now have a much higher dc loop gain and a much lower dc output error.

To avoid degrading ϕ_m, it is good practice[1] to position the second breakpoint of the $|1/\beta|$ curve about a decade below f_2. To find the required value of C_c, we observe that at this breakpoint we have $|Z_{C_c}| = R_c$, or $1/(2\pi C_c f_2/10) = R_c$. Solving gives

$$C_c = \frac{5}{\pi R_c f_2} \tag{8.31}$$

where R_c is given in Eq. (8.30). Again, if $\phi_m \neq 45°$ is desired, replace f_2 with $f_{\phi_m - 180°}$.

EXAMPLE 8.12. (a)–(d) Repeat Example 8.11, but using input-lag compensation. (e) Estimate the actual value of ϕ_m after compensation. (f) Confirm with PSpice.

Solution.

(a) We have $R_c = 447.4\ \Omega$ (use 430 Ω) and $C_c = 5/(\pi 447.4 \times 3 \times 10^6) = 1.186$ nF (use 1.2 nF).

(b) The dc noise gain is now $1/\beta_0 = 1 + R_2/R_1 = 11$ V/V, and $a_0\beta_0 = 10^5/11 = 9091$. Hence, $\epsilon_0 \cong -100/9091 = -0.011\%$.

(c) $E_0 = 11E_I = 11$ mV—quite an improvement!

(d) $f_{-3\,\text{dB}} = 3$ MHz, as before.

(e) As we approach the crossover frequency, $1/\beta$ behaves like a high-pass function, so $1/\beta(jf) \cong 244(jf/0.1f_2)/(1 + jf/0.1f_2)$, with $0.1f_2 = 300$ kHz. The loop gain $T = a\beta$ is then

$$T = \frac{410[1 + jf/(3 \times 10^5)]}{[1 + jf/10^4][1 + jf/(3 \times 10^6)][1 + jf/(3 \times 10^7)][jf/(3 \times 10^5)]}$$

Following Example 8.1, we find that $|T| = 1$ for $f = 2.94$ MHz, where $\angle T = -145.6°$. Thus, $\phi_m = 34.4°$, a reasonable value. If desired, it can be increased by reducing R_c and raising C_c.

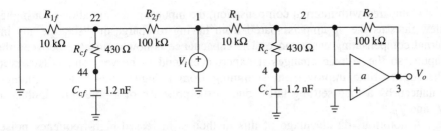

FIGURE 8.26
PSpice circuit of Example 8.12.

(*f*) With reference to Fig. 8.26, we write the following file.

```
Input-lag compensation:
Vi 1 0 ac 1V
*Main circuit:
R1 1 2 10k
R2 2 3 100k
Rc 2 4 430
Cc 4 0 1.2nF
*a0 = 100V/mV, f1 = 10kHz, f2 = 3MHz, f3 = 30MHz:
ea 3 0 Laplace {V(0,2)} = {1E5/((1+s/628E2)*(1+s/188E5)*(1+s/188E6))}
*Circuit to plot 1/beta:
R2f 1 22 100k
R1f 22 0 10k
Rcf 22 44 430
Ccf 44 0 1.2nF
.ac dec 10 1kHz 100MegHz
.probe ;a = V(3)/V(0,2), 1/beta = V(1)/V(22), A = V(3)/V(1)
.end
```

The results of the simulation are shown in Fig. 8.27.

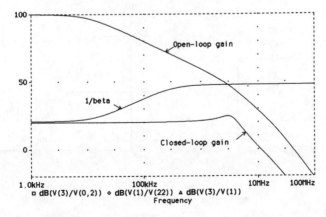

FIGURE 8.27
Frequency plots for the circuit of Fig. 8.26.

Compared with internal compensation, the input-lag method allows for higher slew rates as the op amp is spared from having to charge or discharge any internal compensating capacitance. The capacitance is now connected between the inputs, so the voltage changes it experiences tend to be very small. However, the settling-time improvement stemming from a higher slew rate is counter-balanced by a long settling tail[15] due to the presence of a pole-zero doublet at f_z and f_p.

A notorious disadvantage of this method is increased high-frequency noise, since the noise-gain curve is raised significantly in the vicinity of the crossover frequency f_x. Another disadvantage is a much lower closed-loop differential input impedance Z_d, since z_d is now in parallel with $Z_c = R_c + 1/j2\pi f C_c$, and Z_c is much smaller than z_d. Though this is inconsequential in inverting configurations, it may cause intolerable high-frequency input loading and feedthrough in noninverting configurations.

Input-lag compensation is nevertheless popular. It is also used in connection with constant-GBP op amps as an alternative to the capacitive-load isolation technique discussed in Section 8.2. We still apply Eqs. (8.30) and (8.31), but with f_2 replaced by $f_p = 1/2\pi r_o C_L$. An additional application of the input-lag method is the stabilization of decompensated op amps, discussed at the end of this section.

Feedback-Lead Compensation

This technique[1] uses a feedback capacitance C_f to create phase lead in the feedback path. This lead is designed to occur in the vicinity of the crossover frequency f_x, which is where ϕ_m needs to be boosted. Alternatively, we can view this method as a reshaping of the $|1/\beta|$ curve near f_x to reduce the rate of closure ROC. Referring to Fig. 8.28a and assuming $r_d = \infty$ and $r_o = 0$, we have $1/\beta = 1 + Z_2/R_1$,

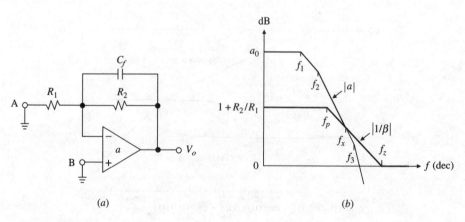

(a) (b)

FIGURE 8.28
Feedback-lead compensation.

$Z_2 = R_2 \parallel (1/j2\pi f C_f)$. Expanding, we can write

$$\frac{1}{\beta(jf)} = \left(1 + \frac{R_2}{R_1}\right) \frac{1 + jf/f_z}{1 + jf/f_p} \qquad (8.32)$$

where $f_p = 1/2\pi R_2 C_f$ and $f_z = (1 + R_2/R_1)f_p$. As depicted in Fig. 8.28b, $|1/\beta|$ has the low- and high-frequency asymptotes $|1/\beta_0| = 1 + R_2/R_1$ and $|1/\beta_\infty| = 0$ dB, and two breakpoints at f_p and f_z.

The phase lag provided by $1/\beta(jf)$ is maximum[1] at the geometric mean of f_p and f_z, so the optimum value of C_f is the one that makes this mean coincide with the crossover frequency, or $f_x = \sqrt{f_p f_z} = f_p\sqrt{1 + R_2/R_1}$. Under such a condition we have $|a(jf_x)| = \sqrt{1 + R_2/R_1}$, which can be used to find f_x via trial and error. Once f_x is known, we find $C_f = 1/2\pi R_2 f_p$, or

$$C_f = \frac{\sqrt{1 + R_2/R_1}}{2\pi R_2 f_x} \qquad (8.33)$$

The closed-loop bandwidth is $1/2\pi R_2 C_f$. Moreover, C_f helps combat the effect of the inverting-input stray capacitance C_n.

One can readily verify that at the geometric mean of f_p and f_z we have $\sphericalangle(1/\beta) = 90° - 2\tan^{-1}\sqrt{1 + R_2/R_1}$, so the larger the value of $1 + R_2/R_1$, the greater the contribution of $1/\beta$ to ϕ_m. For example, with $1 + R_2/R_1 = 10$ we get $\sphericalangle(1/\beta) = 90° - 2\tan^{-1}\sqrt{10} \cong -55°$, which yields $\sphericalangle T = \sphericalangle a - (-55°) = \sphericalangle a + 55°$. We observe that for this compensation scheme to work with a given ϕ_m, the open-loop gain must satisfy $\sphericalangle a(jf_x) \geq \phi_m - 90° - 2\tan^{-1}\sqrt{1 + R_2/R_1}$.

EXAMPLE 8.13. (a) Using an op amp with $a_0 = 10^5$ V/V, $f_1 = 1$ kHz, $f_2 = 100$ kHz, and $f_3 = 5$ MHz, design a noninverting amplifier with $A_0 = 20$ V/V. Hence, verify that the circuit needs compensation. (b) Stabilize it with the feedback-lead method, and find ϕ_m. (c) Find the closed-loop bandwidth.

Solution.

(a) For $A_0 = 20$ V/V use $R_1 = 1.05$ kΩ and $R_2 = 20.0$ kΩ. Then $\beta_0 = 1/20$ V/V, and $a_0\beta_0 = 10^5/20 = 5000$. Thus, without compensation we have

$$T(jf) = \frac{5000}{[1 + jf/10^3][1 + jf/10^5][1 + jf/(5 \times 10^6)]}$$

Using trial and error as in Example 8.1, we find that $|T| = 1$ for $f = 700$ kHz, and that $\sphericalangle T(j700 \text{ kHz}) = -179.8°$. So, $\phi_m = 0.2°$, indicating a circuit in bad need of compensation.

(b) Using again trial and error we find that $|a| = \sqrt{20}$ V/V for $f = 1.46$ MHz, and $\sphericalangle a(j1.46 \text{ MHz}) = -192.3°$. Letting $f_x = 1.46$ MHz in Eq. (8.33) yields $C_f = 24.3$ pF. Moreover, $\phi_m = 180° + \sphericalangle a - (90° - 2\tan^{-1}\sqrt{20}) = 180° + (-192.3°) - (90° - 2 \times 77.4°) = 52.5°$.

(c) $f_{-3\text{dB}} = 1/2\pi R_2 C_f = 327$ kHz.

We observe that feedback-lead compensation does not enjoy the slew-rate advantages of input-lag compensation; however, it provides better filtering capabilities for internally generated noise. These are some of the factors the user needs to consider when deciding which method is best for a given application.

Decompensated Op Amps

These op amps are compensated for unconditional stability only when used with $1/\beta$ above a specified value, such as $1/\beta \geq (1/\beta)_{min} = 5$ V/V, or $\beta \leq \beta_{max} = 0.2$ V/V. Consequently, they provide a constant GBP only for $|a| \geq (1/\beta)_{min}$. Being less conservatively compensated, decompensated op amps offer higher GBPs and SRs. For instance, the fully compensated LF356 op amp uses $C_c \cong 10$ pF to provide GBP $= 5$ MHz and SR $= 12$ V/μs for any $|a| \geq 1$ V/V. The LF357, its decompensated version, uses $C_c \cong 3$ pF to provide GBP $= 20$ MHz and SR $= 50$ V/μs, but only for $|a| \geq 5$ V/V.

We observe that the constraint $1/\beta \geq (1/\beta)_{min}$ need be satisfied only in the vicinity of the crossover frequency; elsewhere we can shape the $1/\beta$ curve as we please. For instance, we can use input-lag compensation to operate a decompensated op amp at values of $1/\beta$ below $(1/\beta)_{min}$ while retaining the high-speed advantages of decompensation. To this end, we still use Eqs. (8.30) and (8.31), but with $|a(jf_2)|$ replaced by $(1/\beta)_{min}$, f_2 replaced by $\beta_{max} \times$ GBP, and R_2 replaced by R_f.

> **EXAMPLE 8.14.** Figure 8.29 shows a common way of configuring a decompensated op amp as a voltage follower. It is apparent that at low frequencies, where C_c acts as an open circuit, we have $A_0 = 1$ V/V. (a) Given that the 357 op amp is compensated for $(1/\beta)_{min} = 5$ V/V, specify suitable components to stabilize the circuit. (b) Find $A(jf)$.
>
> **Solution.**
>
> (a) By Eq. (8.30), $R_c = R_f/(5 - 1 - R_f/\infty) = R_f/4$. Let $R_c = 3$ kΩ and $R_f = 12$ kΩ. Also, $f_x \cong \beta_{max} \times$ GBP $= (1/5) \times 20 = 4$ MHz, so $C_c = 5/(\pi \times 3 \times 10^3 \times 4 \times 10^6) \cong 133$ pF (use 130 pF).
>
> (b) $A(jf) \cong 1/[1 + jf/(4 \text{ MHz})]$ V/V.

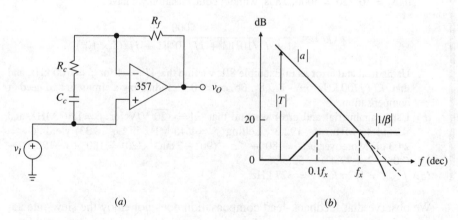

(a)   (b)

FIGURE 8.29
Configuring a decompensated op amp as a unity-gain voltage follower.

8.5

STABILITY IN CFA CIRCUITS[16]

381

SECTION 8.5
Stability in CFA
Circuits

The open-loop response $z(jf)$ of a current-feedback amplifier (CFA) is dominated by a single pole only over a designated frequency band. Beyond this band,
higher-order roots come into play, which increase the overall phase shift. When
frequency-independent feedback is applied around a CFA, the latter will offer unconditional stability with a specified phase margin ϕ_m only as long as $1/\beta \geq (1/\beta)_{min} =
|z(jf_{\phi_m-180°})|$, where $f_{\phi_m-180°}$ is the frequency at which $\sphericalangle z = \phi_m - 180°$. Lower-
ing the $1/\beta$ curve below $(1/\beta)_{min}$ would increase the phase shift, thus eroding ϕ_m
and inviting instability. This behavior is similar to that of decompensated op amps.
The value of $(1/\beta)_{min}$ can be found from the data-sheet plots of $|z(jf)|$ and $\sphericalangle z(jf)$.
As with voltage-feedback amplifiers (VFAs), instability in CFA circuits may also
stem from feedback phase lag due to external reactive elements.

Effect of Feedback Capacitance

To investigate the effect of feedback capacitance, refer to Fig. 8.30a. At low frequencies, C_f acts as an open circuit, so we can apply Eq. (6.58) and write $1/\beta_0 =
R_2 + r_n(1 + R_2/R_1)$. At high frequencies, R_2 is shorted out by C_f, so $1/\beta_\infty =
1/\beta_0|_{R_2 \to 0} = r_n$. Since $1/\beta_\infty \ll 1/\beta_0$, the crossover frequency f_x is pushed into
the region of greater phase shift, as shown in Fig. 8.30b. If this shift reaches $-180°$,
the circuit will oscillate.

We thus conclude that *direct capacitive feedback must be avoided in CFA circuits.* In particular, the familiar inverting or Miller integrator is not amenable to CFA
implementation, unless suitable measures are taken to stabilize it (see Problem 8.44).
However, the noninverting or Deboo integrator is acceptable because β in the vicinity of f_x is still controlled by the resistance in the negative-feedback path. Likewise,
we can readily use CFAs in those filter configurations that do not employ any direct
capacitance between the output and the inverting input, such as *KRC* filters.

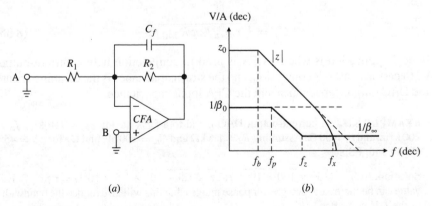

(a)   (b)

FIGURE 8.30
A large feedback capacitance C_f tends to destabilize a CFA circuit.

Stray Input Capacitance Compensation

In Fig. 8.31a C_n appears in parallel with R_1. Replacing R_1 with $R_1 \parallel (1/j2\pi f C_n)$ in Eq. (6.58) yields, after minor algebra,

$$\frac{1}{\beta} = \frac{1}{\beta_0}(1 + jf/f_z) \tag{8.34a}$$

$$\frac{1}{\beta_0} = R_2 + r_n\left(1 + \frac{R_2}{R_1}\right) \qquad f_z = \frac{1}{2\pi(R_1 \parallel R_2 \parallel r_n)C_n} \tag{8.34b}$$

As shown in Fig. 8.31b, the $1/\beta$ curve starts to rise at f_z, and if C_n is sufficiently large to make $f_z < f_x$, the circuit will become unstable.

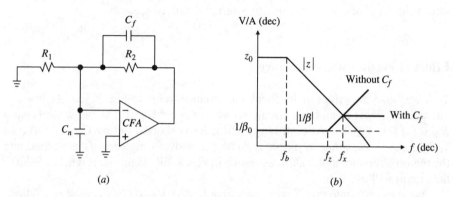

FIGURE 8.31
Input stray capacitance compensation in CFA circuits.

Like a VFA, a CFA is stabilized by counteracting the effect of C_n with a small feedback capacitance C_f. Together with R_2, C_f creates a pole frequency for $1/\beta$ at $f_p = 1/2\pi R_2 C_f$. For $\phi_m = 45°$, impose $f_p = f_x$. We observe that f_x is the geometric mean of f_z and $\beta_0 z_0 f_b$. Letting $1/2\pi R_2 C_f = \sqrt{\beta_0 z_0 f_b f_z}$ and solving, we get

$$C_f = \sqrt{r_n C_n/2\pi R_2 z_0 f_b} \tag{8.35}$$

A typical application is when a CFA is used in conjunction with a current-output DAC to perform fast I-V conversion, and the stray capacitance is the combined result of the DAC output capacitance and the CFA input capacitance.

EXAMPLE 8.15. A current-output DAC is fed to a CFA having $z_0 = 750\,k\Omega$, $f_b = 200\,kHz$, and $r_n = 50\,\Omega$. Assuming $R_2 = 1.5\,k\Omega$ and $C_n = 100\,pF$, find C_f for $\phi_m = 45°$. Verify with PSpice.

Solution. $C_f = \sqrt{50 \times 100 \times 10^{-12}/(2\pi \times 1.5 \times 10^3 \times 1.5 \times 10^{11})} = 1.88\,pF$. This value can be increased for a greater phase margin, but this will also reduce the bandwidth of the I-V converter.

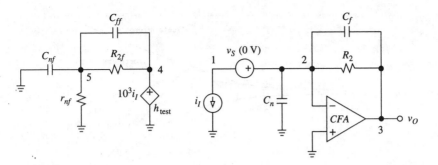

FIGURE 8.32
PSpice circuit of Example 8.15.

Referring to Fig. 8.32 and using the CFA subcircuit of Example 6.17, we write the following circuit file.

```
I-V converter using a CFA:
.subckt CFA vP vN vO
ein 1 0 vP 0 1
rn 1 2 50
vS 2 vN dc 0
fCFA 0 3 vS 1
Req 3 0 750k
Ceq 3 0 1.061pF
eout vO 0 3 0 1
.ends CFA
*Main circuit:
iI 1 0 ac 1mA pulse(0 1mA 0 0.1ns 0.1ns 50ns 100ns)
vS 2 1 dc 0
Cn 2 0 100pF
R2 2 3 1.5k
Cf 2 3 1.88pF
X1 0 2 3 CFA
*Circuit to plot 1/beta:
htest 4 0 vs 1k
R2f 4 5 1.5k
Cff 4 5 1.88pF
Cnf 5 0 100pF
rnf 5 0 50
*Circuit to plot z:
X2 4 6 7 CFA
Rs 6 0 100
RL 7 0 1Meg
.ac dec 100 1MegHz 1GHz
.tran 1ns 50ns
.probe ;A = V(3)/I(Vs), z = V(7)/I(Rs), 1/beta = V(4)/I(rnf), vO = v(3)
.end
```

The results of the simulation are shown in Fig. 8.33.

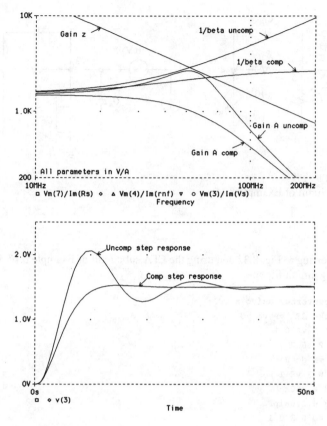

FIGURE 8.33
Frequency and transient responses of the circuit of Fig. 8.32.

8.6
COMPOSITE AMPLIFIERS

Two or more op amps can be combined to achieve improved overall performance.[17] The designer need be aware that when an op amp is placed within the feedback loop of another, stability problems may arise. In the following we shall designate the gains of the individual op amps as a_1 and a_2, and the gain of the composite device as a.

Increasing the Loop Gain

Two op amps, usually from a dual-op-amp package, can be connected in cascade to create a composite amplifier with a gain $a = a_1 a_2$ much higher than the individual gains a_1 and a_2. We expect the composite device to provide a much greater loop gain, and thus a much lower gain error. However, if we denote the individual unity-gain frequencies as f_{t1} and f_{t2}, we observe that at high frequencies, where $a = a_1 a_2 \cong (f_{t1}/jf)(f_{t2}/jf) = -f_{t1} f_{t2}/f^2$, the phase shift of the composite response approaches $-180°$, thus requiring frequency compensation.

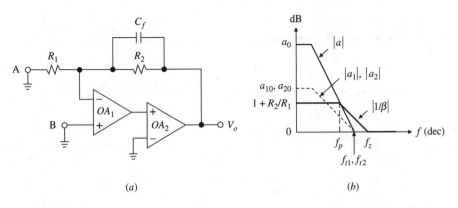

FIGURE 8.34
Composite amplifier with feedback-lead compensation.

In applications with sufficiently high closed-loop dc gains, the composite amplifier can be stabilized via the feedback-lead method[18] shown in Fig. 8.34a. As usual, the circuit can be either an inverting or a noninverting amplifier, depending on whether we insert the input source at node A or B. The decibel plot of $|a|$ is obtained by adding together the individual decibel plots of $|a_1|$ and $|a_2|$. This is illustrated in Fig. 8.34b for the case of matched op amps, or $a_1 = a_2$.

As we know, the $1/\beta$ curve has a pole frequency at $f_p = 1/2\pi R_2 C_f$ and a zero frequency at $f_z = (1 + R_2/R_1)f_p$. For ROC = 30 dB/dec, or $\phi_m = 45°$, we place f_p right on the $|a|$ curve. This yields $1 + R_2/R_1 = |a(jf_p)| = f_{t1}f_{t2}/f_p^2$. Solving for f_p and then letting $C_f = 1/2\pi R_2 f_p$ gives

$$C_f = \sqrt{(1 + R_2/R_1)/f_{t1}f_{t2}}/2\pi R_2 \qquad (8.36)$$

The closed-loop bandwidth is $f_B = f_p$. It can be shown (see Problem 8.46) that increasing C_f by the factor $(1 + R_2/R_1)^{1/4}$ will make the crossover frequency f_x coincide with the geometric mean $\sqrt{f_p f_z}$ and thus maximize ϕ_m; however, this will also decrease the closed-loop bandwidth in proportion.

EXAMPLE 8.16. (a) The circuit of Fig. 8.34a is to be used as a noninverting amplifier with $R_1 = 1$ kΩ and $R_2 = 99$ k. (a) Assuming op amps of the 741 type, find C_f for $\phi_m = 45°$. Then compare ϕ_m, T_0, and f_B with the case of a single-op-amp realization. (b) Find C_f for the maximum phase margin. What are the resulting values of ϕ_m and f_p? (c) What happens if C_f is increased above the value found in (b)?

Solution.

(a) Insert the input source at node B. Letting $f_{t1} = f_{t2} = 1$ MHz in Eq. (8.36) gives $C_f = 16.1$ pF for $\phi_m = 45°$. Moreover, $T_0 = a_0^2/100 = 4 \times 10^8$, and $f_B = f_p = 100$ kHz. Had a single op amp been used, then $\phi_m = 90°$, $T_0 = a_0/100 = 2 \times 10^3$, and $f_B = 10^6/100 = 10$ kHz.

(b) $C_f = (100)^{1/4} \times 16.1 = 50.8$ pF, $f_p = 31.62$ kHz, $\phi_m = 180° + \sphericalangle a - \sphericalangle(1/\beta) \cong 180° - 180° - [\tan^{-1}(f_x/f_z) - \tan^{-1}(f_x/f_p)] = -(\tan^{-1} 0.1 - \tan^{-1} 10) = 78.6°$.

(c) Increasing C_f above 50.8 pF will reduce ϕ_m until eventually $\phi_m \to 0°$, indicating that overcompensation is detrimental.

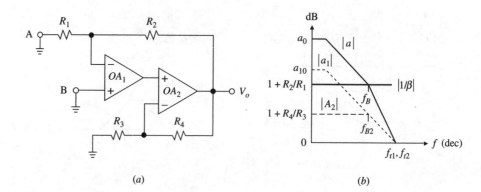

FIGURE 8.35
Composite amplifier with compensation provided by OA_2.

In Fig. 8.34 we have stabilized the composite amplifier by acting on its feedback network. An alternative[19] type of compensation is by controlling the pole of the second op amp using local feedback, in the manner depicted in Fig. 8.35. The composite response $a = a_1 A_2$ has the dc gain $a_0 = a_{01}(1 + R_4/R_3)$, and two pole frequencies at f_{b1} and at $f_{B2} = f_{t2}/(1 + R_4/R_3)$. Without the second amplifier, the closed-loop bandwidth would be $f_{B1} = f_{t1}/(1+R_2/R_1)$. With the second amplifier in place, the bandwidth is expanded to $f_B = (1+R_4/R_3)f_{B1} = f_{t1}(1+R_4/R_3)/(1 + R_2/R_1)$. It is apparent that if we align f_B and f_{B2}, then ROC = 30 dB/dec, or $\phi_m = 45°$. Thus, imposing $f_{t1}(1 + R_4/R_3)/(1 + R_2/R_1) = f_{t2}/(1 + R_4/R_3)$ yields

$$1 + R_4/R_3 = \sqrt{(f_{t2}/f_{t1})(1 + R_2/R_1)} \qquad (8.37)$$

We observe that for the benefits of using OA_2 to be significant the application must call for a sufficiently high closed-loop gain.

EXAMPLE 8.17. (a) Assuming op amps of the 741 type in the circuit of Fig. 8.35a, specify suitable components for operation as an inverting amplifier with a dc gain of -100 V/V. Compare with a single-op-amp realization.

Solution. Insert the input source at node A and let $R_1 = 1\,\text{k}\Omega$ and $R_2 = 100\,\text{k}\Omega$. Then, $R_4/R_3 = \sqrt{101} - 1 = 9.05$. Pick $R_3 = 2\,\text{k}\Omega$ and $R_4 = 18\,\text{k}\Omega$. The dc loop gain is $T_0 = a_{10}(1 + R_4/R_3)/(1 + R_2/R_1) \cong 2 \times 10^4$, and the closed-loop bandwidth is $f_B \cong f_t/10 = 100\,\text{kHz}$. If only one op amp had been used, then $\phi_m \cong 90°$, $T_0 \cong 2 \times 10^3$ and $f_B \cong 10\,\text{kHz}$, indicating an order-of-magnitude improvement brought about by the second op amp.

Optimizing dc and ac Characteristics

There are applications in which it is desirable to combine the dc characteristics of a low-offset, low-noise device, such as a bipolar voltage-feedback amplifier (VFA), with the dynamics of a high-speed device, such as a current-feedback amplifier (CFA). The two sets of technologically conflicting specifications can be met with a composite amplifier. In the topology of Fig. 8.36a we use a CFA with local feedback

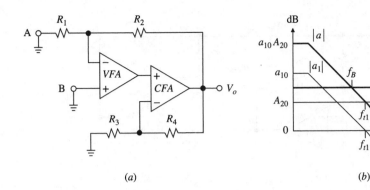

(a)   (b)

FIGURE 8.36
VFA-CFA composite amplifier.

to shift the $|a_1|_{dB}$ curve upward by the amount $|A_2|_{dB}$, and thus improve the dc loop gain by the same amount. As long as $f_{B2} \gg f_{t1}$, the phase shift due to the pole frequency at $f = f_{B2}$ will be insignificant at $f = f_{t1}$, indicating that we can operate the VFA with a feedback factor of unity, or at the maximum bandwidth f_{t1}. Imposing

$$1 + R_4/R_3 = 1 + R_2/R_1 \qquad (8.38)$$

will maximize also the closed-loop bandwidth f_B of the composite device, which is now $f_B = f_{t1}$.

The composite topology offers important advantages other than bandwidth. Since the CFA is operated within the feedback loop of the VFA, its generally poorer input dc and noise characteristics become insignificant when referred to the input of the composite device, where they are divided by a_1. Moreover, with most of the signal swing being provided by the CFA, the slew-rate requirements of the VFA are significantly relaxed, thus ensuring high full-power bandwidth (FPB) capabilities for the composite device. Finally, since the VFA is spared from having to drive the output load, self-heating effects such as thermal feedback become insignificant, so the composite device retains optimum input-drift characteristics.

There are practical limitations to the amount of closed-loop gain achievable with a CFA. Even so, it pays to use a CFA as part of a composite amplifier. For instance, suppose we need an overall dc gain $A_0 = 10^3$ V/V, but using a CFA having only $A_{20} = 50$ V/V. Clearly, the VFA will now have to operate with a gain of $A_0/50 \doteq 20$ V/V and a bandwidth $f_{t1}/20$. This is still 50 times better than if the VFA were to operate alone, not to mention the slew-rate and thermal-drift advantages.

In the arrangement of Fig. 8.36a the composite bandwidth is set by the VFA, so the amplification provided by the CFA above this band is in effect wasted. The alternative topology of Fig. 8.37 exploits the dynamics of OA_2 to their fullest extent by allowing it to participate directly in the feedback mode, but only at high frequencies. The circuit works as follows.

At dc, where the capacitances act as opens, the circuit reduces to that of Fig. 8.34a, so $a_0 = a_{10}a_{20}$. Clearly, the dc characteristics are set by OA_1, which provides OA_2 with whatever drive is needed to force $V_n \to V_{OS1}$. Moreover, any gross bias current at the inverting input of OA_2 is prevented from disturbing node V_n because of the dc blocking action by C_2.

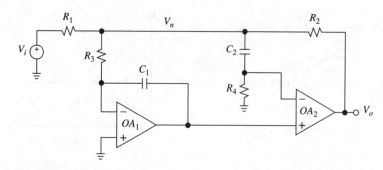

FIGURE 8.37
Composite amplifier enjoying the dc characteristics of OA_1 and the ac characteristics of OA_2.

As we increase the operating frequency, we witness a gradual decrease in OA_1's gain $A_1 = -1/(jf/f_1)$, $f_1 = 1/2\pi R_3 C_1$, while the crossover network $C_2 R_4$ gradually changes the mode of operation of OA_2 from open-loop to closed-loop. Above the crossover network frequency $f_2 = 1/2\pi R_4 C_2$ we can write $V_o \cong a_2(A_1 V_n - V_n)$, or

$$V_o \cong -\frac{a_{20}}{1 + jf/f_{b2}} \frac{1 + jf/f_1}{jf/f_1} V_n$$

It is apparent that if we impose $f_1 = f_{b2}$, or $R_3 C_1 = 1/2\pi f_{b2}$, then we obtain a *pole-zero cancellation* and $V_o = -a V_n$, $a = a_{20}/(jf/f_1) = a_{20} f_{b2}/jf \cong a_2$, indicating that the high-frequency dynamics are fully controlled by OA_2.

In a practical realization the zero-pole cancellation is difficult to maintain because f_{b2} is an ill-defined parameter. Consequently, in response to an input step, the composite device will not completely stabilize until the integrator loop has settled to its final value. The resulting settling tail may be of concern in certain applications.

Improving Phase Accuracy

As we know, a single-pole amplifier exhibits an error function of the type $1/(1 + 1/T) = 1/(1 + jf/f_B)$, whose phase error is $\epsilon_\phi = -\tan^{-1}(f/f_B)$, or $\epsilon_\phi \cong -f/f_B$ for $f \ll f_B$. This error is intolerable in applications requiring high phase accuracy. In the composite arrangement[20] of Fig. 8.38, OA_2 provides active feedback around OA_1 to maintain a low phase error over a much wider bandwidth than in the uncompensated case. This is similar to the active compensation of integrators of Section 6.5.

To analyze the circuit, let $\beta = R_1/(R_1 + R_2)$ and $\alpha = R_3/(R_3 + R_4)$. We note that OA_2 is a noninverting amplifier with gain $A_2 = (1/\beta)/(1 + jf/\beta f_{t2})$. Consequently, the feedback factor around OA_1 is $\beta_1 = \beta \times A_2 \times \alpha = \alpha/(1 + jf/\beta f_{t2})$.

The closed-loop gain of the composite device is $A = A_1 = a_1/(1 + a_1 \beta_1)$, where we are using the fact that OA_1 too is operating in the noninverting mode. Substituting $a_1 \cong f_{t1}/jf$ and $\beta_1 = \alpha/(1 + jf/\beta f_{t2})$, and letting $f_{t1} = f_{t2} = f_t$, we obtain, for $\alpha = \beta$,

$$A(jf) = A_0 \frac{1 + jf/f_B}{1 + jf/f_B - (f/f_B)^2} \tag{8.39}$$

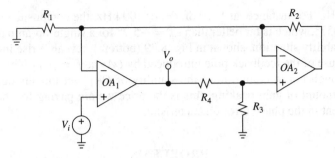

FIGURE 8.38
Composite amplifier with high phase accuracy.

where $A_0 = 1 + R_2/R_1$ and $f_B = f_t/A_0$. As discussed in Section 6.5, this error function offers the advantage of a very small phase error, namely, $\epsilon_\phi = -\tan^{-1}(f/f_B)^3$, or $\epsilon_\phi \cong -(f/f_B)^3$ for $f \ll f_B$.

Figure 8.39 (top) shows the results of the PSpice simulation of a composite amplifier with $A_0 = 10$ V/V using a matched pair of 10-MHz op amps, so that

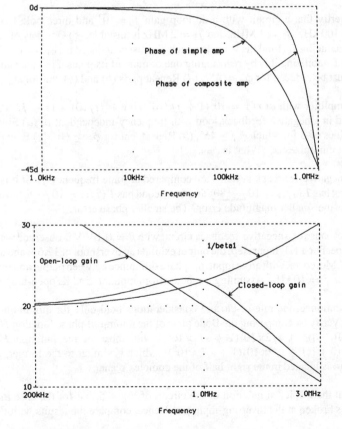

FIGURE 8.39
Frequency plots of the circuit of Fig. 8.38.

$f_B = 1$ MHz. For instance, at $1/10$ of f_B, or 100 kHz, the composite circuit gives $\epsilon_\phi = -0.057°$, which is far better than $\epsilon_\phi = -5.7°$ for a single-op-amp realization.

The stability situation, shown in Fig. 8.39 (bottom), reveals a rise in the $|1/\beta_1|$ curve because of the feedback pole introduced by OA_2 at $f = \beta f_{t2}$. This frequency is high enough not to compromise the stability of OA_1, yet low enough to cause a certain amount of gain peaking: this is the price we are paying for the dramatic improvement in the phase-error characteristic!

PROBLEMS

8.1 The stability problem

8.1 An op amp with $a_0 = 10^3$ V/V and two pole frequencies at $f_1 = 100$ kHz and $f_2 = 2$ MHz is connected as a unity-gain voltage follower. Find ϕ_m, ζ, Q, GP, OS, and $A(jf)$. Would you have much use for this circuit?

8.2 An amplifier has three identical pole frequencies so that $a(jf) = a_0/(1 + jf/f_1)^3$, and is placed in a negative-feedback loop with a frequency-independent feedback factor β. Find an expression for $f_{-180°}$ as well as the corresponding value of T.

8.3 (a) Verify that a circuit with a dc loop gain $T_0 = 10^2$ and three pole frequencies at $f_1 = 100$ kHz, $f_2 = 1$ MHz, and $f_3 = 2$ MHz is unstable. (b) One way of stabilizing it is by reducing T_0. Find the value to which T_0 must be reduced for $\phi_m = 45°$. (c) Another way of stabilizing it is by rearranging one or more of its poles. Find the value to which f_1 must be reduced for $\phi_m = 45°$. (d) Repeat parts (b) and (c), but for $\phi_m = 60°$.

8.4 An amplifier with $a(jf) = 10^5(1 + jf/10^4)/[(1 + jf/10) \times (1 + jf/10^3)]$ V/V is placed in a negative-feedback loop with frequency-independent β. (a) Find the range of values of β for which $\phi_m \geq 45°$. (b) Repeat, but for $\phi_m \geq 60°$. (c) Find the value of β that minimizes ϕ_m. What is $\phi_{m(min)}$?

8.5 Two negative-feedback systems are compared at some frequency f_1. If it is found that the first has $T(jf_1) = 10 \underline{/-180°}$ and the second has $T(jf_1) = 10 \underline{/-90°}$, which system enjoys the smaller magnitude error? The smaller phase error?

8.6 The response of a negative-feedback circuit with $\beta = 0.1$ V/V is observed with the oscilloscope. For a 1-V input step, the output exhibits an overshoot of 12.6% and a final value of 9 V. Moreover, with an ac input, the phase difference between output and input reaches 90° for $f = 10$ kHz. Assuming a 2-pole error amplifier, find its open-loop response.

8.7 As mentioned, the rate-of-closure considerations hold only for minimum-phase systems. Verify by comparing the Bode plots of the minimum-phase function $H(s) = (1 + s/2\pi 10^3)/[(1 + s/2\pi 10)(1 + s/2\pi 10^2)]$ with those of the function $H(s) = (1 - s/2\pi 10^3)/[(1 + s/2\pi 10)(1 + s/2\pi 10^2)]$, which is similar to the former, except that its zero is located in the right half of the complex plane.

8.8 Repeat the PSpice simulation of the circuit of Fig. 8.5, but for the case in which the loop is broken at the inverting-input pin. Hence, compare the results with Fig. 8.6.

8.9 Assuming ideal op amp, derive an expression for the loop gain of the equal-component *KRC* filter of Example 3.8. Hence, discuss the stability of the circuit. What is its gain margin?

8.10 The response of an unconditionally stable op amp can be approximated with a dominant pole frequency f_1 and a single high-frequency pole f_2 to account for the phase shift due to its higher-order roots. (a) Assuming $a_0 = 10^5$ V/V, $f_1 = 10$ Hz, and $\beta = 1$ V/V, find the actual bandwidth f_B and phase margin ϕ_m if $f_2 = 1$ MHz. (b) Find f_2 for $\phi_m = 60°$; what is the value of f_B? (c) Repeat (b), but for $\phi_m = 45°$.

8.11 An op amp with $a(jf) = 10^5/(1 + jf/10)$ is placed in a negative-feedback loop with $\beta(jf) = \beta_0/(1 + jf/10^5)^2$. Find the values of β_0 corresponding to (a) the onset of oscillatory behavior, (b) $\phi_m = 45°$, and (c) GM $= 20$ dB.

8.12 A Howland current pump is implemented with a constant-GBP op amp and four identical resistances. Using rate-of-closure reasoning, show that as long as the load is resistive or capacitive the circuit is stable, but can become unstable if the load is inductive. How would you compensate it?

8.13 Specify R_s in the differentiator of Example 8.2 for $\phi_m = 60°$. Hence, derive an expression for $H(jf)$. What is the value of Q?

8.14 An alternative frequency compensation method for the differentiator of Fig. 8.8a is by means of a suitable feedback capacitance C_f in parallel with R. Assuming $C = 10$ nF, $R = 78.7$ kΩ, and GBP $= 1$ MHz, specify C_f for $\phi_m = 45°$.

8.15 The noninverting differentiator of Fig. P3.2 uses an op amp with GBP $= 1$ MHz. If $R = 78.7$ kΩ and $C = 10$ nF, verify that the circuit needs compensation. How would you stabilize it?

8.16 (a) Show that the circuit of Fig. 8.10a gives $A = -R_2/R_1 \times H_{LP}$, where H_{LP} is the standard second-order low-pass response defined in Eq. (3.44) with $f_0 = \sqrt{\beta_0 f_t f_z}$ and $Q = \sqrt{\beta_0 f_t/f_z}/(1 + \beta_0 f_t/f_p)$. (b) Find Q in the circuit of Example 8.3 before compensation. (c) Compensate the circuit for $\phi_m = 45°$, and find Q after compensation.

8.17 In the circuit of Example 8.3 find C_f for $\phi_m = 60°$; hence, exploit Problem 8.16 to find $A(jf)$, GP, and OS.

8.18 An alternative way of stabilizing a circuit against stray input capacitance C_n is by scaling down all resistances to raise f_z until $f_z \geq f_x$. (a) Scale the resistances of the circuit of Example 8.3 so that with $C_f = 0$ the circuit yields $\phi_m = 45°$. (b) Repeat, but for $\phi_m = 60°$. (c) What is the main advantage and disadvantage of this technique?

8.19 The high-sensitivity I-V converter of Fig. 2.2 uses $R = 1$ MΩ, $R_1 = 1$ kΩ, $R_2 = 10$ kΩ, and the LF351 JFET-input op amp, which has GBP $= 4$ MHz. (a) Assuming an overall input stray capacitance $C_n = 10$ pF, show that the circuit does not have enough phase margin. (b) Find a capacitance C_f that, when connected between the output and the inverting input, will provide neutral compensation. What is the closed-loop bandwidth of the compensated circuit?

8.20 Using the op amp data of Example 8.5, find the maximum C_L that can be connected to the output of the circuit of Fig. 8.14a and still allow for $\phi_m \geq 45°$ if (a) $R_1 = R_2 = 20$ kΩ, (b) $R_1 = 2$ kΩ, $R_2 = 18$ kΩ, (c) $R_1 = \infty$, $R_2 = 0$. (d) Repeat (c), but for $\phi_m \geq 60°$.

8.21 Using PSpice, check the frequency and transient response of the circuit of (a) Example 8.4 and (b) Example 8.5.

8.22 Using the op amp data of Example 8.5, design an amplifier with $A_0 = +10$ V/V, under the constraint that the sum of all resistances be 200 kΩ, and that it be capable of driving a 10-nF load. Then use PSpice to verify its frequency and transient responses.

8.23 Modify the circuit of Example 8.5 for unity-gain voltage-follower operation. Then use PSpice to find GP and OS.

8.24 Assuming constant-GBP op amps, use linearized Bode plots to investigate the stability of (a) the wideband band-pass filter of Fig. 3.11, (b) the multiple-feedback low-pass filter of Fig. 3.32, and (c) the -KRC low-pass filter of Problem 3.27.

8.25 (a) Assuming the op amp has a constant GBP of 1 MHz, discuss the stability of the multiple-feedback band-pass filter of Fig. 3.31, and verify with PSpice. (b) Repeat, but for the -KRC band-pass filter of Problem 3.28 for the case $R_1 = R_2 = 1.607$ kΩ, $kR_2 = 1.445$ MΩ, and $C_1 = C_2 = 3.3$ nF.

8.3 Internal frequency compensation

8.26 Find f_d to stabilize the $\mu A702$ op amp of Example 8.6 for a noninverting gain of 10 V/V with (a) $\phi_m = 60°$, (b) GM = 12 dB, (c) GP = 2 dB, (d) OS = 5%.

8.27 A voltage comparator is a high-gain amplifier intended for open-loop operation. Figure P8.27 shows a way of configuring such a device as a voltage follower. (a) Assuming a two-pole device with $a_0 = 10^3$ V/V, $f_1 = 1$ MHz, and $f_2 = 10$ MHz, use rate-of-closure reasoning to show that the circuit can be stabilized by making the product RC sufficiently large. (b) Assuming a FET-input device, specify R and C for $\phi_m = 45°$. (c) Estimate the small-signal bandwidth.

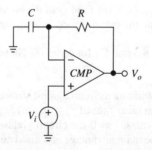

FIGURE P8.27

8.28 An amplifier has $a_0 = 10^4$ V/V, a dominant-pole frequency $f_1 = 1$ kHz, and an adjustable higher-order pole frequency f_2. Find β and f_2 for a maximally flat closed-loop response with a dc gain of 60 dB. What is the -3-dB frequency?

8.29 In Fig. P8.29 three CMOS inverters are cascaded to create a rudimentary op amp, which, in turn, is configured as an ac-coupled inverting amplifier with a closed-loop gain of -100 V/V. (a) Assuming $a_1 = a_2 = a_3 = -10^2/(1 + jf/10^5)$, show that with $R_c = C_c = 0$ the circuit is unstable. (b) Specify suitable values for R_c and C_c to provide dominant-pole stabilization with $\phi_m = 45°$.

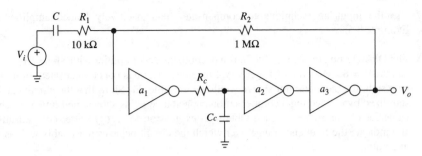

FIGURE P8.29

8.30 Referring to Fig. 8.20a, apply KCL at nodes V_1 and V_2, and then eliminate V_1 to find an expression for the transfer function V_2/V_d. Hence, prove Eqs. (8.24) and (8.25). *Hint:* Given two characteristic frequencies f_1 and f_2 such that $f_1 \ll f_2$, you can approximate $(1 + jf/f_1)(1 + jf/f_2) \cong 1 + jf/f_1 - f^2/f_1f_2$.

8.31 For the op amp of Example 8.9a calculate the actual values of f_x and ϕ_m after compensation. Then verify that the effect of the zero f_z is to *reduce* the phase margin by 9°.

8.32 (a) An op amp has a dominant pole at $s = -2\pi f_1$, and two additional poles at $s = -2\pi f_2$ and $s = -2\pi f_3$, $f_3 = 10 \times \text{GBP}$. Show that for $\phi_m \geq 60°$ we must have $f_2 \geq 2.2 \times \text{GBP}$. (b) An op amp has a dominant pole at $s = -2\pi f_1$, a second pole at $s = -2\pi f_2$, and a zero at $s = +2\pi f_z$, $f_z = 10 \times \text{GBP}$. Show that for $\phi_m \geq 45°$ we must have $f_2 \geq 1.2 \times \text{GBP}$.

8.33 Prove Eqs. (8.26) and (8.27). Use the hint of Problem 8.30.

8.34 Use PSpice to verify the pole-zero compensation scheme of Example 8.10. Show both the frequency and transient responses.

8.4 External frequency compensation

8.35 The op amp of Example 8.11 is configured as a unity-gain inverting amplifier with two 100-kΩ resistances. Use input-lag compensation to stabilize it for $\phi_m = 45°$. Hence, find $A(jf)$.

8.36 In Fig. P8.36 let $R_1 = R_2 = R_4 = 100$ kΩ, $R_3 = 10$ kΩ, and let the op amp have $a_0 = 10^5$ V/V, $f_1 = 10$ kHz, $f_2 = 200$ kHz, and $f_3 = 2$ MHz. (a) Verify that the circuit is unstable. (b) Use input-lag compensation to stabilize it for $\phi_m = 45°$. (c) Find the closed-loop bandwidth after compensation.

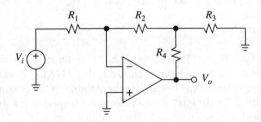

FIGURE P8.36

8.37 Use the input-lag technique to compensate the capacitively loaded amplifier of Example 8.5.

8.38 The OPA637 op amp of Fig. P8.38 is a decompensated amplifier with SR = 135 V/μs and GBP = 80 MHz for $1/\beta \geq 5$ V/V. Since the op amp is not compensated for unity-gain stability, the integrator shown would be unstable. (*a*) Show that the circuit can be stabilized by connecting a compensation capacitance C_c as shown, and find a suitable value for C_c for $\phi_m = 45°$. (*b*) Obtain an expression for $H(jf)$ after compensation and indicate the frequency range over which the circuit behaves reasonably well as an integrator.

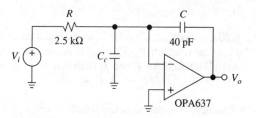

FIGURE P8.38

8.39 An op amp with GBP = 6 MHz and $r_o = 30 \ \Omega$ is to operate as a unity-gain voltage follower with an output load of 5 nF. Design an input-lag network to stabilize it. Then verify its frequency and transient responses via PSpice.

8.40 Using a decompensated op amp with GBP = 80 MHz and $\beta_{max} = 0.2$ V/V, design a unity-gain inverting amplifier, and find $A(jf)$.

8.41 Using an LF357 decompensated op amp, which has GBP = 20 MHz and $\beta_{max} = 0.2$ V/V, design an *I-V* converter with a sensitivity of 0.1 V/μA under the following constraints: (*a*) no compensation capacitances are allowed, and (*b*) the closed-loop bandwidth must be maximized. Then find an expression for $A(jf)$.

8.42 An op amp with $a_0 = 10^6$ V/V and two coincident pole frequencies $f_1 = f_2 = 10$ Hz is configured as an inverting amplifier with $R_1 = 1 \ k\Omega$ and $R_2 = 20 \ k\Omega$. (*a*) Use feedback-lead compensation to stabilize it for $\phi_m = 45°$; then find $A(jf)$. (*b*) Find the value of C_f that will maximize ϕ_m; next find ϕ_m as well as the corresponding closed-loop bandwidth.

8.43 The wideband band-pass filter of Example 3.5 is implemented with an op amp having $a_0 = 10^5$ V/V and two pole frequencies $f_1 = 10$ Hz and $f_2 = 2$ MHz. Sketch the Bode plots of $|a|$ and $|1/\beta|$ in the vicinity of f_x and find ϕ_m.

8.5 Stability in CFA circuits

8.44 The CFA integrator of Fig. P8.44 uses a series resistance R_2 between the summing junction and the inverting-input pin to ensure $1/\beta \geq (1/\beta)_{min}$ over frequency and thus avoid instability problems. (*a*) Investigate the stability of the circuit using Bode plots. (*b*) Assuming the CFA parameters of Problem 6.57, specify suitable components for $f_0 = 1$ MHz. (*c*) List possible disadvantages of this circuit.

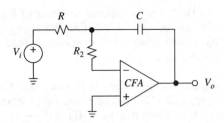

FIGURE P8.44

8.45 The CFA of Problem 6.57 is to be used to design a Butterworth band-pass filter with $f_0 = 10$ MHz and $H_{0BP} = 0$ db, and two alternatives are being considered, namely, the multiple-feedback and the *KRC* designs. Which configuration are you choosing, and why? Show the final circuit.

8.46 (*a*) Show that without C_f the CFA *I-V* converter of Fig. 8.32 yields $V_o/I_i = R_2 H_{LP}$, where H_{LP} is the standard second-order low-pass response defined in Eq. (3.44) with $f_0 = (z_0 f_b/2\pi r_n R_2 C_n)^{1/2}$ and $Q = z_0 f_b/(r_n + R_2) f_0$. (*b*) Predict the GP and OS for the circuit of Example 8.15 before compensation.

8.47 A certain CFA has $r_n = 50\ \Omega$ and an open-loop dc gain of 1 V/μA, and its frequency response can be approximated with two pole frequencies, one at 100 kHz and the other at 100 MHz. The CFA is to be used as a unity-gain voltage follower. (*a*) Find the feedback resistance needed for a phase margin of 45°; what is the closed-loop bandwidth? (*b*) Repeat, but for a 60° margin.

8.6 Composite amplifiers

8.48 (*a*) With reference to the circuit of Fig. 8.34*a*, show that ϕ_m is maximized for $C_f = (1 + R_2/R_1)^{3/4}/[2\pi R_2(f_{t1} f_{t2})^{1/2}]$. (*b*) Show that for $\phi_{m(max)} \geq 45°$ we must have $1 + R_2/R_1 \geq \tan^2 67.5° = 5.8$. (*c*) Assuming 741 op amps, specify suitable component values for operation as an inverting amplifier with $A_0 = -10$ V/V and maximum phase margin. Hence, find the actual values of ϕ_m and $A(jf)$.

8.49 (*a*) Compare the circuit of Example 8.16 with a circuit implemented by cascading two amplifiers with individual dc gains $A_{10} = A_{20} = \sqrt{|A_0|}$ V/V. (*b*) Repeat, but for the circuit of Example 8.17.

8.50 An alternative to Eq. (8.37) is $1 + R_4/R_3 = \sqrt{(1 + R_2/R_1)/2}$, where we have assumed $f_{t1} = f_{t2}$. (*a*) Verify that this alternative yields $\phi_m \cong 65°$. (*b*) Apply it to the design of a composite amplifier with dc gain $A_0 = -50$ V/V. (*c*) Assuming $f_{t1} = f_{t2} = 4.5$ MHz, find $A(jf)$.

8.51 In the composite amplifier of Fig. 8.37 assume OA_1 has $a_{10} = 100$ V/mV, $f_{t1} = 1$ MHz, $V_{OS1} \cong 0$, and $I_{B1} \cong 0$, and OA_2 has $a_{20} = 25$ V/mV, $f_{t2} = 500$ MHz, $V_{OS2} = 5$ mV, and $I_{B2} = 20\ \mu$A. Specify suitable components for $A_0 = -10$ V/V, under the constraint $f_2 = 0.1 f_1$. What is the output dc error E_O and the closed-loop bandwidth f_B?

8.52 For the circuit of Problem 8.51 find the total rms output noise E_{no} if $e_{n1} = 2$ nV/$\sqrt{\text{Hz}}$, $i_{n1} = 0.5$ pA/$\sqrt{\text{Hz}}$, $e_{n2} = 5$ nV/$\sqrt{\text{Hz}}$, and $i_{n2} = 5$ pA/$\sqrt{\text{Hz}}$. Ignore $1/f$ noise. Can you reduce E_{no}?

8.53 (*a*) Find ϕ_m, GP, and OS for the composite amplifier of Fig. 8.38. (*b*) Find its $1°$ phase-error bandwidth, and compare it with that of a single-op-amp realization with the same value of A_0, as well as with that of the cascade realization of two amplifiers with individual dc gains $\sqrt{A_0}$.

8.54 The active-compensation scheme of Fig. P8.54 (see *IEEE Trans. Circuits Syst.*, vol. CAS-26, Feb. 1979, pp. 112–117) works for both the inverting and the noninverting mode of operation of OA_1. Show that $V_o = [(1/\beta)V_2 + (1 - 1/\beta)V_1]/(1 + 1/T)$, $1/(1 + 1/T) = (1 + jf/\beta_2 f_{t2})/(1 + jf/\beta f_{t1} - f^2/\beta f_{t1}\beta_2 f_{t2})$, $\beta = R_1/(R_1 + R_2)$, $\beta_2 = R_3/(R_3 + R_4)$.

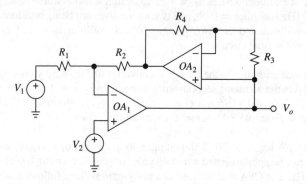

FIGURE P8.54

8.55 Apply the scheme of Problem 8.54 to the design of a high-phase-accuracy (*a*) voltage follower, (*b*) I-V converter with a sensitivity of 10 V/mA, and (*c*) difference amplifier with a dc gain of 100 V/V. Assume matched op amps with $f_t = 10$ MHz.

REFERENCES

1. J. K. Roberge, *Operational Amplifiers: Theory and Practice,* John Wiley & Sons, New York, 1975.
2. R. C. Dorf, *Modern Control Systems,* Addison-Wesley, Reading, MA, 1967.
3. S. Rosenstark, *Feedback Amplifier Principles,* Macmillan, New York, 1986.
4. J. G. Graeme, "Phase Compensation Counteracts Op Amp Input Capacitance," *EDN,* January 6, 1994, pp. 97–104.
5. S. Franco, "Simple Techniques Provide Compensation for Capacitive Loads," *EDN,* June 8, 1989, pp. 147–149.
6. J. Graeme, "Phase Compensation Extends Op Amp Stability and Speed," *EDN,* September 16, 1991, pp. 181–192.
7. J. Williams, "High-Speed Amplifier Techniques," Application Note AN-47, *Linear Applications Handbook Volume II,* Linear Technology, Milpitas, CA, 1993.
8. A. P. Brokaw, "An IC Amplifiers User's Guide to Decoupling, Grounding, and Making Things Go Right for a Change," Application Note AN-202, *Applications Reference Manual,* Analog Devices, Norwood, MA, 1993.
9. A. P. Brokaw, "Analog Signal Handling for High Speed and Accuracy," Application Note AN-342, *Applications Reference Manual,* Analog Devices, Norwood, MA, 1993.
10. J.-H. Broeders, M. Meywes, and B. Baker, "Noise and Interference," *1996 Design Seminar,* Burr-Brown, Tucson, AZ, 1996.

11. P. R. Gray and R. G. Meyer, *Analysis and Design of Analog Integrated Circuits,* 3d ed., John Wiley & Sons, New York, 1993.

12. J. E. Solomon, "The Monolithic Operational Amplifier: A Tutorial Study," *IEEE J. Solid-State Circuits,* vol. SC-9, Dec. 1974, pp. 314–332.

13. J. G. Graeme, G. E. Tobey, and L. P. Huelsman, *Operational Amplifiers: Design and Applications,* McGraw-Hill, New York, 1971.

14. R. J. Widlar, "Feedforward Compensation Speeds Op Amp," Linear Brief LB-2, *Linear Applications Handbook,* National Semiconductor, Santa Clara, CA, 1994.

15. J. Dostál, *Operational Amplifiers,* 2d ed., Butterworth-Heinemann, Stoneham, MA, 1993.

16. Based on the author's article "Current-Feedback Amplifiers Benefit High-Speed Designs," *EDN,* January 5, 1989, pp. 161–172. ©Cahners Publishing Company, 1997, a Div. of Reed Elsevier Inc.

17. J. Williams, "Composite Amplifiers," Application Note AN-21, *Linear Applications Handbook Volume I,* Linear Technology, Milpitas, CA, 1990.

18. J. Graeme, "Phase Compensation Perks Up Composite Amplifiers," *Electronic Design,* August 19, 1993, pp. 64–78.

19. J. Graeme, "Composite Amplifier Hikes Precision and Speed," *Electronic Design Analog Applications Issue,* June 24, 1993, pp. 30–38.

20. J. Wong, "Active Feedback Improves Amplifier Phase Accuracy," *EDN,* September 17, 1987.

9

NONLINEAR CIRCUITS

9.1 Voltage Comparators
9.2 Comparator Applications
9.3 Schmitt Triggers
9.4 Precision Rectifiers
9.5 Analog Switches
9.6 Peak Detectors
9.7 Sample-and-Hold Amplifiers
     Problems
     References

All circuits encountered so far are designed to behave linearly. Linearity is achieved by (*a*) using negative feedback to force the op amp to operate within its linear region and (*b*) implementing the feedback network with linear elements.

Using a high-gain amplifier with positive feedback, or even with no feedback at all, causes the device to operate primarily in saturation. This bistable behavior is highly nonlinear and forms the basis of voltage-comparator and Schmitt-trigger circuits.

Nonlinear behavior can also be achieved by implementing the feedback network with nonlinear elements, such as diodes and analog switches. Common examples include precision rectifiers, peak detectors, and sample-and-hold amplifiers. Another class of nonlinear circuits exploits the predictable exponential characteristic of the BJT to achieve a variety of nonlinear transfer characteristics, such as logarithmic amplification and analog multiplication. This category of nonlinear circuits will be investigated in Chapter 13.

9.1
VOLTAGE COMPARATORS

399

SECTION 9.1
Voltage
Comparators

The function of a voltage comparator is to compare the voltage v_P at one of its inputs against the voltage v_N at the other, and output either a low voltage V_{OL} or a high voltage V_{OH} according to

$$v_O = V_{OL} \qquad \text{for } v_P < v_N \qquad (9.1a)$$

$$v_O = V_{OH} \qquad \text{for } v_P > v_N \qquad (9.1b)$$

As shown in Fig. 9.1a, the symbolism used for comparators is the same as for op amps. We observe that while v_P and v_N are *analog* variables because they can assume a continuum of values, v_O is a *binary* variable because it can assume only one of two values, V_{OL} or V_{OH}. It is fair to view the comparator as a one-bit analog-to-digital converter.

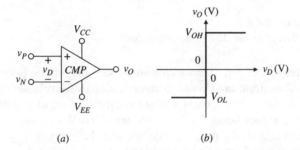

(a)   (b)

FIGURE 9.1
Voltage-comparator symbolism and ideal VTC. (All node
voltages are referenced to ground.)

Introducing the differential input voltage $v_D = v_P - v_N$, the above equations can also be expressed as $v_O = V_{OL}$ for $v_D < 0$ V, and $v_O = V_{OH}$ for $v_D > 0$ V. The voltage transfer curve (VTC), shown in Fig. 9.1b, is a nonlinear curve. At the origin, the curve is a vertical segment, indicating an infinite gain there, or $v_O/v_D = \infty$. A practical comparator can only approximate this idealized VTC, with actual gains being typically in the range from 10^3 to 10^6 V/V. Away from the origin, the VTC consists of two horizontal lines positioned at $v_O = V_{OL}$ and $v_O = V_{OH}$. These levels need not necessarily be symmetric, though symmetry may be desirable in certain applications. All that matters is that the two levels be sufficiently far apart to make their distinction reliable. For example, digital applications require $V_{OL} \cong 0$ V and $V_{OH} \cong 5$ V.

The Response Time

In high-speed applications it is of interest to know how rapidly a comparator responds as the input state changes from $v_P < v_N$ to $v_P > v_N$, and vice versa. Comparator speed is characterized in terms of the *response time,* also called the *propagation delay* t_{PD}, defined as the time it takes for the output to accomplish 50% of its transition in response to a predetermined voltage step at the input. Figure 9.2 illustrates the setup

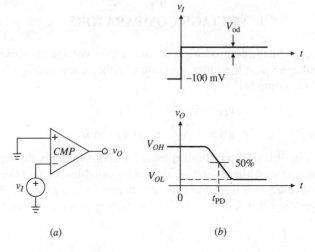

FIGURE 9.2
The response time of a comparator.

for the measurement of t_{PD}. Though the input step magnitude is typically on the order of 100 mV, its limits are chosen to barely exceed the level required to cause the output to switch states. This excess voltage is called the *input overdrive* V_{od}, with typical overdrive values being 1 mV, 5 mV, and 10 mV. In general t_{PD} decreases with V_{od}. Depending on the particular device and the value of V_{od}, t_{PD} can range from a few microseconds to a few nanoseconds.

The Op Amp as a Voltage Comparator

When speed is not critical, an op amp can make an excellent comparator,[1] especially in view of the extremely high gains and low input offsets available from many popular op amp families. The VTC of a practical op amp was depicted in Fig. 1.39, where we expressed v_D in microvolts in order to be able to visualize the slope of the VTC in the linear region. In comparator applications v_D can be a hefty signal, so it is more appropriate express it in volts than in microvolts. If we do so, the horizontal scale undergoes so much compression that the linear-region portion of the VTC coalesces with the vertical axis, resulting in a curve of the type of Fig. 9.1*b*.

The circuit of Fig. 9.3*a* uses a 301 op amp to compare v_I against some voltage threshold, V_T. When $v_I < V_T$ the circuit gives $v_O = -V_{sat} \cong -13$ V, and when $v_I > V_T$ it gives $v_O = +V_{sat} \cong +13$ V. This is illustrated in the figure via both the VTC and the voltage waveforms. Since v_O goes high whenever v_I rises above V_T, the circuit is aptly called a *threshold detector*. If $V_T = 0$ V, the circuit is referred to as a *zero-crossing detector*.

It is important to realize that when used as a comparator, the op amp has no control over v_N due to the absence of feedback. The amplifier now operates in the open-loop mode and, because of its extremely high gain, it spends most of its time in saturation. Clearly, v_N no longer tracks v_P!

Though the output transitions in Fig. 9.3*c* have been shown as instantaneous, we know that in practice they take some time due to slew-rate limiting. Had we

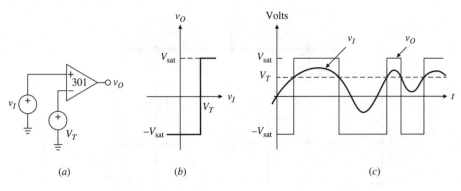

FIGURE 9.3
Threshold detector.

used a 741 op amp, the time to accomplish 50% of the output transition would have been $t_R = V_{sat}/SR = (13 \text{ V})/(0.5 \text{ V}/\mu s) = 26 \ \mu s$, an intolerably long time in many applications. The reason for using the 301 op amp is that it comes without the internal frequency-compensation capacitance C_c, so it slews more rapidly than the 741 op amp. Frequency compensation is indispensable in negative-feedback applications but is superfluous in open-loop applications, where it only slows down the comparator unnecessarily.

Whether internally compensated or not, op amps are intended for negative-feedback operation, so their dynamics are not necessarily optimized for open-loop operation. Moreover, their output saturation levels are generally awkward to interface to digital circuitry. These and other needs peculiar of the voltage-comparison operation have provided the motivation for developing a category of high-gain amplifiers specifically optimized for this operation and thus called voltage comparators.

General-Purpose IC Comparators

Figure 9.4 depicts one of the earliest and most popular voltage comparators, the LM311 (National Semiconductor). The input stage consists of the *pnp* emitter followers Q_1 and Q_2 driving the differential pair Q_3-Q_4. The output of this pair is further amplified by the Q_5-Q_6 pair and then by the Q_7-Q_8 pair, from which it emerges as a single-ended current drive for the base of the output transistor Q_O. Circuit operation is such that for $v_P < v_N$, Q_8 sources substantial current to the base of Q_O, keeping it in heavy conduction; for $v_P > v_N$, the base drive is removed and Q_O is thus in cutoff. Summarizing,

$$Q_O = \text{Off} \qquad \text{for } v_P > v_N \qquad (9.2a)$$

$$Q_O = \text{On} \qquad \text{for } v_P < v_N \qquad (9.2b)$$

The function of Q_9 and R_5 is to provide overload protection for Q_O, in the manner discussed in Section 5.7 for op amps. The reason for using *pnp* input transistors is to allow for the input voltage range as defined in Section 5.7 to extend all the way down to V_{EE}, and also to sustain a high differential input voltage.

When on, Q_O can draw up to 50 mA of current. When off, it draws a negligible leakage current of 0.2 nA typical. Both the collector and the emitter terminals (ignoring R_5) are externally accessible to allow for custom biasing of Q_O. The

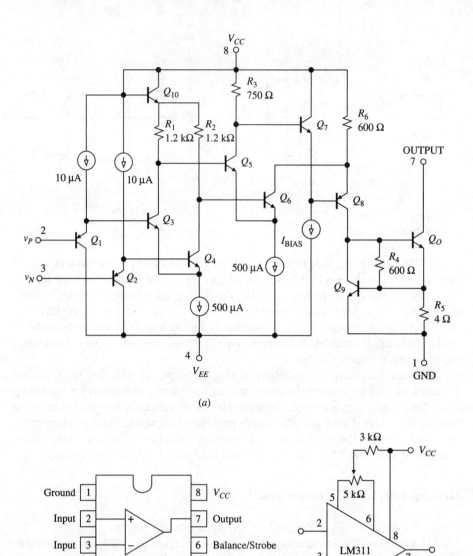

FIGURE 9.4
The LM311 voltage comparator: (*a*) simplified circuit diagram, (*b*) pinout, and (*c*) offset nulling. (Courtesy of National Semiconductor.)

most common biasing scheme involves a mere pullup resistance R_C, as shown in Fig. 9.5*a*. For $v_P < v_N$, Q_O saturates and is thus modeled with a source $V_{CE(\text{sat})}$ as in Fig 9.5*b*. So, $v_O = V_{EE(\text{logic})} + V_{CE(\text{sat})}$. Typically $V_{CE(\text{sat})} \cong 0.1$ V, so we can approximate

$$v_O = V_{OL} \cong V_{EE(\text{logic})} \qquad \text{for } v_P < v_N \qquad (9.3a)$$

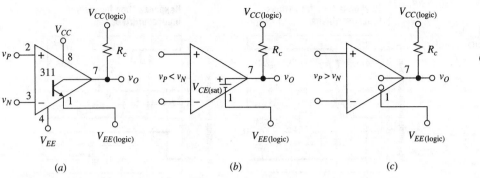

(a) (b) (c)

FIGURE 9.5
(a) Biasing the LM311 output stage with a pullup resistance R_C. Equivalent circuits for the (a) "output low" and (b) "output high" states.

For $v_P > v_N$, Q_O is in cutoff and is modeled with an open circuit as in Fig. 9.5c. By the pullup action of R_c we can write

$$v_O = V_{OH} \cong V_{CC(\text{logic})} \qquad \text{for } v_P > v_N \tag{9.3b}$$

The above expressions indicate that the output logic levels are under the control of the user. For example, letting $V_{CC(\text{logic})} = 5$ V and $V_{EE(\text{logic})} = 0$ V provides TTL and CMOS compatibility. Letting $V_{CC(\text{logic})} = 15$ V and $V_{EE(\text{logic})} = -15$ V yields ± 15-V output levels, but without the notorious uncertainties of op amp saturation voltages. The 311 can also operate from a single 5-V logic supply if we let $V_{CC(\text{logic})} = V_{CC} = 5$ V and $V_{EE(\text{logic})} = V_{EE} = 0$ V. In fact, in the single-supply mode the device is rated to function all the way up to $V_{CC} = 36$ V.

Figure 9.6a shows another popular biasing scheme, which uses a pulldown resistance R_E to operate Q_O as an emitter follower. This alternative is useful

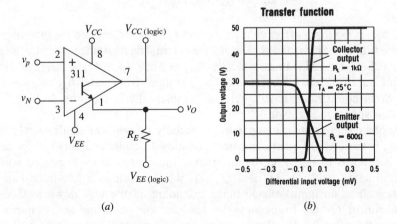

(a) (b)

FIGURE 9.6
(a) Biasing the LM311 output stage with a pulldown resistance R_E. (b) VTC comparison for pullup and pulldown biasing. (Courtesy of National Semiconductor.)

**Response time for various
input overdrives**

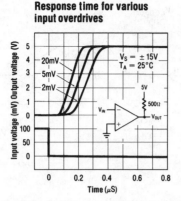

**Response time for various
input overdrives**

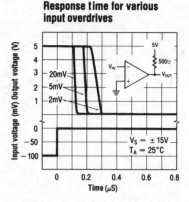

FIGURE 9.7
Typical response times of the LM311 comparator. (Courtesy of National
Semiconductor.)

when interfacing to grounded loads such as silicon controlled rectifiers (SCRs),
an example of which will be discussed in Section 11.5. The VTCs for the two
biasing schemes are shown in Fig. 9.6*b*. Note the opposing polarities of the two
curves.

Figure 9.7 shows the response times of the 311 for various input overdrives.
The responses corresponding to $V_{od} = 5$ mV are often used for comparing different devices. Based on the diagrams, we can characterize the 311 as basically a
200-ns comparator when used with a pullup resistor on the order of a few kiloohms.

Like their op amp cousins, voltage comparators suffer from dc input errors
whose effect is to shift the input tripping point by an error

$$E_I = V_{OS} + R_n I_N - R_p I_P \tag{9.4}$$

where V_{OS} is the input offset voltage, I_N and I_P the currents into the inverting- and
noninverting-input pins, and R_n and R_p the external dc resistances seen by the same
pins. At 25 °C, the LM311 has, typically, $V_{OS} = 2$ mV, $I_B = (I_P + I_N)/2 = 100$ nA
(flowing out of the device because of the *pnp* input BJTs), and $I_{OS} = I_P - I_N =$
6 nA. Some comparators have provisions for internal offset nulling. Nulling for the
LM311 is shown in Fig. 9.4*c*.

Another very popular comparator, especially in low-cost single-supply applications, is the LM339 quad comparator (National Semiconductor) and its derivatives. As shown in Fig. 9.8*a*, its differential input stage is implemented with the
pnp Darlington pairs Q_1-Q_2 and Q_3-Q_4, which result in a low-input-bias current as well as an input voltage range extending all the way down to 0 V. The
current mirror Q_5-Q_6 forms an active load for this stage and also converts to a
single-ended drive for Q_7. This transistor provides additional gain as well as the
base drive for the open-collector output transistor Q_O. The state of Q_O is controlled by v_P and v_N according to Eq. (9.2). Open-collector output stages are
suited to wired-OR operation, just like open-collector TTL gates. When on, Q_O

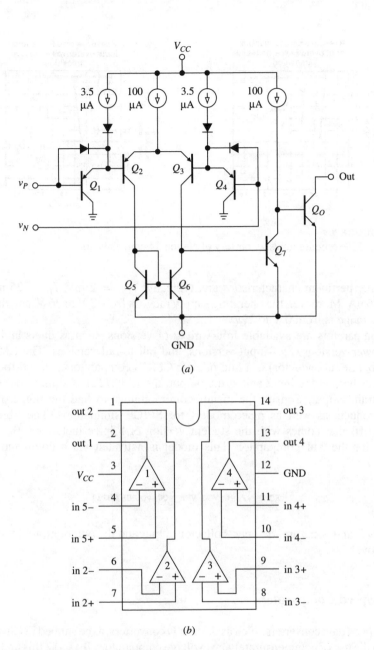

FIGURE 9.8
Simplified circuit diagram and pinout of the LM339 quad comparator.
(Courtesy of National Semiconductor.)

can sink 16 mA typical, 6 mA minimum; when off, its collector leakage is typically
0.1 nA.

The waveforms of Fig. 9.9, obtained with a 5.1-kΩ pullup resistance, reveal
that for a given input overdrive, the circuit takes longer to swing from V_{OL} to V_{OH}
than from V_{OH} to V_{OL}. This dissymmetry is due to charge-storage effects in Q_O.

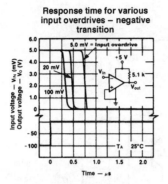
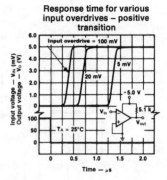

FIGURE 9.9
LM339 response times. (Courtesy of National Semiconductor.)

The other pertinent characteristics are, typically, $V_{OS} = 2$ mV, $I_B = 25$ nA, and $I_{OS} = 5$ nA. Moreover, the operating supply range is from 2 V to 36 V, and the input voltage range is from 0 V to $V_{CC} - 1.5$ V.

Comparators are available in a variety of versions, such as duals and quads, low-power versions, FET-input versions, and rail-to-rail versions. The LMC7211 (National Semiconductor) is a micropower CMOS comparator with rail-to-rail capabilities both at the input and at the output; the LMC7221 is similar, but with an open-drain output. Consult the manufacturer catalogs to find the range of available products as well as macromodels for SPICE simulations. The library file EVAL.LIB that comes with the student version of PSpice includes a Boyle-type model for the 311 comparator. This model is activated via a command of the type

```
XCMP vP vN VCC VEE vOC vOGND LM111
```

where voc and vogND are the open-collector output and the output ground terminals, respectively.

High-Speed Comparators

High-speed data converters, such as flash A-D converters, to be studied in Chapter 12, rely on the use of commensurately fast voltage comparators. To serve this and similar needs, very high-speed comparators are available with response times on the order of 10 ns or less. Such speeds are achieved through circuit techniques and fabrication processes similar to those of the faster logic families such as Schottky TTL and ECL. Moreover, to fully realize these capabilities, suitable circuit construction techniques and power-supply bypass are mandatory on the part of the user.[2]

These comparators are often equipped with output latch capabilities, which allow freezing the output state in a latch flip-flop and holding it indefinitely until the arrival of a new latch-enable command. This feature is especially useful in flash A-D

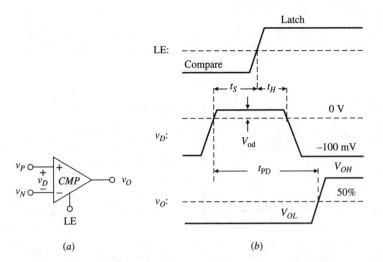

FIGURE 9.10
Comparator with latch enable, and waveforms.

converters. The symbolism and timing for these comparators are shown in Fig. 9.10. To guarantee proper output data, v_D must be valid at least t_S ns before the latch-enable command is asserted, and must remain valid for at least t_H ns thereafter, where t_S and t_H represent, respectively, the *setup* and *hold* times. Popular examples of latch comparators are the CMP-05 (Analog Devices) and LT1016 (Linear Technology). The latter has $t_S = 5$ ns, $t_H = 3$ ns, and $t_{PD} = 10$ ns.

Another useful feature available in some comparators is the strobe control, which disables the device by forcing its output stage into a high-impedance state. This feature is designed to facilitate bus interfacing in microprocessor applications. Finally, for increased flexibility, some comparators provide the output both in true (Q) and in negated ($\overline{Q}$) form.

9.2
COMPARATOR APPLICATIONS

Comparators are used in various phases of signal generation and transmission, as well as in automatic control and measurement. They appear both alone or as part of systems, such as A-D converters, switching regulators, function generators, V-F converters, power-supply supervisors, and a variety of others.

Level Detectors

The function of a level detector, also called a *threshold detector,* is to monitor a physical variable that can be expressed in terms of a voltage, and signal whenever

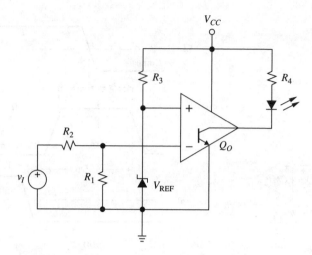

FIGURE 9.11
Basic level detector with optical indicator.

the variable rises above (or drops below) a prescribed value called the *set value*. The detector output is then used to undertake a specific action as demanded by the application. Typical examples are the activation of a warning indicator, such as a light-emitting diode (LED) or a buzzer, the turning on of a motor or heater, or the generation of an interrupt to a microprocessor.

As shown in Fig. 9.11, the basic components of a level detector are: (*a*) a voltage reference V_{REF} to establish a stable threshold, (*b*) a voltage divider R_1 and R_2 to scale the input v_I, and (*c*) a comparator. The latter trips whenever v_I is such that $[R_1/(R_1 + R_2)]v_I = V_{REF}$. Denoting this special value of v_I as V_T, we get

$$V_T = (1 + R_2/R_1)V_{REF} \tag{9.5}$$

For $v_I < V_T$, Q_O is off and so is the LED. For $v_I > V_T$, Q_O saturates and makes the LED glow, thus providing an indication of when v_I rises above V_T. Interchanging the input pins will make the LED glow whenever v_I drops below V_T. The function of R_3 is to bias the reference diode, and that of R_4 is to set the LED current.

EXAMPLE 9.1. In the circuit of Fig. 9.11 let $V_{REF} = 2.0$ V, $R_1 = 20$ kΩ, and $R_2 = 30$ kΩ. Assuming a 339 comparator with $V_{OS} = 5$ mV (maximum) and $I_B = 250$ nA (maximum), estimate the worst-case error of your circuit.

Solution. In this circuit, I_P has no effect because $R_p \cong 0$. Since I_N flows out of the comparator, it raises the inverting-input voltage by $(R_1 \parallel R_2)I_N = 3$ mV (maximum) when the comparator is about to trip. The worst-case scenario occurs when V_{OS} adds in the same direction, for a net inverting-input rise of $V_{OS} + (R_1 \parallel R_2)I_N = 5 + 3 = 8$ mV (maximum). This has the same effect as an 8-mV drop in V_{REF}, giving $V_T = (1 + 30/20)(2 - 0.008) = 4.98$ V instead of $V_T = 5.00$ V.

If v_I is V_{CC} itself, the circuit will monitor its own power supply and function as an *overvoltage* indicator. If the input pins are interchanged with each other so that $v_N = V_{REF}$ and $v_P = v_I/(1 + R_2/R_1)$, then we have an *undervoltage* indicator.

EXAMPLE 9.2. Using comparators of the 339 type, an LM385 2.5-V reference diode ($I_R \cong 1$ mA), and two HLMP-4700 LEDs ($I_{LED} \cong 2$ mA and $V_{LED} \cong 1.8$ V), design a circuit that monitors a 12-V car battery and causes the first LED to glow whenever the battery voltage rises above 13 V, and causes the second LED to glow whenever it drops below 10 V.

Solution. We need two comparators, one for overvoltage and the other for undervoltage sensing. The comparators share the same reference diode, and in both cases v_I is the battery voltage V_{CC}. For the overvoltage circuit we need $13 = (1 + R_2/R_1)2.5$ and $R_4 = (13 - 1.8)/2$; use $R_1 = 10.0$ kΩ and $R_2 = 42.2$ kΩ, both 1%, and $R_4 = 5.6$ kΩ. For the undervoltage circuit we interchange the input pins and we impose $10 = (1 + R_2/R_1)2.5$ and $R_4 = (10 - 1.8)/2$; use $R_1 = 10.0$ kΩ and $R_2 = 30.1$ kΩ, both 1%, and $R_4 = 3.9$ kΩ. To bias the reference diode, use $R_3 = (12 - 2.5)/1 \cong 10$ kΩ.

On-Off Control

Level detection can be applied to any physical variable that can be expressed in terms of a voltage via a suitable transducer. Typical examples are temperature, pressure, strain, position, fluidic level, and light or sound intensity. Moreover, the comparator can be used not only to monitor the variable, but also to control it.

Figure 9.12 shows a simple temperature controller, or thermostat. The comparator, a 339 type, uses the LM335 temperature sensor to monitor temperature, and the LM395 high-beta power transistor to switch a heater on and off in order to keep temperature at the setpoint established via R_2. The LM335 is an active reference diode designed to produce a temperature-dependent voltage according to $V(T) = T/100$, where T is absolute temperature, in kelvins. The purpose of R_5 is to bias the sensor. For the circuit to work over a wide range of supply voltages, the transducer-bridge voltage must be stabilized. This function is provided by the LM329 6.9-V reference diode, which is biased via R_4.

The circuit operates as follows. As long as temperature is above the setpoint, we have $v_N > v_P$; Q_O saturates and keeps the LM395-heater combination off. If, however, temperature drops below the setpoint, then $v_N < v_P$; Q_O is now in cutoff, thus diverting the current supplied by R_6 to the base of the LM395 transistor. The latter then saturates, turning the heater fully on.

Both the sensor and the heater are placed inside an oven and can be used, for instance, to thermostat a quartz crystal. This also forms the basis of *substrate thermostating,* a technique often used to stabilize the characteristics of voltage references and log/antilog amplifiers. We will see examples in Chapter 11 and 13.

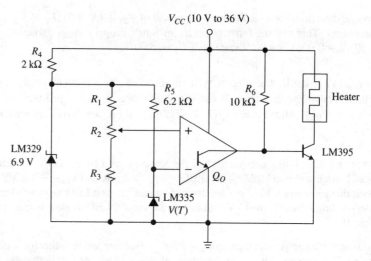

FIGURE 9.12
On-off temperature controller.

EXAMPLE 9.3. In the circuit of Fig. 9.12 specify suitable resistances so that the set-point can be adjusted anywhere between 50 °C and 100 °C by means of a 5-kΩ potentio-meter.

Solution. Since $V(50 \text{ °C}) = (273.2 + 50)/100 = 3.232$ V, and $V(100 \text{ °C}) = 3.732$ V, the current through R_2 is $(3.732 - 3.232)/5 = 0.1$ mA. Consequently, $R_3 = 3.232/0.1 = 32.3$ kΩ (use 32.4 kΩ, 1%), and $R_1 = (6.9 - 3.732)/0.1 = 31.7$ kΩ (use 31.6 kΩ, 1%).

Window Detectors

The function of a window detector, also called a *window comparator,* is to indicate when a given voltage falls within a specified *band,* or *window.* This function is implemented with a pair of level detectors, whose thresholds V_{TL} and V_{TH} define the lower and upper limits of the window. Referring to Fig. 9.13a, we observe that as long as $V_{TL} < v_I < V_{TH}$, both Q_{O1} and Q_{O2} are off, so R_c pulls v_O to V_{CC} to yield a high output. Should, however, v_I fall outside the range, the output BJT of one of the comparators will go on (Q_{O1} for $v_I > V_{TH}$, Q_{O2} for $v_I < V_{TL}$) and bring v_O near 0 V. Figure 9.13b shows the resulting VTC.

If R_c is replaced by an LED in series with a suitable current-limiting resistor, the LED will glow whenever v_I falls outside the window. If we wish the LED to glow whenever v_I falls inside the window, then we must insert an inverting stage between the comparators and the LED-resistor combination. An inverter example is offered by the 2N2222 BJT of Fig. 9.14.

The window detector shown monitors whether its own supply voltage is within tolerance. The top comparator pulls the base of the 2N2222 BJT low whenever V_{CC}

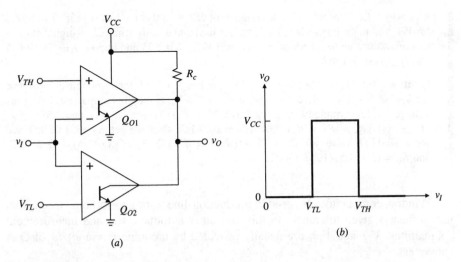

FIGURE 9.13
Window detector and its VTC.

drops below a given lower limit, and the bottom comparator pulls the base low whenever V_{CC} rises above a given upper limit; in either case the LED is off. For V_{CC} within tolerance, however, the output BJTs of both comparators are off, letting R_4 turn on the 2N2222 BJT and thus causing the LED to glow.

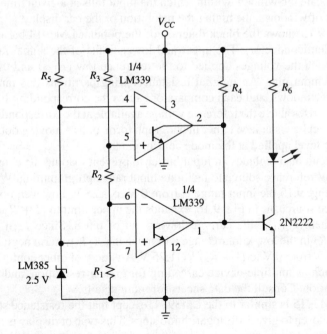

FIGURE 9.14
Power-supply monitor; LED glows as long as V_{CC} is within specification.

EXAMPLE 9.4. Specify suitable component values so that the LED of Fig. 9.14 glows for V_{CC} within the band 5 V $\pm 5\%$, which is the band usually required by digital circuits to work according to specification. Assume $V_{LED} \cong 1.5$ V, and impose $I_{LED} \cong 10$ mA and $I_{B(2N2222)} \cong 1$ mA.

Solution. For $V_{CC} = 5 + 5\% = 5.25$ V we want $v_N = 2.5$ V for the bottom comparator; for $V_{CC} = 5 - 5\% = 4.75$ V we want $v_P = 2.5$ V for the top comparator. Using the voltage divider formula twice gives $2.5/5.25 = R_1/(R_1 + R_2 + R_3)$, and $2.5/4.75 = (R_1 + R_2)/(R_1 + R_2 + R_3)$. Let $R_1 = 10.0$ kΩ; then we get $R_2 = 1.05$ kΩ and $R_3 = 10.0$ kΩ. Moreover, $R_4 = (5 - 0.7)/1 = 4.3$ kΩ, $R_5 = (5 - 2.5)/1 \cong 2.7$ kΩ, and $R_6 = (5 - 1.5)/10 \cong 330$ Ω.

Window comparators are used in production-line testing to sort out circuits that fail to meet a given tolerance. In this and other automatic test and measurement applications, V_{TL} and V_{TH} are usually provided by a computer via a pair of D-A converters.

Bar Graph Meters

A bar graph meter provides a visual indication of the input signal level. The circuit is a generalization of the window detector in that it partitions the input signal range into a string of consecutive windows, or steps, and uses a string of comparator-LED pairs to indicate the window within which the input falls at a given time. The larger the number of windows, the higher the resolution of the bar display.

Figure 9.15 shows the block diagram of the popular LM3914 bar graph meter (National Semiconductor). The upper and lower limits of the signal range are set by the user via the voltages applied to the reference low (R_{LO}) and the reference high (R_{HI}) input pins. An internal resistance string partitions this range into ten consecutive windows, and each comparator causes the corresponding LED to glow whenever v_I rises above the reference voltage available at the corresponding tap. The input level can be visualized either in bar graph form, or as a moving dot, depending on the logic level applied at the mode control pin 9.

The circuit also includes an input buffer to prevent loading the external source and a 1.25-V reference source to facilitate input range programming. With the connection of Fig. 9.15 the input range is from 0 V to 1.25 V; however, bootstrapping the reference source, as in Fig. 9.16, expands the upper limit to $(1 + R_2/R_1)1.25 + R_2 I_{ADJ}$, where I_{ADJ} is the current flowing out of pin 8. Since $I_{ADJ} \cong 75$ μA, specifying R_2 in the low-kilohm range will make the $R_2 I_{ADJ}$ term negligible, so the input range is from 0 V to $(1 + R_2/R_1)1.25$ V. A variety of other configurations are possible, such as multiple-device cascading for greater resolution, and zero-center meter operation. Consult the data sheets for more details.

The LM3915 is similar to the LM3914, except that the resistance string values have been chosen to give 3-dB logarithmic steps. This type of display is intended for signals with wide dynamic ranges, such as audio level, power, and light intensity. The LM3916 is similiar to the LM3915, except that the steps are chosen to configure the device for VU meter readings, the type of readings commonly used in audio and radio applications.

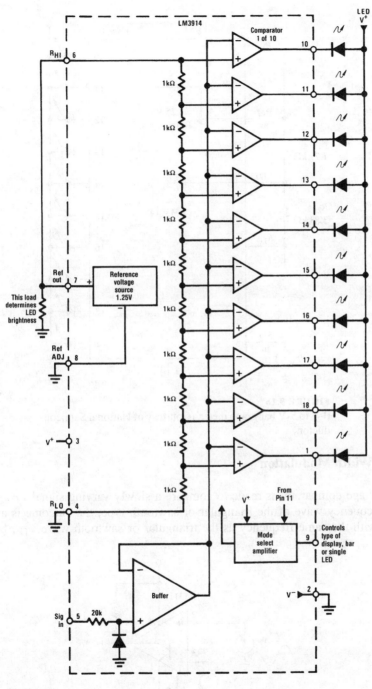

FIGURE 9.15
The LM3914 dot/bar display driver. (Courtesy of National Semiconductor.)

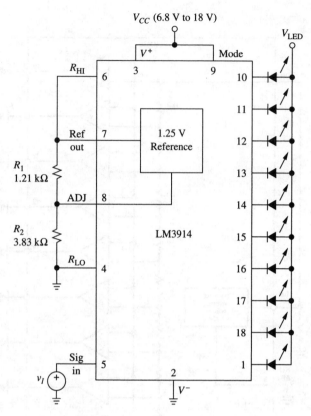

FIGURE 9.16
0-V to 5-V bar graph meter. (Courtesy of National Semiconductor.)

Pulse-Width Modulation

If a voltage comparator is made to compare a slowly varying signal v_I against a high-frequency wave of the triangular or sawtooth type, the outcome is a square wave with the same frequency as the triangular or sawtooth wave v_{TR}, but with

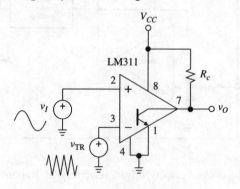

FIGURE 9.17
Modulating a high-frequency triangular wave v_{TR} with a low-frequency signal v_I.

its symmetry controlled by v_I in linear fashion. This is illustrated in Fig. 9.17 for the case of a sinusoidal wave v_I and a triangular wave v_{TR}. Exploiting the 311 macromodel available in the EVAL.LIB file of PSpice, we use the following circuit file to visualize v_O for the case of v_{TR} alternating between 0 and $V_m = 10$ V at 1 kHz, and v_I alternating between 0.5 V and 9.5 V at 100 Hz.

```
PWM Circuit:
vI 2 0 sin (5V 4.5V 100Hz)
vTR 3 0 pulse (0V 10V -0.25ms 0.5ms 0.5ms 1us 1ms)
VCC 8 0 dc 12V
.lib eval.lib
XCMP 2 3 8 0 7 0 LM111
Rc 8 7 3.3k
.tran 100us 10ms 0ms 100us
.probe
.end
```

The waveforms are shown in Fig. 9.18.

The degree of symmetry of v_O is expressed via the *duty cycle*

$$D(\%) = 100\frac{T_H}{T_L + T_H} \tag{9.6}$$

where T_L and T_H denote, respectively, the times spent by v_O in the low and the high state within a given cycle of v_{TR}. For instance, if v_O is high for 0.75 ms and low for 0.25 ms, then $D(\%) = 100 \times 0.75/(0.25 + 0.75) = 75\%$. It is readily seen that for the example illustrated we have

$$D(\%) = 100\frac{v_I}{V_m} \tag{9.7}$$

indicating that varying v_I over the range $0 < v_I < V_m$ varies D over the range $0\% < D < 100\%$. We can regard v_O as a train of pulses whose widths are controlled, or modulated, by v_I. *Pulse-width modulation* (PWM) finds application in signal transmission and power control.

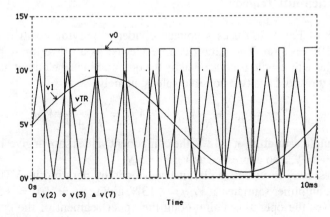

FIGURE 9.18
PWM waveforms.

9.3
SCHMITT TRIGGERS

Having investigated the behavior of high-gain amplifiers with no feedback, we now turn to amplifiers with positive feedback, also known as *Schmitt triggers*. While negative feedback tends to keep the amplifier within the linear region, positive feedback forces it into saturation. The two types of feedback are compared in Fig. 9.19. At power turn-on, both circuits start out with $v_O = 0$. However, any input disturbance that might try to force v_O away from zero will elicit opposite responses. The amplifier with negative feedback will tend to neutralize the perturbation and return to the equilibrium state $v_O = 0$. Not so in the case of positive feedback, for now the reaction is in the same direction as the perturbation, indicating a tendency to reinforce rather than neutralize it. The ensuing regenerative effect will drive the amplifier into saturation, indicating two stable states, namely, $v_O = V_{OH}$ and $v_O = V_{OL}$.

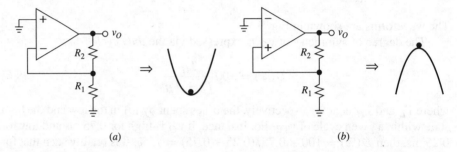

(a)   (b)

FIGURE 9.19
Mechanical models of (*a*) negative and (*b*) positive feedback.

In Fig. 9.19 negative feedback is likened to a ball at the bottom of a bowl, and positive feedback to a ball at the top of a dome. If we shake the bowl to simulate electronic noise, the ball will eventually return to its equilibrium position at the bottom, but shaking the dome will cause the ball to fall to either side.

Inverting Schmitt Trigger

The circuit of Fig. 9.20*a* uses a voltage divider to provide positive dc feedback around a 301 op amp. The circuit can be viewed as an inverting-type threshold detector whose threshold is controlled by the output. Since the output has two stable states, this threshold has two possible values, namely,

$$V_{TH} = \frac{R_1}{R_1 + R_2} V_{OH} \qquad V_{TL} = \frac{R_1}{R_1 + R_2} V_{OL} \qquad (9.8)$$

With the output saturating at ± 13 V, the component values shown give $V_{TH} = +5$ V and $V_{TL} = -5$ V, also expressed as $V_T = \pm 5$ V.

The best way to visualize circuit behavior is by deriving its VTC. Thus, for $v_I \ll 0$, the amplifier saturates at $V_{OH} = +13$ V, giving $v_P = V_{TH} = +5$ V. Increasing v_I moves the operating point along the upper segment of the curve until v_I reaches V_{TH}. At this juncture the regenerative action of positive feedback causes v_O

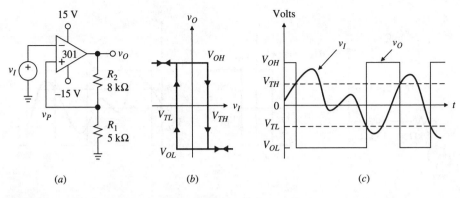

(a)   (b)   (c)

FIGURE 9.20
Inverting Schmitt trigger, VTC, and sample waveforms.

to snap from V_{OH} to V_{OL} as fast as the amplifier can swing. This, in turn, causes v_P to snap from V_{TH} to V_{TL}, or from $+5$ V to -5 V. If we wish to change the output state again, we must now lower v_I all the way down to $v_P = V_{TL} = -5$ V, at which juncture v_O will snap back to V_{OH}. In summary, as soon as $v_N = v_I$ approaches $v_P = V_T$, v_O and, hence, v_P, snap away from v_N. This behavior is opposite to that of negative feedback, where v_N tracks v_P!

Looking at the VTC of Fig. 9.20b, we observe that when coming from the left, the threshold is V_{TH}, and when coming from the right it is V_{TL}. This can also be appreciated from the waveforms of Fig. 9.20c, where it is seen that during the times of increasing v_I the output snaps when v_I crosses V_{TH}, but during the times of decreasing v_I it snaps when v_I crosses V_{TL}. Note also that the horizontal portions of the VTC can be traveled in either direction, under external control, but the vertical portions can be traveled only *clockwise,* under the regenerative effect of positive feedback.

A VTC with two separate tripping points is said to exhibit *hysteresis.* The hysteresis *width* is defined as

$$\Delta V_T = V_{TH} - V_{TL} \tag{9.9}$$

and in the present case can be expressed as

$$\Delta V_T = \frac{R_1}{R_1 + R_2}(V_{OH} - V_{OL}) \tag{9.10}$$

With the component values shown, $\Delta V_T = 10$ V. If desired, ΔV_T can be varied by changing the ratio R_1/R_2. Decreasing this ratio will bring V_{TH} and V_{TL} closer together until, in the limit $R_1/R_2 \to 0$, the two vertical segments coalesce at the origin. The circuit is then an inverting zero-crossing detector.

Noninverting Schmitt Trigger

The circuit of Fig. 9.21a is similar to that of Fig. 9.20a, except that v_I is now applied at the noninverting side. For $v_I \ll 0$, the output will saturate at V_{OL}. If we want v_O

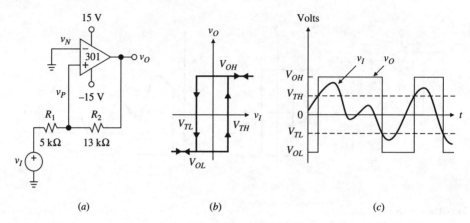

FIGURE 9.21
Noninverting Schmitt trigger, VTC, and sample waveforms.

to switch state, we must raise v_I to a high enough value to bring v_P to cross $v_N = 0$, since this is when the comparator trips. This value of v_I, aptly denoted as V_{TH}, must be such that $(V_{TH} - 0)/R_1 = (0 - V_{OL})/R_2$, or

$$V_{TH} = -\frac{R_1}{R_2}V_{OL} \qquad (9.11a)$$

Once v_O has snapped to V_{OH}, v_I must be lowered if we want v_O to snap back to V_{OL}. The tripping voltage V_{TL} is such that $(V_{OH} - 0)/R_2 = (0 - V_{TL})/R_1$, or

$$V_{TL} = -\frac{R_1}{R_2}V_{OH} \qquad (9.11b)$$

The resulting VTC, shown in Fig. 9.21b, differs from that of Fig. 9.20b in that the vertical segments are traveled in the *counterclockwise* direction. The output waveform is similar to that of the inverting Schmitt trigger, except for a reversal in polarity. The hysteresis width is now

$$\Delta V_T = \frac{R_1}{R_2}(V_{OH} - V_{OL}) \qquad (9.12)$$

and it can be varied by changing the ratio R_1/R_2. In the limit $R_1/R_2 \to 0$ we obtain a noninverting zero-crossing detector.

VTC Offsetting

In single-supply Schmitt triggers the need arises to offset the VTC so that it lies entirely within the first quadrant. The circuit of Fig. 9.22a achieves the positive offset depicted in Fig. 9.22b by using a pullup resistance R_2. To find suitable design

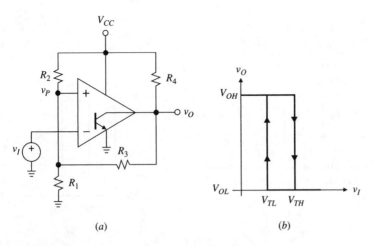

FIGURE 9.22
Single-supply inverting Schmitt trigger.

equations, we apply the superposition principle and write

$$v_P = \frac{R_1 \parallel R_3}{(R_1 \parallel R_3) + R_2} V_{CC} + \frac{R_1 \parallel R_2}{(R_1 \parallel R_2) + R_3} v_O$$

As we know, the circuit gives $V_{OL} \cong 0$ V. To achieve $V_{OH} \cong V_{CC}$, we specify $R_4 \ll R_3 + (R_1 \parallel R_2)$. Then, imposing $v_P = V_{TL}$ for $v_O = V_{OL} = 0$, and $v_P = V_{TH}$ for $v_O = V_{OH} = V_{CC}$, we get

$$V_{TL} = \frac{R_1 \parallel R_3}{(R_1 \parallel R_3) + R_2} V_{CC} \qquad V_{TH} = \frac{R_1}{R_1 + (R_2 \parallel R_3)} V_{CC}$$

Rearranging gives

$$\frac{1}{R_2} = \frac{V_{TL}}{V_{CC} - V_{TL}} \left(\frac{1}{R_1} + \frac{1}{R_3} \right) \qquad \frac{1}{R_1} = \frac{V_{CC} - V_{TH}}{V_{TH}} \left(\frac{1}{R_2} + \frac{1}{R_3} \right) \quad (9.13)$$

Since we have two equations and four unknown resistances, we fix two, say, R_4 and $R_3 \gg R_4$, and then solve for the other two.

EXAMPLE 9.5. Let the comparator of Fig. 9.22a be the LM339 type with $V_{CC} = 5$ V. Specify suitable resistances for $V_{OL} = 0$ V, $V_{OH} = 5$ V, $V_{TL} = 1.5$ V, and $V_{TH} = 2.5$ V.

Solution. Let $R_4 = 2.2$ kΩ (a reasonable value) and let $R_3 = 100$ kΩ (which is much greater than 2.2 kΩ). Then, $1/R_2 = (1.5/3.5)(1/R_1 + 1/100)$ and $1/R_1 = 1/R_2 + 1/100$. Solving yields $R_1 = 40$ kΩ (use 39 kΩ) and $R_2 = 66.7$ kΩ (use 68 kΩ).

Figure 9.23a shows the noninverting realization of the single-supply Schmitt trigger. Here, the function of R_1 and R_2 is to provide a suitable bias for v_N. Imposing $R_5 \ll R_3 + R_4$ to ensure $V_{OH} \cong V_{CC}$, and following a similar line of reasoning,

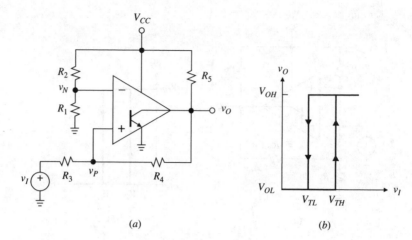

(a)

(b)

FIGURE 9.23
Single-supply noninverting Schmitt trigger.

one can readily show (see Problem 9.10) that

$$\frac{R_3}{R_4} = \frac{V_{TH} - V_{TL}}{V_{CC}} \qquad \frac{R_2}{R_1} = \frac{V_{CC} - V_{TL}}{V_{TH}} \qquad (9.14)$$

These equations are used to achieve the desired V_{TL} and V_{TH}.

Eliminating Comparator Chatter

When processing slowly varying signals, comparators tend to produce multiple output transitions, or bounces, as the input crosses the threshold region. Figure 9.24 shows an example. Referred to as *comparator chatter*, these bounces are due to ac noise invariably superimposed on the input signal, especially in industrial environments. As this signal crosses the threshold region, noise is amplified with the

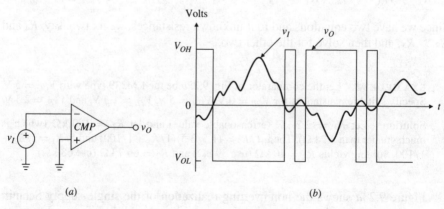

(a)

(b)

FIGURE 9.24
Comparator chatter.

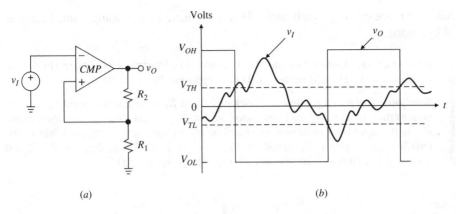

FIGURE 9.25
Using hysteresis to eliminate chatter.

full open-loop gain, causing output chatter. For instance, the LM311 comparator, whose gain is typically 200 V/mV, requires an input noise spike of only $(5/200,000) = 25\ \mu V$ to cause a 5-V output swing. Chatter is unacceptable in counter-based applications.

The problem is eliminated with the help of hysteresis, as shown in Fig. 9.25. In this case, as soon as v_I crosses the present threshold, the circuit snaps and activates the other threshold, so v_I must swing back to the new threshold in order to make v_O snap again. Making the hysteresis width greater than the maximum peak-to-peak amplitude of noise prevents spurious output transitions.

Even in situations where the input signal is relatively clean, it always pays to introduce a small amount of hysteresis, say, a few millivolts, to stave off potential oscillations due to stray ac feedback caused by parasitic capacitances and the distributed impedances of the power-supply and ground busses. This stabilization technique is particularly important in flash A-D converters.

Hysteresis in On-Off Controllers

Hysteresis is used in on-off control to avoid overfrequent cycling of pumps, furnaces, and motors. Consider, for instance, the temperature controller discussed in connection with Fig. 9.12. We can easily turn it into a home thermostat by having the comparator drive a power switch like a relay or a triac to turn a home furnace on or off. Starting with temperatures below the setpoint, the comparator will activate the furnace and cause temperature to rise. This rise is monitored by the temperature sensor and conveyed to the comparator in the form of an increasing voltage. As soon as the temperature reaches the setpoint, the comparator will trip and shut off the furnace. However, the smallest temperature drop following furnace shut off will suffice to trip the comparator back to the active state. As a result, the furnace will be cycled on and off at a rapid pace, a very taxing affair.

In general, temperature need not be regulated to such a sharp degree. Allowing a hysteresis of a few degrees will still ensure a comfortable environment and yet reduce furnace cycling significantly. This we achieve by providing a small amount of hysteresis.

EXAMPLE 9.6. Modify the temperature controller of Example 9.3 to ensure a hysteresis of about $\pm 1\,°C$. The LM395 power BJT has typically $V_{BE(\text{on})} = 0.9$ V.

Solution. Connect a positive-feedback resistance R_F between the output v_O and the noninverting input v_P of the comparator, so that $\Delta v_P = \Delta v_O R_W/(R_W + R_F)$, where R_W is the equivalent resistance presented to R_F by the wiper. With the wiper in the middle, $R_W = (R_1 + R_2/2)\|(R_3 + R_2/2) = 17.2$ kΩ. Using $\Delta v_O = 0.9$ V and $\Delta v_P = \pm 1 \times 10$ mV $= 20$ mV, and solving, we get $R_F \cong 750$ kΩ.

9.4
PRECISION RECTIFIERS

A *half-wave rectifier* (HWR) is a circuit that passes only the positive (or only the negative) portion of a wave, while blocking out the other portion. The transfer characteristic of the positive HWR, pictured in Fig. 9.26a, is

$$v_O = v_I \qquad \text{for } v_I > 0 \tag{9.15a}$$

$$v_O = 0 \qquad \text{for } v_I < 0 \tag{9.15b}$$

A *full-wave rectifier* (FWR), besides passing the positive portion, inverts and then passes also the negative portion. Its transfer characteristic, depicted in Fig. 9.26b, is $v_O = v_I$ for $v_I > 0$, and $v_O = -v_I$ for $v_I < 0$, or, more concisely,

$$v_O = |v_I| \tag{9.16}$$

An FWR is also referred to as an *absolute-value circuit*.

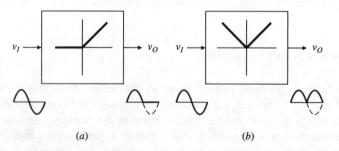

FIGURE 9.26
Half-wave rectifier (HWR) and full-wave rectifier (FWR).

Rectifiers are implemented using nonlinear devices such as diodes. The nonzero forward-voltage drop $V_{D(\text{on})}$ of a practical diode may cause intolerable errors in low-level signal rectification. As we shall see, this shortcoming is avoided by placing the diode inside the negative-feedback path of an op amp.

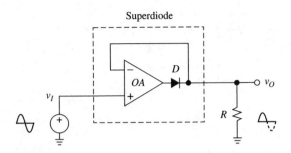

FIGURE 9.27
Basic half-wave rectifier.

Half-Wave Rectifiers

The analysis of the circuit of Fig. 9.27 is facilitated if we consider the cases $v_I > 0$ and $v_I < 0$ separately.

1. $v_I > 0$: In response to a positive input, the op amp output v_{OA} will also swing positive, turning on the diode and thus creating the negative-feedback path shown in Fig. 9.28a. This allows us to apply the virtual-short principle and write $v_O = v_I$. We observe that to make v_O track v_I, the op amp rides its output a diode drop above v_O, that is, $v_{OA} = v_O + V_{D(on)} \cong v_O + 0.7$ V. Placing the diode within the feedback loop in effect eliminates any errors due to its forward-voltage drop. To emphasize this dramatic effect of negative feedback, the diode-op amp combination is referred to as a *superdiode*.

2. $v_I < 0$: Now the op amp output swings negative, turning the diode off and thus causing the current through R to go to zero. Hence, $v_O = 0$. As pictured in Fig. 9.28b, the op amp is now operating in the open-loop mode, and since $v_P < v_N$, the output saturates at $v_{OA} = V_{OL}$. With $V_{EE} = -15$ V, $v_{OA} \cong -13$ V.

A disadvantage of this circuit is that when v_I changes from negative to positive, the op amp output has to come out of saturation and then swing all the way from $v_{OA} = V_{OL} \cong -13$ V to $v_{OA} \cong v_I + 0.7$ V in order to close the feedback loop. All this takes time, and if v_I has changed appreciably meanwhile, v_O may exhibit

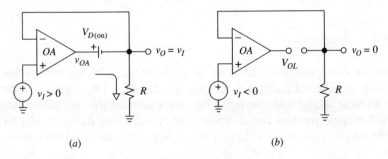

(a) (b)

FIGURE 9.28
Equivalent circuits of the basic HWR for (a) positive and (b) negative inputs.

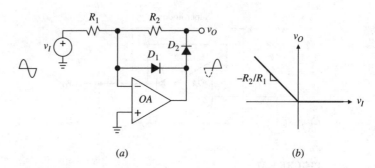

FIGURE 9.29
Improved HWR and its VTC.

intolerable distortion. The improved HWR of Fig. 9.29a alleviates this inconvenience by using a second diode to clamp the negative saturation level just a diode drop below ground. Proceeding as usual, we identify two cases:

1. $v_I > 0$: A positive input causes D_1 to conduct, thus creating a negative-feedback path around the op amp. By the virtual-ground principle we have $v_N = 0$, indicating that D_1 now clamps the op amp output at $v_{OA} = -V_{D1(on)}$. Moreover, D_2 is off, so no current flows through R_2 and, hence, $v_O = 0$.

2. $v_I < 0$: A negative input causes the op amp output to swing positive, thus turning D_2 on. This creates an alternative negative-feedback path via D_2 and R_2, which still ensures $v_N = 0$. Clearly, D_1 is now off, so the current sourced by the op amp to R_2 must equal the current sunk by v_I from R_1, or $(v_O - 0)/R_2 = (0 - v_I)/R_1$. This gives $v_O = (-R_2/R_1)v_I$. Moreover, $v_{OA} = v_O + V_{D2(on)}$.

Circuit behavior is summarized as

$$v_O = 0 \qquad \text{for } v_I > 0 \qquad (9.17a)$$

$$v_O = -(R_2/R_1)v_I \qquad \text{for } v_I < 0 \qquad (9.17b)$$

and the VTC is shown in Fig. 9.29b. In words, the circuit acts as an inverting HWR with gain. The op amp output v_{OA} still rides a diode drop above v_O when $v_O > 0$; however, when $v_O = 0$, v_{OA} is clamped at about -0.7 V, that is, within the linear region. Consequently, the absence of saturation-related delays and the reduced output voltage swing result in much improved dynamics.

Full-Wave Rectifiers

One way of synthesizing the absolute value of a signal is by combining the signal itself with its inverted half-wave rectified version in a 1-to-2 ratio, as shown in Fig. 9.30. Here OA_1 provides inverting half-wave rectification, and OA_2 sums v_I and the HWR output v_{HW} in a 1-to-2 ratio to give $v_O = -(R_5/R_4)v_I - (R_5/R_3)v_{HW}$. Considering that $v_{HW} = -(R_2/R_1)v_I$ for $v_I > 0$, and $v_{HW} = 0$ for $v_I < 0$, we can write

$$v_O = A_p v_I \qquad \text{for } v_I > 0 \text{ V} \qquad (9.18a)$$

$$v_O = -A_n v_I \qquad \text{for } v_I < 0 \text{ V} \qquad (9.18b)$$

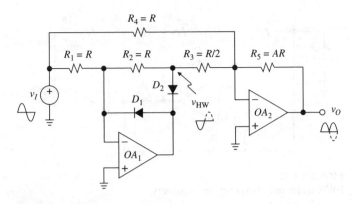

FIGURE 9.30
Precision FWR, or absolute-value circuit.

where

$$A_n = \frac{R_5}{R_4} \qquad A_p = \frac{R_2 R_5}{R_1 R_3} - A_n \qquad (9.19)$$

We want both halves of the input wave to be amplified by the same gain $A_p = A_n = A$, for then we can write $v_O = A v_I$ for $v_I > 0$ and $v_O = -A v_I$ for $v_I < 0$, or, concisely,

$$v_O = A|v_I| \qquad (9.20)$$

One way of achieving this goal is by imposing $R_1 = R_2 = R_4 = R$, $R_3 = R/2$, and $R_5 = AR$, as shown; then, $A = R_5/R$.

Because of resistance tolerances, A_p and A_n will generally differ from each other. Their difference

$$A_p - A_n = \frac{R_2 R_5}{R_1 R_3} - 2\frac{R_5}{R_4}$$

is maximized when R_2 and R_4 are maximized and R_1 and R_3 are minimized. (R_5 can be ignored since it appears in both terms.) Denoting percentage tolerance as p and substituting $R_2 = R_4 = R(1 + p)$ and $R_1 = 2R_3 = R(1 - p)$ gives

$$|A_p - A_n|_{\max} = 2A \left(\frac{1 + p}{(1 - p)^2} - \frac{1}{1 + p} \right)$$

where $A = R_5/R$. For $p \ll 1$ we can ignore higher-order powers of p and use the approximations $(1 \pm p)^{-1} \cong (1 \mp p)$. This allows us to estimate the *maximum percentage difference* between A_p and A_n as

$$100 \left| \frac{A_p - A_n}{A} \right|_{\max} \cong 800p$$

For instance, with 1% resistances, A_p and A_n may differ from each other by as much as about $800 \times 0.01 = 8\%$. To minimize this error, we can either use more precise resistors, such as laser-trimmed IC resistor arrays, or trim one of the first four resistors, say, R_2.

The alternative FWR realization of Fig. 9.31 requires only two matched resistors. For $v_I > 0$, D_1 is on, allowing OA_1 to keep its inverting input at virtual ground.

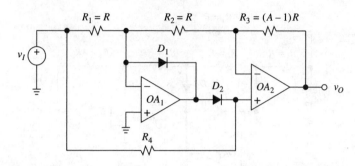

FIGURE 9.31
FWR using only two matched resistors.

With the output of OA_1 clamped at $-V_{D_1(\text{on})}$, D_2 is off, allowing R_4 to transmit v_I to OA_2. The latter, acting as a noninverting amplifier, gives $v_O = A_p v_I$,

$$A_p = 1 + \frac{R_3}{R_2}$$

For $v_I < 0$, D_1 is off and D_2 is forward biased by R_4. OA_1 still keeps its inverting input at virtual ground, but via the feedback path $D_2\text{-}OA_2\text{-}R_3\text{-}R_2$. By KCL, $(0-v_I)/R_1 = (v_O - 0)/(R_2 + R_3)$, or $v_O = -A_n v_I$,

$$A_n = \frac{R_2 + R_3}{R_1}$$

Imposing $A_p = A_n = A$ allows us to write concisely $v_O = A|v_I|$. This condition is met by imposing $R_1 = R_2 = R$ and $R_3 = (A-1)R$, as shown. Clearly, only two matched resistances are needed.

Ac-dc Converters

The most common application of precision absolute-value circuits is ac-dc conversion, that is, the generation of a dc voltage proportional to the amplitude of a given ac wave. To accomplish this task, the ac signal is first full-wave rectified, and then low-pass filtered to synthesize a dc voltage. This voltage is the *average* of the rectified wave,

$$V_{\text{avg}} = \frac{1}{T} \int_0^T |v(t)| \, dt$$

where $v(t)$ is the ac wave and T is its period. Substituting $v(t) = V_m \sin 2\pi f t$, where V_m is the peak amplitude and $f = 1/T$ is the frequency, gives

$$V_{\text{avg}} = (2/\pi)V_m = 0.637 V_m$$

An ac-dc converter is calibrated so that when fed with an ac signal it gives its *root-mean-square (rms)* value,

$$V_{\text{rms}} = \left(\frac{1}{T} \int_0^T v^2(t) \, dt \right)^{1/2}$$

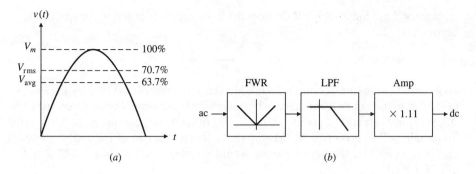

FIGURE 9.32
(a) Relationship between V_{rms} and V_m, and between V_{avg} and V_m. (b) Block diagram of an ac-dc converter.

Substituting $v(t) = V_m \sin 2\pi ft$ and integrating gives

$$V_{\text{rms}} = V_m / \sqrt{2} = 0.707 V_m$$

The relationships between average and rms values and peak value are depicted in Fig. 9.32a. These relationships, which hold for sinusoidal waves but not necessarily for other waveforms, indicate that in order to obtain V_{rms} from V_{avg}, we need to multiply the latter by $(1/\sqrt{2})/(2/\pi) = 1.11$. The complete block diagram of an ac-dc converter is thus as in Fig. 9.32b.

Figure 9.33 shows a practical ac-dc converter implementation. The gain of 1.11 V/V is adjusted by means of the 50-kΩ pot, and the capacitance provides low-pass filtering with cutoff frequency $f_0 = 1/2\pi R_5 C$, where R_5 is the net resistance in parallel with C, or $1.11 \times 200 = 222$ kΩ. Hence, $f_0 = 0.717$ Hz. Using the LT1122 fast-settling JFET-input op amps allow the circuit to process a 10-V peak-to-peak ac signal with a 2-MHz bandwidth.

The capacitance must be sufficiently large to keep the residual output ripple within specified limits. This requires that f_0 be well below the minimum operating

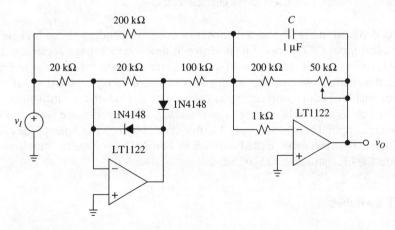

FIGURE 9.33
Wideband ac-dc converter.

frequency f_{min}. Since an FWR doubles the frequency, the criterion for specifying C becomes

$$C \gg \frac{1}{4\pi R_5 f_{min}}$$

As a conservative rule of thumb, C should exceed the right-hand term by the *inverse* of the *fractional ripple error* that can be tolerated at the output. For instance, for a 1% ripple error, C should be about $1/0.01 = 100$ times as large as the right-hand term. To remain within this error all the way down to the low end of the audio range, so that $f_{min} = 20$ Hz, the above circuit would require $C = 100/(4\pi \times 222 \times 10^3 \times 20) \cong 1.8\ \mu F$.

9.5
ANALOG SWITCHES

Many circuits require electronic switches, that is, switches whose state is voltage-programmable. Chopper amplifiers, D-A converters, function generators, S/H amplifiers, and switching power supplies are common examples. Switches are also used to route signals in data acquisition systems, and to reconfigure circuits in programmable instrumentation.

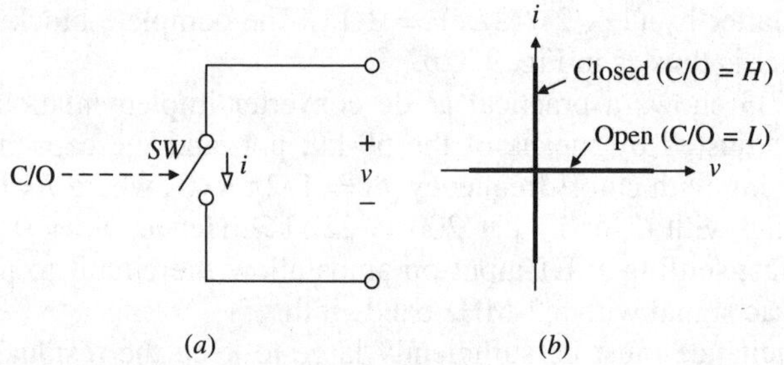

FIGURE 9.34
Ideal switch and its i-v characteristics.

As depicted in Fig. 9.34a, SW closes or opens, depending on the logic level at the control input C/O. When SW is closed it drops zero voltage regardless of the current, and when SW is open it draws zero current regardless of the voltage, thus giving the characteristic of Fig. 9.34b. This behavior can be approximated by any device capable of high on-off resistance ratios, such as field-effect transistors (FETs). An FET acts as a variable resistor called *channel,* whose resistance is controlled by the voltage applied between a control terminal called *gate* G and one of the channel terminals. These terminals, called *source* S and *drain* D, are usually interchangeable because the FET structure is symmetric.

JFET Switches

Figure 9.35 shows the characteristics of the n-channel JFET, or n-JFET for short. Each curve represents the i-v characteristic of the channel for a different value of

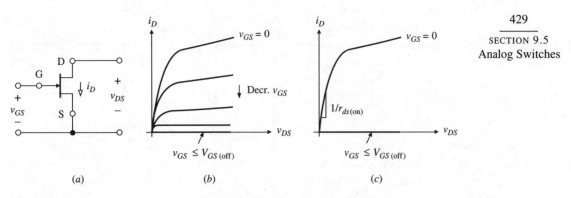

FIGURE 9.35
The *n*-channel JFET ($V_{GS(\text{off})} < 0$), and its *i-v* characteristics.

the control voltage v_{GS} applied between gate and source. For $v_{GS} = 0$ the channel is highly conductive, this being the reason why JFETs are said to be *normally on* devices. Making v_{GS} progressively more negative reduces channel conductivity until a cutoff threshold $V_{GS(\text{off})} < 0$ is reached, such that for $v_{GS} \leq V_{GS(\text{off})}$ conductivity drops to zero and the channel acts as an open circuit. $V_{GS(\text{off})}$ is typically in the range of -0.5 V to -10 V, depending on the device.

In switch applications we are interested only in two curves, the ones corresponding to $v_{GS} = 0$ and $v_{GS} \leq V_{GS(\text{off})}$. The former is highly nonlinear; however, when the channel is used as a closed switch, its operation is near $v_{DS} = 0$ V, where the curve is fairly steep and linear. The slope is inversely proportional to a resistance $r_{ds(\text{on})}$ called the *dynamic resistance* of the channel,

$$\frac{di_D}{dv_{DS}} = \frac{1}{r_{ds(\text{on})}} \tag{9.21}$$

For ideal switch operation, this resistance should be zero; in practice, it is typically in the range of 10^2 Ω or less, depending on the device type.

When the channel is off, its resistance is virtually infinite. The only currents of potential concern in this case are the leakage currents, namely, the *drain cutoff current* $I_{D(\text{off})}$ and the *gate reverse current* I_{GSS}. At ambient temperature these currents are typically in the picoampere range; however, they double with about every 10 °C increase. This can be of concern in certain applications, as we shall see.

A popular *n*-JFET switch is the 2N4391 (Siliconix), whose room-temperature ratings are: -4 V $\leq V_{GS(\text{off})} \leq -10$ V, $r_{ds(\text{on})} \leq 30$ Ω, $I_{D(\text{off})} \leq 100$ pA, $I_{GSS} \leq 100$ pA flowing out of the gate, turn-on delay ≤ 15 ns, and turn-off delay ≤ 20 ns. Figure 9.36 illustrates a typical switch application. The function of the switch is to provide a make/break connection between a source v_I and a load R_L, whereas the function of the switch driver is to translate the TTL-compatible logic command O/C to the proper gate drive.

With O/C low ($\cong 0$ V), the E-B junction of Q_1 is off, so both Q_1 and Q_2 are off. By the pullup action of R_2, D_1 is reverse-biased, allowing R_1 to keep the gate at the same potential as the channel. We thus have $v_{GS} = 0$ regardless of v_I, so the switch is heavily on.

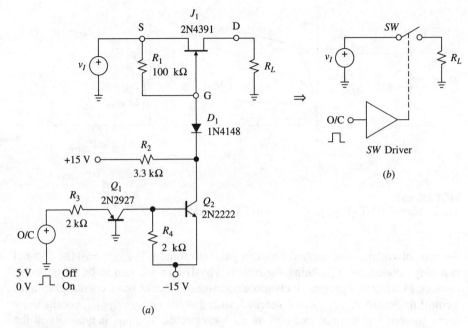

FIGURE 9.36
The n-channel JFET as a switch.

With O/C high ($\cong 5$ V), Q_1 conducts and forces Q_2 to saturate, thus pulling the gate close to -15 V. With a gate voltage this negative, the switch is off. To prevent J_1 from inadvertently going on, we must limit v_I in the negative direction,

$$v_{I(\text{min})} = V_{EE} + V_{CE2(\text{sat})} + V_{D1(\text{on})} - V_{GS(\text{off})} \tag{9.22}$$

For instance, with $V_{GS(\text{off})} = -4$ V we obtain $v_{I(\text{min})} \cong -15 + 0.1 + 0.7 - (-4) \cong -10$ V, indicating that the circuit will operate properly only as long as the input is above -10 V.

For high-speed operation,[2] connect a 100-pF capacitor between the control input and the base of Q_2 to speed up the turn-on and turn-off times of Q_2, and an HP2810 Schottky diode between the base and collector of Q_2 (anode at the base) to eliminate the storage delay of Q_2. JFET drivers as well as JFET-driver combinations are available in IC form from a variety of manufacturers.

The configuration of Fig. 9.36 requires a dedicated driver because the switch must ride with the signal v_I. If the switch is allowed to remain at a nearly constant potential, such as the virtual-ground potential of an op amp, then the driver can be simplified or even eliminated, as the configuration of Fig. 9.37 shows. Referred to as *analog-ground-switch* or *current switch*, this configuration uses a p-JFET designed for direct compatibility with standard logic levels. The p-JFET is similar to the n-JFET, except that the cutoff voltage is now positive, or $V_{GS(\text{off})} > 0$. Moreover, the fabrication of p-JFETs is compatible with low-cost bipolar technology. The switch operates as follows.

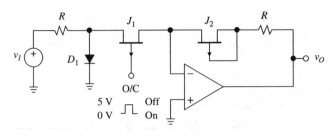

FIGURE 9.37
Analog-ground-switch using p-channel JFETs.

When the control input O/C is low we have $v_{GS1} \cong 0$, indicating that J_1 is heavily on. To compensate for the presence of $r_{ds1(on)}$, a dummy JFET J_2 is used in the feedback path of the op amp with the gate and source tied together to keep it permanently on. J_1 and J_2 are matched devices to ensure $r_{ds2(on)} = r_{ds1(on)}$ and, hence, $v_O/v_I = -1$ V/V.

When O/C is high, or $v_{GS1} > V_{GS1(off)}$, J_1 is off and signal propagation is thus inhibited, so now $v_O/v_I = 0$. D_1 provides a clamping function to prevent the channel from inadvertently turning on during the positive alternations of v_I. Summarizing, the circuit provides unity gain when O/C is low, and zero gain when O/C is high.

The principle of Fig. 9.37 is especially useful in summing-amplifier applications. Replicating the input resistor-diode-switch combination k times gives a *k-channel analog multiplexer*, a device widely used in data acquisition and audio signal switching. The AH5010 quad switch (National Semiconductor) consists of four p-FET switches and relative diode clamps plus a dummy FET in the same package. With an external op amp and five resistors, one can implement a four-channel multiplexer, and by cascading multiple AH5010s one can expand to virtually any number of channels.

MOSFET Switches

Since MOS technology forms the basis of digital VLSI, MOS switches are particularly attractive when analog and digital functions must coexist on the same chip. MOSFETs are available both in normally on, or *depletion,* versions, and in normally off, or *enhancement,* versions. The latter are by far the most common, since they form the basis of CMOS technology.

Figure 9.38 shows the characteristics of the enhancement n-channel MOSFET, or n-MOSFET, for short. Its behavior is similar to that of the n-JFET, except that with $v_{GS} = 0$ the device is off. To make the channel conductive, v_{GS} must be raised above some threshold $V_{GS(on)} > 0$; the greater v_{GS} compared to $V_{GS(on)}$, the more conductive the channel. When operated in a virtual-ground arrangement of the type of Fig. 9.37, an n-MOSFET opens when the gate voltage is low, and closes when the gate voltage is high.

If the n-MOSFET is connected in a floating arrangement of the type of Fig. 9.36, the on-state conductivity is no longer uniformly high, but varies with v_I since v_{GS} itself is a function of v_I. The channel is much less conductive during positive

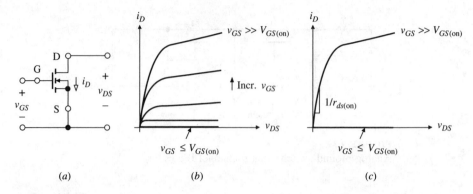

FIGURE 9.38
The enhancement n-channel MOSFET ($V_{GS(\text{on})} > 0$), and its i-v characteristics.

than during negative alternations of v_I, and for sufficiently positive values of v_I it may actually turn off. These drawbacks are eliminated by using a pair of complementary MOS (CMOS) FETs, one handling the negative and the other the positive alternations of v_I. The former is an enhancement n-MOSFET, and the latter an enhancement p-MOSFET, whose characteristics are similar to those of the n-MOSFET, except that the turn-on threshold is now negative. So, to make a p-MOSFET conductive, we need $v_{GS} < V_{GS(\text{on})} < 0$; the lower v_{GS} compared to $V_{GS(\text{on})}$, the more conductive the channel. For proper operation, the p-MOSFET must be driven in antiphase with respect to the n-MOSFET. As shown in Fig. 9.39a for the case of symmetric power supplies, this drive is provided by an ordinary CMOS inverter.

When C/O is high, the gate of n-MOSFET M_n is high and that of p-MOSFET M_p is low, turning both devices on. As depicted in Fig. 9.39b, M_n offers low resistance only over the lower portion of the signal range, and M_p only over the upper portion.

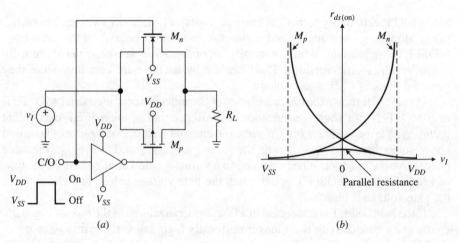

FIGURE 9.39
CMOS transmission gate, and its dynamic resistance as a function of v_I.

However, as a team, they offer a combined parallel resistance that is reasonably low throughout the entire range $V_{SS} \leq v_I \leq V_{DD}$. Finally, when C/O is low, both FETs are off and signal transmission is inhibited.

Also called a *transmission gate,* the basic configuration of Fig. 9.39a is available in a variety of versions and performance ratings. Two of the oldest examples are the CD4066 quad bilateral switch and the CD4051 eight-channel multiplexer/demultiplexer, originally introduced by RCA. The 4051 also provides logic-level translation to allow the switches to work with bipolar analog signals while accepting unipolar logic levels. A wide variety of other MOS-switch products can be found by consulting the data books.

9.6
PEAK DETECTORS

The function of a *peak detector* is to capture the peak value of the input and yield $v_O = v_{I(\text{peak})}$. To achieve this goal, v_O is made to track v_I until the peak value is reached. This value is then held until a new, larger peak comes along, in which case the circuit will update v_O to the new peak value. Figure 9.40a shows an example of input and output waveforms. Peak detectors find application in test and measurement instrumentation.

From the above description we identify the following four blocks: (a) an analog memory to hold the value of the most recent peak—this is the capacitor, whose ability to store charge makes it act as a voltage memory, as per $V = Q/C$; (b) a unidirectional current switch to further charge the capacitor when a new peak comes along: this is the diode; (c) a device to force the capacitance voltage to track the input voltage when a new peak comes along: this is the voltage follower; (d) a switch to periodically reinitialize v_O to zero: this is accomplished with a FET discharge switch in parallel with the capacitor.

In the circuit of Fig. 9.40b the above tasks are performed, respectively, by C_H, D_2, OA_1, and SW. The function of OA_2 is to buffer the capacitor voltage to prevent discharge by R and by any external load. Moreover, D_1 and R prevent OA_1 from

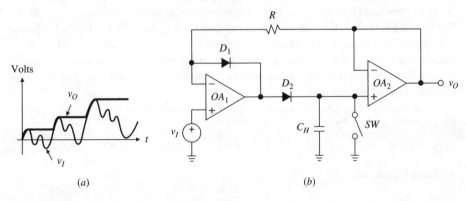

(a)   (b)

FIGURE 9.40
Peak-detector waveforms and circuit diagram.

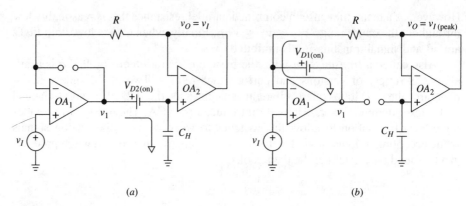

FIGURE 9.41
Peak-detector equivalents during (a) the track mode, and (b) the hold mode.

saturating after a peak has been detected, and thus speed up recovery when a new peak comes along. The circuit operates as follows.

With the arrival of a new peak, OA_1 swings its output v_1 positive, turning D_1 off and D_2 on as shown in Fig. 9.41a. OA_1 uses the feedback path D_2-OA_2-R to maintain a virtual short between its inputs. Since no current flows through R, the result is that v_O will track v_I. During this mode, aptly called the *track mode*, OA_1 sources current to charge C_H via D_2, and its output rides a diode drop above v_O, or $v_1 = v_O + V_{D2(on)}$.

After peaking, v_I starts to decrease, causing the output of OA_1 also to decrease. Consequently, D_2 goes off and D_1 goes on, thus providing an alternative feedback path for OA_1, as depicted in Fig. 9.41b. By the virtual-short concept, the output of OA_1 now rides a diode drop below v_I, or $v_1 = v_I - V_{D1(on)}$. During this mode, called the *hold mode,* the capacitor voltage remains constant, and the function of R is to provide a current path for D_1.

We observe that placing D_2 and OA_2 within the feedback path of OA_1 eliminates any errors due to the voltage drop across D_2 and the input offset voltage of OA_2. All that is required at the input of OA_2 is a sufficently low input bias current to minimize capacitance discharge between peaks. The requirements of OA_1 are a sufficiently low dc input error, and a sufficiently high output-current capability to charge C_H during fast peaks. Moreover, OA_1 may need to be stabilized against the feedback-loop pole introduced by r_{o1} and C_H, and that introduced by OA_2. This is usually achieved by connecting suitable compensation capacitors in parallel with D_1 and R. Typically, R is on the order of a few kiloohms, and the compensation capacitances on the order of a few tens of picofarads.

It is readily seen that reversing the diode directions causes the circuit to detect the negative peaks of v_I.

Voltage Droop and Sagback

During the hold mode, v_O should remain rigorously constant. In practice, because of leakage currents, the capacitor will slowly charge or discharge, depending on leakage

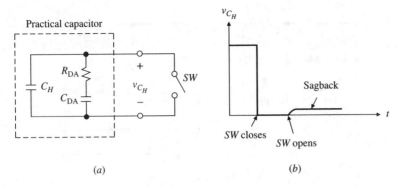

FIGURE 9.42
(*a*) Circuit model for dielectric absorption, and (*b*) the sagback effect.

polarity. Leakage stems from various sources, namely, from diode, capacitor, and reset-switch leakage; printed-circuit board leakage; and the input bias current of OA_2. Using the capacitance law $i = C\,dv/dt$ and denoting the net capacitance leakage as I_L, we define the *voltage droop rate* as

$$\frac{dv_0}{dt} = \frac{I_L}{C_H} \tag{9.23}$$

For instance, a 1-nA leakage current through a 1-nF capacitance produces a voltage droop rate of $10^{-9}/10^{-9} = 1$ V/s $= 1$ mV/ms. Droop is minimized by reducing the individual leakage components.

The most crucial limitations of a practical capacitor in analog memory applications are *leakage* and *dielectric absorption*. Leakage causes the device to slowly discharge when in the hold mode; dielectric absorption causes the new voltage to creep back toward the previous voltage after the capacitance is subjected to a rapid voltage change. This sagback effect, stemming from charge storage phenomena in the bulk of the dielectric, can be modeled with a series of internal R-C stages, each in parallel with C_H. Referring to the first-order model of Fig. 9.42*a*, we observe that even though C_H discharges almost instantaneously when SW is closed, C_{DA} will retain some charge because of the series resistance R_{DA}. After SW is opened, C_{DA} will transfer part of its charge back to C_H to achieve equilibrium, thus causing the sagback effect depicted in Fig. 9.42*b*. Though more than one time constant may intervene in the sag, a single time constant is often sufficient to characterize the sag, with C_{DA} typically one or more orders of magnitude smaller than C_H, and a time constant ranging from fractions of a millisecond to fractions of a second. Capacitor types are available with low leakage and low dielectric absorption. These include polystyrene, polypropylene, and Teflon types.[3]

Printed-circuit board leakage is minimized by input guarding techniques of the type discussed in Section 5.3. In the present circuit the ring is driven by v_O and is made to surround all traces associated with the noninverting input of OA_2, as shown in the practical example of Fig. 9.43.

A FET-input op amp is often chosen for OA_2 to take advantage of its low-input-bias-current characteristics. However, this current doubles with about every 10 °C

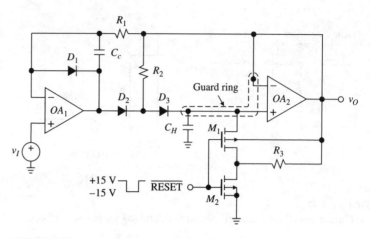

FIGURE 9.43
Peak detector for extended hold.

increase, so if an extended range of operating temperatures is anticipated, a BJT-input op amp with ultralow-input bias current may be preferable.

When reverse biased, a diode draws a leakage current that also doubles with every $10\,°C$ increase. The circuit of Fig. 9.43 eliminates the effect of diode leakage by using a third diode D_3 and the pullup resistance R_2. During the track mode, the D_2-D_3 pair acts as a unidirectional switch, but with a voltage drop twice as large. During the hold mode, R_2 pulls the anode of D_3 to the same potential as the cathode, thus eliminating D_3's leakage; the reverse bias is sustained solely by D_2.

A similar technique can be used to minimize reset-switch leakage. In the example shown, this switch is implemented with two 3N163 enhancement p-MOSFETs (Siliconix). Applying a negative pulse to their gates turns both FETs on and also discharges C_H. Upon pulse removal both FETs go off; however, with R_3 pulling the source of M_1 to the same potential as the drain, M_1's leakage is eliminated; the switch voltage is sustained solely by M_2. If TTL compatibility is desired, one can use a suitable voltage-level translator, such as the DH0034 (National Semiconductor).

A good choice for the op amps of Fig. 9.43 is a dual JFET-input device such as the precision, high-speed OP-249 op amp (Analog Devices). The diodes can be any general-purpose devices, such as the 1N914 or 1N4148 types, and suitable values for the various resistances are in the 10-kΩ range. The purpose of C_c, typically in the range of a few tens of picofarads, is to stabilize the capacitively loaded op amp OA_1 during the track mode. C_H should be large enough to reduce the effect of leakage, yet small enough to allow for its rapid charge during fast peaks. A reasonable compromise is typically in the 1-nF range.

Speed Limitations

Peak-detector speed is limited by the slew rates of its op amps as well as the maximum rate at which OA_1 can charge or discharge C_H. The latter is I_{sc1}/C_H, where

I_{sc1} is the short-circuit output current of OA_1. For instance, with $C_H = 0.5$ nF, an op amp having $SR_1 = 30$ V/μs and $I_{sc1} = 20$ mA gives $I_{sc1}/C_H = 40$ V/μs, indicating that speed is limited by SR_1. However, with $C_H = 1$ nF, we get $I_{sc1}/C_H = 20$ V/μs, indicating that speed is now limited by I_{sc1}. The output current drive of OA_1 can be boosted by replacing D_3 with the *B-E* junction of an *npn* BJT, whose collector is returned to V_{CC} via a series resistance on the order of $10^2 \Omega$ to limit current spikes below a proper safety level.

9.7
SAMPLE-AND-HOLD AMPLIFIERS

It is often necessary to capture the value of a signal in response to a suitable logic command, and hold it until the arrival of a new capture command. We have been exposed to this concept in Chapter 5 in connection with autozeroing amplifiers, where the signal in question is an offset-nulling voltage. Other examples will be encountered in Chapter 12 in connection with A-D and D-A converters.

A *sample-and-hold amplifier* (SHA) is a circuit in which the value of the input signal is captured instantaneously, as shown in Fig. 9.44*a*. Though mathematically convenient in sampled-data theory, instantaneous capture is unfeasible because of inherent dynamic limitations of physical circuits. Rather, a practical circuit is made to track the input for a prescribed time interval, and then hold its most recent value for the remainder of the cycle. The timing of the *track-and-hold amplifier* (THA) is shown in Fig. 9.44*b*. In spite of the obvious differences between the diagrams, engineers use the designations SHA and THA interchangeably.

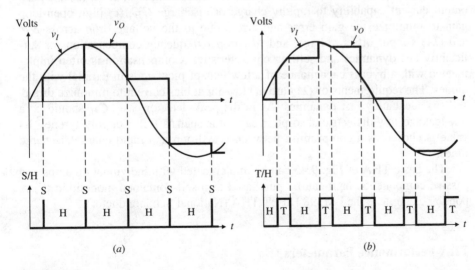

FIGURE 9.44
Idealized responses of (*a*) the sample-and-hold amplifier (SHA), and (*b*) the track-and-hold amplifier (THA).

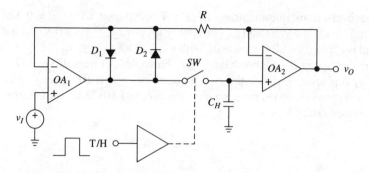

FIGURE 9.45
Basic track-and-hold amplifier.

Figure 9.45 shows one of the most popular THA topologies. The circuit is reminiscent of the peak detector, except for the replacement of the diode switch with an externally controlled bidirectional switch to charge as well as discharge C_H, depending on the case. The circuit operates as follows.

During the *track* mode, SW is closed to create the feedback path SW-OA_2-R around OA_1. Due to the low voltage drop across SW, both diodes are off, indicating a 0-V drop across R. OA_1 thus acts as a voltage follower, providing C_H with whatever current it takes to make v_O track v_I.

During the *hold* mode, SW is opened, allowing C_H to retain whatever voltage it had at the instant of switch aperture; OA_2 then buffers this voltage to the outside. The function of D_1 and D_2 is to prevent OA_1 from saturating, and thus facilitate OA_1's recovery when a new track command is received.

The switch is usually a JFET, a MOSFET, or a Schottky diode bridge, and is equipped with a suitable driver to make the T/H command TTL- or CMOS-compatible. The main requirements of OA_1 are (a) low-input dc error, (b) adequate output current capability to rapidly charge or discharge C_H, (c) high open-loop gain to minimize the gain error and errors due to the voltage drop across SW and OA_2's input offset voltage, and (d) proper frequency compensation for sufficiently fast dynamics and settling characteristics. Compensation is often implemented with a bypass capacitance of a few tens of picofarads in parallel with the diodes. The requirements of OA_2 are (a) low-input bias current to minimize droop, and (b) adequately fast dynamics. As in the peak-detector case, C_H should be a low-leakage, low-dielectric-absorption capacitor, such as Teflon or polystyrene.[3] Its value is chosen as a compromise between low droop and rapid charge/discharge times.

The basic THA of Fig. 9.45 can be implemented with individual op amps and passive components, or it can be purchased as a self-contained monolithic IC. A popular example is the LM398 BiFET THA (National Semiconductor).

THA Performance Parameters

In the track mode, a THA is designed to behave like an ordinary amplifier, so its performance is characterized in terms of the dc and gain errors, the dynamics, and

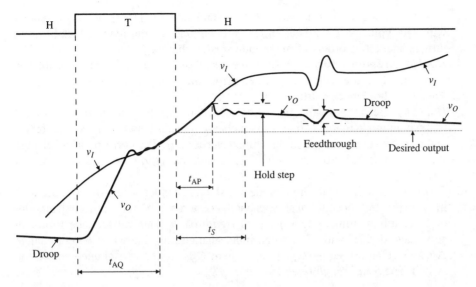

FIGURE 9.46
THA terminology.

other parameters peculiar to amplifiers. However, during the transition from the track to the hold mode and vice versa, as well as during the hold mode itself, performance is characterized by specifications peculiar to THAs. The following list uses the expanded timing diagram of Fig. 9.46 as a guideline.

1. *Acquisition Time (t_{AQ}).* Following the track command, v_O starts slewing toward v_I, and t_{AQ} is the time it takes for v_O to begin tracking v_I within a specified error band after the inception of the track command. This includes propagation delays through the switch driver and the switch, and delays due to slew-rate limiting and settling times of the op amps. The acquisition time increases with the step magnitude as well as the narrowness of the error band. Usually t_{AQ} is specified for a 10-V step and for error bands of 1%, 0.1%, and 0.01% of full scale. The input must be fully acquired before switching to the hold mode.
2. *Aperture Time (t_{AP}).* Because of propagation delays through the driver and the switch, v_O will cease tracking v_I some time after the inception of the hold command. This is the aperture time. The hold command would have to be advanced by t_{AP} for precise timing.
3. *Aperture Uncertainty (Δt_{AP}).* Also called *aperture jitter,* it represents the variation in aperture time from sample to sample. If t_{AP} is compensated for by advancing the hold command by t_{AP}, then Δt_{AP} establishes the ultimate timing error and, hence, the maximum sampling frequency for a given resolution. Aperture jitter results in an output error $\Delta v_O = (dv_I/dt)\Delta t_{AP}$, indicating that the actual sampled waveform can be viewed as the sum of an ideally sampled waveform and a noise component. The signal-to-noise ratio of an otherwise ideal sampling circuit with a sinusoidal input of frequency f_i is[4]

$$\text{SNR} = -20\log_{10}(2\pi f_i \Delta t_{AP(\text{rms})}) \tag{9.24}$$

where $\Delta t_{AP(rms)}$ is the rms value of Δt_{AP}, and the latter is assumed uncorrelated with v_I. Typically, Δt_{AP} is an order of magnitude smaller than t_{AP}, and t_{AP}, in turn, is one to two orders of magnitude smaller than t_{AQ}.

4. *Hold Mode Settling Time (t_S)*. After the inception of the hold command, it takes some time to settle within a specified error band, such as 1%, 0.1%, or 0.01%. This is the hold mode settling time.

5. *Hold Step*. Because of parasitic switch capacitances, when the circuit is switched to the hold mode there is an unwanted charge transfer between the switch driver and C_H, causing a change in the voltage across C_H. The corresponding change Δv_O is referred to as *hold step, pedestal error*, or *sample-to-hold offset*.

6. *Feedthrough*. When in the hold mode, v_O should be independent of any variations in v_I. In practice, because of stray capacitance across SW, there is a small amount of ac coupling from v_I to v_O called *feedthrough*. This capacitance forms an ac voltage divider with C_H, so an input change Δv_I causes an output change $\Delta v_O = [C_{SW}/(C_{SW} + C_H)]\Delta v_I$, where C_{SW} is the capacitance across the switch. The *feedthrough rejection ratio*

$$\text{FRR} = 20\log\frac{\Delta v_I}{\Delta v_O} \tag{9.25}$$

gives an indication of the amount of stray coupling. For example, if FRR = 80 dB, a hold mode change $\Delta v_I = 10$ V results in $\Delta v_O = \Delta v_I/10^{80/20} = 10/10^4 = 1$ mV.

7. *Voltage Droop*. THAs are subject to the same droop limitations as peak detectors. Droop is of special concern when C_H must be kept low to ensure a fast acquisition.

In the case of a JFET switch, feedthrough is due to the drain-source capacitance C_{ds}, and the hold step is due to the gate-drain capacitance C_{gd}. (For discrete devices, these capacitances are typically in the picofarad range.) As the driver pulls the gate from near v_O to near V_{EE}, it removes the charge $\Delta Q \cong C_{gd}(V_{EE} - v_O)$ from C_H, causing a hold step

$$\Delta v_O \cong \frac{C_{gd}}{C_{gd} + C_H}(V_{EE} - v_O) \tag{9.26}$$

This step varies with v_O. For example, with $C_H = 1$ nF and $V_{EE} = -15$ V, the hold step for every picofarad of C_{gd} is about -15 mV/pF for $v_O = 0$, -20 mV/pF for $v_O = 5$ V, and -10 mV/pF for $v_O = -5$ V. A C_{gd} of just a few picofarads can cause intolerable errors!

There are various techniques for minimizing the signal-dependent hold step. One such technique is to implement the switch with the CMOS transmission gate of Fig. 9.39a. Since the two FETs are driven in antiphase, one FET will inject and the other will remove charge, and if their geometries are properly scaled, the two charges will cancel each other out.

An alternative technique[5] is depicted in Fig. 9.47. As the circuit goes into hold, OA_4 produces a positive-going output swing that, by the superposition principle, depends on v_O as well as on the negative step on the switch gate. This swing is

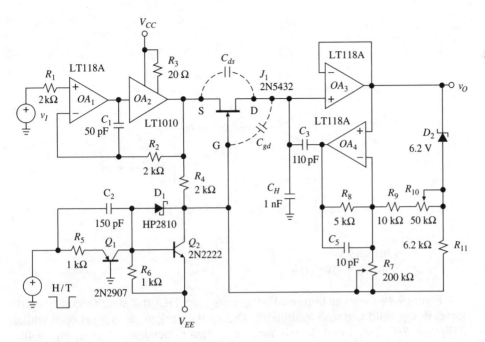

FIGURE 9.47
A 5-MHz THA with charge-transfer compensation to minimize the hold step. (Courtesy of
Linear Technology.)

designed to inject into C_H, via C_3, a charge packet of magnitude equal to that
removed via C_{gd}, thus resulting in a net charge transfer of zero. The hold step is
made independent of v_O with R_7, and adjusted to zero via R_{10}. To calibrate the
circuit, adjust R_7 for equal hold steps with $v_I = \pm 5$ V; then, null the residual offset
via R_{10}.

To achieve high speed, the circuit utilizes fast op amps and boosts OA_1 with the
LT1010 fast power buffer to rapidly charge and discharge C_H in the track mode.
Moreover, by using local feedback around the OA_1-OA_2 pair, the settling dynamics
of the input and output stages are kept separate and simpler. With OA_3 no longer
inside the control loop, its input offset voltage is no longer irrelevant; however, its
offset as well as that of the input buffer are automatically compensated for during
calibration of R_{10}. For long hold periods, OA_3 can be replaced by a FET-input device
such as the LF356 to reduce droop.

Charge compensation can be simplified considerably if the switch is operated in
a virtual-ground arrangement. This is the case with THAs of the *integrating type*,[4,6]
so-called because the holding capacitor is placed in the feedback path of the output
amplifier, as exemplified in Fig. 9.48. Since the switch always sees a virtual ground,
the charge removed from the summing junction via C_{gd} is constant regardless of v_O.
Consequently, the hold step appears as a constant offset that can easily be nulled by
standard techniques, such as adjusting the offset of OA_1, as shown. With an easier-
to-compensate hold step, the holding capacitance can be reduced significantly to
achieve faster acquisition times. The HA-5330 high-speed monolithic THA (Harris)
uses a 90-pF holding capacitance to achieve $t_{AQ} = 400$ ns to 0.01%.

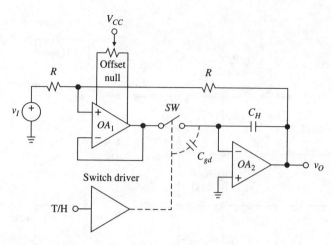

FIGURE 9.48
Integrating-type THA.

Figure 9.49 shows an improved integrating-type THA that simultaneously optimizes droop, hold step, and feedthough. During the track mode, SW_3 is open while SW_1 and SW_2 are closed. In this mode the circuit operates just as in Fig. 9.48, with v_O slewing toward $-v_I$. During the hold mode, SW_1 and SW_2 are open while SW_3 is closed, causing the circuit to hold whatever voltage it acquired during sampling. Note, however, that by grounding the input to the buffer via SW_3, any variations in v_I are muzzled, thus improving the FRR significantly. Moreover, since both SW_1 and SW_2 experience a voltage drop very close to zero, switch leakage is virtually eliminated. The main source of leakage is now the input bias current of OA. However, returning its noninverting input to a dummy capacitance C of size equal to C_H produces a hold step and a droop that, to a first approximation, will cancel out the hold step and droop of C_H. An example of a THA utilizing this technique is the SHC803/804 (Burr-Brown), whose typical ambient-temperature ratings are: $t_{AQ} = 250$ ns and $t_S = 100$ ns, both to 0.01%; $t_{AP} = 15$ ns; $\Delta t_{AP} = \pm 15$ ps; FRR $= \pm 0.005\%$, or 86 dB; hold mode offset $= \pm 2$ mV; droop rate $= \pm 0.5\ \mu V/\mu s$.

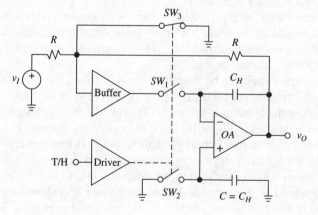

FIGURE 9.49
Improved THA. (Switch settings shown for the hold mode.)

THAs are available from a variety of sources and in a wide range of perfor- mance specifications and prices. Consult the catalogs to familiarize yourself with the available products.

PROBLEMS

9.1 Voltage comparators

9.1 (a) Using a 311 comparator powered from ±15-V regulated supplies, design a threshold detector such that $v_O \cong 0$ V for $v_I > 1$ V, and $v_O \cong 5$ V for $v_I < 1$ V. (b) Repeat, but with $v_O \cong -15$ V for $v_I > 5$ V, and $v_O \cong 0$ V for $v_I < 5$ V.

9.2 Comparator applications

9.2 The thermal characteristic of a certain class of thermistors can be expressed as $R(T) = R(T_0)\exp[B(1/T - 1/T_0)]$, where T is absolute temperature, T_0 is some reference temperature, and B is a suitable constant, all three parameters being in kelvins. Using a single comparator of the 339 type and a thermistor having $R(25\,°C) = 100$ kΩ and $B = 4000$ K, design a bridge comparator circuit that gives $v_O = V_{OH}$ for $T > 100\,°C$, and $v_O = V_{OL}$ for $T < 100\,°C$. Assuming 10% component tolerances, make provision for the exact adjustment of the setpoint, and outline the calibration procedure.

9.3 Using an op amp, two comparators of the 339 type, a 2N2222 *npn* BJT, and resistors as needed, design a circuit that accepts a data input v_I and a control input $V_T \geq 0$, and causes a 10-mA, 1.5-V LED to glow whenever $-V_T < v_I < V_T$. Assume ±15-V regulated supplies.

9.4 Using two comparators of the 339 type and a thermistor of the type of Problem 9.2 with $R(25\,°C) = 10$ kΩ and $B = 4000$ K, design a circuit that yields $v_O \cong 5$ V for $0\,°C \leq T \leq 5\,°C$, and $v_O \cong 0$ V otherwise. Assume a single 5-V regulated supply.

9.5 Show that the window detector of Fig. P9.5 has a window whose center is controlled by v_1 and whose width is controlled by v_2; then sketch and label the VTC if $v_1 = 3$ V and $v_2 = 1$ V.

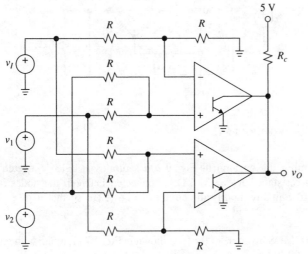

FIGURE P9.5

9.6 Using three comparators of the 339 type, an LM385 2.5-V reference diode, an HLMP-4700 LED of the type of Example 9.2, and resistors as needed, design a circuit that monitors a 15-V ± 5% power supply and causes the LED to glow whenever the supply is within range.

9.7 Using an LM385 2.5-V reference diode, an LM339 quad comparator, and four HLMP-4700 LEDs of the type of Example 9.2 design a 0-V to 4-V bar graph meter. The circuit must have an input impedance of at least 100 kΩ and must be powered from a single 5-V supply.

9.8 Using a 311 comparator powered from ±15-V regulated supplies, design a circuit that accepts a triangular wave with peak values of ±10 V, and generates a square wave with peak values of ±5 V and duty cycle D variable from 5% to 95% by means of a 10-kΩ potentiometer.

9.3 Schmitt triggers

9.9 In the circuit of Fig. 9.20a let v_I be a triangular wave of ±10 V peak values and let $\pm V_{\text{sat}} = \pm 13$ V. Modify the circuit so that the phase of its square-wave output, relative to that of the input, is variable from 0° to 90° by means of a 10-kΩ potentiometer. Show the input and output waveforms when the wiper is in the middle.

9.10 (a) Derive Eq. (9.14). (b) Specify suitable resistances in the circuit of Fig. 9.23 to achieve $V_{OL} = 0$ V, $V_{OH} = 5$ V, $V_{TL} = 1.5$ V, and $V_{TH} = 2.5$ V with $V_{CC} = 5$ V. Try minimizing the effect of the input bias current.

9.11 Assuming $V_{D(\text{on})} = 0.7$ V and $\pm V_{\text{sat}} = \pm 13$ V, sketch and label the VTC of the inverting Schmitt trigger of Fig. P9.11.

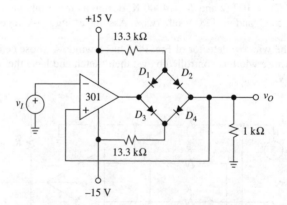

FIGURE P9.11

9.12 (a) Assuming the op amp of Fig. 9.20a saturates at ±13 V, sketch and label the VTC if a resistance $R_3 = 33$ kΩ is connected between the nodes labeled v_P and −15 V. (b) Suitably modify the circuit of Fig. 9.21a so that it gives $V_{TL} = 1$ V and $V_{TH} = 2$ V.

9.13 (a) Using CMOS inverters of the type shown in Fig. 10.11, along with resistances in the 10-kΩ to 100-kΩ range, design a noninverting Schmitt trigger with $V_{TL} = (1/3)V_{DD}$ and

$V_{TH} = (2/3)V_{DD}$; assume $V_T = 0.5V_{DD}$. (b) Modify the circuit so that $V_{TL} = (1/5)V_{DD}$ and $V_{TH} = (1/2)V_{DD}$. (c) How would you turn the preceding circuits into Schmitt triggers of the inverting type?

9.14 Suitably modify the circuit of Problem 9.2 to ensure a hysteresis of $\pm0.5\,°C$. Outline the calibration procedure.

9.15 In the Schmitt trigger of Fig. 9.20a let the input v_I be applied to the inverting-input pin via a voltage divider made up of two 10-kΩ resistances, let R_1 be replaced by the series combination of two 4.3-V Zener diodes connected back to back anode with anode, and let the output v_O be obtained at the node where R_2 joins the Zener network. Draw the circuit. Hence, assuming a 0.7-V forward-bias diode voltage drop, sketch and label the VTC.

9.16 In the circuit of Fig. P1.18 let the source be a variable source, denoted as i_I, and let the op amp saturate at ±10 V. (a) Sketch and label v_O versus i_I for i_I variable over the range -1 mA $\leq i_I \leq 1$ mA. (b) Repeat, but with a 2-kΩ resistor in parallel with i_I, and for i_I variable over the range -2 mA $\leq i_I \leq 2$ mA. *Hint:* Take into account the considerations made in connection with Eq. (1.76).

9.17 A circuit consists of a 311 comparator and three equal resistors, $R_1 = R_2 = R_3 = 10$ kΩ. The 311 is powered between 15 V and ground, and has $V_{EE(logic)} = 0$. Moreover, R_1 is connected between the 15-V supply and the noninverting-input pin, R_2 between the noninverting-input pin and the open-collector output pin, and R_3 between the open-collector output pin and ground. Moreover, the input v_I is applied to the comparator's inverting-input pin. Draw the circuit, and sketch and label its VTC if the output v_O is obtained from: (a) the node where R_1 joins R_2; (b) the node where R_2 joins R_3.

9.18 Consider the circuit obtained by removing R, C, and OA from Fig. 10.19a. What is left then is a noninverting Schmitt trigger, whose input is the node labeled as v_{TR}, and whose output is the node labeled as v_{SQ}. Sketch and label its VTC if $R_1 = 10$ kΩ, $R_2 = 13$ kΩ, $R_3 = 4.7$ kΩ, and the Zener diode is a 5.1-V device; assume forward-bias diode voltage drops of 0.7 V.

9.4 Precision rectifiers

9.19 Sketch and label the VTC of the circuit of Fig. 9.29a if $R_2 = 2R_1$ and the noninverting input of the op amp is lifted off ground and returned to a -5-V reference voltage. Next, sketch and label v_O if v_I is a triangular wave with peak values of ±10 V.

9.20 Sketch and label the VTC of the circuit of Fig. 9.29a if $R_1 = R_2 = 10$ kΩ, and a third resistance $R_3 = 150$ kΩ is connected between the $+15$-V supply and the inverting-input pin of the op amp. (b) Repeat, but with the diode polarities reversed.

9.21 One side of a 10-kΩ resistance is driven by a source v_I, and the other side is left floating. Denoting the voltage at the floating side as v_O, use a superdiode circuit to implement a *variable precision clamp,* that is, a circuit that gives $v_O = v_I$ for $v_I \leq V_{clamp}$ and $v_O = V_{clamp}$ for $v_I \geq V_{clamp}$, where V_{clamp} is a continuously adjustable voltage from 0 to 10 V by means of a 100-kΩ pot. Assume ±15-V regulated supplies. List advantages and drawbacks of your circuit.

9.22 Suitably modify the FWR of Fig. 9.30 so that, when fed with a triangular wave of ±5-V peak values, it gives a triangular wave of ±5-V peak values, but twice the frequency. Assume regulated ±15-V supplies.

9.23 Assuming $R_1 = R_2 = R_4 = 10$ kΩ and $R_3 = 20$ kΩ in the FWR of Fig. 9.31, find all node voltages for $v_I = 10$ mV, 1 V, and -1 V. For a forward-biased diode, assume $v_D = (26$ mV$) \ln[i_D/(20$ fA$)]$.

9.24 Discuss the effect of resistance mismatches in the FWR of Fig. 9.31, and derive an expression for $100|(A_p - A_n)/A|$. Compare with the FWR of Fig. 9.30, and comment.

9.25 Consider the circuit obtained from that of Fig. 9.31 by grounding the left terminals of R_1 and R_4, lifting the noninverting input of OA_1 off ground and driving it with source v_I. (a) Show that the modified circuit gives $v_O = A_p v_I$ for $v_I > 0$ and $v_O = -A_n v_I$ for $v_I < 0$, where $A_p = 1 + (R_2 + R_3)/R_1$ and $A_n = R_3/R_2$. (b) Specify component values for $v_O = 5|v_I|$. List advantages and disadvantages of this circuit.

9.26 Consider the circuit obtained from that of Fig. 9.31 by removing R_1, grounding the left terminal of R_4, lifting the noninverting input of OA_1 off ground and driving it with source v_I. Analyze the modified circuit if $R_2 = R_3 = R$. Afterward, discuss the implications of mismatched resistances.

9.27 (a) Find the VTC of the circuit of Fig. P9.27. (b) Assuming $\pm V_{sat} = \pm 13$ V and $V_{D(on)} = 0.7$ V, show all node voltages for $v_I = +3$ V and $v_I = -5$ V. (c) List advantages and disadvantages of this circuit.

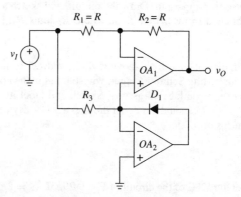

FIGURE P9.27

9.28 The circuit of Fig. 9.30 can be turned into a high-input-impedance FWR by lifting both noninverting inputs off ground, tying them together, and driving them with a common input v_I; moreover, R_4 is removed, and the left terminal of R_1 is grounded. (a) Assuming $R_1 = R_2 = R_3 = R$ and $R_5 = 2R$, find the VTC of the modified circuit. (b) Assuming $V_{D(on)} = 0.7$ V, show all node voltages for $v_I = +2$ V and $v_I = -3$ V. (c) Investigate the effect of mismatched resistances.

9.29 (a) Find the VTC of the circuit of Fig. P9.29; then, assuming $V_{D(on)} = 0.7$ V, show all node voltages for $v_I = +1$ V and $v_I = -3$ V. (b) Suitably modify the circuit so that it accepts two inputs v_1 and v_2, and gives $v_O = |v_1 + v_2|$.

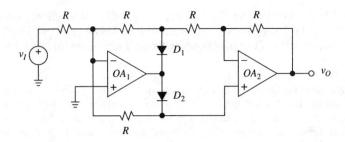

FIGURE P9.29

9.30 Investigate the effect of the input offset voltages V_{OS1} and V_{OS2} of OA_1 and OA_2 in the FWR of Fig. 9.30.

9.5 Analog switches

9.31 Using a 311 comparator, a 2N4391 n-JFET, and a 741 op amp, design a circuit that accepts an analog signal v_I and two control signals v_1 and v_2, and yields a signal v_O such that $v_O = 10v_I$ for $v_1 > v_2$, and $v_O = -10v_I$ for $v_1 < v_2$. Assume ±15-V supplies.

9.32 For small values of $|v_{DS}|$, the channel resistance of a MOSFET can be found as $1/r_{ds(on)} \cong k(|v_{GS}| - |V_{GS(on)}|)$, where k is called the *device transconductance parameter,* in amperes per square volt. Assuming ±5-V supplies in the transmission gate of Fig. 9.39a, and truly complementary FETS with $k = 100\ \mu A/V^2$ and $|V_{GS(on)}| = 2.5$ V, find the net switch resistance for $v_I = \pm5$ V, ±2.5 V, and 0 V. What are the corresponding values of v_O if $R_L = 100$ kΩ?

9.6 Peak detectors

9.33 Consider the circuit obtained from that of Fig. 9.40b by returning the noninverting input of OA_1 to ground, and applying the source v_I to the inverting input of OA_1 via a series resistance having the same value as the feedback resistance R. Discuss how the modified circuit operates, and show its response to a sinusoidal input of increasing amplitude.

9.34 Design a peak-to-peak detector, that is, a circuit that gives $v_O = v_{I(max)} - v_{I(min)}$.

9.35 Using the circuit of Fig. 9.29a as a starting point, design a circuit to provide the *magnitude peak-detector* function, $v_O = |v_I|_{max}$.

9.36 Three superdiodes of the type of Fig. 9.27 are driven by three separate sources v_1, v_2, and v_3, and their outputs are tied together and returned to -15 V via a 10-kΩ resistor. What function does the circuit provide? What happens if the diode polarities are reversed? If the node common to the outputs is returned to the node common to the inverting inputs via a voltage divider?

9.7 Sample-and-hold amplifiers

9.37 Suitably modify the THA of Fig. 9.45 for a gain of 2 V/V. What is the main disadvantage of the modified circuit, and how would you take care of it?

9.38 In the THA of Fig. 9.48 let $C_{gd} = 1$ pF, $C_H = 1$ nF, and let the net leakage current through C_H be 1 nA, flowing from right to left. Assuming $v_I = 1.000$ V, find v_O (a) shortly after the circuit is switched to the hold mode, and (b) 50 ms later.

9.39 The THA of Fig. P9.39 uses a feedback capacitor $C_F = C_H$ to provide a first-order compensation for the droop due to leakage in C_H. (a) Explain how the circuit works. What are the functions of the p-channel JFET J_1 and the n-channel JFETs J_2 and J_3? (b) Assuming an average leakage of 1 nA in each capacitor and a leakage mismatch of 5%, estimate the voltage droop for the case $C_F = C_H = 1$ nF. What would be the leakage if C_F were absent and replaced with a wire?

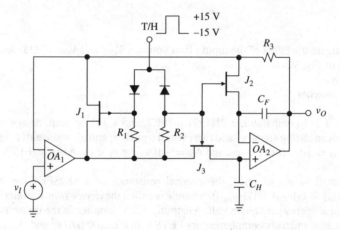

FIGURE P9.39

REFERENCES

1. J. Sylvan, "High-Speed Comparators Provide Many Useful Circuit Functions When Used Correctly," *Analog Dialogue,* vol. 23, no. 4, Analog Devices, Norwood, MA, 1989.
2. J. Williams, "High-Speed Comparator Techniques," Application Note AN-13, *Linear Applications Handbook Volume I,* Linear Technology, Milpitas, CA, 1990.
3. S. Guinta, "Capacitance and Capacitors," *Analog Dialogue,* vol. 30, no. 2, Analog Devices, Norwood, MA, 1996.
4. B. Razavi, *Principles of Data Conversion System Design,* IEEE Press, Piscataway, NJ, 1995.
5. R. J. Widlar, "Unique IC Buffer Enhances Op Amp Designs, Tames Fast Amplifiers," Application Note AN-16, *Linear Applications Handbook Volume I,* Linear Technology, Milpitas, CA, 1990.
6. D. A. Johns and K. W. Martin, *Analog Integrated Circuit Design,* John Wiley and Sons, New York, 1997.

10

SIGNAL GENERATORS

10.1 Sine Wave Generators
10.2 Multivibrators
10.3 Monolithic Timers
10.4 Triangular Wave Generators
10.5 Sawtooth Wave Generators
10.6 Monolithic Waveform Generators
10.7 *V-F* and *F-V* Converters
      Problems
      References

The circuits investigated so far can be categorized as processing circuits because they operate on existing signals. We now wish to investigate the class of circuits used to generate the signals themselves. Though signals are sometimes obtained from transducers, in most cases they need to be synthesized within the system. The generation of clock pulses for timing and control, signal carriers for information transmission and storage, sweep signals for information display, test signals for automatic test and measurement, and audio signals for electronic music and speech synthesis are some of the most common examples.

The function of a signal generator is to produce a waveform of prescribed characteristics such as frequency, amplitude, shape, and duty cycle. Sometimes these characteristics are designed to be externally programmable via suitable control signals, the voltage-controlled oscillator being the most typical example. In general, signal generators employ some form of feedback together with devices possessing time-dependent characteristics, such as capacitors. The two main categories of signal generators that we shall investigate are sinusoidal oscillators and relaxation oscillators.

Sinusoidal Oscillators

These oscillators employ concepts from systems theory to create a pair of conjugate poles right on the imaginary axis of the complex plane to maintain sustained sinusoidal oscillation. The specter of instability that was of so much concern in Chapter 8 is now exploited on purpose to achieve predictable oscillation.

The sinusoidal purity of a periodic wave is expressed via its *total harmonic distortion*

$$\text{THD } (\%) = 100\sqrt{D_2^2 + D_3^2 + D_4^2 + \cdots} \tag{10.1}$$

where D_k $(k = 2, 3, 4, \ldots)$ is the ratio of the amplitude of the kth harmonic to that of the fundamental in the Fourier series of the given wave. For instance, the triangular wave, for which $D_k = 1/k^2$, $k = 3, 5, 7, \ldots$, has $\text{THD} = 100 \times \sqrt{1/3^4 + 1/5^4 + 1/7^4 + \cdots} \cong 12\%$, indicating that as a crude approximation to a sine wave, a triangular wave has a THD of 12%. On the other hand, a pure sine wave has all harmonics, except for the fundamental, equal to zero, so $\text{THD} = 0\%$ in this case. Clearly, the objective of a sine wave generator is to achieve a THD as low as possible.

Relaxation Oscillators

These oscillators employ bistable devices, such as switches, Schmitt triggers, logic gates, and flip-flops, to repeatedly charge and discharge a capacitor. Typical waveforms obtainable with this method are the triangular, sawtooth, exponential, square, and pulse waves. As we proceed, we shall often need to find the time Δt it takes to charge (or discharge) a capacitance by a given amount Δv. The two most common forms of charge/discharge are linear and exponential.

When driven with a constant current I, a capacitance C charges or discharges at a constant rate, yielding a *linear transient* or *ramp* of the type of Fig. 10.1a. Engineers

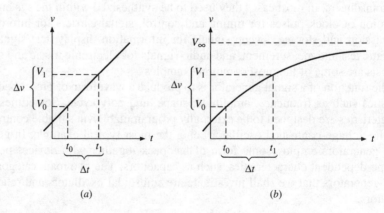

(a)   (b)

FIGURE 10.1
Linear and exponential waveforms.

often describe this ramp via the easy-to-remember relationship

$$C \Delta v = I \Delta t$$

or "cee delta vee equals aye delta tee." This allows us to estimate the time it takes to effect a constant-rate change Δv as

$$\Delta t = \frac{C}{I} \Delta v \qquad (10.2)$$

An *exponential transient* occurs when C is charged or discharged via a series resistance R. With reference to Fig. 10.1b, the instantaneous capacitance voltage is

$$v(t) = V_\infty + (V_0 - V_\infty) \exp[(t - t_0)/\tau]$$

where V_0 is the initial voltage, V_∞ is the steady-state voltage that would be reached in the limit $t \to \infty$, and $\tau = RC$ is the time constant governing the transient. This equation holds regardless of the values and polarities of V_0 and V_∞. The transient reaches a specified intermediate value V_1 at an instant t_1 such that $V_1 = V_\infty + (V_0 - V_\infty) \exp[(t_1 - t_0)/\tau]$. Taking the natural logarithm of both sides and solving for $\Delta t = t_1 - t_0$ allows us to estimate the time it takes to charge or discharge C from V_0 to V_1 as

$$\Delta t = \tau \ln \frac{V_\infty - V_0}{V_\infty - V_1} \qquad (10.3)$$

As we proceed, we shall make frequent use of these equations.

10.1
SINE WAVE GENERATORS

The sine wave is certainly one of the most fundamental waveforms—both in a mathematical sense, since any other waveform can be expressed as a Fourier combination of basic sine waves, and in a practical sense, since it finds extensive use as a test, reference, and carrier signal. In spite of its simplicity, its generation can be a challenging task if near-purity is sought. The op amp circuits that have gained the most prominence in sine wave generation are the *Wien-bridge oscillator* and the *quadrature oscillator*, to be discussed next. Another technique, based on the conversion of the triangular to the sine wave, will be discussed in Section 10.4.

Basic Wien-Bridge Oscillator

The circuit of Fig. 10.2a uses both negative feedback, via R_2 and R_1, and positive feedback, via the series and parallel RC networks. Circuit behavior is strongly affected by whether positive or negative feedback prevails. The components of the RC networks need not be equal-valued; however, making them so simplifies analysis as well as inventory.

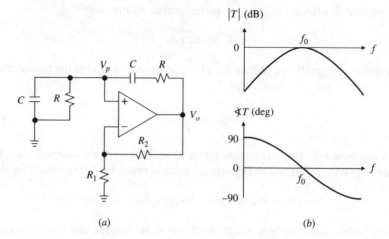

FIGURE 10.2
Wien-bridge circuit and its loop gain $T(jf)$ for the case $R_2/R_1 = 2$.

The circuit can be viewed as a noninverting amplifier that amplifies V_p by the amount

$$A = \frac{V_o}{V_p} = 1 + \frac{R_2}{R_1} \tag{10.4}$$

where we are assuming an ideal op amp for simplicity. In turn, V_p is supplied by the op amp itself via the two RC networks as $V_p = [Z_p/(Z_p + Z_s)]V_o$, where $Z_p = R \parallel (1/j2\pi fC)$ and $Z_s = R + 1/j2\pi fC$. Expanding, we get

$$B(jf) = \frac{V_p}{V_o} = \frac{1}{3 + j(f/f_0 - f_0/f)} \tag{10.5}$$

where $f_0 = 1/2\pi RC$. The overall gain experienced by a signal in going around the loop is $T(jf) = AB$, or

$$T(jf) = \frac{1 + R_2/R_1}{3 + j(f/f_0 - f_0/f)} \tag{10.6}$$

This is a band-pass function since it approaches zero at both high and low frequencies. Its peak value occurs at $f = f_0$ and is

$$T(jf_0) = \frac{1 + R_2/R_1}{3} \tag{10.7}$$

The fact that $T(jf_0)$ is real indicates that a signal of frequency f_0 will experience a net phase shift of zero in going around the loop. Depending on the magnitude of $T(jf_0)$, we have three distinct possibilities:

1. $T(jf_0) < 1$, that is, $A < 3$ V/V. Any disturbance of frequency f_0 arising at the input of the op amp is first amplified by $A < 3$ V/V, and then by $B(jf_0) = \frac{1}{3}$ V/V, for a net gain of less than unity. Intuition tells us that this disturbance lessens each time it goes around the loop until it eventually decays to zero. We can state that negative feedback (via R_2 and R_1) prevails over positive feedback (via Z_s

and Z_p), resulting in a stable system. Consequently, the circuit poles lie in the left half of the complex plane.

2. $T(jf_0) > 1$, that is, $A > 3$ V/V. Now positive feedback prevails over negative feedback, indicating that a disturbance of frequency f_0 will be amplified regeneratively, causing the circuit to break out into oscillations of growing magnitude. The circuit is now *unstable,* and its poles lie in the right half of the complex plane. As we know, the oscillations build up until the saturation limits of the op amp are reached. Thereafter, v_O will appear as a clipped sine wave when observed with the oscilloscope or visualized via PSpice.

3. $T(jf_0) = 1$, or $A = 3$ V/V *exactly,* a condition referred to as *neutral stability* because positive and negative feedback are now applied in equal amounts. Any disturbance of frequency f_0 is first amplified by 3 V/V and then by $\frac{1}{3}$ V/V, indicating that once started, it will be sustained indefinitely. As we know, this corresponds to a pole pair right on the $j\omega$ axis. The conditions $\angle T(jf_0) = 0°$ and $|T(jf_0)| = 1$ are together referred to as the *Barkhausen criterion* for oscillation at $f = f_0$. The band-pass nature of $T(jf)$ allows for oscillation to occur only at $f = f_0$; any attempt to oscillate at other frequencies is naturally discouraged because $\angle T \neq 0°$ and $|T| < 1$ there. By Eq. (10.7), neutral stability is achieved with

$$\frac{R_2}{R_1} = 2 \qquad (10.8)$$

It is apparent that when this condition is met, the components around the op amp form a *balanced bridge* at $f = f_0$.

In a real-life circuit, component drift makes it difficult to keep the bridge exactly balanced. Moreover, provisions must be made so that (*a*) oscillation starts spontaneously at power turn-on, and (*b*) its amplitude is kept below the op amp saturation limits to avoid excessive distortion. These objectives are met by making the ratio R_2/R_1 *amplitude-dependent* such that at low signal levels it is slightly greater than 2 to ensure oscillation start-up, and that at high signal levels it is slightly less than 2 to limit amplitude. Then, once the oscillation has started, it will grow and automatically stabilize at some intermediate level where $R_2/R_1 = 2$ *exactly.*

Amplitude stabilization takes on many forms, all of which use nonlinear elements to either decrease R_2 or increase R_1 with signal amplitude. To provide an intuitive basis for our discussion, we shall continue using the function $T(jf)$, but in an incremental sense because of the nonlinearity now present in the circuit.

Automatic Amplitude Control

The circuit of Fig. 10.3*a* uses a simple diode-resistor network to control the effective value of R_2. At low signal levels the diodes are off, so the 100-kΩ resistance has no effect. We thus have $R_2/R_1 = 22.1/10.0 = 2.21$, or $T(jf_0) = (1 + 2.21)/3 = 1.07 > 1$, indicating oscillation buildup. As the oscillation grows, the diodes are gradually brought into conduction on alternate half-cycles. In the limit of heavy diode conduction, R_2 would effectively change to $(22.1 \parallel 100) = 18.1$ kΩ, giving $T(jf_0) = 0.937 < 1$. However, before this limiting condition is reached, amplitude

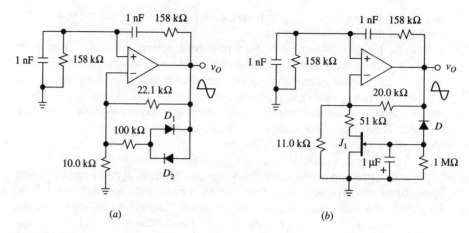

(a)

(b)

FIGURE 10.3
Practical Wien-bridge oscillators.

will automatically stabilize at some intermediate level of diode conduction where $R_2/R_1 = 2$ exactly, or $T(jf_0) = 1$. The process can be visualized via PSpice using the following file.

```
Wien-Bridge Oscillator:
Cp 3 0 1nF IC=0V
Rp 3 0 158k
Cs 3 36 1nF IC=0V
Rs 36 6 158k
R1 2 0 10k
R2 2 6 22.1k
R3 2 26 100k
D1 26 6 D1N4148
D2 6 26 D1N4148
.model D1N4148 D(Is=0.1p Rs=16 CJO=2p Tt=12n Bv=100 Ibv=0.1p)
.lib eval.lib
XOA 3 2 7 4 6 ua741
VCC 7 0 dc 15V
VEE 4 0 dc -15V
.tran 50us 15ms 0ms 50us UIC
.probe
.end
```

As shown in Fig. 10.4, the output stabilizes automatically at a peak amplitude $V_{om} \cong$ 1.5 V.

A disadvantage of the above circuit is that V_{om} is quite sensitive to variations in the diode-forward voltage drops. The circuit of Fig. 10.3b overcomes this drawback by using an n-JFET as the stabilizing element.[1] At power turn-on, when the 1-μF capacitance is still discharged, the gate voltage is near 0 V, indicating a low channel resistance. The JFET effectively shorts the 51-kΩ resistance to ground to give $R_2/R_1 \cong 20.0/(11.0 \| 51) \cong 2.21 > 2$, so oscillation starts to build up. The diode and the 1-μF capacitance form a negative peak detector whose voltage becomes progressively more negative as the oscillation grows. This gradually reduces the

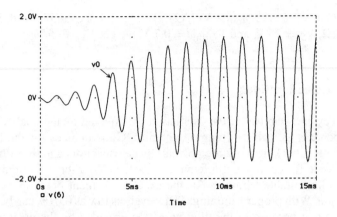

FIGURE 10.4
Using PSpice to display the output of the circuit of Fig. 10.3a.

conductivity of the JFET until, in the limit of complete cutoff we would have $R_2/R_1 = 20.0/11.0 = 1.82 < 2$. However, amplitude stabilizes automatically at some intermediate level where $R_2/R_1 = 2$ exactly. Denoting the corresponding gate-source voltage as $V_{GS(\text{crit})}$, and the output peak amplitude as V_{om}, we have $-V_{om} = V_{GS(\text{crit})} - V_{D(\text{on})}$. For instance, with $V_{GS(\text{crit})} = -4.3$ V we get $V_{om} \cong 4.3 + 0.7 = 5$ V.

Figure 10.5 shows yet another popular amplitude-stabilization scheme,[2] this time using a diode limiter for easier programming of amplitude. As usual, for low output levels the diodes are biased in cutoff, yielding $R_2/R_1 = 2.21 > 2$. The oscillation grows until the diodes become conductive on alternate output peaks. Thanks to the symmetry of the clamping network, these peaks are likewise symmetric, or $\pm V_{om}$. To estimate V_{om}, consider the instant when D_2 starts to conduct. Assuming the current through D_2 is still negligible, and denoting the voltage at the anode of D_2 as V_2, we use KCL to write $(V_{om} - V_2)/R_3 \cong [V_2 - (-V_S)]/R_4$, where $V_2 = V_n + V_{D2(\text{on})} \cong V_{om}/3 + V_{D2(\text{on})}$. Eliminating V_2 and solving gives

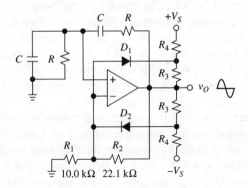

FIGURE 10.5
Wien-bridge oscillator using a limiter for amplitude stabilization.

$V_{om} \cong 3[(1 + R_4/R_3)V_{D2(on)} + V_S]/(2R_4/R_3 - 1)$. For example, with $R_3 = 3$ kΩ, $R_4 = 20$ kΩ, $V_S = 15$ V, and $V_{D(on)} = 0.7$ V, we get $V_{om} \cong 5$ V.

Practical Considerations

The accuracy and stability of the oscillation are affected by the quality of the passive components as well as op amp dynamics. Good choices for the elements in the positive-feedback network are polycarbonate capacitors and thin-film resistors. To compensate for component tolerances, practical Wien-bridge circuits are often equipped with suitable trimmers for the exact adjustment of f_0 as well as THD minimization. With proper trimming, THD levels as low as 0.01% can be achieved.[1] We observe that because of the filtering action provided by the positive-feedback network, the sine wave v_P available at the noninverting input is generally purer than v_O. Consequently, it may be desirable to use v_P as the output, though a buffer would be needed to avoid perturbing circuit behavior.

To avoid slew-rate limiting effects for a given output peak-amplitude V_{om}, the op amp should have SR $> 2\pi V_{om} f_0$. Once this condition is met, the limiting factor becomes the finite GBP, whose effect is a downshift in the actual frequency of oscillation. It can be proved[2] that to contain this shift within 10% when a constant-GBP op amp is used, the latter should have GBP $\geq 43 f_0$. To compensate for this downshift, one can suitably predistort the element values of the positive-feedback network, in a manner similar to the filter predistortion techniques of Section 6.6.

The low end of the frequency range depends on how large the components in the reactive network can be made. Using FET-input op amps to minimize input-bias-current errors, the value of R can easily be increased to the range of tens of megohms. For instance, using $C = 1$ μF and $R = 15.9$ MΩ gives $f_0 = 0.01$ Hz.

Quadrature Oscillators

We can generalize the above ideas and make an oscillator out of any second-order filter that is capable of giving $Q = \infty$ as well as $Q < 0$. To this end, we first ground the input, since it is no longer necessary; then, we design for an initially negative Q to force the poles in the right half of the complex plane and thus ensure oscillation startup; finally, we include a suitable amplitude-dependent network to automatically pull the poles back to the $j\omega$ axis and give $Q = \infty$, or sustained oscillation.

Of special interest are filter topologies of the dual-integrator-loop type, since they provide two oscillations in quadrature, that is, with a relative phase shift of 90°. Figure 10.6 shows how a biquad filter can be turned into a quadrature oscillator. To save an op amp, OA_2 is a noninverting, or Deboo, integrator with $f_0 = 1/2\pi RC$, and it is adjusted to make it slightly regenerative to ensure oscillation startup. At low signal levels OA_1 is a lossless integrator with $f_0 = 1/2\pi RC$. However, as soon as signal amplitude has grown enough to activate the diode limiter, OA_1 becomes lossy. Thereafter, the loss due to OA_1 will compensate for the regeneration due to OA_2,

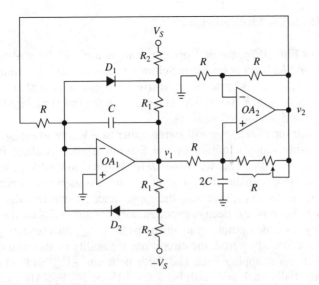

FIGURE 10.6
Quadrature oscillator.

thus sustaining oscillation at $f_0 = 1/2\pi RC$. To estimate V_{om}, consider the instant at which v_1 reaches its positive peak. Retracing familiar steps, we use KCL to write $(V_{om} - V_{D2(on)})/R_1 \cong [V_{D2(on)} - (-V_S)]/R_2$, or $V_{om} \cong V_{D(on)} + (R_1/R_2)(V_S + V_{D2(on)})$. The THD of v_1 is typically of the order of 1%; however, that of v_2 is lower, thanks to the additional filtering provided by OA_2.

10.2
MULTIVIBRATORS

Multivibrators are regenerative circuits intended especially for timing applications. Multivibrators are classified as bistable, astable, and monostable.

In a *bistable multivibrator* both states are stable, so external commands are needed to force the circuit to a given state. This is the popular *flip-flop,* which in turn takes on different names, depending on the way in which the external commands are effected.

An *astable multivibrator* toggles spontaneously between one state and the other, without any external commands. Also called a *free-running multivibrator,* its timing is set by a suitable network, usually comprising a capacitor or a quartz crystal.

A *monostable multivibrator,* also called a *one-shot,* is stable only in one of its two states. If forced into the other state via an external command called a *trigger,* it returns to its stable state spontaneously, after a delay set by a suitable timing network.

Here we are interested in astable and monostable multivibrators. These circuits are implemented with voltage comparators or with logic gates, especially CMOS gates.

Basic Free-Running Multivibrator

In the circuit of Fig. 10.7a, the 301 op amp comparator and the positive-feedback resistances R_1 and R_2 form an inverting Schmitt trigger. Assuming symmetric output saturation at $\pm V_{\text{sat}} = \pm 13$ V, the Schmitt-trigger thresholds are also symmetric at $\pm V_T = \pm V_{\text{sat}} R_1 / (R_1 + R_2) = \pm 5$ V. The signal to the inverting input is provided by the op amp itself via the RC network.

At power turn-on ($t = 0$) v_O will swing either to $+V_{\text{sat}}$ or to $-V_{\text{sat}}$, since these are the only stable states admitted by the Schmitt trigger. Assume it swings to $+V_{\text{sat}}$, so that $v_P = +V_T$. This will cause R to charge C toward V_{sat}, leading to an exponential rise in v_N with the time constant $\tau = RC$. As soon as v_N catches up with $v_P = V_T$, v_O snaps to $-V_{\text{sat}}$, reversing the capacitance current and also causing v_P to snap to $-V_T$. So, now v_N decays exponentially toward $-V_{\text{sat}}$ until it catches up with $v_P = -V_T$, at which point v_O again snaps to $+V_{\text{sat}}$, thus repeating the cycle. It is apparent that once powered, the circuit has the ability to start and then sustain oscillation, with v_O snapping back and forth between $+V_{\text{sat}}$ and $-V_{\text{sat}}$, and v_N slewing exponentially back and forth between $+V_T$ and $-V_T$. After the power-on cycle, the waveforms become periodic.

We are interested in the frequency of oscillation, which is found from the period T as $f_0 = 1/T$. Thanks to the symmetry of the saturation levels, v_O has a duty cycle of 50%, so we only need to find $T/2$. Applying Eq. (10.3) with $\Delta t = T/2$, $\tau = RC$, $V_\infty = V_{\text{sat}}$, $V_0 = -V_T$, and $V_1 = +V_T$, we get

$$\frac{T}{2} = RC \ln \frac{V_{\text{sat}} + V_T}{V_{\text{sat}} - V_T}$$

Substituting $V_T = V_{\text{sat}} / (1 + R_2/R_1)$ and simplifying finally gives

$$f_0 = \frac{1}{T} = \frac{1}{2RC \ln(1 + 2R_1/R_2)} \tag{10.9}$$

With the components shown, $f_0 = 1/(1.62RC)$. If we use the ratio $R_1/R_2 = 0.859$, then $f_0 = 1/2RC$.

We observe that f_0 depends only on the external components. In particular, it is unaffected by V_{sat}, which is known to be an ill-defined parameter since it varies

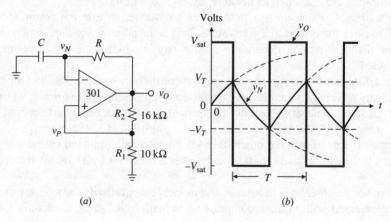

(a) $\qquad\qquad\qquad\qquad\qquad\qquad$ (b)

FIGURE 10.7
Basic free-running multivibrator.

from one op amp to another and also depends on the supply voltages. Any variation in V_{sat} will cause V_T to vary in proportion, thus ensuring the same transition time and, hence, the same oscillation frequency.

The maximum operating frequency is determined by the comparator speed. With the 301 op amp as a comparator, the circuit yields a reasonably good square wave up to the 10-kHz range. This can be extended significantly by using a faster device. At higher frequencies, however, the stray capacitance of the noninverting input toward ground becomes a limiting factor. This can be compensated by using a suitable capacitance in parallel with R_2.

The lowest operating frequency depends on the practical upper limits of R and C, as well as the net leakage at the inverting input node. FET-input comparators may be a good choice in this case.

Although f_0 is unaffected by uncertainties in V_{sat}, it is often desirable to stabilize the output levels for a cleaner and more predictable square-wave amplitude. This is readily achieved with a suitable voltage-clamping network. If it is desired to vary f_0, a convenient approach is to use an array of decade capacitances and a rotary switch for decade selection, and a variable resistance for continuous tuning within the selected decade.

EXAMPLE 10.1. Design a square-wave generator meeting the following specifications: (a) f_0 must be variable in decade steps from 1 Hz to 10 kHz; (b) f_0 must be variable continuously within each decade interval; (c) amplitude must be ± 5 V, stabilized. Assume ± 15-V poorly regulated supplies.

Solution. To ensure stable ± 5 V output levels, use a diode-bridge clamp as in Fig. 10.8. When the op amp saturates at $+13$ V, current flows through the path R_3-D_1-D_5-D_4, thus clamping v_O at $V_{D1(\text{on})} + V_{Z5} + V_{D4(\text{on})}$. To clamp at 5 V, use $V_{Z5} = 5 - 2V_{D(\text{on})} = 5 - 2 \times 0.7 = 3.6$ V. When the op amp saturates at -13 V, current flows through the path D_3-D_5-D_2-R_3, clamping v_O at -5 V.

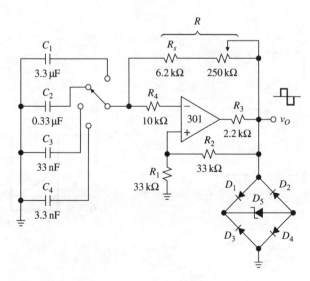

FIGURE 10.8
Square-wave generator of Example 10.1.

To vary f_0 in decade steps, use the four capacitances and rotary switch shown. To vary f_0 within a given decade, implement R with a pot. To cope with component tolerances, ensure an adequate amount of overlap between adjacent decade intervals. To be on the safe side, impose a range of continuous variability from 0.5 to 20, that is, over a 40-to-1 range. We then have $R_{pot} + R_s = 40R_s$, or $R_{pot} = 39R_s$. To keep input-bias-current errors low, impose $I_{R(min)} \gg I_B$, say, $I_{R(min)} = 10 \ \mu A$. Moreover, let $R_1 = R_2 = 33 \ k\Omega$, so that $V_T = 2.5 \ V$. Then, $R_{max} = (5 - 2.5)/(10 \times 10^{-6}) = 250 \ k\Omega$. Since $R_s \ll R_{pot}$, use a 250-kΩ pot. Then, $R_s = 250/39 = 6.4 \ k\Omega$ (use 6.2 kΩ).

To find C_1, impose $f_0 = 0.5 \ Hz$ with the pot set to its maximum value. By Eq. (10.9), $C_1 = 1/[2 \times 0.5 \times (250 + 6.2) \times 10^3 \times \ln 3] = 3.47 \ \mu F$. The closest standard value is $C_1 = 3.3 \ \mu F$. Then, $C_2 = 0.33 \ \mu F$, $C_3 = 33 \ nF$, and $C_4 = 3.3 \ nF$.

The function of R_4 is to protect the comparator input stage at power turn-off, when the capacitors may still be charged, and that of R_3 is to supply current to the bridge, R_2, R, and to the external load, if any. The maximum current drawn by R is when $v_O = +5 \ V$, $v_N = -2.5 \ V$, and the pot is set to zero. This current is $[5 - (-2.5)]/6.2 = 1.2 \ mA$. We also have $I_{R_2} = 5/66 = 0.07 \ mA$. Imposing a bridge current of 1 mA and allowing for a maximum load current of 1 mA, we have $I_{R_3(max)} = 1.2 + 0.07 + 1 + 1 \cong 3.3 \ mA$. Hence, $R_3 = (13 - 5)/3.3 = 2.4 \ k\Omega$ (use 2.2 kΩ to be safe). For the diode bridge use a CA3039 array (Harris).

Figure 10.9 shows a multivibrator designed for single-supply operation. By using a fast comparator, the circuit can operate well into the hundreds of kilohertz. As we know, the circuit gives $V_{OL} \cong 0$ and, if $R_4 \ll R_3 + (R_1 \parallel R_2)$, it gives $V_{OH} \cong V_{CC}$. At power turn-on ($t = 0$), when C is still discharged, v_O is forced high, causing C to charge toward V_{CC} via R. As soon as v_N reaches V_{TH}, v_O snaps low, causing C to discharge toward ground. Henceforth, the oscillation becomes periodic with duty cycle $D(\%) = 100T_H/(T_L + T_H)$ and $f_0 = 1/(T_L + T_H)$. Applying Eq. (10.3)

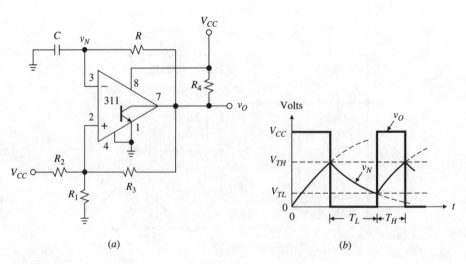

(a)   (b)

FIGURE 10.9
Single-supply free-running multivibrator.

twice, first with $\Delta t = T_L$, $V_\infty = 0$, $V_0 = V_{TH}$, and $V_1 = V_{TL}$, then with $\Delta t = T_H$, $V_\infty = V_{CC}$, $V_0 = V_{TL}$, and $V_1 = V_{TH}$, we get, after combining terms,

$$f_0 = \frac{1}{RC \ln \left(\dfrac{V_{TH}}{V_{TL}} \times \dfrac{V_{CC} - V_{TL}}{V_{CC} - V_{TH}} \right)} \qquad (10.10)$$

To simply inventory and achieve $D = 50\%$, it is customary to impose $R_1 = R_2 = R_3$, after which $f_0 = 1/(RC \ln 4) = 1/1.39RC$. Oscillators of this type can easily achieve stabilities approaching 0.1% with initial predictability of the order of 5% to 10%.

EXAMPLE 10.2. In the circuit of Fig. 10.9 specify components for $f_0 = 1$ kHz, and verify with PSpice for $V_{CC} = 5$ V.

Solution. Use $R_1 = R_2 = R_3 = 33$ kΩ, $R_4 = 2.2$ kΩ, $C = 10$ nF, and $R = 73.2$ kΩ. The input file is:

```
Astable Multivibrator:
VCC 8 0 dc 5V
.lib eval.lib
XCMP 2 3 8 0 7 0 LM111
C 0 3 10nF IC=0
R 3 7 73.2k
R1 2 0 33k
R2 2 8 33k
R3 2 7 33k
R4 7 8 2.2k
.tran 10us 2ms 0 10us uic
.probe
.end
```

The waveforms are shown in Fig. 10.10.

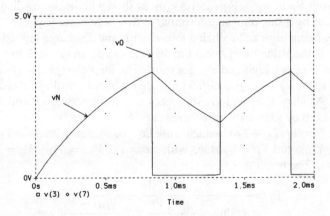

FIGURE 10.10
Waveforms for the circuit of Example 10.2.

Free-Running Multivibrator Using CMOS Gates

CMOS logic gates are particularly attractive when analog and digital functions must coexist on the same chip. A CMOS gate enjoys an extremely high input impedance, a rail-to-rail input range and output swing, extremely low power consumption, and the speed and low cost of logic circuitry. The simplest gate is the inverter depicted in Fig. 10.11. This gate can be regarded as an inverting-type threshold detector giving $v_O = V_{OH} = V_{DD}$ for $v_I < V_T$, and $v_O = V_{OL} = 0$ for $v_I > V_T$. The threshold V_T is the result of internal transistor operation, and is nominally halfway between V_{DD} and 0, or $V_T \cong V_{DD}/2$. The protective diodes, normally in cutoff, prevent v_I from rising above $V_{DD} + V_{D(on)}$ or dropping below $-V_{D(on)}$, and thus protect the FETs against possible electrostatic discharge.

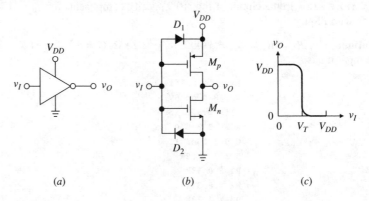

(a)   (b)   (c)

FIGURE 10.11
CMOS inverter: logic symbol, internal circuit diagram, and VTC.

In the circuit of Fig. 10.12a assume at power turn-on ($t = 0$) v_2 goes high. Then, by I_2's inverting action, v_O remains low, and C starts charging toward $v_2 = V_{DD}$ via R. The ensuing exponential rise is conveyed to I_1 via R_1 as signal v_1. As soon as v_1 rises to V_T, I_1 changes state and pulls v_2 low, forcing I_2 to pull v_O high. Since the voltage across C cannot change instantaneously, the step change in v_O causes v_3 to change from V_T to $V_T + V_{DD} \cong 1.5V_{DD}$, as shown in the timing diagram. These changes occur by a snapping action similar to that of Schmitt triggers.

With v_3 being high and v_2 being low, C will now discharge toward $v_2 = 0$ via R. As soon as the value of v_3 decays to V_T, the circuit snaps back to the previous state; that is, v_2 goes high and v_O goes low. The step change in v_O causes v_3 to jump from V_T to $V_T - V_{DD} \cong -0.5V_{DD}$, after which v_3 will again charge toward $v_2 = V_{DD}$. As shown, v_2 and v_O snap back and forth between 0 and V_{DD}, but in antiphase, and they snap each time v_3 reaches V_T.

To find $f_0 = 1/(T_H + T_L)$, we again use Eq. (10.3), first with $\Delta t = T_H$, $V_\infty = 0$, $V_0 = V_T + V_{DD}$, and $V_1 = V_T$, then with $\Delta t = T_L$, $V_\infty = V_{DD}$, $V_0 = V_T - V_{DD}$, and $V_1 = V_T$. The result is

$$f_0 = \cfrac{1}{RC \ln \left(\cfrac{V_{DD} + V_T}{V_T} \times \cfrac{2V_{DD} - V_T}{V_{DD} - V_T} \right)} \tag{10.11}$$

For $V_T = V_{DD}/2$ we get $f_0 = 1/(RC \ln 9) = 1/2.2RC$ and $D(\%) = 50\%$. In practice, due to production variations, there is a spread in the values of V_T. This, in turn,

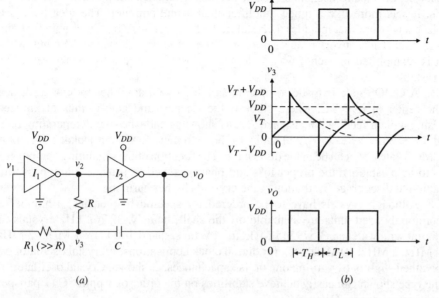

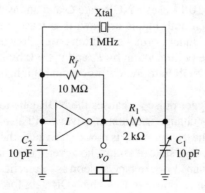

FIGURE 10.12
CMOS-gate free-running multivibrator.

affects f_0, thus limiting the circuit to applications where frequency accuracy is not of primary concern.

We observe that if v_3 were applied to I_1 directly, the input protective diodes of I_1 would clamp v_3 and alter the timing significantly. This is avoided by using the decoupling resistance $R_1 \gg R$ (in practice, $R_1 \cong 10R$ will suffice.)

CMOS Crystal Oscillator

In precise timekeeping applications, frequency must be much more accurate and stable than that afforded by simple RC oscillators. These demands are met with crystal oscillators, an example of which is shown in Fig. 10.13. Since the circuit

FIGURE 10.13
CMOS-gate crystal oscillator.

exploits the electromechanical-resonance characteristics[3] of a quartz crystal to set f_0, it acts more like a tuned amplifier than a multivibrator. The idea here is to place a network that includes a crystal in the feedback loop of a high-gain inverting amplifier. This network routes a portion of the output signal back to the input, where it is reamplified in such a way as to sustain oscillation at a frequency set by the crystal.

A CMOS gate is made to operate as a high-gain amplifier by biasing it near the center of its VTC, where slope is the steepest and gain is thus maximized. Using a plain feedback resistance R_f, as shown, establishes the dc operating point at $V_O = V_I = V_T \cong V_{DD}/2$. Thanks to the extremely low input leakage current of CMOS gates, R_f can be made quite large. The function of the remaining components is to help establish the proper loss and phase, as well as provide a low-pass filter action to discourage oscillation at the crystal's higher harmonics.

Although crystals have to be ordered for specific frequencies, a number of commonly used units are available off the shelf, namely, 32.768 kHz crystals for digital wristwatches, 3.579545 MHz for TV tuners, and 100 kHz, 1 MHz, 2 MHz, 4 MHz, 5 MHz, 10 MHz, etc., for digital clock applications. A crystal oscillator can be tuned slightly by varying one of its capacitances, as shown. Crystal oscillators of the type shown can easily achieve stabilities on the order of 1 ppm/°C (1 part-per-million per degree Celsius).[4]

The duty cycle of clock generators is not necessarily 50%. Applications requiring perfect square-wave symmetry are easily accommodated by feeding the oscillator to a toggle flip-flop. The latter then produces a square wave with $D(\%) = 50\%$, but with half the frequency of the oscillator. To achieve the desired frequency we simply use a crystal with a frequency rating twice as high.

Monostable Multivibrator

On receiving a trigger pulse at the input, a monostable multivibrator or one-shot produces a pulse of a specified duration T. This duration can be generated digitally, by counting a specified number of pulses from a clock source, or in analog fashion, by using a capacitor for time-out control. One-shots are used to generate strobe commands and delays, and in switch debouncing.

The circuit of Fig. 10.14 uses a NOR gate G and an inverter I. The NOR yields a high output only when both inputs are low; if at least one of the inputs is high, the output will be low. Under normal conditions, v_I is low and C is in steady state, so $v_2 = V_{DD}$ due to the pullup action by R, and $v_O = 0$ by inverter action. Further, since both inputs to the NOR gate are low, its output is high, or $v_1 = V_{DD}$, indicating zero voltage across C.

The arrival of a trigger pulse v_I causes the NOR gate to pull v_1 low. Since the voltage across C cannot change instantaneously, v_2 will also go low, causing in turn v_O to go high. Even if the trigger pulse is now deactivated, the NOR gate will keep v_1 low because v_O is high. This state of affairs, however, cannot last indefinitely because R is now charging C toward V_{DD}. In fact, as soon as v_2 reaches V_T, the inverter snaps, forcing v_O back low. In response to this, the NOR gate forces v_1 high, and C then transmits this step to the inverter, thus reinforcing its initial snap in Schmitt-trigger

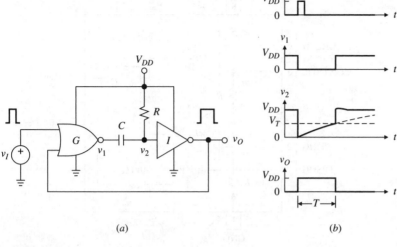

FIGURE 10.14
CMOS-gate one shot.

fashion. Even though v_2 tries to swing from V_T to $V_T + V_{DD} \cong 1.5V_{DD}$, the internal protective diode D_1 of the inverter, shown explicitly in Fig. 10.11b, will clamp v_2 near V_{DD}, thus discharging C. The circuit is now back in the stable state preceding the arrival of the trigger pulse. The timeout T is found via Eq. (10.3) as

$$T = RC \ln \frac{V_{DD}}{V_{DD} - V_T} \qquad (10.12)$$

For $V_T = V_{DD}/2$, this reduces to $T = RC \ln 2 = 0.69RC$.

A *retriggerable* one-shot begins a new cycle each time the trigger is activated, including activation during T. By contrast, a *nonretriggerable* one-shot is insensitive to triggering during T.

10.3
MONOLITHIC TIMERS

The need for the astable and monostable functions arises so often that special circuits,[4] called *IC timers*, are available to satisfy these needs. Among the variety of available products, the one that has gained the widest acceptance in terms of cost and versatility is the 555 timer. Another popular product is the 2240 timer, which combines a timer with a programmable counter to provide additional timing flexibility.

The 555 Timer

As shown in Fig. 10.15, the basic blocks of the 555 timer are: (*a*) a trio of identical resistors, (*b*) a pair of voltage comparators, (*c*) a flip-flop, and (*d*) a BJT switch Q_O. The resistances set the comparator thresholds at $V_{TH} = (2/3)V_{CC}$ and $V_{TL} = (1/3)V_{CC}$.

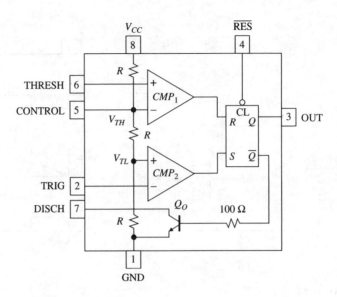

FIGURE 10.15
The 555 timer block diagram.

For additional flexibility, the upper threshold node is externally accessible via pin 5 so that the user can modulate the value of V_{TH}. Regardless of the value of V_{TH}, we always have $V_{TL} = V_{TH}/2$.

The state of the flip-flop is controlled by the comparators as follows: (*a*) Whenever the voltage at the trigger input (TRIG) drops below V_{TL}, CMP_2 fires and sets the flip-flop, forcing Q high and $\overline{Q}$ low; with a low voltage at its base, Q_O is in cutoff. (*b*) Whenever the voltage at the threshold input (THRESH) rises above V_{TH}, CMP_1 fires and clears the flip-flop, forcing Q low and $\overline{Q}$ high. With a high voltage applied to its base via the 100-Ω resistance, Q_O is now heavily on. Summarizing, lowering TRIG below V_{TL} turns Q_O off, and raising THRESH above V_{TH} turns Q_O heavily on. The flip-flop includes a reset input ($\overline{\text{RES}}$) to force Q low and turn Q_O on regardless of the conditions at the inputs of the comparators.

The 555 is available in both bipolar and CMOS versions. The bipolar versions operate over a wide range of supply voltages, typically 4.5 V $\leq V_{CC} \leq$ 18 V, and are capable of sourcing and sinking output currents of 200 mA. The TLC555 (Texas Instruments), which is a popular CMOS version, is designed to operate over a power-supply range of 2 V to 18 V, and has output current sinking and sourcing capabilities of 100 mA and 10 mA, respectively. The transistor switch is an enhancement-type *n*-MOSFET. The advantages of CMOS timers are low power consumption, very high input impedances, and a rail-to-rail output swing.

The 555 as an Astable Multivibrator

Figure 10.16 shows how the 555 is configured for astable operation using just three external components. To understand circuit operation, refer also to the internal diagram of Fig. 10.15.

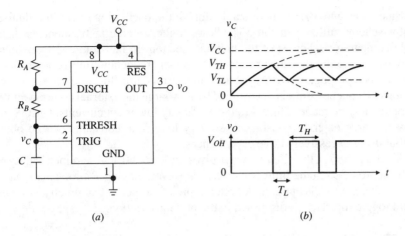

(a)   *(b)*

FIGURE 10.16
The 555 timer as an astable multivibrator.

At power turn-on ($t = 0$), when the capacitor is still discharged, the voltage at the TRIG input is less than V_{TL}. This forces Q high and keeps the BJT in cutoff, thus allowing C to charge toward V_{CC} via the series $R_A + R_B$. As soon as v_C reaches V_{TH}, CMP_1 fires and forces Q low. This turns on Q_O, which then pulls the DISCH pin to $V_{CE(\text{sat})} \cong 0$ V. Consequently, C now discharges toward ground via R_B. As soon as v_C reaches V_{TL}, CMP_2 fires, forcing Q high and turning off Q_O. This reestablishes the conditions for a new cycle of astable operation.

The time intervals T_L and T_H are found via Eq. (10.3). During T_L the time constant is $R_B C$, so $T_L = R_B C \ln(V_{TH}/V_{TL}) = R_B C \ln 2$; during T_H the time constant is $(R_A + R_B)C$, so $T_H = (R_A + R_B)C \ln[(V_{CC} - V_{TL})/(V_{CC} - V_{TH})]$. Consequently,

$$T = T_L + T_H = R_B C \ln 2 + (R_A + R_B)C \ln \frac{V_{CC} - V_{TH}/2}{V_{CC} - V_{TH}} \qquad (10.13)$$

Substituting $V_{TH} = (2/3)V_{CC}$ and solving for $f_0 = 1/T$ and $D(\%) = 100 T_H / (T_L + T_H)$ gives

$$f_0 = \frac{1.44}{(R_A + 2R_B)C} \qquad D(\%) = 100\frac{R_A + R_B}{R_A + 2R_B} \qquad (10.14)$$

We observe that the oscillation characteristics are set by the external components and are independent of V_{CC}. To prevent power-supply noise from causing false triggering when v_C approaches either threshold, use a 0.01-μF bypass capacitor between pin 5 and ground: this will clean V_{TH} as well as V_{TL}. The timing accuracy[4] of the 555 astable approaches 1%, with a temperature stability of 0.005%/°C and a power-supply stability of 0.05%/V.

EXAMPLE 10.3. In the circuit of Fig. 10.16 specify suitable components for $f_0 = 50$ kHz and $D(\%) = 75\%$.

Solution. Let $C = 1$ nF, so that $R_A + 2R_B = 1.44/f_0 C = 28.85$ kΩ. Imposing $(R_A + R_B)/(R_A + 2R_B) = 0.75$ gives $R_A = 2R_B$. Solving gives $R_A = 14.4$ kΩ (use 14.3 kΩ) and $R_B = 7.21$ kΩ (use 7.15 kΩ).

Since V_{TL} and V_{TH} remain stable during the oscillation cycle, the dual-comparator scheme utilized in the 555 allows higher operating frequencies than the single-comparator schemes of the previous section. In fact some 555 versions can easily operate to the megahertz range. The upper frequency limit is determined by the combined propagation delays of the comparators, flip-flop, and transistor switch. The lower frequency limit is determined by how large the external component values can practically be made. Thanks to the extremely low input currents, CMOS timers allow for large external resistances, so very long time constants can be obtained without using excessively large capacitances.

Since $T_H > T_L$, the circuit always gives $D(\%) > 50\%$. A symmetric duty cycle can be approached in the limit $R_A \ll R_B$; however, making R_A too small may lead to excessive power dissipation. A better approach to perfect symmetry is to use an output toggle flip-flop, as discussed in the previous section.

The 555 as a Monostable Multivibrator

Figure 10.17 shows the 555 connection for monostable operation. Under normal conditions, the TRIG input is held high, and the circuit is in the stable state represented by Q low. Moreover, the BJT switch Q_O is closed, keeping C discharged, or $v_C \cong 0$.

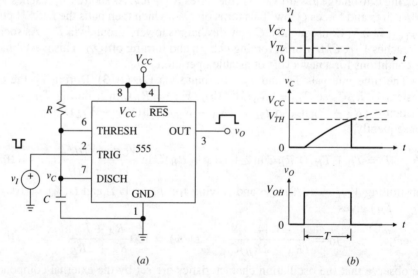

(a)   (b)

FIGURE 10.17
The 555 timer as a monostable multivibrator.

The circuit is triggered by lowering the TRIG input below V_{TL}. When this is done, CMP_2 sets the flip-flop, forcing Q high and turning off Q_O. This frees C to charge toward V_{CC} via R. However, as soon as v_C reaches V_{TH}, the upper comparator clears the flip-flop, forcing Q low and turning Q_O heavily on. The capacitance is rapidly discharged, and the circuit returns to the stable state preceding the arrival of the trigger pulse.

The pulse width T is readily found via Eq. (10.3) as

$$T = RC \ln \frac{V_{CC}}{V_{CC} - V_{TH}} \tag{10.15}$$

Letting $V_{TH} = (2/3)V_{CC}$ gives $T = RC \ln 3$, or

$$T = 1.10RC \qquad (10.16)$$

Note once again the independence of V_{CC}. To enhance noise immunity, connect a $0.01\text{-}\mu\text{F}$ capacitor between pin 5 and ground (see Fig 10.15).

Voltage Control

If desired, the timing characteristics of the 555 can be modulated via the CONTROL input. Changing V_{TH} from its nominal value of $(2/3)V_{CC}$ will result in longer or shorter capacitance charging times, depending on whether V_{TH} is increased or decreased.

When the timer is configured for astable operation, modulating V_{TH} varies T_H while leaving T_L unchanged, as indicated by Eq. (10.13). Consequently, the output is a train of constant-width pulses with a variable repetition rate. This is referred to as *pulse-position modulation* (PPM).

When the timer is configured for monostable operation, modulating V_{TH} varies T, as per Eq. (10.15). If the monostable is triggered by a continuous pulse train, the output will be a pulse train with the same frequency as the input but with the pulse width modulated by V_{TH}. We now have *pulse-width modulation* (PWM).

PPM and PWM represent two common forms of information encoding for storage and transmission. Note that once V_{TH} is overridden externally, V_{TH} and V_{CC} are no longer related; hence, the timing characteristics are no longer independent of V_{CC}.

EXAMPLE 10.4. Assuming $V_{CC} = 5$ V in the multivibrator of Example 10.3, find the range of variation of f_0 and $D(\%)$ if the voltage at the CONTROL input is modulated by ac coupling to it an external sine wave with a peak amplitude of 1 V.

Solution. The range of variation of V_{TH} is $(2/3)5 \pm 1$ V, or between 4.333 V and 2.333 V. Substituting into Eq. (10.13) gives $T_L = 4.96 \,\mu\text{s}$ and $7.78 \,\mu\text{s} \le T_H \le 31.0 \,\mu\text{s}$, so we have 27.8 kHz $\le f_0 \le$ 78.5 kHz, and 61.1% $\le D(\%) \le$ 86.2%.

Timer/Counter Circuits

In applications requiring very long delays, the values of the timing components can become impractically large. This drawback is overcome by using components of manageable size and then stretching the multivibrator time scale with a binary counter. This concept is exploited in the popular 2240 timer/counter circuit, as well as other similar devices. As shown in Fig. 10.18, the basic elements of the 2240 are a time-base oscillator (TBO), an 8-bit ripple counter, and a control flip-flop (FF). The TBO is similar to the 555 timer, except that R_B has been eliminated to reduce the external component count, and the comparator thresholds have been changed to $V_{TL} = 0.27V_{CC}$ and $V_{TH} = 0.73V_{CC}$ to make the value of the logarithm in Eq. (10.13) exactly unity. Thus, the time-base is

$$T = RC \qquad (10.17)$$

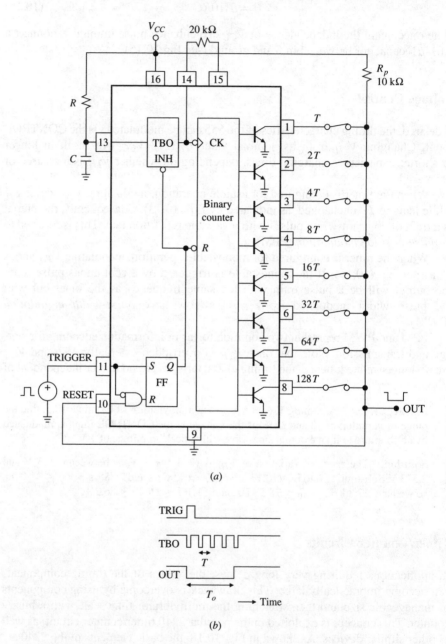

FIGURE 10.18
(a) Programmable delay generator using the XR-2240 timer/counter. (b) Timing diagram. (Courtesy of Exar.)

The binary counter consists of eight toggle flip-flops that are buffered by open-collector BJTs. The desired amount of time stretching is programmed by connecting a suitable combination of counter outputs to a common pullup resistor R_p in a wired-OR configuration. Once a particular combination is selected, the output will be low as long as any one of the selected outputs is low. For instance, connecting only pin 5 to the pullup resistor gives $T_o = 16T$, while connecting pins 1, 3, and 7 gives $T_o = (1+4+64)T = 69T$, where T_o is the duration of the output timing cycle. By suitable choice of the connection pattern, one can program T_o anywhere over the range $T \leq T_o \leq 255T$.

The purpose of the control flip-flop is to translate the external TRIGGER and RESET commands to the proper controls for the TBO and the counter. At power turn-on the circuit comes up in the reset state, where the TBO is inhibited and all open-collector outputs are high. On receiving an external trigger pulse, the control flip-flop goes high and initiates a timing cycle by enabling the TBO and forcing the common output node of the counter low. The TBO will now run until the count programmed by the wired-OR pattern is reached. At this point the output goes high, resetting the control flip-flop and stopping the TBO. The circuit is now in the reset state, awaiting the arrival of the next trigger pulse.

Cascading the counter stages of two or more 2240s makes it possible to achieve truly long delays. For instance, cascading two 8-bit counters yields an effective counter length of 16 bits, which allows T_o to be programmed anywhere in the range from T to over $65 \times 10^3 T$. In this manner, delays of hours, days, or months can be generated using relatively small timing component values. Since the counters do not affect the timing accuracy, the accuracy of T_o depends only on that of T, which is typically around 0.5%. T can be fine-tuned by adjusting R.

10.4
TRIANGULAR WAVE GENERATORS

Triangular waves are generated by alternately charging and discharging a capacitor with a constant current. In the circuit of Fig. 10.19a the current drive for C is provided by OA, a JFET-input op amp functioning as a floating-load V-I converter. The converter receives a two-level drive from a 301 op amp comparator configured as a Schmitt trigger. Because of the inversion introduced by OA, the Schmitt trigger must be of the noninverting type. Also shown is a diode clamp to stabilize the Schmitt-trigger output levels at $\pm V_{\text{clamp}} = \pm(V_{Z5} + 2V_{D(\text{on})})$. Consequently, the Schmitt-trigger input thresholds are $\pm V_T = \pm(R_1/R_2)V_{\text{clamp}}$.

Circuit behavior is visualized in terms of the waveforms of Fig. 10.19b. Assume at power turn-on ($t = 0$) CMP swings to $+V_{\text{sat}}$ so that $v_{\text{SQ}} = +V_{\text{clamp}}$. OA converts this voltage to a current of value V_{clamp}/R entering C from the left. This causes v_{TR} to ramp downward. As soon as v_{TR} reaches $-V_T$, the Schmitt trigger snaps and v_{SQ} switches from $+V_{\text{clamp}}$ to $-V_{\text{clamp}}$. OA converts this new voltage to a capacitance current of the same magnitude but opposite polarity. Consequently, v_{TR} will now ramp upward. As soon as v_{TR} reaches $+V_T$, the Schmitt trigger snaps again, thus repeating the cycle. Figure 10.19b shows also the waveform v_1 at the noninverting input of CMP. By the superposition principle, this waveform is a linear combination of v_{TR} and v_{SQ}, and it causes the Schmitt trigger to snap whenever it reaches 0 V.

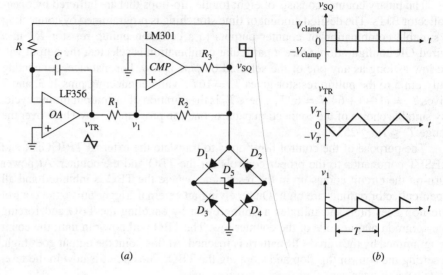

FIGURE 10.19
Basic triangular/square-wave generator.

By symmetry, the time taken by v_{TR} to ramp from $-V_T$ to $+V_T$ is $T/2$. Since the capacitor is operated at constant current, we can apply Eq. (10.2) with $\Delta t = T/2$, $I = V_{clamp}/R$ and $\Delta v = 2V_T = 2(R_1/R_2)V_{clamp}$. Letting $f_0 = 1/T$ gives

$$f_0 = \frac{R_2/R_1}{4RC} \tag{10.18}$$

indicating that f_0 depends only on external components, a desirable feature indeed. As usual, f_0 can be varied continuously by means of R, or in decade steps by means of C. The operating frequency range is limited at the upper end by the SR and GBP of OA as well as the speed of response of CMP; at the lower end by the size of R and C, as well as the input bias current of OA and capacitor leakage. A FET-input op amp is usually a good choice for OA, while CMP should be an uncompensated op amp or, better yet, a high-speed voltage comparator.

EXAMPLE 10.5. In the circuit of Fig. 10.19a specify suitable components for a square wave with peak values of ± 5 V, a triangular wave with peak values of ± 10 V, and f_0 continuously variable from 10 Hz to 10 kHz.

Solution. We need $V_{Z5} = V_{clamp} - 2V_{D(on)} = 5 - 2 \times 0.7 = 3.6$ V, and $R_2/R_1 = V_{clamp}/V_T = 5/10 = 0.5$ (use $R_1 = 20$ kΩ, $R_2 = 10$ kΩ). Since f_0 must be variable over a 1000 : 1 range, implement R with a pot and a series resistance R_s such that $R_{pot} + R_s = 1000R_s$, or $R_{pot} \cong 10^3 R_s$. Use $R_{pot} = 2.5$ MΩ and $R_s = 2.5$ kΩ. For $R = R_{min} = R_s$ we want $f_0 = f_{0(max)} = 10$ kHz. By Eq. (10.18), $C = 0.5/(10^4 \times 4 \times 2.5 \times 10^3) = 5$ nF. The function of R_3 is to provide current to R, R_2, the diode bridge, and the output load under all operating conditions. Now, $I_{R(max)} = V_{clamp}/R_{min} = 5/2.5 = 2$ mA, and $I_{R_2(max)} = V_{clamp}/R_2 = 5/10 = 0.5$ mA. Imposing a bridge current of 1 mA and allowing for a maximum load current of 1 mA yields $I_{R_3(max)} = 2 + 0.5 + 1 + 1 = 4.5$ mA. Then, $R_3 = (13 - 5)/4.5 = 1.77$ kΩ (use 1.5 kΩ to be safe). For the diode bridge, use a CA3039 diode array (Harris).

Slope Control

With the modification of Fig. 10.20a, the charge and discharge times can be adjusted independently to generate asymmetric waves. With $v_{SQ} = +V_{clamp}$, D_3 is on and D_4 is off, so the discharge current is $I_H = [V_{clamp} - V_{D(on)}]/(R_H + R)$. With $v_{SQ} = -V_{clamp}$, D_3 is off and D_4 is on, and the charge current is $I_L = [V_{clamp} - V_{D(on)}]/(R_L + R)$. The charge and discharge times are found as $C \times 2V_T = I_L T_L$ and $C \times 2V_T = I_H T_H$, respectively. The function of D_1 and D_2 is to compensate for the $V_{D(on)}$ term due to D_3 and D_4. With D_1 and D_2 in place we now have $V_T/R_1 = [V_{clamp} - V_{D(on)}]/R_2$. Combining all the above information yields

$$T_L = 2\frac{R_1}{R_2}C(R_L + R) \qquad T_H = 2\frac{R_1}{R_2}C(R_H + R) \qquad (10.19)$$

The frequency of oscillation is $f_0 = 1/(T_H + T_L)$. Note that if one of the slopes is made much steeper than the other, v_{TR} will approach a sawtooth and v_{SQ} a train of narrow pulses.

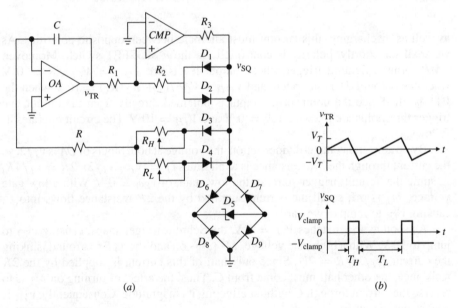

FIGURE 10.20
Triangular wave generator with independently adjustable slopes.

Voltage-Controlled Oscillator

Many applications require that f_0 be programmable automatically, for instance, via a control voltage v_I. The required circuit, known as a *voltage-controlled oscillator* (VCO), is designed to give $f_0 = kv_I$, $v_I > 0$, where k is the sensitivity of the VCO, in hertz per volt.

Figure 10.21 shows a popular VCO realization. Here *OA* is a *V-I* converter that forces C to conduct a current linearly proportional to v_I. To ensure capacitor charging

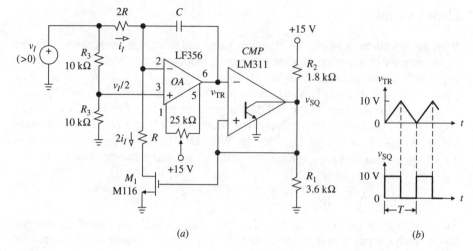

FIGURE 10.21
Voltage-controlled triangular/square-wave oscillator. (Power supplies are ±15 V.)

as well as discharging, this current must alternate between opposite polarities. As we shall see shortly, polarity is controlled via the *n*-MOSFET switch. Moreover, *CMP* forms a Schmitt trigger whose output levels are $V_{OL} = V_{CE(\text{sat})} \cong 0$ V when the output BJT is saturated, and $V_{OH} = V_{CC}/(1 + R_2/R_1) = 10$ V when the BJT is off. Since the noninverting input is obtained directly from the output, the trigger thresholds are likewise $V_{TL} = 0$ V and $V_{TH} = 10$ V. The circuit operates as follows.

By op amp and voltage-divider action, the voltage at both inputs of *OA* is $v_I/2$, so the current through the $2R$ resistance is at all times $i_I = (v_I - v_I/2)/2R = v_I/4R$. Assume the Schmitt trigger starts in the low state, or $v_{SQ} \cong 0$ V. With a low gate voltage, M_1 is off, so all the current supplied by the $2R$ resistance flows into C, causing v_{TR} to ramp downward.

As soon as v_{TR} reaches $V_{TL} = 0$ V, the Schmitt trigger snaps, causing v_{SQ} to jump to 10 V. With a high gate voltage, M_1 turns on and shorts R to ground, sinking the current $(v_I/2)/R = 2i_I$. Since only half of this current is supplied by the $2R$ resistance, the other half must come from C. Thus, the effect of turning on M_1 is to reverse the current through C without affecting its magnitude. Consequently, v_{TR} is now ramping upward.

As soon as v_{TR} reaches $V_{TH} = 10$ V, the Schmitt trigger snaps back to 0 V, turning off M_1 and reestablishing the conditions of the previous half-cycle. The circuit is therefore oscillating. Using Eq. (10.2) with $\Delta t = T/2$, $I = v_I/4R$, and $\Delta v = V_{TH} - V_{TL}$, and then solving for $f_0 = 1/T$ gives

$$f_0 = \frac{v_I}{8RC(V_{TH} - V_{TL})} \tag{10.20}$$

With $V_{TH} - V_{TL} = 10$ V we get $f_0 = kv_I$, $k = 1/80RC$. Using, for example, $R = 10$ kΩ, $2R = 20$ kΩ, and $C = 1.25$ nF gives a sensitivity $k = 1$ kHz/V. Then, varying v_I over the range of 10 mV to 10 V sweeps f_0 over the range of 10 Hz to 10 kHz.

The accuracy of Eq. (10.20) is limited at high frequencies by the dynamics of OA, CMP, and M_1, and at low frequencies by the input bias current and offset voltage of OA. To null the latter, set v_I to a low value, say, 10 mV, and then adjust the offset-nulling pot for a 50% duty cycle. Another source of error is the channel resistance $r_{ds(on)}$ of the FET switch. The data sheets of the M116 FET (Siliconix) give $r_{ds(on)} = 100\,\Omega$ typical. With $R = 10\,k\Omega$, this represents an error of only 1%; if desired, this can be eliminated by reducing R from $10\,k\Omega$ to $10\,k\Omega - 100\,\Omega = 9.9\,k\Omega$.

Triangular-to-Sine Wave Conversion

If a triangular wave is passed through a circuit exhibiting a sinusoidal VTC, as shown in Fig. 10.22a, the result is a sine wave. Since nonlinear wave shaping is independent of frequency, this form of sine wave generation is particularly convenient when used in connection with triangular-output VCOs, since the latter offer much wider tuning ranges than Wien-bridge oscillators. Practical wave shapers approximate a sinusoidal VTC by exploiting the nonlinear characteristics of diodes or transistors.[4]

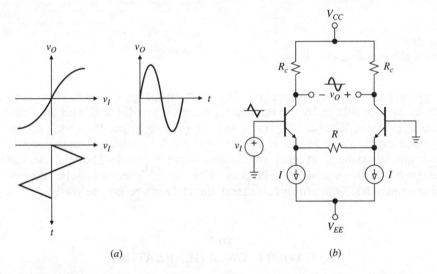

FIGURE 10.22
(a) VTC of a triangular-to-sine wave converter. (b) Logarithmic wave shaper.

In the circuit of Fig. 10.22b a sinusoidal VTC is approximated by suitably over-driving an emitter-degenerated differential pair. Near the zero-crossings of the input, the gain of the pair is approximately linear; however, as either peak is approached, one of the BJTs is driven to the verge of cutoff, where the VTC becomes logarithmic and produces a gradual rounding of the triangular wave. The THD of the output is minimized[4] at about 0.2% for $RI \cong 2.5V_T$ and $V_{im} \cong 6.6V_T$, where V_{im} is the peak amplitude of the triangular wave and V_T is the thermal voltage ($V_T \cong 26\,mV$ at room temperature). This translates to $RI \cong 65\,mV$ and $V_{im} \cong 172\,mV$, indicating that the triangular wave must be properly scaled to fit the requirements of the wave shaper.

Figure 10.23 shows a practical wave shaper realization. The shaping function is performed by the LM394 matched BJT pair, whose output is converted to a

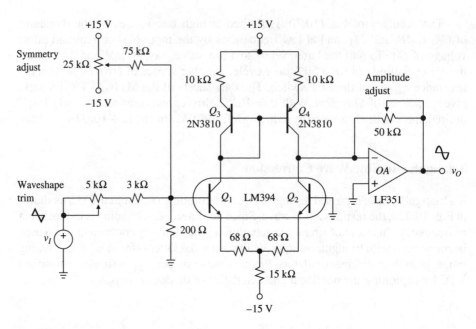

FIGURE 10.23
Practical logarithmic wave shaper.

single-ended current with the help of the current mirror Q_3-Q_4; this current is then converted to a voltage by the op amp. The circuit is calibrated with the help of an oscilloscope/spectrum analyzer as follows: (*a*) first, adjust the 25-kΩ pot for a symmetrical output; (*b*) next, adjust the 5-kΩ pot for minimum output distortion; (*c*) finally, adjust the 50-kΩ pot for the desired output amplitude. The input attenuator, designed for triangular waves with peak values of ± 5 V, can easily be adapted to other amplitudes. With proper calibration, the THD can be kept below 1%.

10.5
SAWTOOTH WAVE GENERATORS

A sawtooth cycle is generated by charging a capacitor at a constant rate and then rapidly discharging it with a switch. Figure 10.24 shows a circuit utilizing this principle. The current drive for C is provided by *OA*, a floating-load *V-I* converter. In order for v_{ST} to be a positive ramp, i_I must always flow out of the summing junction, or $v_I < 0$. R_2 and R_3 establish the threshold $V_T = V_{CC}/(1 + R_2/R_3) = 5$ V.

At power turn-on ($t = 0$), when C is still discharged, the 311 comparator inputs are $v_P = 0$ V and $v_N = 5$ V, indicating that the output BJT is in saturation and $v_{PULSE} \cong -15$ V. With a gate voltage this low, the *n*-JFET J_1 is in cutoff, allowing C to charge. As soon as the ensuing ramp v_{ST} reaches V_T, the comparator output BJT goes off, allowing the 2-kΩ resistor to pull v_{PULSE} to ground. This change of state takes place in a snapping fashion because of the positive-feedback action provided by C_1. Since now $v_{GS} = 0$ V, the JFET switch closes and rapidly discharges C, bringing v_{ST} to 0 V.

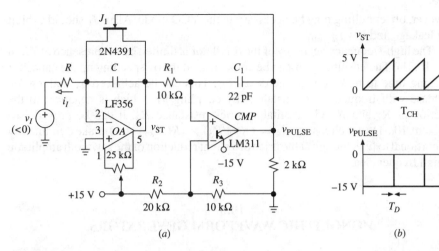

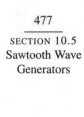

FIGURE 10.24
Voltage-controlled sawtooth/pulse-wave oscillator.

The comparator is prevented from responding immediately to this change in v_{ST} because of the charge accumulated in C_1 during the transition of v_{PULSE} from -15 V to 0 V. This one-shot action, whose duration T_D is proportional to $R_1 C_1$, is designed to ensure that C undergoes complete discharge. With the component values shown, $T_D < 1$ μs. After timing out, v_{PULSE} returns to -15 V, turning J_1 off again and allowing C to resume charging. The cycle, therefore, repeats itself.

The charging time T_{CH} is found using Eq. (10.2) with $\Delta t = T_{CH}$, $I = |v_I|/R$, and $\Delta v = V_T$. Letting $f_0 = 1/(T_{CH} + T_D)$, we obtain

$$f_0 = \frac{1}{RCV_T/|v_I| + T_D} \tag{10.21}$$

As long as $T_D \ll T_{CH}$, this simplifies to

$$f_0 = \frac{|v_I|}{RCV_T} \tag{10.22}$$

indicating that f_0 is linearly proportional to the control voltage v_I. With $R = 90.9$ kΩ and $C = 2.2$ nF, $f_0 = k|v_I|$, $k = 1$ kHz/V, so varying v_I from -10 mV to -10 V will sweep f_0 from 10 Hz to 10 kHz. The circuit can also function as a *current-controlled oscillator* (CCO) if we drive it directly with a current sink i_I. Then, $f_0 = i_I/CV_T$. A common application of sawtooth CCOs is found in electronic music, where the control current is provided by an exponential *V-I* converter designed for a sensitivity of 1 octave per volt over a 10-decade frequency range, typically from 16.3516 Hz to 16.744 kHz.

Practical Considerations

A good choice for *OA* is a FET-input op amp combining low input bias current, which is critical at the low end of the control range, with good slew-rate performance, which is critical at the high end. The input offset voltage is not critical in the CCO mode;

however, offset nulling may be necessary in the VCO mode. Also, J_1 should exhibit low leakage and low $r_{ds(\text{on})}$.

The high-frequency accuracy of the oscillator is limited by the presence of T_D in Eq. (10.21). The ensuing error can be compensated for by speeding up the capacitor charging time to make up for the delay T_D. This can be achieved by making V_T decrease with frequency, for instance, by coupling v_I, which is negative, to the junction of R_2 and R_3 via a suitable series resistance R_4. It can be proved (see Problem 10.31) that choosing $R_4 = (R_2 \parallel R_3) \times (RC/T_D - 1)$ makes f_0 linearly proportional to $|v_I|$, though at the price of a slight reduction of the sawtooth amplitude at high frequencies.

10.6
MONOLITHIC WAVEFORM GENERATORS

Also called *function generators,* these circuits are designed to provide the basic waveforms with a minimum of external components. The heart of a waveform generator is a VCO that generates the triangular and square waves. Passing the triangular wave through an on-chip wave shaper yields the sine wave, whereas configuring the oscillator for a highly asymmetric duty cycle gives the sawtooth and pulse-train waves. The two most frequent VCO configurations are the *grounded capacitor* and the *emitter-coupled types,*[4] both of which are available either as stand-alone units or as part of complex systems, such as phase-locked loops (PLLs), tone decoders, *V-F* converters, and PWM controllers.

Grounded-Capacitor VCOs

These circuits are based on the principle of charging and discharging a grounded capacitor at rates controlled by programmable current generators. With reference to Fig. 10.25a, we note that when the switch *SW* is in the up position, C charges at a rate set by the current source i_H. Once v_{TR} reaches the upper threshold V_{TH}, the Schmitt trigger changes state and flips *SW* to the down position, causing C to discharge at

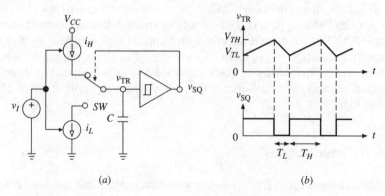

(a)   (b)

FIGURE 10.25
Grounded-capacitor VCO.

a rate set by the current sink i_L. Once v_{TR} reaches V_{TL}, the trigger changes state again, flipping SW to the up position and repeating the cycle.

To allow for automatic frequency control, i_H and i_L are made programmable via an external control voltage v_I. If the magnitudes of i_L and i_H are equal, the output waveform will be symmetric. Conversely, if one of the currents is made much larger than the other, v_{TR} will approach a sawtooth.

The grounded-capacitor configuration is used in the design of temperature-stable VCOs with operating frequencies up to tens of megahertz. Popular products utilizing this configuration are the NE566 function generator (Signetics) and the ICL8038 precision waveform generator (Harris).

The ICL 8038 Waveform Generator

In the circuit[5] of Fig. 10.26, Q_1 and Q_2 form two programmable current sources whose magnitudes are set by the external resistors R_A and R_B. The drive for Q_1 and Q_2 is provided by the emitter follower Q_3, which also compensates for their base-emitter voltage drops to yield $i_A = v_I/R_A$ and $i_B = v_I/R_B$, with v_I being referenced to V_{CC} as shown. While i_A is fed to C directly, i_B is diverted to the current mirror Q_4-Q_5-Q_6 where it undergoes polarity reversal as well as amplification by 2 due to the combined action of Q_5 and Q_6. The result is a current sink of magnitude $2i_B$.

The Schmitt trigger is similar to that of the 555 timer, with $V_{TL} = (1/3)V_{CC}$ and $V_{TH} = (2/3)V_{CC}$. When the flip-flop output Q is high, Q_7 saturates and pulls the

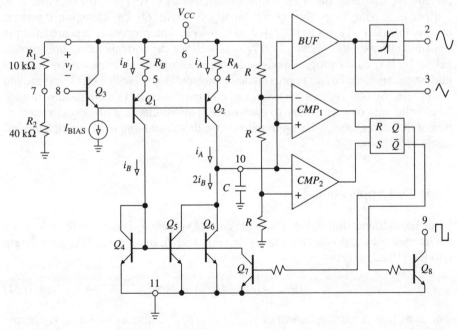

FIGURE 10.26
Simplified circuit diagram of the ICL8038 waveform generator. (Courtesy of Harris Semiconductor.)

bases of Q_5 and Q_6 low, shutting off the current sink. Consequently, C charges at a rate set by $i_H = i_A$. Once the capacitance voltage reaches V_{TH}, CMP_1 fires and clears the flip-flop, turning Q_7 off and enabling the current mirror. The net current out of C is now $i_L = 2i_B - i_A$; as long as $2i_B > i_A$, this current will cause C to discharge. Once V_{TL} is reached, CMP_2 fires and sets the flip-flop, thus repeating the cycle. It can be shown (see Problem 10.32) that

$$f_0 = 3 \left(1 - \frac{R_B}{2R_A}\right) \frac{v_I}{R_A C V_{CC}} \qquad D(\%) = 100 \left(1 - \frac{R_B}{2R_A}\right) \qquad (10.23)$$

With $R_A = R_B = R$ the circuit yields symmetric waveforms with $f_0 = kv_I$, $k = 1.5/RCV_{CC}$. As shown in the figure, the device is also equipped with a unity-gain buffer to isolate the waveform developed across C, a wave shaper to convert the triangular wave to a low-distortion sine wave, and an open-collector transistor (Q_8) to provide, with the help of an external pullup resistor, a square-wave output.

Figure 10.27 shows the wave shaper[5] utilized in the 8038. The circuit is known as a *breakpoint wave shaper* because it uses a set of breakpoints at designated signal levels to fit a nonlinear VTC by a piecewise linear approximation. The circuit, designed to process triangular waves alternating between $(1/3)V_{CC}$ and $(2/3)V_{CC}$, uses the resistive strings shown at the right to establish two sets of breakpoint voltages symmetric about the midrange value of $(1/2)V_{CC}$. These voltages are then buffered by the even-numbered emitter-follower BJTs. The circuit works as follows.

For v_I near $(1/2)V_{CC}$, all odd-numbered BJTs are off, giving $v_O = v_I$. Consequently, the initial slope of the VTC is $a_0 = \Delta v_O/\Delta v_I = 1$ V/V. As v_I is increased to the first breakpoint, the common-base BJT Q_1 goes on and loads down the source, changing the VTC slope from a_0 to $a_1 = 10/(1 + 10) = 0.909$ V/V. Further increasing v_I to the second breakpoint turns Q_3 on, changing the slope to $a_2 = (10 \| 2.7)/[1 + (10 \| 2.7)] = 0.680$ V/V. The process is repeated for the remaining breakpoints above $(1/2)V_{CC}$ as well for the corresponding breakpoints below $(1/2)V_{CC}$. By progressively reducing the slope as v_I moves away from its midrange value, the circuit approximates a sinusoidal VTC with THD levels around 1% or less. We observe that the even- and odd-numbered BJTs associated with each breakpoint are complementary to each other. This results in a first-order cancellation of the corresponding base-emitter voltage drops, yielding more predictable and stable breakpoints.

Basic 8038 Applications[5]

In the basic connection of Fig. 10.28 the control voltage v_I is derived from V_{CC} via the internal voltage divider R_1 and R_2 (see Fig. 10.26), so $v_I = (1/5)V_{CC}$. Inserting into Eq. (10.23) gives

$$f_0 = \frac{0.3}{RC} \qquad D(\%) = 50\% \qquad (10.24)$$

indicating that f_0 is independent of V_{CC}, a desirable feature as we know. By proper choice of R and C, the circuit can be made to oscillate at any frequency from 0.001 Hz to 1 MHz. The thermal drift of f_0 is typically 50 ppm/°C. For optimum performance, confine i_A and i_B within the 1-μA to 1-mA range.

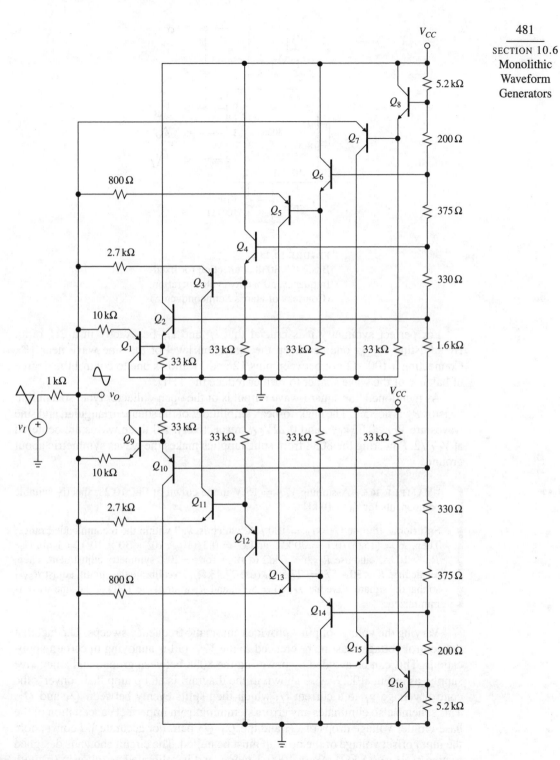

FIGURE 10.27
Breakpoint wave shaper. (Courtesy of Harris Semiconductor.)

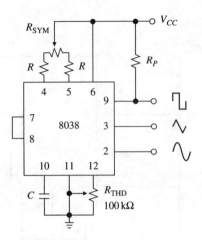

FIGURE 10.28
Basic ICL8038 connection for fixed-
frequency, 50% duty cycle operation.
(Courtesy of Harris Semiconductor.)

For perfect symmetry it is crucial that i_L and i_H be exactly in a 2:1 ratio.
By adjusting R_{SYM} one can keep the distortion level of the sine wave near 1%.
Connecting a 100-kΩ pot between pins 12 and 11 allows one to control the degree
of balance of the wave shaper to further reduce the THD.

As mentioned, the square-wave output is of the open-collector type, so a pullup
resistor R_p is needed. The peak-to-peak amplitudes of the square, triangular, and sine
waves are V_{CC}, $0.33V_{CC}$, and $0.22V_{CC}$, respectively. All three waves are centered
at $V_{CC}/2$. Powering the 8038 from split supplies makes the waves symmetric about
ground.

EXAMPLE 10.6. Assuming $V_{CC} = 15$ V in the circuit of Fig. 10.28, specify suitable
components for $f_0 = 10$ kHz.

Solution. Impose $i_A = i_B = 100$ μA, which is well within the recommended range.
Then, $R = (15/5)/0.1 = 30$ kΩ, and $C = 0.3/(10 \times 10^3 \times 30 \times 10^3) = 1$ nF. Use
$R_p = 10$ kΩ, and use $R_{\text{SYM}} = 5$ kΩ to allow for a $\pm20\%$ symmetry adjustment. Then,
recalculate R as $30 - 5/2 = 27.5$ kΩ (use 27.4 kΩ). To calibrate the circuit, adjust R_{SYM}
so that the square wave has $D(\%) = 50\%$, and R_{THD} until the THD of the sine wave is
minimized.

Varying the voltage of pin 8 provides automatic frequency sweeps. The fact that
the control voltage must be referenced to the V_{CC} rail is annoying in certain applications. This can be avoided by powering the 8038 between ground and a negative
supply, as in Fig. 10.29. Also shown in the diagram is an op amp that converts the
control voltage v_I to a current i_I, which then splits evenly between Q_1 and Q_2.
This scheme also eliminates any errors stemming from imperfect cancellation of the
base-emitter voltage drops of Q_3 and the Q_1-Q_2 pair. For accurate V-I conversion,
the input offset voltage of the op amp must be nulled. The circuit shown is designed
to give $i_I = v_I/(5$ k$\Omega)$ over a 1000:1 range, and is calibrated as follows: (a) with
$v_I = 10.0$ V and the wiper of R_3 set in the middle, adjust R_2 for $D(\%) = 50\%$; hence,

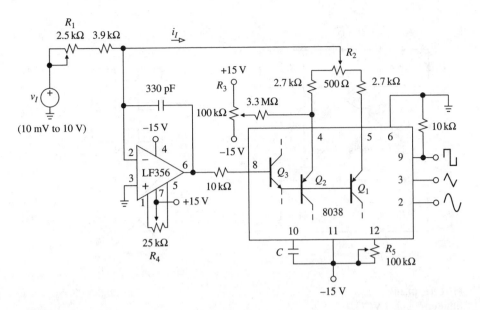

FIGURE 10.29
The ICL8038 as a linear voltage-controlled oscillator. (Courtesy of Harris Semiconductor.)

adjust R_1 for the desired full-scale frequency f_{FS}; (b) with $v_I = 10.0$ mV, adjust R_4 for $f_0 = f_{FS}/10^3$; hence, adjust R_3 for $D(\%) = 50\%$; repeat the adjustment of R_4, if necessary; (c) with $v_I = 1$ V, adjust R_5 for minimum THD.

Emitter-Coupled VCOs

These VCOs use a pair of cross-coupled Darlington stages and an emitter-coupling timing capacitor, as shown[4] in Fig. 10.30a. The two stages are biased with matched emitter currents, and their collector swings are constrained to just one diode voltage drop by clamps D_1 and D_2.

The cross-coupling between the two stages ensures that either Q_1-D_1 or Q_2-D_2 (but not both) are conducting at any given time. This bistable behavior is similar to cross-coupled inverters in flip-flop realizations. Unlike flip-flops, however, the capacitive coupling between the emitters causes the circuit to alternate between its two states in astable-multivibrator fashion. During any half cycle, the capacitor plate connected to the stage that is on remains at a constant potential, while the plate connected to the stage that is off ramps downward at a rate set by i_I. As the ramp approaches the emitter conduction threshold of the corresponding BJT, the latter goes on, forcing the other BJT to go off because of the positive-feedback action stemming from cross-coupling. Thus, C is alternately charged and discharged at a rate set by i_I.

Circuit operation is better visualized by tracing through the waveforms of Fig. 10.30b. Note that the emitter waveforms are identical except for a half-cycle delay. Feeding them to a high input-impedance difference amplifier yields a symmetric triangular wave with a peak-to-peak amplitude of two base-emitter voltage drops.

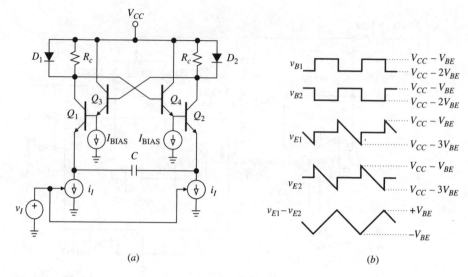

FIGURE 10.30
Emitter-coupled VCO.

The frequency of oscillation is found via Eq. (10.2) with $\Delta t = T/2$ and $\Delta v = 2V_{BE}$. Letting $f_0 = 1/T$ gives

$$f_0 = \frac{i_I}{4CV_{BE}} \tag{10.25}$$

indicating the CCO capability of the circuit.

The emitter-coupled oscillator enjoys a number of advantages: (a) it is simple and symmetric, (b) it lends itself to automatic frequency control, and (c) it is inherently capable of high-frequency operation since it consists of nonsaturating *npn*-BJTs. In its basic form of Fig. 10.30a, however, it suffers from a major drawback, namely, the thermal drift of V_{BE}, which is typically -2 mV/°C. There are various methods[4] of stabilizing f_0 with temperature. One method makes i_I proportional to V_{BE} to render their ratio temperature-independent. Popular devices utilizing this technique are the PLLs of the NE560 (Signetics) and XR-210/15 (Exar) types. Other methods modify the basic circuit to eliminate the V_{BE} term altogether. Though the increased circuit complexity lowers the upper end of the usable frequency range, these methods achieve thermal drifts as low as 20 ppm/°C. Popular products using this approach are the XR-2206/07 monolithic function generators (Exar) and the AD537 *V-F* converter (Analog Devices).

The XR-2206 Function Generator

This device uses an emitter-coupled CCO to generate the triangular and square waves, and a logarithmic wave shaper to convert the triangle to the sine wave.[4] The CCO parameters are designed so that when the circuit is connected in the basic

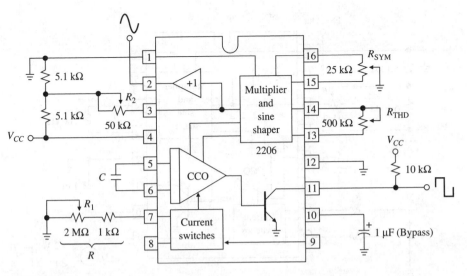

FIGURE 10.31
Basic XR-2206 connection for low-distortion sine wave generation. (Courtesy of Exar.)

configuration of Fig. 10.31, the frequency of oscillation is

$$f_0 = \frac{1}{RC} \tag{10.26}$$

The operating frequency range is from 0.01 Hz to more than 1 MHz, with a typical thermal stability of 20 ppm/°C. The recommended range for R is from 1 kΩ to 2 MΩ, and the optimum range is 4 kΩ to 200 kΩ. Varying R with a pot, as shown, allows for a 2000 : 1 sweep of f_0. Symmetry and distortion adjustments are provided, respectively, by R_{SYM} and R_{THD}. With proper calibration the circuit can achieve THD $\cong 0.5\%$.

The amplitude and offset of the sine wave are set by the resistive network external to pin 3. Denoting the equivalent resistance seen by this pin as R_3, the peak amplitude is approximately 60 mV for every kilohm of R_3. For instance, with the wiper of R_2 set in the middle, the peak amplitude of the sine wave is $[25 + (5.1 \parallel 5.1)] \times (60\,\text{mV}) \cong 1.65$ V. The sine wave offset is the same as the dc voltage established by the external network. With the components shown, this is $V_{CC}/2$.

Open circuiting pins 13 and 14 disables the rounding action by the wave shaper so that the output waveform becomes triangular. Its offset is the same as that of the sine wave; however, its peak amplitude is approximately twice as large. The square-wave output is of the open-collector type, hence, a pullup resistor is required.

Figure 10.32 shows another widely used 2206 configuration, which exploits the device's ability to operate with two separate timing resistances R_1 and R_2. With control pin 9 open-circuited or driven high, only R_1 is active and the circuit oscillates at $f_1 = 1/R_1C$; similarly, with pin 9 driven low, only R_2 is active and the circuit oscillates at $f_2 = 1/R_2C$. Thus, frequency can be keyed between two levels, often referred to as *mark* and *space frequencies,* whose values are set independently by R_1 and R_2. *Frequency shift keying* (FSK) is a widely used method of transmitting

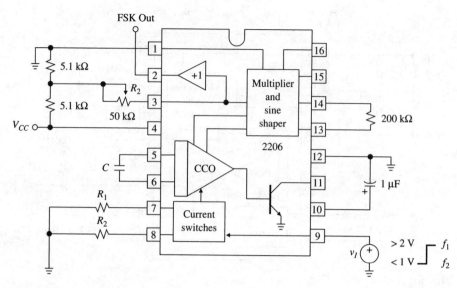

FIGURE 10.32
Sinusoidal FSK generator. (Courtesy of Exar.)

data over telecommunication links. If the FSK control signal is obtained from the square-wave output, R_1 and R_2 will be active on alternate half-cycles of oscillation. This feature can be exploited to configure the 2206 as a sawtooth/pulse generator.

10.7
V-F AND F-V CONVERTERS

The function of a *voltage-to-frequency converter* (VFC) is to accept an analog input v_I and generate a pulse train with frequency

$$f_O = kv_I \qquad (10.27)$$

where k is the VFC *sensitivity,* in hertz per volt. As such, the VFC provides a simple form of analog-to-digital conversion. The primary reason for this type of conversion is that a pulse train can be transmitted and decoded much more accurately than an analog signal, especially if the transmission path is long and noisy. If electrical isolation is also desired, it can be accomplished without loss of accuracy using inexpensive optocouplers or pulse transformers. Moreover, combining a VFC with a binary counter and digital readout provides a low-cost digital voltmeter.[6]

VFCs usually have more stringent performance specifications than VCOs. Typical requirements are (*a*) wide dynamic range (four decades or more), (*b*) the ability to operate to relatively high frequencies (hundreds of kilohertz, or higher), (*c*) low linearity error (less than 0.1% deviation from the straight line going from zero to the full scale), (*d*) high scale-factor accuracy and stability with temperature and supply voltage. The output waveform, on the other hand, is of secondary concern as long as its levels are compatible with standard logic signals. VFCs fall into two categories: *wide-sweep multivibrators* and *charge-balancing VFCs*.[4]

Wide-Sweep Multivibrator VFCs

These circuits are essentially voltage-controlled astable multivibrators designed with VFC performance specifications in mind. The multivibrator is usually a temperature-stabilized version of the basic CCO concept of Fig. 10.30. A popular product[7] in this category is the AD537 (Analog Devices) shown in Fig. 10.33. The op amp and Q_1 form a buffer *V-I* converter that converts v_I to the current drive i_I for the CCO according to $i_I = v_I/R$. The CCO parameters have been chosen so that $f_O = i_I/10C$, or

$$f_O = \frac{v_I}{10RC} \tag{10.28}$$

This relationship holds fairly accurately over a dynamic range of at least four decades, up to a full-scale current of 1 mA and a full-scale frequency of 100 kHz. For instance, with $C = 1$ nF, $R = 10$ kΩ and $V_{CC} = 15$ V, varying v_I from 1 mV to 10 V varies i_I from 0.1 μA to 1 mA and f_O from 10 Hz to 100 kHz. To minimize the *V-I* conversion error at the low end of the range, the op amp input offset error is nulled internally via R_{OS}. With a capacitor of suitable quality (polystyrene or NPO ceramic for low thermal drift and low dielectric absorption), the linearity error ratings are 0.1% typical for $f_O \leq 10$ kHz, 0.15% typical for $f_O \leq 100$ kHz.

Though the figure shows the connection for $v_I > 0$, we can easily configure the device for $v_I < 0$ by grounding the noninverting input of the op amp, lifting the left terminal of R off ground, and applying v_I there. The device can also function as a current-to-frequency converter (CFC) if we make the control current flow out of the inverting input node. For instance, grounding pin 5 and replacing R by a photodetector diode current sink will convert light intensity to frequency.

The AD537 also includes an on-chip precision voltage reference to stabilize the CCO scale factor. This yields a typical thermal stability of 30 ppm/°C. To further enhance the versatility of the device, two nodes of the reference circuitry are made available to the user, namely, V_R and V_T. Voltage V_R is a stable 1.00-V voltage reference. Obtaining v_I from pin 7 in Fig. 10.33 yields $f_O = 1/10RC$, and if R is

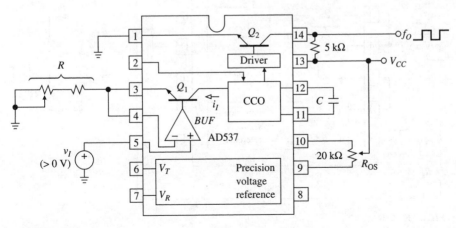

FIGURE 10.33
The AD537 voltage-to-frequency converter. (Courtesy of Analog Devices.)

a resistive transducer, such as a photoresistor or a thermistor, it will convert light or temperature to frequency.

Voltage V_T is a voltage linearly proportional to absolute temperature T as $V_T = (1\ mV/K)T$. For instance, at $T = 25\ °C = 273.2\ K$ we have $V_T = 298.2\ mV$. If v_I is derived from pin 6 in Fig. 10.33, then $f_O = T/(RC \times 10^4K)$, indicating that the circuit converts absolute temperature to frequency. For instance, with $R = 10\ k\Omega$ and $C = 1\ nF$, the sensitivity is 10 Hz/K. Other temperature scales, such as Celsius and Fahrenheit, can be accommodated by suitably offsetting the input range with the help of V_R.

> **EXAMPLE 10.7.** In the circuit of Fig. 10.34 specify suitable components to yield Celsius-to-frequency conversion with a sensitivity of 10 Hz/°C; then outline the calibration procedure.
>
> **Solution.** For $T = 0\ °C = 273.2\ K$ we have $V_T = 0.2732\ V$ and we want $f_O = 0$. Thus, R_3 must develop a 0.2732-V drop. Imposing $0.2732/R_3 = (1.00 - 0.2732)/R_2$ yields $R_2 = 2.66R_3$. For a sensitivity of 10 Hz/°C we want $10 = 1/10^4RC$, where $R = R_1 + (R_2 \parallel R_3)$ is the effective resistance seen by Q_1. Let $C = 3.9\ nF$; then $R = 2.564\ k\Omega$. Let $R_3 = 2.74\ k\Omega$; then $R_2 = 2.66 \times 2.74 = 7.29\ k\Omega$ (use 6.34 kΩ in series with a 2-kΩ pot). Finally, $R_1 = 2.564 - (2.74 \parallel 7.29) = 572\ \Omega$ (use 324 Ω in series with a 500-Ω pot).
>
> To calibrate, place the IC in a 0 °C environment and adjust R_2 so that the circuit is barely oscillating, say, $f_O \cong 1\ Hz$. Then move the IC to a 100 °C environment and adjust R_1 for $f_O = 1.0\ kHz$.

Figure 10.34 shows another useful feature of the AD537, namely, the ability to transmit information over a twisted pair. This pair serves the dual purpose of supplying power to the device and carrying frequency data in the form of current

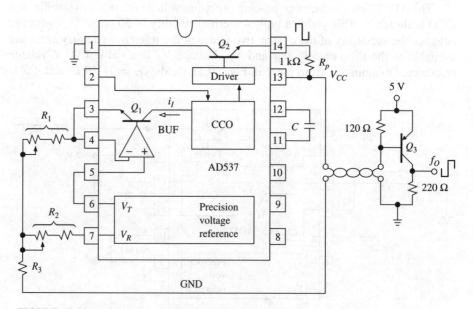

FIGURE 10.34
AD537 application as a temperature-to-frequency converter with two-wire transmission. (Courtesy of Analog Devices.)

modulation. With the parameter values shown, the current drawn by the AD537 alternates between about 1.2 mA during the half-cycle in which Q_2 is off, and $1.2 + [5 - V_{EB3(sat)} - V_{CE2(sat)}]/R_p \cong 1.2 + (5 - 0.8 - 0.1)/1 = 5.3$ mA during the half-cycle in which Q_2 is on. This current difference is sensed by Q_3 as a voltage drop across the 120-Ω resistance. This drop is designed to be low enough to keep Q_3 in cutoff when the current is 1.2 mA, yet large enough to drive Q_3 in saturation when the current is 5.3 mA. Consequently, Q_3 reconstructs a 5-V square wave at the receiving end. The ripple of about 0.5 V appearing across the 120-Ω resistance does not affect the performance of the AD537, thanks to its high PSRR.

Charge-Balancing VFCs

The charge-balancing technique[8] supplies a capacitor with continuous charge at a rate that is linearly proportional to the input voltage v_I, while simultaneously pulling discrete charge packets out of the capacitor at a rate f_O such that the net charge flow is always zero. The result is $f_O = kv_I$. Figure 10.35 illustrates the principle using the VFC32 *V-F* converter (Burr-Brown).

OA converts v_I to a current $i_I = v_I/R$ flowing into the summing junction; the value of R is chosen such that we always have $i_I < 1$ mA. With *SW* open, i_I flows into C_1 and causes v_1 to ramp downward. As soon as v_1 reaches 0 V, *CMP* fires and triggers a precision one-shot that closes *SW* and turns on Q_1 for a time interval

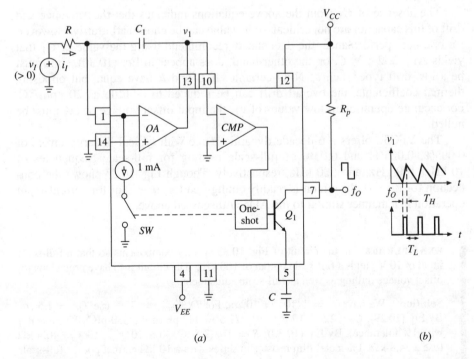

(a)   (b)

FIGURE 10.35
The VFC32 voltage-to-frequency converter. (Courtesy of Burr-Brown.)

T_H set by C. The one-shot, whose details have been omitted for simplicity, uses a threshold of 7.5 V and a charging current of 1 mA to give

$$T_H = \frac{7.5 \text{ V}}{1 \text{ mA}} C \qquad (10.29)$$

The closure of SW causes a net current of magnitude $(1 \text{ mA} - i_I)$ to flow out of the summing junction of OA. Consequently, during T_H, v_1 ramps upward by an amount $\Delta v_1 = (1 \text{ mA} - i_I) T_H / C_1$. After the one-shot times out, SW is opened and v_1 resumes ramping downward at a rate again set by i_I. The time T_L it takes for v_1 to return to zero is such that $T_L = C_1 \Delta v_1 / i_I$. Eliminating Δv_1 and letting $f_O = 1/(T_L + T_H)$ gives, with the help of Eq. (10.29),

$$f_O = \frac{v_I}{7.5RC} \qquad (10.30)$$

where f_O is in hertz, v_I in volts, R in ohms, and C in farads. As desired, f_O is linearly proportional to v_I. Moreover, the duty cycle $D(\%) = 100 \times T_H/(T_H + T_L)$ is readily found to be

$$D(\%) = 100 \frac{v_I}{R \times 1 \text{ mA}} \qquad (10.31)$$

and it is also proportional to v_I. For best linearity, the data sheets recommend designing for a maximum duty cycle of 25%, which corresponds to $i_{I(\max)} = 0.25$ mA.

The absence of C_1 from the above equations indicates that the tolerance and drift of this capacitor are not critical, so its value can be chosen arbitrarily. However, for optimum performance, the data sheets recommend using the value of C_1 that yields $\Delta v_1 \cong 2.5$ V. C, on the other hand, does appear in Eq. (10.30), so it must be a low-drift type, such as NPO ceramic. If C and R have equal but opposing thermal coefficients, the overall drift can be reduced to as little as 20 ppm/°C. For accurate operation to low values of v_I, the input offset voltage of OA must be nulled.

The VFC32 offers a 6-decade dynamic range with typical linearity errors of 0.005%, 0.025%, and 0.05% of full-scale reading for full-scale frequencies of 10 kHz, 100 kHz, and 500 kHz, respectively. Though Fig. 10.35 shows the connection for $v_I > 0$, the circuit is readily configured for $v_I < 0$ or for current-input operation in a manner similar to the AD537 discussed above.

EXAMPLE 10.8. In the circuit of Fig. 10.35 specify components so that a full-scale input of 10 V yields a full-scale output of 100 kHz. The circuit is to have provisions for offset voltage nulling as well as full-scale adjustment.

Solution. We have $T = 1/10^5 = 10 \ \mu s$. For $D(\%)_{\max} = 25\%$ use $T_H = 2.5 \ \mu s$. By Eq. (10.29), $C = 2.5 \times 10^{-6} \times 10^{-3}/7.5 = 333$ pF (use a 330-pF NPO capacitor with 1% tolerance). By Eq. (10.30), $R = 10/(7.5 \times 330 \times 10^{-12} \times 10^5) = 40.4$ kΩ (use a 34.8-kΩ, 1% metal-film resistor in series with a 10-kΩ cermet pot for full-scale adjustment). Imposing $\Delta v_{1(\max)} = 2.5$ V yields $C_1 = (10^{-3} \times 2.5 \times 10^{-6})/2.5 = 1$ nF.

To null the input offset voltage of OA, use the scheme of Fig. 5.19b with $R_A = 62 \ \Omega$, $R_B = 150$ kΩ, and $R_C = 100$ kΩ. The calibration is similar to that of Example 10.7.

The *frequency-to-voltage converter* (FVC) performs the inverse operation, namely, it accepts a periodic waveform of frequency f_I and yields an analog output voltage

$$v_O = kf_I \qquad (10.32)$$

where k is the FVC sensitivity, in volts per hertz. FVCs find application as tachometers in motor speed control and rotational measurements. Moreover, they are used in conjunction with VFCs to convert the transmitted pulse train back to an analog voltage.

A charge-balancing VFC can easily be configured as an FVC by applying the periodic input to the comparator and deriving the output from the op amp, which now has the resistance R in the feedback path (see Fig. 10.36). The input signal usually requires proper conditioning to produce a voltage with reliable zero-crossings for *CMP*. Shown in the figure is a high-pass network to accommodate inputs of the TTL and CMOS type. On each negative spike of v_1, *CMP* triggers the one-shot, closing *SW* and pulling 1 mA out of C_1 for a duration T_H as given in Eq. (10.29). In response to this train of current pulses, v_O builds up until the current pulled out of the summing junction of *OA* in 1-mA packets is exactly counterbalanced by that injected by v_O via R continuously, or $f_I \times 10^{-3} \times T_H = v_O/R$. Solving for v_O and using Eq. (10.29) gives

$$v_O = 7.5RCf_I \qquad (10.33)$$

The value of C is determined on the basis of a maximum duty cycle of 25%, as

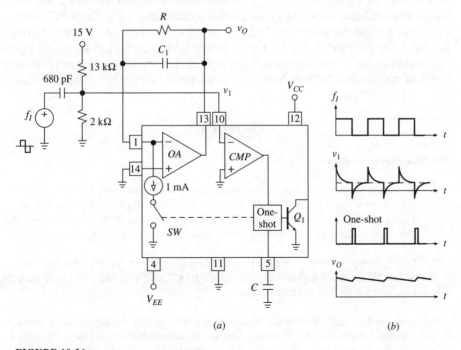

(a)

(b)

FIGURE 10.36
VFC connection for frequency-to-voltage conversion, and corresponding waveforms. (Courtesy of Burr-Brown.)

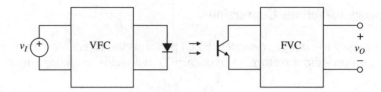

FIGURE 10.37
Transmission of analog information in isolated form.

discussed earlier, while R now establishes the full-scale value of v_O. As in the VFC case, the input offset voltage of OA should be nulled to avoid degrading the conversion accuracy at the low end of the range.

Between consecutive closures of SW, R will cause C_1 to discharge somewhat, resulting in output ripple. This can be objectionable, especially at the low end of the conversion range where the ripple-to-signal ratio is the worst. The maximum ripple is $V_{r(\max)} = (1 \text{ mA})T_H/C_1$. Using Eq. (10.29), we get

$$V_{r(\max)} = \frac{C}{C_1}7.5 \text{ V} \tag{10.34}$$

indicating that the ripple can be reduced by making C_1 suitably large. Too large a capacitance, however, slows down the response to a rapid change in f_I since this response is governed by the time constant $\tau = RC_1$. The optimum value of C_1 is, therefore, a compromise between the two opposing demands.

Figure 10.37 shows, in block diagram form, a typical VFC-FVC arrangement for transmitting analog information in isolated form. Here v_I is usually a transducer signal that has been amplified by an instrumentation amplifier. The VFC converts v_I to a train of current pulses for the LED, the phototransistor reconstructs the pulse train at the receiving end, and the FVC converts frequency back to an analog signal v_O. The example shown utilizes an opto-isolator; however, other forms of isolated coupling are possible, such as fiber optic links, pulse transformers, and RF links.

PROBLEMS

10.1 Sine wave generators

10.1 Show that for arbitrary component values in its positive-feedback network, the Wien-bridge circuit of Fig. 10.2a gives $B(jf_0) = 1/(1 + R_s/R_p + C_p/C_s)$ and $f_0 = 1/2\pi\sqrt{R_s R_p C_s C_p}$, where R_p and C_p are the parallel and R_s and C_s the series elements. Hence, verify that neutral stability requires $R_2/R_1 = R_s/R_p + C_p/C_s$.

10.2 Disregarding the limiter in Fig. 10.3a, obtain expressions for $T(s)$ for the cases in which the feedback resistance is 22.1 kΩ, 20.0 kΩ, and 18.1 kΩ. Then, find the pole locations for each of the three cases.

10.3 Problem 10.1 indicates that the frequency of a Wien-bridge oscillator can be varied by varying, for instance, R_p. However, to maintain neutral stability, we must also vary R_s in such a way as to keep the ratio R_s/R_p constant. This awkward constraint is avoided by the circuit[9] of Fig. P10.3. (a) Show that f_0 is still as in Problem 10.1, but neutral stability now requires $(R_2/R_1)(1 + R_3/R_p) = R_s/R_p + C_p/C_s$. (b) Verify that if we let $R_2/R_1 = C_p/C_s$, this condition simplifies to $R_3 = (R_1/R_2)R_s$. (c) Assuming sufficiently fast JFET-input of amps in the design shown, find the range of variability of f_0.

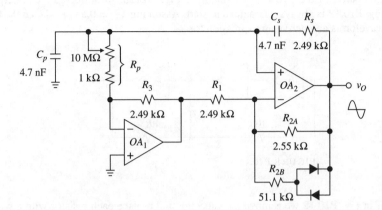

FIGURE P10.3

10.4 In the quadrature oscillator of Fig. 10.6 specify suitable components for $f_0 = 10$ kHz and $V_{om} = 5$ V. Hence, assuming 1N4148 diodes and 741 op amps, verify with PSpice.

10.5 In the quadrature oscillator of Fig. 10.6 let the variable resistance be adjusted to $R(1 - \epsilon)$, $\epsilon \ll 1$. Show that at power turn-on the poles are located in the right half of the s plane at $s = (\epsilon/4 \pm j)/RC$.

10.6 (a) Assuming $R_1 = 20$ kΩ, $R_2 = 10$ kΩ, $C_1 = 20$ nF, and $C_2 = 10$ nF in the low-pass KRC filter of Fig. 3.23, show a design to turn it into a sine wave oscillator without changing the values or the topology of the elements given here. (b) Find f_0.

10.2 Multivibrators

10.7 In the circuit of Fig. 10.7a let $R = 330$ kΩ, $C = 1$ nF, $R_1 = 10$ kΩ, $R_2 = 20$ kΩ. Assuming ± 15-V supplies, find f_0 and $D(\%)$ if a third resistance $R_3 = 30$ kΩ is connected between the noninverting-input pin of the 301 and the -15-V supply.

10.8 In the circuits of Fig. 10.7a let $R_1 = R_2 = 10$ kΩ, and suppose a control source v_I is connected to the noninverting input of the comparator via a 10-kΩ series resistance. Sketch the modified circuit, and show that it allows for automatic duty-cycle control. What are the expressions for $D(\%)$ and f_0 in terms of v_I? What is the permissible range for v_I?

10.9 In the circuits of Fig. 10.9a and Fig. 10.12a specify suitable components for $f_0 = 100$ kHz. The circuits must have provision for the exact adjustment of f_0.

10.10 (a) Using a 339 comparator, design a single-supply astable multivibrator with $f_0 = 10$ kHz and $D(\%) = 60\%$. (b) Repeat (a), but with $D(\%) = 40\%$.

10.11 The inverters of Fig. 10.12 have the following threshold ratings at $V_{DD} = 5$ V:

$V_T = 2.5$ V typical, 1.1 V minimum, and 4.0 V maximum. (a) Specify suitable components for $f_0 = 100$ kHz typical. (b) Find the percentage spread of f_0 due to the spread of V_T.

10.12 Compared to the two-gate oscillator of Fig. 10.12a, the three-gate counterpart of Fig. P10.12 is always guaranteed to start. Assuming $V_T = 0.5V_{DD}$, sketch the timing waveforms and derive an expression for f_0.

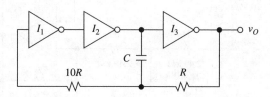

FIGURE P10.12

10.13 If in Fig. P10.12 we remove the capacitor and replace each resistor with a wire, the resulting circuit is called a *ring oscillator* and is often used to measure the propagation delays of logic gates. (a) Sketch the voltages at the gate outputs versus time; then derive a relationship between the average gate propagation delay t_P and the frequency of oscillation f_0. (b) Can this technique be extended to four gates within the loop? Explain.

10.14 Assuming the threshold spread specifications of Problem 10.11, find suitable components for $T = 10$ μs (typical) in the one-shot of Fig. 10.14a; then find the percentage spread of T.

10.15 Design a one-shot using two CMOS NAND gates. Next, explain how it works, show its waveforms, and derive an expression for T. (Recall that the output of a NAND gate goes low only when both inputs are high.)

10.16 Consider the circuit obtained from the one-shot of Fig. 10.14a by connecting the output of G to the input of I directly, inserting a resistance R between the lower input of G and ground, and returning the output of I to the lower input of G via a series capacitance C. Draw the modified circuit; then, sketch and label its waveforms, and find T if $R = 100$ kΩ, $C = 220$ pF, and $V_T = 0.4V_{DD}$.

10.3 Monolithic timers

10.17 Let the 555 astable multivibrator of Fig. 10.16a be modified as follows: R_B is shorted out, and the wire connecting the bottom node of R_A to pin 7 is cut to allow for the insertion of a series resistance R_C. (a) Sketch the modified circuit and show that choosing $R_C = R_A/2.362$ gives $D(\%) = 50\%$. (b) Specify suitable components for $f_0 = 10$ kHz and $D(\%) = 50\%$.

10.18 (a) Verify that if the THRESHOLD and TRIGGER terminals of the TLC555 CMOS timer are tied together to form a common input, then the device forms an inverting Schmitt trigger with $V_{TL} = (1/3)V_{DD}$, $V_{TH} = (2/3)V_{DD}$, $V_{OL} = 0$ V, and $V_{OH} = V_{DD}$, where V_{DD} is the supply voltage. (b) Using just one resistor and one capacitor, configure the device as a 100-kHz free-running multivibrator, and verify that its duty cycle is 50%.

10.19 Design a 555 one-shot whose pulse width can be varied anywhere from 1 ms to 1 s by means of a 1-MΩ pot.

10.20 A 10-μs 555 one-shot is powered from $V_{CC} = 15$ V. What voltage must be applied to the CONTROL input to stretch T from 10 μs to 20 μs? To shrink T from 10 μs to 5 μs?

10.21 Using a 555 timer powered from $V_{CC} = 5$ V, design a voltage-controlled astable multivibrator whose frequency of oscillation is $f_0 = 10$ kHz when $V_{TH} = (2/3)V_{CC}$, but can be varied over the range 5 kHz $\leq f_0 \leq$ 20 kHz by externally varying V_{TH}. What are the values of V_{TH} and $D(\%)$ corresponding to the extremes of the above frequency range?

10.22 In the circuit of Fig. 10.18 specify suitable components and output interconnections for $T = 1$ s and $T_o = 3$ min.

10.4 Triangular wave generators

10.23 In the circuit of Fig. 10.19a let the noninverting input of OA be lifted off ground and returned to a $+3$-V source. Draw the modified circuit; then, sketch and label its waveforms and find f_0 and $D(\%)$ if $R = 30$ kΩ, $C = 1$ nF, $R_1 = 10$ kΩ, $R_2 = 13$ kΩ, $R_3 = 2.2$ kΩ, and D_5 is a 5.1-V reference diode.

10.24 In the circuit of Fig. 10.19a let $R_1 = R_2 = R = 10$ kΩ, $R_3 = 3.3$ kΩ, $V_{D(on)} = 0.7$ V, $V_{Z5} = 3.6$ V, and suppose a control source v_I is connected to the inverting input of OA via a 10-kΩ series resistance. Sketch the modified circuit, and show that it allows for automatic duty-cycle control. What are the expressions for $D(\%)$ and f_0 in terms of v_I? What is the permissible range for v_I?

10.25 In the circuit of Fig. 10.20a specify suitable components so that both waves have peak amplitudes of 5 V and T_L and T_H are independently adjustable from 50 μs to 50 ms.

10.26 Using a CMOS op amp connected as a Deboo integrator, and a CMOS 555 timer connected as a Schmitt trigger in the manner of Problem 10.18, design a single-supply triangular wave generator. Then, show its waveforms and derive an expression for f_0.

10.27 The effect of component tolerances in the VCO of Fig. 10.21a can be compensated for by inserting a variable resistance R_s in series between the control source v_I and the rest of the circuit, and suitably decreasing the nominal value of C to allow for the adjustment of k in both directions. Design a VCO with $k = 1$ kHz/V, k adjustable over a range of $\pm 25\%$.

10.28 Shown in Fig. P10.28 is another popular VCO. Sketch and label its waveforms, and find an expression for f_0 in terms of v_I.

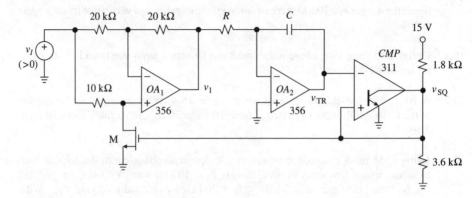

FIGURE P10.28

10.29 Design a wave shaper circuit that accepts the triangular output of the VCO of Fig. 10.21 and converts it to a sine wave of variable amplitude and offset. Amplitude and offset must be separately adjustable over the ranges 0 to 5 V and -5 V to $+5$ V, respectively.

10.30 Figure P10.30 shows a crude triangular-to-sine wave converter. R_1 and R_2 are found by imposing that v_{SINE} and $v_{TR}/(1 + R_2/R_1)$ have (a) identical slopes at the zero-crossings, and (b) peak values equal to $V_{D(on)}$. Assuming $V_{D(on)} = 0.7$ V at $I_D = 1$ mA, find R_1 and R_2 if v_{TR} has peak values of ± 5 V; then use PSpice to plot v_{TR} and v_{SINE} versus time.

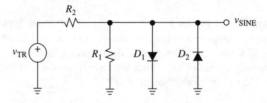

FIGURE P10.30

10.5 Sawtooth wave generators

10.31 (a) Show that connecting a resistance R_4 between the source v_I and the inverting-input pin of the 311 in Fig. 10.24a gives $V_T = V_{T0} - k|v_I|$, $V_{T0} = V_{CC}/[1 + R_2/(R_3 \parallel R_4)]$, $k = 1/[1 + R_4/(R_2 \parallel R_3)]$. ($b$) Verify that letting $R_4 = (R_2 \parallel R_3)(RC/T_D - 1)$ gets rid of the T_D term in Eq. (10.21) and gives $f_0 = |v_I|/RCV_{T0}$. (c) Assuming $T_D \cong 0.75 \mu s$, specify suitable components for a sensitivity of 2 kHz/V and a low-frequency sawtooth amplitude of 5 V. Your circuit is to be compensated against the error due to T_D.

10.6 Monolithic waveform generators

10.32 Derive Eq. (10.23).

10.33 Assuming $V_{CC} = 15$ V, design an ICL8038 sawtooth generator with $f_0 = 1$ kHz and $D(\%) = 99\%$. The circuit must have provision for frequency adjustment over a $\pm 20\%$ range.

10.34 Specify C for a 20-kHz full-scale frequency in the VCO of Fig. 10.29.

10.35 Assuming $V_{CC} = 15$ V, design an XR-2206 sawtooth generator with $f_0 = 1$ kHz, $D(\%) = 99\%$, and sawtooth peaks of 5 V and 10 V.

10.7 *V-F and F-V converters*

10.36 (*a*) Using the AD537 VFC, design a circuit that accepts a voltage in the range -10 V $<$ $v_S < 10$ V and converts it to a frequency in the range 0 Hz $< f_O < 20$ kHz. The circuit is to be powered from ± 15-V poorly regulated supplies. (*b*) Repeat, but for the case of an input 4 mA $< i_S < 20$ mA and an output range $0 < f_O < 100$ kHz.

10.37 Repeat Example 10.7, but for the Fahrenheit scale.

10.38 The circuit of Fig. P10.38 allows for the VFC32 to work with bipolar inputs. (*a*) Analyze the circuit for both $v_I > 0$ and $v_I < 0$, and find a condition for the resistances that will ensure $f_O = k|v_I|$. (*b*) Specify suitable components for a VFC sensitivity of 10 kHz/V.

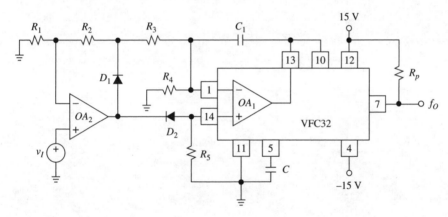

FIGURE P10.38

10.39 Specify suitable component values so that the FVC of Fig. 10.36 yields a full-scale output of 10 V for a full-scale input of 100 kHz with a maximum ripple of 10 mV. Then, estimate how long it takes for the output to settle within 0.1% of the final value for a full-scale change in f_I.

10.40 Using a 4N28 optocoupler, design an external resistive network to provide an optocoupled link between the VFC of Example 10.8 and the FVC of Problem 10.39. The transistor of the 4N28 gives $I_{C(\min)} = 1$ mA with a diode forward current $I_D = 10$ mA. Assume ± 15-V supplies.

REFERENCES

1. "Sine Wave Generation Techniques," Application Note AN-263, *Linear Applications Handbook,* National Semiconductor, Santa Clara, CA, 1994.

2. E. J. Kennedy, *Operational Amplifier Circuits: Theory and Applications,* Oxford University Press, New York, 1988.

3. J. Williams, "Circuit Techniques for Clock Sources," Application Note AN-12, *Linear Applications Handbook Volume I,* Linear Technology, Milpitas, CA, 1990.

4. A. B. Grebene, *Bipolar and MOS Analog Integrated Circuit Design,* John Wiley & Sons, New York, 1984.

5. *Linear & Telecom ICs for Analog Signal Processing Applications,* Harris Semiconductor, Melbourne, FL, 1993–1994, pp. 7-120–7-129.

6. P. Klonowski, "Analog-to-Digital Conversion Using Voltage-to-Frequency Converters," Application Note AN-276, *Applications Reference Manual,* Analog Devices, Norwood, MA, 1993.

7. B. Gilbert and D. Grant, "Applications of the AD537 IC Voltage-to-Frequency Converter," Application Note AN-277, *Applications Reference Manual,* Analog Devices, Norwood, MA, 1993.

8. J. Williams, "Design Techniques Extend V/F Converter Performance," *EDN,* May 16, 1985, pp. 153–164.

9. P. Brokaw, "FET Op Amps Add New Twist to an Old Circuit," *EDN,* June 5, 1974, pp. 75–77.

11

VOLTAGE REFERENCES AND REGULATORS

11.1 Performance Specifications
11.2 Voltage References
11.3 Voltage-Reference Applications
11.4 Linear Regulators
11.5 Linear-Regulator Applications
11.6 Switching Regulators
11.7 Monolithic Switching Regulators
    Problems
    References

The function of a voltage reference/regulator is to provide a stable dc voltage V_O starting from a less stable power source V_I. The general setup is depicted in Fig. 11.1.

In the case of a regulator, V_I is usually a poorly specified voltage, such as the crudely filtered output of a transformer and diode rectifier. The regulated output V_O is then used to power other circuits, collectively referred to as the *load* and characterized by the current I_O that the load draws from the regulator.

In the case of a voltage reference, V_I is already regulated to some degree, so the function of the reference is to produce an even more stable voltage V_O to serve as a standard for other circuits. The role of a reference is similar to that of a tuning fork for a musical ensemble. For example, the full-scale accuracy of a digital multimeter is set by an internal voltage reference of suitable quality. Similarly, power supplies; A-D, D-A, *V-F*, and *F-V* converters; transducer circuits; VCOs; log/antilog amplifiers; and a variety of other circuits and systems require some kind of reference standard, or yardstick, to function with the desired degree of accuracy. The primary requirements of a voltage reference are thus *accuracy* and *stability*. Typical stability requirements are on the order of 100 ppm/°C (parts per million per degree Celsius) or better. To minimize errors due to self-heating, voltage references come with modest output-current capabilities, usually on the order of a few milliamperes.

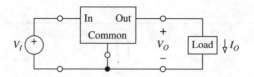

FIGURE 11.1
Basic connection of a voltage reference/ regulator.

Traditionally, the standard of voltages has been the Weston cell, an electrochemical device that, at 20 °C, yields a reproducible voltage of 1.018636 V with a thermal coefficient of 40 ppm/°C. Solid-state references are now available with far better stability. Even though semiconductor devices are strongly affected by temperature, clever compensating techniques have been devised to achieve thermal coefficients below 1 ppm/°C! These techniques are also exploited in the synthesis of voltages or currents with predictable thermal coefficients for use in temperature-sensing applications. This forms the basis of a variety of monolithic temperature transducers and signal conditioners.

The performance parameters of voltage regulators are similar to those of voltage references, except that the requirements are less stringent and the output current capabilities are much higher. Depending on the regulator type, the output current rating may range from as low as 100 mA to 10 A or higher.

In this chapter we discuss two popular categories, namely, *linear regulators* and *switching regulators*. Linear regulators control V_O by continuously adjusting a power transistor connected in series between V_I and V_O. The simplicity of this scheme comes at the price of poor efficiency because of the power dissipated in the transistor.

Switching regulators improve efficiency by operating the transistor as a high-frequency switch, which inherently dissipates less power than a transistor operating in the continuous mode. Moreover, unlike their linear counterparts, switching regulators can generate outputs that are higher than the unregulated input or even of the opposite polarity; they can provide multiple outputs, isolated outputs, and can be made to run directly off the ac power line, with no need for bulky power transformers. The price for these advantages is the need for coils, capacitors, and more complex control circuitry, along with much noisier behavior. Nonetheless, switching regulators are widely used to power computers and portable equipment. Even in power-supply design for analog systems it is common to exploit the efficiency and the light-weight advantages of switching regulators to generate preregulated— if noisy—voltages and then use linear regulators to provide cleaner postregulated voltages for critical analog circuitry.[1]

11.1
PERFORMANCE SPECIFICATIONS

The ability of a voltage reference or regulator to maintain a constant output under varying external conditions is characterized in terms of performance parameters such as *line* and *load regulation,* and the *thermal coefficient.* In the case of voltage references, *output noise* and *long-term stability* are also significant.

Line and Load Regulation

Line regulation, also called *input,* or *supply regulation,* gives a measure of the circuit's ability to maintain the prescribed output under varying input conditions. In the case of voltage references, the input is typically an unregulated voltage or, at best, a regulated voltage of lower quality than the reference itself. In the case of voltage regulators, the input is usually derived from the 60-Hz line via a step-down transformer, a diode-bridge rectifier, and a capacitor filter and is afflicted by significant ripple. With reference to the symbolism of Fig. 11.1, we define

$$\text{Line regulation} = \frac{\Delta V_O}{\Delta V_I} \qquad (11.1a)$$

where ΔV_O is the output change resulting from a change ΔV_I at the input. Line regulation is expressed in millivolts or microvolts per volt, depending on the case. An alternative definition is

$$\text{Line regulation } (\%) = 100\frac{\Delta V_O/V_O}{\Delta V_I} \qquad (11.1b)$$

with the units being percent per volt. As you consult the catalogs, you will find that both forms are in use.

A related parameter is the *ripple rejection ratio* (RRR), expressed in decibels as

$$\text{RRR}_{dB} = 20\log_{10}\frac{V_{ri}}{V_{ro}} \qquad (11.2)$$

where V_{ro} is the output ripple resulting from a ripple V_{ri} at the input. The RRR is used especially in connection with voltage regulators to provide an indication of the amount of ripple (usually 120-Hz ripple) feeding through to the output.

Load regulation gives a measure of the circuit's ability to maintain the prescribed output voltage under varying load conditions, or

$$\text{Load regulation} = \frac{\Delta V_O}{\Delta I_O} \qquad (11.3a)$$

Both voltage references and voltage regulators should behave like ideal voltage sources, delivering a prescribed voltage regardless of the load current. The *i-v* characteristic of such a device is a vertical line positioned at $v = V_O$. A practical reference or regulator exhibits a nonzero output impedance whose effect is a slight dependence of V_O on I_O. This dependence is expressed via the load regulation, in millivolts per milliampere or per ampere, depending on the output current capabilities. The alternative definition

$$\text{Load regulation } (\%) = 100\frac{\Delta V_O/V_O}{\Delta I_O} \qquad (11.3b)$$

expresses the above dependence in percent per milliampere or per ampere.

EXAMPLE 11.1. The data sheets of the μA7805 5-V voltage regulator (Fairchild) indicate that V_O typically changes by 3 mV when V_I is varied from 7 V to 25 V, and by 5 mV when I_O is varied from 0.25 A to 0.75 A. Moreover, $\text{RRR}_{dB} = 78$ dB at 120 Hz. (a) Estimate the typical line and load regulation of this device. What is the output impedance of the regulator? (b) Estimate the amount of output ripple V_{ro} for every volt of V_{ri}.

Solution.

(a) Line regulation $= \Delta V_O/\Delta V_I = 3 \times 10^{-3}/(25 - 7) = 0.17$ mV/V. Alternatively, line regulation $= 100(0.17 \text{ mV/V})/(5 \text{ V}) = 0.0033\%$/V. Load regulation $= \Delta V_O/\Delta I_O = 5 \times 10^{-3}/[(750 - 250)10^{-3}] = 10$ mV/A. Alternatively, load regulation $= 100(10 \text{ mV/A})/(5 \text{ V}) = 0.2\%$/A. The output impedance is $\Delta V_O/\Delta I_O = 0.01 \ \Omega$.

(b) $V_{ro} = V_{ri}/10^{78/20} = 0.126 \times 10^{-3} \times V_{ri}$. Thus, a 1-V, 120-Hz ripple at the input will result in an output ripple of 0.126 mV.

Thermal Coefficient

The *thermal coefficient* of V_O, denoted as $\text{TC}(V_O)$, gives a measure of the circuit's ability to maintain the prescribed output voltage V_O under varying thermal conditions. It is defined in two forms,

$$\text{TC}(V_O) = \frac{\Delta V_O}{\Delta T} \tag{11.4a}$$

in which case it is expressed in millivolts or microvolts per degree Celsius, or

$$\text{TC}(V_O) \ (\%) = 100\frac{\Delta V_O/V_O}{\Delta T} \tag{11.4b}$$

in which case it is expressed in percent per degree Celsius. Replacing 100 by 10^6 gives the TC in parts per million per degree Celsius. Good voltage references have TCs on the order of a few parts per million per degree Celsius.

EXAMPLE 11.2. The data sheets of the REF101KM 10-V precision voltage reference (Burr-Brown) give a typical line regulation of 0.001%/V, a typical load regulation of 0.001%/mA, and a maximum TC of 1 ppm/°C. Find the variation in V_O brought about by: (a) a change of V_I from 13.5 V to 35 V; (b) a ±10-mA change in I_O; (c) a temperature change from 0 °C to 70 °C.

Solution.

(a) By Eq. (11.1b), 0.001%/V $= 100(\Delta V_O/10)/(35 - 13.5)$, or $\Delta V_O = 2.15$ mV typical.

(b) By Eq. (11.3b), 0.001%/mA $= 100(\Delta V_O/10)/(\pm10 \text{ mA})$, or $\Delta V_O = \pm1$ mV typical.

(c) By Eq. (11.4b), 1 ppm/°C $= 10^6(\Delta V_O/10)/(70 \ °\text{C})$, or $\Delta V_O = 0.7$ mV maximum. You will agree that these are rather small variations for a 10-V source!

In the case of voltage references, *output noise* and *long-term stability* are also important. The data sheets of the aforementioned REF101 give a typical output noise of 6 μV peak-to-peak from 0.1 Hz to 10 Hz, and a typical long-term stability of 50 ppm/(1000 hours). This means that over a period of 1000 hours (about 42 days) the reference output may typically change by $(50 \times 10^{-6})10 \text{ V} = 0.5$ mV.

Let us apply the above concepts to the analysis of the classical shunt regulator of Fig. 11.2. The input is a raw voltage assumed to lie within known limits, or $V_{I(\min)} \leq V_I \leq V_{I(\max)}$. The goal is to produce an output V_O that is as insensitive as possible to both input and load variations. This is achieved by exploiting the nearly vertical i-v characteristic of a Zener diode. As depicted in Fig. 11.3a, this characteristic can be approximated with a straight line having a slope of $1/r_z$ and a v-axis intercept at $-V_{Z0}$, so the coordinates V_Z and I_Z of an arbitrary operating point down the curve are related as $V_Z = V_{Z0} + r_z I_Z$. The resistance r_z, called the *dynamic resistance* of the Zener diode, is typically in the range of a few ohms to several hundreds of ohms, depending on the diode. Zener diodes are specified at the point corresponding to 50% of the power rating. Thus, a 6.8-V, 0.5-W, 10-Ω Zener diode has, at the 50% power point, $I_Z = (P_Z/2)/V_Z = (500/2)/6.8 \cong 37$ mA. Moreover, $V_{Z0} = V_Z - r_z I_Z = 6.8 - 10 \times 37 \times 10^{-3} = 6.43$ V.

It is apparent that a Zener diode can be modeled with a voltage source V_{Z0} and a series resistance r_z, so the circuit of Fig. 11.2b can be redrawn as in Fig. 11.3b. To function as a regulator, the diode must operate well within the breakdown region under all possible line and load conditions. In particular, I_Z must never be allowed to

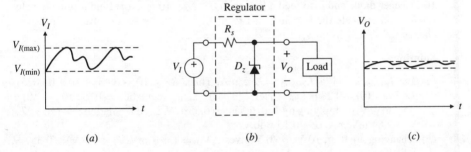

| | | |
| (a) | (b) | (c) |

FIGURE 11.2
The Zener diode as a shunt regulator.

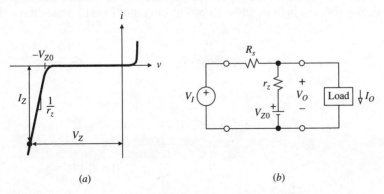

| | |
| (a) | (b) |

FIGURE 11.3
Breakdown diode characteristic, and equivalent circuit of the shunt regulator.

drop below some safety value $I_{Z(\min)}$. Simple analysis reveals that R_s must satisfy

$$R_s \leq \frac{V_{I(\min)} - V_{Z0} - r_z I_{Z(\min)}}{I_{Z(\min)} + I_{O(\max)}} \tag{11.5}$$

The value of $I_{Z(\min)}$ is chosen as a compromise between the need to ensure proper worst-case operation and the need to avoid excessive power wastage. A reasonable compromise is $I_{Z(\min)} \cong (1/4)I_{O(\max)}$.

We are now ready to estimate the line and load regulation. Applying the superposition principle, we readily find

$$V_O = \frac{r_z}{R_s + r_z}V_I + \frac{R_s}{R_s + r_z}V_{Z0} - (R_s \parallel r_z)I_O \tag{11.6}$$

Only the second term on the right-hand side is a desirable one. The other two indicate dependence on line and load as

$$\text{Line regulation} = \frac{r_z}{R_s + r_z} \tag{11.7a}$$

$$\text{Load regulation} = -(R_s \parallel r_z) \tag{11.7b}$$

Multiplying by $100/V_O$ gives the regulations in percentage form.

EXAMPLE 11.3. A raw voltage $10\text{ V} \leq V_I \leq 20\text{ V}$ is to be stabilized by a 6.8-V, 0.5-W, 10-Ω Zener diode and is to feed a load with $0 \leq I_O \leq 10$ mA. (a) Find a suitable value for R_s, and estimate the line and load regulation. (b) Estimate the effect of full-scale changes of V_I and I_O on V_O.

Solution.

(a) Let $I_{Z(\min)} = (1/4)I_{O(\max)} = 2.5$ mA. Then, $R_s \leq (10 - 6.43 - 10 \times 0.0025)/(2.5 + 10) = 0.284$ kΩ (use 270 Ω). Line regulation $= 10/(270 + 10) = 35.7$ mV/V; multiplying by $100/6.5$ gives 0.55%/V. Load regulation $= -(10 \parallel 270) = -9.64$ mV/mA, or -0.15%/mA.

(b) Changing V_I from 10 V to 20 V gives $\Delta V_O = (35.7\text{ mV/V}) \times (10\text{ V}) = 0.357$ V, which represents a 5.5% change in V_O. Changing I_O from 0 to 10 mA gives $\Delta V_O = -(9.64\text{ mV/mA}) \times (10\text{ mA}) = -0.096$ V, which represents a -1.5% change.

The modest line and load regulation capabilities of a diode can be improved dramatically with the help of an op amp. The circuit of Fig. 11.4 uses the artifice of

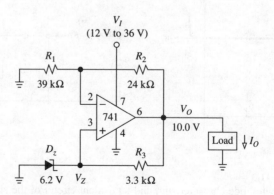

FIGURE 11.4
Self-regulated 10-V reference.

powering the diode from V_O, that is, from the very voltage we are trying to regulate. The result is a far more stable voltage V_Z, which the op amp then amplifies to give

$$V_O = \left(1 + \frac{R_2}{R_1}\right)V_Z \qquad (11.8)$$

This artifice, aptly referred to as *self-regulation,* shifts the burden of line and load regulation from the diode to the op amp. As an additional advantage, V_O is now adjustable, for instance, via R_2. Moreover, R_3 can now be raised to avoid unnecessary power wastage and self-heating effects.

By inspection, we now have

$$\text{Load regulation} \cong -\frac{z_o}{1 + a\beta} \qquad (11.9)$$

where a and z_o are the open-loop gain and output impedance, and $\beta = R_1/(R_1+R_2)$. To find the line regulation, we observe that because of single-supply operation, a 1-V change in V_I is perceived by the op amp both as a 1-V supply change and as a 0.5-V input common-mode change. This results in a worst-case input offset voltage change $\Delta V_{OS} = \Delta V_I (1/\text{PSRR} + 1/2\text{CMRR})$ appearing in series with V_Z. The op amp then gives $\Delta V_O = (1 + R_2/R_1)\Delta V_{OS}$, so

$$\text{Line regulation} = \left(1 + \frac{R_2}{R_1}\right) \times \left(\frac{1}{\text{PSRR}} + \frac{0.5}{\text{CMRR}}\right) \qquad (11.10)$$

We observe that since z_o, a, PSRR, and CMRR are frequency-dependent, so are the line and load regulation. In general, both parameters tend to degrade with frequency.

EXAMPLE 11.4. Assuming typical 741 dc parameters, find the line and load regulation of the circuit of Fig. 11.4.

Solution. Load regulation $= -75/[1 + 2 \times 10^5 \times 39/(39 + 24)] = -0.6\ \mu\text{V/mA} = -0.06\ \text{ppm/mA}$. Using $1/\text{PSRR} = 30\ \mu\text{V/V}$ and $1/\text{CMRR} = 10^{-90/20} = 31.6\ \mu\text{V/V}$, we get line regulation $= (1 + 24/39) \times (30 + 15.8)10^{-6} = 74\ \mu\text{V/V} = 7.4\ \text{ppm/V}$. They represent dramatic improvements over the circuit of Example 11.3.

Dropout Voltage

The circuit of Fig. 11.4 will work properly as long as V_I does not drop too low to cause the op amp to saturate. This holds for voltage references and regulators in general, and the minimum difference between V_I and V_O for which the circuit still functions properly is called the *dropout voltage* V_{DO}. In the example of Fig. 11.4 the 741 requires that V_{CC} be at least a couple of volts higher than V_O, so in this case $V_{DO} \cong 2$ V. Moreover, since the maximum supply rating of the 741 is 36 V, it follows that the permissible input voltage range for the circuit is 12 V $< V_I <$ 36 V.

Start-up Circuitry

In the self-regulated circuit of Fig. 11.4, V_O depends on V_Z, and V_Z, in turn, depends on V_O being greater than V_Z to keep the diode reverse biased. If at power turn-on V_O fails to swing to a value greater than V_Z, the diode will never turn on, making positive feedback via R_3 prevail over negative feedback via R_2 and R_1. The result is a Schmitt trigger latched in the undesirable state $V_O = V_{OL}$. The possibility for this undesirable behavior is common in most self-biased circuits, and is avoided by using suitable circuitry, known as *start-up circuitry,* to override the amplifier and prevent it from latching in this undesirable state when power is first applied.

The particular implementation of Fig. 11.4 will start properly because of the internal nature of the op amp being used. With reference to Fig. 5.1, we observe that at power turn-on, when v_P and v_N are still zero, the first two stages of the 741 remain off, allowing I_B to turn on the output stage. Consequently, V_O will swing positive until the Zener diode turns on and the circuit stabilizes at $V_O = (1 + R_2/R_1)V_Z$. However, if another op amp type is used, the circuit may never be able to properly bootstrap itself, thus requiring start-up circuitry. We shall see an example in Section 11.4.

11.2
VOLTAGE REFERENCES

Besides line and load regulation, thermal stability is the most demanding performance requirement of voltage references due to the tendency of IC components to be strongly influenced by temperature.[2] For example, consider the silicon *pn* junction, which forms the basis of diodes and BJTs. Its forward-bias voltage V_D and current I_D are related as $V_D = V_T \ln(I_D/I_s)$, where V_T is the *thermal voltage* and I_s the *saturation current.* Their expressions are

$$V_T = kT/q \qquad (11.11a)$$

$$I_s = BT^3 \exp(-V_{G0}/V_T) \qquad (11.11b)$$

where $k = 1.381 \times 10^{-23}$ is Boltzmann's constant, $q = 1.602 \times 10^{-19}$ C is the electron charge, T is absolute temperature, B is a proportionality constant, and $V_{G0} = 1.205$ V is the *bandgap voltage* for silicon.

The TC of the thermal voltage is

$$\text{TC}(V_T) = k/q = 0.0862 \text{ mV/}^\circ\text{C} \qquad (11.12)$$

The TC of the junction voltage drop V_D at a given bias I_D is $\text{TC}(V_D) = \partial V_D/\partial T = (\partial V_T/\partial T)\ln(I_D/I_s) + V_T \partial[\ln(I_D/I_s)]/\partial T = V_D/T - V_T \partial(3\ln T - V_{G0}/V_T)/\partial T$. The result is

$$\text{TC}(V_D) = -\left(\frac{V_{G0} - V_D}{T} + \frac{3k}{q}\right) \qquad (11.13)$$

Assuming $V_D = 650$ mV at 25 °C, we get $\text{TC}(V_D) \cong -2.1$ mV/°C. Engineers remember this by saying that the forward drop of a silicon junction decreases by about 2 mV for every degree Celsius increase. Equations (11.12) and (11.13) form the basis of two common approaches to thermal stabilization, namely, *thermally compensated Zener diode references* and *bandgap references.* Equation 11.12 forms also the basis of solid-state *temperature sensors.*

Thermally Compensated Zener Diode References

The thermal stability of V_O in the self-regulated reference of the previous section can be no better than that of V_Z itself. As depicted in Fig. 11.5a, TC(V_Z) is a function of V_Z as well as I_Z. There are two different mechanisms by which the *i-v* characteristic breaks down: *field emission breakdown,* which dominates below 5 V and produces negative TCs, and *avalanche breakdown,* which dominates above 5 V and produces positive TCs. The idea behind thermally compensated Zener diodes is to connect a forward-biased diode in series with a Zener diode having an equal but opposing TC, and then fine-tune I_Z to drive the TC of the composite device to zero.[3] This is illustrated in Fig. 11.5b for the compensated diodes of the popular 1N821-9 series (Motorola). The composite device, whose voltage we relabel as $V_Z = 5.5 + 0.7 = 6.2$ V, uses $I_Z = 7.5$ mA to minimize TC(V_Z). This TC ranges from 100 ppm/°C (1N821) to 5 ppm/°C (1N829).

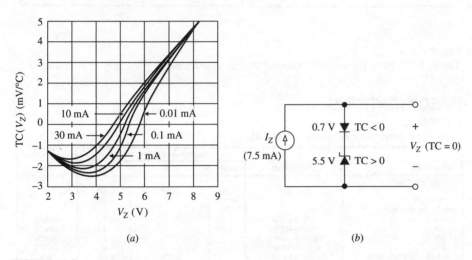

(a) (b)

FIGURE 11.5
(a) TC(V_Z) as a function of V_Z and I_Z. (b) Thermally compensated breakdown diode. (Courtesy of Motorola, Inc.)

Self-regulated references based on thermally compensated Zener diodes are available in monolithic form. An example is the REF101 10-V precision reference (Burr-Brown) depicted in Fig. 11.6a. The device includes also a pair of matched 20-kΩ resistances to facilitate applications. The typical drift curve of Fig. 11.6b indicates a maximum output change of 0.7 mV for a thermal excursion of 0 °C to 70 °C. Other specifications are shown in Fig. 11.7.

Another popular device[4] is the LM329 precision reference diode (National Semiconductor) shown in Fig. 11.8 (bottom). This device uses Zener diode Q_3 in series with the BE junction of Q_{13} to achieve TCs ranging from 100 ppm/°C to 6 ppm/°C, depending on the version. The device uses also active feedback circuitry to lower the effective dynamic resistance to $r_z = 0.6$ Ω typical, 1 Ω maximum. Except for its much greater stability and much lower dynamic resistance, it acts as an ordinary Zener diode, and it is biased via a series resistor to provide shunt regulation. The bias current may be anywhere between 0.6 mA and 15 mA.

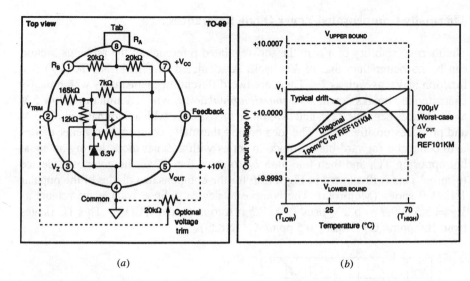

(a)   (b)

FIGURE 11.6
The REF101 10-V voltage reference and its drift characteristic. (Courtesy of Burr-Brown.)

SPECIFICATIONS

ELECTRICAL

At T_A = +25°C and +15VDC power supply, unless otherwise noted.

| PARAMETER | CONDITIONS | REF101JM, KM, RM, SM | | | UNITS |
		MIN	TYP	MAX	
OUTPUT VOLTAGE					
Initial	T_A = +25°C	9.995	10.000	10.005	V
Trim Range[1]		−0.10(		+0.250	V
vs Temperature[2]					
KM	0°C to +70°C			1	ppm/°C
JM	0°C to +70°C			2	ppm/°C
SM	−55°C to +125°C			3	ppm/°C
RM	−55°C to +125°C			6	ppm/°C
vs Supply (line regulation)	V_{CC} = 13.5 to 35V		0.001	0.002	%/V
vs Output Current					
(load regulation)	I_L = 0 to ±10mA		0.001	0.002	%/mA
vs Time	T_A = +25°C		50		ppm/1000hrs
NOISE	0.1Hz to 10Hz		6	25	µVp-p
OUTPUT CURRENT	Source or Sink	±10			mA
INPUT VOLTAGE RANGE		13.5		35	V
QUIESCENT CURRENT	I_{OUT} = 0		4.5	6	mA
WARM-UP TIME	To 0.1%		10		µs
UNCOMMITTED RESISTORS					
Resistance			20		kΩ
Match			±0.01	±0.05	%
TCR			50		ppm/°C
TCR Tracking			2		ppm/°C
TEMPERATURE RANGE					
Specification					
JM, KM		0		+70	°C
RM, SM		−55		+125	°C
Operating					
JM, KM		−25		+85	°C
RM, SM		−55		+125	°C
Storage		−65		+125	°C

NOTES : (1) Trimming the offset voltage will affect the drift slightly. See Installation and operating Instructions for details. (2) The "box method" is used to specify output voltage drift vs temperature. See the Discussion of Performance section.

FIGURE 11.7
REF101 10-V voltage reference specifications. (Courtesy of Burr-Brown.)

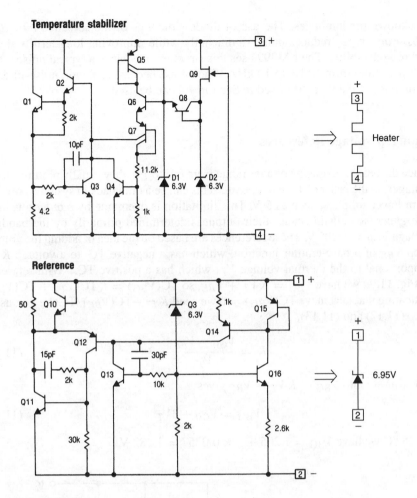

FIGURE 11.8
Circuit diagram of the LM399 6.95-V thermally stabilized reference. (Courtesy of National Semiconductor.)

Thermal stability can be improved further via substrate thermostating.[4] The LM399 stabilized reference of Fig. 11.8 uses the aforementioned LM329 active diode (shown at the bottom) to provide the reference proper, and suitable stabilizing circuitry (shown at the top) to sense the substrate temperature and hold it at some set value above the maximum expected ambient temperature. Thermal sensing is done via the BE junction of Q_4, and substrate heating via the power-dissipating transistor Q_1. At power turn-on, Q_1 heats the substrate to 90 °C, where it is then maintained within less than 2 °C over ambient variations from 0 °C to 70 °C. The result is a typical TC of 0.3 ppm/°C. Another thermally stabilized reference is the LTZ1000 Super Zener (Linear Technology). An obvious drawback of these devices is the additional power required to heat the chip. For instance, at 25 °C, the LM399 dissipates 300 mW. An LM399 application will be shown in Fig. 11.11.

A notorious problem with breakdown diodes is noise, especially avalanche noise, which plagues devices with breakdown voltages above 5 V, where avalanche breakdown predominates. The use of diode structures of the so-called *buried*, or *subsurface*, type[4] reduces noise significantly while improving long-term stability and reproducibility. The LM399 uses this structure to achieve a typical noise rating of 7 μV (rms) from 10 Hz to 10 kHz. When noise becomes a factor, noise-filtering techniques of the type discussed in Section 7.4 can be used.

Bandgap Voltage References

Since the best breakdown voltages range from 6 V to 7 V, they usually require supply voltages on the order of 10 V to operate. This can be a drawback in systems powered from lower supplies, such as 5 V. This limitation is overcome by *bandgap voltage references*, so called because their output is determined primarily by the bandgap voltage $V_{G0} = 1.205$ V. These references are based on the idea of adding the voltage drop V_{BE} of a base-emitter junction, which has a negative TC, to a voltage $K V_T$ proportional to the thermal voltage V_T, which has a positive TC.[2] With reference to Fig. 11.9a we have $V_{BG} = K V_T + V_{BE}$, so $\text{TC}(V_{BG}) = K\text{TC}(V_T) + \text{TC}(V_{BE})$, indicating that to achieve $\text{TC}(V_{BG}) = 0$ we need $K = -\text{TC}(V_{BE})/\text{TC}(V_T)$ or, using Eqs. (11.12) and (11.13),

$$K = \frac{V_{G0} - V_{BE}}{V_T} + 3 \tag{11.14}$$

Substituting into $V_{BG} = K V_T + V_{BE}$ gives

$$V_{BG} = V_{G0} + 3V_T \tag{11.15}$$

At 25 °C we have $V_{BG} = 1.205 + 3 \times 0.0257 = 1.282$ V.

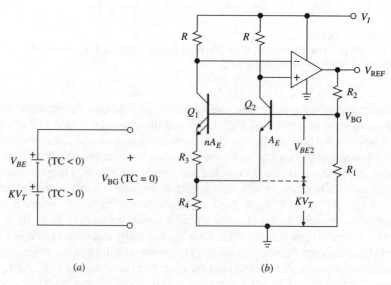

(a)

(b)

FIGURE 11.9
Bandgap voltage reference.

Figure 11.9*b* shows one of several popular bandgap-cell realizations. Known as the *Brokaw cell* for its inventor,[5] the circuit is based on two BJTs of different emitter areas. The emitter area of Q_1 is n times as large as the emitter area A_E of Q_2, so the saturation currents satisfy $I_{s1}/I_{s2} = n$, by Eq. (5.32). With identical collector resistances, the collector currents are also identical, by op amp action. Ignoring base currents, we have $KV_T = R_4(I_{C1} + I_{C2}) = 2R_4 I_{C1}$, or

$$KV_T = 2R_4 \frac{V_{BE2} - V_{BE1}}{R_3} = \frac{2R_4}{R_3} V_T \ln \frac{I_{C2} I_{s1}}{I_{s2} I_{C1}} = \frac{2R_4}{R_3} V_T \ln n$$

indicating that

$$K = 2\frac{R_4}{R_3} \ln n \tag{11.16}$$

This constant can be fine-tuned by adjusting the ratio R_4/R_3. The op amp raises the cell's voltage to $V_{REF} = (1 + R_2/R_1)V_{BG}$.

> **EXAMPLE 11.5.** Assuming $n = 4$ and $V_{BE2}(25°C) = 650\,\text{mV}$ in the circuit of Fig. 11.9*b* specify R_4/R_3 for TC(V_{BG}) = 0 at 25 °C, and R_2/R_1 for $V_{REF} = 5.0$ V.
>
> **Solution.** By Eq. (11.14), $K = (1.205 - 0.65)/0.0257 + 3 = 24.6$. Then, $R_4/R_3 = K/(2\ln 4) = 8.87$. Moreover, imposing $5.0 = (1 + R_2/R_1)1.282$ gives $R_2/R_1 = 2.9$.

Thanks to their ability to operate with low supply voltages, bandgap references (see also the alternative realizations[2] of Problems 11.5 and 11.6) find wide application as part of systems such as voltage regulators; D-A, A-D, *V-F*, and *F-V* converters; bar graph meters; and power-supply supervisory circuits. They are also available as stand-alone products, either as *two-terminal* or as *three-terminal references,* and sometimes they come with provisions for external trimming.

An example of a two-terminal reference is the already familiar LM385 2.5-V micropower reference diode (National Semiconductor). Besides the bandgap cell, the device includes circuitry to minimize its dynamic resistance as well as raise the cell voltage to 2.5 V. Typically, it has a TC of 20 ppm/°C and a dynamic resistance of 0.4 Ω. It is biased with a plain series resistance, and its operating current may be anywhere between 20 μA and 20 mA.

An example of a three-terminal reference is the REF-05 5-V precision reference (Analog Devices). Its output, rated at 5.00 V ± 30 mV, can be adjusted externally over a ±300-mV range. The REF-05A version has, typically, TC = 3 ppm/°C for $-55\,°C \le T \le 125\,°C$, line regulation = 0.006%/V for $8\,\text{V} \le V_I \le 33\,\text{V}$, load regulation = 0.005%/mA for $0 \le I_O \le 10$ mA, output noise = 10 μV peak-to-peak from 0.1 Hz to 10 Hz, and long-term stability = 65 ppm/1000 hours.

Monolithic Temperature Sensors

The voltage KV_T arising in bandgap cells is linearly *proportional to absolute temperature* (PTAT). As such it forms the basis for a variety of monolithic temperature sensors[6] known as *VPTATs* and *IPTATs,* depending on whether they produce a PTAT voltage or a PTAT current. These sensors enjoy the low-cost advantages of IC fabrication and do not require the costly linearization circuitry common to other sensors,

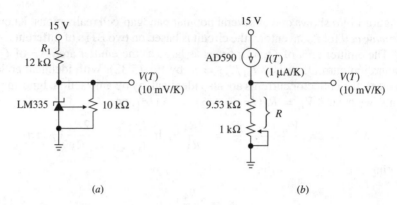

(a)   (b)

FIGURE 11.10
Basic temperature sensors using the LM335 VPTAT and the AD590 IPTAT.

such as thermocouples, RTDs, and thermistors. Besides temperature measurement and control, common applications include fluid-level detection, flow-rate measurement, anemometry, PTAT circuit biasing, and thermocouple cold-junction compensation. Moreover, IPTATs are used in remote-sensing applications because of their insensitivity to voltage drops over long wire runs.

A popular VPTAT is the LM335 precision temperature sensor (National Semiconductor). As shown in Fig. 11.10a, this device acts as a reference diode, except that its voltage is PTAT with $TC(V) = 10$ mV/K. Thus, at room temperature it gives $V(25 \, °C) = (10 \text{ mV/K}) \times (273.2 + 25)K = 2.982$ V. The device is also equipped with a third terminal for the exact adjustment of its TC. The LM335A version comes with an initial room-temperature accuracy of ± 1 °C. After calibration at 25 °C, its typical accuracy is ± 0.5 °C for $-40 \, °C \leq T \leq 100$ °C. Its operating current may be anywhere between 0.5 mA and 5 mA, and its dynamic resistance is less than 1 Ω.

A popular IPTAT is the AD590 two-terminal temperature transducer (Analog Devices). To the user this device appears as a high-impedance current source providing 1 μA/K. Terminating it on a grounded resistance as in Fig. 11.10b gives a VPTAT with a sensitivity of $R \times (1 \, \mu\text{A/K})$. The AD590M version comes with a room-temperature accuracy of ± 0.5 °C maximum. After calibration at 25 °C, the accuracy is ± 0.3 °C maximum for $-55 \, °C \leq T \leq 150$ °C. The device operates properly as long as the voltage across its terminals is between 4 V and 30 V.

Additional temperature-processing devices include Celsius and Fahrenheit sensors, and thermocouple signal conditioners. Consult the manufacturer catalogs to see what is available.

11.3
VOLTAGE-REFERENCE APPLICATIONS

When applying voltage references, care must be exercised to prevent the external circuitry and wiring interconnections from degrading the performance of the reference. This may require the use of precision op amps and low-drift resistors, along

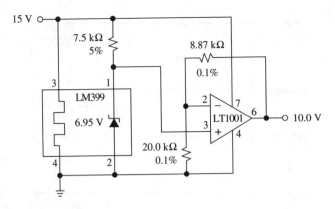

FIGURE 11.11
Buffered 10-V reference.

with special wiring and circuit-construction techniques. As an example, consider the circuit[7] of Fig. 11.11, which uses a precision op amp to raise the output of a thermally stabilized reference to 10.0 V. We wish to assess the impact of op amp and wiring nonidealities. The LM399 data sheets give $TC_{max} = 2$ ppm/°C and $r_{z(max)} = 1.5\,\Omega$, and the LT1001 data sheets give $TC(V_{OS})_{max} = 1\,\mu V/°C$, $TC(I_B) \cong 4$ pA/°C, $CMRR_{dB(min)} = 106$ dB, and $PSRR_{dB(min)} = 103$ dB.

The maximum drift due to the LM399 is $2 \times 10^{-6} \times 6.95 = 13.9\,\mu V/°C$, and that due to the overall input error of the LT1001 is $1 \times 10^{-6} + (20 \parallel 8.87)10^{3} \times 4 \times 10^{-12} \cong 1\,\mu V/°C$; consequently, the worst-case output drift is $(1 + 8.87/20) \times (13.9 + 1) = 1.44 \times 14.9 = 21.5\,\mu V/°C$. The worst-case line regulation due to the LM399 is $1.5/(1.5 + 7500) = 200\,\mu V/V$, and that due to the LT1001 is $10^{-103/20} + 0.5 \times 10^{-106/20} = (7.1 + 2.5) = 9.6\,\mu V/V$; consequently, the overall worst-case line regulation is $1.44(200 + 9.6) = 303\,\mu V/V$. To give an idea, a 1-V power-supply change has the same effect as a temperature change of $303/21.5 \cong 14$ °C. It is apparent that the use of a precision op amp causes negligible degradation in the present example.

However, when the circuit is fabricated, its drift may be compromised by thermocouple effects arising from thermal gradients across dissimilar metals. The kovar leads of the ICs form thermocouple junctions with the copper traces of the printed-circuit board. A gradient of just 1 °C between the leads of the chip-heated LM399 will generate an error on the order of 50 μV. Thermal gradients are reduced by using equal-size pads and traces to ensure equal amounts of heat dissipation at the two junctions, and by paying attention to other sources of heat, such as power stages.

Even after all the above sources of error have been minimized, special attention must be paid to wiring and interconnections, since voltage drops across stray resistances may degrade performance significantly. For instance, a copper trace with a stray resistance of 1 Ω develops an error of 1 mV/mA and introduces a TC of 4 $(\mu V/mA)/°C$ (the TC of copper is 0.004%/°C). For a 10-V reference, this corresponds to an accuracy degradation of 0.01% and a TC of 0.4 ppm/°C.

An effective technique for combating stray-resistance errors, especially in high-current applications, is *remote sensing,* as already illustrated in Fig. 2.22 in

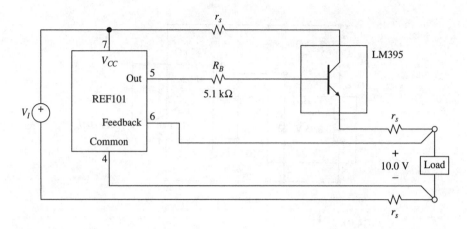

FIGURE 11.12
Remote sensing to eliminate the effect of unwanted voltage drops due to the stray wire
resistances r_s.

connection with instrumentation amplifiers. The technique is shown in Fig. 11.12
for a REF101 reference whose output current capability is boosted with an LM395
high-gain power transistor. To prevent the voltage losses across the stray resistances
r_s from degrading the voltage received by the load, the feedback and common pins
are connected to the load by a separate pair of wires, thus ensuring that the 10.0-V
voltage appears *directly across the load,* regardless of the offending voltage drops.
The stray resistance of this additional set of wires is less critical due to the much
lower currents involved.

Voltage Sources

Voltage references can readily be used as the basis for a variety of precision voltage
sources. The circuit of Fig. 11.13 utilizes the matched resistance pair inside the
REF101 10-V reference of Fig. 11.6a to provide a variable voltage source. When
the wiper is at the bottom, the op amp acts as a unity-gain inverting amplifier and
gives $V_O = -10$ V; when the wiper is at the top, it acts as a unity-gain buffer and

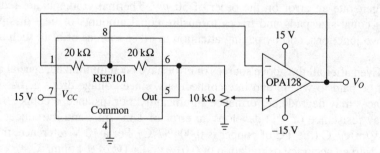

FIGURE 11.13
Variable reference over the range -10 V $\leq V_O \leq 10$ V.

gives $V_O = +10$ V. Consequently, varying the wiper from end to end varies the output over the range -10 V $\leq V_O \leq +10$ V. With imagination, a variety of other useful circuits can easily be devised[7,8] (see also the end-of-chapter problems).

Current Sources

A voltage reference can readily be turned into a current reference[9] by bootstrapping its common terminal with a voltage follower, as in Fig. 11.14. By op amp action, the voltage across R is always V_{REF}, so the circuit gives

$$I_O = \frac{V_{REF}}{R} \tag{11.17}$$

regardless of the voltage V_L developed by the load, provided no saturation effects occur. The permissible range of values of V_L is called the *voltage compliance* of the current source.

EXAMPLE 11.6. The circuit of Fig. 11.14 uses a 5-V reference with TC $= 20\ \mu\text{V}/^\circ\text{C}$, line regulation $= 50\ \mu\text{V/V}$, and dropout voltage $V_{DO} = 3$ V, and a JFET-input op amp with TC(V_{OS}) $= 5\ \mu\text{V}/^\circ\text{C}$ and CMRR$_{dB} = 100$ dB. (*a*) Specify R for $I_O = 10$ mA. (*b*) Find the worst-case values of TC(I_O) and of the resistance R_o seen by the load. (*c*) Assuming ± 15-V supplies, find the voltage compliance.

Solution.

(*a*) $R = 5/10 = 500\ \Omega$ (use 499 Ω, 1%).
(*b*) A 1 °C change in T causes a worst-case change in the voltage across R of $20 + 5 = 25\ \mu\text{V}/^\circ\text{C}$; the corresponding change in I_O is $25 \times 10^{-6}/500 = 50$ nA/°C. A 1-V change in V_L causes a 50-μV/V change in V_{REF} and a $10^{-100/20} = 10\ \mu\text{V/V}$ change in V_{OS}, for a worst-case change in I_O of $(50 + 10)10^{-6}/500 = 120$ nA/V. Thus, $R_{o(min)} = (1\text{ V})/(120\text{ nA}) = 8.33$ MΩ.
(*c*) $V_L \leq V_{CC} - V_{DO} - V_{REF} = 15 - 3 - 5 = 7$ V.

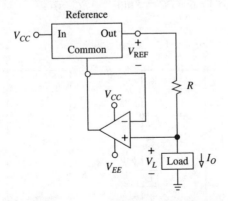

FIGURE 11.14
Turning a voltage reference into a current source.

The bootstrapping principle can readily be applied to the case of diode references to implement either current sources or current sinks. This is shown in Fig. 11.15,

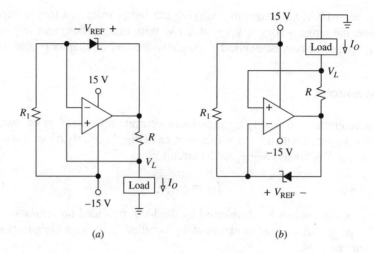

FIGURE 11.15
Using a reference diode to implement a current source and a current sink.

where $I_O = V_{REF}/R$ for both circuits. The function of R_1 is to bias the diode. If an LM385 reference diode is used, imposing a bias current of 100 μA when $V_L = 0$ yields $R_1 = 150$ kΩ. The voltage compliance of the source is $V_L \leq V_{OH} - V_{REF}$, and that of the sink is $V_L \geq V_{OL} + V_{REF}$. If a 741 op amp and a 2.5-V diode are used, then $V_L \leq 10.5$ V for the source, and $V_L \geq -10.5$ V for the sink.

When the circuits just discussed fail to meet load-current demands, we can use current-boosting transistors. The circuit of Fig. 11.16a uses a *pnp* BJT to source current. By op amp action, the voltage across the current-setting resistance R is V_{REF}, so the current entering the emitter is $I_E = V_{REF}/R$. The current leaving the collector is $I_C = [\beta/(\beta + 1)]I_E$, so $I_O = [\beta/(\beta + 1)]V_{REF}/R \cong V_{REF}/R$. The voltage compliance is $V_L \leq V_{CC} - V_{REF} - V_{EC(sat)}$.

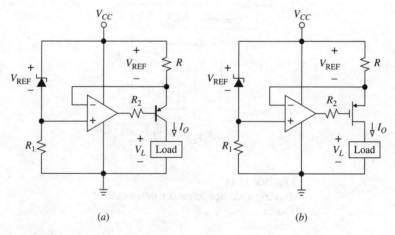

FIGURE 11.16
Current sources with current-boosting transistors.

EXAMPLE 11.7. Let the circuit of Fig. 11.16a use a 741 op amp with $V_{CC} = 15$ V, an LM385 2.5-V diode with a bias current of 0.5 mA, and a 2N2905 BJT with $R_2 = 1$ kΩ. (a) Specify R and R_1 for $I_O = 100$ mA. (b) Assuming typical BJT parameters, find the voltage compliance of the source, and check that the 741 is operating within specifications.

Solution.

(a) We have $R = 2.5/0.1 = 25$ Ω (use 24.9 Ω, 1%), and $R_1 = (15 - 2.5)/0.5 = 25$ kΩ (use 24 kΩ).

(b) $V_L \leq 15 - 2.5 - 0.2 = 12.3$ V. The 741 inputs are at $15 - 2.5 = 12.5$ V, which is within the input voltage range specifications. Assuming $\beta = 100$ so that $I_B = 1$ mA, we find that the 741 output is at $V_{CC} - V_{REF} - V_{EB(on)} - R_2 I_B = 15 - 2.5 - 0.7 - 1 \times 1 = 10.8$ V (which is below $V_{OH} = 13$ V), and sinks a current of 1 mA (which is below $I_{sc} = 25$ mA). Consequently, the 741 is operating within specifications.

For higher output currents, the transistor can be replaced by a power *pnp* Darlington, or by a power enhancement *p*-MOSFET as in Fig. 11.16b. In these cases, heat-sinking, to be discussed in Section 11.5, may be required.

Temperature-Sensor Applications

In thermometer applications it is desirable that $V(T)$ and $I(T)$ be calibrated in degrees Celsius or Fahrenheit rather than in kelvins. If a VPTAT or an IPTAT is used, then suitable conditioning circuitry is required.[6]

The circuit of Fig. 11.17 senses temperature via the AD590 IPTAT, whose current can be expressed as $I(T) = 273.2 \ \mu\text{A} + (1 \ \mu\text{A}/°\text{C})T$, T in degrees Celsius. By the superposition principle,

$$V_O(T) = R_2(273.2 + T)10^{-6} - 10R_2/R_1$$

It is apparent that letting $R_1 = 10/(273.2 \times 10^{-6}) = 36.6$ kΩ will cause a cancellation and leave $V_O(T) = R_2 10^{-6}T$, T in degrees Celsius. For a sensitivity of 100 mV/°C, use $R_2 = (100 \ \text{mV})/(1 \ \mu\text{A}) = 100$ kΩ. To compensate for the various tolerances, implement R_1 with a 35.7-kΩ resistor in series with a 2-kΩ pot, and R_2 with a 97.6-kΩ resistor in series with a 5-kΩ pot. To calibrate, (a) place the IPTAT in an ice bath ($T = 0$ °C) and adjust R_1 for $V_O(T) = 0$ V; (b) place the IPTAT in boiling water ($T = 100$ °C) and adjust R_2 for $V_O(T) = 10.0$ V.

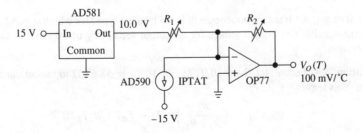

FIGURE 11.17
Celsius sensor.

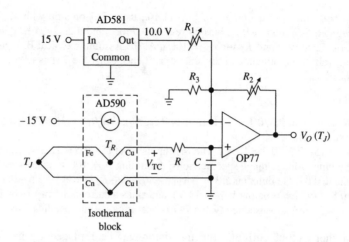

FIGURE 11.18
Thermocouple cold-junction compensation using the AD590
IPTAT.

Another popular application of temperature sensors is cold-junction compensation in thermocouple measurements.[6] A thermocouple is a temperature sensor consisting of two wires of dissimilar metals and producing a voltage of the type

$$V_{TC} = \alpha(T_J - T_R)$$

where T_J is the temperature at the measurement or *hot* junction; T_R is the temperature at the reference or *cold* junction, formed where the thermocouple is connected to the leads (usually of copper) of the measuring device; α is the *Seebeck coefficient*. For example, Type J thermocouples are made up of iron and constantan (55% Cu and 45% Ni), and give $\alpha = 52.3\ \mu V/°C$.

It is apparent that a thermocouple inherently provides only *relative* temperature information. If we want to measure T_J regardless of T_R, we must use another sensor to measure T_R, as exemplified in Fig. 11.18. Using again the superposition principle,

$$V_O = \left(1 + \frac{R_2}{R_1 \parallel R_3}\right)\alpha(T_J - T_R) + R_2(273.2 + T_R)10^{-6} - 10R_2/R_1$$

where both T_J and T_R are in degrees Celsius. As before, we select R_1 to cancel out the 273.2 term, R_3 to cancel out T_R, and R_2 to achieve the desired output sensitivity.

EXAMPLE 11.8. If the thermocouple of Fig. 11.18 is a type J for which $\alpha = 52.3\ \mu V/°C$, specify suitable component values for an output sensitivity of 10 mV/°C. Outline its calibration.

Solution. As before, let $R_1 = 10/(273.2 \times 10^{-6}) = 36.6\ k\Omega$ to cancel out the 273.2 term. This leaves

$$V_O = \left(1 + \frac{R_2}{R_1 \parallel R_3}\right)\alpha(T_J - T_R) + R_2T_R10^{-6}$$

Next, impose $[1 + R_2/(R_1 \parallel R_3)]\alpha = R_210^{-6} = 10$ mV/°C to cancel out T_R as well as achieve the desired output sensitivity. The results are $R_2 = 10.0\ k\Omega$ and $R_3 = 52.65\ \Omega$.

In practice we would use $R_3 = 52.3$ Ω, 1%, and make R_1 and R_2 adjustable as follows: (*a*) place the hot junction in an ice bath and adjust R_1 for $V_O(T_J) = 0$ V; (*b*) place the hot junction in a hot environment of known temperature and adjust R_2 for the desired output (the second adjustment can also be performed with the help of a thermocouple voltage simulator).

To suppress noise pickup by the thermocouple wires, use an *RC* filter as shown, say $R = 10$ kΩ and $C = 0.1$ μF.

Thermocouple cold-junction compensators are also available as self-contained IC modules. Two examples are the AD594/5/6/7 series (Analog Devices) and the LT1025 (Linear Technology).

11.4
LINEAR REGULATORS

As shown in Fig. 11.19, a voltage regulator uses the Darlington pair Q_1 and Q_2, also called the *series-pass element,* to transfer power from an unregulated input source V_I to a load at a prescribed regulated voltage V_O. The feedback network R_1 and R_2 samples V_O and feeds a portion thereof to the error amplifier *EA* for comparison against a reference V_{REF}. The amplifier provides the series-pass element with whatever drive it takes to force the error close to zero. The regulator is a classic example of series-shunt feedback, and it can be viewed as a noninverting op amp that has been equipped with a Darlington current booster to give

$$V_O = \left(1 + \frac{R_2}{R_1}\right) V_{\text{REF}} \tag{11.18}$$

Since the error amplifier provides currents on the order of milliamperes and the load may draw currents on the order of amperes, a current gain on the order of 10^3 A/A is required. A single power BJT is usually insufficient, so a Darlington pair is used instead, whose overall current gain is $\beta \cong \beta_1 \times \beta_2$. We observe that for an *npn* BJT to work in the forward-active region, where $I_C = \beta I_B$, the conditions $v_{BE} = V_{BE(\text{on})}$ and $v_{CE} \geq V_{CE(\text{sat})}$ must hold. A low-power BJT has typically

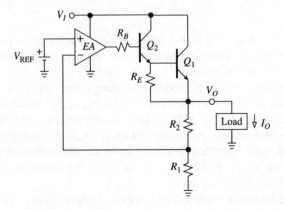

FIGURE 11.19
Basic series voltage regulator.

$\beta \cong 100$, $V_{BE(on)} \cong 0.7$ V, and $V_{CE(sat)} \cong 0.1$ V; a power BJT may have $\beta \cong 20$, $V_{BE(on)} \cong 1$ V, and $V_{CE(sat)} \cong 0.5$ V.

Because of wide thermal excursions due to self-heating, and voltage errors due to stray resistances in the wiring system, the accuracy and stability of voltage regulators are less stringent than those of voltage references. The source V_{REF} is usually of the bandgap type, and the regulator is configured for the desired V_O by proper selection of the ratio R_2/R_1.

The *efficiency* of the regulator is given by the ratio of the average power delivered to the load to that absorbed from the source, or $\eta = P_O/P_I$. Since $P_O = V_O I_O$ and $P_I \cong V_I I_O$, we get

$$\eta(\%) \cong 100 \frac{V_O}{V_I} \qquad (11.19)$$

where we have ignored the currents drawn by the reference, amplifier, and feedback network compared to I_O.

EXAMPLE 11.9. Let $R_B = 510$ Ω and $R_E = 3.3$ kΩ in the regulator of Fig. 11.19. Assuming a bandgap reference and typical BJT parameters, find (a) R_2/R_1 for $V_O = 5.0$ V, (b) the error-amplifier output drive needed to provide $I_O = 1$ A, (c) the dropout voltage V_{DO} if the error amplifier saturates at $V_{OH} = V_I - 0.5$ V, and (d) the maximum efficiency attainable for the given I_O.

Solution.

(a) Imposing $5 = (1 + R_2/R_1)1.282$ gives $R_2/R_1 = 2.9$.

(b) For $I_O = 1$ A we have $I_{B1} = I_{E1}/(\beta_1 + 1) \cong 1/21 \cong 48$ mA, and $I_{E2} = I_{B1} + V_{BE1(on)}/R_E \cong 48$ mA. The error amplifier must thus source $I_{OA} = I_{B2} = I_{E2}/(\beta_2 + 1) \cong 48/101 \cong 0.47$ mA; moreover, $V_{OA} = V_{R_B} + V_{BE2(on)} + V_{BE1(on)} + V_O \cong 0.51 \times 0.47 + 0.7 + 1 + 5 \cong 7$ V.

(c) For the circuit to work properly we need $v_{OA} \le V_{OH}$ and $v_{CE} \ge V_{CE(sat)}$ for both BJTs. It is readily seen that these conditions are met if $V_I \ge 7.5$ V. Hence, $V_{DO} = 7.5 - 5 = 2.5$ V.

(d) Since $V_I \ge 7.5$ V, $\eta(\%) \le 5/7.5 \cong 67\%$.

Protections

The reliable performance of a power BJT is critically affected by power dissipation capabilities, current and voltage ratings, maximum junction temperature, and second breakdown, a phenomenon resulting from the formation of hot spots within the BJT, which cause uneven sharing of the total load among different regions of the device.[10] The above factors define a restricted region of the i_C-v_{CE} characteristic, known as the *safe operating area* (SOA), within which the device can be operated without the risk of failure or performance degradation. Figure 11.20 shows typical SOA data for the case of continuous operation. Note, for instance, that while the BJT can draw a current of 10 A up to $v_{CE} \cong 12$ V, at $v_{CE} = 100$ V it can only handle 1 A without risking second-breakdown failure.

Voltage regulators are equipped with special circuitry to protect the power stage against *current overload, second breakdown,* and *thermal overload.* Each circuit is designed to be inactive under normal operating conditions, but to become active as soon as an attempt is made to exceed the corresponding safety limits.

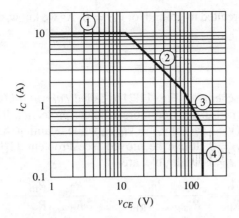

FIGURE 11.20
Typical power BJT safe operating area
(SOA): (1) bonding-wire limited, (2) thermally limited, (3) second-breakdown limited, and (4) voltage-rating limited.

Current overload protection is dictated by maximum power-rating considerations. Since the power dissipated by the series-pass BJT is $P \cong (V_I - v_O)i_O$, we
must ensure $i_O \leq P_{max}/(V_I - v_O)$ for safe operation. The protection scheme of
Fig. 11.21a, similar to that discussed at the end of Chapter 5 for op amps, uses a
brute-force approach to keep i_O below the limit $I_{sc} = P_{max}/V_I$, which occurs when

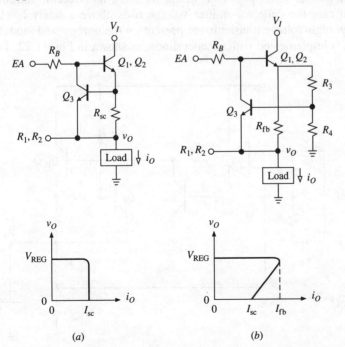

FIGURE 11.21
Output overload protection: (a) short-circuit protection, and (b)
current fold-back protection.

the output is short-circuited to ground, or $v_O = 0$. As we know, the resulting design equation is

$$R_{sc} = \frac{V_{BE3(on)}}{I_{sc}} \qquad (11.20)$$

The alternative scheme of Fig. 11.21b, called *current fold-back* for the shape of its curve, is designed to provide more efficient protection by raising the upper limit to $I_{fb} = P_{max}/(V_I - v_O)$ at $v_O = V_{REG}$, while retaining the short-circuit limit $I_{sc} = P_{max}/V_I$ at $v_O = 0$. It can be proved (see Problem 11.15) that the design equations, assuming I_{B3} is negligible, are

$$\frac{1}{R_{fb}} = \frac{1}{R_{sc}} - \frac{I_{fb} - I_{sc}}{V_{REG}} \qquad \frac{R_3}{R_4} = \frac{R_{fb}}{R_{sc}} - 1 \qquad (11.21)$$

EXAMPLE 11.10. A 5-V regulator with $V_I = 8$ V uses a 12-W series-pass BJT. (a) Assuming typical BJT parameters, specify suitable components for output short-circuit protection. (b) Repeat, but for fold-back protection.

Solution.

(a) $I_{sc} = 12/8 = 1.5$ A; $R_{sc} = 0.7/1.5 = 0.47$ Ω.

(b) $I_{fb} = 12/(8 - 5) = 4$ A; $R_{fb} = [1/0.47 - (4 - 1.5)/5]^{-1} = 0.61$ Ω; $R_3/R_4 = 0.61/0.47 - 1 = 0.3$. For $v_O = 0$, impose $V_{BE3(on)}/(R_3 + R_4) \cong 10I_{B3}$. Assuming $I_{B3} = 0.1$ mA, we get $R_3 \cong 160$ Ω and $R_4 \cong 540$ Ω.

To confine the series-pass BJT within its SOA, its collector current must be reduced in case the collector-emitter voltage rises above a safety level, a likely event when high-voltage transients are present on the unregulated input line. This protection is implemented with a Zener diode, as shown in Fig. 11.22. This diode,

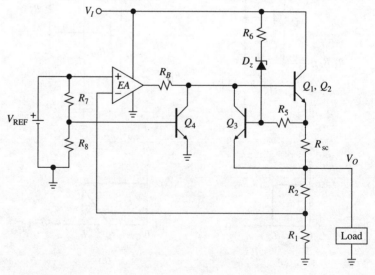

FIGURE 11.22
Positive regulator with overload, SOA, and thermal protection.

normally in cutoff, is designed to turn on as soon as V_I rises above a safety level. The current supplied by D_z will then turn on Q_3 and divert current away from the base of the series-pass BJT, as in the case of current overload. The function of R_5 is to decouple the base of Q_3 from the low-impedance emitter of the power BJT, and that of R_6 is to limit the current through D_z, particularly in the presence of large noise spikes on the input line.

Excessive self-heating may cause permanent damage to BJTs, unless junction temperatures are kept from rising above a safety level, usually 175 °C or less. The series-pass BJT is protected by sensing its instantaneous temperature and reducing its collector current in case of thermal overload. In the circuit of Fig. 11.22 this protection is provided by Q_4, a BJT mounted in close thermal coupling with the series-pass element. Temperature is sensed by exploiting the negative TC of V_{BE4}. This BJT is designed to be in cutoff during acceptable thermal conditions, but to turn on as soon as the temperature approaches 175 °C. Once in conduction, Q_4 will divert current away from the base of the series-pass BJT, reducing its conduction to the point of even shutting it off until the temperature drops to a more tolerable level.

EXAMPLE 11.11. Assuming V_{BE4} (25 °C) = 700 mV, find R_7 and R_8 to cause thermal shutdown at 175 °C if V_{REF} is a bandgap reference.

Solution. The voltage required to turn on Q_4 can be estimated as V_{BE4} (175 °C) = V_{BE4} (25 °C) + TC(V_{BE4})(175 − 25) °C ≅ 700 mV + (−2 mV/°C)150 °C ≅ 400 mV. Ignoring I_{B4} and imposing $0.4 = [R_8/(R_8 + R_7)]1.282$ gives $R_7/R_8 = 2.2$. Assuming $I_{B4} = 0.1$ mA and imposing $V_{REF}/(R_7 + R_8) \cong 10I_{B4}$ gives $R_7 = 880\ \Omega$ and $R_8 = 400\ \Omega$.

Monolithic Voltage Regulators

Manufacturers' data books report a wide variety of monolithic regulators. For reasons of space, we limit ourselves to a few examples.

Two of the earliest products to gain wide popularity were the μA7800 series of positive regulators and the μA7900 series of negative regulators (Fairchild). Figure 11.23 depicts the 7800 series, where we identify the following functional blocks.

1. Q_{16} and Q_{17} form the series-pass element.
2. Q_{15}, D_2, and Q_{14} provide, respectively, output short-circuit protection, SOA protection, and thermal shutdown.
3. Q_1 through Q_7 form a combined bandgap-reference/error-amplifier designed to keep the base of Q_6 at 5 V via negative feedback.
4. R_{19} and R_{20} form a feedback network designed to give

$$V_O = \left(1 + \frac{R_{20}}{R_{19}}\right)5\ \text{V} \tag{11.22}$$

V_O is factory-programmed for a variety of different values by selecting the proper tap on R_{20} during fabrication. For instance, with $R_{20} = 0$ the device is configured

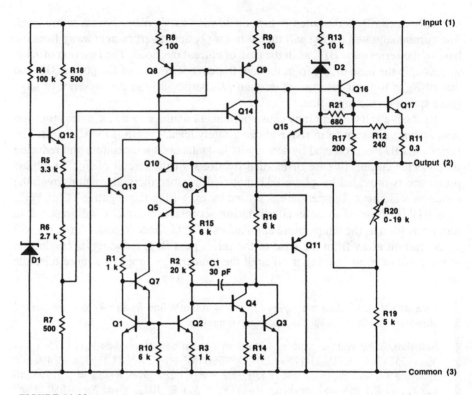

FIGURE 11.23
The μA7800 series of three-terminal positive voltage regulators. (Copyright, Fairchild Semiconductor Corporation, 1982. Used by permission.)

for $V_O = 5$ V and is called 7805; likewise, $R_{20} = 10$ kΩ yields the 7815 15-V regulator, and $R_{20} = 7$ kΩ yields the 7812 12-V regulator.

5. Q_{13}, along with the biasing network consisting of D_1 and Q_{12}, functions as the start-up circuit. At power turn-on, Q_{13} brings up the voltage-reference section and also turns on the series element Q_{16}-Q_{17} via the current mirror Q_8-Q_9. This causes V_O to swing positive, until negative feedback takes over and turns off Q_{13}, which thus remains inactive during normal operation.

Figure 11.24 shows the electrical characteristics of the 7805.

The μA78G is similar to the 7800, except that R_{19} and R_{20} are omitted and the base of Q_6, referred to as the *control pin,* is made accessible to the user for the external setting of V_O. Called a *four-terminal adjustable regulator,* the device is especially useful in remote sensing. As depicted in Fig. 11.25, mounting the feedback network right across the load and equipping it with separate returns will ensure a regulated voltage $V_{REG} = (1 + R_2/R_1)5$ V *right at the load,* irrespective of any voltage drops across the stray resistances r_s of the wires. The four-terminal version of the 7900 negative regulators is called the μA79G.

Another popular class of products is offered by *three-terminal adjustable regulators,* of which the LM317 positive regulator and the LM337 negative regulator

Absolute Maximum Ratings

Input Voltage (5 V through 18 V)	35 V
(24 V)	40 V
Internal Power Dissipation	Internally Limited
Storage Temperature Range	−65°C to +150°C
Operating Junction	
Temperature Range	
μA7800	−55°C to +150°C
μA7800C	0°C to +125°C

μA7805C
Electrical Characteristics $V_{IN} = 10$ V, $I_{OUT} = 500$ mA, $0°C \leq T_J \leq 125°C$, $C_{IN} = 0.33$ μF, $C_{OUT} = 0.1$ μF, unless otherwise specified.

Characteristic	Condition (Note)		Min	Typ	Max	Unit
Output Voltage	$T_J = 25°C$		4.8	5.0	5.2	V
Line Regulation	$T_J = 25°C$	7 V $\leq V_{IN} \leq 25$ V		3	100	mV
		8 V $\leq V_{IN} \leq 12$ V		1	50	mV
Load Regulation	$T_J = 25°C$	5 mA $\leq I_{OUT} \leq 1.5$ A		15	100	mV
		250 mA $\leq I_{OUT} \leq 750$ mA		5	50	mV
Output Voltage	7 V $\leq V_{IN} \leq 20$ V 5 mA $\leq I_{OUT} \leq 1.0$ A $P \leq 15$ W		4.75		5.25	V
Quiescent Current	$T_J = 25°C$			4.2	8.0	mA
Quiescent Current Change	with line	7 V $\leq V_{IN} \leq 25$ V			1.3	mA
	with load	5 mA $\leq I_{OUT} \leq 1.0$ A			0.5	mA
Output Noise Voltage	$T_A = 25°C$, 10 Hz $\leq f \leq 100$ kHz			40		μV
Ripple Rejection	$f = 120$ Hz, 8 V $\leq V_{IN} \leq 18$ V		62	78		dB
Dropout Voltage	$I_{OUT} = 1.0$ A, $T_J = 25°C$			2.0		V
Output Resistance	$f = 1$ kHz			17		mΩ
Short-Circuit Current	$T_J = 25°C$, $V_{IN} = 35$ V			750		mA
Peak Output Current	$T_J = 25°C$			2.2		A
Average Temperature Coefficient of Output Voltage	$I_{OUT} = 5$ mA, $0°C \leq T_J \leq 125°C$			1.1		mV / °C

FIGURE 11.24
Electrical characteristics of the μA7805 voltage regulator. (Copyright, Fairchild Semiconductor Corporation, 1982. Used by permission.)

(National Semiconductor) are among the most widely known examples. In the LM317 functional diagram[11] of Fig. 11.26a, the diode is a 1.25-V bandgap reference biased at 50 μA. The error amplifier provides whatever drive it takes to keep the voltage at the output pin 1.25 V higher than the voltage at the adjustment pin. Thus, connecting the device as in Fig. 11.26b gives $V_O = V_{ADJ} + 1.25$ V. By the

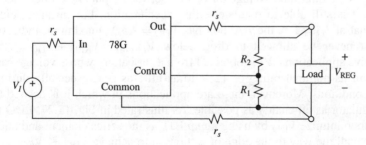

FIGURE 11.25
Adjustable regulator with remote sensing.

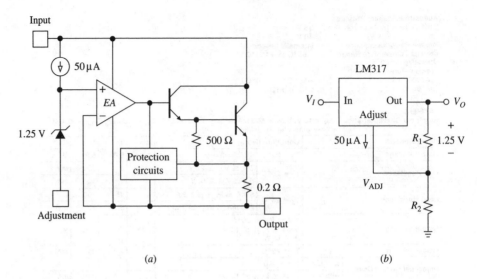

(a)   (b)

FIGURE 11.26
Functional diagram and typical connection of the LM317 three-terminal adjustable regulator.
(Courtesy of National Semiconductor.)

superposition principle, $V_{ADJ} = V_O/(1+R_1/R_2)+(R_1 \parallel R_2)(50\,\mu A)$. Eliminating V_{ADJ} gives

$$V_O = \left(1 + \frac{R_2}{R_1}\right)1.25\text{ V} + R_2(50\,\mu A) \tag{11.23}$$

The purpose of R_1 and R_2, besides setting the value of V_O, is to provide a conductive path toward ground for the quiescent current of the error amplifier and the remaining circuitry in the absence of a load. The data sheets recommend imposing a current of 5 mA through R_1 to meet this requirement. One can then verify that the effect of the 50-μA current becomes negligible, so $V_O = (1 + R_2/R_1)1.25$ V. By varying R_2, V_O can be adjusted anywhere between 1.25 V and 35 V.

Lastly, we mention *low-dropout* (LDO) regulators. As we know, the dropout voltage V_{DO} is the minimum voltage difference between input and output under which the circuit is still able to regulate within specification. For instance, Fig. 11.24 shows that at $I_O = 1$ A the μA7805 has $V_{DO} = 2.5$ V maximum, indicating that V_I must never be allowed to drop below $V_{I(min)} = V_{REG} + V_{DO} = 7.5$ V. In automotive applications, V_I is obtained from a car battery whose voltage can easily drop from its nominal rating of 12 V to as little as 6 V, especially during "cold crank" conditions. Moreover, there are applications in which it is desired to operate a regulator as efficiently as possible. As illustrated in Fig. 11.27, LDO positive regulators minimize V_{DO} by using a *pnp* BJT as the series element and allowing it to operate all the way to the edge of saturation to achieve $V_{DO} \cong V_{EC(sat)}$, which is usually on the order of a few tenths of a volt. To avoid using R_{sc}, which would increase V_{DO}, the *pnp* BJT is equipped with an additional small-area collector to provide collector-current sensing information for the overload protection circuitry.

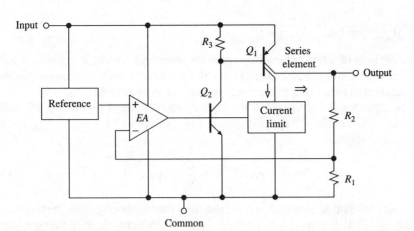

FIGURE 11.27
Block-diagram of a low-dropout (LDO) regulator.

LDOs are often used to provide postregulation of the noisier outputs of switching regulators.

11.5
LINEAR-REGULATOR APPLICATIONS

The primary application of voltage regulators is in power supplies, especially distributed supplies, where the unregulated voltage is brought to different subsystems to be treated locally by dedicated regulators. Aside from a few simple requirements, a linear regulator is generally easy to use. As exemplified in Fig. 11.28, the device should always be equipped with an input capacitor to reduce the effects of stray inductance in the input wires, especially if the regulator is located away from the unregulated source, and an output capacitor to help improve the response to sudden load-current changes. For best results, use thick wires and traces, keep the leads short, and mount both capacitors as close as possible to the regulator. Depending on the case, heat-sinking may be required to keep the internal temperature within tolerable levels.

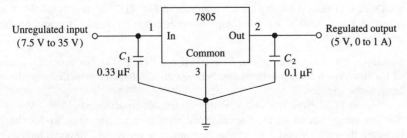

FIGURE 11.28
Typical circuit connection of the μA7805 voltage regulator. (Copyright, Fairchild Semiconductor Corporation, 1982. Used by permission.)

Power Sources

With the help of a few external components, a voltage regulator can, like a voltage reference, by configured for a variety of voltage source or current source applications, the main difference lying in the much higher currents available.

A regulator is configured for a higher output voltage by raising its common terminal to a suitable voltage pedestal. In Fig. 11.29a we have $V_O = V_{REG} + R_2 \times V_O/(R_1 + R_2)$, or

$$V_O = \left(1 + \frac{R_2}{R_1}\right) V_{REG} \tag{11.24}$$

The role of the op amp, which is powered from the regulated output to eliminate any PSRR and CMRR errors, is to prevent the feedback network from being loaded by the common terminal. However, if the current of this terminal is sufficiently small, as in the case of adjustable regulators such as the LM317 and LM337 types, then we can do without the op amp and the circuit simplifies to the familiar form of Fig. 11.26b.

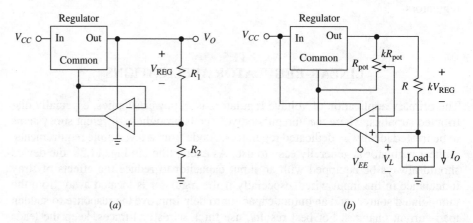

FIGURE 11.29
Configuring a regulator (a) as a power voltage source, and (b) as an adjustable power current source.

EXAMPLE 11.12. Assuming a 7805 regulator in Fig. 11.29a, specify suitable components for $V_O = 15.0$ V. What is the permissible range for V_{CC}? Comment on the line and load regulation.

Solution. Use a 741 op amp with $R_1 = 10\,k\Omega$ and $R_2 = 20\,k\Omega$. For the exact adjustment of V_O, interpose a 1-$k\Omega$ potentiometer between R_1 and R_2, and connect the noninverting input to the wiper.

Figure 11.24 gives $V_{DO} = 2$ V, so the permissible input range is $17\,V \leq V_{CC} \leq 35\,V$. The percentage values of the line and load regulation are the same as for the 7805; however, their mV/V and mV/A values are now $1 + R_2/R_1 = 3$ times as large.

In Fig. 11.29b the op amp bootstraps the regulator's common terminal with the voltage V_L developed by the output load, and the regulator keeps the voltage across R at kV_{REG}, where k represents the fraction of the potentiometer between the wiper and the regulator's output, $0 \leq k \leq 1$. Consequently, the circuit gives

$$I_O = k\frac{V_{REG}}{R} \tag{11.25}$$

regardless of V_L, provided no saturation effects occur. We thus have an adjustable current source, and its voltage compliance is $V_L \leq V_{CC} - V_{DO} - kV_{REG}$. If a current sink is needed, then we can use a negative regulator. To maximize the compliance for a given V_{CC}, use a regulator with low V_{DO} and V_{REG}. An adjustable regulator of the 317 or 337 type is a good choice.

> **EXAMPLE 11.13.** The circuit of Fig. 11.29b uses an LM317 1.25-V regulator, whose ratings are $V_{DO} = 2$ V and line regulation = 0.07%/V maximum. Assuming a 10-kΩ potentiometer, an op amp with $CMRR_{dB} \geq 70$ dB, and ±15-V supplies, specify R for an adjustable current from 0 to 1 A; next, find the voltage compliance and the minimum equivalent resistance seen by the load for the case $k = 1$.
>
> **Solution.** $R = 1.25\ \Omega$, 1.25 W (use 1.24 Ω, 2 W). $V_L \leq 15 - 2 - 1.25 = 11.75$ V. A 1-V change in V_L causes a worst-case change in I_O of $(1.25 \times 0.07/100 + 10^{-70/20})/1.25 = 0.953$ mA, so $R_{o(min)} = (1\ \text{V})/(0.953\ \text{mA}) = 1.05\ \text{k}\Omega$.

Thermal Considerations

The power dissipated in the base-collector junction of the series-pass BJT is converted into heat, which raises the junction temperature T_J. To prevent permanent damage to the BJT, T_J must be kept within a safe limit. For silicon devices, this limit[10] is in the range of 150 °C to 200 °C. To avoid excessive temperature buildup, heat must be expelled from the silicon chip to the surrounding package structure and from there to the ambient. At thermal equilibrium, the temperature rise of a constant-power dissipating BJT with respect to the ambient can be expressed as

$$T_J - T_A = \theta_{JA}P_D \tag{11.26}$$

where T_J and T_A are the junction and ambient temperatures, P_D is the dissipated power, and θ_{JA} is the *junction-to-ambient thermal resistance,* in degrees Celsius per watt. This resistance, representing the amount of temperature rise per unit of dissipated power, is given in the data sheets. For instance, for $\theta_{JA} = 50$ °C/W the chip temperature rises above the ambient temperature by 50 °C for every watt of dissipated power. If $T_A = 25$ °C and $P_D = 2$ W, then $T_J = T_A + \theta_{JA}P_D = 25 + 50 \times 2 = 125$ °C. We can also regard θ_{JA} as a measure of a device's ability to expel heat. The lower θ_{JA}, the smaller the temperature rise for a given P_D. It is apparent that θ_{JA} and $T_{J(max)}$ set an upper limit on P_D for a given $T_{A(max)}$.

The heat-transfer process can be modeled with an electrical-conduction analog where power corresponds to current, temperature to voltage, and thermal resistance to ohmic resistance. This analogy is illustrated in Fig. 11.30 for the case of free-air

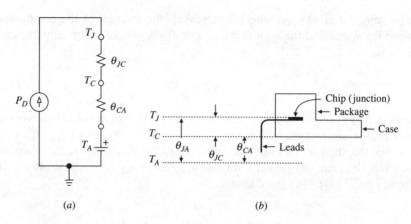

(a)   (b)

FIGURE 11.30
(a) Electrical analog of heat flow. (b) Typical package structure operating in free air.

operation, that is, with no provisions for cooling. The thermal resistance θ_{JA} consists of two components,

$$\theta_{JA} = \theta_{JC} + \theta_{CA} \tag{11.27}$$

where θ_{JC} is the thermal resistance from *junction to case*, and θ_{CA} that from *case to ambient*. Using Ohm's law and KVL, we can find the temperature at any point of the heat-flow path once the other parameters are known. If the path involves more than one resistance, the net resistance is the sum of the individual resistances.

The component θ_{JC} is set by device layout and packaging. To help reduce θ_{JC}, the device is encapsulated in a suitably large case, and the collector region, where most of the heat is dissipated, is placed in direct contact with the case. Figure 11.31 shows two popular packages, along with their thermal ratings for the case of the μA7800 and μA7900 series. Data sheets usually give only θ_{JC} and θ_{JA}; then, we can compute $\theta_{CA} = \theta_{JA} - \theta_{JC}$.

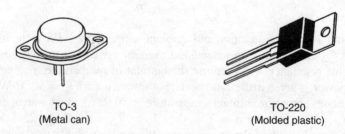

TO-3
(Metal can)

TO-220
(Molded plastic)

FIGURE 11.31
Two popular power packages. For the μA7800 series, the typical (maximum) thermal-resistance ratings are: TO-3: $\theta_{JC} = 3.5(5.5)\ °C/W$, $\theta_{JA} = 40\ (45)\ °C/W$; TO-220: $\theta_{JC} = 3.0\ (5.0)\ °C/W$, $\theta_{JA} = 60\ (65)\ °C/W$.

EXAMPLE 11.14. (*a*) According to Fig. 11.24, $T_{J(max)} = 150$ °C for the $\mu A7805$. Assuming $T_{A(max)} = 50$ °C, find the maximum power that a TO-220 package operating in free air can dissipate. What is the corresponding case temperature T_C? (*b*) Find the maximum current that can be drawn from the device if $V_I = 8$ V.

Solution.

(*a*) $P_{D(max)} = (T_{J(max)} - T_{A(max)})/\theta_{JA} = (150 - 50)/60 = 1.67$ W. By KVL, $T_C = T_J - \theta_{JC} P_D = 150 - 3 \times 1.67 = 145$ °C.

(*b*) Ignoring the current at the common terminal, we have $P_D \cong (V_I - V_O)I_O$, so $I_O \leq 1.67/(8 - 5) = 0.556$ A.

In the case of free-air operation, heat encounters much more resistance in propagating from case to ambient than from junction to case. The user can reduce θ_{CA} significantly by means of a heatsink. This is a metal structure, usually with fins, that is bonded, clipped, or clamped to the device package to facilitate heat flow from case to ambient. The effect of a heatsink is illustrated in Fig. 11.32. While θ_{JC} remains the same, θ_{CA} is altered significantly as

$$\theta_{CA} = \theta_{CS} + \theta_{SA} \tag{11.28}$$

where θ_{CS} is the thermal resistance of the mounting surface and θ_{SA} is that of the heatsink. The mounting surface is usually a thin insulating washer of mica or fiberglass to provide electrical isolation between the case, which is internally connected to the collector, and the sink, which is often bonded to the chassis. Usually smeared with heatsink grease to ensure intimate thermal contact, the mounting surface has a typical thermal resistance of less than 1 °C/W.

Heatsinks are available in a variety of shapes and sizes, with thermal resistances ranging from about 30 °C/W for the smaller types to as little as 1 °C/W or less for

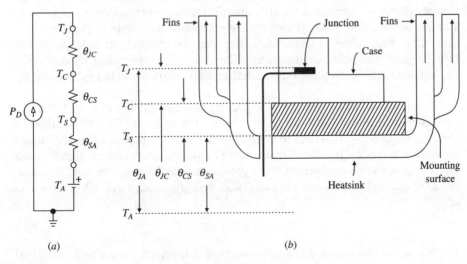

(*a*)   (*b*)

FIGURE 11.32
Heat-flow electrical analog of a package mounted on a heatsink.

the truly massive units. Thermal resistance is specified for the case of a heatsink mounted with fins vertical and with unobstructed airflow. Forced air cooling reduces thermal resistance further. In the limiting case of infinite heatsinking and a thermally perfect mounting surface, θ_{CA} would approach zero and the device's ability to expel heat would be limited only by θ_{JC}. The package-heatsink combination best suited to a given application is determined on the basis of the maximum expected power dissipation, the maximum allowable junction temperature, and the maximum anticipated ambient temperature.

EXAMPLE 11.15. A μA7805 regulator is to meet the following requirements: $T_{A(\text{max})} = 60\,°C$, $I_{O(\text{max})} = 0.8\,A$, $V_{I(\text{max})} = 12\,V$, and $T_{J(\text{max})} = 125\,°C$. Select a suitable package-heatsink combination.

Solution. $\theta_{JA(\text{max})} = (125 - 60)/[(12 - 5)0.8] = 11.6\,°C/W$. Use the TO-220 package, which is cheaper and offers better thermal resistance. Then, $\theta_{CA} = \theta_{JA} - \theta_{JC} = 11.6 - 5 = 6.6\,°C/W$. Allowing 0.6 °C/W for the thermal resistance of the mounting surface, we are left with $\theta_{SA} = 6\,°C/W$. According to the catalogs, a suitable heatsink example is the IERC HP1 series, whose θ_{SA} rating is in the range of 5 °C/W to 6 °C/W.

Power-Supply Supervisory Circuits

The forms of protection discussed in Section 11.4 safeguard the regulator. A well-designed power-supply system will also include circuitry to safeguard the load and to monitor satisfactory power-supply performance. The functions typically required are *over-voltage* (OV) protection, *under-voltage* (UV) sensing, and *ac line loss* detection. The MC3425 (Motorola) is one of a variety of dedicated circuits known as *power-supply supervisory circuits* designed to assist the designer in this task.

As shown in Fig. 11.33, the circuit consists of a 2.5-V bandgap reference and two comparator channels, one for OV protection and the other for UV detection. The input comparators CMP_1 and CMP_3 have open-collector outputs with 200-μA active pullups. These outputs are externally accessible to allow independent adjustment of the response delays of the two channels in order to prevent false triggering in noisy environments. The delays are established by connecting two capacitors between these outputs and ground, as shown in the subsequent figures.

Under normal conditions these outputs are low. Should, however, an OV or UV condition arise, either CMP_1 or CMP_3 will switch its output BJT off to allow the corresponding delay capacitor to charge by the 200-μA pullup. Once the capacitor voltage reaches V_{REF}, the corresponding output comparator fires, signaling that the emergency condition persisted for the entire delay of that channel. The delay of either channel is obtained via Eq. (10.2) as $T_{\text{DLY}} = C_{\text{DLY}}(2.5\,\text{V})/(200\,\mu\text{A})$, or

$$T_{\text{DLY}} = 12,500 C_{\text{DLY}} \tag{11.29}$$

where C_{DLY} is in farads and T_{DLY} in seconds. For instance, using $C_{\text{DLY}} = 0.01\,\mu\text{F}$ yields $T_{\text{DLY}} = 125\,\mu\text{s}$.

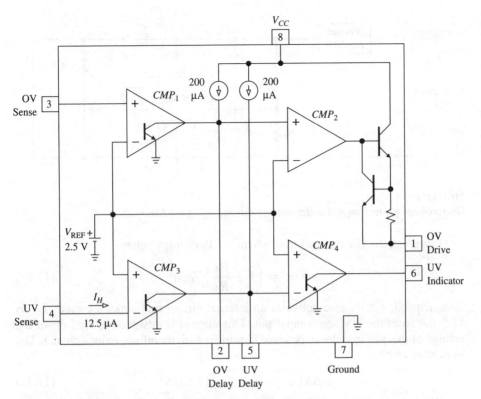

FIGURE 11.33
Simplified diagram of the MC3425 power-supply supervisory/over-under-voltage protection
circuit. (Courtesy of Motorola, Inc.)

Whereas the UV comparator CMP_4 has an open-collector output, the OV comparator CMP_2 has an overload-protected output booster to drive an external silicon
controlled rectifier (SCR) crowbar for emergency power shutdown.

OV/UV Sensing and Line-Loss Detection

Figure 11.34 shows a typical 3425 connection for OV protection and UV sensing.
The OV channel trips whenever V_{CC} tries to rise above a level V_{OV} such that
$V_{OV}/(1 + R_2/R_1) = V_{REF}$, or

$$V_{OV} = \left(1 + \frac{R_2}{R_1}\right) V_{REF} \tag{11.30}$$

If the OV condition persists for the entire delay T_{OV} as set by C_{OV}, the MC3425
fires the SCR, which in turn shorts out the voltage regulator and blows the fuse, thus
protecting the load against prolonged over-voltage and the unregulated input source
against prolonged overload.

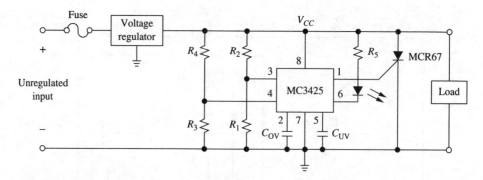

FIGURE 11.34
Over-voltage protection and under-voltage sensing using the MC3425.

Likewise, the UV channel trips whenever V_{CC} drops below

$$V_{UV} = \left(1 + \frac{R_4}{R_3}\right) V_{REF} \tag{11.31}$$

Once tripped, CMP_3 also activates an internal circuit that sinks a current $I_H = 12.5\ \mu A$ from the UV sense input pin. This current is designed to load down the voltage of this pin in order to produce *hysteresis* and, therefore, reduce chatter. The hysteresis width is

$$\Delta V_{UV} = (R_3 \parallel R_4)(12.5\ \mu A) \tag{11.32}$$

Thus, once CMP_3 fires as a result of V_{CC} dropping below V_{UV}, it remains in that state until V_{CC} rises above $V_{UV} + \Delta V_{UV}$. Unless this happens within the delay T_{UV} as set by C_{UV}, CMP_4 also fires and causes the LED to glow. Once V_{CC} returns above $V_{UV} + \Delta V_{UV}$, CMP_3 returns to the original state and deactivates I_H.

EXAMPLE 11.16. In Fig. 11.34 specify suitable components for an OV trip level of 6.5 V with a 100-μs delay, and a UV trip level of 4.5 V with a 0.25-V hysteresis and a 500-μs delay.

Solution. The above equations give $C_{OV} = 8$ nF, $R_2/R_1 = 1.6$, $R_4/R_3 = 0.8$, $R_3 \parallel R_4 = 20$ kΩ, $C_{UV} = 40$ nF. Use $C_{OV} = 8.2$ nF, $C_{UV} = 43$ nF, $R_1 = 10.0$ kΩ, $R_2 = 16.2$ kΩ, $R_3 = 45.3$ kΩ, $R_4 = 36.5$ kΩ.

In microprocessor-based systems, ac line loss, whether total (*blackout*) or partial (*brownout*), must be detected in time to allow the salvage of vital status information in nonvolatile memory, as well as disable any devices that might be adversely affected by underpowered operation, such as motors and pumps. The circuit of Fig. 11.35*a* monitors the ac line via a center-tapped transformer (which can be the very transformer involved in the generation of the unregulated input to the voltage regulator) and uses the UV channel to detect line loss. Circuit operation is best understood with the help of the waveforms of Fig. 11.35*b*.

The delay capacitor C_{UV} is chosen to be large enough so that, under normal line conditions, it does not have enough time between consecutive ac peaks to charge past 2.5 V. This is also referred to as *retriggerable one-shot operation*. However, should the line drop to the extent of causing the peaks at the UV sense pin 4 to drop

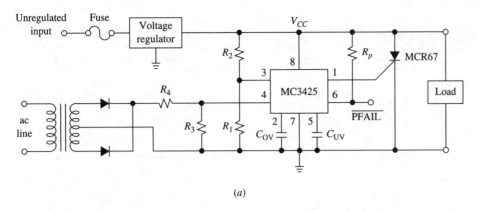

(a)

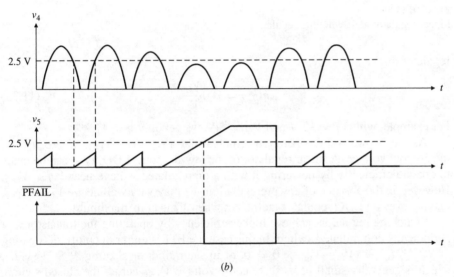

(b)

FIGURE 11.35
Over-voltage protection with ac line-loss detection circuit, and typical waveforms.

below the 2.5-V threshold, C_{UV} will fully charge and trigger CMP_4, thus issuing a $\overline{PFAIL}$ command. This can be used to interrupt the microprocessor and initiate appropriate power-fail routines.

11.6
SWITCHING REGULATORS

As we know, in a linear regulator the series-pass transistor transfers power from V_I to V_O continuously. As depicted in Fig. 11.36a, the BJT operates in the forward-active region, where it acts as a controlled current source dissipating the power $P = V_{CE}I_C + V_{BE}I_B$. Ignoring the base current and the current drawn by the control circuitry compared to the load current I_O, we can write $P \cong (V_I - V_O)I_O$. As already seen, it is precisely this dissipation that limits the efficiency of a linear

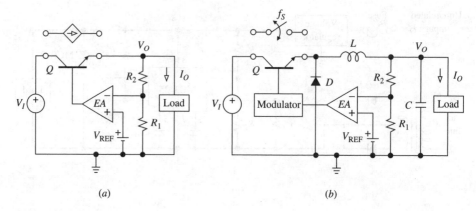

FIGURE 11.36
Linear regulator and switching regulator.

regulator to

$$\eta(\%) = 100\frac{V_O}{V_I} \qquad (11.33)$$

For example, with $V_I = 12$ V and $V_O = 5$ V, we get only $\eta = 41.7\%$.

As we know, proper operation requires that $V_I \geq V_O + V_{DO}$, where V_{DO} is the dropout voltage. A linear regulator of the low-dropout (LDO) type can be made to operate efficiently by powering it with a preregulated voltage near $V_O + V_{DO}$. However, in the absence of any preregulation, V_I may vary well above $V_O + V_{DO}$, making even an LDO regulator inefficient when V_I is at its maximum.

Switching regulators achieve higher efficiency by operating the transistor as a periodically commutated switch. In this case the BJT is either in cutoff, dissipating $P \cong V_{CE}I_C \cong (V_I - V_O) \times 0 = 0$, or in saturation, dissipating $P \cong V_{SAT}I_C$, which is generally small because so is the voltage V_{SAT} across the closed switch. Thus, a switched BJT dissipates much less power than a forward-active BJT. The price for switch-mode operation is the need for a coil to provide a high-frequency transfer of energy packets from V_I to V_O, and a smoothing capacitor to ensure a low output ripple. However, L and C manipulate energy without dissipating any power, at least ideally. Consequently, the combination of switches and low-loss reactive elements makes switching regulators inherently more efficient than their linear counterparts.

Switch-mode regulation is effected by adjusting the duty cycle D of the switch, defined as

$$D = \frac{t_{ON}}{t_{ON} + t_{OFF}} = \frac{t_{ON}}{T_S} = f_S t_{ON} \qquad (11.34)$$

where t_{ON} and t_{OFF} are the time intervals during which the transistor is on and off; $T_S = t_{ON} + t_{OFF}$ is the duration of a switch cycle; and $f_S = 1/T_S$ is the operating frequency of the switch. There are two ways of adjusting the duty cycle: (*a*) in *pulse-width modulation* (PWM), f_S is kept fixed and t_{ON} is adjusted; and (*b*) in *pulse-frequency modulation* (PFM), t_{ON} (or t_{OFF}) is fixed and f_S is adjusted. It is apparent that switching regulators require more complex control circuitry than their linear counterparts.

Basic Topologies

If we view the switch-coil-diode combination as a T structure, then, depending on which leg is occupied by the coil, we have the three topologies of Fig. 11.37, called, for reasons to be justified shortly, the *buck, boost,* and *buck-boost* topologies; clearly, the circuit of Fig. 11.36*b* is a buck circuit. Though the topologies are shown for operation with $V_I > 0$, they can readily be configured for $V_I < 0$ by proper reversal of the switch and diode polarities. Moreover, a wide range of variants[12,13] can be obtained by suitable modification of the coil and switch structures. To gain more insight, we focus on the buck topology, though similar analysis can be applied also to the other topologies. Assuming $V_I > V_O$, we can describe buck operation as follows.

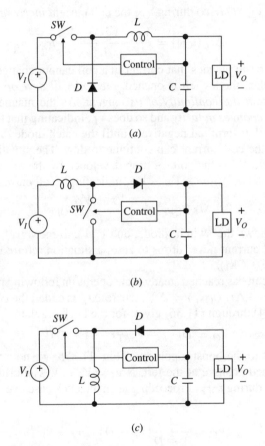

FIGURE 11.37
Basic switching-regulator topologies: (*a*) buck,
(*b*) boost, and (*c*) buck-boost.

During t_{ON} the switch closes and connects the coil to V_I. The diode is off, so the situation is as in Fig. 11.38*a*, where V_{SAT} is the voltage drop developed by the closed switch. During this time, current and magnetic energy build up in the coil according to the familiar laws $di_L/dt = v_L/L$ and $w_L = (1/2)Li_L^2$. If V_I and V_O do not change appreciably during a switch cycle, the coil voltage v_L remains relatively constant at $v_L = V_I - V_{SAT} - V_O$. We can replace differentials with finite differences

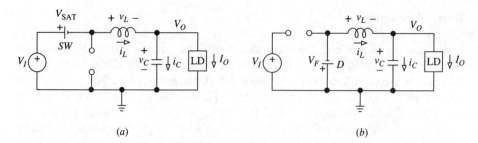

(a)   (b)

FIGURE 11.38
Equivalent circuits of the buck regulator when SW is (a) closed and (b) open.

and write $\Delta i_L = v_L \Delta t / L$, so during t_{ON} the coil current *increases* by

$$\Delta i_L(t_{ON}) = \frac{V_I - V_{SAT} - V_O}{L} t_{ON} \qquad (11.35)$$

Recall from basic physics that current in a coil cannot change instantaneously. Consequently, when the switch is opened, *the coil will develop whatever voltage it takes to maintain the continuity of its current*. As the magnetic field starts to collapse, di_L/dt *changes polarity* and so does v_L, indicating that the coil will swing the voltage of its left terminal negatively until the catch diode turns on to provide a path in which the coil current can continue to flow. The situation is depicted in Fig. 11.38b, where V_F is the voltage drop developed by the forward-biased diode. The coil voltage is now $v_L = -V_F - V_O$, indicating a coil current *decrease*

$$\Delta i_L(t_{OFF}) = -\frac{V_F + V_O}{L} t_{OFF} \qquad (11.36)$$

Figure 11.39a shows the switch, diode, and coil current waveforms for the case in which the coil current never drops to zero, a situation referred to as *continuous conduction mode* (CCM).

Once the circuit has reached steady-state operation following power turn-on, we have $\Delta i_L(t_{ON}) = -\Delta i_L(t_{OFF}) = \Delta i_L$, where Δi_L is called the *coil current ripple*. Using Eqs. (11.34) through (11.36) gives, for the buck regulator,

$$V_O = D(V_I - V_{SAT}) - (1 - D)V_F \qquad (11.37)$$

Turning next to the boost topology of Fig. 11.37b, we note that the coil voltage, again assumed positive at the left, is $v_L = V_I - V_{SAT}$ during t_{ON}, and $v_L = V_I - (V_F + V_O)$ during t_{OFF}. Proceeding as in the buck case, we find, for the boost regulator,

$$V_O = \frac{1}{1 - D}(V_I - DV_{SAT}) - V_F \qquad (11.38)$$

Likewise, the coil voltage in Fig. 11.37c, assumed positive at the top, is $v_L = V_I - V_{SAT}$ during t_{ON}, and $v_L = V_O - V_F$ during t_{OFF}. Consequently, we have, for the buck-boost regulator,

$$V_O = -\frac{D}{1 - D}(V_I - V_{SAT}) + V_F \qquad (11.39)$$

In the ideal limits $V_{SAT} \to 0$ and $V_F \to 0$ the above equations simplify, respectively, to the following *lossless* characteristics.

$$V_O = DV_I \qquad V_O = \frac{1}{1 - D}V_I \qquad V_O = -\frac{D}{1 - D}V_I \qquad (11.40)$$

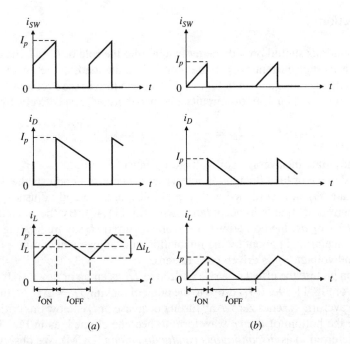

FIGURE 11.39
Current waveforms for the three basic topologies: (*a*) continuous conduction mode (CCM), and (*b*) discontinuous conduction
mode (DCM).

Given that $0 < D < 1$, the buck regulator yields $V_O < V_I$ and the boost regulator
$V_O > V_I$, these being the reasons for their names. By analogy with transformers,
the buck and boost circuits are also referred to as *step-down* and *step-up* regulators. In
the buck-boost circuit the output magnitude can be smaller or greater than the input
magnitude, depending on whether $D < 0.5$ or $D > 0.5$; moreover, the output polarity
is opposite to that of the input, so this regulator is also called an *inverting* regulator.
Note that boost and polarity inversion are not possible with linear regulators!

In the ideal limit of lossless components and zero power dissipation by the
control circuitry, a switching regulator would be 100% efficient, giving $P_O = P_I$,
or $V_O I_O = V_I I_I$. Writing

$$I_I = (V_O/V_I)I_O \qquad (11.41)$$

provides an estimate for the current drawn from the input source.

EXAMPLE 11.17. Given a buck regulator with $V_I = 12$ V and $V_O = 5$ V, find D if
(*a*) the switch and diode are ideal, and (*b*) $V_{SAT} = 0.5$ V and $V_F = 0.7$ V. (*c*) Repeat (*a*)
and (*b*) if 8 V $\leq V_I \leq 16$ V.

Solution.

(*a*) By Eq. (11.40), $D = 5/12 = 41.7\%$.
(*b*) By Eq. (11.37), $D = 46.7\%$.
(*c*) The same equations give, for the two cases, $31.2\% \leq D \leq 62.5\%$, and $35.2\% \leq$
$D \leq 69.5\%$.

Coil Selection

Two observations should provide better insight into the role of L: (a) The coil must carry some average current $I_L \neq 0$ in order to feed the load; in fact, with reference to the continuous mode shown in Fig. 11.39, one can prove (see Problem 11.31) that the buck, boost, and buck-boost circuits are characterized, respectively, by

$$I_L = I_O \qquad I_L = \frac{V_O}{V_I} I_O \qquad I_L = \left(1 - \frac{V_O}{V_I}\right) I_O \qquad (11.42)$$

(b) In steady state the average coil voltage V_L must be zero.

Should a line or load fluctuation intervene, the controller adjusts the duty cycle D to regulate V_O in accordance with Eq. (11.40), and the coil adjusts I_L to meet the load-current demands in accordance with Eq. (11.42). By the inductance law $i_L = (1/L) \int v_L \, dt$, the coil adjusts its average current I_L by integrating the voltage imbalance brought about by the fluctuation; this adjustment continues until the average coil voltage V_L is driven back to zero.

We can picture the effect of a rise or drop in I_O as an up or down shift of the i_L waveform of Fig. 11.39a. If I_O drops to the point of making $I_L = \Delta i_L/2$, the bottom of the i_L waveform reaches zero. Any further decrease of I_O below this critical value will cause the bottom of the i_L waveform to become clipped, as in Fig. 11.39b, a situation referred to as *discontinuous conduction mode* (DCM). We observe that in CCM V_O depends only on D and V_I, regardless of I_O. By contrast, in DCM V_O depends also on I_O, so D will have to be reduced accordingly by the controller; failing to do so would cause, in the limit of an open-circuited output, $V_O \to V_I$ for the buck, $V_O \to \infty$ for the boost, and $V_O \to -\infty$ for the buck-boost regulators.

To estimate a suitable value of L, it is convenient to assume $V_{SAT} = V_F = 0$. Then, for a buck regulator in steady state, Eqs. (11.35) and (11.36) give $t_{ON} = L\Delta i_L/(V_I - V_O)$ and $t_{OFF} = L\Delta i_L/V_O$. Letting $t_{ON} + t_{OFF} = 1/f_S$ gives, for the buck regulator,

$$L = \frac{V_O(1 - V_O/V_I)}{f_S \Delta i_L} \qquad (11.43)$$

Proceeding in similar manner, we find, for the boost regulator,

$$L = \frac{V_I(1 - V_I/V_O)}{f_S \Delta i_L} \qquad (11.44)$$

and for the buck-boost regulator,

$$L = \frac{V_I/(1 - V_I/V_O)}{f_S \Delta i_L} \qquad (11.45)$$

The choice of L is usually a tradeoff between maximum output power with minimum output ripple, and small physical size with fast transient response.[13] Moreover, increasing L for a given I_O will cause the system to go from DCM to CCM. A good starting point is to choose the current ripple Δi_L, and then use the proper equation to estimate L.

There are various criteria for specifying Δi_L. One possibility[13] is to let $\Delta i_L = 0.2 I_{L(max)}$, where $I_{L(max)}$ is dictated either by the maximum output current rating of the regulator, as per Eq. (11.42), or by the maximum peak-current rating of the switch, as per $I_p = I_L + \Delta i_L/2$. The switch rating becomes important especially in step-up situations, where I_L may be considerably larger than I_O. Alternatively, to avoid discontinuous operation, we can let $\Delta i_L = 2I_{O(\min)}$, where $I_{O(\min)}$ is the minimum anticipated load current. Other criteria[13,14] are possible, depending on the type of regulation as well as the goals of the given application.

Once the value of L has been chosen, a coil must be found that can handle both the peak and rms values of i_L. The peak value is limited by core saturation, for if the coil were to saturate, its inductance would drop abruptly, causing an inordinate rise in i_L during t_{ON}. The rms value is limited by losses in the windings and the core. Though the coil has traditionally been perceived as a very intimidating issue, modern switching-regulator data sheets provide a wealth of useful information to ease coil selection, including coil manufacturers' addresses and specific part numbers.

EXAMPLE 11.18. Specify a coil for a boost regulator with $V_I = 5$ V, $V_O = 12$ V, $I_O = 1$ A, and $f_S = 100$ kHz. What is the minimum load current $I_{O(\min)}$ for continuous operation?

Solution. At full load, $I_L = (12/5)1 = 2.4$ A. Let $\Delta i_L = 0.2I_L = 0.48$ A. Then, Eq. (11.44) gives $L = 61$ μH. At full load the coil must withstand $I_p = I_L + \Delta i_L/2 = 2.64$ A, and $I_{\text{rms}} = [I_L^2 + (\Delta i_L/\sqrt{12})^2]^{1/2} \cong I_L = 2.4$ A. Moreover, $I_{O(\min)} = 0.1$ A.

Capacitor Selection

To estimate a suitable value of C in the buck topology of Fig. 11.37a, we observe that the coil current splits between the capacitor and the load as $i_L = i_C + i_O$. In steady state the average capacitance current is zero and the load current is relatively constant. We can therefore write $\Delta i_C = \Delta i_L$, indicating that the i_C waveform is similar to the i_L waveform, except that i_C is centered about zero. The i_C ripple causes a voltage ripple $\Delta v_C = (1/C) \int i_C \, dt$, where integration is from $t_{\text{ON}}/2$ (where v_C reaches its minimum) to $t_{\text{OFF}}/2$ (where v_C reaches its maximum). We easily find the area as $\int i_C \, dt = 1/2 \times (t_{\text{ON}}/2 + t_{\text{OFF}}/2) \times \Delta i_L/2 = \Delta i_L/8 f_S$. This gives, for the buck regulator,

$$C = \frac{\Delta i_L}{8 f_S \Delta v_C} \qquad (11.46)$$

In the boost topology of Fig. 11.37b the coil is disconnected from the output during t_{ON}, so the load current during this time is supplied by the capacitor. Using Eq. (10.2), we estimate the ripple as $\Delta v_C = I_O t_{\text{ON}}/C$. But, $t_{\text{ON}} = D/f_S$ and $D = 1 - V_I/V_O$, so we have, for the boost regulator,

$$C = \frac{I_O(1 - V_I/V_O)}{f_S \Delta v_C} \qquad (11.47)$$

Similar considerations hold for the buck-boost topology, so

$$C = \frac{I_O(1 - V_I/V_O)}{f_S \Delta v_C} \qquad (11.48)$$

The above equations give C for a specific ripple Δv_C. Practical capacitors exhibit a small *equivalent series resistance* (ESR) and a small *equivalent series inductance* (ESL), as modeled in Fig. 11.40. The ESR contributes an output ripple term of the type $\Delta v_{\text{ESR}} = \text{ESR} \times \Delta i_C$, where Δi_C is the capacitor ripple current, indicating

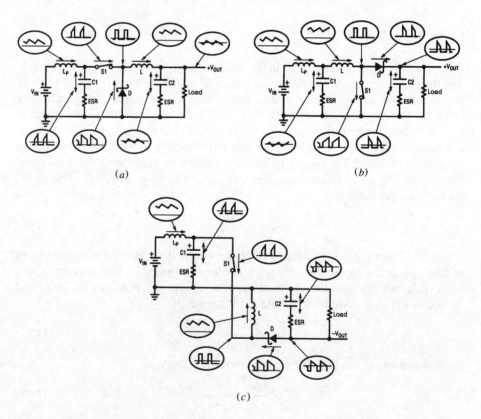

FIGURE 11.40
A practical capacitor has
an equivalent series resistance ESR and inductance
ESL.

the need for low-ESR capacitors. The ripple Δv_C across C in Fig. 11.40 and the ripple Δv_{ESR} across ESR combine to give an overall ripple V_{ro} at the output. For an estimation of the maximum allowed ESR, a reasonable approach[13] is to allow $\frac{1}{3}$ of V_{ro} to come from Δv_C, and $\frac{2}{3}$ of V_{ro} from Δv_{ESR}.

EXAMPLE 11.19. In the boost regulator of Example 11.18, specify a capacitor for an output ripple $V_{ro} \cong 100$ mV.

Solution. At full load and with $\Delta v_C \cong (1/3)V_{ro} \cong 33$ mV, Eq. (11.47) gives $C = 177$ μF. For the boost regulator we have $\Delta i_C = \Delta i_D = I_p$, so at full load $\Delta i_C = 2.64$ A. Then, ESR = $(67$ mV$)/(2.64$ A$) \cong 25$ mΩ.

(a) $\qquad$ (b)

(c)

FIGURE 11.41
Typical waveforms for the (a) buck, (b) boost, and (c) buck-boost regulators. (Courtesy of Linear Technology.)

The C and ESR requirements may be difficult to meet simultaneously, so we can either increase the size of the capacitor, since larger capacitors tend to have smaller ESRs, or we can filter out the existing ripple with an additional LC stage at the output.

A well-constructed switching regulator will include an LC filter also at the input, both to ease the output-impedance requirements of the source V_I and to prevent the injection of electromagnetic interference (EMI) upstream of the regulator. This is illustrated[15] in Fig. 11.41 for the three basic topologies operating in CCM (the waveforms pointing to arrows are element currents, the others are node voltages). We observe that the most taxing situation for a capacitor is when it is in series with either the switch or the diode. When it is in series with the coil, as at the input of the boost or at the output of the buck topology, the filtering action provided by the coil itself results in a smoother waveform. It follows that the buck regulator enjoys the lowest output ripple of the three topologies.

Efficiency

The efficiency of a switching regulator is found as

$$\eta(\%) = \frac{P_O}{P_O + P_{\text{diss}}} \tag{11.49}$$

where $P_O = V_O I_O$ is the power delivered to the load, and

$$P_{\text{diss}} = P_{SW} + P_D + P_{\text{coil}} + P_{\text{cap}} + P_{\text{controller}} \tag{11.50}$$

is the sum of the losses in the switch, the diode, the coil, the capacitor, and the switch controller.

Switch loss is the sum of a *conduction* component and a *switching* component, or $P_{SW} = V_{SAT}I_{SW} + f_S W_{SW}$. The conduction component is due to the nonzero voltage drop V_{SAT}; for the case of a saturating BJT switch this component is found as $V_{CE(\text{sat})}I_{SW(\text{avg})}$, and for the case of a FET switch as $r_{DS(\text{on})}I_{SW(\text{rms})}^2$. The switching component is due to the nonzero rise and fall times of the voltage and current waveforms of the switch; the resulting waveform overlap causes the per-cycle dissipation of an energy packet[14] $W_{SW} \cong 2\Delta v_{SW}\Delta i_{SW}t_{SW}$, where Δv_{SW} and Δi_{SW} are the switch voltage and current changes, and t_{SW} is the effective overlap time.

Diode dissipation is likewise[13] $P_D = V_F I_{F(\text{avg})} + f_S W_D$, $W_D \cong V_R I_F t_{\text{RR}}$, where V_R is the diode reverse voltage, I_F the forward current at turn-off, and t_{RR} the reverse recovery time. Schottky diodes are good choices because of their inherently lower voltage drop V_F and the absence of charge-storage effects.

Capacitor loss is $P_{\text{cap}} = \text{ESR}I_{C(\text{rms})}^2$. Coil loss consists of two terms, namely, the copper loss $R_{\text{coil}}I_{L(\text{rms})}^2$ in the coil resistance, and core losses, which depend on the coil current as well as f_S. Finally, the controller contributes $V_I I_Q$ where I_Q is the average current it draws from V_I, exclusive of the switch.

EXAMPLE 11.20. A buck regulator with $V_I = 15$ V, $V_O = 5$ V, $I_O = 3$ A, $f_S = 50$ kHz, and $I_Q = 10$ mA, uses a switch with $V_{SAT} = 1$ V and $t_{SW} = 100$ ns, a diode with $V_F = 0.7$ V and $t_{RR} = 100$ ns, a coil with $R_{coil} = 50$ mΩ and $\Delta i_L = 0.6$ A, and a capacitor with ESR $= 100$ mΩ. Assuming core losses of 0.25 W, find η and compare with a linear regulator.

Solution. Eq. (11.37) gives $D = 38.8\%$. Then, $P_{SW} \cong V_{SAT} D I_O + 2 f_S V_I I_O t_{SW} = 1.16 + 0.45 = 1.61$ W; $P_D \cong V_F(1 - D)I_O + f_S V_I I_O t_{RR} = 1.29 + 0.22 = 1.51$ W; $P_{cap} = ESR(\Delta i_L / \sqrt{12})^2 = 3$ mW; $P_{coil} = R_{coil} \times (\Delta i_L / \sqrt{12})^2 + 0.25$ W $\cong 0.25$ W; $P_{controller} = 15 \times 10 = 0.15$ W; $P_O = 5 \times 3 = 15$ W; $P_{diss} = 3.52$ W; $\eta = 81\%$.

A linear regulator would have $\eta = 5/15 = 33\%$, indicating that to deliver 15 W of useful power it would dissipate 30 W, while the switching regulator of our example dissipates only 3.52 W.

11.7
MONOLITHIC SWITCHING REGULATORS

Monolithic switching regulators are available in a wide range of performance specifications. For switch currents of up to a few amperes, the switch is usually provided on-chip, along with the control circuitry. All the user needs to provide, then, is the coil, the output filter capacitor, the input bypass capacitor, and the catch diode, usually a Schottky type. When higher currents are called for, the switch is provided externally by the user, and it may be either a power BJT or a power MOSFET. FETs are generally preferable because the absence of second-breakdown limitations and charge-storage effects allows for higher switching frequencies, and hence, smaller energy-storage elements, particularly smaller coils. To minimize power loss, use a FET with a suitably low $r_{DS(on)}$.

Component layout and orientation are extremely critical in switching regulators, so manufacturers provide printed-circuit-board layouts and component-stuffing diagrams. Moreover, to foster switching regulator applications, computer programs are available, such as the *SwitcherCAD* program by Linear Technology[15] and the *Switchers Made Simple* program by National Semiconductor.[16]

Though the market offers both pulse-width modulation (PWM) and pulse-frequency modulation (PFM) controllers, the majority of regulators at present are PWM controllers operating at a fixed frequency f_S in the range of 10^4 to 10^6 Hz. This frequency is chosen as a compromise between small coil/capacitor size on the one hand, and low switching losses and reduced EMI and RFI on the other. There are two types of PWM control, namely, voltage-mode and current-mode control.

Voltage-Mode Control

Voltage-mode control,[17] exemplified in Fig. 11.42 for the buck topology operating in the continuous conduction mode (CCM), controls t_{ON} by modulating a sawtooth waveform v_S of frequency f_S with the error-amplifier output v_C. To gain a better feel for the various issues involved, refer to the simplified equivalent of Fig. 11.43. If the switching frequency f_S is high enough that PWM can be regarded as a continuous

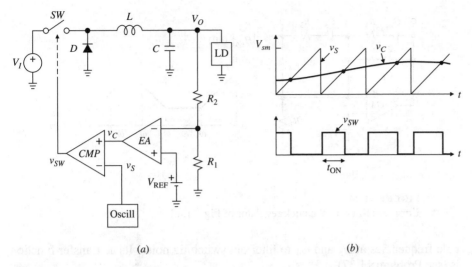

FIGURE 11.42
Voltage-mode control, and typical waveforms.

process over the frequency range of interest, the *control-to-output transfer function* is (see Problem 11.36)

$$H_{CO} = \frac{V_o}{V_c} = \frac{V_I}{V_{sm}} \times \frac{1 + j\omega/\omega_z}{1 - (\omega/\omega_0)^2 + (j\omega/\omega_0)/Q} \qquad (11.51)$$

$$\omega_0 = \frac{1}{\sqrt{LC}} \qquad \omega_z = \frac{1}{ESRC} \qquad Q = \frac{1}{(R_{coil} + ESR)\sqrt{C/L}} \qquad (11.52)$$

where V_{sm} is the peak value of the sawtooth. We observe that the presence of L and C within the loop results in a complex pole pair, and the presence of ESR results in a zero.

The function of the error amplifier is to ensure high-loop gain for good regulation, and adequate phase margin for stability. The error amplifier exemplified in Fig. 11.44 has a pole frequency at the origin to ensure high dc gain, two zero frequencies at ω_1 and ω_2 to provide phase lead in the vicinity of the crossover frequency, and two

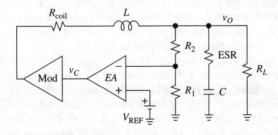

FIGURE 11.43
Equivalent circuit of a buck regulator operating in CCM with voltage-mode control.

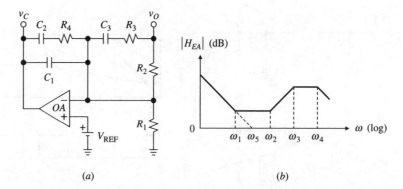

FIGURE 11.44
Error amplifier for the buck regulator of Fig. 11.43.

pole frequencies at ω_3 and ω_4 to filter out switching noise. Its ac transfer function is (see Problem 11.37)

$$H_{EA} = \frac{V_c}{V_o} = -\frac{(1 + j\omega/\omega_1)(1 + j\omega/\omega_2)}{(j\omega/\omega_5)(1 + j\omega/\omega_3)(1 + j\omega/\omega_4)} \quad (11.53)$$

The circuit is implemented with $C_2 \gg C_1$ and $R_3 \ll R_2$, in which case its characteristic frequencies simplify as

$$\omega_1 = \frac{1}{R_4 C_2} \quad \omega_2 = \frac{1}{R_2 C_3} \quad \omega_3 = \frac{1}{R_3 C_3} \quad \omega_4 = \frac{1}{R_4 C_1} \quad \omega_5 = \frac{1}{R_2 C_2} \quad (11.54)$$

The overall loop gain is $T = -H_{EA} H_{CO}$. For a fast response, the crossover frequency f_x should be specified as high as possible, a common choice being[18] $f_x \cong f_S/5$. As we know, the output-shunt feedback action of the regulator will result in a low output impedance only over the frequency range of substantial loop gain. Past f_x, the output impedance reduces to that of the capacitor in parallel with the coil.

EXAMPLE 11.21. Specify suitable components in the error amplifier of Fig. 11.44 for a CCM buck regulator with $V_I = 12$ V, $f_S = 100$ kHz, $V_{sm} = 1$ V, $L = 100$ μH, $C = 300$ μF, ESR $= 0.05$ Ω, and $R_{coil} \ll$ ESR.

Solution. We readily find that H_{CO} has a dc gain of 12 V/V, $f_0 = 920$ Hz, $f_z = 10.6$ kHz, and $Q = 11.5$. Moreover, let $f_x = 100/5 = 20$ kHz.

A common design strategy[18] is to make the loop gain T roll off with an average slope of -20 dB/dec up to f_x. This requires imposing that the zeros of the error amplifier satisfy $f_1 = f_2 = f_0$, and the first pole satisfy $f_3 = f_z$; moreover, to maintain a good phase margin, we position the second pole at $f_4 = 2f_x$.

To fully specify H_{EA} we need one additional parameter, namely, the unity-gain frequency f_5 associated with the pole at the origin. Calculating $|H_{CO}(jf_x)| = 1/18.5$ V/V indicates that we need $|H_{EA}(jf_x)| = 18.5$ V/V in order to make $|T| = 1$ V/V at the specified crossover frequency. Applying Eq. (8.10) twice, we find $f_5 = 1.47$ kHz. Then, using Eq. (11.54) and arbitrarily imposing $R_2 = 10$ kΩ, we find $R_3 = 867$ Ω, $R_4 = 16.0$ kΩ, $C_1 = 240$ pF, $C_2 = 10.8$ nF, and $C_3 = 17.3$ nF, all of which can readily be changed to the nearest standard values. Moreover, $R_1 = R_2/(V_O/V_{REF} - 1)$.

The results of a PSpice simulation, shown in Fig. 11.45, reveal the actual values $f_x \cong 18$ kHz and $\phi_m \cong 60°$.

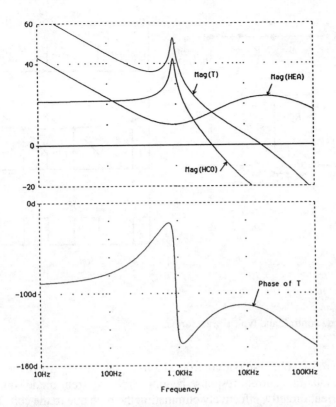

FIGURE 11.45
PSpice frequency plots for the buck regulator of Example
11.21.

We observe that while f_0 is known within the tolerances of L and C, f_z is less predictable because ESR varies with capacitor technology, temperature, and aging. Moreover, Eq. (11.51) reveals that gain depends on V_I, indicating that an increase in V_I will increase f_x, possibly upsetting ϕ_m. An obvious remedy is to make V_{sm} proportional to V_I to ensure a constant ratio between the two. In general, it can be said that the roots of H_{CO} tend to be affected both by the input and by the load, and that they change dramatically as the operating mode changes from CCM to DCM.[12] Clearly, the subject of stability in switching regulators can be a complex one, often requiring cut-and-try techniques for an optimal solution.[13]

Current-Mode Control

Current-mode control,[17] exemplified in Fig. 11.46 for a boost regulator operating in CCM, uses the oscillator only to turn on the switch. A small series resistance R_s senses the coil current, and the switch is turned off when this current reaches a peak controlled by the error-amplifier output v_C. (*Peak current control* would be a more accurate designation.) In spite of its name, this scheme uses current as well as voltage information, as confirmed by the existence of an inner feedback loop due to current sampling by R_s, and an outer feedback loop due to voltage sampling by R_1 and R_2.

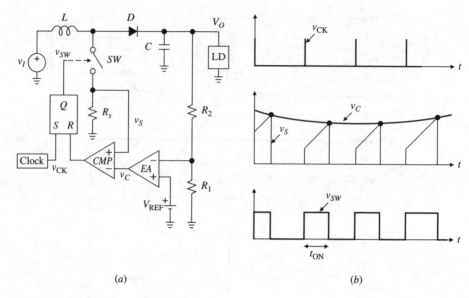

(a)   (b)

FIGURE 11.46
Current-mode control, and typical waveforms.

Compared to voltage-mode control, which emphasizes control of the coil voltage and thus results in a current response lagging by 90°, current-mode control acts on the coil current directly, effectively eliminating the pole due to the coil. The advantages of current-mode control are thus a faster response to line and load variations, along with simplified frequency-compensation requirements (see Problem 11.38); furthermore, overload current protection is inherently provided on a pulse-by-pulse basis.

The LT1070 Monolithic Switching Regulator

Figure 11.47 shows the block diagram of a well-documented[13] monolithic regulator, the LT1070 (Linear Technology). The circuit operates at $f_S = 40$ kHz ($f_S = 100$ kHz in the LT1170 version) and uses current-mode control. The switch is an *npn* BJT with suitable antisaturation circuitry to minimize charge-storage effects and thus reduce switching losses. The switch current is sensed by a 20-mΩ series resistance. The error amplifier is a transconductance amplifier (voltage in, current out) with typical transconductance gain $g_m = 4.4$ mA/V. Its voltage gain is set by an external frequency-compensation network Z_c as $H_{EA} \cong g_m Z_c$. All internal circuitry is powered from an on-chip 2.3-V LDO regulator, which allows the LT1070 to operate over the range 3 V $\leq V_I \leq$ 60 V.

Besides including the various protections discussed in Section 11.4, switching regulators are equipped with provisions to avoid excessive current surges at power turn-on. Referred to as *soft start*, this provision is implemented by limiting the duty cycle D as the regulator builds up its output from zero. In the LT1070, soft start is provided by the capacitor of the external frequency-compensation network.

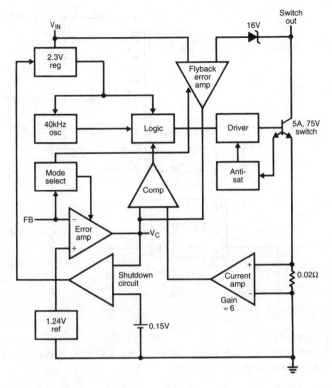

FIGURE 11.47
The LT1070 switching regulator. (Courtesy of Linear Technology.)

Figure 11.48 shows a typical LT1070 application. The output voltage is programmed via R_1 and R_2 as

$$V_O = \left(1 + \frac{R_2}{R_1}\right) V_{REF} \qquad (11.55)$$

where $V_{REF} = 1.24$ V is an internally generated bandgap reference voltage. The

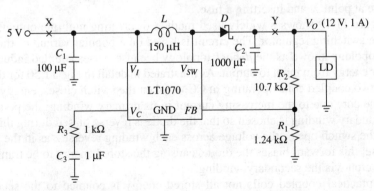

FIGURE 11.48
The LT1070 as a boost regulator. (Courtesy of Linear Technology.)

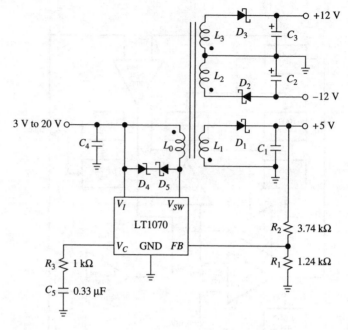

FIGURE 11.49
Triple-output flyback regulator using the LT1070. (Courtesy of
Linear Technology.)

RC network associated with the node labeled V_C is the frequency-compensation
network recommended in the data sheets. The diode is a Schottky type, such as the
1N5822 (Motorola).

The output ripple can be reduced further by breaking the circuit at point Y and
inserting an LC filter consisting of a series 10-μH inductor and a shunt
100-μF capacitor. Unlike the other topologies, the boost regulator is not short-circuit
protected because of the diode connecting the input to the output; this also causes
inrush current at power turn-on. A simple protection can be provided by breaking
the wire at point X and inserting a fuse.

Figure 11.49 shows a widely used method of creating multiple outputs using
just one switching regulator. The circuit is based on a popular variant of the buck-
boost topology known as the *flyback* topology because it uses coupled inductors to
transfer energy from input to output. As illustrated in detail in Fig. 11.50 for the case
of just two coupled coils operating in CCM, when the switch closes, energy builds
up in the core due to the increasing current in its primary winding; the polarity of
the secondary winding is chosen so that the diode is reverse biased during this time.
When the switch opens, the voltage across each winding reverses, as in the single-
coil case; this forward biases the diode, causing the stored energy to be transferred
to the output via the secondary winding.

In practical coupled coils not all stored energy is coupled to the secondary
winding(s); the fraction left in the primary winding leakage inductance LL causes

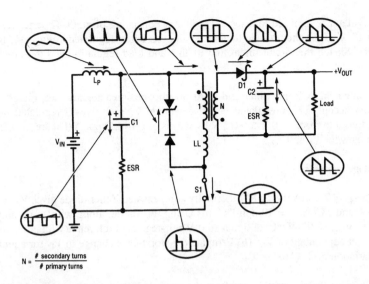

FIGURE 11.50
Typical waveforms for a flyback regulator. (Courtesy of Linear Technology.)

a positive voltage spike across the switch as the latter is opened. To prevent damage to the base-collector junction of the BJT, a voltage clamp is used, consisting of a Zener diode and a rectifier diode, as shown. This clamp provides a current path for the leakage inductance spike, and once the corresponding energy has been fully dissipated in the clamp, the switch voltage settles to its normal flyback value, which is $V_{SW} = V_I + V_O/N$, N being the turns ratio.

Energy transfer can be optimized by suitable choice of the turn ratios of the coils. Moreover, the coupled-coil structure allows for multiple as well as isolated outputs, if desired. In the example of Fig. 11.49, regulation is provided only for the 5-V output, and isolation only for the ±12-V outputs. The ±12-V outputs are scaled to the 5-V output by suitable choice of the turns ratios, and the extent to which they will track the regulated output, also referred to as *cross regulation,* depends on how tight the magnetic coupling of the windings is. If needed, they can be regulated further with the help of individual LDO regulators. Moreover, if isolation is required also for the 5-V output, the feedback signal can be obtained via suitable optocoupler circuitry.[18]

Using the aforementioned SwitcherCAD program, one can find the coil values required for a given set of specifications, such as $V_{O1} = 5$ V at 3 A, and $V_{O2,3} = \pm12$ V at 0.5 A with $V_I = 12$ V.

PROBLEMS

11.1 Performance specifications

11.1 An unregulated voltage $V_I = (26 \pm 2)$ V is applied to a shunt regulator consisting of a series 200-Ω resistor and an 18-V, 20-Ω shunt diode. The output of this regulator is then fed to a second regulator consisting of a 300-Ω series resistor and a 12-V, 10-Ω shunt diode to achieve an even better regulated voltage V_O for a load R_L. Sketch the circuit; then, find its line and load regulation and the minimum R_L allowed.

11.2 Using a 6.2-V Zener diode and a 741 op amp, design a negative self-biased reference that accepts an unregulated negative voltage V_I and gives a regulated output V_O adjustable from -10 V to -15 V by means of a 10-kΩ pot. What are the permissible ranges for V_I and I_O?

11.2 Voltage references

11.3 At $I_Z = 7.5$ mA the 1N827 thermally compensated Zener diode gives $V_Z = 6.2$ V $\pm$ 5% and TC(V_Z) = 10 ppm/$^\circ$C. (*a*) Using this diode, along with an op amp having TC(V_{OS}) = 5 μV/$^\circ$C, design a 10.0-V self-regulated reference with provision for the exact adjustment of V_O. (*b*) Estimate the worst-case change in V_O for a temperature variation of 0 $^\circ$C to 70 $^\circ$C.

11.4 Consider the circuit obtained from the self-regulated reference of Fig. 11.4 by lifting the left terminals of R_1 and D_z off ground, connecting them together, and then returning the resulting common node to ground via a variable resistance R. (*a*) Show that this modification allows us to vary V_O without altering the diode current. (*b*) Obtain a relationship between V_O and V_Z. (*c*) Specify standard components for a variable reference from 10 V to 20 V using the 1N827 diode of Problem 11.3 as the reference element.

11.5 (*a*) Assuming matched BJTs in the alternative bandgap cell[2] of Fig. P11.5, show that $V_{REF} = V_{BE1} + KV_T$, $K = (R_2/R_3)\ln(R_2/R_1)$. (*b*) Assuming I_s (25 $^\circ$C) = 5×10^{-15} A for both BJTs, specify suitable components for TC(V_{REF}) = 0 at 25 $^\circ$C.

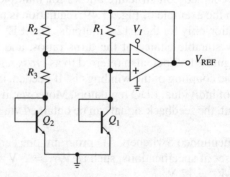

FIGURE P11.5

11.6 The alternative bandgap reference[2] of Fig. P11.6 is known as the *Widlar bandgap cell* for its inventor. (*a*) Assuming matched BJTs with negligible base currents, show that $V_{REF} = V_{BE3} + KV_T$, $K = (R_2/R_3)\ln(I_{C1}/I_{C2})$. (*b*) Specify suitable components for TC(V_{REF}) = 0 at 25 $^\circ$C if I_s (25 $^\circ$C) = 2×10^{-15} A for all BJTs, $I_{C1} = I_{C3} = 0.2$ mA, and $I_{C2} = I_{C1}/5$.

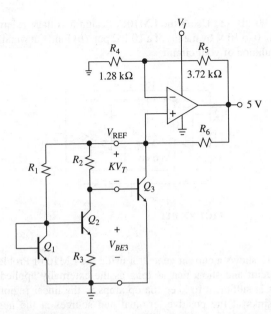

FIGURE P11.6

11.3 Voltage-reference applications

11.7 Using the REF101 10-V reference of Fig. 11.6a and an external op amp, but no additional components, design a circuit that gives (a) +10 V and −10 V, and is powered from ±15-V supplies; (b) +10 V and +5 V, and is powered from a single 15-V supply; (c) +5 V and −5 V, and is powered from ±9-V supplies; (d) +10 V and +20 V, and is powered from a single 24-V supply.

11.8 The LT1029 is a 5-V reference diode that operates with any current between 0.6 mA and 10 mA, and has a maximum TC of 20 ppm/°C. Using the LT1029 and a JFET-input op amp with TC(V_{OS}) = 6 μV/°C, design ±2.5-V split references and estimate their worst-case thermal drifts. Assume ±5-V power supplies.

11.9 (a) Using the REF101 10-V reference of Fig. 11.6a and an external JFET-input op amp, but no additional components, design a 1-mA current source. (b) Assuming ±15-V supplies and TC(V_{OS}) = 1 μV/°C, use the data of Fig. 11.7 to estimate the voltage compliance and the worst-case TC of your source. (c) Find the range of variability of the source if the optional voltage trim connection of Fig. 11.6a is used.

11.10 Assuming ±15-V supplies and using an LM385 2.5-V diode with a bias current of 100 μA as a voltage reference, design a current generator whose output is variable over the range −1 mA ≤ I_O ≤ 1 mA by means of a 10-kΩ pot.

11.11 The LM10 (National Semiconductor) consists of two op amps and a 200-mV reference internally connected as in Fig. P11.11. The op amps have rail-to-rail output swing capability, and the device draws a maximum quiescent current of 0.5 mA from a supply voltage anywhere between 1.1 V and 40 V. The LM10C version has TC(V_{REF}) = 0.003 %/°C, TC(V_{OS}) = 5 μV/°C, line regulation = 0.0001 %/V, and CMRR$_{dB}$ ≅

$PSRR_{dB} \cong 90$ dB. (*a*) Using the LM10C, design a voltage reference continuously variable from 0 to 10 V by means of a 10-kΩ pot. (*b*) Find the worst-case thermal drift and line regulation of your circuit.

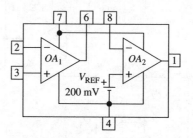

FIGURE P11.11

11.12 Figure P11.12 shows a current generator using the LM10 of Problem 11.11. (*a*) Analyze the circuit and show that as long as the externally applied voltage between its terminals is sufficient to keep the op amps in the linear region, the current that the circuit sinks at the positive terminal and sources at the negative terminal is $I_O = (1 + R_2/R_3)V_{REF}/R_1$. (*b*) Specify suitable components for $I_O = 5$ mA. (*c*) What is the range of external voltages over which your circuit will operate properly?

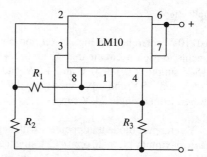

FIGURE P11.12

11.13 Design a circuit that senses the temperatures T_1 and T_2 at two different sites, and yields $V_O = (0.1 \text{ V})(T_2 - T_1)$, T_1 and T_2 in degrees Celsius. The circuit uses two matched diodes with I_s (25 °C) = 2 fA as temperature sensors, and two potentiometers for its calibration. Describe the calibration procedure.

11.14 Specify suitable components in the circuit of Fig. 11.17 for a Fahrenheit sensor with a sensitivity of 10 mV/°F. Outline its calibration.

11.4 Linear regulators

11.15 Obtain expressions for R_{fb} and R_3/R_4 in terms of R_{sc}, I_{sc}, and I_{fb} in the current fold-back scheme of Fig. 11.21*b*, assuming I_{B3} is negligible.

11.16 Using a 741 op amp, an LM385 2.5-V reference diode, and *pnp* BJTs, design an overload-protected negative regulator with $V_O = -12$ V and $I_{O(max)} = 100$ mA.

11.17 Using the LM10 of Problem 11.11 and two *npn* BJTs, design a 100-mA overload-protected voltage regulator whose output can be varied from 0 to 15 V by means of a 10-kΩ pot. Show how you power your circuit, and estimate the lowest permissible supply voltage.

11.18 Figure P11.18 shows a high-voltage regulator based on the LM10 of Problem 11.11. Since the LM10 is powered by three V_{BE} drops, the high-voltage capabilities of the circuit are limited only by the external components. (*a*) Analyze the circuit and find V_O in terms of V_{REF}. (*b*) Specify R_1 and R_2 for $V_O = 100$ V. (*c*) Assuming typical BJT parametrs, estimate V_{DO} for $I_O = 1$ A.

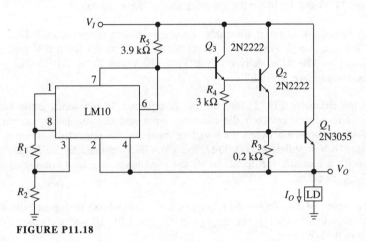

FIGURE P11.18

11.5 Linear-regulator applications

11.19 The LM338 is a 1.2-V, 5-A, adjustable regulator having $V_{DO} = 2.5$ V, a maximum input-output differential voltage of 35 V, and an adjustment-pin current of 45 μA. Using the LM338, design a 5-A regulator whose output can be varied from 0 V to 5 V via a 10-kΩ pot. What are the power-supply requirements of your circuit?

11.20 Using the LT1029 reference diode of Problem 11.8 and the LM338 voltage regulator of Problem 11.19, design a minimum-component circuit for the simultaneous generation of a 5-V reference voltage and a 15-V, 5-A supply voltage. What is the permissible unregulated input voltage range of your circuit?

11.21 Using a μA7805 5-V regulator and 0.25-W (or less) resistances, design a 1-A current source. What is its voltage compliance as a function of the supply voltage?

11.22 In Fig. 11.29*a* let the common terminal of the regulator be connected directly to the node shared by R_1 and R_2 to save the op amp. Assuming a μA7805 5-V regulator, whose specifications are given in Fig. 11.24, find suitable resistances for $V_O = 12$ V; then find the permissible range for V_{CC}, as well as the load and line regulation.

11.23 In the circuit of Fig. 11.26*b* let $V_I = 25$ V and $R_1 = 2.5$ Ω, and let R_2 be an arbitrary load. Find the Norton equivalent of the circuit seen by the load, along with its voltage compliance, given the following LM317 specifications: $V_{DO} \cong 2$ V, line regulation = 0.07%/V maximum, and $\Delta I_{ADJ} = 5$ μA maximum for 2.5 V $\leq (V_I - V_O) \leq 40$ V.

11.24 The LT337 is a -1.25-V, 1.5-A, adjustable negative regulator with $\Delta V_{REG}/\Delta(V_I - V_O) = 0.03\%$/V maximum, and $\Delta I_{ADJ}/\Delta(V_I - V_O) = 0.135$ μA/V maximum. Using this device, design a 500-mA current sink; next, find its Norton equivalent.

11.25 Using an LM317 1.25-V, adjustable positive regulator, and an LM337 -1.25-V, adjustable negative regulator, design a dual-tracking bench power supply whose outputs are adjustable from ± 1.25 V to ± 20 V by means of a single 10-kΩ pot. See Problems 11.23 and 11.24 for the specifications of these regulators.

11.26 (a) Find the maximum allowable operating ambient temperature if $T_{J(max)} = 190$ °C, $P_{D(max)} = 1$ W, and $\theta_{JA} = 60$ °C/W. (b) Find θ_{JA} for a 5-V regulator with $T_{J(max)} = 150$ °C to deliver 1 A at $V_I = 10$ V and $T_A = 50$ °C. Can a μA7805 operating in free air do it?

11.27 In the circuit of Fig. 11.29b the pot is replaced by the series combination of a 2-kΩ resistance between the inverting input and the regulator's output, and an 18-kΩ resistance between the inverting input and the regulator's common. Assuming ± 18-V supplies, $R = 1.00$ Ω, and a μA7805 regulator in the TO-220 package, specify a heatsink for operation all the way down to a load voltage of 0 V with $T_{A(max)} = 60$ °C.

11.28 Using the LM10 of Problem 11.11 and a 1.5-V, 2-mA LED, design an indicator circuit that monitors its own power supply and turns the LED off whenever the supply drops below 4.75 V.

11.29 Specify components in the circuit of Fig. 11.35a to provide OV protection when V_{CC} tries to rise above 6.5 V, and issue a $\overline{PFAIL}$ command when the 120-V (rms), 60-Hz ac line tries to drop below 80% of its nominal value.

11.6 Switching regulators

11.30 The switched coil of Fig. 11.37c bears some similarity to the switched capacitor of Fig. 4.23a. (a) Assuming $V_{SAT} = V_F = 0$, compare the two arrangements and point out similarities as well as fundamental differences. (b) Assuming the coil current waveform of Fig. 11.39, show that the power transferred by the coil from V_I to V_O is $P = f_S W_{cycle}$, where $W_{cycle} = L I_L \Delta i_L$ is the energy packet transferred during each cycle.

11.31 (a) Derive Eq. (11.42). Then, assuming $I_O = 1$ A and $\Delta i_L = 0.2$ A, estimate I_p as well as the minimum value of I_O for continuous operation for the case of (b) a buck regulator with $V_I = 12$ V and $V_O = 5$ V, (c) a boost regulator with $V_I = 5$ V and $V_O = 12$ V, and (d) an inverting regulator with $V_I = 5$ V and $V_O = -15$ V.

11.32 An inverting regulator with 5 V $\leq V_I \leq 10$ V is to deliver $V_O = -12$ V at a full load of 1 A. Assuming continuous operation with $V_{SAT} = V_F = 0.5$ V, find the required range for D, as well as the maximum value of I_I.

11.33 A buck-boost regulator is powered from $+15$ V and operates at 150 kHz. Specify L, C, and ESR for $V_O = -15$ V, $V_{ro(max)} = 150$ mV, and continuous-mode operation over the range 0.2 A $\leq I_O \leq 1$ A.

11.34 A buck regulator has $V_I = 20$ V, $V_O = 5$ V, $f_S = 100$ kHz, $L = 50$ μH, and $C = 500$ μF. Assuming $V_{SAT} = V_F = 0$ and ESR $= 0$ sketch and label i_{SW}, i_D, i_L, i_C, and the voltage v_X at the left terminal of L for the case of (a) continuous-mode operation with $I_O = 3$ A, and (b) discontinuous-mode operation with $t_{ON} = 2$ μs.

11.35 Discuss how η in the regulator of Example 11.20 is affected by (a) doubling V_I, and (b) doubling f_S.

11.7 Monolithic switching regulators

11.36 Find the control-to-output transfer function of the circuit of Fig. 11.43; next, verify that if R_{coil} and ESR are much smaller than the load R_L and the gain-setting network R_1 and R_2, then Eqs. (11.51) and (11.52) result. *Hint:* Considering that $v_O/v_I = D$ and $D = v_C/V_{sm}$, the gain of the Mod block is obtained by differentiating v_O with respect to v_C and letting $v_I = V_I$.

11.37 In the error amplifier of Fig. 11.44a, R_1 and R_2 set the value of the feedback factor; however, for small-signal analysis purposes, V_{REF} is set to zero, and R_1 has thus no effect. Assuming ideal op amp, obtain expressions for ω_1 through ω_5; then verify that for $R_3 \ll R_2$ and $C_2 \gg C_1$ they simplify as in Eq. (11.54).

11.38 Assuming $g_m \cong 4.4$ mA/V, $R_o \cong 180$ kΩ, and $C_o \cong 3$ pF in the transconductance-type error amplifier of the LT1070, find the voltage gain $H_{EA}(jf)$ when the amplifier is terminated on the frequency-compensation network shown in Fig. 11.48. Next, sketch its Bode plots, and comment on your results.

REFERENCES

1. Analog Devices Engineering Staff, *Practical Design Techniques for Power and Thermal Management,* Analog Devices, Norwood, MA, 1998.
2. P. R. Gray and R. G. Meyer, *Analysis and Design of Analog Integrated Circuits,* 3d ed., John Wiley & Sons, New York, 1993.
3. R. Knapp, "Selection Criteria Assist in Choice of Optimum Reference," *EDN,* February 18, 1988, pp. 183–192.
4. "IC Voltage Reference Has 1 ppm per Degree Drift," Application Note AN-161, *Linear Applications Handbook,* National Semiconductor, Santa Clara, CA, 1994.
5. A. P. Brokaw, "A Simple Three-Terminal IC Bandgap Reference," *IEEE Journal of Solid-State Circuits,* vol. SC-9, no. 6, December 1974, pp. 388-393.
6. Analog Devices Engineering Staff, *Practical Design Techniques for Sensor Signal Conditioning,* Analog Devices, Norwood, MA, 1999.
7. B. Huffman, "Voltage Reference Circuit Collection," Application Note AN-42, *Linear Applications Handbook Volume II,* Linear Technology, Milpitas, CA, 1993.
8. J. Graeme, "Op Amps Expand Voltage Reference Options," *Electronic Design,* April 16, 1992, pp. 75–90.
9. J. Graeme, "Precision DC Current Sources," Part 1 and Part 2, *Electronic Design,* April 26, 1990, pp. 191–198 and 201–206.
10. A. B. Grebene, *Bipolar and MOS Analog Integrated Circuit Design,* John Wiley & Sons, New York, 1984.
11. "Applications for an Adjustable IC Power Regulator," Application Note AN-178, *Linear Applications Handbook,* National Semiconductor, Santa Clara, CA, 1994.

12. R. W. Erickson, *Fundamentals of Power Electronics,* Chapman & Hall, International Thompson Publishing, New York, 1997.

13. C. Nelson, "LT1070 Design Manual," Application Note AN-19, *Linear Applications Handbook Volume I,* Linear Technology, Milpitas, CA, 1990.

14. C. Nelson, "LT1074/LT1076 Design Manual," Application Note AN-44, *Linear Applications Handbook Volume II,* Linear Technology, Milpitas, CA, 1993.

15. B. Huffman and C. Nelson, *SwitcherCAD User's Manual,* Linear Technology, Milpitas, CA, 1992.

16. *National Power IC's Databook,* National Semiconductor, Santa Clara, CA, 1995.

17. R. Mammano, "The Pros and Cons of Voltage-Mode and Current-Mode Controllers," *Electronic Design Analog Applications Issue,* June 27, 1994, pp. 53–54.

18. G. C. Chryssis, *High-Frequency Switching Power Supplies: Theory and Design,* 2d ed., McGraw-Hill, New York, 1989.

12

D-A AND A-D CONVERTERS

12.1 Performance Specifications
12.2 D-A Conversion Techniques
12.3 Multiplying DAC Applications
12.4 A-D Conversion Techniques
12.5 Oversampling Converters
  Problems
  References

In their natural state, information-carrying variables—such as voltage, current, charge, temperature, and pressure—are in analog form. However, for processing, transmission, and storage purposes, it is often more convenient to represent information in digital form. Consider, for instance, an op amp circuit that is required to put out a signal v in the range of 0 V to 1 V with an accuracy of 1 mV, or 0.1%. Given the effects of component nonidealities, drift, aging, noise, and imperfect wires and interconnections, even an accuracy requirement this moderate may be difficult to meet.

The demands on circuit performance can be relaxed significantly if information is represented in digital form. For instance, in decimal form, which is the most familiar form to humans, the above signal would be expressed as $v = 0.d_1 d_2 \ldots d_n$, where $d_1, d_2, \ldots, d_n$ are decimal digits between 0 and 9. For a 1-mV resolution over the range $0.000 \text{ V} \leq v < 0.999 \text{ V}$, three such digits are needed. This, in turn, requires three separate circuits to hold the individual digit values; however, the performance requirements are now much more relaxed because each digit-circuit needs to resolve only 10 voltage levels instead of 1000. Individual accuracies of $\pm 5\%$ are sufficient for this task.

Expressing signals digitally, while easing one problem, creates another, namely, the need to convert from analog to digital (A-D) and from digital to analog (D-A).

For instance, a decimal D-A converter for our example would have to determine the values of d_1, d_2, and d_3 as provided by the corresponding circuits (an easy task), and then synthesize the analog signal $v = d_1 10^{-1} + d_2 10^{-2} + d_3 10^{-3}$ with a 1-mV accuracy (an inherently difficult task).

Though convenient for humans, decimal representation does not relax circuit performance requirements to the maximum extent. Rather, this is done by allowing digits to take on just two values, namely, 0 and 1. If we represent these values with sufficiently different voltages, such as 0 V and 5 V, then even the crudest circuit will be able to resolve them. Binary digits, or *bits,* form the basis of digital systems precisely because of this. Bits are held and manipulated by binary circuits such as switches, logic gates, and flip-flops.

Figure 12.1 depicts the most general context[1] within which A-D and D-A conversion is used. An analog input signal, after suitable conditioning, is A-D converted to be processed or perhaps just transmitted or recorded in digital form by the digital signal processor (DSP) block. Once processed, received, or retrieved, it is D-A converted to be reused in analog form, possibly after additional output conditioning.

The A-D converter (ADC) is operated at a rate of f_S samples per second. To avoid any aliasing phenomena,[1,2] the analog input must be band-limited so that its highest frequency component is less than $f_S/2$; antialiasing filters were addressed in Chapter 4. ADCs usually require that the input be held constant during the conversion process, indicating that the ADC must be preceded by an SHA to freeze the band-limited signal just prior to each conversion; SHAs were addressed in Chapter 9. The D-A converter (DAC) is usually operated at the same rate f_S as the ADC and, if the application demands so, it is equipped with appropriate circuitry to remove any output glitches arising in connection with input code changes. The resulting staircase-like signal is finally passed through a smoothing filter to ease the effects of quantization noise.

The scheme of Fig. 12.1 is found, either in full or in part, in countless applications. Digital signal processing (DSP), direct digital control (DDC), digital audio mixing, recording and playback, pulse-code modulation (PCM) communication, data acquisition, computer music and video synthesis, and digital-multimeter instrumentation are only some examples.

This chapter, after introducing converter terminology and performance parameters, discusses the most common D-A and A-D conversion techniques and applications, including $\Sigma\text{-}\Delta$ converters.

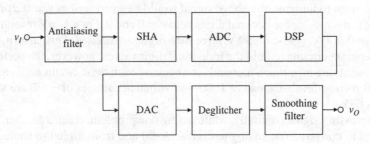

FIGURE 12.1
Sampled-data system.

12.1
PERFORMANCE SPECIFICATIONS

561

SECTION 12.1
Performance
Specifications

A string of n bits, $b_1 b_2 b_3 \ldots b_n$, forms an n-bit word. Bit b_1 is called the *most significant bit* (MSB) and bit b_n the *least significant bit* (LSB). The quantity

$$D = b_1 2^{-1} + b_2 2^{-2} + b_3 2^{-3} + \cdots + b_n 2^{-n} \tag{12.1}$$

is called the *fractional binary value*. Depending on the bit pattern, D can assume 2^n equally spaced values from 0 to $1 - 2^{-n}$. The lower limit is reached when all bits are 0, the upper limit when all bits are 1, and the spacing between adjacent values is 2^{-n}.

D-A Converters (DACs)

A DAC accepts an n-bit input word $b_1 b_2 \ldots b_n$ with fractional binary value D_I, and produces an analog output proportional to D_I. Figure 12.2a depicts a voltage-output DAC, for which we have

$$v_O = K V_{\text{REF}} D_I = V_{\text{FSR}}(b_1 2^{-1} + b_2 2^{-2} + \cdots + b_n 2^{-n}) \tag{12.2}$$

where K is a *scale factor*; V_{REF} is a *reference voltage*; b_k ($k = 1, 2, \ldots, n$) is either 0 or 1, depending on the logic level at the corresponding input; $V_{\text{FSR}} = K V_{\text{REF}}$ is the *full-scale range*. Frequently used values for V_{FSR} are 2.5 V, 5.0 V, and 10.0 V. Though our discussion will focus on voltage-output DACs, the results are readily extended to current-output DACs, characterized by $i_O = K I_{\text{REF}} D_I = I_{\text{FSR}} D_I$. A typical I_{FSR} value is 1.0 mA.

We observe that the DAC output is the result of multiplying the analog signal V_{REF} by the digital variable D_I. A DAC that allows for V_{REF} to vary all the way down to zero is called a *multiplying DAC* (MDAC).

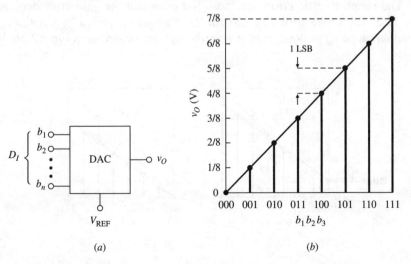

(a)

(b)

FIGURE 12.2
DAC diagram, and ideal transfer characteristic for $n = 3$ and $V_{\text{FSR}} = 1$ V.

Depending on the input bit pattern, v_O can assume 2^n different values ranging from 0 to the *full-scale value* $V_{FSV} = (1 - 2^{-n})V_{FSR}$. The MSB contribution to v_O is $V_{FSR}/2$, and the LSB contribution is $V_{FSR}/2^n$. The latter is called the *resolution,* or simply the *LSB.* Note that V_{FSV} is always 1 LSB short of V_{FSR}. The quantity $DR = 20 \log_{10} 2^n$ is called the *dynamic range* of the DAC. Thus, a 12-bit DAC with $V_{FSR} = 10.000$ V has LSB $= 2.44$ mV, $V_{FSV} = 9.9976$ V, and DR $= 72.25$ dB.

Since there are only 2^n possible input codes, the transfer characteristic of a DAC is a set of points whose envelope is a straight line with end points at $(b_1 b_2 \ldots b_n, v_O) = (00 \ldots 0, 0 \text{ V})$ and $(11 \ldots 1, V_{FSV})$. Figure 12.2b shows the characteristic of a DAC with $n = 3$ and $V_{FSR} = 1.0$ V. The graph consists of $2^3 = 8$ bars ranging in height from 0 to $V_{FSV} = \frac{7}{8}$ V with a resolution of 1 LSB $= \frac{1}{8}$ V. If we drive a DAC with a uniformly clocked n-bit binary counter and observe v_O with the oscilloscope, the waveform will be a staircase. The higher n, the finer the resolution and the closer the staircase to a continuous ramp. DACs are available in word lengths ranging from 6 bits to 20 bits or more. While DACs with 6, 8, 10, 12, and 14 bits are common and economical, DACs with $n > 14$ become progressively more expensive and require the utmost care to realize their full precision.

DAC Specifications[3]

The internal circuitry of a DAC is subject to component mismatches, drift, aging, noise, and other sources of error, whose effect is to degrade conversion performance. The maximum deviation of the actual output from the ideal value predicted by Eq. (12.2) is called the *absolute accuracy* and is expressed in fractions of 1 LSB. Clearly, if an n-bit DAC is to retain its credibility down to its LSB, its absolute accuracy must never be worse than $\frac{1}{2}$ LSB. DAC errors are classified as *static* and *dynamic*.

The simplest static errors are the *offset* error and the *gain* error depicted in Fig. 12.3. The offset error ($+1$ LSB in the example) is nulled by translating the actual envelope up or down until it goes through the origin, as in Fig. 12.3b. What

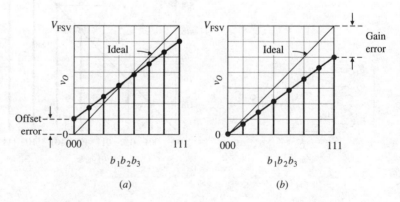

FIGURE 12.3
DAC offset error and gain error.

is left, then, is the gain error (-2 LSB in the example), which is nulled by adjusting the scale factor K.

Even after both errors have been nulled, the actual envelope is likely to deviate from the straight line passing through the end points. The maximum deviation is called the *integral nonlinearity* (INL), or also the *relative accuracy*, and is expressed in fractions of 1 LSB. Ideally, the difference in height between adjacent bars is 1 LSB; the maximum deviation from this ideal value is called the *differential nonlinearity* (DNL). If DNL < -1 LSB, the transfer characteristic becomes *nonmonotonic;* that is, for certain input code transitions v_O will decrease with the input code, rather than increase. A nonmonotonic characteristic is especially undesirable in control, where it may cause oscillations, and in successive-approximation ADCs, where it may lead to missing codes. An example will better clarify these concepts.

EXAMPLE 12.1. Find the INL and DNL of the 3-bit DAC of Fig. 12.4. Comment on your results.

Solution. By inspection, the individual-code integral and differential nonlinearities, in fractions of 1 LSB, are found to be

k:	000	001	010	011	100	101	110	111
INL_k:	0	0	$-1/2$	$1/2$	-1	$1/2$	$-1/2$	0
DNL_k:	0	0	$-1/2$	1	$-3/2$	$3/2$	-1	$1/2$

The maxima of INL_k and DNL_k are, respectively, INL $= 1$ LSB and DNL $= 1\frac{1}{2}$ LSB. We observe a nonmonotonicity as the code changes from 011 to 100, where the step size is $-\frac{1}{2}$ LSB instead of $+1$ LSB; hence, $DNL_{100} = -\frac{1}{2} - (+1) = -\frac{3}{2}$ LSB < -1 LSB. The fact that $DNL_{101} = \frac{3}{2}$ LSB > 1 LSB, though undesirable, does not cause nonmonotonicity.

Remark. Note that $INL_k = \sum_{i=0}^{k} DNL_i$. Can you provide an intuitive justification?

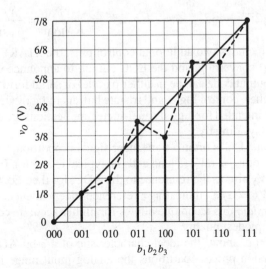

FIGURE 12.4
Example of actual DAC characteristic after the offset and gain errors have been nulled.

DAC performance changes with temperature, age, and power-supply variations; hence, all relevant performance parameters such as offset, gain, INL and DNL, and monotonicity must be specified over the full temperature and power-supply ranges.

The most important dynamic parameter is the *settling time* t_S. This is the time it takes for the output to settle within a specified band (usually $\pm\frac{1}{2}$ LSB) of its final value following a code change at the input (usually a full-scale change). Typically, t_S ranges from under 10 ns to over 10 μs, depending on word length as well as circuit architecture and technology.

Another potential source of concern is the presence of output spikes in connection with major input-code transitions. Called *glitches,* these spikes are due to the internal circuitry's nonuniform response to input bit changes as well as poor synchronization of the bit changes themselves. For instance, if during the center-scale transition from $01\ldots1$ to $10\ldots0$ the MSB is perceived as going on before (or after) all other bits go off, the output will momentarily swing to full scale (or to zero), causing a positive-going (or negative-going) output spike, or glitch.

Glitches are of particular concern in CRT display applications. They can be minimized by synchronizing the input bit changes with a high-speed parallel latch register, or by processing the DAC output with a THA. The THA is switched to the hold mode just prior to the input code change, and is returned to the track mode only after the DAC has recovered from the glitch and settled to its new level.

A-D Converters (ADCs)

An ADC provides the inverse function of a DAC. As shown in Fig. 12.5a, it accepts an analog input v_I and produces an output word $b_1 b_2 \ldots b_n$ of fractional value D_O such that

$$D_O = b_1 2^{-1} + b_2 2^{-2} + \cdots + b_n 2^{-n} = \frac{v_I}{K V_{\text{REF}}} = \frac{v_I}{V_{\text{FSR}}} \qquad (12.3)$$

Usually, an ADC includes two additional control pins: the START input, to tell the ADC when to start converting, and the EOC output, to announce when conversion is complete. The output code can be in either parallel or serial form. ADCs are often equipped with latches, control logic, and tristate buffers to facilitate microprocessor interfacing. ADCs intended for digital panel-meter applications are designed to drive LCD or LED displays directly.

The input to an ADC is often a transducer signal proportional to the transducer supply voltage V_S, or $v_I = \alpha V_S$ (a load cell is a typical example). In these cases it is convenient to use V_S also as the reference to the ADC, for then Eq. (12.3) simplifies as $D_O = \alpha V_S / K V_S = \alpha/K$, indicating a reference-independent conversion. Called *ratiometric conversion,* this technique allows for highly accurate conversions using references of only modest quality.

Figure 12.5b, top, shows the ideal characteristic of a 3-bit ADC with $V_{\text{FSR}} = 1.0$ V. The conversion process partitions the analog input range into 2^n intervals called *code ranges,* and all values of v_I within a given code range are represented by the same code, namely, that corresponding to the midrange value. For example, code 011, corresponding to the midrange value $v_I = \frac{3}{8}$ V, actually represents all

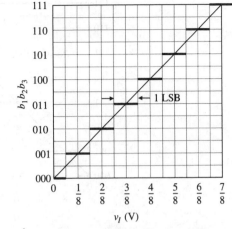

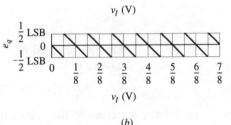

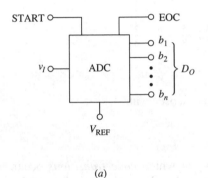

(a)   (b)

FIGURE 12.5
ADC diagram, and ideal transfer characteristic and quantization noise for $n = 3$ and $V_{FSR} = 1$ V.

inputs within the range $\frac{3}{8} \pm \frac{1}{16}$ V. Due to the inability by the ADC to distinguish among different values within this range, the output code can be in error by as much as $\pm\frac{1}{2}$ LSB. This uncertainty, called *quantization error*, or also *quantization noise* e_q, is an inherent limitation of any digitization process. An obvious way to improve it is by increasing n.

As shown in Fig. 12.5b, bottom, e_q is a sawtooth-like variable with a peak value of $\frac{1}{2}$ LSB $= V_{FSR}/2^{n+1}$. Its rms value is readily found to be $E_q = (\frac{1}{2}$ LSB$)/\sqrt{3}$ or

$$E_q = \frac{V_{FSR}}{2^n \sqrt{12}} \tag{12.4}$$

If v_I is a sinusoidal signal, the signal-to-noise ratio is maximized when v_I has a peak amplitude of $V_{FSR}/2$, or an rms value of $(V_{FSR}/2)/\sqrt{2}$. Thus, SNR$_{max}$ = $20 \log_{10}[(V_{FSR}/2\sqrt{2})/(V_{FSR}/2^n\sqrt{12})]$, or

$$\text{SNR}_{max} = 6.02n + 1.76 \text{ dB} \tag{12.5}$$

Increasing n by 1 cuts E_q in half and increases SNR$_{max}$ by 6.02 dB.

ADC Specifications[3]

Similar to the case of DACs, ADC performance is characterized in terms of *offset* and *gain errors*, *differential* and *integral nonlinearity*, and *stability*. However, ADC

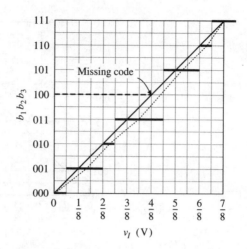

FIGURE 12.6
Example of actual ADC characteristic with
missing code.

errors are defined in terms of the values of v_I at which *code transitions* occur. Ideally, these transitions occur at odd multiples of $\frac{1}{2}$ LSB, as shown in Fig. 12.5*b*. In particular, the first transition (000 → 001) occurs at $v_I = \frac{1}{2}$ LSB $= \frac{1}{16}$ V, and the last (110 → 111) at $v_I = V_{\text{FSV}} - \frac{1}{2}$ LSB $= V_{\text{FSR}} - \frac{3}{2}$ LSB $= \frac{13}{16}$ V.

The *offset error* is the difference between the actual location of the first code transition and $\frac{1}{2}$ LSB, and the *gain error* is the difference between the actual locations of the last and first transition, and the ideal separation of $V_{\text{FSR}} - 2$ LSB. Even after both errors have been nulled, the locations of the remaining code transitions are likely to deviate from their ideal values, as exemplified in Fig. 12.6.

The dotted curve, representing the locus of the midpoints of the actual code ranges, is called the *code center line*. Its maximum deviation from the straight line passing through the end points after the offset and gain errors have been nulled is called the *integral nonlinearity* (INL). Ideally, code transitions are 1 LSB apart. The maximum deviation from this ideal value is called the *differential nonlinearity* (DNL). If the DNL exceeds 1 LSB, some codes may be skipped at the output. Missing codes are undesirable in digital control, where they may lead to instability.

In the example shown, the INL error is maximized in connection with the 011 code range, where this error is $-\frac{1}{2}$ LSB. This range also maximizes the DNL error. The range width of 2 LSB indicates that DNL $= (2 - 1)$ LSB $= 1$ LSB. Not suprisingly, there is a missing code. As you investigate INL and DNL errors, make sure you measure them along the horizontal (or the vertical) axis, not as geometric distances! As a check, you can use the relationship $\text{INL}_k = \sum_{i=0}^{k} \text{DNL}_i$, which holds also for ADCs.

An A-D conversion takes a certain amount of time to complete. Called the *conversion time,* it typically ranges from less than 10 ns to tens of milliseconds, depending on the conversion method, resolution, and technology.

A practical ADC will produce noise in excess of the theoretical quantization noise of Eq. (12.4). It will also introduce distortion due to transfer-characteristic nonlinearities. The effective number of bits is then[4]

$$\text{ENOB} = \frac{S/(N+D) - 1.76 \text{ dB}}{6.02} \qquad (12.6)$$

where $S/(N+D)$ is the actual signal-to-noise-plus-distortion ratio, in decibels.

EXAMPLE 12.2. A 10-bit ADC with $V_{\text{FSR}} = 10.24$ V is found to have $S/(N+D) = 56$ dB. Find E_q, SNR_{max}, and ENOB.

Solution. Using Eqs. (12.4) through (12.6) gives $E_q = 2.89$ mV, $\text{SNR}_{\text{max}} = 61.97$ dB, and ENOB = 9.01, indicating nine effective bits. In other words, the given 10-bit ADC yields the same performance as an ideal 9-bit ADC.

12.2
D-A CONVERSION TECHNIQUES

DACs are available in a variety of architectures and technologies.[2-4] In this section we examine the most common examples.

Weighted-Resistor DACs

Equation (12.2) indicates that the functions required to implement an n-bit DAC are n switches and n binary-weighted variables to synthesize the terms $b_k 2^{-k}$, $k = 1, 2, \ldots, n$; moreover, we need an n-input summer, and a reference. The DAC of Fig. 12.7 uses an op amp to sum n binary-weighted currents derived from V_{REF} via the current-scaling resistances $2R, 4R, 8R, \ldots, 2^n R$. Whether the current $i_k = V_{\text{REF}}/2^k R$ appears in the sum depends on whether the corresponding switch is closed ($b_k = 1$) or open ($b_k = 0$). Writing $v_O = -R_f i_O$ gives

$$v_O = (-R_f/R)V_{\text{REF}}(b_1 2^{-1} + b_2 2^{-2} + \cdots + b_n 2^{-n}) \qquad (12.7)$$

indicating that $K = -R_f/R$. The offset error is nulled by trimming V_{OS}, and the gain error by adjusting R_f. Since the switches are of the virtual-ground type, they can be implemented with p-channel JFETs in the manner of Fig. 9.37.

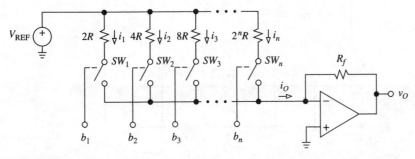

FIGURE 12.7
Weighted-resistor DAC.

The conceptual simplicity of the weighted-resistor DAC is offset by two draw-backs, namely, the nonzero resistances of the switches, and a spread in the current-setting resistances that increases exponentially with n. The effect of switch resistances is to disrupt the binary-weighted relationships of the currents, particularly in the most significant bit positions, where the current-setting resistances are smaller. These resistances can be made sufficiently large to swamp the switch resistances; however, this may result in unrealistically large resistances in the least significant positions. For instance, an 8-bit DAC requires resistances ranging from $2R$ to $256R$. The difficulty in ensuring accurate ratios over a range this wide, especially in mono-lithic form, restricts the practicality of resistor-weighted DACs below 6 bits.

Weighted-Capacitor DACs

Complex MOS ICs such as CODECS and microcomputers require on-chip data conversion capabilities using only MOSFETs and capacitors, which are the natural components of this technology. The DAC of Fig. 12.8 can be viewed as the switched-capacitor counterpart of the weighted-resistor DAC just discussed. Its heart is an array of binary-weighted capacitances plus a terminating capacitance equal in value to the LSB capacitance. Circuit operation alternates between two cycles called the *reset* and *sample* cycles.

During the *reset cycle,* shown in the figure, all switches are connected to ground to completely discharge all capacitors. During the *sample cycle,* SW_0 is opened while each of the remaining switches is either left at ground or connected to V_{REF}, depending on whether the corresponding input bit is 0 or 1, respectively. This results in a redistribution of charge whose effect is to yield a code-dependent output.

Using elementary capacitor-divider principles, we readily find $v_O = V_{REF}\, C_r/C_t$, where C_r represents the sum of all capacitances connected to V_{REF}, and C_t the total capacitance of the array. We can write $C_r = b_1 C + b_2 C/2 + \cdots + b_n C/2^{n-1}$;

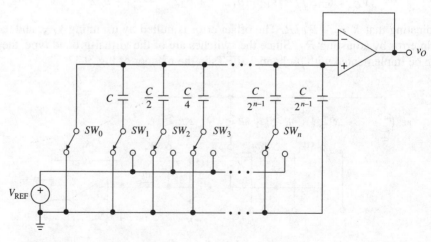

FIGURE 12.8
Weighted-capacitor DAC.

moreover, $C_t = C + C/2 + \cdots + C/2^{n-1} + C/2^{n-1} = 2C$. Substituting gives

$$v_O = V_{REF}(b_1 2^{-1} + b_2 2^{-2} + \cdots + b_n 2^{-n}) \qquad (12.8)$$

indicating that the sample cycle provides an n-bit D-A conversion with $V_{FSR} = V_{REF}$.

By the artifice of switching the bottom plates, as shown, the bottom-plate parasitic capacitances are connected either to ground or to V_{REF}, without affecting charge distribution in the active capacitances. Since MOS capacitance ratios are easily controlled to 0.1% accuracies, the weighted-capacitor scheme is suitable for $n \le 10$. As with weighted-resistor DACs, the main drawback of this scheme is an exponentially increasing capacitance spread.

Potentiometric DACs

It is not difficult to imagine the impact that component mismatches in the most significant bit positions of the previous DACs may have on differential nonlinearity and monotonicity. A *potentiometric* DAC achieves inherent monotonicity by using a string of 2^n resistors to partition V_{REF} into 2^n identical intervals. As depicted in Fig. 12.9 for $n = 3$, a binary tree of switches then selects the tap corresponding to the given input code and connects it to a high-input-impedance amplifier with gain $K = 1 + R_2/R_1$.

No matter how mismatched the resistors, v_O will always increase as the amplifier is switched from one tap to the next, up the ladder, hence the inherent monotonicity. Another advantage is that if the top and bottom nodes of the resistive string are

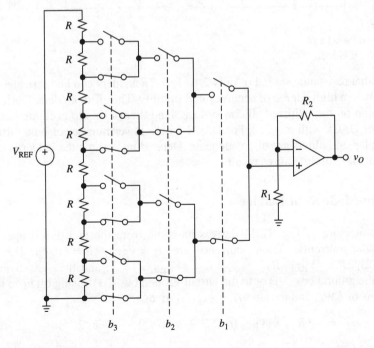

FIGURE 12.9
Potentiometric DAC.

biased at some arbitrary voltages V_H and V_L, the DAC will interpolate between V_L and V_H with a resolution of 2^n steps. However, the large number of resistors (2^n) and switches ($2^{n+1} - 2$) required limits practical potentiometric DACs to $n \leq 8$, even though the switches can be fabricated very efficiently in MOS technology.

R-2R Ladders

Most DACs' architectures are based on the popular R-$2R$ ladder depicted in Fig. 12.10. Starting from the right and working toward the left, one can readily prove that the equivalent resistance to the right of each labeled node equals $2R$. Consequently, the current flowing downward, away from each node, is equal to the current flowing toward the right; moreover, twice this current enters the node from the left. The currents and, hence, the node voltages are binary-weighted,

$$i_{k+1} = \tfrac{1}{2}i_k \qquad v_{k+1} = \tfrac{1}{2}v_k \qquad (12.9)$$

$k = 1, 2, \ldots, n-1$. (Note that the rightmost $2R$ resistance serves a purely terminating function.)

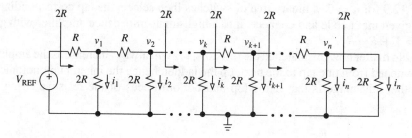

FIGURE 12.10
R-$2R$ ladder.

With a resistance spread of only 2-to-1, R-$2R$ ladders can be fabricated monolithically to a high degree of accuracy and stability. Thin-film ladders, fabricated by deposition on the oxidized silicon surface, lend themselves to accurate laser trimming for DACs with $n \geq 12$. For DACS with a lower number of bits, diffused or ion-implanted ladders are often adequate. Depending on how the ladder is utilized, different DAC architectures result.

Current-Mode R-2R Ladder

The architecture of Fig. 12.11 derives its name from the fact that it operates on the ladder currents. These currents are $i_1 = V_{REF}/2R = (V_{REF}/R)2^{-1}$, $i_2 = (V_{REF}/2)/2R = (V_{REF}/R)2^{-2}, \ldots, i_n = (V_{REF}/R)2^{-n}$, and they are diverted either to the ground bus ($\overline{i_O}$) or to the virtual-ground bus (i_O). Using bit b_k to identify the status of SW_k, and letting $v_O = -R_f i_O$ gives

$$v_O = -(R_f/R)V_{REF}(b_1 2^{-1} + b_2 2^{-2} + \cdots + b_n 2^{-n}) \qquad (12.10)$$

indicating that $K = -R_f/R$. Since $i_O + \overline{i_O} = (1 - 2^{-n})V_{REF}/R$ regardless of the input code, $\overline{i_O}$ is said to be *complementary* to i_O. An important advantage of the

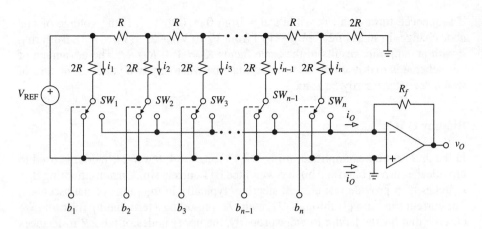

FIGURE 12.11
DAC using a current-mode R-$2R$ ladder.

current mode is that the voltage change across each switch is minimal, so charge injection is virtually eliminated and switch-driver design is made simpler.

We observe that the potential of the i_O bus must be sufficiently close to that of the $\overline{i_O}$ bus; otherwise linearity errors will occur. Thus, in high-resolution DACs, it is crucial that the overall input offset error of the op amp be nulled and have low drift.

Voltage-Mode R-$2R$ Ladder

In the alternative mode of Fig. 12.12, the $2R$ resistances are switched between V_L and V_H, and the output is obtained from the leftmost ladder node. As the input code

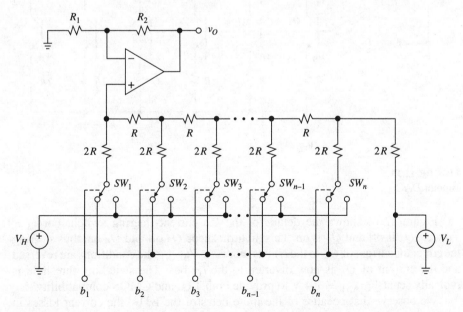

FIGURE 12.12
DAC using a voltage-mode R-$2R$ ladder.

is sequenced through all possible states from $0\ldots0$ to $1\ldots1$, the voltage of this node changes in steps of $2^{-n}(V_H - V_L)$ from V_L to $V_H - 2^{-n}(V_H - V_L)$. Buffering it with an amplifier results in the scale factor $K = 1 + R_2/R_1$. The advantage of this scheme is that it allows us to interpolate between any two voltages, neither of which need necessarily be zero.

Bipolar DACs

In the architecture exemplified in Fig. 12.13 for $n = 4$, the R-$2R$ ladder is used to provide the current bias for n binary-weighted BJT current sinks; n nonsaturating BJT switches then provide fast current steering, typically in the range of nanoseconds. The current sinks are Q_1 through Q_4, with Q_{4t} providing a terminating function. We observe that for the ladder to work properly, the upper nodes of the $2R$ resistances must be equipotential. The voltages at these nodes are set by the emitters of the current sinks. Since the corresponding currents are in 2:1 ratios, the emitter areas must be scaled accordingly as $1A_E$, $2A_E$, $4A_E$, and $8A_E$ to ensure identical V_{BE} drops and, hence, equipotential emitters.

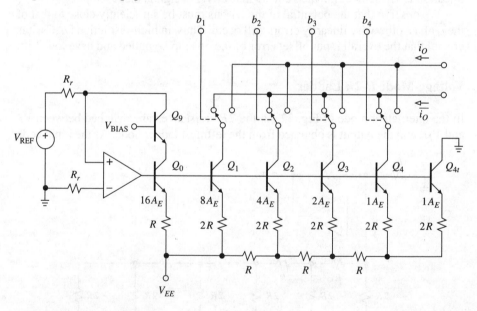

FIGURE 12.13
Bipolar DAC.

Figure 12.14 shows the details of the kth current-steering switch. For $v_k > V_{BIAS1}$, Q_1 is off and Q_2 is on. This, in turn, keeps Q_3 off and Q_4 on, thus steering the collector current of Q_k to the i_O bus. For $v_k < V_{BIAS1}$, the conditions are reversed and the current of Q_k is now diverted to the $\overline{i_O}$ bus. The switching threshold is typically set at $V_{BIAS1} \cong 1.4$ V to provide both TTL and CMOS compatibility.

We observe that because of the finite betas of the BJTs, the current losses in the bases introduce errors. The circuit of Fig. 12.13 uses Q_0 to compensate for the base losses of the current sinks, and Q_9 to compensate for the base losses of

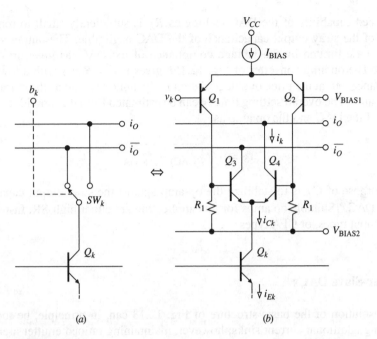

FIGURE 12.14
High-speed current switch.

the switches. The circuit works as follows: by op amp action, $i_{C9} = V_{REF}/R_r$. Using the BJT relationship $i_C = \alpha i_E$, and assuming the same α throughout, we have $i_{E0} = i_{C0}/\alpha = i_{E9}/\alpha = (i_{C9}/\alpha)/\alpha = (V_{REF}/R_r)/\alpha^2$. By ladder action, the emitter current of the kth sink is $i_{Ek} = i_{E0}2^{-k}$. The kth current reaching the i_O bus is $i_k = \alpha i_{Ck} = \alpha(\alpha i_{Ek}) = \alpha^2 i_{E0}2^{-k} = (V_{REF}/R_r)2^{-k}$, indicating the disappearance of base current errors. Summing the various currents on the i_O bus gives

$$i_O = I_{REF}(b_1 2^{-1} + b_2 2^{-2} + b_3 2^{-3} + b_4 2^{-4}) \qquad (12.11)$$

where $I_{REF} = V_{REF}/R_r$.

Figure 12.15 shows the two most common ways of converting i_O to a voltage. The purely resistive termination of Fig. 12.15a, giving $v_O = -R_L i_O$, realizes the

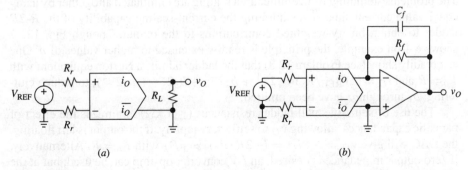

FIGURE 12.15
Bipolar DAC output conditioning.

full-speed capability of the DAC as long as R_L is sufficiently small to render the effect of the stray output capacitance of the DAC negligible. The output swing is in this case limited by the voltage compliance of the DAC, as given in the data sheets. The op amp converter of Fig. 12.15b gives $v_O = R_f i_O$ with a low output impedance, but at the price of a degradation in dynamics as well as the extra cost of the op amp. The overall settling time t_S can be estimated from the individual settling times of the DAC and the op amp as

$$t_S = \sqrt{t_{S(\text{DAC})}^2 + t_{S(\text{OA})}^2} \qquad (12.12)$$

The purpose of C_f is to stabilize the op amp against the stray output capacitance of the DAC.[5] Suitable op amps for this application are either high-SR, fast-settling JFET-input types, or CFA types.

Master-Slave DACs

The resolution of the basic structure of Fig. 12.13 can, in principle, be increased by using additional current sinks; however, maintaining ratioed emitter areas soon leads to extravagant BJT geometries. The architecture of Fig. 12.16 eases the geometry requirements by combining two DACs of the type just discussed in a master-slave configuration in which the current of the terminating BJT Q_{4t} of the master DAC is used to bias the slave DAC. This current, representing 1 LSB of the master DAC, is partitioned by the slave DAC into four additional binary-weighted currents, with Q_{8t} now providing the required termination. The result is an 8-bit DAC with $I_{\text{REF}} = V_{\text{REF}}/R_r$ and a resolution of $I_{\text{REF}}/2^8$. Popular master-slave DACs are the DAC-08 (8-bit) and the DAC-10 (10-bit) (Analog Devices), both of which settle within $\pm\frac{1}{2}$ LSB in 85 ns (typical) and provide output voltage compliance down to -10 V.

Current-Driven R-$2R$ Ladder

The problems stemming from emitter area scaling are eliminated altogether by using equal-value current sinks and exploiting the current-scaling capability of the R-$2R$ ladder to obtain binary-weighted contributions to the output. Though Fig. 12.17 shows a 4-bit example, the principle is readily extended to higher values of n. One can readily show (see Problem 12.8) that the ladder admits a Norton equivalent with $R_o = R$ and $i_O = (2V_{\text{REF}}/R_r)(b_1 2^{-1} + b_2 2^{-2} + b_3 2^{-3} + b_4 2^{-4})$; to reduce cluttering, b_1 through b_4 have been omitted.

The use of suitably small ladder resistances (≤ 1 kΩ) minimizes the effect of parasitic capacitances, allowing v_O to settle very rapidly. If the output is left floating, the DAC will give $v_O = -Ri_O = (-2R/R_r)V_{\text{REF}}D_I$ with $R_o = R$. Alternatively, if zero output impedance is desired, an I-V converter op amp can be used, but at the price of a longer settling time as per Eq. (12.12).

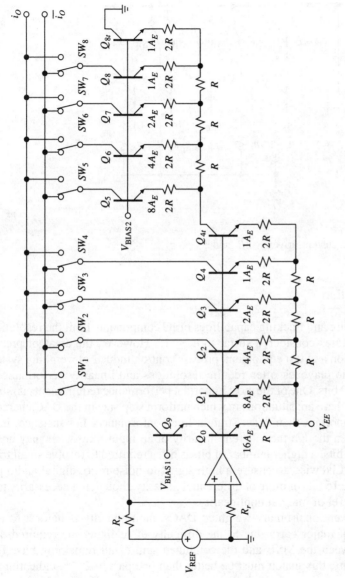

FIGURE 12.16
Master-slave DAC.

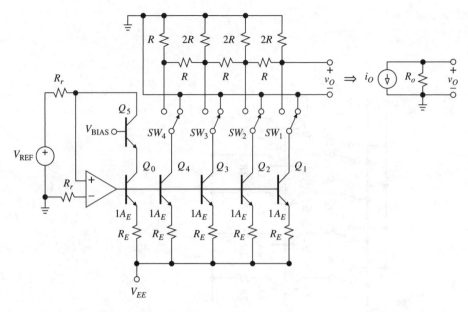

FIGURE 12.17
DAC using a current-driven R-$2R$ ladder.

Segmentation

The matching and tracking capabilities of IC components limit the resolution of the DAC structures considered so far to $n \leq 12$. However, the areas of precision instrumentation and test equipment, process control, industrial weighing systems, and digital audio playback often require resolutions and linearity performance well in excess of 12 bits. One of the most important performance requirements is *monotonicity*. In fact, there are situations in which uniform step size in the DAC characteristics is more important than exact straight-line conformance. For instance, in process control, even though the inherent linearity of an input transducer may not surpass 0.1% or 10 bits, a higher number of bits is often required to resolve small transducer variations. Likewise, to ensure a high signal-to-noise ratio, digital audio playback systems use 16 bits or more of differential linearity, though not necessarily providing the same level of integral nonlinearity.

In conventional binary-weighted DACs, monotonicity is hardest to realize at the point of major carry due to the difficulty of realizing the required degree of match between the MSB and the combined sum of all remaining bits. To ensure monotonicity, this match must be better than one part in 2^{n-1}, indicating that difficulty increases exponentially with n. High-resolution DACs achieve monotonicity by a technique known as *segmentation*. Here the reference range is partitioned into a sufficiently large number of contiguous segments, and a DAC of lesser resolution is then used to interpolate between the extremes of the selected segment. We shall now discuss this technique for both voltage-mode and current-mode DACs.

Voltage-Mode Segmentation

Figure 12.18 illustrates the segmentation technique utilized by the AD7846 16-bit DAC (Analog Devices). The four MS input bits are decoded to select, via switches SW_0 through SW_{16}, one of sixteen voltage segments available along the resistor string. The selected segment is then buffered by the voltage followers and used as a reference voltage of nominal value $V_{REF}/16$ to drive a 12-bit voltage-mode R-$2R$

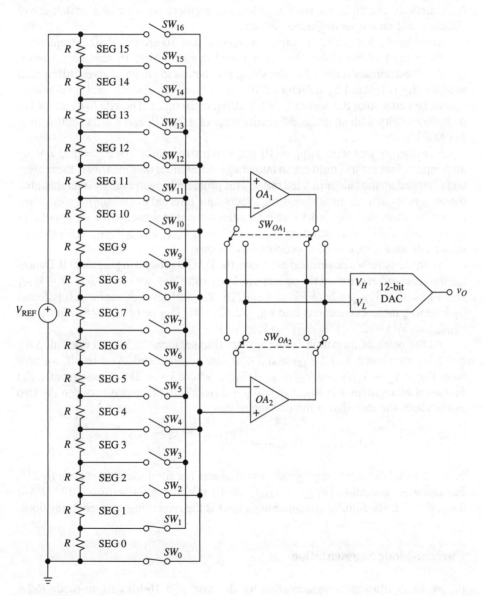

FIGURE 12.18
Simplified diagram of the AD7846 16-bit segmented DAC. (Courtesy of Analog Devices.)

DAC. The latter, in turn, partitions the selected segment into $2^{12} = 4096$ smaller steps, starting at the bottom of the segment and ending one step short of the top, to give

$$v_O = V_L + D_{12}(V_H - V_L) \qquad (12.13)$$

where V_H and V_L are, respectively, the top and the bottom of the selected segment, and D_{12} is the fractional value of the lower 12-bit code. Omitted from the figure for simplicity are an input latch register, the segment decoder and switch-driver circuitry, and an output deglitcher switch.

Since the 65,536 possible output levels consist of 16 groups of 4096 steps each, the major carry of the 12-bit DAC is repeated in each of the 16 segments. Consequently, the accuracy required of the string resistances to ensure a given differential nonlinearity is relaxed by a factor of 16. Note, however, that integral nonlinearity cannot be better than the accuracy of the string resistances. The AD7846 offers 16-bit monotonicity with an integral linearity error of ± 2 LSB, and a 9-μs settling time to 0.0003%.

Considering that with $V_{\text{REF}} = 10$ V the step size is only $10/2^{16} = 152$ μV, op amp input offset errors could cause intolerable differential nonlinearity if the buffers were stepped up the ladder in fixed order. This problem is overcome by interchanging the buffers at each segment transition, a technique referred to as *leapfrogging*. This, in turn, requires that V_H and V_L also be interchanged to preserve the input polarity to the 12-bit DAC. This function is provided by SW_{OA_1} and SW_{OA_2}. The effect of buffer interchanging can be appreciated as follows.

With the switches positioned as shown, the DAC is processing segment 0. Denoting the input offset errors of the op amps as V_{OS1} and V_{OS2}, we have $V_H = V_1 + V_{OS1}$ and $V_L = 0 + V_{OS2}$, where $V_1 = V_{\text{REF}}/16$. The last level of segment 0 is found by inserting these expressions into Eq. (12.13) with $D_{12} = (1 - 2^{-12})$. This gives $v_{O(\text{last})} = V_1(1 - 2^{-12}) + V_{OS1} - (V_{OS1} - V_{OS2})2^{-12}$.

At the point of transition from segment 0 to segment 1, SW_0 is opened, SW_1 and SW_2 are closed, and SW_{OA_1} and SW_{OA_2} are commutated. As a result, we now have $V_H = V_2 + V_{OS2}$ and $V_L = V_1 + V_{OS1}$, where $V_2 = 2V_1$. Consequently, the first level of segment 1 is $v_{O(\text{first})} = V_1 + V_{OS1}$. The difference between the two levels yields the step size at the first major carry,

$$v_{O(\text{first})} - v_{O(\text{last})} = \frac{V_{\text{REF}}}{2^{16}} + \frac{V_{OS2} - V_{OS1}}{2^{12}}$$

indicating that the leapfrogging technique reduces the combined offset error by 2^{12}. For instance, assuming $|V_{OS2} - V_{OS1}| \cong 10$ mV, the error term is $10^{-2}/2^{12} = 2.4$ $\mu V \ll 1$ LSB. Similar considerations hold at the remaining segment transitions.

Current-Mode Segmentation

Figure 12.19 illustrates segmentation for the case of a 16-bit current-mode R-$2R$ DAC. The resistances at the left establish 15 current segments of value V_{REF}/R, so the contribution of each segment to the output is $-(R_f/R)V_{\text{REF}}$. The decode logic examines the 4 MS input bits and diverts to the i_O bus 8 such segments for

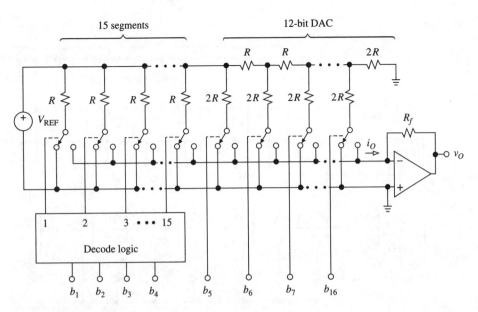

FIGURE 12.19
16-bit segmented DAC using a 12-bit current-mode R-$2R$ ladder.

b_1, 4 segments for b_2, 2 segments for b_3, and 1 segment for b_4. The remaining resistances form an ordinary 12-bit current-mode R-$2R$ DAC, whose contribution to the output is given by Eq. (12.10). Using the superposition principle, we thus have $v_O = -(R_f/R)V_{REF} \times (8b_1 + 4b_2 + 2b_3 + b_4 + b_5 2^{-1} + b_6 2^{-2} + \cdots + b_{16} 2^{-12})$, or

$$v_O = -16\frac{R_f}{R}V_{REF}(b_1 2^{-1} + b_2 2^{-2} + \cdots + b_{16} 2^{-16}) \qquad (12.14)$$

indicating a 16-bit conversion with $V_{FSR} = -16(R_f/R)V_{REF}$. We observe that the segment resistances, like the ladder resistances, need only be accurate to 12 bits to ensure monotonicity at the 16-bit level. An example of a DAC using this principle is the MP7616 16-bit CMOS DAC (Micro Power Systems).

Figure 12.20 shows a 16-bit segmented DAC using the current-driven ladder architecture. Here Q_1 through Q_7 provide 7 current segments of value $V_{REF}/4R_r = 0.25$ mA, which a decoder (not shown for simplicity) steers either to the i_O bus or to ground, depending on the 3 MS bits. Steered to the i_O bus are 4 segments for b_1, 2 segments for b_2, and 1 segment for b_3. Moreover, Q_8 through Q_{20}, along with the R-$2R$ ladder, form a 13-bit current-driven DAC. Proper scaling requires an additional R resistance between the 13-bit DAC and the i_O bus. Consequently, the Norton resistance is now $R_o = 2R$. By the superposition principle, $i_O = (V_{REF}/4R_r)(4b_1 + 2b_2 + b_3 + b_4 2^{-1} + b_5 2^{-2} + \cdots + b_{16} 2^{-12})$, or

$$i_O = 2\frac{V_{REF}}{R_r}(b_1 2^{-1} + b_2 2^{-2} + \cdots + b_{16} 2^{-16}) \qquad (12.15)$$

indicating a 16-bit conversion with $I_{FSR} = 2$ mA. Two popular examples of 16-bit monolithic DACs utilizing this architecture are the PCM52/53 (Burr-Brown) and HI-DAC16 (Harris).

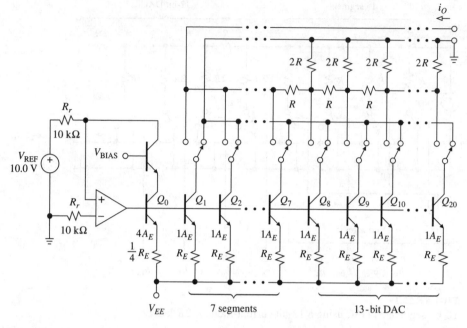

FIGURE 12.20
16-bit segmented DAC using a 13-bit current-driven R-$2R$ ladder.

12.3
MULTIPLYING DAC APPLICATIONS

The R-$2R$ ladder DACs of Figs. 12.11 and 12.12 are especially suited to monolithic fabrication in CMOS technology.[6] The switches are implemented with CMOS transistors, and the ladder and the feedback resistor $R_f = R$ are fabricated by thin-film deposition on the CMOS die. Because of process variations, the resistances, though highly matched, are not necessarily accurate. For instance, a ladder with a nominal rating of 10 kΩ may in practice lie in the range of 5 kΩ to 20 kΩ.

Figure 12.21 shows the circuit diagram of the kth switch, $k = 1, 2, \ldots, n$. The switch proper consists of the n-MOS pair M_8-M_9, while the remaining FETs accept TTL- and CMOS-compatible logic inputs to provide antiphase gate drives for M_8 and M_9. When the logic input is high, M_8 is off and M_9 is on, so i_k is diverted to the i_O bus. When the input is low, M_8 is on, M_9 is off, and i_k is now diverted to the $\overline{i_O}$ bus.

The nonzero resistance $r_{ds(\text{on})}$ of the switches tends to disrupt the 2:1 ratio of the ladder resistances and degrade performance. Since $r_{ds(\text{on})}$ is proportional to the ratio of the channel length L to the channel width W, it could be minimized by fabricating M_8 and M_9 with $L/W \ll 1$; this, however, would lead to extravagant device geometries. A common technique for overcoming this drawback is to taper switch geometries to achieve, at least in the MS bit positions, binary-weighted switch resistances such as $r_{ds1(\text{on})} = 20\ \Omega$, $r_{ds2(\text{on})} = 40\ \Omega$, $r_{ds3(\text{on})} = 80\ \Omega$, and so on. Since the currents halve as the switch resistances double, the product $r_{dsk(\text{on})} \times i_k$ remains constant throughout the tapered bit positions, causing a systematic switch voltage drop, whose value is typically 10 mV. Since this drop is effectively being subtracted from V_{REF}, the result is a gain error that is readily trimmed by adjusting R_f.

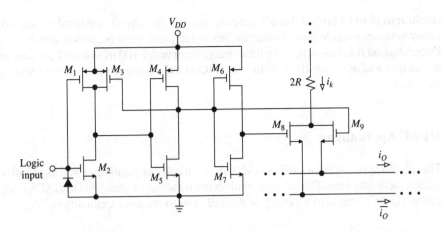

FIGURE 12.21
CMOS switch for R-$2R$ ladder.

EXAMPLE 12.3. A CMOS DAC with $n = 12$ is operated in the current mode depicted in Fig. 12.11. If $V_{REF} = 10.0$ V and the DAC is calibrated at 25 °C, specify TC(V_{REF}) and TC(V_{OS}) so that the individual drift errors contributed by the reference and the op amp are less than $\pm\frac{1}{4}$ LSB over the operating range of 0 °C to 70 °C.

Solution. We have $\frac{1}{4}$ LSB = $10.0/2^{14}$ = 0.61 mV. Since the maximum temperature excursion from the point of calibration is $70° - 25° = 45$ °C, the individual drifts must not exceed $\pm0.61 \times 10^{-3}/45 \cong \pm13.6$ μV/°C. This gives $TC_{max}(V_{REF}) = \pm1.36$ ppm/°C. Moreover, using a conservative estimate of 2 V/V for the noise gain of the op amp, we have $TC_{max}(V_{OS}) \cong \pm13.6/2 = \pm6.8$ μV/°C.

In the following we shall use the functional diagram of Fig. 12.22 to represent a CMOS DAC. This structure is available from various manufacturers in a range of

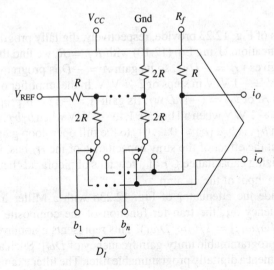

FIGURE 12.22
Functional diagram of a multiplying DAC.

resolutions (8 to 14 bits) and configurations (single, dual, quad, and octal packages). Many versions include input buffer latches to facilitate microprocessor interfacing. Depending on resolution, settling times range from under 100 ns to over 1 μs. One of the earliest and most popular families of CMOS DACs is the AD7500 series (Analog Devices).

MDAC Applications

The reference voltage of a CMOS DAC can be varied over positive as well as negative values, including zero. This inherent multiplicative ability makes CMOS DACs, aptly called *MDACs*, suited to a variety of digitally programmable applications.[6]

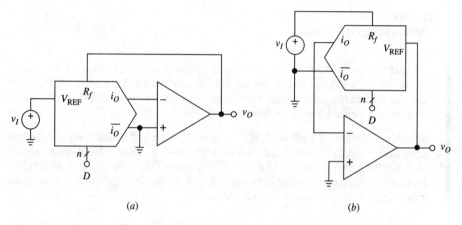

(a) (b)

FIGURE 12.23
(a) Digitally programmable attenuator: $v_O = -Dv_I$; (b) digitally programmable amplifier: $v_O = (-1/D)v_I$.

The circuits of Fig. 12.23 provide, respectively, digitally programmable attenuation and amplification. Using Eq. (12.10) with $R_f = R$, we find that the attenuator of Fig. 12.23a gives $v_O = -Dv_I$, so its gain $A = -D$ is programmable from 0 to $-(1-2^{-n})$ V/V $\cong -1$ V/V in steps of 2^{-n} V/V. In the amplifier of Fig. 12.23b we have $v_I = -Dv_O$, or $v_O = (-1/D)v_I$. Its gain $A = -1/D$ is programmable from $-1/(1-2^{-n}) \cong -1$ V/V when all bits are 1, to -2 V/V when $b_1b_2\ldots b_n = 10\ldots0$, to 2^n V/V when $b_1\ldots b_{n-1}b_n = 0\ldots01$, to the full open-loop gain a when all bits are 0. To combat the effect of the stray capacitance of the i_O bus, it is advisable to connect a stabilizing capacitance C_f of a few tens of picofarads between the output and the inverting input of the op amp.[5]

If we cascade the attenuator of Fig. 12.23a with a Miller integrator having unity-gain frequency ω_1, the transfer function of the composite circuit is $H = (-D) \times [-1/(j\omega/\omega_1)] = 1/(j\omega/D\omega_1)$. This represents a noninverting integrator with a digitally programmable unity-gain frequency of $D\omega_1$. Such an integrator can be used to implement a digitally programmable filter. The filter example of Fig. 12.24 is a state-variable topology of the type encountered in Fig. 4.37, so we can reuse

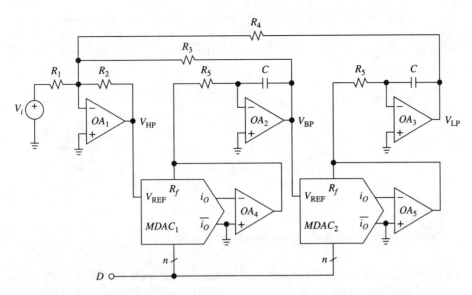

FIGURE 12.24
Digitally programmable filter.

Eq. (4.34) and write

$$\omega_0 = D\sqrt{R_2/R_4}/R_5C \qquad Q = R_3/\sqrt{R_2 R_4} \qquad (12.16a)$$

$$H_{0HP} = -R_2/R_1 \qquad H_{0BP} = -R_3/R_1 \qquad H_{0LP} = -R_4/R_1 \qquad (12.16b)$$

indicating that we can program ω_0 digitally from $2^{-n}\sqrt{R_2/R_4}/R_5C$ to $(1 - 2^{-n})$ $\sqrt{R_2/R_4}/R_5C$. Once we have a digitally programmable filter, we can readily turn it into a digitally programmable oscillator by letting $Q \to \infty$ (see Problem 12.12).

EXAMPLE 12.4. In the circuit of Fig. 12.24 specify suitable components for $Q = 1/\sqrt{2}$, $H_{0BP} = -1$ V/V, and f_0 digitally programmable in 10-Hz steps by means of 10-bit MDACs.

Solution. Impose $R_2 = R_4 = 10.0$ kΩ, and let $C = 1.0$ nF. Then, the full-scale range is $f_{0(FSR)} = 2^{10} \times 10 = 10.24$ kHz, so $R_5 = 1/(2\pi 10, 240 \times 10^{-9}) = 15.54$ kΩ (use 15.4 kΩ, 1%).

Use fast op amps with low-input-offset error and noise characteristics and wide dynamics, such as the OPA627 JFET-input op amps (Burr-Brown). To avoid high-frequency Q enhancement, phase-error compensation may be required, as discussed in Section 6.5.

Figure 12.25 shows a digitally programmable waveform generator. The circuit is similar to that encountered in Fig. 10.19a, except for the use of an MDAC to control the rate of capacitance charge/discharge digitally. To avoid the uncertainties of the ladder resistances, the MDAC is current-driven using the REF200 100-μA current source (Burr-Brown). When v_{SQ} is high, I_{REF} enters the MDAC; when v_{SQ} is low, I_{REF} exits the MDAC. In either case the MDAC divides this current to give

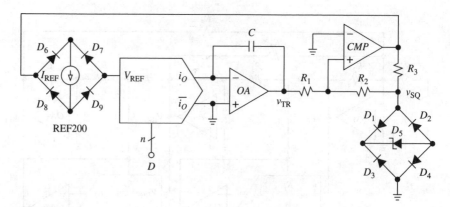

FIGURE 12.25
Digitally programmable triangular/square-wave oscillator.

$i_O = \pm D I_{REF}$. To find the frequency of oscillation f_0, apply Eq. (10.2) with $\Delta t = 1/2 f_0$, $I = D I_{REF}$, and $\Delta v = 2 V_T = 2 (R_1/R_2) V_{clamp}$, where $V_{clamp} = 2 V_{D(on)} + V_{Z5}$. The result is

$$f_0 = D \frac{(R_2/R_1) I_{REF}}{4 C V_{clamp}} \qquad (12.17)$$

indicating that f_0 is linearly proportional to D.

> **EXAMPLE 12.5.** In the circuit of Fig. 12.25 specify suitable components for 5-V wave-form amplitudes and f_0 digitally programmable in 1-Hz steps by means of a 12-bit MDAC.
>
> **Solution.** For $V_{clamp} = 5$ V, use $V_{Z5} = 3.6$ V. Moreover, use $R_1 = R_2 = 20$ kΩ and $R_3 = 6.2$ kΩ. The full-scale range is $f_{0(FSR)} = 2^{12} \times 1 = 4.096$ kHz, so Eq. (12.17) gives $C = 100 \times 10^{-6}/(20 \times 4096) = 1.22$ nF (use 1.0 nF, which is more easily available, and raise R_1 to 24.3 kΩ, 1%). Use a low-offset JFET-input op amp for *OA*, and a high slew-rate op amp for *CMP*.

12.4
A-D CONVERSION TECHNIQUES

This section discusses popular ADC techniques, such as DAC-based ADCs, flash ADCs, integrating ADCs, and variants thereof.[2–4] A more recent technique, known as sigma-delta (Σ-Δ) conversion, is addressed in the next section.

DAC-Based A-D Conversion

A-D conversion can be accomplished by using a DAC and a suitable register to adjust the DAC's input code until the DAC's output comes within $\pm\frac{1}{2}$ LSB of the analog input. The code that achieves this is the desired ADC output $b_1 \ldots b_n$. As shown in Fig. 12.26, this technique requires suitable logic circuitry to direct the register

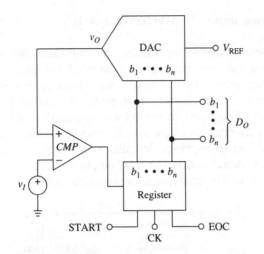

FIGURE 12.26
Functional diagram of a DAC-based ADC.

to perform the code search on the arrival of the START command, and a voltage comparator to announce when v_O has come within $\pm\frac{1}{2}$ LSB of v_I and thus issue an end-of-conversion (EOC) command. Moreover, to center the analog range properly, the DAC output must be offset by $+\frac{1}{2}$ LSB, per Fig. 12.5b.

The simplest code search is a *sequential search,* obtained by operating the register as a binary counter. As the counter steps through consecutive codes starting from $0\ldots0$, the DAC produces an increasing staircase, which the comparator then compares against v_I. As soon as this staircase reaches v_I, *CMP* fires and stops the counter. This also serves as an EOC command to notify that the desired code is sitting in the counter. The counter must be stepped at a low enough frequency to allow for the DAC to settle within each clock cycle. Considering that a conversion can take as many as $2^n - 1$ clock periods, this technique is limited to low-speed applications. For example, a 12-bit ADC with a 1-MHz counter clock will take $(2^{12} - 1)\ \mu s = 4.095$ ms to convert a full-scale input.

A better approach is to allow the counter to start counting from the most recent code rather than restarting from zero. If v_I has not changed drastically since the last conversion, fewer counts will be needed for v_O to catch up with v_I. Also referred to as a *tracking* or a *servo converter,* this scheme uses the register as an up/down counter with the count direction controlled by the comparator: counting will be up when $v_O < v_I$, and down when $v_O > v_I$. Whenever v_O crosses v_I, the comparator changes state and this is taken as an EOC command. Clearly, conversions will be relatively fast only as long as v_I does not change too rapidly between consecutive conversions. For a full-scale change, the conversion will still take $2^n - 1$ clock periods.

The fastest code-search strategy uses binary search techniques to complete an n-bit conversion in just n clock periods, regardless of v_I. Following is a description of two implementations: the *successive-approximation* and the *charge-redistribution* ADCs.

Successive-Approximation Converters (SA ADCs)

This technique uses the register as a *successive-approximation register* (SAR) to find each bit by trial and error. Starting from the MSB, the SAR inserts a trial 1 and then interrogates the comparator to find whether this causes v_O to rise above v_I. If it does, the trial bit is changed back to 0; otherwise it is left as 1. The procedure is then repeated for all subsequent bits, one bit at a time, in a way similar to a chemist's balance. Figure 12.27 illustrates how a 10.8-V input is converted to a 4-bit code with $V_{FSR} = 16$ V. The analog range, in volts, is at the left, and the digital codes at the right. To ensure correct results, the DAC output must be offset by $-\frac{1}{2}$ LSB, or -0.5 V in our example. The conversion takes place as follows.

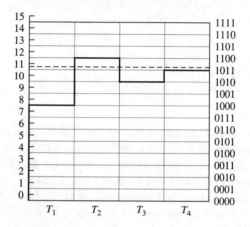

FIGURE 12.27
Idealized DAC output for the 4-bit successive-approximation conversion of $v_I = 10.8$ V with $V_{FSR} = 16$ V.

Following the arrival of the START command, the SAR sets b_1 to 1 with all remaining bits at 0 so that the trial code is 1000. This causes the DAC to output $v_O = 16(1 \times 2^{-1} + 0 \times 2^{-2} + 0 \times 2^{-3} + 0 \times 2^{-4}) - 0.5 = 7.5$ V. At the end of clock period T_1, v_O is compared against v_I, and since $7.5 < 10.8$, b_1 is left at 1.

At the beginning of T_2, b_2 is set to 1, so the trial code is now 1100 and $v_O = 16(2^{-1} + 2^{-2}) - 0.5 = 11.5$ V. Since $11.5 > 10.8$, b_2 is changed back to 0 at the end of T_2.

At the beginning of T_3, b_3 is set to 1, so the trial code is 1010 and $v_O = 10 - 0.5 = 9.5$ V. Since $9.5 < 10.8$, b_3 is left at 1.

At the beginning of T_4, b_4 is set to 1, so the trial code is 1011 and $v_O = 11 - 0.5 = 10.5$ V. Since $10.5 < 10.8$, b_4 is left at 1. Thus, when leaving T_4, the SAR has generated the code 1011, which ideally corresponds to 11 V. Note that any voltage in the range 10.5 V $< v_I <$ 11.5 V would have led to the same code.

Since the entire conversion takes a total of n clock cycles, a SA ADC offers a major speed improvement over a sequential-search ADC. For instance, a 12-bit SA ADC with a clock frequency of 1 MHz will complete a conversion in 12 μs.

Figure 12.28 shows an actual implementation[7] using the Am2504 SAR and the Am6012 bipolar DAC (Advanced Micro Devices), whose settling time is 250 ns,

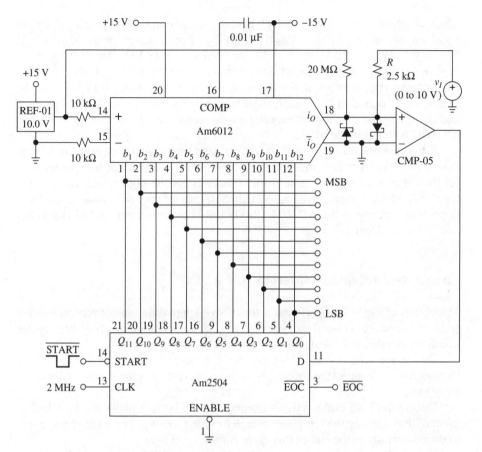

FIGURE 12.28
12-bit, 6-μs successive-approximation ADC.

along with the CMP-05 comparator, whose response time to a 1.2-mV overdrive ($\frac{1}{2}$ LSB) is 125 ns maximum. The desired output code is available both in parallel form from Q_0 through Q_{11}, or in serial form at the data pin D.

To take full advantage of the bipolar DAC speed, i_O is converted to a voltage for the comparator via simple resistive termination. Since its input is $v_D = v_I - Ri_O$, the comparator is in effect comparing i_O against v_I/R. The function of the 20-MΩ resistance is to provide the required $-\frac{1}{2}$-LSB shift, and that of the Schottky diodes is to limit the voltage swing at the comparator input in order to reduce delays caused by the stray output capacitance of the DAC.

The primary factors affecting the speed of a SA ADC are the settling time of the DAC and the response time of the comparator. The conversion time can be further reduced by a number of ingenious techniques,[7] such as comparator speed-up techniques, or variable-clock techniques, which exploit the faster settling times in the least significant bit positions.

The resolution of a SA ADC is limited by the resolution and linearity of the DAC, and the gain of the comparator. A crucial requirement is that the DAC be monotonic to prevent the occurrence of missing codes. The comparator, besides adequate speed, must provide enough gain to magnify an LSB step to a full output logic swing, or $a \geq (V_{OH} - V_{OL})/(V_{FSR}/2^n)$. For instance, with $V_{OH} = 5$ V, $V_{OL} = 0$ V, $V_{FSR} = 10$ V, and $n = 12$, we need $a \geq 2048$ V/V. Another important requirement is that during conversion v_I remain constant within $\pm\frac{1}{2}$ LSB; otherwise an erroneous code may result. For instance, if v_I were to rise above 11.5 V after the second clock period in Fig. 12.27, there would be no way for the SAR to go back and change b_2, so a wrong output code would result. This is avoided by preceding the ADC with a suitable SHA.

SA ADCs are available from a variety of sources and in a wide range of performance characteristics and prices. Conversion times typically range from under 1 μs for the faster 8-bit units to tens of microseconds for the high-resolution ($n \geq 14$) types. SA ADCs equipped with an on-chip SHA are referred to as *sampling* ADCs. A popular example is the AD1674 12-bit, 100-kilosamples per second (ksps) SA ADC (Analog Devices).

Charge-Redistribution Converters (CR ADCs)

The circuit of Fig. 12.29 performs a successive-approximation conversion using a weighted-capacitor DAC of the type of Fig. 12.8. Its operation involves three cycles called the *sample, hold,* and *redistribution* cycles.[2]

During the sample cycle, SW_0 grounds the top-plate bus while SW_i and SW_1 through SW_{n+1} connect the bottom plates to v_I, thus precharging the entire capacitor array to v_I.

During the hold cycle, SW_0 is opened and the bottom plates are switched to ground, thus causing the top-plate voltage to swing to $-v_I$. The voltage presented to the comparator at the end of this cycle is thus $v_P = -v_I$.

During the redistribution cycle, SW_0 is still open, SW_i is connected to V_{REF}, and the remaining switches are sequentially flipped from ground to V_{REF}, and

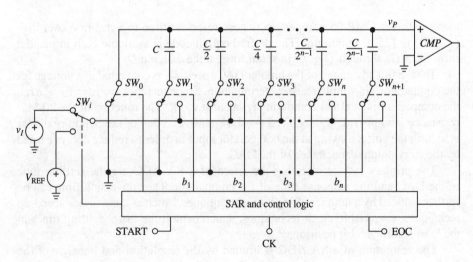

FIGURE 12.29
Charge-redistribution ADC.

possibly back to ground, to perform a successive-approximation search for the desired code.

Flipping a given switch SW_k from ground to V_{REF} causes v_P to increase by the amount $V_{REF}(C/2^{k-1})/C_t = V_{REF}2^{-k}$. If it is found that this increase causes the comparator to change state, then SW_k is returned to ground; otherwise it is left at V_{REF} and the next switch is tried. This procedure is repeated at each bit position, starting from the MSB and progressing down to the LSB (excluding the terminating capacitor switch, which is left permanently grounded). It is readily seen that at the end of the search the voltage presented to the comparator is

$$v_P = -v_I + V_{REF}(b_1 2^{-1} + b_2 2^{-2} + \cdots + b_n 2^{-n})$$

and that v_P is within $\pm\frac{1}{2}$ LSB of 0 V. Thus, the final switch pattern provides the desired output code.

Because of the exponential increase of capacitance spread with n, practical CR ADCs are limited to $n \leq 10$. One way to increase resolution is to combine charge redistribution with potentiometric techniques,[2] as exemplified in Fig. 12.30. Here a resistor string partitions V_{REF} into 2^{n_H} inherently monotonic voltage segments, and an n_L-bit weighted-capacitor DAC interpolates within the selected segment. As long as the capacitances are ratio-accurate to n_L bits, the composite DAC will retain monotonicity to $n = n_H + n_L$ bits, so using it as part of an SA conversion will avoid missing codes. A conversion proceeds as follows.

Initially, SW_f is closed to autozero the comparator, and the bottom plates are connected via the L bus and SW_L to the analog input v_I. This precharges the capacitor

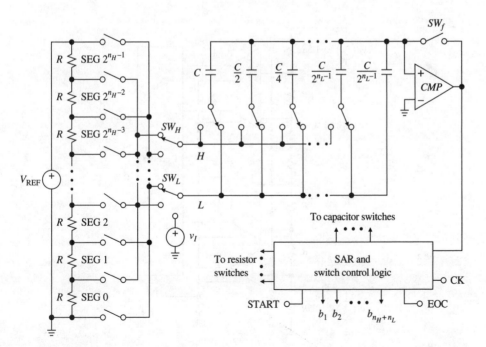

FIGURE 12.30
High-resolution charge-redistribution ADC.

array to v_I minus the comparator's threshold voltage, thus removing this threshold as a possible source of error.

Next, SW_f is opened, and an SA search among the resistor string taps is performed to find the segment within which the voltage held in the capacitor array lies. The outcome of this search is the n_H-bit portion of the desired code.

Once the segment has been found, the H and L busses are connected to the extremes of the corresponding resistor, and a second SA search is performed to find the individual bottom-plate switch settings that make the comparator input converge to its threshold. The outcome of this search is the n_L-bit portion of the desired code. For instance, with $n_H = 4$ and $n_L = 8$, the circuit provides 12 bits of resolution without excessive demands in terms of circuit complexity or capacitance spread and matching.

Flash Converters

The circuit of Fig. 12.31 uses a resistor string to create $2^n - 1$ reference levels separated from each other by 1 LSB, and a bank of $2^n - 1$ high-speed latched comparators to simultaneously compare v_I against each level. Note that to position the analog signal range properly, the top and bottom resistors must be $1.5R$ and $0.5R$, as shown. As the comparators are strobed by the clock, the ones whose reference levels are below v_I will output a logic 1, and the remaining ones a logic 0. The result, referred to as a *bar graph,* or also as a *thermometer* code, is then converted to the desired output code $b_1 \ldots b_n$ by a suitable decoder, such as a priority encoder.

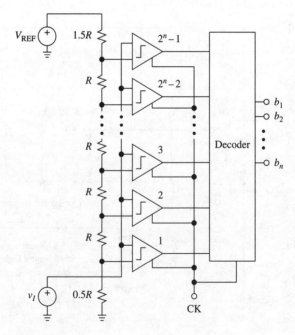

FIGURE 12.31
n-bit flash converter.

Since input sampling and latching take place during the first phase of the clock period, and decoding during the second phase, the entire conversion takes only one clock cycle, so this ADC is the fastest possible. Aptly called a *flash converter,* it is used in high-speed applications, such as video and radar signal processing, where conversion rates on the order of millions of samples per second (Msps) are required, and SA ADCs are generally not fast enough.

The high-speed and inherent-sampling advantages of flash ADCs are offset by the fact that $2^n - 1$ comparators are required. For instance, an 8-bit converter requires 255 comparators. The exponential increase with n in die area, power dissipation, and stray input capacitance makes flash converters impractical for $n > 10$. Flash ADCs are available in bipolar or in CMOS technology, with resolutions of 6, 8, and 10 bits, sampling rates of tens to hundreds of Msps, depending on resolution, and power dissipation ratings on the order of 1 W or less. Consult the catalogs to familiarize yourself with the range of available products.

Subranging Converters

Subranging ADCs trade speed for circuit complexity by splitting the conversion into two subtasks, each requiring less complex circuitry. Also called a *two-step,* or a *half-flash, converter,* this architecture uses a coarse flash ADC to provide an *n*-bit accurate digitization of the n_H most-significant bits. These bits are then fed to a high-speed, *n*-bit accurate DAC to provide a coarse approximation to the analog input. The difference between this input and the DAC output, called the *residue,* is magnified by 2^{n_H} V/V by an amplifier called the *residue amplifier* (RA), and finally fed to a fine flash ADC for the digitization of the n_L least-significant bits of the *n*-bit code, where $n = n_H + n_L$. Note that the half-flash requires an SHA to hold the value of v_I during the digitization of the residue.

Figure 12.32 exemplifies an 8-bit converter with $n_L = n_H = 4$. Besides the SHA, the DAC, and the RA, the circuit uses $2(2^4 - 1) = 30$ comparators, indicating a substantial saving compared to the 255 comparators required by a full-flash. (This saving is even more dramatic for $n \geq 10$.) The main price for this saving is a longer conversion time, with the first phase comprising the conversion time of the coarse ADC, the acquisition time of the SHA, and the settling time of the DAC-subtractor-RA block, and the second phase comprising the conversion time of the

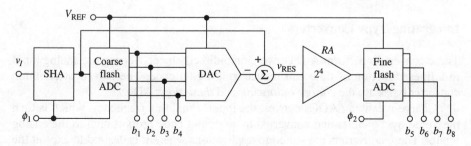

FIGURE 12.32
8-bit subranging ADC. (Note that DAC must be 8-bit accurate.)

fine ADC. Moreover, the requirement that the DAC be n-bit accurate may be a heavy requirement.

Subranging ADCs, though not as fast as full-flash ADCs, are still comparably faster than SA ADCs, so the subranging architecture, or variants[4] thereof, is used in a number of high-speed ADC products.

Pipelined Converters

Pipelined ADCs break down the conversion task into a sequence of N serial subtasks, and use SHA interstage isolation to allow for the individual subtasks to proceed concurrently to achieve high throughput rates. With reference to Fig. 12.33, each subtask stage consists of an SHA, an ADC, a DAC, a subtractor, and an RA, with some or even all functions often combined in one circuit.[4] The first stage samples v_I, digitizes k bits, and uses a DAC-subtractor-RA circuit to create a residue for the next stage in the pipeline. The next stage samples the incoming residue and performs a similar sequence of operations while the previous stage begins processing the next sample. The ability of the various stages to operate concurrently makes the conversion rate depend on the speed of only one stage, usually the first stage. Pipelined structures are used in a variety of formats, including the case $k = 1$, which results in the simplest per-stage circuitry, though n such stages are needed. However, if stages are reused, considerable savings in die area can be achieved.

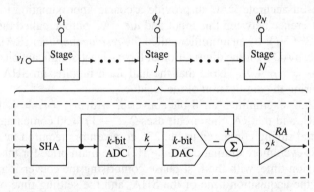

FIGURE 12.33
Pipeline ADC architecture.

Integrating-Type Converters

These converters perform A-D conversion indirectly by converting the analog input to a linear function of time and thence to a digital code. The two most common converter types are the *charge-balancing* and *dual-slope* ADCs.

Charge-balancing ADCs convert the input signal to a frequency, which is then measured by a counter and converted to an output code proportional to the analog input.[8] These converters are suited to applications where it is desired to exploit the ease with which a frequency is transmitted in noisy environments or in isolated form, such as telemetry. However, as seen in Section 10.7, the transfer characteristic of a

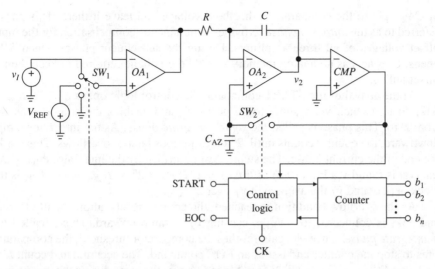

FIGURE 12.34
Functional diagram of a dual-slope ADC.

VFC depends on an RC product whose value is not easily maintained with temperature and time. This drawback is ingeniously overcome by dual-slope converters.

As shown in the functional diagram of Fig. 12.34, a dual-slope ADC, also called a *dual-ramp* ADC, is based on a high-input-impedance buffer, a precision integrator, and a voltage comparator. The circuit first integrates the input signal v_I for a fixed duration of 2^n clock periods, and then it integrates an internal reference V_{REF} of opposite polarity until the integrator output is brought back to zero. The number N of clock cycles required to return to zero is proportional to the value of v_I averaged over the integration period. Consequently, N represents the desired output code. With reference to the waveform diagram of Fig. 12.35, following is a detailed description of how the circuit operates.

Prior to the arrival of the START command, SW_1 is connected to ground and SW_2 closes a loop around the integrator-comparator combination. This forces the autozero capacitance C_{AZ} to develop whatever voltage is needed to bring the output

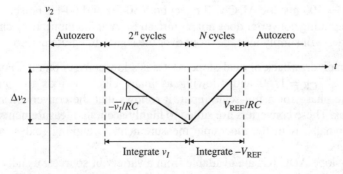

FIGURE 12.35
Dual-slope waveform.

of OA_2 right to the comparator's threshold voltage and leave it there. This phase, referred to as the *autozero phase,* provides simultaneous compensation for the input offset voltages of all three amplifiers. During the subsequent phases, when SW_2 opens, C_{AZ} acts as an analog memory to hold the voltage required to keep the net offset nulled.

At the arrival of the START command, the control logic opens SW_2, connects SW_1 to v_I (which we assume to be positive), and enables the counter, starting from zero. This phase is called the *signal integrate phase*. As the integrator ramps downward, the counter counts until, 2^n clock periods later, it overflows. This marks the end of the current phase. The swing Δv_2 described by the integrator during this interval is found via Eq. (10.2) as $C\Delta v_2 = (\overline{v_I}/R) \times 2^n \times T_{CK}$, where T_{CK} is the clock period, and $\overline{v_I}$ the average of v_I over $2^n T_{CK}$.

As the overflow condition is reached, the counter resets automatically to zero and SW_1 is connected to $-V_{REF}$, causing v_2 to ramp upward. This is called the *deintegrate phase*. Once v_2 again reaches the comparator threshold, the comparator fires to stop the counter and issues an EOC command. The accumulated count N is such that $C\Delta v_2 = (V_{REF}/R)NT_{CK}$. Since $C\Delta v_2$ is the same during the two phases, we get

$$N = 2^n \frac{\overline{v_I}}{V_{REF}} \tag{12.18}$$

We make a number of important observations.

1. The conversion accuracy is independent of R, C, T_{CK}, and the input offset voltage of the three amplifiers. As long as these parameters remain stable over the conversion period, they affect the two integration phases equally, so long-term drifts are automatically eliminated.
2. An integrating ADC offers excellent linearity and resolution, and virtually zero differential nonlinearity. With an integrator of suitable quality, nonlinearity errors can be kept below 0.01%, and resolution can be pushed above 20 bits. Moreover, since v_2 is a continuous function of time, differential nonlinearity, within the limits of clock jitter, is virtually absent, so there are no missing codes.
3. A dual-slope ADC provides excellent rejection of ac noise components with frequencies that are integral multiples of $1/(2^n T_{CK})$. For instance, if we specify T_{CK} so that $2^n T_{CK}$ is a multiple of $1/60 = 16.67$ ms, then any 60-Hz pickup noise superimposed on the input signal will be averaged to zero. In particular, if $2^n T_{CK} = 100$ ms, the ADC will reject both 50-Hz and 60-Hz noise.
4. An integrating converter does not require an SHA at the input. If v_I changes, the converter will simply average it out over the signal-integrate period.

The main drawback of dual-slope ADCs is a low conversion rate. For instance, imposing $2^n T_{CK} = 1/60$ and allowing as many clock periods to complete the deintegrate phase for a full-scale input, it follows that the conversion rate is less than 30 sps. These converters are suited to highly accurate measurements of slowly varying signals, as in thermocouple measurements, weighing scales, and digital multimeters.

Dual-slope ADC ICs are available from a variety of sources, usually in CMOS technology. Besides autozero capabilities, they offer automatic input polarity sensing and reference polarity switching to provide sign and magnitude information.

Moreover, they are available both in microprocessor-compatible and in display-oriented versions. The latter provide the output code in a format suitable for driving decimal LCD or LED displays, and their resolution is expressed in terms of decimal digits rather than bits. Since the leftmost digit is usually allowed to run only to unity, it is counted as $\frac{1}{2}$ digit. Thus, a $4\frac{1}{2}$-digit sign-plus-magnitude ADC having $V_{FSR} = 200$ mV yields all decimal codes within the range of ± 199.99 mV and with a resolution of 10 μV. An example is the ICL7129 $4\frac{1}{2}$-digit ADC (Harris), which, with the help of suitable support circuitry, is easily turned into a full-fledged multimeter to measure both dc and ac voltages and currents, as well as resistances.

We are now able to compare the circuit complexity and the required clock-cycles for the architectures discussed so far:

	Flash	Pipeline	SA	Integrating
Complexity :	2^n	n	1	1
Conversion :	1	1	n	2^n

12.5
OVERSAMPLING CONVERTERS

It is apparent that the most critical part of a data converter is its analog circuitry. Because of component mismatches and nonlinearities, drift and aging, noise, dynamic limitations and parasitics, resolution and speed can be pushed only so far. Oversampling converters ease analog-circuitry requirements at the expense of more complex digital circuitry. These converters are ideal for mixed-mode IC fabrication processes, where fast digital-processing circuitry is far more easily implemented than precise analog circuitry. The principal benefits of oversampling followed by digital filtering are *relaxed analog-filter requirements* and *quantization-noise reduction.* Sigma-delta (Σ-Δ) converters combine with these benefits the additional benefit of *noise shaping* to achieve truly high resolutions ($\geq$ 16 bits) with the simplest analog circuitry (1-bit digitizers).

Before embarking on the study of oversampling and noise shaping, we need to examine in greater detail conventional sampling, also referred to as *Nyquist-rate sampling.*

Nyquist-Rate Sampling

The digitization process, depicted in Fig. 12.36a, has a profound impact on the frequency spectrum of the input signal. We are primarily interested in the situation from dc to the sampling frequency f_S. As depicted in Fig. 12.36b, this range consists of two zones, namely, zone I extending from dc to $f_S/2$, and zone II extending from $f_S/2$ to f_S. Zone I is also called the *baseband,* and $f_S/2$ is called the *Nyquist bandwidth.* The effects of digitization are twofold:[1,2]

1. Digitization, viewed as *discretization in time,* creates additonal spectral components, called *images,* at locations symmetric about the midpoint $f_S/2$; for instance, a spectral component of v_I at $f = f_I$ results in an image at $f = f_S - f_I$, as shown in Fig. 12.36b, top.

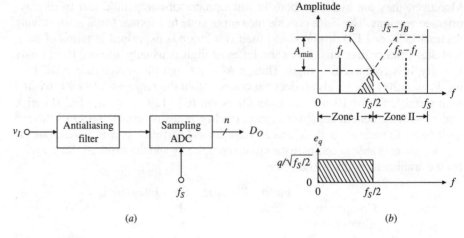

FIGURE 12.36
Nyquist sampling with analog filtering.

2. Digitization, viewed as *discretization in amplitude,* introduces quantization noise, as discussed in Section 12.1. The noise power of v_I folds into the baseband, in the manner depicted in Fig. 12.36b, bottom.

If v_I is a relatively active or busy signal, its quantization noise can, under certain conditions,[9,10] be treated as white noise with spectral density

$$e_q = \frac{q}{\sqrt{f_S/2}} \tag{12.19}$$

where $q = V_{FSR}/2^n \sqrt{12}$. The rms value is

$$E_q = \left(\int_0^{f_S/2} e_q^2 \, df \right)^{1/2} = q \tag{12.20}$$

or $E_q = V_{FSR}/2^n \sqrt{12}$, in accordance with Eq. (12.4). As we know, this results in

$$\text{SNR}_{\text{max}} = 6.02n + 1.76 \text{ dB} \tag{12.21}$$

With reference to Fig. 12.36b, top, we observe that as long as all spectral components of v_I lie within zone I, their images will be confined within zone II. Processing the spectrum of the digitized signal with a low-pass filter having a cutoff frequency of $f_S/2$ will pass the baseband components and block their images, thus allowing for the full recovery of the spectrum of v_I. This spectrum can, in turn, be used to reconstruct v_I itself. However, should v_I possess spectral components in zone II, their images will creep into zone I, overlapping the legitimate components there and causing nonlinear distortion. This phenomenon, referred to as *aliasing,* introduces an ambiguity that prevents the recovery of the spectrum of v_I. *Nyquist's criterion* states that if we want to recover or reconstruct a signal of a given bandwidth f_B from its digitized version, the sampling rate f_S must be such that

$$f_S > 2f_B \tag{12.22}$$

where $2f_B$ is called the *Nyquist rate*. This requirement can be met either by band-limiting v_I below $f_S/2$, or by raising f_S above the Nyquist rate.

A familiar aliasing example is offered by the spoked wheels of a stagecoach in a 16-mm, 24-frames-per-second Western. As long as the coach travels slowly enough relative to the camera's sampling rate of 24 frames per second, its wheels will appear to be turning correctly. However, as the coach speeds up, a point is reached where the wheels will appear to be slowing down, indicating an alias, or unwanted frequency, near the upper end of the baseband. Speeding up further will lower the alias frequency until it reaches dc, where the wheels will appear to be still. Any speed increase beyond this point will result in a negative alias frequency, making the wheels appear to be turning backward! These aliasing effects could be avoided either by limiting the filming only to slow scenes, or by increasing the number of frames per second.

In practical ADCs, to avoid wasting digital data rate, f_S is usually specified not far above the Nyquist rate of $2f_B$. For example, digital telephony, where the band of interest is $f_B = 3.2$ kHz and thus $2f_B = 6.4$ kHz, uses $f_S = 8$ kHz. Likewise, compact-disc audio, where $f_B = 20$ kHz and $2f_B = 40$ kHz, uses $f_S = 44.1$ kHz. Even though f_S is not strictly equal to $2f_B$, these converters are loosely referred to as *Nyquist-rate* converters.

It is apparent that in order to prevent any noise or spurious input spectral components above $f_S/2$ from folding into the baseband, an antialiasing filter is required. Such a filter must provide a flat response up to f_B and must roll off rapidly enough thereafter to provide the desired amount of suppression at $f_S/2$ and beyond. The shaded area of Fig. 12.36b, top, represents the baseband aliases of the unsuppressed signal and noise components above $f_S/2$. The contribution from these aliases must be kept below $\frac{1}{2}$ LSB by suitable choice of A_{min}. Such a choice, in turn, depends on the noise distribution and the spectral makeup of v_I for $f \geq f_S/2$. It is apparent that the performance requirements of the antialiasing filter can be quite stringent. Elliptic filters are a common choice for this task because of their sharp cutoff rate, if at the price of a nonlinear phase response.

Oversampling

Consider now the effect of speeding up the sampling rate by a factor of k, $k \gg 1$. This is shown in Fig. 12.37a. The ensuing benefits, illustrated in Fig. 12.37b, are twofold:

1. The transition band of the analog filter preceding the digitizer is now much wider, providing an opportunity for a drastic reduction in circuit complexity. In fact, in oversampling converters of the Σ-Δ type, this filter can be as simple as a mere *RC* stage!
2. The quantization noise is now spread over a wider band, or

$$e_q = \frac{q}{\sqrt{kf_S/2}} \tag{12.23}$$

indicating a spectral-density reduction by $\sqrt{k}$.

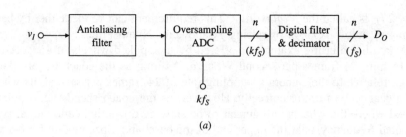

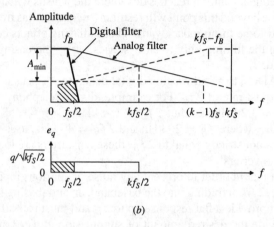

FIGURE 12.37
Oversampling with analog and digital filtering.

The price for the preceding benefits is the need for a *digital filter* at the output of the digitizer to (a) suppress any spectral components and noise above $f_S/2$, and (b) reduce the data rate from kf_S back to f_S, a process known as *decimation*. Though digital filters/decimators are beyond the scope of this book, it must be said that they can be designed for very sharp cutoff characteristics with good phase response. Moreover, they are far more easily implemented and maintained with temperature and time than their analog counterparts, and they can readily be reprogrammed in the software, if needed.

We observe that the rms noise at the output of the digitizer is still $V_{FSR}/2^n\sqrt{12}$; however, only the shaded portion will make it past the filter/decimator, so the rms noise at the output is

$$E_q = \left(\int_0^{f_S/2} \frac{q^2}{kf_S/2}\,df\right)^{1/2} = q/\sqrt{k} \qquad (12.24)$$

or $E_q = V_{FSR}/2^n\sqrt{12k}$. Expressing k in the form $k = 2^m$, we now have

$$\text{SNR}_{max} = 6.02(n + 0.5m) + 1.76 \text{ dB} \qquad (12.25)$$

indicating a $\frac{1}{2}$-bit improvement for every octave of oversampling.

EXAMPLE 12.6. An audio signal is oversampled with a 12-bit ADC. Find the oversampling frequency needed to achieve a 16-bit resolution. What is the corresponding SNR_{max}?

Solution. To gain $16 - 12 = 4$ bits of resolution we need to oversample by $m = 4/(1/2) = 8$ octaves, so the oversampling frequency must be $2^8 \times 44.1$ kHz $= 11.29$ MHz. Moreover, $SNR_{max} = 98.09$ dB.

Remark. Oversampling, while increasing resolution, does not improve linearity: the integral nonlinearity of the final 16-bit conversion cannot be better than that of the 12-bit ADC used!

Noise Shaping and Σ-Δ Converters

It is instructive to develop an intuitive feel for quantization-noise reduction. To this end, refer back to the 3-bit ADC example of Fig. 12.5, and suppose we apply a constant input V_I lying somewhere between $\frac{3}{8}$ V and $\frac{4}{8}$ V. The ADC will yield either $D_O = 011$ or $D_O = 100$, depending on whether V_I is closer to $\frac{3}{8}$ V or to $\frac{4}{8}$ V. Moreover, only *one sample* needs be taken to find D_O. An ingenious way to increase resolution above 3 bits is to add a Gaussian-noise dither $e_n(t)$ to V_I, and take *multiple samples* of the resulting signal $v_I(t) = V_I + e_n(t)$. Because of the fluctuations of $v_I(t)$, the samples will form a Gaussian distribution about some mean value, which we can easily compute by taking the average of our multiple readings. The result gives a more accurate estimate of V_I! In fact, Eq. (12.25) indicates that we need four samples to increase resolution by 1 bit, sixteen samples to increase by 2 bits, sixty-four samples to increase by 3 bits, and so forth.

Σ-Δ ADCs use feedback for the double purpose of (*a*) generating dither to keep the input busy, and (*b*) reshaping the noise spectrum to reduce the amount of oversampling required. In its simplest form[1] depicted in Fig. 12.38*a*, a Σ-Δ ADC consists of a 1-bit digitizer or modulator to convert v_I to a high-frequency serial data stream v_O, followed by a digital filter/decimator to convert this stream to a sequence of n-bit words of fractional binary value D_O at a lower rate of f_S words per second. The modulator is made up of a latched comparator acting as a 1-bit ADC, a 1-bit DAC, and an integrator to integrate (Σ) the difference (Δ) between v_I and the DAC output; hence the name Σ-Δ ADC. The comparator is strobed at a rate of kf_S sps, where k, usually a power of 2, is called the *oversampling ratio*.

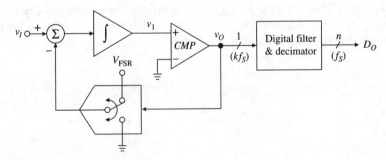

FIGURE 12.38
First-order Σ-Δ ADC.

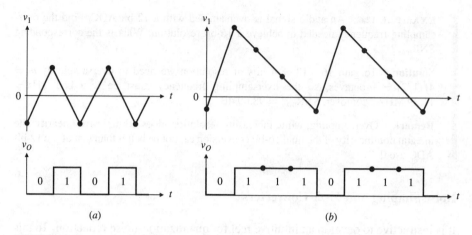

FIGURE 12.39
Integrator and comparator outputs for (a) $v_I = 0.5V_{FSR}$, and (b) $v_I = 0.75V_{FSR}$.

Figure 12.39 shows the integrator and comparator outputs for two representative input conditions (the dots mark the instants in which *CMP* is strobed). In (a) v_I is set at midrange, so the serial stream contains an equal number of 0s and 1s. To decode this stream with a 2-bit resolution, we pass it through a digital filter which computes its average over four samples. The result is the fractional binary value $D_O = 10$, corresponding to $(\frac{1}{2} + \frac{0}{4})V_{FSR}$, or $0.5V_{FSR}$. In (b) v_I is set at $\frac{3}{4}$ of the range, so the serial stream contains three 1s for every 0. After averaging, this gives $D_O = 11$, corresponding to $(\frac{1}{2} + \frac{1}{4})V_{FSR}$, or $0.75V_{FSR}$. It is apparent that the distribution of 0s and 1s in the serial stream depends on the value of v_I within the range of 0 to V_{FSR}.

To understand how noise shaping comes about, refer to Fig. 12.40, where the quantization error is modeled additively via the noise process $e_{qi}(jf) = q/\sqrt{kf_S/2}$. By inspection, the various Fourier transforms are related as $V_o = e_q + H \times (V_i - V_o)$, or

$$V_o(jf) = \frac{1}{1 + 1/H(jf)} V_i(jf) + \frac{1}{1 + H(jf)} e_{qi}(jf) \qquad (12.26)$$

Choosing $H(jf)$ such that its magnitude is sufficiently large over the frequency band of interest will provide the simultaneous benefits of (a) making V_o *closely track* V_i over the given band and (b) *drastically reducing* quantization noise over the same band. This is not surprising for the observant reader who has already noted the similarity of Fig. 12.40 to Fig. 1.25, or the similarity of Eq. (12.26) to Eq. (1.53) with H playing the role of T and e_{qi} that of x_3.

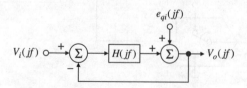

FIGURE 12.40
Linear system model of a Σ-Δ ADC.

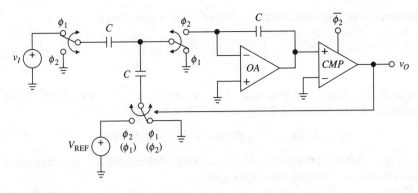

FIGURE 12.41
Switched-capacitor implementation of a first-order modulator. Bottom-switch
phase is (ϕ_1, ϕ_2) for $v_O =$ high, and (ϕ_2, ϕ_1) for $v_O =$ low.

For frequency bands extending down to dc, $H(jf)$ is usually implemented
with integrators; however, depending on the application, other filter types may be
more efficient, such as band-pass filters in telecommunications.[11] In mixed-mode IC
processes, $H(jf)$ is implemented using switched-capacitor techniques. Figure 12.41
shows an SC realization[11] of the 1-bit modulator. Using Eq. (4.22) with $C_1 = C_2$,
$\omega = 2\pi f$, and $T_{CK} = 1/kf_S$, we can express the SC integrator transfer function as
$H(jf) = 1/[\exp(j2\pi f/kf_S) - 1]$. Substituting into Eq. (12.26) gives

$$V_o(jf) = V_i(jf)e^{-j2\pi f/kf_s} + e_{qo}(jf) \qquad (12.27)$$

$$e_{qo}(jf) = (1 - e^{-j2\pi f/kf_s})e_{qi}(jf) \qquad (12.28)$$

By the well-known Fourier-transform property that multiplying by $\exp(-j\omega T)$ in
the frequency domain is equivalent to delaying by T in the time domain, Eq. (12.27)
indicates that v_O is simply v_I delayed by $1/kf_S$. Moreover, applying Euler's identity
to Eq. (12.28), we can write

$$|e_{qo}(jf)| = 2\sin(\pi f/kf_S)|e_{qi}(jf)| \qquad (12.29)$$

The plot of Fig. 12.42 reveals that the modulator shifts most of the noise energy
toward higher frequencies. Only the shaded portion will make it past the filter/

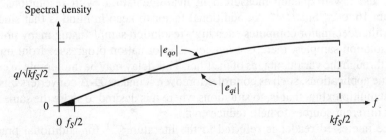

FIGURE 12.42
First-order noise shaping ($k = 16$).

decimator, so the corresponding rms output noise is obtained as

$$E_q = \left(\int_0^{f_S/2} |e_{qo}(jf)|^2 df \right)^{1/2} \tag{12.30}$$

For $k \gg \pi$, we obtain (see Problem 12.22) $E_q = \pi q/\sqrt{3k^3} = \pi V_{FSR}/(2^n \sqrt{36k^3})$. Expressing k in the form $k = 2^m$ gives, for a first-order Σ-Δ ADC,

$$\text{SNR}_{\max} = 6.02(n + 1.5m) - 3.41 \text{ dB} \tag{12.31}$$

indicating a 1.5-bit improvement for every octave of oversampling; this is better than the 0.5-bit improvement without noise shaping.

The benefits of noise shaping can be enhanced further by using higher-order modulators. For instance, suitably cascading[11] two subtractor-integrator blocks gives a second-order Σ-Δ ADC with

$$|e_{qo}(jf)| = [2 \sin(\pi f/k f_S)]^2 |e_{qi}(jf)| \tag{12.32}$$

Substituting into Eq. (12.30), we obtain (see Problem 12.22), for $k \gg \pi$, $E_q = \pi^2 q/\sqrt{5k^5} = \pi^2 V_{FSR}/(2^n \sqrt{60k^5})$. This yields, for a second-order Σ-Δ ADC,

$$\text{SNR}_{\max} = 6.02(n + 2.5m) - 11.14 \text{ dB} \tag{12.33}$$

indicating a 2.5-bit improvement for every octave of oversampling.

> **EXAMPLE 12.7.** Find k for $\text{SNR}_{\max} \geq 96$ dB (or ≥ 16 bits) using (a) a first-order and (b) a second-order Σ-Δ ADC.
>
> **Solution.**
>
> (a) Imposing $6.02(1 + 1.5m) - 3.41 \geq 96$ gives $m \geq 10.3$, or $k \geq 2^{10.3} \cong 1261$.
> (b) Similarly, $k \geq 2^{6.7} \cong 105$.

Besides offering the aforementioned advantages of undemanding and mixed-mode-compatible analog circuitry, 1-bit quantizers are inherently linear: since only two output levels are provided, a straight characteristic results, with no need for trimming or calibration as in multilevel quantizers. Moreover, the presence of the integrator makes the input SHA unnecessary—if at the price of more stringent input-drive requirements due to charge injection effects.[12]

Practical upper limits on sampling rates currently restrict Σ-Δ ADCs to moderate-speed but high-resolution applications, such as digital audio, digital telephony, and low-frequency measurement instrumentation, with resolutions ranging from 16 to 24 bits.[12-14] An additional factor to keep in mind is that since the digital filter/decimator computes each high-resolution sample using many previous low-resolution samples, there is a *latency* as information progresses from input to output through the various stages of the filter. This delay may be intolerable in certain real-time applications, such as control. Moreover, it makes Σ-Δ converters unsuited to input multiplexing, that is, to situations where it is desired to share the same ADC among different sources to help reduce cost.

The interested reader is referred to the literature[9-11] for additional practical issues such as stability and idle tones, system architectures, and the fascinating subject of digital filtering and decimation.

12.1 Performance specifications

12.1 A 3-bit DAC designed for $V_{FSR} = 3.2$ V is sequenced through all input codes from 000 to 111, and the actual output values are found to be $v_O = 0.2, 0.5, 1.1, 1.4, 1.7, 2.0,$ 2.6, and 2.9, all in V. Find the offset error, the gain error, the INL, and the DNL, in fractions of 1 LSB.

12.2 A full-scale sinusoid is applied to a 12-bit ADC. If the digital analysis of the output reveals that the fundamental has a normalized power of 1 W while the remaining power is 0.6 μW, find the effective number of bits of this ADC. What is the SNR if the input sinusoid is reduced to 1/100th of full scale?

12.2 D-A conversion techniques

12.3 A 6-bit weighted-resistor DAC of the type of Fig. 12.7 is implemented with $V_{REF} =$ 1.600 V, but with $R_f = 0.99R$ instead of $R_f = R$, and a low-quality op amp having $V_{OS} = 5$ mV and $a = 200$ V/V. Find the offset and gain errors of this DAC, in fractions of 1 LSB. What is the worst-case value of the output when all bits are set to 1?

12.4 A 4-bit weighted-resistor DAC of the type of Fig. 12.7 is implemented with $V_{REF} =$ -3.200 V and a high-quality op amp, but gross resistor values, namely, $R_f = 9.0$ kΩ instead of 10 kΩ, $2R = 22$ kΩ instead of 20 kΩ, $4R = 35$ kΩ instead of 40 kΩ, $8R =$ 50 kΩ instead of 80 kΩ, and $16R = 250$ kΩ instead of 160 kΩ. Find the gain error, along with the integral and differential nonlinearities. Comment on your findings.

12.5 The AH5010 quad switch (National Semiconductor) consists of four analog-ground p-FET switches and relative diode clamps of the type of Fig. 9.37, plus a fifth dummy FET for $r_{ds(on)}$ compensation. (a) Using an LM385 2.5-V reference diode, an AH5010 quad switch ($r_{ds(on)} \cong 100$ Ω), and a JFET-input op amp with ±15-V supplies, design a 4-bit weighted-resistor DAC with $V_{FSR} = +10.0$ V. (b) Compute v_O for each input code. (c) Repeat if the op amp has $V_{OS} = 1$ mV. What are the offset and gain errors of your DAC?

12.6 One way of curbing excessive resistance spread in an 8-bit weighted-resistor DAC is by combining the outputs of two 4-bit DACs as $v_O = v_{O(MS)} + 2^{-4}v_{O(LS)}$, where $v_{O(MS)}$ is the output of the DAC using the four MSBs of the 8-bit code, and $v_{O(LS)}$ that of the DAC using the four LSBs. Using components of the type of Problem 12.5, design one such 8-bit DAC.

12.7 (a) Using an 8-bit R-$2R$ ladder with $R = 10$ kΩ, an LM385 2.5-V reference diode, and a 741 op amp, design an 8-bit voltage-mode DAC with $V_{FSR} = 10$ V. (b) Modify your circuit so that v_O is offset by -5 V. Assume ±15-V regulated supplies.

12.8 (a) Derive expressions for the element values in the Norton equivalent of the current-driven R-$2R$ ladder DAC of Fig. 12.17. (b) Suppose $V_{REF}/R_r = 1$ mA, $R = 1$ kΩ, and the output of the DAC is fed to a simple I-V converter op amp with a feedback resistance of 1 kΩ. If the I-V converter introduces an offset error of $\frac{1}{4}$ LSB and a gain error of $-\frac{1}{2}$ LSB, find the I-V converter output for $b_1b_2b_3b_4 = 0000, 0100, 1000,$ 1100, and 1111. (c) Find the closed-loop small-signal bandwidth if the op amp has a constant GBP of 50 MHz.

12.3 Multiplying DAC applications

12.9 The programmable attenuator of Fig. 12.23a can be turned into a programmable attenuator/amplifier by using a T-network of the type of Fig. 2.2 in the feedback path. This is achieved by interposing a voltage divider between the op amp output and the R_f pin of the DAC (see Analog Devices Application Note AN-137). Using a 12-bit MDAC with $R_f = 10$ kΩ, design a circuit whose gain can be varied from $\frac{1}{64}$ V/V to 64 V/V as the input code is sequenced from $0\ldots01$ to $1\ldots11$.

12.10 Consider the circuit obtained from the biquad filter of Fig. 3.36 by replacing the inverting amplifier (OA_3 plus the R_3 resistances) with the programmable attenuator of Fig. 12.23a. Find an expression for the band-pass response, and verify that both f_0 and Q are proportional to $\sqrt{D}$, indicating a digitally programmable, constant-bandwidth band-pass filter.

12.11 Consider the circuit obtained from Fig. 12.24 by removing R_4, $MDAC_2$, OA_5, and the OA_3 integrator. (a) Sketch the reduced circuit, and show that now OA_1 and OA_2 provide, respectively, the first-order high-pass and low-pass responses. (b) Specify suitable components so that the low-pass response has a dc gain of 20 dB, the high-pass response has a high-frequency gain of 0 dB, and the characteristic frequency is digitally programmable in 5-Hz steps by means of a dual 10-bit MDAC.

12.12 Modify the quadrature oscillator of Fig. 10.6 for peak amplitudes of 5 V and f_0 digitally programmable in 10-Hz steps by means of a dual 10-bit MDAC.

12.13 Using a 12-bit MDAC and an AD537 wide-sweep CCO (see Fig. 10.33), design a triangular wave generator with peak values of ±5 V and f_0 digitally programmable in 10-Hz steps. The circuit is to have provision for both frequency and amplitude calibration. Assume the triangular wave available across the timing capacitor of the AD537 has a peak-to-peak amplitude of $\frac{5}{3}$ V.

12.14 Using an 8-bit CMOS DAC of the type of Fig. 12.11, an LM385 2.5-V reference diode, and an LM317 regulator of the type of Fig. 11.26, along with other components as needed, design a 1-A power supply digitally programmable over the range 0.0 V to 10.0 V. Assume ±15-V supplies.

12.4 A-D conversion techniques

12.15 As we know, a SA ADC must usually be preceded by a THA. However, if the input is sufficiently slow to change by less than $\pm\frac{1}{2}$ LSB during the conversion cycle, then the THA is unnecessary. (a) Show that a full-scale sine wave input can be converted without the need for a THA, provided its frequency is below $f_{max} = 1/2^n\pi t_{SAC}$, where t_{SAC} is the time it takes for the SA ADC to complete a conversion. (b) Find f_{max} for an 8-bit SA ADC operating at the rate of 10^6 conversions per second. How does f_{max} change if the SA ADC is preceded by an ideal SHA?

12.16 Discuss the general requirements on the reference, DAC, and comparator of an 8-bit SA ADC for a conversion time of 1 μs over the range $0\,°C \le T \le 50\,°C$ with an accuracy of $\pm1/2$ LSB, if $V_{FSR} = 10$ V.

12.17 Consider a charge-redistribution ADC of the type of Fig. 12.29 with $n = 4$, $V_{REF} = 3.0$ V, and $C = 8$ pF. Assuming node v_P has a parasitic capacitance of

4 pF toward ground, find the intermediate values of v_P during the conversion of $v_I = 1.00$ V.

12.18 Assume the 8-bit subranging ADC of Fig. 12.32 has $V_{REF} = 2.560$ V. (a) Find the total number of comparators, their voltage reference levels, and the maximum level tolerances allowed for a $\pm\frac{1}{2}$ LSB accuracy. (b) Find $b_1 \ldots b_8$, v_{RES}, and the quantization error for $v_I = 0.5$ V, 1.054 V, and 2.543 V.

12.19 Show that if the input to the dual-slope ADC of Fig. 12.34 contains an unwanted ac component of the type $v_i = V_m \cos(\omega t + \theta)$, then the result of integrating it over the interval $T = 2^n T_{CK}$ is proportional to the *sampling function* $\text{Sa}(\omega T) = \sin(\omega T)/(\omega T)$. Plot $|\text{Sa}(\omega T)|_{dB}$ vs. ωT, and verify that this type of ADC inherently rejects all unwanted ac components whose frequencies are integral multiples of $1/T$.

12.20 The integrator of a dual-slope ADC is implemented with an op amp having gain $a = 10^3$ V/V. (a) Assuming its output $v_O(t)$ is initially zero, find $v_O(t \geq 0)$ if the input is $v_I = 1$ V. (b) Find the minimum value of RC so that $v_O(t = 100$ ms) is afflicted by an error of less than 1 mV.

12.21 A 14-bit dual-slope ADC of the type of Fig. 12.34 is to be designed so that it rejects the 60-Hz power-line interference frequency and harmonics thereof. (a) What is the required clock frequency f_{CK}? What is the time required to convert a full-scale input? (b) If $V_{REF} = 2.5$ V and the input is in the range 0 to 5 V, what is the value of RC for a peak value of 5 V at the integrator's output for a full-scale input? (c) If component aging causes R to change by $+5\%$ and C by -2%, what is the effect upon the integrator's output for the case of a full-scale input? Upon the conversion's accuracy?

12.5 Oversampling converters

12.22 (a) Plot $|e_{qo}(jf)|$, $0 \leq f \leq k f_S/2$ for the second-order Σ-Δ ADC, and compare with the first order. (b) Show that the rms noise before digital filtering is $\sqrt{2}q$ for the first-order modulator, and $\sqrt{8}q$ for the second-order modulator. (c) Using the approximation $\sin x \cong x$ for $x \ll 1$, show that the rms noise after digital filtering is, for $k \gg \pi$, $\pi q/\sqrt{3k^3}$ for the first-order modulator, and $\pi^2 q/\sqrt{5k^5}$ for the second-order modulator. (d) Find the rms noise percentage removed by the digital filter for both orders if $k = 16$.

12.23 Compare the sampling rates needed for a 16-bit audio ADC using a 1-bit ADC with (a) straight oversampling, (b) first-order noise shaping, and (c) second-order noise shaping.

12.24 An 8-bit ADC that is linear to 12 bits is used to perform conversions over a 100-kHz signal bandwidth. (a) Find the sampling rate required to achieve 12 bits of accuracy using straight oversampling. (b) Repeat, but for the case in which the above ADC is placed inside a first-order Σ-Δ modulator. (c) Repeat, but for a second-order modulator.

12.25 An oversampling audio ADC with $n = 16$, $V_{FSR} = 2$ V, $f_S = 48$ kHz, and $k f_S = 64 f_S$ uses a simple RC network as the input antialiasing filter. (a) Specify RC for a maximum attenuation of 0.1 dB for $0 \leq f \leq 20$ kHz. (b) Assuming the spectral makeup of v_I within the first image band $k f_S \pm 20$ kHz is just white noise with spectral density e_{nw}, find the maximum allowed value of e_{nw} so that the corresponding base-band rms noise is less than $\frac{1}{2}$ LSB.

REFERENCES

1. Analog Devices Engineering Staff, *Mixed-Signal Design Seminar,* Analog Devices, Norwood, MA, 1991.

2. A. B. Grebene, *Bipolar and MOS Analog Integrated Circuit Design,* John Wiley & Sons, New York, 1984.

3. Analog Devices Engineering Staff, *Analog-Digital Conversion Handbook,* 3d ed., Prentice-Hall, Englewood Cliffs, NJ, l986.

4. B. Razavi, *Principles of Data Conversion System Design,* IEEE Press, Piscataway, NJ, 1995.

5. A. P. Brokaw, "Analog Signal-Handling for High Speed and Accuracy," Application Note AN-342, *Applications Reference Manual,* Analog Devices, Norwood, MA, 1993.

6. J. Wilson, G. Whitmore, and D. Sheingold, *Application Guide to CMOS Multiplying D/A Converters,* Analog Devices, Norwood, MA, 1978.

7. J. Williams, "Build Your Own A/D Converter for Optimum Performance," *EDN,* March 20, 1986, pp. 191–198.

8. P. Klonowski, "Analog-to-Digital Conversion Using Voltage-to-Frequency Converters," Application Note AN-276, *Applications Reference Manual,* Analog Devices, Norwood, MA, 1993.

9. M. W. Hauser, "Principles of Oversampling A/D Conversion," *J. Audio Eng. Soc.,* January/February 1991, pp. 3–26.

10. J. C. Candy and G. C. Temes, editors, *Oversampling Delta-Sigma Data Converters,* IEEE Press, Piscataway, NJ, 1992.

11. D. A. Johns and K. W. Martin, *Analog Integrated Circuit Design,* John Wiley & Sons, New York, 1997.

12. Analog Devices Engineering Staff, *Practical Analog Design Techniques,* Analog Devices, Norwood, MA, 1995.

13. Burr-Brown Engineering Staff, *Burr-Brown Design Seminar Manual,* Burr-Brown, Tucson, AZ, 1996.

14. Analog Devices Engineering Staff, *High-Speed Analog Design Techniques,* Analog Devices, Norwood, MA, 1996.

13

NONLINEAR AMPLIFIERS AND PHASE-LOCKED LOOPS

13.1 Log/Antilog Amplifiers
13.2 Analog Multipliers
13.3 Operational Transconductance Amplifiers
13.4 Phase-Locked Loops
13.5 Monolithic PLLs
    Problems
    References

The highly predictable characteristic of the bipolar junction transistor is exploited in the realization of some very useful nonlinear functions, such as logarithmic conversion and variable transconductance multiplication. These functions, in turn, provide the basis for a variety of other analog operations, such as antilogarithmic amplification, true rms conversion, analog division and square-root computation, various forms of linearization, and voltage-controlled amplification, filtering, and oscillation. These precise building blocks simplify analog design considerably while broadening the scope of practical analog circuits to applications where considerations of speed or cost require implementation in analog rather than digital form.

Another important class of nonlinear circuits is provided by phase-locked loops. Though unrelated to those just mentioned, PLLs encompass many of the important topics that we have studied so far. We thus find it appropriate to conclude the book with this subject.

608

CHAPTER 13
Nonlinear
Amplifiers and
Phase-Locked
Loops

13.1
LOG/ANTILOG AMPLIFIERS

A logarithmic amplifier—also called *log amp,* or *logger*—is an *I-V* converter with a transfer characteristic of the type

$$v_O = V_o \log_b \frac{i_I}{I_i} \tag{13.1}$$

where V_o is called the *output scale factor,* I_i the *input reference current,* and b is the *base,* usually 10 or 2. V_o represents the sensitivity of the log amp, in volts per decade (or per octave), and I_i is the value of i_I for which $v_O = 0$. Note that for proper operation we must always have $i_I/I_i > 0$. The quantity

$$DR = \log_b \frac{|i_I|_{max}}{|i_I|_{min}} \tag{13.2}$$

is called the *dynamic range* and is expressed in decades or in octaves, depending on b. For instance, a logger designed to operate over the range $1 \text{ nA} \le i_I \le 1 \text{ mA}$ has $DR = \log_{10}(10^{-3}/10^{-9}) = 6$ decades, or $DR = \log_2 10^6 \cong 20$ octaves.

Plotting Eq. (13.1) on semilog paper with i_I/I_i on the logarithmic axis and v_O on the linear axis, as in Fig. 13.1a, yields a straight line with a slope of V_o V/dec. Any departure of the actual characteristic from the best-fit straight line is called the *log conformity error* e_O. Though this error can only be observed at the output, it is convenient to refer it to the input because of the unique log-function property that equal percentage errors at the input produce equal incremental errors at the output, regardless of the point on the curve. Indeed, denoting the percentage input error as p, we have $e_O = v_{O(actual)} - v_{O(ideal)} = V_o \log_b[(1+p)(i_I/I_i)] - V_o \log_b[i_I/I_i]$, or

$$e_O = V_o \log_b(1+p) \tag{13.3}$$

For instance, with $b = 10$ and $V_o = 1$ V/dec, a 1% input error corresponds to an output error $e_O = 1 \log_{10}(1 + 0.01) = 4.32$ mV. Conversely, $e_O = 10$ mV corresponds to a percentage error p such that $10 \text{ mV} = 1 \log_{10}(1 + p)$, or $p = 2.33\%$.

The main application of log amps is data compression. As an example, consider the digitization of a photodetector current over the range $10 \text{ nA} \le i_I \le 100 \mu A$ with an error of less than 1% of its actual value. Since we have a four-decade range, the required resolution is $0.01/10^4 = 1/10^6$, or 1 ppm. Since $10^6 \cong 2^{20}$, this requires

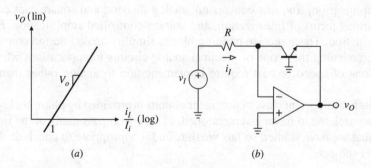

(a)   (b)

FIGURE 13.1
Logarithmic characteristic and the transdiode configuration.

a 20-bit A/D converter, which can be a challenging and expensive proposition. Consider now the effect of compressing the input with a log amp before digitizing. Letting, for instance, $b = 10$, $V_o = 1$ V/dec, and $I_i = 10$ nA, the current range is now compressed to a 0 to 4-V voltage range. Since a 1% current accuracy corresponds to a 4.32-mV voltage interval, the required resolution is now $(4.32 \times 10^{-3})/4 \cong 1/926 \cong 1/2^{10}$, or 10 bits. This represents a substantial reduction in cost and circuit complexity!

The inverse function of logarithmic compression is exponential expansion. This is provided by the *antilogarithmic amplifier* (antilog amp), whose transfer characteristic is

$$i_O = I_o b^{v_I/V_i} \tag{13.4}$$

where I_o is the *output reference current* and V_i the *input scale factor*, in volts per decade or per octave. The output of an antilog amp can be converted to a voltage by means of an op amp *I-V* converter. When plotted on semilog paper with v_I on the linear axis and i_O/I_o on the logarithmic axis, Eq. (13.4) also yields a straight line. The above log conformity error considerations still hold, but with the input and output errors interchanged.

The Transdiode Configuration

Log/antilog amplifiers exploit the exponential characteristic of a forward-active BJT. By Eq. (5.3), this characteristic can be written as $v_{BE} = V_T \ln(i_C/I_s)$. Practical logging BJTs conform to this equation remarkably well over a range of at least six decades,[1] typically for 0.1 nA $\leq i_C \leq$ 0.1 mA. The heart of log/antilog amps is the circuit of Fig. 13.1b, known as the *transdiode configuration*. The op amp converts v_I to the current $i_I = v_I/R$, and then forces the BJT in its feedback path to respond with a logarithmic base-emitter voltage drop to yield

$$v_O = -V_T \ln \frac{v_I}{RI_s} \tag{13.5}$$

If we take into account also the input bias voltage V_{OS} and bias current I_B, then the collector current becomes $i_C = (v_I - V_{OS})/R - I_B$, so the transfer characteristic takes on the more realistic form

$$v_O = -V_T \ln \frac{v_I - V_{OS} - RI_B}{RI_s} \tag{13.6}$$

The input offset error $(V_{OS} + RI_B)$ sets the ultimate limit on the range of inputs that can be processed within a given log conformity error. Wide-dynamic-range loggers use op amps with ultra-low V_{OS} and I_B to approach the ideal characteristic of Eq. (13.5). The ultimate limit is then posed by drift and noise. If the transdiode is driven directly with a current source i_I, Eq. (13.5) reduces to $v_O = -V_T \ln(i_I/I_s)$, and the ultimate limit is now set by the input bias current of the op amp or by the low-end log conformity error of the BJT, whichever is higher. In general, current-driven loggers offer a wider dynamic range than voltage-driven loggers.

610

CHAPTER 13
Nonlinear
Amplifiers and
Phase-Locked
Loops

Stability Considerations

Transdiode circuits are notorious for their tendency to oscillate due to the presence of an active gain element inside the feedback loop. As shown in Fig. 13.2a, the transdiode is stabilized[1] by using an emitter-degeneration resistance R_E to reduce the feedback factor β, and a feedback capacitance C_f to provide feedback lead. To investigate stability we need to find the feedback factor β. To this end, refer to the ac model of Fig. 13.2b, where the BJT has been replaced by its common-base small-signal model.[2] The BJT parameters r_e and r_o depend on the operating current I_C as

$$r_e = \frac{\alpha V_T}{I_C} \qquad r_o = \frac{V_A}{I_C} \qquad (13.7)$$

where V_T is the thermal voltage and V_A is the so-called Early voltage. Typically, $\alpha \cong 1$ and $V_A \cong 100$ V. The base-collector junction capacitance C_μ and the inverting-input stray capacitance C_n are typically on the order of a few picofarads.

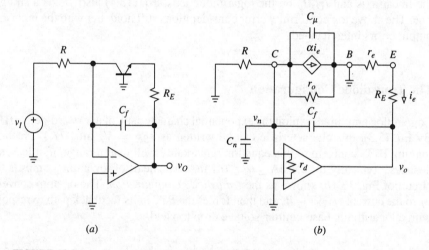

(a)   (b)

FIGURE 13.2
Transdiode circuit with frequency compensation, and its incremental model.

Circuit analysis is facilitated by the introduction of

$$R_a = R \parallel r_o \parallel r_d \qquad R_b = r_e + R_E \qquad (13.8)$$

Applying KCL at the summing junction gives

$$v_n[1/R_a + j\omega(C_n + C_\mu)] + \alpha i_e + j\omega C_f(v_n - v_o) = 0$$

Letting $i_e = -v_o/R_b$, rearranging, and solving for $\beta = v_n/v_o$ gives, for $\alpha \cong 1$,

$$\frac{1}{\beta} = \frac{R_b}{R_a}\frac{1 + jf/f_z}{1 + jf/f_p} \qquad (13.9)$$

where the zero and pole frequencies are

$$f_z = \frac{1}{2\pi R_a(C_n + C_\mu + C_f)} \qquad f_p = \frac{1}{2\pi R_b C_f} \qquad (13.10)$$

The $|1/\beta|$ curve has the low-frequency asymptote $1/\beta_0 = R_b/R_a$, the high-frequency asymptote $1/\beta_\infty = 1 + (C_n + C_\mu)/C_f$, and two breakpoints at f_z and f_p. While f_z and $1/\beta_\infty$ are relatively constant, f_p and $1/\beta_0$ depend on the operating current as per Eq. (13.7), so they can vary over a wide range of values, as exemplified in Fig. 13.3. The hardest condition to compensate is when i_C is maximized, since this minimizes $1/\beta_0$ and maximizes f_p, leading to the highest rate of closure. As a rule of thumb,[1] R_E is chosen so that, when i_C is maximized, $1/\beta_0 \cong 0.5$ V/V and $f_p \cong 0.5 f_x$, where f_x is the crossover frequency.

EXAMPLE 13.1. In the circuit of Fig. 13.2a let $R = 10$ kΩ, 1 mV $< v_I < 10$ V, $C_n + C_\mu = 20$ pF, $V_A = 100$ V, $r_d = 2$ MΩ, and $f_t = 1$ MHz. Find suitable values for R_E and C_f.

Solution. At the upper end of the range, where $i_C = (10 \text{ V})/(10 \text{ k}\Omega) = 1$ mA, we have $r_e \cong 26 \ \Omega$, $r_o \cong 100$ kΩ, and $R_a \cong 9$ kΩ. Imposing $(26 + R_E)/9000 = 0.5$ gives $R_E = 4.47$ kΩ (use 4.3 kΩ).

Next, find f_x using the definition $|a(jf_x)| \times \beta_\infty = 1$. Letting $|a(jf_x)| \cong f_t/f_x$ and using $1/\beta_\infty = 1 + (C_n + C_\mu)/C_f$ we get $f_x = f_t/[1 + (C_n + C_\mu)/C_f]$. Imposing $f_p = 0.5 f_x$ and simplifying, we finally obtain

$$C_f = \frac{1 + (C_n + C_\mu)/C_f}{\pi R_b f_t}$$

Substituting the given parameter values, along with $R_b \cong 4.3$ kΩ, gives $C_f = 90$ pF (use 100 pF).

Figure 13.3 indicates that at low values of v_I the response is dominated by f_p, thus resulting in slow dynamics. This is not surprising, since at low current levels it takes longer to charge or discharge the various capacitances. At low currents we have $r_e \gg R_E$, so $f_p \cong 1/2\pi r_e C_f$, indicating a time constant $\tau \cong r_e C_f \cong (V_T/I_C)C_f = (V_T/v_I)RC_f$. For instance, with $C_f = 100$ pF, at $I_C = 1$ nA we have $\tau \cong (0.026/10^{-9})10^{-10} = 2.6$ ms, so we must be prepared for slow dynamics near the low end of the range.

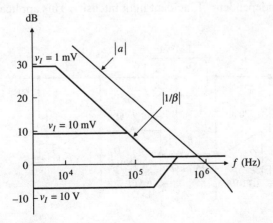

FIGURE 13.3
Bode plots for the transdiode circuit of Example 13.1.

612

CHAPTER 13
Nonlinear
Amplifiers and
Phase-Locked
Loops

Practical Log/Antilog Circuits[3]

Both the output scale factor and the input reference term in Eq. (13.5) depend on temperature. The circuit of Fig. 13.4 overcomes this serious drawback by using a matched BJT pair to eliminate I_s, and a temperature-sensitive voltage divider to compensate for $TC(V_T)$. The op amps force the BJTs to develop $v_{BE1} = V_T \ln(i_I/I_{s1})$ and $v_{BE2} = V_T \ln(I_{REF}/I_{s2})$, where $I_{REF} = V_{REF}/R_r$. By the voltage divider formula, $v_{B2} = v_O/(1 + R_2/R_1)$. But, by KVL, $v_{B2} = v_{BE2} - v_{BE1} = V_T \ln[(I_{REF}/I_{s2})(I_{s1}/i_I)]$. Eliminating v_{B2} and using the property $\ln x = 2.303 \log_{10} x$, we get

$$v_O = V_o \log_{10} \frac{i_I}{I_i} \tag{13.11}$$

$$V_o = -2.303 \frac{R_1(T) + R_2}{R_1(T)} V_T \qquad I_i = \frac{V_{REF}}{R_r} \times \frac{I_{s2}}{I_{s1}} \tag{13.12}$$

For matched BJTs $I_{s2}/I_{s1} = 1$, so we get the temperature-independent expression $I_i = V_{REF}/R_r$. Moreover, for $R_2 \gg R_1$, we can approximate $V_o \cong -2.303 R_2 V_T/R_1(T)$, indicating that V_o can be thermally stabilized by using a resistance $R_1(T)$ with $TC(R_1) = TC(V_T) = 1/T = 3660$ ppm/°C. A suitable resistor is the Q81 (Tel Labs), which must be mounted in close thermal coupling with the BJT pair. The function of D_1 is to protect the BJTs against inadvertent reverse bias. The use of the LT1012 picoampere-input-current, microvolt-offset, low-noise op amp (Linear Technology) allows for a voltage-logging range of $4\frac{1}{2}$ decades. With the given component values, $V_o = -1$ V/dec and $I_i = 10$ μA, so $v_O = -(1$ V/dec$) \log_{10}[v_I/(0.1$ V$)]$. V_o and I_i are calibrated via R_2 and R_r.

If the input reference current I_i is allowed to vary, the log amp is called a *log ratio amplifier* and finds application in wide-dynamic-range ratiometric measurements where the unknown signal is measured against a reference signal that is itself variable. Typical examples are absorbance measurements in medicine and pollution control, where light transmitted through a specimen is measured against incident light, and the result must be independent of incident light intensity. This application is illustrated

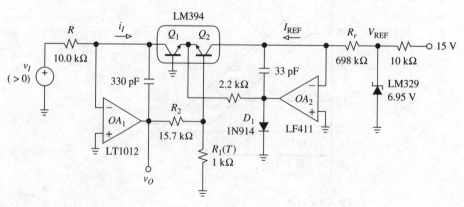

FIGURE 13.4
Logarithmic amplifier.

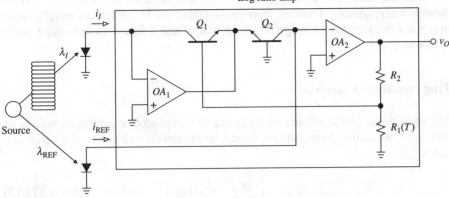

FIGURE 13.5
Log ratio amplifier for absorbance measurements.

in Fig. 13.5, where frequency compensation and reverse-bias protection have been omitted for simplicity. The transmitted light λ_I and the incident light λ_{REF} are converted to the proportional currents i_I and i_{REF} by a pair of matched photodiodes operating in the photovoltaic mode. Then, the circuit computes the log ratio $v_O = V_o \log_{10}(i_I/i_{REF}) = V_o \log_{10}(\lambda_I/\lambda_{REF})$, where V_o is given by Eq. (13.12).

Figure 13.6 shows how the log amp can be rearranged to implement an exponential amp. It is left as an exercise for the reader (see Problem 13.4) to prove that the circuit gives

$$i_O = I_0 10^{v_I/V_i} \tag{13.13}$$

$$I_0 = \frac{V_{REF}}{R_r} \times \frac{I_{s2}}{I_{s1}} \qquad V_i = -2.303 \frac{R_1(T) + R_2}{R_1(T)} V_T \tag{13.14}$$

With the given component values, $I_o = 0.1$ mA and $V_i = -1$ V/dec. It is important that the collector of Q_2 be returned to a 0-V node, such as the virtual-ground node of the I-V converter OA_2, in order to nullify the collector-base leakage current of Q_2. Otherwise, this current may degrade log conformity at the low end of the range.

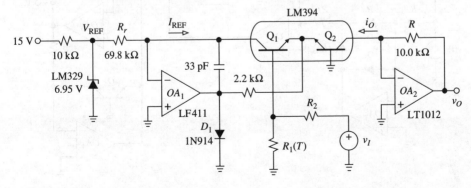

FIGURE 13.6
Antilog amplifier.

614

CHAPTER 13
Nonlinear
Amplifiers and
Phase-Locked
Loops

Log, log-ratio, and antilog amplifiers are available in IC form from various manufacturers (Analog Devices, Burr-Brown, Harris). These devices usually work over a six-decade current range (1 nA to 1 mA) and a four-decade voltage range (1 mV to 10 V).

True rms-to-dc Converters

The logarithmic characteristics of BJTs are also exploited to perform a variety of slide-rule-like analog computations. A popular example is *true rms-to-dc conversion,* defined as

$$V_{rms} = \left(\frac{1}{T} \int_0^T v^2(t)\, dt \right)^{1/2} \tag{13.15}$$

V_{rms} gives a measure of the energy content of $v(t)$, so it provides the basis for accurate and consistent measurements, especially in the case of ill-defined waveforms, such as noise (electronic noise, switch contact noise, acoustical noise), mechanical transducer outputs (stress, vibration, shock, bearing noise), SCR waveforms, low-repetition-rate pulse trains, and other waveforms carrying information on the average energy generated, transmitted, or dissipated.

Equation (13.15) can be mechanized by performing the operations of squaring, averaging, and square rooting. Referred to as *explicit* rms computation, this scheme places severe demands on the dynamic output range of the squarer, which must be twice as wide as the input range. This drawback is overcome by *implicit* rms computation, in which the gain of the squarer is made inversely proportional to V_{rms} to make the output dynamic range comparable to the input range.

A common implementation of this principle is shown in Fig. 13.7, where frequency compensation and reverse-bias protection have been omitted for simplicity.

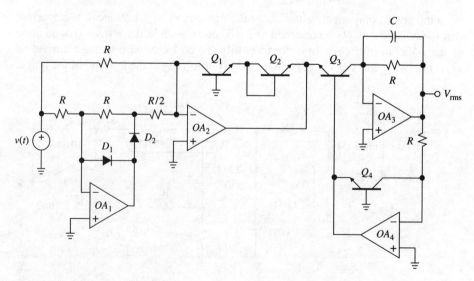

FIGURE 13.7
True rms converter.

OA$_1$ and the associated circuitry convert $v(t)$ to a full-wave rectified current $i_{C1} = |v|/R$ flowing into Q_1. By KVL, $v_{BE3} + v_{BE4} = v_{BE1} + v_{BE2}$, or $V_T \ln[(i_{C3}/I_{s3}) \times (i_{C4}/I_{s4})] = V_T \ln[(i_{C1}/I_{s1}) \times (i_{C2}/I_{s2})]$. Assuming pairwise matched BJTs and ignoring base currents so that $i_{C2} = i_{C1}$, we get

$$i_{C3} = \frac{i_{C1}^2}{i_{C4}}$$

Substituting $i_{C1}^2 = v^2/R^2$ and $i_{C4} = V_{rms}/R$ gives

$$i_{C3} = \frac{v^2}{R V_{rms}}$$

indicating that the scale factor of the squaring function is controlled by V_{rms}, as expected of implicit computation.

OA$_3$ forms a low-pass filter with cutoff frequency $f_0 = 1/2\pi RC$. For signal frequencies sufficiently higher than f_0, OA$_3$ will provide the running average of i_{C3} as $V_{rms} \cong \overline{Ri_{C3}} = \overline{v^2}/V_{rms}$. Making the approximation $\overline{V_{rms}} \cong V_{rms}$, we can write $V_{rms}^2 = \overline{v^2}$, or

$$V_{rms} = (\overline{v^2})^{1/2} \tag{13.16}$$

As a consequence of the approximations made, V_{rms} of Eq. (13.16) will differ from the ideal V_{rms} of Eq. (13.15) by an average (or dc) error as well as an ac (or ripple) error. Both errors can be kept below a specified limit by using a suitably large capacitance.[4] However, too large a capacitance will increase the response time of the circuit, so a compromise must be reached. An effective way of reducing ripple without unduly lengthening the response is to use a *post filter,* such as a low-pass *KRC* type.

The structure of Fig. 13.7 (or improved variations thereof) is available in IC form from various manufacturers. Consult the literature[4] for useful application tips.

13.2
ANALOG MULTIPLIERS

A multiplier produces an output v_O proportional to the product of two inputs v_X and v_Y,

$$v_O = k v_X v_Y \tag{13.17}$$

where k is a scale factor, usually $1/10 \text{ V}^{-1}$. A multiplier that accepts inputs of either polarity and preserves the correct polarity relationship at the output is referred to as a *four-quadrant multiplier.* Both the input and output ranges are usually from -10 V to $+10 \text{ V}$. By contrast, a *two-quadrant* multiplier requires that one of its inputs be unipolar, and a *one-quadrant multiplier* requires that both inputs be unipolar.

Multiplier performance is specified in terms of *accuracy* and *nonlinearity.* Accuracy represents the maximum deviation of the actual output from the ideal value predicted by Eq. (13.17); this deviation is also referred to as the *total error.* Nonlinearity, also referred to as *linearity error,* represents the maximum output deviation from the best-fit straight line for the case where one input is varied from end to

616

CHAPTER 13
Nonlinear
Amplifiers and
Phase-Locked
Loops

end while the other is kept fixed, usually at $+10$ V or -10 V. Both accuracy and nonlinearity are expressed as a percentage of the full-scale output.

Multiplier dynamics are specified in terms of the *small-signal bandwidth,* representing the frequency where the output is 3 dB below its low-frequency value, and the *1% absolute-error bandwidth,* representing the frequency where the output magnitude starts to deviate from its low-frequency value by 1%.

Variable-Transconductance Multipliers

Monolithic four-quadrant multipliers utilize the *variable-transconductance principle*[5] to achieve errors of fractions of 1% over small-signal bandwidths extending well into the megahertz range. This principle is illustrated in Fig. 13.8a. The block uses the differential pair Q_3-Q_4 to provide variable transconductance, and the diode-connected pair Q_1-Q_2 to provide the proper base drive for the former. The following analysis assumes matched BJTs and negligible base currents.

By KVL, $v_{BE1} + v_{BE4} - v_{BE3} - v_{BE2} = 0$, or $v_{BE3} - v_{BE4} = v_{BE1} - v_{BE2}$. Using the logarithmic v-i characteristics of the BJTs, this can be expressed as $V_T \ln(i_3/i_4) = V_T \ln(i_1/i_2)$, or

$$\frac{i_3}{i_4} = \frac{i_1}{i_2}$$

Rewriting as $(i_3 - i_4)/(i_3 + i_4) = (i_1 - i_2)/(i_1 + i_2)$ gives

$$i_3 - i_4 = \frac{(i_1 - i_2) \times (i_3 + i_4)}{i_1 + i_2} \tag{13.18}$$

indicating the circuit's ability to multiply the current difference $(i_1 - i_2)$ by the total emitter current $(i_3 + i_4)$.

To be of practical use, the circuit requires two *V-I* converters to synthesize the terms $(i_1 - i_2)$ and $(i_3 + i_4)$ from the input voltages v_X and v_Y, and an *I-V* converter to convert $(i_3 - i_4)$ to the output voltage v_O. Moreover, provisions must be made to

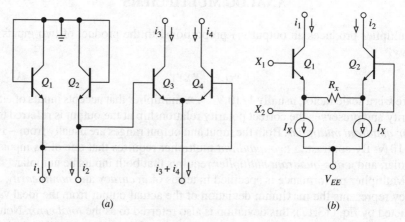

(a)　　　　　　　　　　　　　　(b)

FIGURE 13.8
Linearized transconductance block, and differential *V-I* converter.

ensure four-quadrant operation; as is, the circuit is only two-quadrant because the current $(i_3 + i_4)$ must always flow out of the emitters.

Figure 13.8b shows the circuit used to provide V-I conversion. By KCL, $i_1 = I_X + i_{R_x}$ and $i_2 = I_X - i_{R_x}$, where $i_{R_x} = (v_{E1} - v_{E2})/R_x$ is the current through R_x, assumed to flow from left to right. Consequently,

$$i_1 - i_2 = 2\frac{v_{E1} - v_{E2}}{R_x}$$

By KVL, $v_{E1} - v_{E2} = (v_{X_1} - v_{BE1}) - (v_{X_2} - v_{BE2}) = (v_{X_1} - v_{X_2}) - (v_{BE1} - v_{BE2})$, or

$$v_{E1} - v_{E2} = v_{X_1} - v_{X_2} - V_T \ln\frac{i_1}{i_2}$$

Combining the two equations gives

$$i_1 - i_2 = \frac{2}{R_x}(v_{X_1} - v_{X_2}) - \frac{2V_T}{R_x}\ln\frac{i_1}{i_2} \qquad (13.19)$$

In a well-designed multiplier the last term is on the order of 1% of the other two, so we can ignore it and approximate

$$i_1 - i_2 = \frac{2}{R_x}(v_{X_1} - v_{X_2}) \qquad (13.20)$$

indicating the circuit's ability to provide differential V-I conversion.

Figure 13.9 shows the complete multiplier. Four-quadrant operation is achieved by using two transconductance pairs with the bases driven in antiphase and the

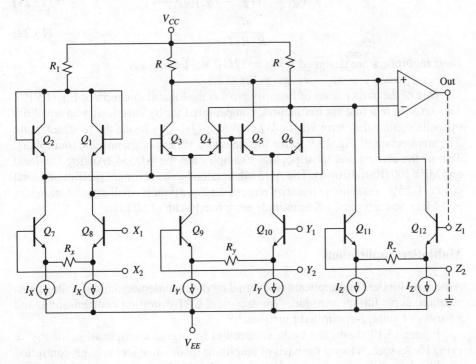

FIGURE 13.9
Four-quadrant analog multiplier.

618

CHAPTER 13
Nonlinear
Amplifiers and
Phase-Locked
Loops

emitters driven by a second *V-I* converter. Substituting Eq. (13.20) into Eq. (13.18) and using the identities $i_1 + i_2 = 2I_X$ and $i_3 + i_4 = i_9$, we obtain

$$i_3 - i_4 = \frac{v_{X_1} - v_{X_2}}{R_x I_X} i_9$$

Likewise, using the identity $i_5 + i_6 = i_{10}$, we obtain

$$i_6 - i_5 = \frac{v_{X_1} - v_{X_2}}{R_x I_X} i_{10}$$

Subtracting the first equation from the second pairwise and using $i_{10} - i_9 = (2/R_y)(v_{Y_1} - v_{Y_2})$, we obtain

$$(i_4 + i_6) - (i_3 + i_5) = \frac{(v_{X_1} - v_{X_2})(v_{Y_1} - v_{Y_2})}{R_x R_y I_X / 2}$$

The output *I-V* converter is made up of the op amp and a third *V-I* converter in its feedback path, namely, Q_{11}-Q_{12}. By KVL, the voltages at the inverting and noninverting inputs are $v_N = V_{CC} - R(i_4 + i_6 + i_{11})$ and $v_P = V_{CC} - R(i_3 + i_5 + i_{12})$. The op amp will provide Q_{12} with whatever drive it takes to make $v_N = v_P$, or $i_4 + i_6 + i_{11} = i_3 + i_5 + i_{12}$, that is,

$$(i_4 + i_6) - (i_3 + i_5) = i_{12} - i_{11} = \frac{2}{R_z}(v_{Z_1} - v_{Z_2})$$

Combining the last two equations, we finally obtain

$$v_{Z_1} - v_{Z_2} = k(v_{X_1} - v_{X_2})(v_{Y_1} - v_{Y_2}) \tag{13.21}$$

$$k = \frac{R_z}{R_x R_y I_X} \tag{13.22}$$

Most multipliers are designed for $k = 1/(10 \text{ V})$. Letting $v_O = v_{Z_1} - v_{Z_2}$, $v_X = v_{X_1} - v_{X_2}$, and $v_Y = v_{Y_1} - v_{Y_2}$ gives Eq. (13.17).

One of the main causes of linearity error is the logarithmic term of Eq. (13.19). This error is, to a first approximation, compensated for by introducing an equal but opposite nonlinearity term via the *V-I* converter Q_{11}-Q_{12} inside the feedback path. The architecture of Fig. 13.9 forms the basis of a variety of monolithic multipliers. Two of the earliest and most popular examples are the AD534 (Analog Devices) and MPY100 (Burr-Brown). The AD534L version has a maximum pretrimmed total error of 0.25%, a maximum linearity error of 0.12%, a typical small-signal bandwidth of 1 MHz, and a typical 1% amplitude error bandwidth of 50 kHz.

Multiplier Applications

Analog multipliers find application in signal modulation/demodulation, analog computation, curve fitting, transducer linearization, CRT distortion compensation, and a variety of voltage-controlled functions.[1,6]

Figure 13.10 shows the basic connection for signal multiplication, or $v_O = v_1 v_2/10$. As such, it forms the basis of amplitude modulation and voltage-controlled amplification. When either input is zero, v_O should also be zero, regardless of the other input. In practice, because of slight component mismatches, a small fraction

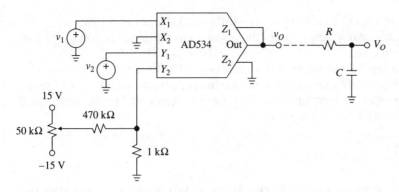

FIGURE 13.10
Basic multiplier connection for $v_O = v_1 v_2/10$. If followed by a low-pass
filter, it can be used for phase detection.

of the nonzero input will feed through to the output, causing an error. In critical
applications such as suppressed-carrier modulation, this error can be minimized by
applying an external trim voltage (± 30-mV range required) to the X_2 or the Y_2 input.

Of particular interest is the case in which the inputs are ac signals, or $v_1 =
V_1 \cos(\omega_1 t + \theta_1)$ and $v_2 = V_2 \cos(\omega_2 t + \theta_2)$, for then their product is, by a well-
known trigonometric identity

$$v_O = \frac{V_1 V_2}{20} \{\cos[(\omega_1 - \omega_2)t + (\theta_1 - \theta_2)] + \cos[(\omega_1 + \omega_2)t + (\theta_1 + \theta_2)]\}$$

indicating that v_O consists of two components, with frequencies equal to the sum
and the difference of the input frequencies. If the input frequencies are the same and
the high-frequency component is suppressed with a low-pass filter, as shown, then
we get

$$v_O = \frac{V_1 V_2}{20} \cos(\theta_1 - \theta_2) \qquad (13.23)$$

In this capacity, the circuit can be used in ac power measurements or as a phase
detector in phase-locked-loop circuits.

Figure 13.11 shows how a multiplier can be configured for two other popular
functions, namely, *analog division* and *square-root extraction*. In Fig. 13.11a we

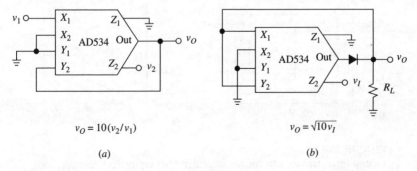

$$v_O = 10(v_2/v_1)$$

(a)

$$v_O = \sqrt{10 v_I}$$

(b)

FIGURE 13.11
Analog divider and square-rooter.

620

CHAPTER 13
Nonlinear
Amplifiers and
Phase-Locked
Loops

have, by Eq. (13.21), $0 - v_2 = (v_1 - 0)(0 - v_O)/10$, or $v_O = 10(v_2/v_1)$. To maximize the denominator range, return the X_2 input to a trimmable voltage (± 3-mV range required).

In Fig. 13.11b we have $0 - v_I = (v_O - 0)(0 - v_O)/10$, or $v_O = \sqrt{10 v_I}$. The function of the diode is to prevent a latching condition, which could arise in the event of the input inadvertently changing polarity. Additional applications are discussed in the end-of-chapter problems.

13.3
OPERATIONAL TRANSCONDUCTANCE AMPLIFIERS

An operational transconductance amplifier (OTA) is a voltage-input, current-output amplifier. Its circuit model is shown in Fig. 13.12a. To avoid loading effects both at the input and at the output, an OTA should have $z_d = z_o = \infty$. The ideal OTA, whose circuit symbol is shown in Fig. 13.12b, gives $i_O = g_m v_D$, or

$$i_O = g_m(v_P - v_N) \tag{13.24}$$

where g_m is the *unloaded transconductance gain,* in amperes per volt.

In its simplest form, an OTA consists of a differential transistor pair with a current-mirror load.[7] We have encountered this configuration when studying op amp input stages, in Chapter 5. In the bipolar example of Fig. 5.1 the OTA consists of the Q_1-Q_2 pair and the Q_3-Q_4 mirror; in the MOS example of Fig. 5.8 it consists of the M_1-M_2 pair and the M_3-M_4 mirror.

Besides serving as building blocks for other amplifiers, OTAs find application in their own right. Since it can be realized with just one stage and it operates on the principle of processing currents rather than voltages, the OTA is an inherently fast device.[8] Moreover, g_m can be varied by changing the bias current of the differential transistor pair, making OTAs suited to electronically programmable functions.

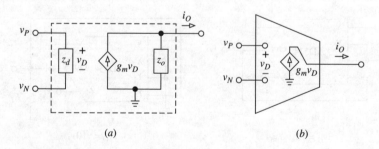

(a)   (b)

FIGURE 13.12
Operational transconductance amplifier: (a) equivalent circuit and (b) ideal model.

g_m-C Filters

A popular OTA application is the realization of fully integrated continuous-time filters, where OTAs have emerged as viable alternatives to traditional op amps.[7-9] OTA-based filters are referred to as g_m-C filters because they use OTAs and capacitors, but no resistors and no inductors. A popular g_m-C filter example is shown in Fig. 13.13a. Its analysis proceeds as follows.

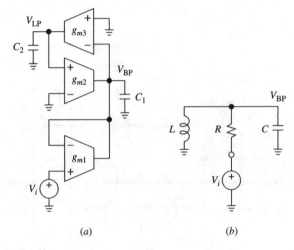

FIGURE 13.13
Second-order g_m-C filter and its *RLC* equivalent.

By Eq. (13.24), $I_1 = g_{m1}(V_i - V_{BP})$, $I_2 = g_{m2}V_{LP}$, and $I_3 = -g_{m3}V_{BP}$. By Ohm's law, $V_{LP} = (1/sC_2)I_3$ and $V_{BP} = (1/sC_1)(I_1 + I_2)$. Combining, we get

$$\frac{V_{BP}}{V_i} = \frac{sC_2 g_{m1}/g_{m2}g_{m3}}{s^2 C_1 C_2/g_{m2}g_{m3} + sC_2 g_{m1}/g_{m2}g_{m3} + 1} \tag{13.25}$$

It is readily seen that this transfer function is the same as that of the *RLC* equivalent of Fig. 13.13b, provided $C = C_1$, $R = 1/g_{m1}$, and $L = C_2/g_{m2}g_{m3}$. Evidently, g_{m1} simulates a resistance, whereas the combination g_{m2}-g_{m3}-C_2 simulates an inductance. Moreover, the circuit provides V_{BP} and V_{LP} simultaneously, a feature not available in its *RLC* counterpart. What is even more important is that we can automatically tune f_0 by varying g_{m2} and g_{m3}, and tune Q by varying g_{m1}.

EXAMPLE 13.2. (a) In the filter of Fig. 13.13a find g_{m1} and $g_{m2} = g_{m3}$ for $\omega_0 = 10^5$ rad/s and $Q = 5$ with $C_1 = C_2 = 100$ pF. (b) What are the values of the simulated resistance and the simulated inductance? (c) The sensitivities of the filter?

Solution.

(a) By inspection, $\omega_0 = \sqrt{g_{m2}g_{m3}/C_1C_2}$ and $Q = \sqrt{C_1/C_2} \times \sqrt{g_{m2}g_{m3}}/g_{m1}$. Substituting the given data, we get $g_{m2} = g_{m3} = 10$ μA/V and $g_{m1} = 2$ μA/V.
(b) $R = 500$ kΩ, and $L = 1$ H.
(c) The sensitivity of Q with respect to g_{m1} is -1; all other sensitivities are either $\frac{1}{2}$ or $-\frac{1}{2}$, which are fairly low.

622

CHAPTER 13
Nonlinear
Amplifiers and
Phase-Locked
Loops

Off-the-Shelf OTAs

Figure 13.14 shows a popular OTA available as an off-the-shelf IC. Its heart is the linearized transconductance multiplier made up of D_1-D_2 and Q_3-Q_4. The remaining blocks, each consisting of a BJT pair and a diode, are high-output-impedance current mirrors of the Wilson type. Denoting the collector current of transistor Q_k as i_k and ignoring base currents, we can describe circuit operation as follows.

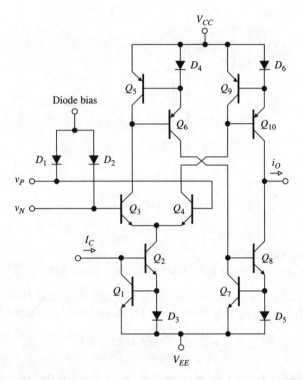

FIGURE 13.14
Bipolar OTA.

The mirror Q_1-D_3-Q_2 accepts the external control current I_C and duplicates it at the emitters of the Q_3-Q_4 pair to give

$$i_3 + i_4 = I_C$$

The mirror Q_5-D_4-Q_6 duplicates i_3 to yield $i_6 = i_3$, and the mirror Q_7-D_5-Q_8 duplicates i_6 to yield $i_8 = i_6$, so $i_8 = i_3$. Likewise, the mirror Q_9-D_6-Q_{10} duplicates i_4 to yield $i_{10} = i_4$. Consequently, KCL gives $i_O = i_{10} - i_8$, or

$$i_O = i_4 - i_3$$

Retracing the reasoning of Section 5.1, we can write

$$i_O = I_C \tanh \frac{v_P - v_N}{2V_T}$$

where V_T is the thermal voltage. As we know, this is a nonlinear characteristic, which, for $|v_P - v_N| \ll 2V_T$, can be approximated as

$$i_O = \frac{I_C}{2V_T}(v_P - v_N) \tag{13.26}$$

indicating that $g_m = I_C/2V_T$.

To accommodate applications requiring a wider linear range at the input, the linearizing diode network of Fig. 13.15 is used. Applying Eq. (13.18), we can write

$$i_O = i_4 - i_3 = \frac{i_2 - i_1}{i_2 + i_1}I_C$$

With $v_1 = v_2$, the bias current provided by R_3 splits evenly between D_1 and D_2, giving $i_2 - i_1 = 0$. Moreover, we can write $i_2 + i_1 = I_{R_3} = (V_{CC} - V_D)/[R_3 + (R_1 \parallel R_2)/2] \cong 1.08$ mA, where we have assumed $V_D \cong 0.7$ V. Making $v_1 \neq v_2$ will unbalance the two halves of the input network and divert a greater portion of I_{R_3} to either D_2 or D_1, depending on the imbalance direction. The voltage variation at the anodes is designed to be negligible compared to V_{CC} over an input range of more than 10 V, so $i_2 + i_1$ can be assumed to be constant. Using simple KCL reasoning, we find that I_{R_3} is completely diverted from one diode to the other when $|v_1 - v_2| \cong (R_1 + R_2)I_{R_3} \cong 11.3$ V. Consequently, for $|v_1 - v_2| \leq 11.3$ V we have $i_2 - i_1 = (v_1 - v_2)/(11.3 \text{ k}\Omega)$. Substituting into the above equation, along with $i_2 + i_1 \cong 1.08$ mA, we obtain

$$i_O = g_m(v_1 - v_2) \tag{13.27a}$$

$$g_m \cong \frac{I_C}{12.2 \text{ V}} \tag{13.27b}$$

The scale factor of approximately $1/(12.2 \text{ V})$ allows for operation over the range -10 V $< (v_1 - v_2) < 10$ V with negligible linearity error and no fear of saturation. Popular OTAs of the type shown are the LM13600 (National Semiconductor) and NE5517 (Signetics). The CA3080 (Harris) comes without the input diodes.

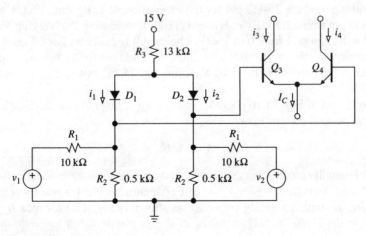

FIGURE 13.15
Input linearization network.

624

CHAPTER 13
Nonlinear
Amplifiers and
Phase-Locked
Loops

Applications with Off-the-Shelf OTAs

Though OTA data sheets propose quite a variety of useful applications, we shall examine a few representative ones, namely, voltage-controlled amplifiers, filters, and oscillators (VCAs, VCFs, and VCOs).

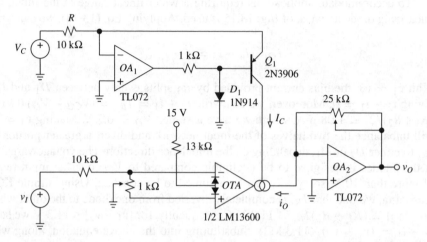

FIGURE 13.16
Voltage-controlled amplifier with linear control.

Figure 13.16 shows a basic VCA (note the alternative symbol for the OTA). Here OA_1 and Q_1 form a *V-I* converter to provide $I_C = V_C/R$, where we are assuming the base current of Q_1 to be negligible. OA_2 converts i_O to a voltage v_O, and since i_O is proportional to the product $I_C \times v_I$, the final result is $v_O = Av_I$,

$$A = kV_C \tag{13.28}$$

where k is a suitable proportionality constant, in V^{-1}. The 1-kΩ pot is used for offset nulling, and the 25-kΩ pot for the calibration of k. By Eq. (13.27), adjusting the 25-kΩ pot near 12.2 kΩ yields $k = 1/(10 \text{ V})$, indicating that varying V_C from 0 to 10 V will change A from 0 to 1 V/V. The circuit is calibrated as follows: (*a*) with $v_I = 0$, sweep V_C from 0 to 10 V, and adjust the 1-kΩ pot for the minimum deviation of v_O from 0 V; (*b*) with $V_C = 10$ V, adjust the 25-kΩ pot so that $v_O = 10$ V for $v_I = 10$ V.

The circuit of Fig. 13.16 provides linear gain control. Audio applications often call for exponential control, or

$$A = A_0 b^{kV_C} \tag{13.29}$$

where b is usually either 10 or 2, k is a proportionality constant in decades or octaves per volt, and A_0 is the gain for $V_C = 0$. Exponential control is readily achieved by generating I_C with an antilog converter, as shown in Fig. 13.17. Since I_C must be sourced to the OTA, the BJTs must be of the *pnp* type. With $V_C = 0$, the circuit gives $I_C = 1$ mA; increasing V_C decreases I_C exponentially with a sensitivity of k dec/V or k oct/V, with k being set by R_2.

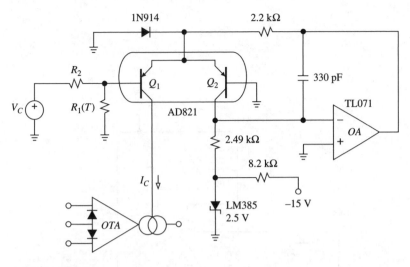

FIGURE 13.17
OTA with exponential control.

OTA-based VCFs and VCOs rely on the integration of the OTA's output current using a capacitor. The example of Fig. 13.18 uses also an op amp to provide low output impedance. Writing $V_o = (-1/sC)I_o = (-1/sC) \times [I_C/(12.2 \text{ V})] \times (V_2 - V_1)$, and letting $s = j2\pi f$, we obtain

$$V_o = \frac{1}{jf/f_0}(V_1 - V_2) \qquad f_0 = \frac{I_C}{2\pi(12.2 \text{ V})C} \qquad (13.30)$$

The circuit integrates the difference $V_1 - V_2$ with a programmable unity-gain frequency f_0. For instance, varying I_C from 1 μA to 1 mA with $C = 652$ pF will sweep f_0 over the entire audio range, from 20 Hz to 20 kHz. I_C can be generated either with a linear V-I converter, as in Fig. 13.16, or with an exponential converter, as in Fig. 13.17.

The circuit of Fig. 13.19 uses two OTA-based integrators to implement a state-variable topology of the type of Fig. 4.37. The output current of the V-I converter, which can be controlled either linearly or exponentially, is split between the two OTAs by the suitably biased AD821 matched BJT pair.

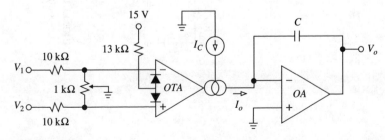

FIGURE 13.18
Current-controlled integrator.

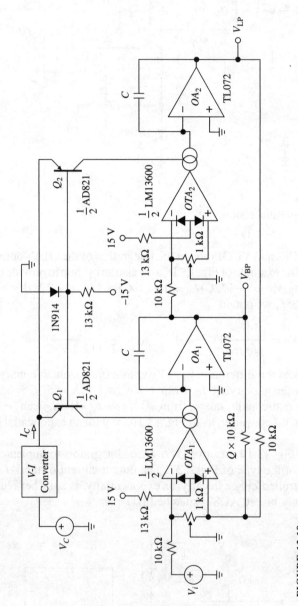

FIGURE 13.19
Voltage-controlled state-variable filter.

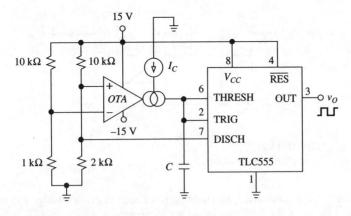

FIGURE 13.20
Current-controlled relaxation oscillator.

Applying Eq. (13.30), we get $V_{\text{BP}} = (V_i - V_{\text{BP}}/Q - V_{\text{LP}})/(jf/f_0)$ and $V_{\text{LP}} = V_{\text{BP}}/(jf/f_0)$. Combining, we obtain

$$\frac{V_{\text{BP}}}{V_i} = QH_{\text{BP}} \qquad \frac{V_{\text{LP}}}{V_i} = H_{\text{LP}} \tag{13.31}$$

where H_{BP} and H_{LP} are the standard second-order band-pass and low-pass functions defined in Section 3.4, and

$$f_0 = \frac{I_C}{4\pi(12.2\text{ V})C}$$

If a resonance gain of unity is desired, then increase the 10-kΩ input resistance by a factor of Q. To reduce Q-enhancement effects, follow the directions of Section 6.5 and use a small phase-lead capacitor in parallel with the 10-kΩ interstage resistance.

In the circuit of Fig. 13.20 the OTA is used to source/sink a current of value I_C, and thus charge/discharge C at a programmable rate. The resulting triangle waveform alternates between 5 V and 10 V, the thresholds of the high-input-impedance CMOS timer. The frequency of oscillation is (see Problem 13.16),

$$f_0 = \frac{I_C}{10C} \tag{13.32}$$

As usual, I_C can be controlled either linearly or exponentially. If the triangular wave is used, a buffer amplifier may be required.

13.4

PHASE-LOCKED LOOPS

A *phase-locked loop* (PLL) is a frequency-selective circuit designed to synchronize with an incoming signal and maintain synchronization in spite of noise or variations in the incoming signal frequency. As depicted in Fig. 13.21, the basic PLL system comprises a *phase detector*, a *loop filter*, and a *voltage-controlled oscillator* (VCO).

The phase detector compares the phase θ_I of the incoming signal v_I against the phase θ_O of the VCO output v_O, and develops a voltage v_D proportional to the difference $\theta_I - \theta_O$. This voltage is sent through a low-pass filter to suppress

628

CHAPTER 13
Nonlinear
Amplifiers and
Phase-Locked
Loops

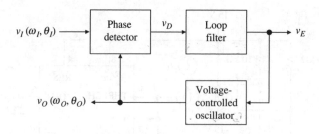

FIGURE 13.21
Basic phase-locked loop.

high-frequency ripple and noise, and the result, called the *error voltage* v_E, is applied to the control input of the VCO to adjust its frequency ω_O.

The VCO is designed so that with $v_E = 0$ it is oscillating at some initial frequency ω_0 called the *free-running frequency,* so its characteristic is

$$\omega_O(t) = \omega_0 + K_o v_E(t) \tag{13.33}$$

where K_o is the *sensitivity* of the VCO, in radians-per-second per volt. If a periodic input is applied to the PLL with frequency ω_I sufficiently close to the free-running frequency ω_0, an error voltage v_E will develop, which will adjust ω_O until v_O becomes synchronized with v_I, that is, until *for every input cycle there is one, and only one, VCO cycle.* At this point the PLL is said to be *locked* on the incoming signal, and it gives $\omega_O = \omega_I$ exactly.

Should ω_I change, the phase shift between v_O and v_I will start to increase, changing v_D and, hence, the control voltage v_E. This change in v_E is designed to adjust the VCO until ω_O is brought back to the same value as ω_I. This self-adjusting ability by the feedback loop allows the PLL, once locked, to track input frequency changes. Since a change in ω_I is ultimately reflected by a change in v_E, we use v_E as the output of the PLL whenever we wish to detect changes in ω_I, as in FM and FSK demodulation.

A PLL can be designed to lock on the incoming signal in spite of noise that might afflict such a signal. A noisy input will generally cause the phase-detector output v_D to jitter around some average value. However, if the filter cutoff frequency is low enough to suppress this jitter, v_E will emerge as a clean signal, in turn resulting in a stable VCO frequency and phase. We thus use ω_O as the output of the PLL whenever we wish to recover a signal buried in noise, and also in frequency-related applications such as frequency synthesis and synchronization.

Lock and Capture

To develop a concrete understanding of PLL operation, consider the case of phase detection being accomplished with a balanced mixer of the type discussed in Section 13.2. As we know, the mixer output contains the sum and difference frequencies $\omega_I \pm \omega_O$. When the loop is locked, the sum is twice ω_I and the difference is zero, or dc. The low-pass filter suppresses the sum but passes the dc component, which thus keeps the loop in lock.

If the loop is not locked, and if the difference frequency falls above the cutoff frequency of the filter, it will be suppressed along with the sum frequency, leaving the loop unlocked and oscillating at its free-running frequency. However, if ω_O is sufficiently close to ω_I to make the difference frequency approach the filter band edge, part of this component is passed, tending to drive ω_O toward ω_I. As the difference $\omega_O - \omega_I$ is reduced, more error signal is transmitted to the VCO, resulting in a constructive effect that ultimately brings the PLL in lock.

The *capture range* is the frequency range $\pm\Delta\omega_C$, centered about ω_0, over which the loop can acquire lock. This range is affected by the filter characteristics, and gives an indication of how close ω_I must be to ω_0 to acquire lock. The *lock range* is the frequency range $\pm\Delta\omega_L$, also centered about ω_0, over which the loop can track the input once lock has been established. The lock range is affected by the operating range of the phase detector and the VCO. The capture process is a complex phenomenon, and the capture range is never greater than the lock range.

The time it takes for a PLL to capture the incoming signal is called the *capture time* or *pull-in time*. This time depends on the initial frequency and phase differences between v_I and v_O, as well as the filter and other loop characteristics. In general, it can be said that reducing the filter bandwidth has the following effects: (*a*) it slows down the capture process, (*b*) it increases the pull-in time, (*c*) it decreases the capture range, and (*d*) it increases the interference-rejection capabilities of the loop.

The PLL in the Locked Condition

When in the locked condition, a PLL can be modeled[10-12] as in Fig. 13.22. This diagram is similar to that of Fig. 13.21, except that we are now working with Laplace transforms of *signal changes* (symbolized by lowercase letters with lowercase subscripts) about some operating point, and *operations* on these changes, both of which are generally functions of the complex frequency s. The phase detector develops the voltage change

$$v_d(s) = K_d\theta_d(s) \tag{13.34a}$$

$$\theta_d(s) = \theta_i(s) - \theta_o(s) \tag{13.34b}$$

where K_d is the phase-detector *sensitivity*, in volts per radian. This voltage is sent through the loop filter, whose transfer function is denoted as $F(s)$, and possibly

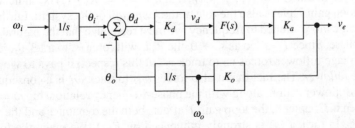

FIGURE 13.22
Block diagram of the basic PLL system in the locked condition.

630

CHAPTER 13
Nonlinear
Amplifiers and
Phase-Locked
Loops

an amplifer with gain K_a, in volts per volt, to produce the error-voltage variation $v_e(s)$. This, in turn, is converted to the frequency variation $\omega_o(s) = K_o v_e(s)$, by Eq. (13.33).

Since the phase detector processes phase, we need a means for converting from frequency to phase. Considering that frequency represents the rate of change of phase with time, or $\omega = d\theta(t)/dt$, we have

$$\theta(t) = \theta(0) + \int_0^t \omega(\xi)\,d\xi \tag{13.35}$$

indicating that frequency-to-phase conversion is inherently an operation of integration. Exploiting the well-known Laplace transform property that integration in the time domain corresponds to division by s in the frequency domain, we use the $1/s$ blocks shown.

If we were to open the loop at the inverting input of the phase comparator, the overall gain experienced by $\theta_i(s)$ in going around the path and emerging as $\theta_o(s)$ is $K_d \times F(s) \times K_a \times K_o \times 1/s$, or

$$T(s) = K_v \frac{F(s)}{s} \tag{13.36}$$

$$K_v = K_d K_a K_o \tag{13.37}$$

where $T(s)$ is the *open-loop gain,* in radians per radian, and K_v is called the *gain factor,* in s^{-1}. With the loop closed, we readily find

$$H(s) = \frac{\theta_o(s)}{\theta_i(s)} = \frac{T(s)}{1 + T(s)} = \frac{K_v F(s)}{s + K_v F(s)} \tag{13.38}$$

Other transfer functions may be of interest, depending on what we consider as input and output. For instance, substituting $\theta_i(s) = \omega_i(s)/s$ and $\theta_o = (K_o/s)v_e(s)$, we readily get

$$\frac{v_e(s)}{\omega_i(s)} = \frac{1}{K_o}H(s) \tag{13.39}$$

which allows us to find the voltage change $v_e(s)$ in response to an input frequency change $\omega_i(s)$, as in FM and FSK demodulation.

Comparing Fig. 13.22 with Fig. 1.21, we observe that a PLL is a negative-feedback system with $x_i = \theta_i$, $x_f = \theta_o$, and $a\beta = T = K_v F(s)/s$, indicating that the open-loop gain T plays also the role of the *loop gain* of the system. Further, even though we are more attuned to frequency, we must recognize that the natural input of a PLL is phase. Since $T \to \infty$ as $s \to 0$, the PLL will force θ_o to track θ_i, just as an op amp voltage follower forces v_o to track v_i. In this respect, it pays to view a PLL as a *phase follower.* The fact that it also forces ω_O to track ω_I is a consequence of this phase-follower action, along with the phase-frequency relationship $\omega = d\theta/dt$.

As seen in Chapter 8, the loop gain T affects both the dynamics and the stability of the PLL. In turn, $T(s)$ is strongly influenced by $F(s)$. We make the following observations: (*a*) The number of poles of $H(s)$ defines the *order* of the loop; (*b*) the number of $1/s$ terms (or integrations) present within the loop defines the *type* of a loop. Because of the $1/s$ function associated with the VCO, a PLL is at least Type I, and its order equals the order of the filter plus 1.

First-Order Loop

Consider the instructive case in which there is no loop filter, or $F(s) = 1$. The result is a *first-order loop,* and the above equations simplify, after the substitution $s \rightarrow j\omega$, as

$$T(j\omega) = \frac{1}{j\omega/K_v} \tag{13.40}$$

$$\frac{v_e(j\omega)}{\omega_i(j\omega)} = \frac{1/K_o}{1 + j\omega/K_v} \tag{13.41}$$

Equation (13.40) indicates a *Type I* loop with crossover frequency $\omega_x = K_v$ and phase margin $\phi_m = 90°$. Equation (13.41) indicates that the loop inherently provides a first-order low-pass response with a dc gain of $1/K_o$ V/(rad/s) and a cutoff frequency of K_v rad/s.

If $\omega_i(t)$ is a step change, the resulting change $v_e(t)$ will be an exponential transient governed by the time constant $\tau = 1/K_v$. If $\omega_i(t)$ is varied sinusoidally with a modulating frequency ω_m, $v_e(t)$ will also vary sinusoidally with the same frequency ω_m; its amplitude is $|v_e| = (1/K_o)|\omega_i|$ at low frequencies, and rolls off with ω_m at the rate of -1 dec/dec past K_v.

EXAMPLE 13.3. A first-order PLL with $K_v = 10^4$ s^{-1} uses a VCO with a free-running frequency of 10 kHz and a sensitivity of 5 kHz/V. (a) What is the control voltage needed to lock the PLL on a 20-kHz input signal? On a 5-kHz input signal? (b) Find the response $v_e(t)$ if the input frequency is changed stepwise as $f_I = [10 + u(t)]$ kHz, where $u(t) = 0$ for $t < 0$ and $u(t) = 1$ for $t > 0$. (c) Repeat if the input frequency is modulated as $f_I = 10(1 + 0.1 \cos 2\pi f_m t)$ kHz, $f_m = 2.5$ kHz.

Solution.

(a) By Eq. (13.33), $v_E = (\omega_O - \omega_0)/K_o$, where $\omega_0 = 2\pi 10^4$ rad/s and $K_o = 2\pi \times 5 \times 10^3 = \pi 10^4$ (rad/s)/V. For $\omega_O = 2\pi \times 20 \times 10^3$ rad/s we get $v_E = 2$ V, and for $\omega_O = 2\pi \times 5 \times 10^3$ rad/s we get $v_E = -1$ V.

(b) The response to a step increase $\omega_i(t) = 2\pi u(t)$ krad/s is an exponential transient with amplitude $|\omega_i(t)|/K_o = 2\pi 10^3/10^4\pi = 0.2$ V, and time constant $1/K_v = 1/10^4 = 100$ μs, so

$$v_e(t) = 0.2[1 - e^{-t/(100 \, \mu s)}]u(t) \text{ V}$$

(c) Now $\omega_i(t) = 2\pi \times 10^4 \times 0.1 \cos 2\pi f_m t = 2\pi 10^3 \cos 2\pi 2500t$ rad/s. Calculating Eq. (13.41) at $j\omega = j\omega_m = j2\pi 2500$ rad/s gives

$$\frac{v_e(j\omega_m)}{\omega_i(j\omega_m)} = \frac{1/10^4\pi}{1 + j2\pi 2500/10^4} = \frac{0.5370}{10^4\pi} \underline{/-57.52°} \text{ V/(rad/s)}$$

Letting $\omega_i(j\omega_m) = 2\pi 10^3 \underline{/0°}$ rad/s gives $v_e(j\omega_m) = 0.1074 \underline{/-57.52°}$ V, so

$$v_e(t) = 0.1074 \cos(2\pi 2500t - 57.52°) \text{ V}$$

The absence of a loop filter drastically limits the selectivity and noise-suppression capabilities of a PLL, so first-order loops are seldom used in practice.

632

CHAPTER 13
Nonlinear
Amplifiers and
Phase-Locked
Loops

Second-Order Loops

Most PLLs utilize a one-pole low-pass filter and are thus *second-order loops*. Such a filter provides a flywheel-like function that allows the VCO to smooth over noise and jumps in the input frequency. As seen in Chapter 8, the presence of a second pole within the loop erodes the phase margin, so care must be exercised to avoid instability. Second-order loops are stabilized by introducing also a filter zero to counterbalance the phase lag due to the filter pole.

A popular loop filter is shown in Fig. 13.23a. Called a *passive lag-lead filter*, it provides the transfer function

$$F(s) = \frac{1 + s/\omega_z}{1 + s/\omega_p} \tag{13.42}$$

where $\omega_z = 1/R_2C$ and $\omega_p = 1/(R_1 + R_2)C$. By Eq. (13.36), the loop gain is now

$$T(j\omega) = \frac{1 + j\omega/\omega_z}{(j\omega/K_v)(1 + j\omega/\omega_p)} \tag{13.43}$$

indicating a *second-order, Type I loop*. This gain is plotted in Fig. 13.23b for the case in which ω_z is positioned at the geometric mean of ω_p and K_v, or $\omega_z = \sqrt{\omega_p K_v}$. The crossover frequency is then ω_z itself, and the phase margin is 45°. Also shown for comparison is the loop gain of the first-order loop, or $F(s) = 1$.

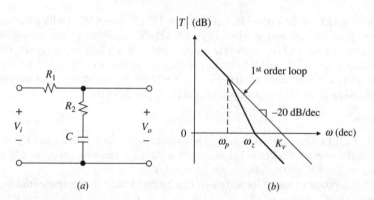

FIGURE 13.23
Passive lag-lead filter, and magnitude plot of the loop gain T.

EXAMPLE 13.4. (a) Given a PLL system with $K_v = 10^4$ s^{-1}, specify a passive lag-lead filter for a crossover frequency $\omega_x = 10^3$ rad/s and a phase margin $\phi_m = 45°$. (b) What are the actual values of ω_x and ϕ_m?

Solution.

(a) For $\phi_m \cong 45°$ we want $\omega_z = \omega_x = 10^3$ rad/s, so $\omega_p = \omega_z^2/K_v = 10^6/10^4 = 100$ rad/s. Let $C = 0.1$ μF. Then, $R_2 = 1/\omega_zC = 10$ kΩ, and $R_1 = 1/\omega_pC - R_2 = 90$ kΩ (use 91 kΩ).

(b) Using Eq. (13.43), along with the trial-and-error technique of Example 8.1, we find the actual values $\omega_x = 1.27$ krad/s, and $\phi_m = 180° + \sphericalangle T(j1.27 \times 10^3) = 56°$.

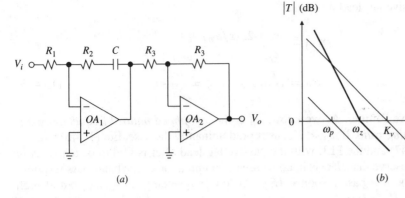

(a)   (b)

FIGURE 13.24
Active PI filter, and magnitude plot of the loop gain T.

Another popular loop filter is the *active PI filter* of Fig. 13.24a, so called because its output is proportional to the input as well as to the integral of the input. The inverting stage OA_2 can be omitted by interchanging the phase-detector inputs, if needed. Assuming ideal op amps, the filter gives

$$F(s) = \frac{1 + s/\omega_z}{s/\omega_p} \tag{13.44}$$

where $\omega_z = 1/R_2 C$ and $\omega_p = 1/R_1 C$. The corresponding loop gain is

$$T(j\omega) = \frac{1 + j\omega/\omega_z}{(j\omega/K_v)(j\omega/\omega_p)} \tag{13.45}$$

indicating a *second-order, Type II loop*. As shown in Fig. 13.24b, the slope is -40 dB/dec below ω_z, and -20 dB/dec above ω_z. Imposing again $\omega_z = \sqrt{\omega_p K_v}$ gives $\omega_x \cong \omega_z$ and $\phi_m \cong 45°$.

Compared to the passive filter, whose dc gain is $F(0) = 1$, the active filter gives $F(0) = \infty$, indicating that the phase error θ_D needed to sustain the control voltage v_E approaches zero when the PI filter is used. In practice, $F(0)$ is limited by the finite dc gain of OA_1; even so, θ_D will still be vanishingly small, implying $\theta_O \cong \theta_I$, or phase coherence at the detector input. The use of the active filter also avoids possible loading effects at the output.

Damping Characteristics

Additional insight is gained by substituting Eqs. (13.42) and (13.44) into Eq. (13.38), and then expressing the latter in the standard form of Eq. (3.40). The results (see Problem 13.20) are

$$H(s) = \frac{(2\zeta - \omega_n/K_v)(s/\omega_n) + 1}{(s/\omega_n)^2 + 2\zeta(s/\omega_n) + 1} \tag{13.46a}$$

$$\omega_n = \sqrt{\omega_p K_v} \qquad \zeta = \frac{\omega_n}{2\omega_z}\left(1 + \frac{\omega_z}{K_v}\right) \tag{13.46b}$$

634

CHAPTER 13
Nonlinear
Amplifiers and
Phase-Locked
Loops

for the passive lag-lead filter, and

$$H(s) = \frac{2\zeta(s/\omega_n) + 1}{(s/\omega_n)^2 + 2\zeta(s/\omega_n) + 1} \tag{13.47a}$$

$$\omega_n = \sqrt{\omega_p K_v} \qquad \zeta = \frac{\omega_n}{2\omega_z} \tag{13.47b}$$

for the active PI filter. As we know, ω_n is the *undamped natural frequency,* and ζ is the *damping ratio.* If $\omega_n \ll K_v$, as is predominantly the case, Eq. (13.46) reduces to Eq. (13.47) and the PLL with the passive lag-lead filter is said to be a *high-gain loop.* We observe that $H(s)$ is in both cases a combination of the band-pass response H_{BP} and the low-pass response H_{LP}. At low frequencies $H \to H_{LP}$, but at high frequencies $H \to H_{BP}$.

Recall that for $\zeta < 1$ the step response exhibits overshoot. To keep the latter within reason, it is customary to design for $0.5 \le \zeta \le 1$. Under this condition, the time constant governing the loop response to small phase or frequency changes is roughly[7]

$$\tau \cong \frac{1}{\omega_n} \tag{13.48}$$

and the loop bandwidth, obtained by imposing $|H(j\omega)| = 1/\sqrt{2}$, is[13]

$$\omega_{-3\,dB} = \omega_n[1 \pm 2\zeta^2 + \sqrt{1 + (1 \pm 2\zeta^2)^2}]^{1/2} \tag{13.49}$$

where the plus (minus) sign holds for high-gain (low-gain) loops.

EXAMPLE 13.5. (*a*) Find ζ, τ, and $\omega_{-3\,dB}$ for the PLL of Example 13.4. (*b*) Find the response $v_e(t)$ to small input changes of the type $\omega_i = |\omega_i|u(t)$ and $\omega_i = |\omega_i|\cos\omega_m t$, $\omega_m = 1$ krad/s.

Solution.

(*a*) By Eq. (13.46b), $\omega_n = \sqrt{10^2 \times 10^4} = 1$ krad/s and $\zeta = [10^3/(2 \times 10^3)](1 + 10^3/10^4) = 0.55$. Using Eq. (13.49) for the high-gain loop case gives $\omega_{-3\,dB} \cong 1.9\omega_n = 1.9$ krad/s. By Eq. (13.48), $\tau \cong 1/10^3 = 1$ ms.

(*b*) Substituting the above data into Eq. (13.46a) gives

$$H(s) = \frac{s/10^3 + 1}{(s/10^3)^2 + 1.1(s/10^3) + 1}$$

This function has a complex pole pair at $s = -550 \pm j835$ complex Np/s, indicating a step response of the type

$$v_e(t) = \frac{|\omega_i|}{K_o}[1 - Ae^{-550t}\cos(835t + \phi)]$$

with A and ϕ suitable constants. Calculating $H(s)$ at $s = j\omega_m$ as in Example 13.3, we find the ac response as

$$v_e(t) = \frac{|\omega_i|}{K_o}1.286\cos(10^3 t - 45°)$$

Filter Design Criteria

In general, ω_n is chosen high enough to ensure satisfactory dynamics, yet low enough to provide sufficient flywheel action for smoothing over undesired frequency jumps or noise. A typical design process proceeds as follows: (a) first, choose ω_n to achieve either the desired $\omega_{-3\,dB}$ or the desired τ, depending on the application; (b) next, using Eq. (13.46b) or (13.47b), specify ω_p for the chosen ω_n; (c) finally, specify ω_z for the desired ζ.

We observe that because of the filter zero, a second-order PLL acts as a first-order loop at high frequencies, indicating a reduced ability to suppress ripple and noise. This drawback can be overcome by adding a capacitance $C_2 \ll C$ in parallel with R_2 in either of the above filters. This creates an additional high-frequency pole and turns the loop into a *third-order loop*. To avoid perturbing the existing values of ω_x and ϕ_m significantly, this pole is positioned about a decade above ω_x by imposing $1/R_2C_2 \cong 10\omega_x$.

> **EXAMPLE 13.6.** Redesign the filter of Example 13.4 for $\omega_{-3\,dB} = 1$ krad/s and $\zeta = 1/\sqrt{2}$. What are the new values of τ and ϕ_m? What value of C_2 would yield a third-order loop without reducing ϕ_m too much?
>
> **Solution.** With $\zeta = 1/\sqrt{2}$ we get $\omega_n \cong \omega_{-3\,dB}/2 = 10^3/2.0 = 500$ rad/s, so $\tau \cong 2$ ms. Equation (13.46b) gives $\omega_p = 25$ rad/s and $\omega_z = 366$ rad/s, which can be realized with $C = 1\ \mu F$, $R_1 = 39\ k\Omega$, and $R_2 = 2.7\ k\Omega$. Proceeding as in Example 13.4, we find $\omega_x \cong 757$ rad/s, and $\phi_m \cong 66°$. Use $C_2 \cong C/10 = 0.1\ \mu F$.

13.5
MONOLITHIC PLLs

Monolithic PLLs are available in various technologies and in a wide range of performance specifications.[12] In the following we discuss the popular 4046 CMOS PLL as a representative example.

The 74HC(T)4046A CMOS PLL

Originally developed by RCA, the 4046 family of CMOS PLLs has gone through a series of improvements, and presently includes the 74HC(T)4046A, the 74HC(T)7046A, and the 74HC(T)9046A (Philips).[13] We select the 4046A version, shown in simplified form in Fig. 13.25, because it includes the three most common phase detector types, known as *Type I (PC_1)*, *Type II (PC_2)*, and *Type III (PC_3)* *phase comparators*. Since the circuit is powered from a single supply (typically $V_{SS} = 0$ V and $V_{DD} = 5$ V), all analog signals are referenced to $V_{DD}/2$, or 2.5 V.

The VCO

The VCO, whose details[7,13] are omitted for brevity, is a current-controlled multivibrator operating on a principle similar to that of the emitter-coupled VCO of

636

CHAPTER 13
Nonlinear
Amplifiers and
Phase-Locked
Loops

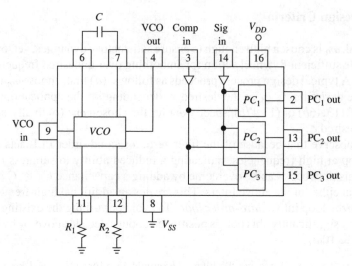

FIGURE 13.25
Simplified block diagram of the 4046A CMOS PLL.

Fig. 10.30. The current for the capacitor is obtained from the control voltage v_E via a *V-I* converter whose sensitivity is set by R_1 and whose output is offset by R_2. The VCO characteristic is of the type

$$f_O = \frac{k_1}{R_1 C} v_E + \frac{k_2}{R_2 C} \tag{13.50}$$

where k_1 and k_2 are suitable circuit constants. As shown in Fig. 13.26, the value of f_O corresponding to $v_E = V_{DD}/2$ is called the *center frequency* f_0. It is apparent that if R_2 is omitted ($R_2 = \infty$), the *frequency offset* $f_{O(off)} = k_2/R_2 C$ becomes zero. The maximum VCO frequency of CMOS PLLs is typically on the order of 10 MHz.

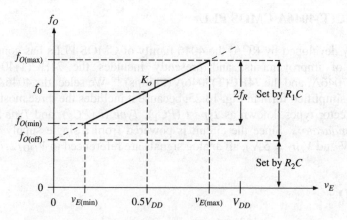

FIGURE 13.26
VCO characteristic and terminology.

The VCO characteristic of Eq. (13.50) holds only as long as v_E is confined within the range $v_{E(min)} \leq v_E \leq v_{E(max)}$. For a 4046A PLL with $V_{DD} = 5$ V, this range is typically[13] 1.1 V $\leq v_E \leq 3.9$ V. The frequency range corresponding to the permissible range of v_E is called the *VCO frequency range* $2f_R$. Outside this range the VCO characteristic depends on the particular 4046 version, and it can be found in the data sheets.

The VCO sensitivity is $K_o = 2f_R/[v_{E(max)} - v_{E(min)}]$. In FM applications it is usually required that the *V-F* characteristic of the VCO be highly linear in order to minimize distortion. However, in such applications as frequency synchronization, synthesis, and reconstruction the linearity requirements are less stringent.

The Type I Phase Comparator

The Type I phase comparator, depicted in Fig. 13.27a, is an exlusive-OR (XOR) gate. This gate outputs $v_D = V_{DD} = 5$ V whenever its input levels disagree with each other, and $v_D = V_{SS} = 0$ whenever they agree. This is exemplified in the timing diagram of Fig. 13.28, where the waveforms have been plotted as a function of $\omega_I t$. It is apparent that if we average out $v_D(t)$ by means of a low-pass filter, the result is $V_D = DV_{DD}$, where D is the duty cycle of v_D. D is minimized when the inputs

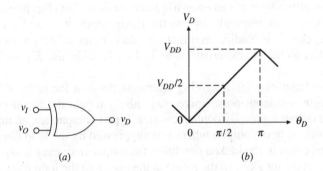

(a)　　　　　　　　(b)

FIGURE 13.27
Type I phase comparator, and its output average V_D as a function of the input phase difference.

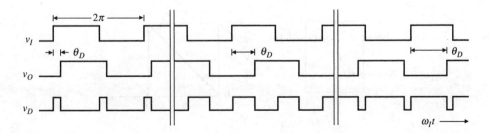

FIGURE 13.28
Typical waveforms for the Type I phase comparator in the locked condition: $\theta_D = \pi/6$ (left), $\theta_D = \pi/2$ (center), and $\theta_D = (5/6)\pi$ (right).

638

CHAPTER 13
Nonlinear
Amplifiers and
Phase-Locked
Loops

are in phase with each other, and maximized when they are in antiphase. If both input waveforms have 50% duty cycles, as shown, then $0 \leq D \leq 1$. Consequently, PC_1 will exhibit the characteristic of Fig. 13.27b, and $K_d = V_{DD}/\pi = 5/\pi = 1.59$ V/rad.

An alternative implementation of the Type I comparator, especially in bipolar PLLs designed to work with low-amplitude inputs, is a four-quadrant multiplier, as discussed in Section 13.2. Also called a *balanced modulator*, the multiplier is implemented with a scale factor high enough to ensure that v_I will typically overdrive the multiplier and thus render the sensitivity K_d independent of the amplitude of v_I.[10]

The Type I comparator requires that both inputs have 50% duty cycles; if at least one input is asymmetrical (see Problem 13.23), the characteristic will generally be clipped, reducing the lock range. Another notorious feature of the Type I comparator is that it may allow the PLL to lock on harmonics of the input signal. Note that if v_I is absent, v_D oscillates at the same frequency as v_O, so the average of v_D is $V_D = 0.5V_{DD}$, and $\omega_O = \omega_0$.

The Type III Phase Comparator

The Type III comparator, shown in simplified form in Fig. 13.29a, overcomes both of the above limitations by using an edge-triggered set-reset (SR) flip-flop. As depicted in Fig. 13.30, v_D now responds only to the rising edges of v_I and v_O, regardless of the duty cycles. It is readily seen that the phase range of PC_3 is twice as large as that of PC_1, so the characteristic is as in Fig. 13.29b, and $K_d = V_{DD}/2\pi = 0.796$ V/rad.

The advantages of edge-triggering operation come at the price of higher sensitivity to noise. An input noise spike may falsely toggle the flip-flop and cause unacceptable output errors. By contrast, with a Type I comparator, an input spike is merely transmitted to the output, where it is suppressed by the loop filter.

We observe that in the locked condition the output frequency is $\omega_D = 2\omega_I$ for PC_1, and $\omega_D = \omega_I$ for PC_3, so the ripple at the output of the loop filter is generally higher with PC_3 than with PC_1. Note that with v_I absent, PC_3 will drive ω_O as low as it can.

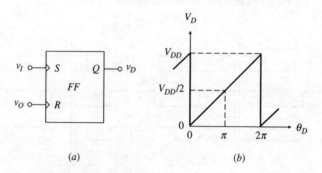

(a)   (b)

FIGURE 13.29
Type III phase comparator, and its output average V_D as a function of the input phase difference.

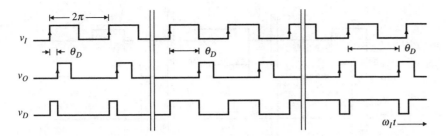

FIGURE 13.30
Typical waveforms for the Type III phase comparator in the locked condition: $\theta_D = \pi/4$ (left), $\theta_D = \pi$ (center), and $\theta_D = (7/4)\pi$ (right).

The Type II Phase Comparator

The Type II comparator differs from PC_1 and PC_3 because its output depends not only on the phase error $\theta_I - \theta_O$, but also on the frequency error $\omega_I - \omega_O$ when the loop has not yet acquired lock. Also called a *phase-frequency detector* (PFD), the circuit is shown in simplified form in Fig. 13.31a.

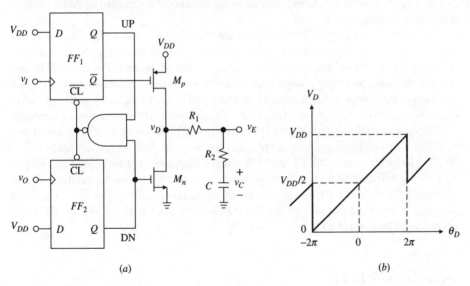

FIGURE 13.31
Type II phase comparator, and its output average V_D as a function of the input phase difference.

With reference to Fig. 13.32, we observe that PC_3 produces UP pulses when the rising edge of v_I leads that of v_O, DN pulses when the rising edge of v_I lags that of v_O, and no pulses when the leading edges are aligned. An UP pulse closes the MOSFET switch M_p and causes the filter capacitance C to charge toward V_{DD} via the series $R_1 + R_2$. A DN pulse closes switch M_n and discharges C toward $V_{SS} = 0$ V. Between pulses, both M_p and M_n are off, providing a high-impedance state to the filter. When PC_2 is in this state, C acts as an analog memory, retaining whatever charge it had accumulated at the end of the last UP or DN pulse.

640

CHAPTER 13
Nonlinear
Amplifiers and
Phase-Locked
Loops

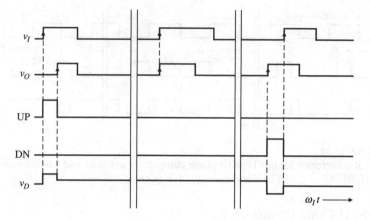

FIGURE 13.32

Typical waveforms for the Type II phase comparator for $\omega_O = \omega_I$: v_O lags v_I (left), v_O is in phase with v_I (center), and v_O leads v_I (right).

Clearly, we now have $v_D = v_E = v_C$. The characteristic is as in Fig. 13.31b, with $K_d = V_{DD}/4\pi = 0.398$ V/rad. For obvious reasons, PC_2 is also called a *charge-pump phase comparator.*

To appreciate its operation, suppose we initially have $\omega_I > \omega_O$. Since v_I generates more rising edges per unit time than v_O, UP will be high most of the time, pumping charge into C and thus raising ω_O. Conversely, when $\omega_I < \omega_O$, DN is high most of the time, pumping charge out of C and lowering ω_O. In either case, PC_2 will keep pumping charge until the inputs become equal *both* in frequency and in phase, or $\omega_O = \omega_I$ and $\theta_O = \theta_I$. We conclude that PC_2 approaches *ideal-integrator* behavior.

It is apparent that a PLL with a Type II comparator will lock under any condition, and that it drives the input-phase error to zero over the full frequency range of the VCO. Moreover, since the UP and DN pulses disappear entirely once the loop is locked, v_E will exhibit no ripple, so there are no unwanted phase modulation effects. The main drawback of PC_2 is its susceptibility to noise spikes, just like PC_3. Even so, PC_2 is the most popular of the three PCs. Note that with v_I absent, PC_2 will drive ω_O as low as it can.

Designing with PLLs

The design process of a PLL-based system involves a number of decisions[12] dictated by the performance specifications of the given application, along with considerations of circuit simplicity and cost. For 4046 PLLs, this process requires (*a*) the specification of the VCO parameters f_0 and $2f_R$, the choice of (*b*) the phase-detector type and (*c*) the filter type, and (*d*) the specification of the filter parameters ω_p and ω_z.

To simplify the process, computer programs are available that accept specifications by the user and translate them into actual resistance and capacitance values to meet the VCO and filter requirements. An example is the *HCMOS Phase-Locked Loop Program,* by Philips Semiconductors (check our Web site at http://www.mhhe.com/franco to find how to download this program), which also provides important data about the loop dynamics and displays the frequency response via Bode plots. Once a PLL system has been designed, it can be simulated by computer,[7,12] for instance, using suitable SPICE macromodels.[14] However, the designer still needs a sound understanding of PLL theory to judge the results of any simulation!

Popular PLL applications[10] include FM, PM, AM, and FSK modulation/demodulation, frequency synchronization and synthesis, clock reconstruction, and motor speed control. Here we discuss two examples, FM demodulation and frequency synthesis. Other examples can be found in the end-of-chapter problems.

EXAMPLE 13.7. An FM signal is being modulated over the range of 1 MHz $\pm$ 10 kHz with a modulating frequency of 1 kHz. Using a 4046A PLL, design a circuit to demodulate such a signal.

Solution. For the VCO we let $f_0 = 1$ MHz, and choose $2f_R$ wide enough to accommodate parameter spread. Thus, let $2f_R = 0.5$ MHz. This gives $K_o = 2\pi \times 0.5 \times 10^6/2.8 = 1.122 \times 10^6$ (rad/s)/V. Using the data sheets or the aforementioned PLL program, we find that a suitable set of VCO components is $R_1 = 95.3$ kΩ, $R_2 = 130$ kΩ, and $C = 100$ pF.

Next, anticipating a noisy input signal, we choose PC_1, so $K_d = 5/\pi$ V/rad and $K_v = K_d K_o = 1.786 \times 10^6$ s^{-1}. To allow for the possibility of a weak input, we take advantage of the fact that the detector input buffers are self-biased near $V_{DD}/2$, where gain is maximized. Consequently, the input signal is ac coupled, as shown in Fig. 13.33.

Finally, to minimize cost, we use a passive lag-lead filter. Impose $\zeta = 0.707$ and choose $f_{-3\,dB} > f_m$, say, $f_{-3\,dB} = 10$ kHz. Proceeding as in Example 13.6, we find $\omega_p = 553$ rad/s and $\omega_z = 22.5$ krad/s, which can be met with the filter components shown in the figure.

Just as inserting a voltage divider within the feedback loop of an op amp increases the output voltage swing, inserting a frequency divider inside the PLL loop downstream of the VCO increases the VCO frequency. A frequency divider is implemented with a counter, and the VCO output frequency becomes $\omega_O = N\omega_I$, where N is the counter modulus. Making the counter programmable allows the synthesis of variable frequencies that are integral multiples of ω_I.

The PLL formalism still holds, but with K_o replaced by K_o/N. We observe that varying N varies also the gain factor K_v, so care must be exercised to ensure that stability and dynamics are maintained over the full range of values of N.

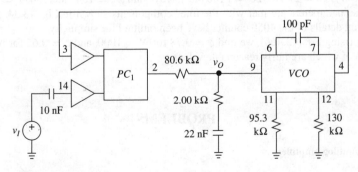

FIGURE 13.33
FM demodulator using the 4046A PLL.

642

CHAPTER 13
Nonlinear
Amplifiers and
Phase-Locked
Loops

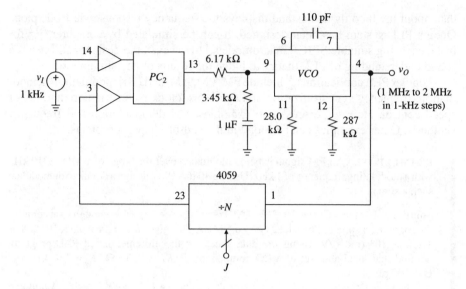

FIGURE 13.34
Frequency synthesizer using the 4046A PLL.

EXAMPLE 13.8. Using a 4046A PLL, design a circuit that accepts a 1-kHz reference frequency and synthesizes all frequencies between 1 MHz and 2 MHz in 1-kHz steps.

Solution. To span the given range, we need a programmable counter between $N_{min} = 10^6/10^3 = 1000$ and $N_{max} = 2000$. Choose, for instance, a 4059 counter, which allows for N to be programmed anywhere from 3 to 15,999 via a set of inputs referred to as the *jam inputs J* in the data sheets.

For the VCO, specify f_0 halfway between the extremes, or $f_0 = 1.5$ MHz, and again choose $2f_R$ wide enough, say, $2f_R = 1.5$ MHz. This gives $K_o = 3.366 \times 10^6$ (rad/s)/V. Using the data sheets or the aforementioned PLL program, we find the VCO component values $R_1 = 28.0$ kΩ, $R_2 = 287$ kΩ, and $C = 110$ pF.

Anticipating relatively clean on-board signals, we choose PC_2, so $K_d = 5/4\pi$ V/rad. Since N is variable, a reasonable approach[12] is to design for the geometric mean of the extremes, or for $N_{mean} = \sqrt{N_{min}N_{max}} = 1414$. The corresponding gain factor is then $K_{v(mean)} = K_dK_o/N_{mean} = 947$ s^{-1}.

We again use a passive lag-lead filter. Imposing $\zeta = 0.707$ and arbitrarily choosing $\omega_n = \omega_I/20 = 2\pi 10^3/20 = \pi 100$ rad/s, we obtain $\omega_p = 104$ rad/s and $\omega_z = 290$ rad/s. These parameters are met with the filter components shown in Fig. 13.34, where the wiring details of the 4059 counter have been omitted for simplicity.

Using Eq. (13.46b), we find $\zeta = 0.78$ for $N = 1000$, and $\zeta = 0.65$ for $N = 2000$, both of which are fairly reasonable values.

PROBLEMS

13.1 Log/antilog amplifiers

13.1 In the transdiode of Fig. 13.2a let $R = 10$ kΩ, $C_n + C_\mu = 20$ pF, $V_A = 100$ V, $r_d = 2$ MΩ, and $f_t = 1$ MHz. If $R_E = 4.3$ kΩ and $C_f = 100$ pF, calculate $1/\beta_0$,

1/β_∞, f_z, and f_p for $v_I = 1$ mV, 10 mV, and 10 V; hence, confirm the linearized plots of Fig. 13.3.

643

Problems

13.2 Find the phase margin of the circuit of Example 13.1.

13.3 Modify the circuit of Fig. 13.4 to yield $v_O = -(2 \text{ V/dec}) \log_{10}[v_I/(1 \text{ V})]$.

13.4 (*a*) Derive Eqs. (13.13) and (13.14). (*b*) Design a circuit that accepts an input voltage $-5 \text{ V} \le v_I \le +5 \text{ V}$, and gives $i_O = (10 \, \mu\text{A})2^{-v_I/(1 \text{ V})}$; this circuit is useful in electronic music. (*c*) Modify the above circuit so that it gives the same output range, but for $0 \text{ V} \le v_I \le 10 \text{ V}$.

13.5 The log conformity error at the upper end of the current range is due primarily to the bulk resistance of the emitter region, which can be modeled with a small resistance r_s in series with the emitter itself. (*a*) Recompute the transfer characteristic of the transdiode of Fig. 13.1*b*, but with r_s in place. If $r_s = 1 \, \Omega$, what is the log conformity error at $i_I = 1$ mA? At $i_I = 0.1$ mA? (*b*) The effect of r_s can be compensated by feeding a small portion of v_I to the base of the BJT. This is achieved by lifting the base off ground, returning it to ground via a resistance R_x, and connecting a second resistance R_y between the source v_I and the base of the BJT. Sketch the modified transdiode, and show that choosing $R_y/R_x = R/r_s - 1$ will eliminate the error due to r_s.

13.6 In the log amp of Fig. 13.4 the bulk-resistance error (see Problem 13.5) can be compensated by connecting a suitable network between the base of Q_2 and the output of OA_2. Such a network consists of a resistance R_c in series with a diode D_c (cathode at the output of OA_2). Likewise, in the antilog amp of Fig. 13.6 the compensation network is connected between the base of Q_1 and the output of OA_1 (cathode at the output of OA_1). Show that the error is nulled when $R_c = (R_1 \| R_2)(2.2 \text{ k}\Omega)/r_s$. What is the required R_c, given that the LM394 has $r_s = 0.5 \, \Omega$?

13.2 Analog multipliers

13.7 A popular multiplier application is *frequency doubling*. One way of configuring the AD534 for this operation is as follows:[6] connect X_2 and Y_1 to ground, connect X_1 and Y_2 together and drive them with a source $v_I = 10 \cos \omega t$ V, connect the Out pin to Z_1 via a 10-kΩ resistor, connect Z_1 to Z_2 via another 10-kΩ resistor, and drive Z_2 with a 10-V reference voltage. (*a*) Sketch the circuit; then, using the identity $\cos^2 \alpha = (1 + \cos 2\alpha)/2$ obtain an expression for the output v_O. (*b*) Assuming well-regulated ± 15-V supplies, design a circuit to generate the 10-V reference for Z_2.

13.8 The AD534 multiplier can be made to approximate the *sine function* within 0.5% of full scale as follows:[6] connect Y_2 to ground, connect Y_1 and Z_2 together and drive them with a source v_I, connect Y_1 to X_2 via a 10-kΩ resistor, connect X_2 to ground via an 18-kΩ resistor, connect the Out pin to Z_1 via a 4.7-kΩ resistor, connect Z_1 to X_1 via a 4.3-kΩ resistor, and connect X_1 to ground via a 3-kΩ resistor. (*a*) Sketch the circuit, derive an expression for the output v_O as a function of v_I, and calculate v_O at some significant points to verify that the circuit approximates the function $v_O = 10 \sin[(v_I/10)90°]$ V. (*b*) Using additional components as needed, design a circuit that accepts a triangular wave with peak values of ± 5 V and gives a sine wave with the same frequency and peak values as the input.

644

CHAPTER 13
Nonlinear
Amplifiers and
Phase-Locked
Loops

13.9 The AD534 multiplier can be configured to yield the percentage deviation between two signals v_1 and v_2 as follows: connect X_1 and Z_1 together and drive them with v_1, connect X_2 and Y_1 to ground, drive Z_2 with v_2, connect the Out pin to Y_2 via a resistance R_1, and connect Y_2 to ground via a resistance R_2. Develop an expression for the output v_O, and specify R_1 and R_2 for $v_O = 100(v_2 - v_1)/v_1$.

13.10 Figure P13.10 shows a transducer-response linearization technique using a four-quadrant multiplier. Derive an expression for V_O as a function of δ and verify that it is linearly proportional to δ in spite of the fact that the voltage across the transducer is a nonlinear function of δ.

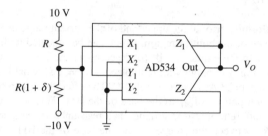

FIGURE P13.10

13.11 Using the AD534 as a voltage-controlled attenuator, design a programmable first-order low-pass filter with a dc gain of 20 dB and $f_0 = kV_C$, $0.1 \text{ V} \le V_C \le 10 \text{ V}$ and $k = 100 \text{ Hz/V}$. *Hint:* See Problem 12.11.

13.3 Operational transconductance amplifiers

13.12 Find the transfer function of the g_m-C filter of Fig. P13.12.

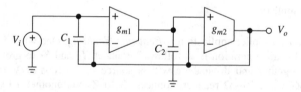

FIGURE P13.12

13.13 Design an exponential VCA such that $A = 2^{-V_C/(1 \text{ V})}$ V/V, $0 \le V_C \le 10$ V. Hence, outline its calibration procedure.

13.14 Design a programmable state-variable filter with $Q = 10$, $H_{OBP} = 1$, and f_0 variable over the audio range by means of control voltage V_C as $f_0 = (20 \text{ kHz})2^{-V_C/(1 \text{ V})}$, $0 \le V_C \le 10$ V.

13.15 The VCA610 (Burr-Brown) is a wideband VCA that accepts two signal inputs v_P and v_N and a control input V_C, and gives $v_O = A(v_P - v_N)$, where $A = 0.01 \times 10^{-V_C/(0.5 \text{ V})}$ V/V for $-2 \text{ V} \le V_C \le 0$. Using a VCA610 and an OPA620 wideband precision op amp, design a first-order low-pass filter with unity dc gain and programmable cutoff frequency from 100 Hz to 1 MHz.

13.16 (a) Sketch and label all relevant waveforms in the CCO of Fig. 13.20, and derive Eq. (13.32). (b) Find C so that $f_0 = 100$ kHz for $I_C = 1$ mA; next, using this CCO as basis, design a VCO such that $f_0 = (100\text{ kHz})10^{-V_C/(2\text{ V})}$, $0 \leq V_C \leq 10$ V. Outline its calibration procedure.

13.4 Phase-locked loops

13.17 Find the phase response $\theta_d(t)$, in degrees, in parts (a) and (b) of Example 13.3.

13.18 If we let $R_2 = 0$ in Fig. 13.23a, the zero is moved to infinity, resulting in a *passive lag filter*. Such a filter finds limited use because it does not allow for ω_x to be specified independently of K_v. (a) Verify that if we let $R_2 = 0$ in the filter of Example 13.4, the phase margin is inadequate. (b) Specify a new set of values for R_1 and C to ensure $\phi_m \cong 45°$ with $R_2 = 0$. What is the corresponding value of ω_x?

13.19 Repeat Example 13.4, but using an active PI filter.

13.20 Prove Eqs. (13.46) through (13.47).

13.21 A PLL has $\omega_0 = 2\pi 10^6$ rad/s, $K_d = 0.2$ V/rad, $K_a = 1$ V/V, and $K_o = \pi 10^6$ (rad/s)/V. Design an active PI filter for a loop time-constant of approximately 100 periods of the free-running frequency and $Q = 0.5$.

13.22 If a 0.1-μF capacitance is connected in parallel with R_2 in the loop filter of Example 13.6, find how it affects ω_x and ϕ_m.

13.5 Monolithic PLLs

13.23 (a) Sketch and label the average V_D versus θ_D for a Type I phase comparator if the duty cycles of v_I and v_O are $D_I = \frac{1}{2}$ and $D_O = \frac{1}{3}$. (b) Repeat, but with $D_I = \frac{1}{3}$ and $D_O = \frac{1}{2}$. Comment.

13.24 Sketch v_I, v_O, UP, DN, and v_D for a Type II detector if (a) ω_I is slightly higher than ω_O, (b) ω_I is slightly lower than ω_O, (c) $\omega_I \gg \omega_O$, and (d) $\omega_I \ll \omega_O$.

13.25 A certain CMOS PLL is powered between 5 V and 0 V, and uses a Type I phase comparator and a VCO with $K_o = 5$ MHz/V and $f_0 = 10$ MHz for $v_E = 2.5$ V. (a) Design a passive lag-lead filter for $\omega_n = 2\pi 5$ krad/s and $Q = 0.5$. (b) Sketch v_I, v_O, v_D, and v_E for the case in which the loop is locked to an input frequency of 7.5 MHz.

13.26 Find $v_e(t)$ in the FM demodulator of Example 13.7.

13.27 Dual-slope ADCs are clocked at a frequency that is locked to the ac line frequency f_line in order to reject line-induced noise. Using a 4046A PLL, design a circuit that accepts f_line (either 60 Hz or 50 Hz) and gives $f_\text{CK} = 2^{16} \times f_\text{line}$. Specify as many parameters and components as you can in your circuit.

13.28 Using a 4046A for phase detection and an 8038 as VCO, design a circuit that generates a 1-kHz sine wave synchronized on a 1-MHz crystal oscillator.

13.29 An FSK signal v_I alternates between $f_L = 1200$ Hz (logic 0) and $f_H = 2400$ Hz (logic 1). One way[13] to decode this signal with a 4046A PLL is to use PC_3, a loop

646

CHAPTER 13
Nonlinear
Amplifiers and
Phase-Locked
Loops

filter consisting of a plain RC stage with $1/2\pi RC = f_H$, the VCO with $f_0 = (f_L + f_H)/2 = 1.8$ kHz and $2f_R = 2$ kHz, and a positive edge-triggered latch flip-flop of the type of Fig. 13.31, with v_I as the D input and the VCO output v_O as the clock; the $\overline{Q}$ output of the flip-flop is the FSK decoder output. Draw the circuit; then sketch and label v_I, the average of v_E, v_O, and $\overline{Q}$ both for $f_I = f_L$ and $f_I = f_H$. What is the distinguishing feature of PC_3 that makes it attractive in this application?

REFERENCES

1. D. H. Sheingold, ed., *Nonlinear Circuits Handbook,* Analog Devices, Norwood, MA, 1974.

2. P. R. Gray and R. G. Meyer, *Analysis and Design of Analog Integrated Circuits,* 3d ed., John Wiley & Sons, New York, 1993.

3. "Theory and Applications of Logarithmic Amplifiers," Application Note AN-311, *Linear Applications Handbook,* National Semiconductor, Santa Clara, CA, 1994.

4. C. Kitchin and L. Counts, *RMS to DC Conversion Application Guide,* Analog Devices, Norwood, MA, 1983.

5. B. Gilbert, "Translinear Circuits—25 Years On," *Electronic Engineering:* Part I, August 1993, pp. 21–24; Part II, September 1993, pp. 51–53; Part III, October 1993, pp. 51–56.

6. D. H. Sheingold, ed., *Multiplier Application Guide,* Analog Devices, Norwood, MA, 1978.

7. D. A. Johns and K. W. Martin, *Analog Integrated Circuit Design,* John Wiley & Sons, New York, 1997.

8. C. Toumazou, F. J. Lidgey, and D. G. Haigh, eds., *Analogue IC Design: The Current-Mode Approach,* IEEE Circuits and Systems Series, Peter Peregrinus Ltd., London, U.K., 1990.

9. R. Schaumann, M. S. Ghausi, and K. R. Laker, *Design of Analog Filters: Passive, Active RC, and Switched Capacitor,* Prentice-Hall, Englewood Cliffs, NJ, 1990.

10. A. B. Grebene, *Bipolar and MOS Analog Integrated Circuit Design,* John Wiley & Sons, New York, 1984.

11. F. M. Gardner, *Phaselock Techniques,* 2d ed., John Wiley & Sons, New York, 1979.

12. R. E. Best, *Phase-Locked Loops: Theory, Design, and Applications,* 3d ed., McGraw-Hill, New York, 1997.

13. *CMOS Phase-Locked Loops,* Philips Semiconductors, Sunnyvale, CA, June 1995.

14. J. A. Connelly and P. Choi, *Macromodeling with SPICE,* Prentice-Hall, Englewood Cliffs, NJ, 1992.

INDEX

A

μA78G, 524
μA79G, 524
μA741 op amp, 1, 2, 212, 215, 216, 249–257, 340
μA7800 series, 523–525, 530
μA7805 regulator, 525, 527
μA7900 series, 523, 530
Absolute accuracy, 562
Absolute maximum ratings, 241
Absolute-value circuit, 422
ac-dc converters, 426–428
ac noise. *See* Noise
Acquisition time, 439
Active compensation of integrators, 285
Active filters, 106–210
    all-pass filters.
        See All-pass filters
    audio filter applications, 121–126. *See also* Audio filter applications
    band-pass filters.
        See Band-pass filters
    band-reject filters.
        See Band-reject filters
    biquad filters, 147, 148
    common frequency responses, 106–108
    first-order filters, 115–121.
        See also First-order active filters
    GBP, and, 289–293
    higher-order filters, 160–210.
        See also Higher-order active filters
    high-pass filters.
        See High-pass filters
    KRC filters, 133–141. *See also* *KRC* filters
    limitation, 109
    low-pass filters. *See* Low-pass filters
    multiple-feedback filters, 141–144
    nature of, 109
    RLC filters, 109
    SC filters. *See* SC filters
    second-order filters, 126–133.
        See also Second-order filters
    sensitivity, 150–152

SV filters, 144–147
transfer function, 109–114
universal filters, 144–150
Active guard drive, 86, 87
Active PI filter, 633
Active tone control, 123–125
AD522, 81
AD534, 618
AD537 V-F converter, 484
AD537 voltage-to-frequency converter, 487–489
AD549, 222
AD590 IPTAT, 518
AD590 two-terminal temperature transducer, 512
AD594/5/6/7 series, 519
AD817, 364
AD7846 16-bit segmented DAC, 577
A-D converters, 559–606.
    See also D-A converters
    CR ADC, 588–590
    DAC-based A-D conversion, 584, 585
    errors, 566
    flash converters, 590, 591
    INL/DNL, 566
    integrating-type converters, 592–595
    performance specifications, 564, 565
    pipelined converters, 592
    SA ADC, 586–588
    subranging converters, 591, 592
ADCs. *See* A-D converters
Aliasing, 596
All-pass filters
    all-pass response, 107, 108
    second-order filters, 132
All-pole ladder, 182
Alternative *D*-element realization, 178
Amount of feedback, 24
AMP-01, 83–85
Amplifier
    bridge, 57. *See also* Transducer bridge amplifiers
    CFAs, 293–303. *See also* Current-feedback amplifiers (CFAs)
    composite, 384–390

current, 71, 72
difference, 19, 20, 73–79.
    See also Difference amplifiers
fundamentals, 2–5
IAs, 79–91. *See also* Instrumentation amplifiers
inverting, 12–15, 266–268
linear, 3
log/antilog, 608–615
main, 234
noninverting, 8–12, 262–264
nonlinear. *See* Nonlinear amplifiers
nulling, 234
op amp. *See* Op amp
OTA, 620–627
photodetector, 63
SHAs, 437–443
THAs, 437–443
transconductance, 63–71.
    See also Voltage-to-current converters
transducer bridge, 91–97
transimpedance, 61, 294
transresistance, 5, 61–63
VFAs, 293, 300–303
voltage, 3
Analog division, 619, 620
Analog-ground-switch, 430
Analog multipliers, 615–620
Analog switches, 428–433
Analog to digital converters. *See* A-D converters
Angular frequency, 109, 258
Antilogarithmic amplifier, 609, 613
Aperture jitter, 439
Aperture time, 439
Aperture uncertainty, 439
Astable multivibrator, 457
Audio filter applications, 121–126
    active tone control, 123–125
    graphic equalizers, 125, 126
    phono preamplifier, 122
    tape preamplifier, 123
Autozero mode, 235
Autozero phase, 594
Autozero (AZ) technique, 234
Avalanche breakdown, 507

Avalanche noise, 324
Averaging-type meters, 314
AZ op amps, 234, 235
AZ technique, 234

B

Balanced bridge, 20, 67
Balanced modulator, 638
Balanced transmission, 86
Bandgap voltage, 506
Bandgap voltage references,
510, 511
Band-pass filters
band-pass response, 107
cascade design, 171–174
direct synthesis, 196–198
KRC filters, 139, 140
multiple-feedback filters,
141, 142
SC filters, 196–198
second-order filters, 130, 131
wideband filters, 120
Band-reject filters
band-reject response, 107
cascade design, 175
KRC filters, 140, 141
multiple-feedback filters,
143, 144
notch response, 107
second-order filters, 131, 132
Bar graph, 590
Bar graph meters, 412–414
Barkhausen criterion, 453
Baseband, 595
Basic free-running multivibrator,
458–461
Basic series voltage
regulator, 519
Basic triangular/square-wave
generator, 472
Basic Wien-bridge oscillator,
451–453
Bass/treble control, 123–125
Bessel filters, 165, 166
Bidirectional, 65
Bipolar DAC output
conditioning, 573
Bipolar DACs, 572–574
Bipolar junction transistor
(BJT), 1
Bipolar op amps, 230–233
Bipolar OTAs, 622
Biquad filter, 147, 148, 152
Bistable multivibrator, 457
BJT, 1
Black, Harold S., 347
Bode plots, 114
Boost circuit, 537
Bootstrapping principle, 515
Breakpoint wave shaper, 480, 481

Bridge amplifier, 57. *See also*
Transducer bridge amplifiers
Bridge calibration, 93, 94
Bridge imbalance
difference amplifiers, 74–77
V-I converters, 68, 69
Bridge legs, 93
Bridge linearization, 97
Broadband, 265
Brokaw cell, 511
Buck-boost circuit, 537
Buck circuit, 537
BUF-03, 12
Buffer, 12
Buffered 10-V reference, 513
Buried diode structure, 510
Butterworth filters, 163, 164
Butterworth low-pass element
values, 183
Butterworth response, 129

C

CA3080, 623
Cancellation. *See also*
Compensation
input-bias-current, 222
pole-zero, 372, 373
zero-pole, 284
Capacitive loading, 363
Capacitive load isolation,
362–364
Capture range, 629
Capture time, 629
Cascade design, 166–175
band-pass filters, 171–174
band-reject filters, 175
high-pass filters, 171
low-pass filters, 168–171
SC filters, 202–204
Case to ambient, 530
Cauer filters, 165
CCM, 538, 539
CCO, 477
Celsius sensor, 517
CF, 313, 314
CFA-derived VFA, 302
CFAs. *See* Current-feedback
amplifiers (CFAs)
Characteristic frequencies, 110
Charge-balancing ADCs, 592
Charge-balancing VFC,
489, 490
Charge-pump phase
comparator, 640
Charge-redistribution converters
(CR ADCs), 588–590
Chatter, 420, 421
Chebyshev filters, 165
Chebyshev low-pass element
values, 183

Chopper stabilization (CS)
technique, 234
Chopper-stabilized op amp
(CSOA), 234
Circuits with resistive feedback,
60–105
current amplifiers, 71, 72
current-to-voltage converters,
61–63
difference amplifiers, 73–79
instrumentation amplifiers,
79–86. *See also*
Instrumentation
amplifiers
instrumentation applications,
86–91
transducer bridge amplifiers,
91–97. *See also*
Transducer bridge
amplifiers
voltage-to-current converters,
63–71. *See also*
Voltage-to-current
converters
Closed-loop gain, 9
Closed-loop response, 263–268
Closed-loop transresistance
gain, 36
Closed-loop voltage gain, 36
CMOS-gate crystal oscillator,
463, 464
CMOS-gate free-running
multivibrator, 463
CMOS-gate one shot, 465
CMOS inverter, 462
CMOS logic gates, 462
CMOS op amps, 45, 341
CMOS transmission gate,
432, 433
CMP-05, 407
CMRR, 75, 76, 227, 228
Code center line, 566
Code ranges, 564
Coil current ripple, 538
Cold-junction compensation, 518
Commercial range, 240
Common-centroid layout, 231
Common-mode gain, 75
Common-mode input
capacitance, 269
Common-mode input
impedance, 269
Common-mode input
resistance, 74
Common-mode rejection ratio
(CMRR), 75, 76, 227, 228
Comparator chatter, 420, 421
Comparators. *See* Voltage
comparators
Compensated differentiator, 358

Compensation
    active, of integrators, 285, 286
    cold junction, 518
    external frequency, 374–380.
        See also External
        frequency compensation
    input offset-error, 235–240
    internal frequency, 365–374.
        See also Internal
        frequency compensation
    neutral, 360
    passive, of integrators, 284
    phase-error, 284
    Q enhancements, 286–289
    stray input-capacitance,
        359–362, 382–384
Complex frequency, 109
Composite amplifiers, 384–390
Constant-GBP op amps, 261
Contact noise, 323
Continuous conduction mode
    (CCM), 538, 539
Continuous-time filters. *See*
    Higher-order active filters
Control-to-output transfer
    function, 545
Conversion time, 566
Converting RC integrator to SC
    integrator, 188
Corner frequency, 316
CR ADCs, 588–590
Crest factor (CF), 313, 314
Critical frequencies, 110
Crossover frequency, 262
Cross regulation, 551
Cross-talk for common return
    impedance, 78
CS op amps, 234, 235
CS technique, 234
Current amplifiers, 4, 71, 72
Current cancellation, 222
Current-controlled integrator, 625
Current-controlled oscillator
    (CCO), 477
Current-controlled relaxation
    oscillator, 627
Current-driven *R-2R* ladders,
    574, 576
Current-feedback amplifiers
    (CFAs), 293–303
    applying CFAs, 299
    closed loop gain, 295, 296
    dynamics, 297, 298
    high-speed voltage-feedback
        amplifiers, 300–303
    noise, 334
    PSpice models, 299, 300
    second-order effects, 298, 299
    simplified circuit diagram, 294
    stability, 381–384

Current flow, 42–44
Current fold-back, 522
Current foldback protection, 521
Current-input IA, 91
Current mirror, 72
Current-mode control, 547, 548
Current mode *R-2R* ladders,
    570, 571
Current-mode segmentation,
    578, 579
Current-output IA, 89–91
Current reverser, 72
Current sources, 515–517
Current switch, 430
Current-to-voltage converters,
    61–63
Cutoff frequency, 106, 129, 161
Cyclical frequency, 258

D

D-element realization, 177, 178
DABP filter, 178, 179
DAC-based A-D conversion,
    584, 585
D-A converters, 559–606.
    See also A-D converters
    bipolar DACs, 572–574
    current-driven *R-2R* ladders,
        574, 576
    current-mode *R-2R* ladders,
        570, 571
    current-mode segmentation,
        578, 579
    errors, 562, 563
    glitches, 564
    INL/DNL, 563
    master-slave DACs, 574, 575
    MDAC applications, 580–584
    performance specification,
        561, 562
    potentiometric DACs, 569, 570
    R-2R ladders, 570–572, 574
    segmentation, 576–579
    voltage-mode *R-2R* ladders,
        571, 572
    voltage-mode segmentation,
        577, 578
    weighted-capacitor DACs,
        568, 569
    weighted-resistor DACs,
        567, 568
DACs. *See* D-A converters
Damping ratio, 127, 634
Darlington pair, 215, 519
dc gain, 118, 259
DCM, 539, 540
dc noise, 311
dc noise gain, 235
dc-offsetting amplifier, 19
Deboo integrator, 117

Decimation, 598
Decompensated op amps, 380
Deintegrate phase, 594
Delay filters, 108. *See also*
    All-pass filters
Delyiannis-Friend filter, 141
Desensitivity factor, 26
Dielectric absorption, 435
Difference amplifier
    calibration, 76
Difference amplifiers,
    19, 20, 73–79
    ground-loop interference
        elimination, 77–79
    resistance mismatches, 74–77
    variable gain, 77
Difference-input,
    difference-output amplifier,
    79
Differential input capacitance,
    269
Differential input impedance, 269
Differential input-pair noise,
    340, 341
Differential input voltage, 6
Differential-mode gain, 75
Differential-mode input
    resistance, 74
Differential nonlinearity (DNL),
    563, 566
Differential V-I converter,
    616, 617
Differentiator, 20, 21, 115, 116
Differentiator circuit, 357, 358
Digitally programmable
    amplifier, 582
Digitally programmable
    attenuator, 582
Digitally programmable
    filter, 583
Digitally programmable
    gain, 87–89
Digitally programmable IA, 88
Digital to analog converters.
    See D-A converters
Direct-design, 181–186
Direct synthesis (design),
    181–186
    high-pass filters, 185, 186
    low-pass filters, 182–185
    SC filters, 196–198
Discontinuous conduction mode
    (DCM), 539, 540
DNL, 563, 566
Dominant-pole compensation,
    367, 368
Dominant-pole frequency, 259
Dominant-pole PSpice
    model, 263
Dominant-pole response, 259

Doubly terminated all-pole RLC ladder, 194
Doubly terminated series-resonant RLC ladder, 182
Drain cutoff current, 429
Dropout voltage, 505
Dual-amplifier band-pass (DABP) filter, 178, 179
Dual-integrator-loop SC filter, 192–194
Dual-op-amp IA, 82, 83
Dual-slope ADC, 593
Dual-slope waveform, 593
Duty cycle, 415
Dynamic op amp limitations, 258–310. *See also* Static op amp limitations
    CFAs, 293–303. *See also* Current-feedback amplifiers (CFAs)
    closed-loop response, 263–268
    finite GBP/filters, 289–293
    finite GBP/integrator circuits, 283–289
    full-power bandwidth, 278, 279
    input/output impedances, 269–275
    open-loop response, 259–263
    rise time, 275, 276
    settling time, 279–281
    slew-rate limiting, 276–278, 281, 282
    transient response, 275–282
Dynamic output range, 45
Dynamic range, 562, 608
Dynamic resistance, 429

E

8-bit subranging ADC, 591
EL2044C low-power/low-voltage 120-MHz unity-gain stable op amp, 303
Element values (Butterworth/Chebyshev low-pass filters), 183
Elliptic filters, 165
Elliptic low-pass response, 182
Emitter-coupled VCO, 483–486
Emitter degeneration, 282
Equal-component *KRC* circuit, 135, 136
Equalized preamplifiers, 121
Equivalent series inductance (ESL), 541, 542
Equivalent series resistance (ESR), 541, 542
Error amplifier, 23
Error function, 24

Error signal, 23
Error voltage, 628
ESL, 541, 542
ESR, 541, 542
EVAL.LIB, 216, 415
Excess noise, 323
Explicit rms computation, 614
Exponential/linear waveforms, 450
Exponential transient, 451
External frequency compensation, 374–380
    decompensated op amps, 380
    feedback-lead compensation, 378–380
    input-lag compensation, 375–378
    reducing loop gain, 374, 375
External (interference) noise, 311, 312
External offset nulling, 238–240

F

False ground, 280
FDNR, 177, 179–181
Feedback, 29–37. *See also* Negative feedback
Feedback factor (β), 39–41
Feedback-lead compensation, 378–380
Feedback pole, 356, 357
Feedforward compensation, 373, 374
Feedthrough, 440
Feedthrough gain, 32, 34
Feedthrough rejection ratio, 440
FET-input op amps, 234
Field-effect transistors (FETs), 428
Field emission breakdown, 507
Fifth-order 0.1/40-dB elliptic high-pass filter, 186
Fifth-order 3-dB responses, 166
Fifth-order SC low-pass filter, 196
FILDES, 163
Filter, 106. *See also* Active filters
Filter approximation, 161–166
Finite open-loop gain, 70
First-order active filters, 115–121
    differentiator, 115, 116
    GBP, and, 290
    high-pass filter with gain, 119, 120
    integrators, 116–118
    low-pass filter with gain, 118, 119
    phase shifters, 121
    wideband band-pass filter, 120
First-order loops, 632, 633

First-order noise shaping, 601
555 timer, 465–469
555 timer block diagram, 466
Flash converters, 590, 591
Flicker noise, 323, 324
Flip-flop, 457
Floating-load current amplifier, 72
Floating-load V-I converters, 64, 65
Flyback regulator, 550, 551
Flying capacitor techniques, 85, 86
Folded-cascode bipolar VFA, 302
4046 CMOS PLL, 635–642
Four-quadrant analog-multiplier, 617
Four-quadrant multiplier, 615
Four-terminal adjustable regulator, 524
Fourth-order, 1-dB, 1-kHz elliptic low-pass filter, 204
Fourth-order, 1-dB, 2-kHz Chebyshev low-pass filter, 203
Fourth-order band-pass filter, 197
FPB, 278, 279
Fractional binary value, 561
Fractional deviation, 92
Fractional elongation, 95
Free-running frequency, 628
Free-running multivibrator, 457–463
Frequency-compensation capacitance, 215
Frequency compensation techniques, 365–380
Frequency doubling, 643
Frequency response, 106, 112
Frequency shift keying (FSK), 485, 486
Frequency-to-voltage converter (FVC), 491, 492
FSK, 485, 486
Full-power bandwidth (FPB), 278, 279
Full-scale range, 561
Full-wave rectifier (FWR), 422, 424–426
Function generators, 478–486
FVC, 491, 492
F-V converters, 491, 492
FWR, 422, 424–426

G

Gain-bandwidth product, 261. *See also* GBP
Gain-bandwidth tradeoff, 265
Gain desensitivity, 25, 26
Gain error, 24, 562, 566

Gain margin (GM), 348, 349
Gain node, 294
Gain peaking (GP), 350, 351
Gate reverse current, 429
GBP, 261
    effect, on filters, 289–293
    effect, on integrator circuits,
        283–289
Generalized impedance converter
    (GIC), 175–181
General-purpose IC comparators,
    401–406
GIC, 175–181
Glitches, 564
GM, 348, 349
g_m-C filters, 621
GP, 350, 351
Graphic equalizers, 125, 126
Grounded-capacitor VCO,
    478–483
Grounded FDNR, 177
Grounded-load current
    amplifier, 72
Grounded-load V-I converters,
    66–68
Grounding, 365
Ground-loop interference, 78
Ground-loop interference
    elimination, 77–79
Guard-ring layout/
    connections, 225
Gyrator, 206, 207

H

HA-2725 programmable op
    amp, 282
HA-5330 high-speed monolithic
    THA, 441
Half-flash converter, 591
Half-wave rectifier (HWR),
    422–424
HCMOS Phase-Locked Loop
    Program, 640
Heatsink, 531
HI-DAC16, 579
Higher-order active filters,
    160–210
    Bessel approximation,
        165, 166
    Butterworth approximation,
        163, 164
    cascade design, 166–175
    Cauer approximation, 165
    Chebyshev approximation, 165
    direct-design, 181–186
    filter approximations, 161–166
    GBP, and, 291–293
    generalized impedance
        converter (GIC), 175–181
    PSpice, 163

High-frequency gain, 120, 129
High-gain loop, 634
High-pass filters
    cascade design, 171
    direct design, 185, 186
    gain, and, 119, 120
    high-pass response, 107
    KRC filters, 138, 139
    SC filters, 185, 186
    second-order filters, 129, 130
High-pass notch, 149
High-resolution
    charge-redistribution
        ADC, 589
High-sensitivity I-V converter, 62
High-speed comparators,
    406, 407
High-speed current switch, 573
High-speed voltage-feedback
    amplifiers, 300–303
Hold mode, 434
Hold mode settling time, 440
Hold step, 440
Howland circuit calibration, 69
Howland current pump, 66–68
Huelsman, L. P., 144
HWR, 422–424
Hysteresis, 421

I

IAs. See Instrumentation
    amplifiers
IC noise densities, 317
IC timers, 465–471
ICL7129 4$^{1}/_{2}$-digit ADC, 595
ICL8038 waveform generator,
    479–483
Ideal, 4
Ideal inverting amplifier, 13–15
Idealized filter responses, 107
Ideal noninverting amplifier, 11
Ideal op amp, 7
Ideal op amp analysis, 15–23
    difference amplifier, 19, 20
    differentiator, 20, 21
    integrator, 21, 22
    negative-resistance converter
        (NIC), 22, 23
    summing amplifier, 17–19
    virtual short, 16
Ideal terminal resistances, 5
Images, 595
Imbalance factor
    difference amplifiers, 75
    V-I converters, 68
Impedance
    input/output, 269–275
    series, 270, 271
    shunt, 272–275
Impedance transformation, 22

Implicit rms computation, 614
Improved Howland current
    pump, 70, 71
Improved THA, 442
INA101, 81
INA105, 76
Inadequate power-supply
    filtering, 364, 365
Inductance simulator, 177
Industrial range, 240
Infinite-gain filters, 141–144
Inherent noise, 312
Initial input offset voltage, 230
INL, 563, 566
Input bias current, 63, 217–221
Input-bias-current cancellation,
    221, 222
Input-bias-current
    characteristics, 224
Input bias-current drift, 224
Input current constraint, 15
Input guarding, 225
Input-lag compensation, 375–378
Input linearization network, 623
Input offset current, 217–221
Input offset-error compensation,
    235–240
Input offset voltage, 225–230
Input/output impedances,
    269–275
Input overdrive, 400
Input-pair load noise, 341
Input reference current, 608
Input-referred errors, 211.
    See also Static op amp
    limitations
Input regulation, 501
Input resistance, 3
Input scale factor, 609
Input-series, 29
Input-shunt, 29
Input shunt topology, 273
Input signal-to-noise ratio, 334
Input stage, 213–216
Input voltage constraint, 15
Input voltage range, 241, 242
Instability. See Stability
Instrumentation amplifiers,
    79–91
    active guard drive, 86, 87
    current-input IA, 91
    current-output IA, 89–91
    digitally programmable gain,
        87–89
    dual-op-amp IA, 82, 83
    flying capacitor techniques,
        85, 86
    monolithic IA, 83–85
    output-offsetting, 89
    triple-op-amp IA, 79–82

Instrumentation applications, 86–91

Integral nonlinearity (INL), 563, 566

Integrated-circuit noise, 316, 317

Integrating-type converters, 592–595

Integrating-type THA, 441, 442

Integration unity-gain frequency, 200

Integrator, 21, 22

Integrator circuits, and finite GBP, 283–289

Integrators, 116–118

Interference noise, 311, 312

Intermediate (second) stage, 215

Internal frequency compensation, 365–374

dominant-pole compensation, 367, 368

feedforward compensation, 373, 374

Miller compensation, 369–372

pole-zero compensation, 372, 373

shunt-capacitance compensation, 368, 369

Internal (inherent) noise, 312

Internally compensated op amps, 259

Internal offset nulling, 237

Internal power dissipation, 42

Inverting amplifier, 12–15, 266–268

Inverting configuration, 34–36

Inverting integrator, 356

Inverting regulator, 539

Inverting Schmitt trigger, 416, 417

Inverting SC integrator, 189

IPTATs, 511, 512

I-V converter, 61–63

J

JFET-input op amps, 222

JFET switches, 428–431

Johnson noise, 322

Junction-to-ambient thermal resistance, 529

Junction to case, 530

K

k-channel analog multiplexer, 431

Kerwin, W. J., 144

KHN filter, 144–147

KRC filters, 133–141

band-pass filters, 139, 140

band-reject filters, 140, 141

equal-component KRC circuit, 135, 136

high-pass filters, 138, 139

low-pass filters, 134, 135

sensitivity, 151

unit-gain KRC circuit, 136–138

L

Ladder simulation

continuous-time filters, 181–186

SC filters, 194, 195

Large-signal conditions, 214

LDO regulators, 526, 527

Leakage, 435, 436

Leapfrogging, 578

Least significant bit (LSB), 561

Level detectors, 407–409

LF356, 441

LF356 biFET op amp, 222, 223

Limitations. See Dynamic op amp limitations; Static op amp limitations

Linear amplifier, 3

Linear/exponential waveforms, 450

Linearity, 398

Linearity error, 615, 618

Linearized bridge circuit, 104

Linearized transconductance block, 616

Linearizing effect of negative feedback, 27

Linear region, 44, 45

Linear regulators, 519–535

monolithic voltage regulators, 523–527

power sources, 528, 529

power-supply supervisory circuits, 532–535

protections, 520–523

thermal considerations, 529–532

uses, 527–535

Line-loss detection, 533–535

Line regulation, 501

LM308 op amp, 221

LM311 voltage comparator, 402–404

LM317 functional diagram, 525, 526

LM317 positive regulator, 524

LM318 op amp, 282

LM329 6.9-V reference diode, 409

LM329 precision reference diode, 507

LM335, 409

LM335 precision temperature sensor, 512

LM337 negative regulator, 524

LM339 quad comparator, 405, 406

LM385 2.5-V micropower reference diode, 511

LM395, 409

LM398 BiFET THA, 438

LM399 6.95-V thermally stabilized reference, 509

LM3914 dot/bar display driver, 413

LM3915, 412

LM3916, 412

LM13600, 623

LMC6464 CMOS op amp, 242

Load, 499

Loading, 3, 4

Load regulation, 501

Lock range, 629

Log amp, 608

Log/antilog amplifiers, 608–615

antilog amp, 609, 612

log amp, 608, 609, 612

log ratio amplifier, 612, 613

practical circuits, 612–614

stability, 610, 611

transdiode configuration, 609

true rms-to-dc converters, 614, 615

Logarithmic amplifier, 608, 609, 612

Logarithmic wave shaper, 475, 476

Log conformity error, 608

Logger, 608

Log ratio amplifier, 612, 613

Loop gain (T)

feedback factor, 39–41

finding it directly, 37, 38

graphical visualization, 262

increasing, 384–386

PSpice, 353–355

reducing, 374, 375

Lossy integrator, 22, 119

Low-distortion sine wave generation, 485

Low-dropout (LDO) regulators, 526, 527

Lower saturation region, 44, 45

Low-input-bias-current op amps, 221–225

Low-input-offset-voltage op amps, 230–235

Low-noise op amps, 339–342

differential input-pair noise, 340, 341

input-pair load noise, 341

second-stage noise, 341

ultralow-noise op amps, 341, 342

Low-pass filters
    cascade design, 168–171
    direct design, 182–185, 196
    FDNR, and, 180, 181
    gain, and, 118, 119
    KRC filters, 134, 135
    low-pass response, 106, 107
    multiple-feedback filters,
      142, 143
    SC filters, 182–185, 196
    second-order filters, 127–129
Low-pass filter tables, 169
Low-pass notch, 149
LSB, 562
LT1010, 441
LT1016, 407
LT1025, 519
LT1028, 340
LT1070 monolithic switching
    regulator, 548–550
LT1360, 364
LT1363 70-MHz, 1000-V/μs op
    amp, 302
LTC1060, 198
LTC1064 series, 198
LTZ1000 Super Zener, 509

M

Macromodel, 216
Main amplifier, 234
Mark frequency, 485
Master-slave DACs, 574, 575
Maximum internal power
    dissipation (P_{max}), 241
Maximum passband ripple, 161
Maximum ratings, 240–244
MC3425, 532–534
MDAC, 561, 580–584
Mean square value, 313
MF10 universal monolithic dual
    SC filter, 199
MF10 universal SC filter,
    199–202
Micromodel, 215
Micropower op amps, 42
Midfrequency gain, 120
Military range, 240
Miller compensation, 369–372
Miller effect, 35
Miller integrator, 116
Monolithic IA, 83–85
Monolithic PLLs, 635–642
    designing, 640–642
    Type I phase comparator,
      637, 638
    Type II phase comparator,
      639, 640
    Type III phase comparator,
      638, 639
    VCO, 635–637

Monolithic switching regulators,
    544–551
    current-mode control,
      547, 548
    flyback regulators, 550, 551
    LT1070, 548–550
    voltage-mode control,
      544–547
Monolithic temperature sensors,
    511, 512
Monolithic timers, 465–471
Monolithic voltage regulators,
    523–527
Monolithic waveform generators,
    478–486
Monostable multivibrator, 457,
    464, 465
MOSFET-input op amps,
    222–224
MOSFET switches, 431–433
Most significant bit (LSB), 561
MP7616 16-bit CMOS DAC, 579
MPY100, 618
Multiple-feedback band-pass
    filter, 141, 142
Multiple-feedback filters,
    141–144, 152
Multiple-feedback low-pass filter,
    142, 143
Multiple-op-amps filters. *See*
    Universal filters
Multipliers, 615–620
Multiplying DAC (MDAC), 561,
    580–584
Multivibrators, 457–465
    basic free-running, 458–461
    classification, 457
    CMOS crystal oscillator,
      463, 464
    free-running, using CMOS
      gates, 462, 463
    monostable, 464, 465

N

NAB equalization curve and tape
    preamplifier, 123
Natural response, 111
n-bit flash converter, 590
n-channel JFET, 429, 430
n-channel MOSFET, 431, 432
NE560, 484
NE5517, 623
NE5533/5534 low-noise audio op
    amp, 340
NEB, 318–320
Negative feedback, 23–29
    block diagram, 23
    disturbances, and, 29
    gain desensitivity, 25, 26
    noise, and, 29

nonlinear distortion reduction,
    26–28
    stability, and, 347
Negative feedback mode, 45
Negative-feedback topologies,
    29, 30
Negative resistance, 22, 23
Negative-resistance converter
    (NIC), 22, 23
Negative saturation region,
    44, 45
Neper frequency, 109
Neutral compensation, 360
Newcomb, R. W., 144
NIC, 22, 23
Noise, 311–346
    avalanche, 324
    BJTs, in, 324, 325
    crest factor, 313, 314
    defined, 311
    dynamics, 317–322
    filtering, 334, 335, 337, 338
    flicker, 323, 324
    IC, 316, 317
    inherent, 312
    interference, 311, 312
    JFETs, in, 325
    low-noise op amps, 339–342
    MOSFETs, in, 325, 326
    NEB, 318–320
    negative feedback, and, 29
    observation/measurement, 314
    $1/f$, 316
    op amp, 328–335. *See also* Op
      amp noise
    photodiode amplifiers, in,
      335–339
    piecewise graphical
      integration, 320
    pink-noise tangent principle,
      321, 322
    PSpice modeling, 326, 327
    rms value, 313
    shot, 323
    SNR, 312
    sources of, 322–327
    spectra, 315
    summation, 314, 315
    thermal, 322
    white, 316
Noise dynamics, 317–322
Noise equivalent bandwidth
    (NEB), 318–320
Noise filtering, 334, 335,
    337, 338
Noise gain, 29, 317
Noise power density, 315
Noise shaping, 599–602
Noise spectra, 315
Noise summation, 314, 315

Noninverting amplifier, 8–12, 264–266
Noninverting configuration, 31–33
Noninverting (Deboo) integrator, 117
Noninverting Schmitt trigger, 417, 418
Noninverting SC integrator, 189, 190
Noninverting SC integrator waveforms, 190
Nonlinear amplifiers, 607–646
  analog multipliers, 615–620
  log/antilog amplifiers, 608–615. *See also* Log/antilog amplifiers
  operational transconductance amplifiers, 620–627
  PLLs, 627–642. *See also* Phase-locked loops (PLLs)
Nonlinear circuits, 398–448
  ac-dc converters, 426–428
  analog switches, 428–431
  FWR, 424–426
  HWR, 422–424
  MOSFET switches, 431–433
  peak detectors, 433–437
  precision rectifiers, 422–428
  sample-and-hold amplifier (SHA), 437–443
  Schmitt triggers, 416–422
  THA, 437–443
  voltage comparators, 399–415. *See also* Voltage comparators
Nonlinear distortion reduction, 26–28
Nonretriggerable one-shot, 465
Notch filters. *See* Band-reject filters
Notch response, 107
Nulling amplifier, 234
Nyquist bandwidth, 595
Nyquist rate, 597
Nyquist-rate converters, 597
Nyquist-rate sampling, 595–597
Nyquist's criterion, 596

O

Octave equalizer, 126
Offset error, 562, 566
Offset nulling, 237–240
Offset-voltage adjustment range, 237
Off-the-shelf OTAs, 622–627
On-chip trimming, 232
1% absolute-error bandwidth, 616

$1/f$ noise, 316, 323
1-mA source, 67
1N821-9 series, 507
1N5822, 550
One-quadrant multiplier, 615
One-shot, 457, 465
On-off controllers, 409, 410, 421, 422
On-off temperature controller, 410
Op amp, 5–8
  feedback, 29–37
  historical overview, 1, 2
  ideal, 7. *See also* op amp analysis
  limitations. *See* Dynamic op amp limitations; Static op amp limitations
  noise. *See* Op amp noise
  powering, 41–47
  practical, 65
  PSpice simulation, 7, 8
  simplified block diagram, 259
  simplified circuit diagram, 212–217
  voltage comparator, as, 400, 401
Op amp circuit, 8
Op amp differentiator, 20, 21
Op amp integrator, 21, 22
Op amp manufacturers, 2
Op amp noise, 328–335
  CFAs, in, 334
  noise filtering, 334, 335
  overall input spectral density, 329, 330
  rms output noise, 330–333
  SNR, 333, 334
Op amp powering, 41–47
  current flow, 42–44
  output saturation, 44–47
  power dissipation, 42–44
OP-07 op amp, 222
OP-27, 232, 233
OPA129, 222
OPA501, 244
OPA627, 341
Open-circuit noise, 330
Open-circuit voltage gain, 3
Open-loop -3-dB frequency, 259
Open-loop bandwidth, 259
Open-loop dc gain, 259
Open-loop gain, 9, 23
Open-loop response, 259–263
Open-loop transimpedance gain, 294
Operating temperatures, 240
Operational transconductance amplifier (OTA), 620–627
OS, 350, 351

Oscillation, 347.
  See also Stability
Oscillators
  CMOS crystal, 463, 464
  quadrature, 456, 457
  relaxation, 450, 451
  ring, 494
  sinusoidal, 450
  VCO, 473–475
  Wien-bridge, 451–456
OTA, 620–627
Output/input impedances, 269–275
Output-offsetting, 89
Output overload protection, 521
Output reference current, 609
Output resistance, 3
Output saturation, 44–47
Output scale factor, 608
Output-series, 30
Output short-circuit current, 242
Output-shunt, 30
Output shunt topology, 272
Output spectral density, 331
Output stage, 215
Output voltage swing, 242
Overall input spectral density, 329, 330
Overdamped, 127
Overload protection, 242–244
Oversampling, 597–599
Oversampling converters, 595–602
  noise shaping, 599–602
  Nyquist-rate sampling, 595–597
  oversampling, 597–599
  sigma-delta converters, 599–602
Overshoot (OS), 350, 351
Overvoltage protection, 533–535
OV protection, 533–535

P

P_{max}, 241
PA04, 244
Passband, 107, 161
Passive band-pass filter prototype, 179
Passive compensation of integrators, 284
Passive lag-lead filter, 632
PCM52/53, 579
Peak current control, 547
Peak detectors, 433–437
Peak frequency, 130
Peaking, 129
Pedestal error, 440
Percentage deviation, 92

Phase accuracy, improving,
388–390
Phase-error compensation, 284
Phase-locked loops (PLLs),
627–642
basic system, 627, 628
designing, 640–642
dumping characteristics, 633
filter design criteria, 635
first-order loop, 631
lock and capture, 628, 629
locked condition, 629–631
monolithic, 635–642. *See also*
Monolithic PLLs
second-order loops, 632, 633
Phase shifters, 121
Phono preamplifier, 122
Photoconductive detectors, 63
Photodetector amplifiers, 63
Photovoltaic detectors, 63
Pink-noise curve, 321
Pink-noise tangent principle,
321, 322
Pipelined converters, 592
PLLs. *See* Phase-locked loops
(PLLs)
Pole splitting, 369, 370
Pole-zero cancellation, 372, 373
Pole-zero circuit, 153, 154
Poor grounding, 364, 365
Positive resistance, 22
Positive saturation region,
44, 45
Post filter, 615
Potentiometric DACs, 569, 570
Power dissipation, 42–44
Power op amps, 244
Power sources, 528, 529
Power-supply busses, 365
Power-supply rejection ratio
(PSRR), 229
Power-supply supervisory
circuits, 532–535
PPM, 469
Practical logarithmic wave
shaper, 476
Practical long/antilog circuits,
612–614
Practical op amp, 65
Practical Wien-bridge
oscillators, 454
Precision integrator, 21
Precision rectifiers, 422–428
Printed-circuit
board leakage, 435
Programmable op amps, 282
Propagation delay, 399
Proportional to absolute
temperature (PTAT), 511
PSpice, 7, 163, 216

PSpice models.
See also SPICE models
CFAs, 299, 300
finding T, 353–355
noise, 326, 327
PSRR, 229
PTAT, 511
Pull-in time, 629
Pulse-position modulation
(PPM), 469
Pulse-width modulation (PWM),
414, 415, 469
Push-pull pair, 215
PWM, 414, 415, 469
PWM waveforms, 415

Q

Q-enhancement compensation,
286–289
Q multiplier, 157
Quadrature oscillators, 456, 457
Quantization error, 565
Quantization noise, 565
Quiescent supply current, 42

R

RA, 591
Ragazzini, John R., 1
Rail-to-rail op amps, 45, 242
Rate of closure (ROC), 352, 353
Ratiometric conversion, 564
RC-SC integrator conversion, 188
REF-05 5-V precision
reference, 511
REF101, 502
REF101 10-V precision
reference, 507, 508
Reference voltage, 561
References/regulators. *See* Volt-
age references/regulators
Regions of operation, 44, 45
Relative accuracy, 563
Relaxation oscillators, 450, 451
Remote sensing, 513, 514
Reset-switch leakage, 436
Residue, 591
Residue amplifier (RA), 591
Resistance mismatches
difference amplifiers, 74–77
V-I converters, 68, 69
Resistance trimming, 69
Resistance values, 58, 59
Resistive feedback. *See* Circuits
with resistive feedback
Resistive transducers, 91
Resolution, 562
Resonance frequency, 130, 131
Resonance gain, 130
Response time, 399, 400
Retriggerable one-shot, 465

Retriggerable one-shot
operation, 534
Return ratio, 24
RIAA playback equalization
curve and phono
preamplifier, 122
Ring oscillator, 494
Ripple band, 161
Ripple rejection ratio (RRR), 501
Rise time, 275, 276
RLC filters, 109
RLC ladder simulation, 181–186
rms output noise, 330–333
rms value, 313
ROC, 352, 353
Root-mean-square (rms)
value, 313
R-$2R$ ladders, 570–572, 574
RRR, 501

S

SA ADCs, 586–588
Safe operating area (SOA),
520, 521
Sagback effect, 435
Sallen-Key filters, 134.
See also KRC filters
Sample-and-hold amplifier
(SHA), 437–443
Sampled-data system, 560
Sample-to-hold offset, 440
Sampling mode, 235
SAR, 586
Saturating amplifiers, 44–47
Saturation current, 506
Sawtooth wave generators,
476–478
Scale factor, 561
Scaling factor, 110
SC biquad filter, 193
SC filters, 192–198
cascade design, 202–204
direct synthesis of band-pass
filters, 196–198
direct synthesis of low-pass
filters, 196
dual-integrator-loop filters,
192–194
ladder simulation, 194, 195
practical limitations, 190–192
universal filters, 198–204
Schmitt triggers, 416–422
SC integrators, 188–190
Second-order, Type I loop, 632
Second-order, Type II loop, 633
Second-order filters, 126–133
all-pass response, 132
band-pass response, 130, 131
filter measurements, 133
high-pass response, 129, 130

Second-order filters—*Cont.*
  low-pass response, 127–129
  notch response, 131, 132
Second-order g_m-C filter, 621
Second-order loops, 632, 633
Second stage, 215
Second-stage noise, 341
Seebeck coefficient, 518
Segmentation, 576–579
Selectivity factor, 161
Self-regulated 10-V
  reference, 504
Self-regulation, 505
Sensitivity, 150–152
Sequential search, 585
Series impedances, 270, 271
Series-pass element, 519
Series-shunt, 31
Servo converter, 585
Settling time, 279–281, 564
Seventh-order 0.28/60-dB elliptic
  low-pass filter, 184
74HC(T)4046A CMOS PLL,
  635–642
741 op amp, 1, 2, 212, 215, 216,
  249–257, 340
741 op amp macromodel, 217
7800 series, 523–525, 530
SHA, 437–443
SHC803/804, 442
Short-circuit current gain, 4
Short-circuit noise, 330
Short-circuit protection, 521
Shot noise, 323
Shunt-capacitance compensation,
  368, 369
Shunt impedances, 272–275
Shunt-shunt, 34, 36
Sigma-delta converters, 599–602
Signal gain, 235
Signal generators, 449–498
  emitter-coupled VCO,
    483–486
  F-V converters, 491, 492
  grounded-capacitor VCO,
    478–483
  logarithmic wave shaper,
    475, 476
  monolithic timers, 465–471
  monolithic waveform
    generators, 478–486
  multivibrators, 457–465.
    See also Multivibrators
  quadrature oscillators,
    456, 457
  relaxation oscillators, 450, 451
  sawtooth wave generators,
    476–478
  sine wave generators, 451–457
  sinusoidal oscillators, 450

  triangular wave generators,
    471–476
  V-F converters, 486–490
  voltage-controlled oscillator
    (VCO), 473–475
  Wien-bridge oscillator,
    451–456
Signal integrate phase, 594
Signal-to-noise ration (SNR),
  312, 333, 334
Silicon photodiode, 63
Simplified op amp circuit
  diagram, 212–217
Simplified op-amp block
  diagram, 259
Sine wave generators, 451–457
Single-op-amp bridge
  amplifier, 96
Single-supply free-running
  multivibrator, 460
Single-supply inverting Schmitt
  trigger, 419
Single-supply noninverting
  Schmitt trigger, 420
Sinusoidal FSK generator, 486
Sinusoidal oscillators, 450
Sixth-order 0.1/40-dB elliptic
  low-pass filter, 172
Sixth-order 1-dB Chebyshev
  low-pass filter, 170
Sixth-order 1.0/40-dB band-pass
  filter, 174
Sixth-order Butterworth
  band-pass filter, 173
Skirt, 161
Slew rate (SR), 276, 277
Slew-rate limiting, 276–278,
  281, 282
Small-signal bandwidth, 616
Small-signal conditions, 214
SNR, 312, 333, 334
SOA, 520, 521
Soft start, 548
Source-free response, 111
Source-to-load gain, 3
Space frequency, 485
Spectral noise densities, 315
SPICE, 2. *See also* PSpice,
  PSpice models
SPICE models, 215–217
Square-root extraction, 619, 620
SR, 276, 277
Stability, 347–397
  capacitive load isolation,
    362–364
  CFA circuits, 381–384
  composite amplifiers, 384–390
  constant-GBP op amp circuits,
    356–365
  differentiator circuit, 357, 358

  external frequency
    compensation, 374–380.
    See also External
    Frequency compensation
  feedback pole, 356, 357
  finding T using PSpice,
    353–355
  gain margin, 348, 349
  grounding, 364, 365
  $H(s)$, and, 111, 112
  internal frequency
    compensation, 365–374.
    See also Internal
    Frequency compensation
  log/antilog amplifiers, 610, 611
  peaking/ringing, 350–352
  power-supply filtering,
    364, 365
  rate of closure, 352, 353
  stray input-capacitance
    compensation, 359–362,
    382–384
Standard resistance values, 58, 59
Start-up circuitry, 506
State-variable (SV) filters,
  144–147
Static op amp limitations,
  211–257. *See also* Dynamic
  op amp limitations
  autozero/chopper-stabilized op
    amps, 234, 235
  bipolar op amps, 230–233
  CMRR, 227, 228
  external nulling, 238–240
  FET-input op amps, 234
  input bias current, 217–221
  input bias-current drift, 224
  input guarding, 225
  input offset current, 217–221
  input offset-error
    compensation, 235–240
  input offset voltage, 225–230
  input-bias-current
    cancellation, 222
  internal nulling, 237
  JFET-input op amps, 222
  low-input-bias-current op
    amps, 221–225
  low-input-offset-voltage op
    amps, 230–235
  maximum ratings, 240–244
  MOSFET-input op amps,
    222–224
  overload protection, 242–244
  PSRR, 229
  superbeta-input op amps,
    221, 222
  thermal drift, 227
Step-down regulator, 539
Step-up regulators, 539

Stopband, 107, 161
Strain gauge bridges, 94–96
Stray input-capacitance
    compensation, 359–362,
    382–384
Subranging converters,
    591, 592
Substrate thermostating, 409
Subsurface diode structure, 510
Successive-approximation
    converters (SA ADCs),
    586–588
Successive-approximation
    register (SAR), 586
Summing amplifier, 17–19
Summing junction, 18
Superbeta-input op amps,
    221, 222
Superbeta input stage, 221
Superbeta transistors, 221
Superdiode, 423
Supply regulation, 501
SV filters, 144–147
Switched capacitor, 187–192
Switched-capacitor filters.
    See SC filters
SwitcherCAD, 544, 551
Switchers Made Simple, 544
Switches, 428–433
Switching regulators, 535–544
    basic topologies, 537–539
    capacitor selection, 541–543
    coil selection, 540, 541
    efficiency, 543, 544
    monolithic, 544–551. *See also*
        Monolithic switching
        regulators
Symmetric notch, 149

T

T. See Loop gain (*T*)
Tape preamplifier, 123
TC, 502
Temperature coefficient, 227
Temperature control
    linear regulators, 529–532
    monolithic temperature
        sensors, 511, 512
    voltage regulators, 517–519
Temperature-to-frequency
    converter, 488
THA, 437–443
THD, 450
-3-dB frequency, 119, 120
Thermal coefficient (TC), 502
Thermal drift, 227
Thermally compensated Zener
    diode references, 507–510
Thermal noise, 322
Thermal voltage, 506

Thermocouple cold-junction
    compensators, 518, 519
Thermometer, 590
Third-order loop, 635
Thomson filters, 166
Three-terminal adjustable
    regulators, 524
Three-terminal reference, 511
Threshold detector, 400,
    401, 407
THS4401 high-speed VFA, 303
Timer/counter circuits, 469–471
Timing diagram, 470
TL080 op amp, 282
TLC279 CMOS op amp, 223
TLE2426 Rail Splitter, 45
T-network photodiode amplifier,
    338, 339
Tone control, 123–125
Total error, 615
Total harmonic distortion
    (THD), 450
Total output rms noise, 317
Total rms input noise, 333
Total rms output noise, 331, 333
Tow-Thomas filter, 147, 148
Track-and-hold amplifier (THA),
    437–443
Tracking converter, 585
Track mode, 434
Transconductance amplifiers, 5,
    63–71. *See also*
        Voltage-to-current
        converters
Transducer bridge amplifiers,
    91–97
    bridge calibration, 93, 94
    bridge linearization, 97
    single-op-amp amplifier, 96
    strain gauge bridges, 94–96
    transducer resistance
        deviation, 92, 93
Transducer resistance deviation,
    92, 93
Transfer curve, 26
Transient response, 275–282
Transimpedance amplifiers,
    61, 294
Transistor noise models, 324
Transition band, 161
Transition frequency, 261
Transmission gate, 432, 433
Transresistance amplifier,
    5, 61–63
Treble/bass control, 123–125
Triangular-to-sine wave
    conversion, 475, 476
Triangular wave generators,
    471–476
Triple-op-amp IA, 79–82

Triple-output flyback
    regulator, 550
True difference amplifier, 20
True rms meters, 314
True rms-to-dc converters,
    614, 615
2240 timer/counter circuit,
    469–471
Two-quadrant multiplier, 615
Two-step converter, 591
Two-terminal reference, 511
Type I phase comparator,
    637, 638
Type II phase comparator,
    639, 640
Type III phase comparator,
    638, 639

U

Ultralow-noise op amps, 341, 342
Unbalanced bridge
    difference amplifiers, 74–77
    V-I converters, 68, 69
Uncompensated
    differentiator, 357
Undamped, 127
Underdamped, 127
Undamped natural frequency,
    127, 634
Undervoltage sensing, 533–535
Unity-gain amplifier, 11
Unity-gain frequency, 116,
    261, 357
Unity-gain *KRC* circuit, 136–138
Universal filters, 144–150
    biquad filters, 147, 148
    notch response, 148–150
    sensitivity, 152
    SV filters, 144–147
Universal SC filters, 198–204
Unloaded current gain, 4
Unloaded transconductance
    gain, 620
Unloaded voltage gain, 3
Upper saturation region, 44, 45
UV sensing, 533–535

V

v_{CM}, 73
v_{DM}, 73
Variable gain, 77
Variable precision damp, 445
Variable-transconductance
    multipliers, 616–618
Variable transconductance
    principle, 616
VCO, 473–475
VFA-CFA composite
    amplifier, 387
VFAs, 293, 300–303

VFC, 486–490
VFC32 voltage-to-frequency
    converter, 489, 490
V-F converters, 486–490
V-I converters. *See*
    Voltage-to-current
    converters
Virtual short, 16
Voltage amplifier, 3
Voltage comparators, 399–415
    bar graph meters, 412–414
    general-purpose IC
        comparators, 401–406
    high-speed comparators,
        406, 407
    level detectors, 407–409
    on-off control, 409, 410
    op amp as, 400, 401
    pulse-width modulation,
        414, 415
    response time, 399, 400
    uses, 407–415
    window detectors, 410–412
Voltage compliance, 64, 515
Voltage-controlled oscillator
    (VCO), 473–475
Voltage-controlled
    sawtooth/pulse-wave
    oscillator, 477
Voltage-controlled state-variable
    filter, 626
Voltage-controlled
    triangular/square-wave
    oscillator, 474
Voltage droop, 434, 435, 440
Voltage droop rate, 435
Voltage-feedback amplifiers
    (VFAs), 293, 300–303
Voltage follower, 11, 12, 356
Voltage gain factor, 3
Voltage-mode control, 544–547

Voltage mode *R*-2*R* ladders,
    571, 572
Voltage-mode segmentation, 577,
    578
Voltage references, 506–519
    bandgap voltage references,
        510, 511
    current sources, 515–517
    monolithic temperature
        sensors, 511, 512
    remote sensing, 513, 514
    temperature-sensor
        applications, 517–519
    thermally compensated
        Zener diode references,
        507–510
    uses, 512–519
    voltage sources, 514, 515
Voltage references/regulators,
    499–558
    basic connection, 500
    linear regulators, 519–535.
        See also Linear regulators
    monolithic switching
        regulators, 544–551
    performance specifications,
        500–506
    switching regulators, 535–551.
        See also Switching
        regulators
    voltage references, 506–519.
        See also Voltage
        references
Voltage sources, 514, 515
Voltage-to-current converters,
    63–71
    finite open-loop gain, 70
    floating-type converters,
        64, 65
    grounded-load converters,
        66–68

Howland current pump, 66–68
improved Howland current
    pump, 70, 71
practical op amp limitations,
    65, 66
resistance mismatches, 68, 69
Voltage-to-frequency converter
    (VFC), 486–490
VPTATs, 511, 512
VTC offsetting, 418–420

W

Weighted-capacitor DACs,
    568, 569
Weighted-resistor DACs,
    567, 568
White noise, 316
White-noise floors, 316
Wideband ac-dc converter, 427
Wideband band-pass filter, 120
Wide-sweep multivibrator VFC,
    487–489
Widlar, Robert J., 1
Wien-bridge oscillator, 451–456
Window detectors, 410–412

X

XR-210/15, 484
XR-2206 function generator,
    484–486
XR-2206/07 monolithic function
    generators, 484
XR-2240 timer/counter, 469–471

Z

Zener diode (as shunt
    regulator), 503
Zener zapping, 232
Zero-crossing detector, 400
Zero-pole cancellation, 284